ENGINEERING MECHANICS

ENGINEERING MECHANICS

Arvind Kumar Dubey
B.Tech., M.Tech.
Director
Arvind's Physics Centre
Motihari, Bihar

Anil Kumar
B.E.
Director
Concept Classes and Educational Consultant
Bangalore (Karnataka)

PUBLISHING FOR ONE WORLD

NEW AGE INTERNATIONAL (P) LIMITED, PUBLISHERS
New Delhi • Bangalore • Chennai • Cochin • Guwahati • Hyderabad
Jalandhar • Kolkata • Lucknow • Mumbai • Ranchi
Visit us at www.newagepublishers.com

Published by New Age International (P) Ltd., Publishers
First Edition: 2009

Branches:
- 36, Malikarjuna Temple Street, Opp. ICWA, Basavanagudi, **Bangalore**. ✆ (080) 26677815
- 26, Damodaran Street, T. Nagar, **Chennai**. ✆ (044) 24353401
- Hemsen Complex, Mohd. Shah Road, Paltan Bazar, Near Starline Hotel, **Guwahati**. ✆ (0361) 2543669
- No. 105, 1st Floor, Madhiray Kaveri Tower, 3-2-19, Azam Jahi Road, Nimboliadda, **Hyderabad**. ✆ (040) 24652456
- RDB Chambers (Formerly Lotus Cinema) 106A,1st Floor, S.N. Banerjee Road, **Kolkata**. ✆ (033) 22275247
- 18, Madan Mohan Malviya Marg, **Lucknow**. ✆ (0522) 2209578
- 142C, Victor House, Ground Floor, N.M. Joshi Marg, Lower Parel, **Mumbai**. ✆ (022) 24927869
- 22, Golden House, Daryaganj, **New Delhi**. ✆ (011) 23262370, 23262368

ISBN : 978-81-224-2396-9

Rs. 295.00

C-08-09-3136

Printed in India at Glorious Printers, Delhi.
Typeset at Kalyani Computer Services, New Delhi.

PUBLISHING FOR ONE WORLD
NEW AGE INTERNATIONAL (P) LIMITED, PUBLISHERS
4835/24, Ansari Road, Daryaganj, New Delhi-110002
Visit us at **www.newagepublishers.com**

Dedicated

To

my mother Smt. Gayatri Devi

&

my father Shri Jatashanker Dubey

— Er. Arvind Kumar Dubey

To

In the loving memory of my beloved mother

Late Smt. Moti Devi whom I miss a lot

&

my father Sri Asta Nand Prasad Gupta

Your incredible contributions to education will not go unnoticed.

— Er. Anil Kumar

Foreword

I am glad to write the foreword of the excellent endeavour on Engineering Mechanics by Er. Arvind Kumar Dubey and Er. Anil Kumar. This textbook is by and large geared to meet the syllabus of Engineering Mechanics across the country.

This textbook is so comprehensive that it addresses the students as if a teacher would address the students in class-room. The mathematical parts have been rendered to pain-free simplicity and the textual matter is accompanied by self explanatory figures, and encourages the students to lean towards the subject. Well chosen exercise problems are also included.

I have great pleasure in recommending the textbook to students and teachers. And I have every hope that this book shall be immensely helpful in aiding the fundamentals of the subject.

Dr. Achintya
Professor, Department of Civil Engineering
Muzaffarpur Institute of Technology (MIT)
Muzaffarpur-842003

• • •

The book titled *"Engineering Mechanics"* by Er. A.K. Dubey and Er. Anil Kumar has been written for first year engineering students. The book concentrates primarily on static mechanics and provides a useful introduction to the fundamental concepts in an important area of mechanical and structural engineering.

The strength of the book is the inclusion of a large number of worked-out and exercise problems of different levels of complexity. Thus not only that the book will be beneficial for the freshman engineering students but it can be also be a valuable reference for students preparing for various engineering entrance examinations across India.

Dr. Anindya Deb
Professor
Centre for Product Design and Manufacturing
Indian Institute of Science
Bangalore

• • •

An excellent book to learn from and use as a reference. The Authors do a fantastic job of teaching the readers about Engineering Mechanics and its principles. The texts and the worked out examples are crystal clear. I specially recommend this book for the first year engineering students of all branches in engineering courses in technical universities in India.

Mr. Shiva Prasad T.S.
(Former HOD, Deptt. of CSE/ISE SVCE, Bangalore)
ERP Consultant,
American Solution Inc.,
Newark, DE, USA

Preface

Engineering Mechanics is that branch of science, which should be taught at an advanced level. To be able to do this, one must have a level of understanding of Mathematics. This book titled *Engineering Mechanics* is designed to meet the requirements of students and teachers who want to study and teach this subject at a progressive level.

This book is divided into eight chapters, covering all the concepts in STATIC mechanics. Each of these chapter has a large number of worked out examples of various types, which will help the readers tackle the exercise problems with more confidence. While dealing with the numerous problems in this book, we have emphasized more on a mathematical approach to solving these problems rather than the conventional approach. The readers will be able to relate this subject to the actual field of engineering mechanics since the exercises in each chapter has problems related to practical applications. In the theory section of each chapter, we have explained the concepts clearly with the aid of neat diagrams that will help perfect the readers' understanding of this subject.

Newton's Laws of Motion is said to be the essence of mechanics, thus in the fourth chapter of the book these three fundamental laws are discussed in detail and they form as the framework for the subsequent chapters. Several problems involving forces are clearly depicted using free body diagrams. An important concept such as frame of reference is discussed in detail as the readers can then relate it to the prime properties of surface like the centroid and the moment of inertia. Pseudo force though quite confusing is dealt with in a methodical manner as it has wide applications in Mechanics. Since it is essential to have a strong knowledge of Mathematics and Vector, we have included one chapter on Mechanics and Mathematics and one chapter is dedicated only to Vector.

In conclusion this book is written to enjoy Mechanics, but it is also very useful for students and teachers under technical universities for the first year engineering courses all over india. Finally, we hope readers will enjoy reading this book as much as we enjoyed writing and completing it. Good Luck !

AUTHORS

Acknowledgements

In writing a book like this we have benefited enormously from the contributions of our friends, they helped us for removing the errors and they took great pains in going through the entire manuscript and made valuable comments and others who gave us advice and where we got time for fruitful discussions. Let us take this opportunity to express our deep sense of gratitude to them. We have been very fortunate in one aspect, that we had studied mechanics under the guidance of a great teacher like Prof. M.M.R. Akhtar. This book is a result of that guidance and our interest in the subject. We have also benefited from Juhi Choudhary and Suraj Mysore L, while we were removing the errors. We would like to express our sincere thanks to Dr. Achintya and Mr. Shiva Prasad T. S., they gave their valuable time for writing foreword to this book. We thank to the editorial and production staffs at New Age International (P) Limited, Publishers, New Delhi for publishing this book. We are also beholden our friends Manoj Shah (Director Satkar Etta Udhyog, Kalaiya), Er. Sunil B. Shegunasi, Er. Arun Kumar, Er. Saroj Kumar Gupta, Dinesh Kumar, Brajesh Kumar, Er. Rajesh Kumar, Er. Saurav Kumar Singh, Er. Pawan Bhatt, Er. Nitesh Kumar, Er. Rajeev Ranjan and Er. Dipankar Chakrabarty and my maternal uncle Je. Sri Ramagya Prasad and Manoj Kumar Shah. They gave us moral supports and all the needs whenever we asked them to do. Finally, the persons responsible for any success in our life are our fathers. They center our universe and has been with us in every endeavour, we have undertaken in our life so far. From more than one decade my (Anil Kumar) elder sisters are giving their valuable supports in my education, for that I cannot express my gratitude in words. Any suggestion for further improvement of the book will be acknowledged and appreciated. Your comments could be sent at engineeringmechanics@gmail.com

AUTHORS

Brief Contents

Contents

1 MATHEMATICS AND MECHANICS

1.1 RELATION BETWEEN MATHEMATICS AND MECHANICS

The description of laws of mechanics becomes easy if we have the freedom to use mathematics. For example, in inertial frame of reference sum of external forces on a body of constant mass is proportional to acceleration. Mathematically, we could represent this as follows:

Sum of external Force ∝ Acceleration

or $\Sigma \vec{F}_{ext.} \propto \vec{a}$

or $\Sigma \vec{F}_{ext.} = m\vec{a}$; where 'm' is proportionality constant.

or $\Sigma F_x = F_{1x} + F_{2x} + ... + F_{nx} = ma_x$;

or $\Sigma F_y = F_{1y} + F_{2y} + ... + F_{ny} = ma_y$;

and $\Sigma F_z = F_{1z} + F_{2z} + ... + F_{nz} = ma_z$;

Further, the division of mathematics such as algebra, trigonometry and calculus could be used to make easy description.

Thus, mathematics is the language of mechanics. It helps to make easy description to discover, understand and explain laws of mechanics. The importance of mathematics in today's world couldnot be disputed.

However, mathematics itself is not mechanics. As we need language to express our idea like Hindi, Marathi, English, C, C++ etc. But idea that we want to express has the main attention. If we are poor at grammar and vocabulary, it would be difficult for us to communicate our feeling that we want to express, suppose you want to go Bangalore from Patna in Rajdhani Express with collection of favourite songs, then the sweet memories of the journey is Rajdhani Express having yours favourite songs is just like mathematics, the mathematics is essential thing in the world of mechanics.

1.2 BASIC TRIGONOMETRICAL FUNCTIONS

1.2.1 Trigonometrical functions of angle θ

$$\sin\theta = \frac{y}{r} \qquad \cos\theta = \frac{x}{r}$$

$$\tan\theta = \frac{y}{x} \qquad \cot\theta = \frac{x}{y}$$

$$\sec\theta = \frac{r}{x} \qquad \text{cosec}\,\theta = \frac{r}{y}$$

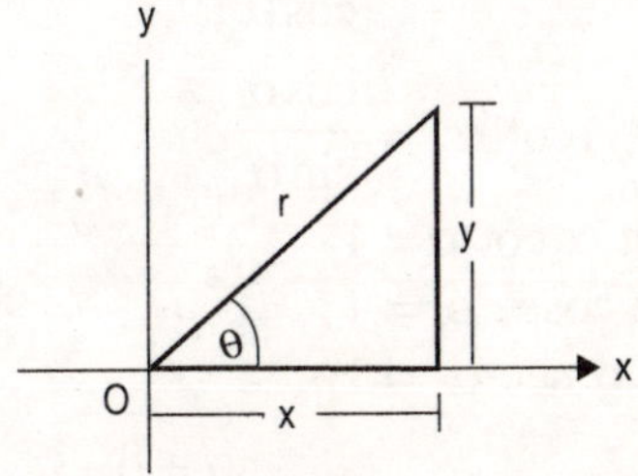

Fig. 1.1

1.2.2 Pythagorean theorem

$$x^2 + y^2 = r^2$$

or $$r = \sqrt{x^2 + y^2}$$

1.2.3 Trigonometrical identities

$$\sin^2\theta + \cos^2\theta = 1$$

$$\sec^2\theta - \tan^2\theta = 1 \quad \text{cosec}^2\theta - \cot^2\theta = 1$$

$$\sin(\alpha \pm \beta) = \sin\alpha\cos\beta \pm \cos\alpha\sin\beta$$

$$\cos(\alpha \pm \beta) = \cos\alpha\cos\beta \mp \sin\alpha\sin\beta$$

$$\tan(\alpha \pm \beta) = \frac{\tan\alpha \pm \tan\beta}{1 \mp \tan\alpha\tan\beta}$$

$$\sin 2\theta = 2\sin\theta\cos\theta$$

$$\tan 2\alpha = \frac{2\tan\alpha}{1 - \tan^2\alpha}$$

$$\cot 2\alpha = \frac{\cot^2\alpha - 1}{2\cot\alpha}$$

$$\cos 2\theta = \cos^2\theta - \sin^2\theta = 2\cos^2\theta - 1 = 1 - 2\sin^2\theta$$

$$\sin\alpha = \frac{1}{\sqrt{1 + \cot^2\alpha}} = \frac{1}{\text{cosec}\,\alpha}$$

$$\cos\alpha = \frac{1}{\sqrt{1 + \tan^2\alpha}} = \frac{1}{\sec\alpha}$$

$$\tan\alpha = \frac{\sin\alpha}{\cos\alpha}$$

$$\cot\alpha = \frac{\cos\alpha}{\sin\alpha}$$

$$\tan\alpha\cot\alpha = 1$$

$$\sin\alpha\,\text{cosec}\,\alpha = 1$$

$$\cos\alpha\sec\alpha = 1$$

$$\sin\alpha + \sin\beta = 2\sin\frac{\alpha+\beta}{2}\cos\frac{\alpha-\beta}{2}$$

$$\cos\alpha + \cos\beta = 2\cos\frac{\alpha+\beta}{2}\cos\frac{\alpha-\beta}{2}$$

$$\sin\alpha - \sin\beta = 2\cos\frac{\alpha+\beta}{2}\sin\frac{\alpha-\beta}{2}$$

$$\cos\alpha - \cos\beta = -2\sin\frac{\alpha+\beta}{2}\sin\frac{\alpha-\beta}{2}$$

$$\tan\alpha \pm \tan\beta = \frac{\sin(\alpha \pm \beta)}{\cos\alpha\cos\beta}$$

$$\cot\alpha \pm \cot\beta = \pm\frac{\sin(\alpha \pm \beta)}{\sin\alpha\sin\beta}$$

$$\sin\frac{\alpha}{2} = \sqrt{\frac{1 - \cos\alpha}{2}}$$

$$\cos\frac{\alpha}{2} = \sqrt{\frac{1 + \cos\alpha}{2}}$$

$$2\sin\alpha\sin\beta = \cos(\alpha - \beta) - \cos(\alpha + \beta)$$

$$2\sin\alpha\cos\beta = \sin(\alpha - \beta) + \sin(\alpha + \beta)$$

$$2\cos\alpha\cos\beta = \cos(\alpha - \beta) + \cos(\alpha + \beta)$$

$$2\cos\alpha\sin\beta = \sin(\alpha - \beta) - \sin(\alpha + \beta)$$

$$\sin x = x - \frac{x^3}{6} + \frac{x^5}{120} - \ldots\ldots \quad -\infty < x < \infty$$

$$\cos x = 1 - \frac{x^2}{2} + \frac{x^4}{24} - \ldots\ldots \quad -\infty < x < \infty$$

$$\tan x = x + \frac{x^3}{3} + \frac{2x^5}{15} + \ldots\ldots \quad -\frac{\pi}{2} < x < \frac{\pi}{2}$$

where x is in radian.

1.2.4 Approximations for α (in radian) and for very small number

$$\sin x \approx x$$

$$\cos x \approx 1 - \frac{x^2}{2}$$

$$\tan x \approx x$$

1.2.5 Triangle: sine rule and cosine rule

Angles of triangle are A, B, C and opposite sides are a, b and c.

Then A + B + C = 180°

$$\frac{\sin A}{a} = \frac{\sin B}{b} = \frac{\sin C}{c}$$

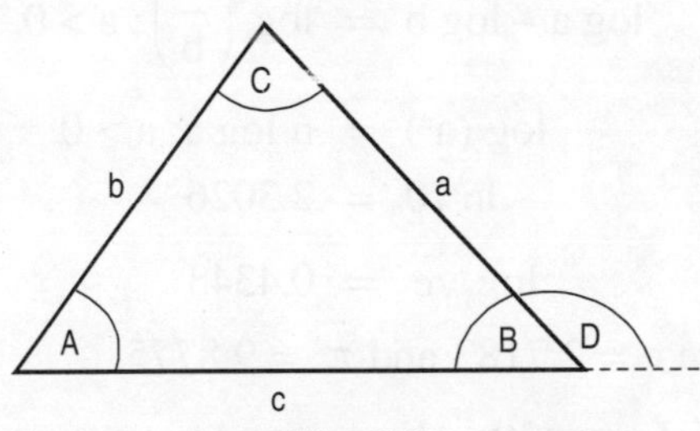

Fig. 1.2

and $c^2 = a^2 + b^2 - 2\,ab\cos C$

Exterior angle D = A + C

$\sin(90 - \theta) = \cos\theta$

$\cos(90 - \theta) = \sin\theta$

$\sin(-\alpha) = -\sin\alpha$

$\cos(-\alpha) = \cos\alpha$

$\sin\theta = 0 \Rightarrow \theta = n\pi;\ n \in z$

$\cos\theta = 0 \Rightarrow \theta = (2n+1)\dfrac{\pi}{2};\ n \in z$

$\tan\theta = 0 \Rightarrow \theta = n\,\pi;\ n \in z$

$\sin\theta = \sin\theta_0 \Rightarrow \theta = n\pi + (-1)^n\theta_0;$

where $n \in I$

$\cos\theta = \cos\theta_0 \Rightarrow \theta = 2n\pi \pm \theta_0;\ n \in I$

$\tan\theta = \tan\theta_0 \Rightarrow \theta = n\pi + \theta_0;\ n \in I$

$e^{\pm i\theta} = \cos\theta \pm i\sin\theta$

$$\sin\theta = \frac{e^{i\theta} - e^{-i\theta}}{2i}$$

$$\cos\theta = \frac{e^{i\theta} - e^{-i\theta}}{2}$$

where $i^2 = -1$.

1.2.6 Sign of trigonometrical ratios

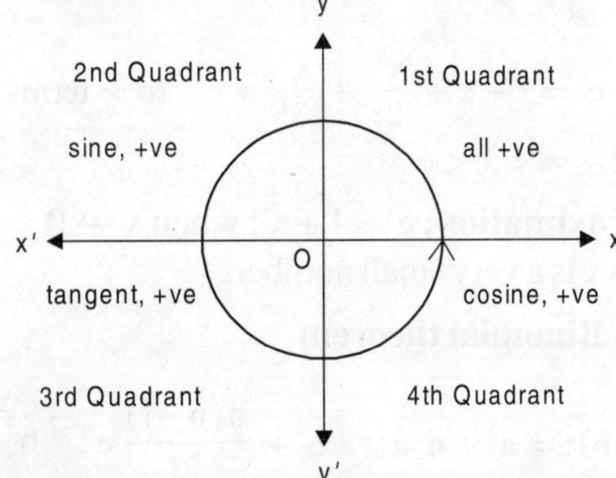

Table 1.1 : Value of trigonometrical ratio for some particular angles

ratio \ θ	0°	30°	45°	60°	90°
sin θ	$\sqrt{\frac{0}{4}} = 0$	$\sqrt{\frac{1}{4}} = \frac{1}{2}$ = 0.5	$\sqrt{\frac{2}{4}} = \frac{1}{\sqrt{2}}$ = 0.707	$\sqrt{\frac{3}{4}} = \frac{\sqrt{3}}{2}$ = 0.866	$\sqrt{\frac{4}{4}} = 1$
cos θ	$\sqrt{\frac{4}{4}} = 1$	$\sqrt{\frac{3}{4}} = \frac{\sqrt{3}}{2}$ = 0.806	$\sqrt{\frac{2}{4}} = \frac{1}{\sqrt{2}}$ = 0.707	$\sqrt{\frac{1}{4}} = \frac{1}{2}$ = 0.5	$\sqrt{\frac{0}{4}} = 0$
tan θ	$\sqrt{\frac{0}{4}} = 0$	$\sqrt{\frac{1}{3}} = \frac{1}{\sqrt{3}}$ = 0.577	$\sqrt{\frac{2}{2}} = 1$	$\sqrt{\frac{3}{1}} = \sqrt{3}$ = 1.732	$\sqrt{\frac{4}{0}}$ = undefined

1.3 ALGEBRA

1.3.1 Quadratic equation

If $ax^2 + bx + c = 0$; $a \neq 0$, then roots of quadratic equation are $x = \dfrac{-b \pm \sqrt{b^2 - 4ac}}{2a}$.

1.3.2 Exponential functions

$$a^{-x} = \frac{1}{a^x}$$

$$a^{x-y} = \frac{a^x}{a^y}$$

$$a^{x+y} = a^x a^y$$

1.3.3 Exponential expansion

$$e^x = 1 + x + \frac{x^2}{2!} + \frac{x^3}{3!} + \ldots\ldots \text{ to } \infty \text{ terms,}$$

where $-\infty < x < \infty$.

Approximation: $e^x \approx 1 + x$; when $x \to 0$ that is x is a very small number.

1.3.4 Binomial theorem

$$(a + b)^n = a^n + n\, a^{n-1} b + \frac{n(n-1)}{2!} a^{n-2} b^2 + \frac{n(n-1)(n-2)}{3!} a^{n-3} b^3 + \ldots\infty.$$

where n is real a number and $a^2 > b^2$.

$$(1 + x)^n = 1 + \frac{nx}{1!} + \frac{n(n-1)}{2!} x^2 + \frac{n(n-1)(n-2)}{3!} x^2 + \ldots\ldots\infty.$$

$n \in \underset{R}{I^+}$ and $x^2 > 1$.

$(1 + x)^n = 1 + nx + {}^nC_2\, x^2 + {}^nC_3\, x^3 + \ldots\ldots + x^n$ contains n + 1 terms where $x^2 < 1$ and $n \in I^+$.

Approximation: $(1 \pm x)^n \approx 1 \pm nx$; where x is a very small number.

1.3.5 Logarithms

$$\log a + \log b = \log (ab);\ a > 0, b > 0$$

$$\log a - \log b = \log \left(\frac{a}{b}\right);\ a > 0, b > 0$$

$$\log (a^n) = n \log a;\ a > 0$$

$$\ln 10 = 2.3026$$

$$\log_{10} e = 0.4343$$

where $e = 2.7183$ and $\pi^2 = 9.8775$.

1.3.6 Logarithmic expansion

$$\ln (1 + x) = \frac{x^1}{1} - \frac{x^2}{2} + \frac{x^3}{3} - \frac{x^4}{4} + \frac{x^5}{5} \ldots\ldots\infty,$$

where $-1 < x < 1$.

Approximation: $\ln (1 + x) \approx x$, when x is a very small number, then the higher powers of x can be neglected.

1.3.7 Taylor's expansion

$$f(x_0 + x) = f(x_0) + x\, f^1(x_0) + \frac{x^2}{2!} f^2(x_0) + \frac{x^3}{3!} f^3(x_0) + \ldots\ldots$$

$$\frac{1}{1+x} = 1 - x + x^2 - x^3 + \ldots\ldots$$

$$\frac{1}{1-x} = 1 + x + x^2 + x^3 + \ldots\ldots$$

$$\sqrt{1+x} = 1 + \frac{x}{2} - \frac{x^2}{8} + \frac{x^3}{16} - \ldots\ldots$$

$$\frac{1}{\sqrt{1+x}} = 1 - \frac{x}{2} + \frac{3x^2}{8} - \frac{5x^3}{16} + \ldots\ldots$$

1.3.8 Determinant

$$\begin{vmatrix} a_1 & a_2 & a_3 \\ b_1 & b_2 & b_3 \\ c_1 & c_2 & c_3 \end{vmatrix}$$

$$= a_1\begin{vmatrix} b_2 & b_3 \\ c_2 & c_3 \end{vmatrix} - a_2\begin{vmatrix} b_1 & b_3 \\ c_1 & c_3 \end{vmatrix} + a_3\begin{vmatrix} b_1 & b_2 \\ c_1 & c_2 \end{vmatrix}$$

1.3.9 Cramer's rule

Let two simultaneous equations with unknowns x and y be

$$a_1x + b_1y = c_1$$

and $$a_2x + b_2y = c_2$$

then the solutions of these simultaneous equations is given by

$$x = \frac{\begin{vmatrix} c_1 & b_1 \\ c_2 & b_2 \end{vmatrix}}{\begin{vmatrix} a_1 & b_1 \\ a_2 & b_2 \end{vmatrix}} \text{ and } y = \frac{\begin{vmatrix} a_1 & c_1 \\ a_2 & c_2 \end{vmatrix}}{\begin{vmatrix} a_1 & b_1 \\ a_2 & b_2 \end{vmatrix}}$$

where $a_1b_2 \neq a_2b_1$.

1.4 VECTOR ALGEBRA

1. Let $\vec{\mathbf{i}}, \vec{\mathbf{j}}, \vec{\mathbf{k}}$ are unit vectors along x, y and z axes respectively.

Then $$\vec{\mathbf{i}}.\vec{\mathbf{i}} = \vec{\mathbf{j}}\cdot\vec{\mathbf{j}} = \vec{\mathbf{k}}\cdot\vec{\mathbf{k}} = 1$$

$$\vec{\mathbf{i}}\cdot\vec{\mathbf{j}} = \vec{\mathbf{j}}.\vec{\mathbf{k}} = \vec{\mathbf{k}}.\vec{\mathbf{i}} = 0$$

$$\vec{\mathbf{i}}\times\vec{\mathbf{j}} = \vec{\mathbf{j}}\times\vec{\mathbf{k}} = \vec{\mathbf{k}}\times\vec{\mathbf{i}} = 0$$

$$\vec{\mathbf{i}}\times\vec{\mathbf{j}} = \vec{\mathbf{k}}, \vec{\mathbf{j}}\times\vec{\mathbf{k}} = \vec{\mathbf{i}}, \vec{\mathbf{k}}\times\vec{\mathbf{i}} = \vec{\mathbf{j}}$$

2. A vector $\vec{a}$ having components (a_x, a_y, a_z) along the x, y, z axes respectively could be given by

$$\vec{a} = a_x\,\vec{i} + a_y\,\vec{j} + a_z\,\vec{k}$$

3. Let $\vec{a}, \vec{b}$ and $\vec{c}$ are an arbitrary vectors of magnitudes a, b and c respectively.

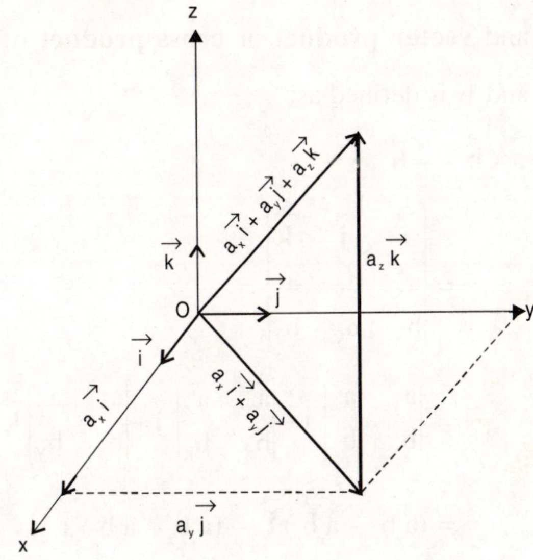

Fig. 1.3

Then $$\vec{a}\cdot\left(\vec{b}+\vec{c}\right) = \vec{a}\cdot\vec{b} + \vec{a}\cdot\vec{c}$$

$$\left(s\,\vec{a}\right)\cdot\vec{b} = s\left(\vec{a}\cdot\vec{b}\right)$$

$$= \vec{a}\cdot\left(s\,\vec{b}\right)$$

where s is a scalar.

$$\vec{a}\times\left(\vec{b}+\vec{c}\right) = \vec{a}\times\vec{b} + \vec{a}\times\vec{c}$$

$$\left(s\,\vec{a}\right)\times\vec{b} = s\left(\vec{a}\times\vec{b}\right) = \vec{b}\times\left(s\,\vec{b}\right)$$

4. Let $\vec{a} = (a_x, a_y, a_z)$, $\vec{b} = (b_x, b_y, b_z)$ and θ be the vectorial angle between $\vec{a}$ and $\vec{b}$. Then, **dot product** or **scalar product** of $\vec{a}$ and $\vec{b}$ is defined as,

$$\vec{a}\cdot\vec{b} = \vec{b}\cdot\vec{a} = a_xb_x + a_yb_y + a_zb_z$$

$$= b_xa_x + b_ya_y + b_za_z$$

$$= ab\cos\theta$$

and **vector product** or **cross product** of $\vec{a}$ and $\vec{b}$ is defined as:

$$\vec{a}\times\vec{b} = -\vec{b}\times\vec{a}$$

$$= \begin{vmatrix} \vec{\mathbf{i}} & \vec{\mathbf{j}} & \vec{\mathbf{k}} \\ a_x & a_y & a_z \\ b_x & b_y & b_z \end{vmatrix}$$

$$= \begin{vmatrix} a_y & a_z \\ b_y & b_z \end{vmatrix}\vec{\mathbf{i}} - \begin{vmatrix} a_x & a_z \\ b_x & b_z \end{vmatrix}\vec{\mathbf{j}} + \begin{vmatrix} a_x & a_y \\ b_x & b_y \end{vmatrix}\vec{\mathbf{k}}$$

$$= (a_y b_z - a_z b_y)\vec{\mathbf{i}} - (a_x b_z - a_z b_x)\vec{\mathbf{j}} + (a_x b_y - a_y b_x)\vec{\mathbf{k}}$$

$= ab \sin\theta\, \hat{n}$ where $\hat{n}$ is an unit vector perpendicular to plane of $\vec{a}$ and $\vec{b}$ and $|\vec{a}\times\vec{b}| = ab\sin\theta$.

***Note:** The constant unit vectors* $\vec{\mathbf{i}}, \vec{\mathbf{j}}, \vec{\mathbf{k}}$ *obey the circular permutation.*

i.e., $\vec{\mathbf{i}}\times\vec{\mathbf{j}} = \vec{\mathbf{k}}, \vec{\mathbf{j}}\times\vec{\mathbf{k}} = \vec{\mathbf{i}}$ and $\vec{\mathbf{k}}\times\vec{\mathbf{i}} = \vec{\mathbf{j}}$.

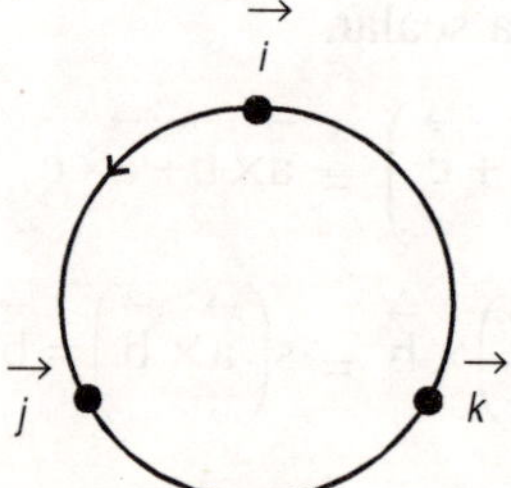

Fig. 1.4

5. $\vec{a}\cdot\left(\vec{b}\times\vec{c}\right) = \vec{b}\cdot\left(\vec{c}\times\vec{a}\right) = \vec{c}\cdot\left(\vec{a}\times\vec{b}\right)$

$$= \left[\vec{a}\,\vec{b}\,\vec{c}\right] = \left[\vec{b}\,\vec{c}\,\vec{a}\right] = \left[\vec{c}\,\vec{a}\,\vec{b}\right]$$

$$= \begin{vmatrix} a_x & a_y & a_z \\ b_x & b_y & b_z \\ c_x & c_y & c_z \end{vmatrix}$$

***Note:** Scalar triple product obeys circular permutation too.*

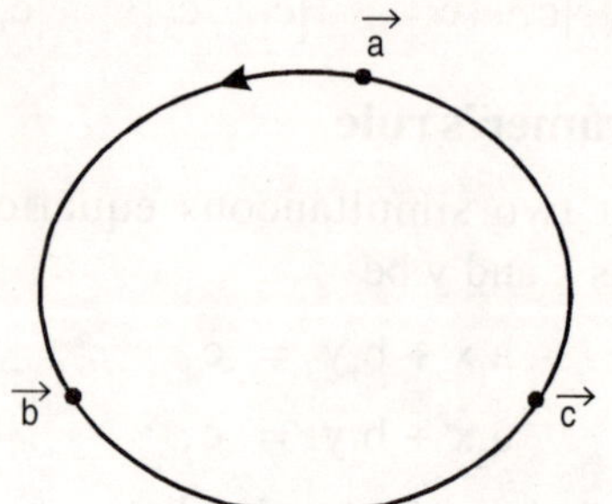

Fig. 1.5

6. $\vec{a}\times\left(\vec{b}\times\vec{c}\right) = \left(\vec{a}\cdot\vec{c}\right)\vec{b} - \left(\vec{a}\cdot\vec{b}\right)\vec{c}$

7. $\dfrac{d}{dt}\left(\vec{a}+\vec{b}\right) = \dfrac{d\vec{a}}{dt} + \dfrac{d\vec{b}}{dt}$

8. $\dfrac{d}{dt}\left(k\vec{a}\right) = k\dfrac{d\vec{a}}{dt}$

9. $\dfrac{d}{dt}\left(\vec{a}\cdot\vec{b}\right) = \vec{a}\cdot\dfrac{d\vec{b}}{dt} + \dfrac{d\vec{a}}{dt}\cdot\vec{b}$

10. $\dfrac{d}{dt}\left(\vec{a}\times\vec{b}\right) = \dfrac{d\vec{a}}{dt}\times\vec{b} + \vec{a}\times\dfrac{d\vec{b}}{dt}$

(maintain the order)

1.5 GEOMETRY

1. Circular area of radius, r = πr^2.
2. Circumference of circle of radius, r = $2\pi r$.
3. Surface area of sphere of radius r, A = $4\pi r^2$.
4. Volume of sphere of radius r, V = $\dfrac{4}{3}\pi r^3$.
5. Area of triangle of base a and altitude h, A = $\dfrac{1}{2}ah$.
6. Volume of right circular cone, V = $\dfrac{1}{3}\pi r^2 l$, where r is base radius and l is slant height of cone.
7. Area of right circular cylinder of radius r and height h, A = $2\pi r^2 + 2\pi rh$ and V = $r\pi^2 h$.

1.6 CARTESIAN COORDINATE GEOMETRY

1. Distance between two 2D points

Let A (x_1, y_1) and B (x_2, y_2) are two points then length between the points is denoted by $\overline{AB}$ and is given by

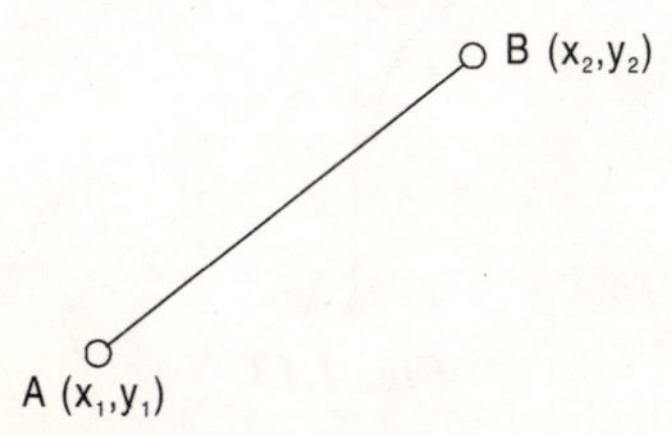

Fig. 1.6

$$\overline{AB} = \sqrt{\Delta x^2 + \Delta y^2}$$

where $\Delta x = x_1 \sim x_2$ and $\Delta y = y_1 \sim y_2$.

Similarly, distance between two 3D points P (x_1, y_1, z_1) and Q (x_2, y_2, z_2), is given by

$$\overline{PQ} = \sqrt{(x_1 \sim x_2)^2 + (y_1 \sim y_2)^2 + (z_1 \sim z_2)^2}$$

$$= \sqrt{\Delta x^2 + \Delta y^2 + \Delta z^2}$$

where $\Delta x = x_1 \sim x_2$, $\Delta y = y_1 \sim y_2$ and $\Delta z = z_1 \sim z_2$

2. Straight line

General equation of straight line is given by $ax + by + c = 0$.

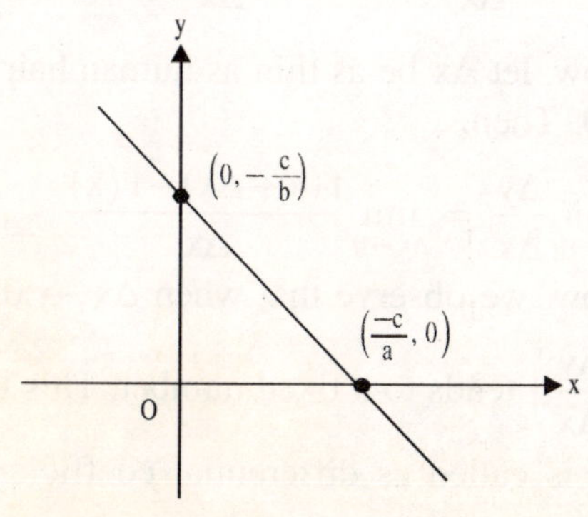

Fig. 1.7

3. Equation of x-axis $y = 0$.

4. Equation of y-axis $x = 0$.

5. Equation of line parallel to x-axis $y = k$, where k is an arbitrary constant.

6. Equation of line parallel to y-axis $x = k$, where k is an arbitrary constant.

7. Equation of straight line that bisects 1st and 3rd quadrant is $y = x$ and that of 2nd and 4th is $y = -x$.

8. Equation of straight line passing through origin is $y = mx$, where m is slope and $m = \tan \theta$.

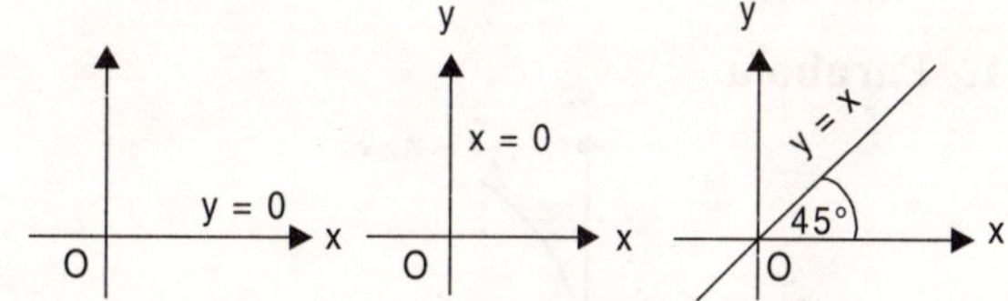

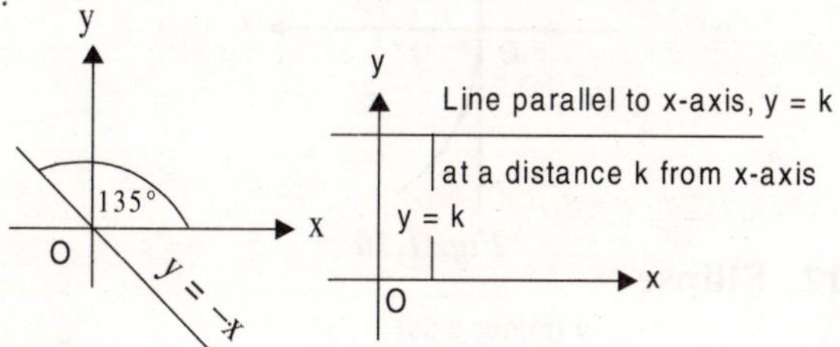

Fig. 1.8a

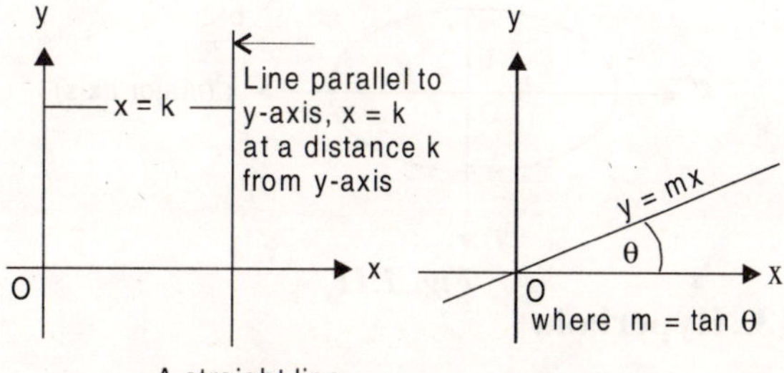

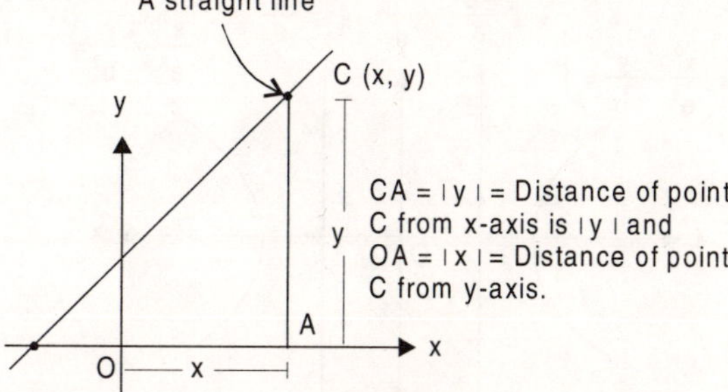

Fig. 1.8b

9. The distance of a point from x-axis is $|y|$ and from y-axis is $|x|$.

All the equations shown in Fig. (1.8a and 1.8b).

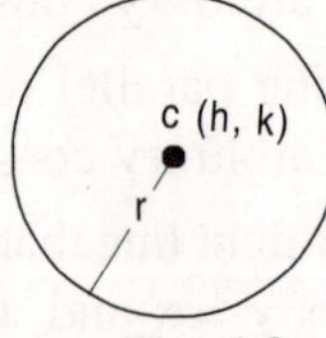

Fig. 1.9

10. Circle

Equation of circle having radius r and centre (h, k) is given by

$$(x - h)^2 + (y - k)^2 = r^2.$$

11. Parabola

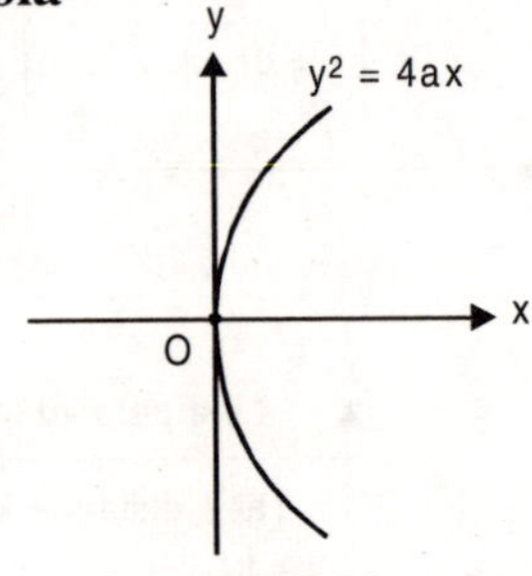

Fig. 1.10

12. Ellipse

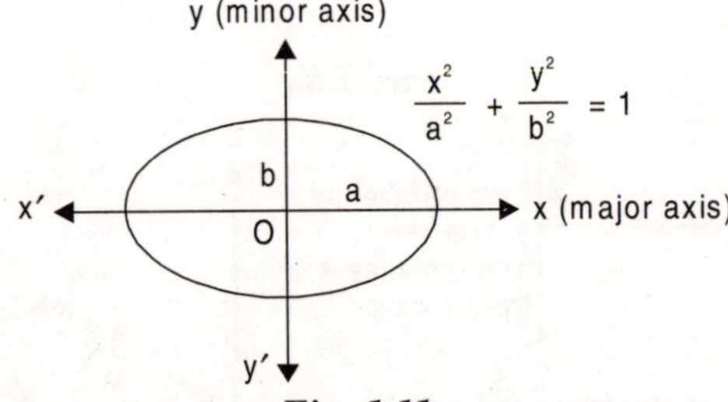

Fig. 1.11

13. Hyperbola

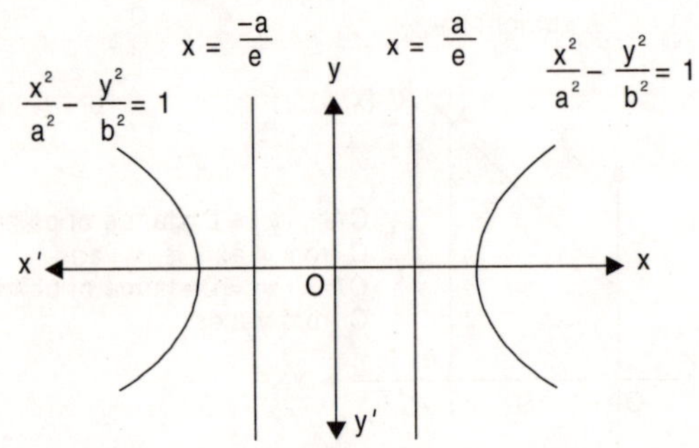

Fig. 1.12

14. Rectangular hyperbola

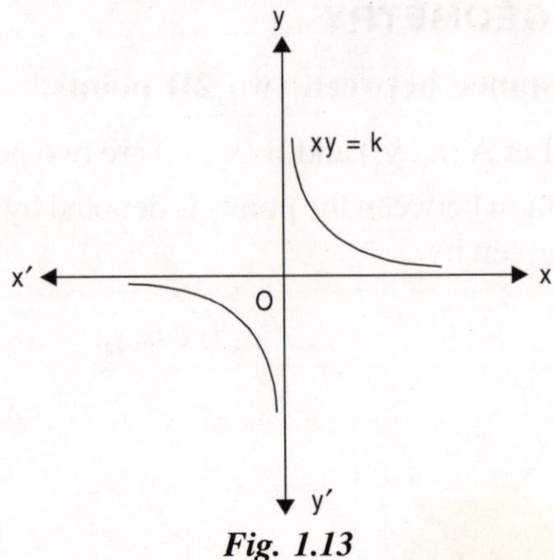

Fig. 1.13

1.7 DIFFERENTIAL COEFFICIENT AND ITS APPLICATIONS IN MECHANICS

Let y = f (x) represents a cartesian curve C as shown in Fig. (1.14). Now, increase x by Δx at this value of x, then change in y or f (x) could be Δy or $f(x + \Delta x) - f(x)$.

$$y = f(x) \qquad \text{....(i)}$$

$$y + \Delta y = f(x + \Delta x) \qquad \text{....(ii)}$$

Subtracting (i) from (ii), we get

$$\Delta y = f(x + \Delta x) - f(x) \qquad \text{....(iii)}$$

Dividing (iii) by Δx, we get

$$\frac{\Delta y}{\Delta x} = \frac{f(x+\Delta x)-f(x)}{\Delta x} \qquad \text{....(iv)}$$

Now, let Δx be as thin as human hair i.e., $\Delta x \to 0$. Then,

$$\lim_{\Delta x \to 0} \frac{\Delta y}{\Delta x} = \lim_{\Delta x \to 0} \frac{f(x+\Delta x)-f(x)}{\Delta x} \qquad \text{....(v)}$$

Now, we observe that when $\Delta x \to 0$, the ratio $\left(\frac{\Delta y}{\Delta x}\right)$ tends to a fixed number. This fixed number is called as differential coefficient of f(x) at x and is denoted by $\frac{dy}{dx}$.

Note: $\frac{dy}{dx}$ *is not ratio of dy and dx that means it is not dy ÷ dx, it is just a notation.*

Thus, $$\frac{dy}{dx} = \lim_{\Delta x \to 0} \frac{\Delta y}{\Delta x}$$

$$= \lim_{\Delta x \to 0} \frac{f(x+\Delta x)-f(x)}{\Delta x} \quad ....(1.1)$$

Let $x + \Delta x = a$ then $\Delta x = a - x$

when $\Delta x \to 0$

then $x \to a$

Thus an another form of $\frac{dy}{dx}$ is

$$\lim_{x \to a} \frac{f(a)-f(x)}{a-x} = \lim_{x \to a} \frac{f(x)-f(a)}{x-a} \quad ....(1.2)$$

is differential coefficient of f(x) at x = a.

1.8 USES OF DIFFERENTIAL COEFFICIENT

1. Although, $\frac{dy}{dx} \neq \frac{\Delta y}{\Delta x}$ but $\lim_{\Delta x \to 0} \frac{\Delta y}{\Delta x}$. But for a small change Δx we could get approximation.

i.e., $$\frac{\Delta y}{\Delta x} \cong \frac{dy}{dx}$$

or $$\Delta y \cong \frac{dy}{dx} \Delta x$$

Then, when we write,

$$\frac{dy}{dx} = \sin x$$

or $$dy = \sin x \, dx$$

or $$\int dy = \int \sin x \, dx$$

or $$y = \int \sin x \, dx$$

Actually we mean that,

$$\frac{dy}{dx} = \frac{\Delta y}{\Delta x} \quad \text{because,}$$

$$\frac{dy}{dx} \neq dy \div dx$$

i.e., $\frac{\Delta y}{\Delta x} = \sin x$, where Δx is infinitely small number.

or $$\Delta y = \sin x \, \Delta x$$

or $$\int \Delta y = \int \sin x \, \Delta x$$

But as we know that, we don't see the things the way they are that means the thing that we are doing is not correct but it is a good approximation so in the mathematical operations $\Delta y = \int \sin x \, \Delta x$ is equal to $dy = \int \sin x \, dx$.

2. Let y = f (x) represents a curve C shown in Fig. (1.14). Take a point P (x, y) on the curve and a point close to point P (x, y) as Q (x + Δx, y + Δy).

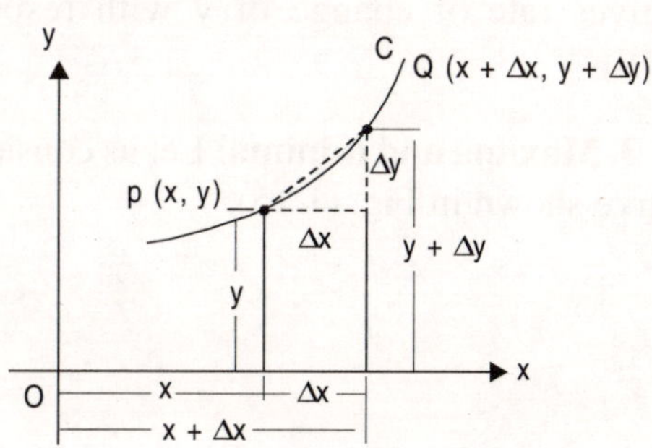

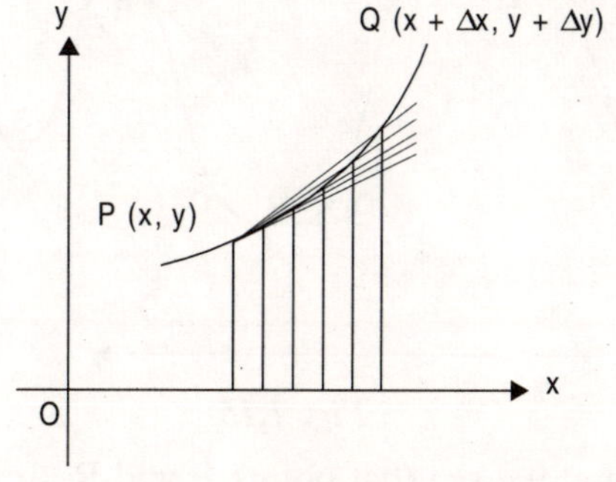

Fig. 1.14

Now, in Δ PQR, we could get

$$\tan \theta = \frac{\Delta y}{\Delta x}$$

We observe that when $\Delta x \to 0$ then $\Delta y \to 0$ and the ratio $\frac{\Delta y}{\Delta x}$ becomes tangent of the angle made by line which is tangent at point P with x-axis.

i.e., $\tan \theta_t = \lim_{\Delta x \to 0} \frac{\Delta y}{\Delta x} = \frac{dy}{dx}$, where θ_t is angle made by line of tangent at the point P on the curve.

Therefore, $\frac{dy}{dx} = \tan \theta_t$ = slope of tangent (m) at point P. ....(1.3)

Thus, geometrically derivative at a particular point of the curve represents slope of tangent at that point and mathematically, it also gives rate of change of y with respect to x.

3. Maxima and minima: Let us consider the curve shown in Fig. (1.15).

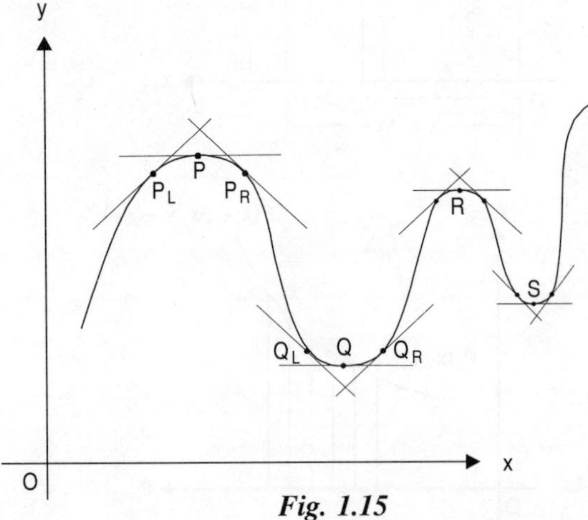

Fig. 1.15

Here, we see that points P and R are two local maxima points for points P_L and P_R points. Q and S are two local minima points for the points Q_L and Q_R. We see that tangents at maxima or minima points are parallel to x-axis i.e., slope of tangents are zero.

$$\text{Thus,} \quad \left.\frac{dy}{dx}\right|_{P,\,Q,\,R,\,S} = 0 \qquad \text{....(1.4)}$$

Therefore, analytically slope at maxima or minima that is $\frac{dy}{dx}$ is equal to zero for any curve (cartesian) and geometrically tangent is parallel to x-axis for curve of the form $y = f(x)$.

Also, we observe that, slopes in neighbourhood of maxima changes positive to negative (from left to right). Thus, rate of change of slopes at maximum points are negative (final – initial < 0).

$$\text{Thus,} \quad \frac{d}{dx}\left(\frac{dy}{dx}\right) = \frac{d^2y}{dx^2} < 0 \text{ at maxima} \qquad \text{....(1.5)}$$

And we also observe that sign of slopes in neighbourhood of minima changes from negative to positive.

Thus, rate of change of slope at minima > 0

$$\text{i.e.,} \quad \frac{d}{dx}\left(\frac{dy}{dx}\right) > 0$$

$$\text{or} \quad \frac{d^2y}{dx^2} > 0$$

$$\text{Therefore,} \frac{d^2y}{dx^2} > 0 \text{ at minima} \qquad \text{....(1.6)}$$

1.9 ROLE OF CALCULUS IN MECHANICS

To realize the importance of calculus, let us take one example like a thin rod of length l has linear charge density $\rho(x) = ax + b$; coulomb/meter, $0 \le x \le l$. Determine the net force applied by rod on a point change having charge 'q' at a distance d from its right end as shown in Fig. (1.16). The rod is placed in vacuum.

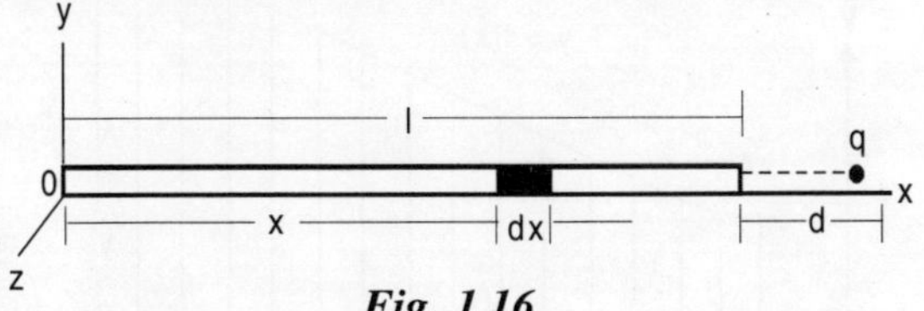

Fig. 1.16

Solution:

For two point charges q_1 and q_2, separated by a distance d then force between them is given by the formula.

$$F_{q_1q_2} = \frac{Kq_1q_2}{d^2}$$

But in this case, however, charge is distributed over the rod whose dimension could not be ignored in comparison with the characteristic distance d considered in the problem. Hence, charge over rod cannot be considered to be a point charge.

Now linear charge density at point which is left in most of the rod is given by,

$\rho(0) = b$ coul/mt. and at right most end

$\rho(l) = (al + b)$ coul/mt.

Let us suppose an intermediate position x, where $\rho(x) = ax + b;\ 0 \le x \le l$.

On increasing this position by a very small element dx over this range, considering dx is very small we can ignore change in linear density.

Since dx is very small (it may be of 10^{-50} order), so change over dx can be treated as a point and change in magnitude is given by,

$dq = (ax + b)dx$ coul

Now, distance between dq and q is,

$d' = l - x + d$ mt.

Therefore, force between dq and q is,

$$dF = \frac{K(dq)q}{(d^1)^2}$$

or $$dF = \frac{K(ax+b)q}{(1-x+d)^2}dx$$

Now, using integration technique we can add dF over whole rod i.e., force applied by the rod is,

$$F = \int_0^F dF = \int_0^l Kq \frac{ax+b}{(1-x+d)^2}dx$$

$$= Kq\int_0^l \frac{ax+b}{(1-x+d)^2}dx$$

where, $ax + b = c_1\frac{d}{dx}(l-x+d)^2 + c_2$ for some constants c_1 and c_2.

$$= Kq\left[\ln(l-l+d)^2 - \ln(l-0+d)^2\right]$$

$$= Kq\left[\ln d^2 - \ln(l+d)^2\right]$$

$$= Kq\ \ln\left(\frac{d}{l+d}\right)^2$$

Therefore, $F = 2Kq\ \ln\left(\frac{1}{l+\frac{1}{d}}\right)$N,

considering above assumptions.

So far we have seen that use of differentiation and integration in "Transition from elementary system to such complicated system is always facilitated by the use of them."

For defining work, calculation of centre of mass, centroid, moment of inertia, etc., we use calculus. Hence, first we will see basic functions, its differentiation and then integration.

1.10 INTEGRAL CALCULUS AND ITS USES IN MECHANICS

Let AB be a curve representing the relation between two quantities x and y (Fig. 1.17). The point A corresponds to x = a and B corresponds to x = b. Draw perpendiculars AM and BN on the x-axis respectively. We are interested in finding the area AMNB. Let us denote the value of y at x by the symbol y = f (x).

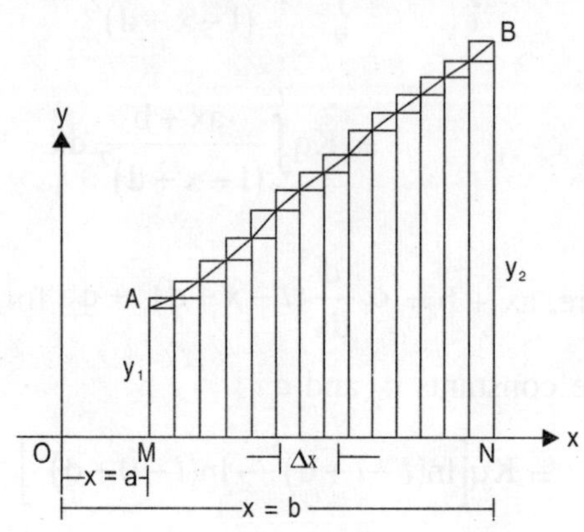

Fig. 1.17

Let, $y_1 = f(a) = \overline{AM}$ and $y_2 = f(b) = \overline{BN}$. Now let us divide the length MN in N equal elements each of length $\Delta x = \dfrac{ON - OM}{N}$ $= \dfrac{b-a}{N}$. From the ends of each small length, we draw lines parallel to y-axis. This constructs the rectangular bars shown below.

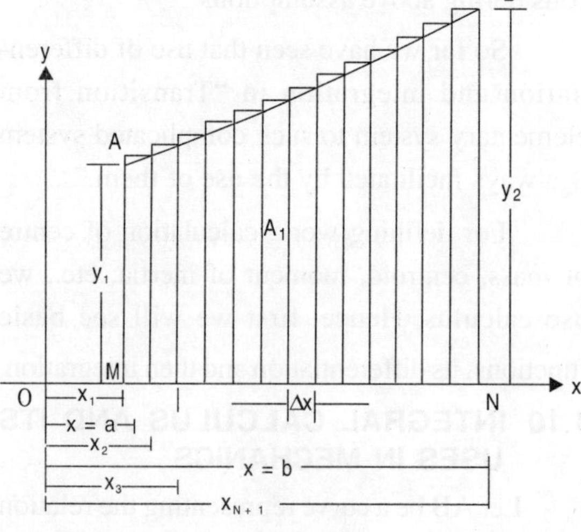

Fig. 1.17(a)

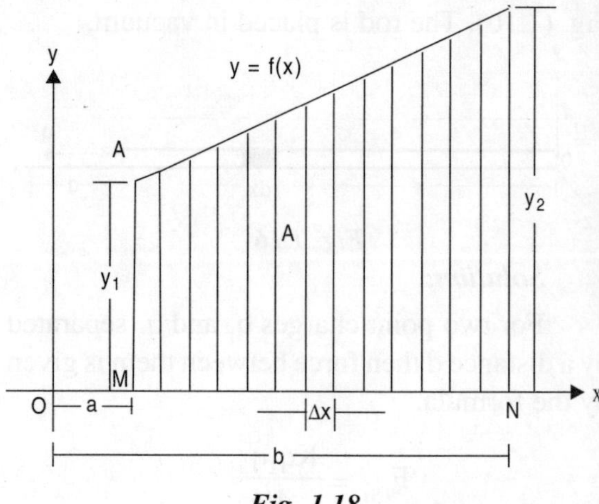

Fig. 1.18

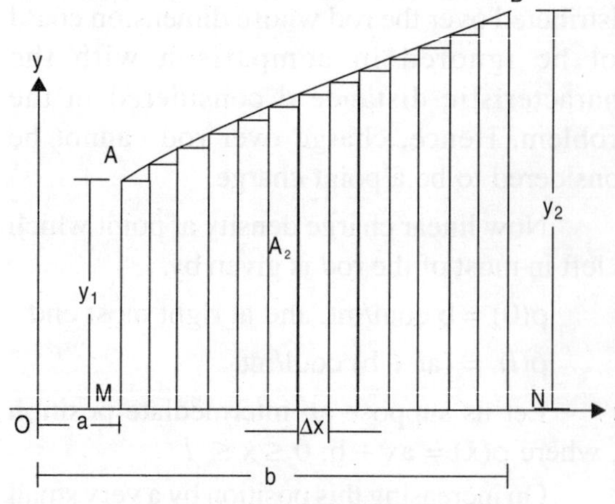

Fig. 1.19

From the above figures we observe that,

$$A_2 < A < A_1$$

$f(a)\,\Delta x + f(a + \Delta x)\,\Delta x + f(a + 2\Delta x)\,\Delta x + ... + f[a + (N-1)\,\Delta x]\,\Delta x < A < f(a + \Delta x)\,\Delta x + f(a + 2\Delta x)\,\Delta x + f(a + 3\Delta x)\,\Delta x + ... + f(a + N\Delta x)\,\Delta x$

This may be written as,

$$\sum_{i=1}^{N} f(x_i)\Delta x < A < \sum_{i=2}^{N+1} f(x_i)\Delta x$$

where x_i takes the values a, a + Δx, a + 2Δx ..... b – Δx and b this inequality could be neglected when the number of intervals (N) increases towards infinity i.e., $N \to \infty$. In this case, vertices of the bars touch the curve AB at more points and the total area of small triangles decreases. As, $N \to \infty$ i.e., $\Delta x \to 0$,

$\because \Delta x = \dfrac{b-a}{\infty} \to 0$, the vertices of the bars touch the curve at infinite number of points and total area of triangle tends to zero. Thus we could write,

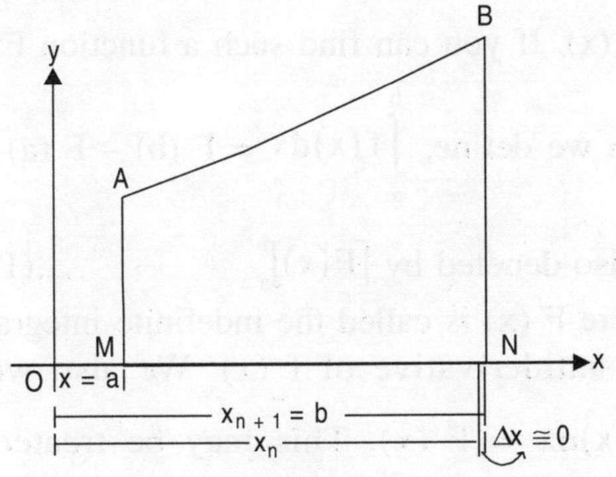

Fig. 1.20

$$A = \lim_{\Delta x \to 0} \sum_{i=1}^{\infty} f(x_i)\Delta x = A_1 = A_2 \quad ....(1.7)$$

when $\Delta x \to \infty$ then, n or n + 1 both tend to ∞ in mathematics, equation (1.7) is denoted by,

$$A = \int_a^b f(x)\,dx = \lim_{\Delta x \to 0} \sum_{i=1}^{\infty} f(x_i)\Delta x \quad ....(1.8)$$

and is read as the integral of f (x) = y, with respect to x within the limits x = a to x = b. Here a is called the lower limit and b upper limit of integration. The integral is the sum of a large number of terms of the type f (x) dx with x continuously varying from a to b and the number of terms tending to infinity. Thus, area under curve is the definite integral of the function from lower limit to upper limit.

Let us use the above method to find the area of the triangle. Let us suppose the line PQ is represented by the equation y = mx, points O and A are on the x-axis representing x = 0 and x = a. We have to find the area of the triangle OAQ.

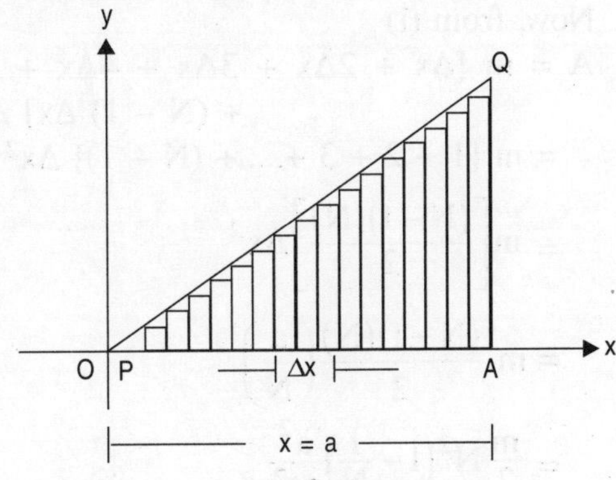

Fig. 1.21

Let us divide the length OA in N-equal intervals, the length of each interval is $\Delta x = \dfrac{a-0}{N} = \dfrac{a}{N}$. The height of the first shaded bar is zero corresponding to x = 0 and of the second it is y = mΔx and that of third bar is y = m 2Δx and so on. The height of the Nth bar is y = m (N – 1) Δx. The width of each bar is Δx. So the total area of the bars is,

$$A = 0 \times \Delta x + m(\Delta x)\Delta x + m(2\Delta x)\Delta x + ...+ m\{(N-1)\Delta x)\}\Delta x$$

$$= [m + 2m + 3m + ... + (N-1)m]\Delta x^2$$

$$= m[\Delta x + 2\Delta x + 3\Delta x + ... + (N-1)\Delta x]\Delta x$$

$$= m\left(\sum_{i=0}^{N-1} x_i \Delta x\right) \quad ....(i)$$

where, $\Delta x = \dfrac{a}{N}$ and x_i = 0, Δx, 2Δx, 3Δx, ...

As $\Delta x \to 0$, the total area of the bar becomes the area of the shaded part O(P)AQ.

Thus the required area under curve is,

$$A = m \lim_{\Delta x \to 0} \sum_{i=0}^{N-1} x_i \, \Delta x$$

$$= m \int_0^a x \, dx \qquad \text{....(ii)}$$

Now, from (i)

$$A = m\,[\Delta x + 2\Delta x + 3\Delta x + 4\Delta x + ... + (N-1)\,\Delta x]\,\Delta x$$

$$= m\,[1 + 2 + 3 + ... + (N-1)]\,\Delta x^2$$

$$= m\left[\frac{(N-1)(N)}{2}\right]\Delta x^2$$

$$= m\,\frac{(N-1)(N)}{2}\left(\frac{a}{N}\right)^2$$

$$= \frac{m}{2}N^2\left(1-\frac{1}{N}\right)\frac{a^2}{N^2}$$

$$= \frac{m}{2}\left(1-\frac{1}{N}\right)a^2$$

$$= \frac{m}{2}\,(1-0)\,a^2 \text{ as } N \to \infty \text{ then } \frac{1}{N} \to 0$$

$$= \frac{1}{2}\,ma^2 = \frac{1}{2}\,a \times ma$$

$$= \frac{1}{2} OA \times AQ$$

$$= \frac{1}{2} \times \text{base} \times \text{height}$$

= area of triangle

$$\text{From (ii) } A = m\int_0^a x\,dx = \left[m\frac{1}{2}x^2\right]_0^a$$

$$= \frac{m}{2}a^2 = \frac{1}{2} \times a \times ma$$

= area of triangle

Thus we see that,

$$\lim_{\Delta x \to 0} \sum_{i=1}^{\infty} f(x_i)\Delta x = \int_a^b f(x)\,dx \qquad \text{....(1.9)}$$

= area of region bounded by curve y = f (x), x-axis and ordinates at x = a and x = b.

This is also known as geometrical meaning of definite integral. In mathematics special method has been developed to find the integration of various functions of x. A very useful method is as follows. Suppose we wish to find $\int_a^b f(x)\,dx = \lim_{\Delta x \to 0} \sum_{i=1}^{N \to \infty} f(x_i)\Delta x$ where

$\Delta x = \dfrac{b-a}{N}$ and x_i = a, a + Δx, a + 2Δx,....., b – Δx and b. Now look at function F (x) such that the derivative of F (x) is f (x) i.e., $\dfrac{dF(x)}{dx}$ = f (x). If you can find such a function F (x), then we define, $\int_a^b f(x)\,dx$ = F (b) – F (a) and is also denoted by $[F(x)]_a^b$ ....(1.10)

where F (x) is called the indefinite integral or the antiderivative of f (x). We also write, $\int f(x)\,dx$ = F (x). This may be treated as another way of writing $\dfrac{dF(x)}{dx}$ = f (x), i.e., anti-integration is differentiation. From this definition we get idea about differentiation.

$$\text{For example, } \frac{d}{dx}\left(\frac{1}{2}x^2\right)$$

$$= \frac{1}{2}\frac{d}{dx}\left(x^2\right) = \frac{1}{2} \times 2x = x$$

$$\text{Thus, } m\int_0^a x\,dx = m\left[\frac{1}{2}x^2\right]_0^a$$

[∵ f (x) = x]

$$= \frac{1}{2}\,ma^2$$

$$= \frac{1}{2} \times a \times ma$$

$$= \frac{1}{2} \times \text{base} \times \text{height.}$$

Table (1.2) shows the list of some important integration formulae.

Table 1.2: Integration formulae

$f(x)$	$F(x) = \int f(x)dx$	$f(x)$	$F(x) = \int f(x)dx$
sin x	– cos x	$x^n\ (n \neq -1)$	$\frac{x^{n+1}}{n+1}$
cos x	sin x	$\frac{1}{x}$	$ln x$
$\sec^2 x$	tan x	$\frac{1}{x^2+a^2}$	$\frac{1}{a}\tan^{-1}\frac{x}{a}$
$\text{cosec}^2 x$	– cot x	$\frac{1}{\sqrt{a^2-x^2}}$	$\sin^{-1}\frac{x}{a}$
sec x tan x	sec x	$\sqrt{x}$	$\frac{2}{3}x^{\frac{3}{2}}$
cosec x cot x	– cosec x	c	cx

Some important rules for integration are as follows:

1. $\int cf(x)dx = c\int f(x)dx$

2. Let $\int f(x)dx = F(x)$

then $\int f(cx)dx = \frac{1}{c}F(cx)$

3. $\int[f(x) \pm g(x)]dx = \int f(x)dx \pm \int g(x)dx$

Example 1

$$\int_2^3 (\sin x + \cos x + x^3)dx$$

$$= \left[-\cos x + \sin x + \frac{x^4}{4}\right]_2^3$$

$$= [\cos 2 - \cos 3] + [\sin 3 - \sin 2] + ¼[3^4 - 2^4]$$

$$= 16.26$$

Some important uses:

Note: *As we denote,* $\lim_{\Delta x \to 0} \sum_{i=1}^{N} f(x_i)\Delta x$

$$= \int_a^b f(x)dx = [F(b) - F(a)] = [F(x)]_a^b$$

where, $\Delta x = \frac{b-a}{N}$ and $\frac{dF(x)}{dx} = f(x)$.

Then from the above definition we observe that instead of $\Delta x \to 0$, we write as dx and $\sum_{i=1}^{N=\infty}$ is written as $\int_a^b$.

4. Finding the centroid of area under the curve y = f (x), between the ordinates at x = a and x = b and x-axis.

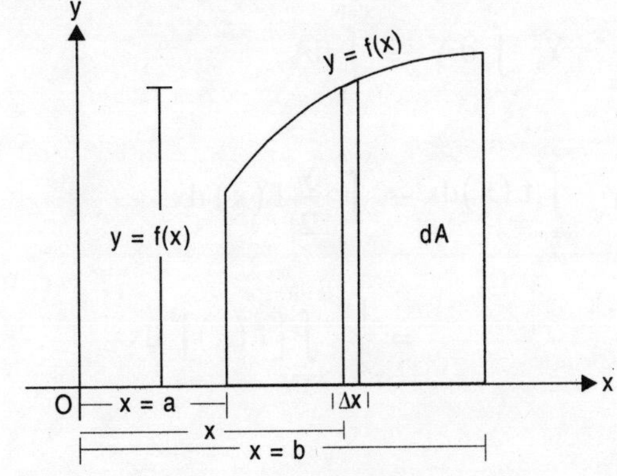

Fig. 1.22

Let us consider a rectangular bar of width dx and height y = f (x).

Therefore, dA = height × width

= y dx

= f (x) dx

As width of bar is dx that is, a very small number that means it could be assumed zero. So, x-cordinate of bar could be strongly approximated by $x_c = x$ then from definition of centroid,

$$X_c \int_{x=a}^{x=b} dA = \int_{x=a}^{x=b} x_c \, dA$$, where X_c is centroid of total area,

$$\text{or } X_c \int_{x=a}^{x=b} f(x)dx = \int_{x=a}^{x=b} x f(x)dx$$

$$\therefore \quad X_c = \frac{\int_{x=a}^{x=b} x f(x)dx}{\int_{x=a}^{x=b} f(x)dx} \quad ....(1.11)$$

As height of bar is y, therefore, y-coordinate of centroid of bar is given by $y_c = \frac{y}{2}$. Thus from definition of centroid, its y-coordinate is,

$$Y_c \int_{x=a}^{x=b} dA = \int_{x=a}^{x=b} dA$$

$$\text{or } Y_c \int_{x=a}^{x=b} f(x)dx = \int_{x=a}^{x=b} \frac{y}{2} f(x)\,dx$$

$$= \frac{1}{2} \int_{x=a}^{x=b} [f(x)]^2 dx$$

$$\text{Therefore, } Y_c = \frac{\frac{1}{2} \int_{x=a}^{x=b} [f(x)]^2 dx}{\int_{x=a}^{x=b} f(x)dx}$$

5. If y = L (x) denotes the intensity of the loading, then we could find the position of net load.

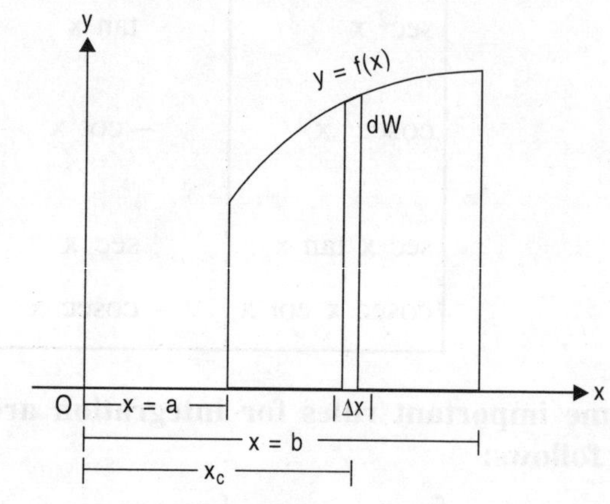

Fig. 1.23

Since load per unit length (intensity) is L(x), therefore for dx net load,

dw = intensity × width length

= L(x) dx

= L(x) dx

Therefore, net load over the interval x = a to x = b.

$$w = \int_w dw$$

$$= \int_{x=a}^{x=b} L(x)dx$$

Let x_c be the position of the load dw, therefore, moment of dw is $x_c$dw. Now, sum of moment is $\int_{x=a}^{x=b} x_c\,dw$ noting that all the moments are in anticlockwise direction. Therefore, we can add them.

Let X_c be the position from where net load passes then we know that,

$$\text{Then, } X_c \int_{x=a}^{x=b} L(x)dx = \int_{x=a}^{x=b} x_c L(x)dx \quad ...(1.12)$$

So, by this way we could find position of net loading.

6. We could find moment of inertia by integration.

If y = f (x) is a curve (C) in cartesian form, we want to find moment of inertia of the area under the curve (C) and x-axis and between ordinates at x = a and x = b.

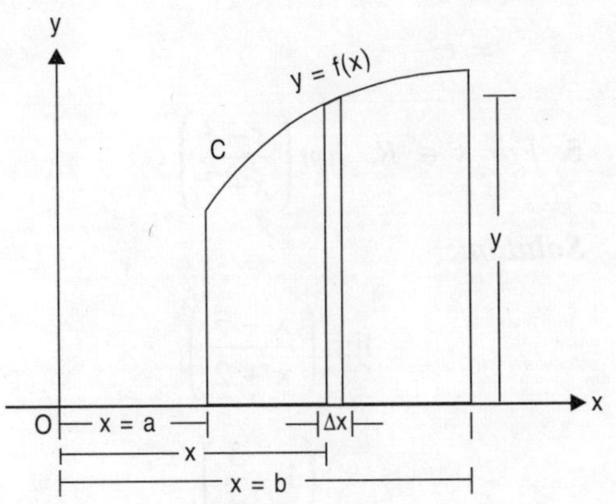

Fig. 1.24

Let width of rectangular stript be dx and height be y = f (x).

If dA is the area of this strip then,

$$\begin{aligned} dA &= \text{width} \times \text{height} \\ &= ydx \\ &= ydx \end{aligned}$$

Now, let I_x = the moment of inertia of area dA about x-axis.

$$\text{Then} \quad I_x = \int_c d^2 (dA)$$

where, d is the distance of the area dA from x-axis. Since, dA is a rectangular bar of height y and with dx, so from definition, centroid the area, dA could be assumed to be concentrated at its centroid. Since, we know that geometrically centroid of the rectangle lies at its geometrical centre, i.e.,

$$C\left(\frac{\text{width}}{2} = x_c, \frac{\text{height}}{2} = y_c\right).$$

Since, y_c represents total area concentrated at a distance $y_c = \frac{\text{height}}{2}$ from x-axis.

$$\text{Therefore, } I_x = \int_c y_c^2 dA$$

$$= \int_c \left(\frac{y}{2}\right)^2 ydx$$

$$= \frac{1}{2}\int_{x=a}^{x=b} y^3 dx = \frac{1}{2}\int_{x=a}^{x=b} [f(x)]^3 dx \quad ....(1.13)$$

$$\text{Similarly, } I_y = \int_c x_c^2\, dA$$

$$= \int_{x=a}^{x=b} x^2 ydx, \text{ where } x_c = x$$

Since dx is almost as thin as human hair, so we can approximate x + dx = x = x_c

$$\text{Therefore, } I_y = \int_{x=a}^{x=b} x^2 f(x)dx \quad ....(1.14)$$

So, for finding moment of inertia we can directly use integrations **1.13** and **1.14.**

WORKED OUT EXAMPLES

1. *Find the value of the parameter a, for which the function f (x) = 1 + ax, a ≠ 0 is the inverse of itself.*

Solution: Let y = f(x)

$\Rightarrow \quad y = 1 + ax$

$\Rightarrow \quad x = \dfrac{y-1}{a}$

$\Rightarrow \quad f^{-1}(x) = \dfrac{x-1}{a}$

Now, $f(x) = f^{-1}(x)$

$\Rightarrow \quad 1 + ax = \dfrac{x-1}{a}$

$\Rightarrow \quad 1 + ax = \dfrac{1}{a}x - \dfrac{1}{a}$

$\Rightarrow \quad a = -1.$

2. *Determine the domain of definition of the function y(x) given by the equation $2^x + 2^y = 2$.*

Solution: As we know that, $2^y > 0, \forall y \in R$

Thus, $2^y = 2 - 2^x < 2, \forall x \in R.$

As 2^x is increasing function, therefore domain of definition is $-\infty < x < 1$.

3. *Find the value of* $\lim_{x\to 0} \dfrac{\int_0^{x^2} \cos t^2\, dt}{x \sin x}$.

Solution: $\lim_{x\to 0} \dfrac{\int_0^{x^2} \cos t^2\, dt}{x \sin x} \qquad \left[\dfrac{0}{0}\text{ form}\right]$

$= \lim_{x\to 0} \dfrac{2x \cos x^4}{\cos x + \sin x}$

[Using L'Hospital's Rule]

$= \lim_{x\to 0} \dfrac{2\cos x^4 - 8x^4 \sin x^4}{2\cos x - x \sin x}$

[Using Hospital's Rule again]

$= \dfrac{2-0}{2-0} = 1$

4. *Find the value of* $\lim_{x\to\infty} \left(\dfrac{x+6}{x+1}\right)^{x+4}$.

Solution:

$$\lim_{x\to\infty} \left(\frac{x+6}{x+1}\right)^{x+4}$$

$$= \lim_{x\to\infty} \left[\left(1 + \frac{1}{\frac{x+1}{5}}\right)^{\frac{x+1}{5}}\right]^{\frac{5(x+4)}{x+1}}$$

$$= e^{\lim_{x\to\infty} \frac{5(x+4)}{(x+1)}}$$

$$= e^{\lim_{x\to\infty} \frac{5\left(1+\frac{4}{x}\right)}{\left(1+\frac{1}{x}\right)}}$$

$$= e^5.$$

5. *For x ∈ R,* $\lim_{x\to y} \left(\dfrac{x-3}{x+2}\right)^x$.

Solution:

$$\lim_{x\to\infty} \left(\frac{x-3}{x+2}\right)^x$$

$$= \lim_{x\to\infty} \frac{\left(1-\frac{3}{x}\right)^x}{\left(1+\frac{2}{x}\right)^x}$$

$$= \frac{\lim_{x\to\infty} \left[\left(1-\frac{3}{x}\right)\right]^{-x/3}}{\lim_{x\to\infty} \left[\left(1+\frac{2}{x}\right)\right]^{x/2}}$$

$$= \frac{e^{-3}}{e^2}$$

$= e^{-5}$.

6. *If* $f(x) = \begin{cases} \dfrac{1-\cos 10x}{x^2}, & x<0 \\ a, & x=0 \\ \dfrac{\sqrt{x}}{\sqrt{625+\sqrt{x}}-25}, & x>0 \end{cases}$

then find the value of a so that f (x) is continuous at x = 0.

Solution: We have, RHL at x = 0

$$= \lim_{h\to 0} f(0+h)$$

$$= \lim_{h\to 0} \frac{\sqrt{h}}{\sqrt{625+\sqrt{h}}-25}$$

$$= \lim_{h\to 0} \frac{\sqrt{h}\left[\sqrt{625+\sqrt{h}}+25\right]}{\sqrt{625+\sqrt{h}-625}}$$

$$\lim_{h\to 0}\left[\sqrt{625+\sqrt{h}}+25\right]$$

$= 25 + 25$

$= 50.$

LHL at x = 0

$$= \lim_{h\to 0} f(0-h)$$

$$= \lim_{h\to 0} \frac{1-\cos 10h}{h^2}$$

$$= \lim_{h\to 0} \frac{2\sin^2 5h}{h^2}$$

$$= 2 \lim_{h\to 0}\left(\frac{\sin 5h}{5h}\right)^2 \times 25$$

$= 50$

and f (0) = a.

Since f(x) is continuous at x = 0

$$\therefore \quad \lim_{h\to 0} f(0+h) \lim_{h\to 0} f(0-h) = f(0)$$

$\Rightarrow$ a = 50.

7. *Let g (x) = x f (x), where*

$f(x) = \begin{cases} x \sin\dfrac{1}{x}, & x\neq 0 \\ 0, & x=0 \end{cases}$, *then find g′ (x)*

Solution :

It is given that

$$g(x) = \begin{cases} x^2 \sin\dfrac{1}{x}, & x\neq 0 \\ 0, & x=0 \end{cases}.$$

Now for $x \neq 0$

$$g'(x) = x^2\left(-\frac{1}{x^2}\right)\cos\frac{1}{x} + 2x\sin\frac{1}{x}$$

$$= -\cos\frac{1}{x} + 2x\sin\frac{1}{x}. \text{ at } \quad x = 0$$

$$g'(0) = \lim_{x\to 0}\frac{g(x)-g(0)}{x-0}$$

$$= \lim_{x\to 0}\frac{x^2\sin\frac{1}{x} - 0}{x}$$

$$= \lim_{x\to 0} x\sin\frac{1}{x}$$

$= 0$

$$\therefore \quad g'(x) = \begin{cases} 2x\sin\dfrac{1}{x} - \cos\dfrac{1}{x}, & x\neq 0 \\ 0, & x=0 \end{cases}$$

$g'(x)$ is not continuous at x = 0 as $\cos\dfrac{1}{x}$ is not continous at x = 0. Also, it is not differentiable at x = 0.

8. *Let* $y^2 = p(x)$ *then find* $\dfrac{d}{dx}\left(2y^3\dfrac{d^2y}{dx^2}\right)$.

Solution:

We have, $y^2 = P(x)$ ....(1)

$$\Rightarrow \quad 2y\frac{dy}{dx} = P'(x) \qquad ....(2)$$

$$\Rightarrow \quad 2\frac{dy}{dx}\cdot\frac{dy}{dx} + 2y.\frac{d^2y}{dx^2} = P''(x)$$

$$\Rightarrow \quad 2y^2\left(\frac{dy}{dx}\right)^2 + 2y^3 . \frac{d^2y}{dx^2} = y^2P''(x)$$

$$\Rightarrow \quad 2y^3 \frac{d^2y}{dx^2} = y^2 P''(x) - 2y^2\left(\frac{dy}{dx}\right)^2$$

$$= y^2 P''(x) - \frac{1}{2}[P'(x)]^2 \text{ [from (2)]}$$

$$\Rightarrow \quad 2\frac{d}{dx}\left(y^3\frac{d^2y}{dx^2}\right)$$

$$= 2y\frac{dy}{dx} P''(x) + y^2 P'''(x)$$

$$- \frac{1}{2} 2P'(x). P''(x)$$

$$= P'(x) P''(x) + y^2 P'''(x) - P'(x) P''(x)$$

$$\left[\because 2y\frac{dy}{dx} = P'(x)\right]$$

$$= y^2 P'''(x)$$

$$= P(x) P'''(x). \qquad [\because y^2 = P'(x)]$$

9. *If $xe^{xy} = y + \sin^2 x$, then at $x = 0$ then, find $\frac{dy}{dx}$.*

Solution: When x = 0, we get y = 0.

Now differentiating both sides of the given equation w.r.t. x, we get

$$e^{xy} + x\,e^{xy}\left(x\frac{dy}{dx} + y\right) = \frac{dy}{dx} + 2\sin x \cos x$$

and putting x = 0, y = 0, we get

$$e^0 + 0 \times e^0\left(0.\frac{dy}{dx}\Big]_{(0,0)} + 0\right) = \frac{dy}{dx}\Big]_{(0,0)} + 2\sin 0 \cos 0$$

$$\Rightarrow \quad \frac{dy}{dx}\Big]_{(0,0)} = 1$$

10. *A curve $y = f(x)$ passes through the point P(1, 1). The slope of the tangent at any point on the curve is proportional to the ordinate of the point, then find the equation of the curve.*

Solution: Let slope of the normal at (1, 1) be $-\frac{1}{a}$.

$\Rightarrow$ Slope of the tangent at (1, 1) is a

i.e., $$\frac{dy}{dx}\bigg|_{(1,1)} = a. \qquad ...(1)$$

It is given that $\frac{dy}{dx} \propto y$

or $\frac{dy}{dx} = ky$, where k is some constant.

or $\frac{dy}{y} = k\,dx$

or $\log|y| = kx + c$, wehre c is a constant.

or $|y| = e^{kx+c}$

or $y = \pm e^c e^{kx} = Ac^{kx}$, where A is constant.

Since the curve passes through (1, 1).

Therefore, $1 = Ae^k$

$\Rightarrow \quad A = e^{-k}$

Therefore, $y = e^{-k}. e^{kx} = e^{k(x-1)}$

$$\Rightarrow \quad \frac{dy}{dx} = ke^{k(x-1)}$$

$$\Rightarrow \quad \frac{dy}{dx}\bigg|_{(1,1)} = k$$

$\Rightarrow \quad a = k$ [Using (1)]

Thus, the required curve is $y = e^{a(x-1)}$.

11. *Find the number of values of x where the function $f(x) = \cos x + \cos(\sqrt{2}x)$ attains its maximum.*

Solution: It is given that,

$$f(x) = \cos x + \cos(\sqrt{2}\,x)$$

$$\Rightarrow \quad |f(x)|$$

$$= \left|\cos x + \cos\sqrt{2}x\right| \le |\cos x| + \left|\cos\sqrt{2}\,x\right|$$

$= 1 + 1 = 2, \forall x \in R.$

$\therefore$ Maximum value of f (x) = 2.

This requires

$\cos x = 1$ and $\cos \sqrt{2}\, x = 1$

$\Rightarrow \quad x = 2n\pi$

and $\quad \sqrt{2}\, x = 2m\,\pi$, n, m $\in$ I.

$\Rightarrow \quad 2n\pi = \dfrac{2m\pi}{\sqrt{2}}$

$\Rightarrow \quad n = \dfrac{m}{\sqrt{2}}.$

This is possible only when n = m = 0.

$\therefore$ There is only one value (x = 0) at which f (x) attains its maximum value.

12. $\int 7^{7^{7^x}}.7^{7^x}.7^x$ *dx is equal to.*

Solution:

Put $7^{7^{7^x}} = t$

$$\Rightarrow 7^{7^{7^x}} \frac{d}{dx}\left(7^{7^x}\right).\log 7 = dt$$

$$\Rightarrow 7^{7^{7^x}}.7^{7^x}(\log 7)^3 dx = dt$$

$$\therefore \int 7^{7^{7^x}}.7^{7^x}.7^x\, dx$$

$$= \frac{1}{(\log 7)^3}\int dt$$

$$= \frac{7^{7^{7^x}}}{(\log 7)^3} + C$$

13. *If* $\int \dfrac{4e^x + 6x^{-x}}{9e^x - 7e^{-x}}\, dx = Ax + V\,(9x^{2xz}$ *– 4) + C, then what is the value of A.*

Solution:

$$\int \frac{4e^x + 6e^{-x}}{9e^x - 4e^{-x}}\, dx$$

$$= \int \frac{4e^{2x} + 6}{9e^{2x} - 4}\, dx$$

$$= \int \frac{35 + 2t}{9t\,(t+4)}\, dt$$

$$\left[\begin{array}{l} \text{Put } 9e^{2x} - 4 = t \quad \Rightarrow e^{2x} = \frac{1}{9}(4+t) \\ \qquad\qquad \Rightarrow 2e^{2x}\, dx = \frac{1}{9}\, dt \\ \qquad\qquad \Rightarrow dx = \frac{dt}{2(+4)} \end{array}\right]$$

$$= \int \left[\frac{35}{36t} - \frac{3}{4(t+4)}\right] dt \text{ [by partial fractions]}$$

$$= \frac{35}{36}\log t - \frac{3}{4}\log(t+4) + C$$

$$= \frac{35}{36}\log(9e^{2x} - 4) - \frac{3}{4}\log(9e^{2x}) + C$$

$$= \frac{35}{36}\log(9e^{2x} - 4) - \frac{3}{4}(\log 9 + 2x) + C$$

$$= \frac{3}{2}x + \frac{35}{36}\log(9e^{2x} - 4) + \left(C - \frac{3}{2}\log 3\right)$$

$$\Rightarrow A = \frac{-3}{2},\ V = \frac{35}{36}\ C = \text{a constant.}$$

14. *Find the value of* $\displaystyle\int_{e-1}^{e^2}\left|\frac{\log_e x}{x}\right| dx.$

Solution: $\displaystyle\int_{e^{-1}}^{e^2}\left|\frac{\log_e x}{x}\right| dx$

$$= \int_{e^{-1}}^{1}\left|\frac{\log_e x}{x}\right| dx + \int_{1}^{e^2}\left|\frac{\log_e x}{x}\right| dx$$

$$= -\int_{e^{-1}}^{1}\frac{\log x}{x}\, dx + \int_{1}^{e^2}\frac{\log x}{x}\, dx$$

$$= -\int_{-1}^{0} z\, dz + \int_{0}^{2} z\, dx$$

$$\left[\text{Putting } \log x = z \text{ so that } \frac{1}{x}\, dx = dx\right]$$

$$= \left[\frac{-z^2}{2}\right]_{-1}^{0} + \left[\frac{z^2}{2}\right]_{0}^{2}$$

$= \frac{1}{2}+2$

$= \frac{5}{2}.$

15. *Find the value of* $\int_{-1}^{2}[x]\,dx$ *where* [x] *represents the greatest integer function of* x.

Solution:

$$\int_{-1}^{2}[x]\,dx = \int_{-1}^{0}[x]\,dx + \int_{0}^{1}[x]\,dx + \int_{1}^{2}[x]\,dx$$

$$= \int_{-1}^{2}(-1)\,dx + \int_{0}^{1}0\,dx + \int_{1}^{2}1\,dx$$

$$= [-x]_{-1}^{0} + [x]_{1}^{2}$$

$= -1 + 2 - 1$

$= 0.$

16. *The slope of the tangent to a curve* $y = f(x)$ *at* $(x, f(x))$ *is* $2x + 1$. *If the curve passes through the point (1,2), then find the area of the region bounded by the curve, the x-axis and the line* $x = 1$.

Solution: Given $\frac{dy}{dx} = 2x + 1$

$\Rightarrow \quad y = x^2 + x + k$

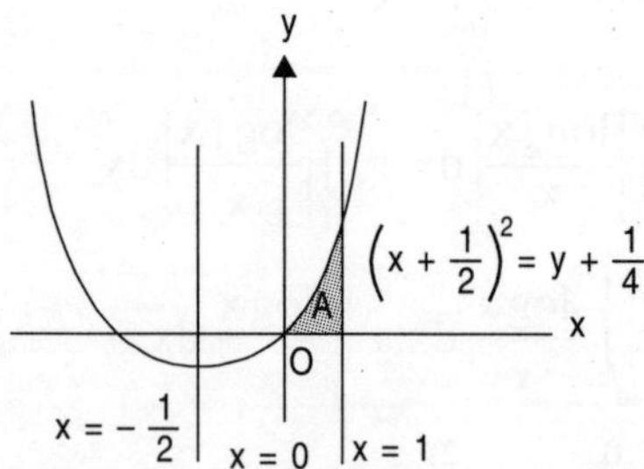

Fig. 1-W1

Since the curve passes through the point (1, 2)

$\therefore \quad 2 = 1 + 1 + k \Rightarrow k = 0$

$\therefore$ The curve is $y = x^2 + x$

So, the required area,

$$A = \int_{0}^{1}(x^2 + x)\,dx$$

$$= \left[\frac{x^3}{3}+\frac{x^2}{2}\right]_0^1$$

$= \frac{1}{3}+\frac{1}{2}$

$= \frac{5}{6}.$

17. *Let PQR be a right-angled isosceles triangle, right angled at P(2, 1). If the equation of the line QR is* $2x + y = 3$, *then find the equation representing the pair of lines PQ and PR.*

Solution: Given line is

$$2x + y = 3 \qquad ...(i)$$

Slope of line (i) = – 2

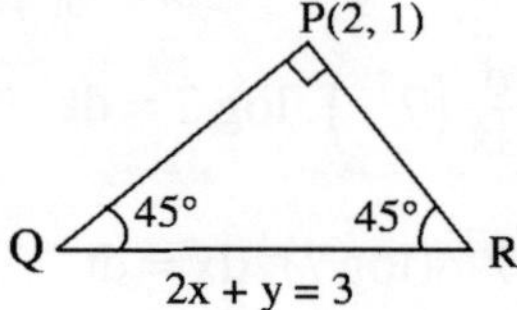

Fig. 1-W2

Let m be the slope of PQ or PR.

Since ∠ PQR = ∠ PRQ = 45°

$$\therefore \quad \frac{m-(-2)}{1+(-2)m} = \pm \tan 45°$$

$$\Rightarrow \quad m = 3, -\frac{1}{3}$$

∴ Equations of lines PQ or PR are

$(y - 1) = 3(x - 2)$ and $(y - 1) = -\frac{1}{3}(x - 2)$

$\Rightarrow 3x - y - 5 = 0$ and $x + 3y - 5 = 0$.

Their combined equation is

$(3x - y - 5)(x + 3y - 5) = 0$

$3x^2 - 3y^2 + 8xy - 20x - 10y + 25 = 0$

18. *If the circle* $x^2 + y^2 + 2x + 6ky + 6 = 0$ *and* $x^2 + y^2 + 2ky + k = 0$ *intersect orthogonally, then find the value of k.*

Solution: Using the condition, $2gg_1 + 2ff_1 = c + c_1$, we get

$$2(1)(0) + 2(k)(k) = 6 + k$$

or $$2k^2 - k - 6 = 0$$

or $$(2k + 3)(k - 2) = 0$$

$$\Rightarrow \quad k = -\frac{3}{2}, 2.$$

19. *If the sum of the roots of the equation $ax^2 + 2x + 3a = 0$, is equal to their product, then find the value of a.*

Solution: Sum of the roots $= \frac{-2}{a}$ and product of the roots $= \frac{3a}{a} = 3$

Given $\frac{-2}{a} = 3 \Rightarrow a = \frac{-2}{3}$.

20. *Let $R = \left(5\sqrt{5} + 11\right)^{2n+1}$ and $f = R\,[R]$ where [ ] denotes the greatest integer function. Then find R f.*

Solution: Here f = R – [R] is the fraction part of R. Thus if I is the integral part of R, then

$$R = I + f = \left(5\sqrt{5} + 11\right)^{2n+1}, \text{ and } 0 < f < 1.$$

Now, $\because 5\sqrt{5} - 11 = 0.18034 < 1$.

$\therefore$ if $f' = (5\sqrt{5} - 11)^{2n+1}$, then $0 < f' < 1$

Now, $I + f - f' = (5\sqrt{5} + 11)^{2n+1} - (5\sqrt{5} - 11)^{2n+1}$

$= 2[{}^{2n+1}C_1\ (5\sqrt{5})^{2n} \times 11 + {}^{2n+1}C_3\ (5\sqrt{5})^{2n-2} \times 11^3 + ...]$ ....(1)

= An even integer.

$\Rightarrow f - f'$ must also be an integer.

$\Rightarrow f - f' = 0, \quad [\because 0 < f < 1, 0 < f' < 1]$

$\Rightarrow f = f'$

$$\therefore Rf = Rf' = (5\sqrt{5} + 11)^{2n+1}(5\sqrt{5} - 11)^{2n+1}$$
$$= (125 - 121)^{2n+1}$$
$$= 4^{2n+1}.$$

21. *If* $f(x) = \begin{vmatrix} \cos x & x & 1 \\ 2\sin x & x^2 & 2x \\ \tan x & x & 1 \end{vmatrix}$ *then find* $\lim_{x \to 0}\left[\frac{f'(x)}{x}\right]$.

Solution:

We have, $f(x) = \begin{vmatrix} \cos & x & 1 \\ 2\sin x & x^2 & 2x \\ \tan x & x & 1 \end{vmatrix}$

$$= \begin{vmatrix} \cos x - \tan x & 0 & 0 \\ 2\sin x & x^2 & 2x \\ \tan x & x & 1 \end{vmatrix}$$

[Applying $R_1 \to R_1 - R_3$]

$= (\cos x - \tan x)(x^2 - 2x^2)$

$= -(\cos x - \tan x)$ [Expanding along R_1]

$\therefore f'(x) = -2x(\cos x - \tan x) - x^2(-\sin x - \sec^2 x)$

$$\therefore \lim_{x \to 0} \frac{f'(x)}{x} = \lim_{x \to 0} -2(\cos x - \tan x) + x(\sin x + \sec^2 x)$$
$$= -2 \times 1$$
$$= -2.$$

22. *Let $f(\theta) = \sin\theta(\sin\theta + \sin 3\theta)$. Then find the minimum value of $f(\theta)$.*

Solution: We have, $f(\theta)$

$$= \sin\theta(\sin\theta + \sin 3\theta)$$
$$= \sin\theta(2\theta \sin 2\theta \cos\theta)$$
$$= \sin^2 2\theta \geq 0, \text{ for all real } \theta.$$

$$\Rightarrow [f(\theta)]_{min} = 0$$

EXERCISE

1. Compute the followings:

(i) $\lim\limits_{x \to \frac{\pi}{2}} \dfrac{\tan 2x}{x - \frac{\pi}{2}}$ (ii) $\lim\limits_{x \to 1} \dfrac{x^3 - 1}{(x-1)^2}$

(iii) $\lim\limits_{x \to 0} \dfrac{x^4 + 5x - 3}{2 - \sqrt{x^2 + 4}}$ (iv) $\lim\limits_{x \to 0} \dfrac{x^3 - 7x}{x^3}$

(v) $\lim\limits_{x \to 0} \dfrac{\cos 2x - 1}{\cos x - 1}$ (vi) $\lim\limits_{x \to 4} \dfrac{3 - \sqrt{x+5}}{x-4}$

(vii) $\lim\limits_{x \to 3} \dfrac{x^4 - 81}{2x^2 - 5x - 3}$

2. Find c of Lagrange's Mean Value Theorem for $f(x) = \sqrt{25 - x^2}$ in [1, 5].

3. Find the critial points for
$$f(x) = (x - 2)^{2/3} \times (2x + 1).$$

4. Let $f(x) = x^3 - 2x^2 + 6$, find points at which f(x) have local maximum and local minimum.

5. $f(x) = \begin{cases} 6 & x \le 1 \\ 7 - x & x > 1 \end{cases}$, then for f(x) at x = 1 discuss maxima and minima.

6. If $ax^2 + bx + c = 0$; a, b, c ∈ R. Find condition that this equation would have at least one root in (0, 1).

7. If at each point of the curve $y = x^3 - ax^2 + x + 1$ tangent is inclined at an acute angle with positive direction of x-axis. Find interval in which a lies.

8. If $f(x) = \log_e x$ and $g(x) = x^2$ and $c \in (4, 5)$, then find value of $c \log \dfrac{4^{25}}{5^{16}}$.

9. Let a + b = 4 and a < 2 and let g (x) be a differentiable function. If $\dfrac{dg}{dx} > 0 \; \forall \; x$ then prove that $\int_0^a g(x)\,dx + \int_0^b g(x)\,dx$ increases as b – a increases.

10. Find the area between $x = y + 3$ and $x = y^2$ from y = –1 to y = 1.

11. Find the equation of the normal to the curve $x^2 = 4y$ passing through the point (1, 2).

12. Find the point (x, y) on the graph of $y = \sqrt{x}$ nearest to point (4, 0).

13. If $e^x + e^y = e^{x+y}$ then prove that $\dfrac{dy}{dx} = -\dfrac{e^x (e^y - 1)}{e^y (e^x - 1)}$.

14. Let f be twice differentiable such that $f'(x) = -f(x)$ and $f^4(x) = g(x)$. If $h(x) = \{f(x)\}^2 + \{g(x)\}^2$ where h (5) = 11. Find h(10).

15. You are standing at the edge of a slow-moving river which is one mile wide and wish to return to your campground on the opposite side of the river. You can swim at 2 kmph and walk at 3 kmph. You must first swim across the river to any point on the opposite bank. From there walk to campground which is one mile from the point directly across the river from where you start your swim. What route will take the least amount of time?

16. Find the points on the curve $xy^2 = 1$ which are nearest to the origin.

17. If a, b, c be non-zero real numbers such that $\int_0^1 (1 + \cos^8 x)(ax^2 + bx + c)\,dx$
$= \int_0^2 (1 + \cos^8 x)(ax^2 + bx + c)\,dx = 0$
then the equation $ax^2 + bx + c = 0$ will have

18. Find the equation of the tangent to the hyperbola $2x^2 - 3y^2 = 6$ which is parallel to the line y = 3x + 4.

19. Find the value of $\int \left(\sqrt{\tan x} + \sqrt{\cot x}\right) dx$.

20. $\int \dfrac{x+1}{x(1 + xe^x)^2}\,dx$ is equal to ....

21. $\int_{-1}^{1} |1-x|\, dx$ equals ....

22. Find the area of the region bounded by the curves $y = e^x \log_e x$ and $y = \frac{\log_e x}{e^x}$.

23. Two vertices of a triangle are (3, –1) and (–2, 3) and its orthocentre is origin, the coordinates of the third vertex are ....

24. Let the algebraic sum of the perpendicular distances from the points A (2, 0), B(0, 2), C (1, 1) to a variable line be zero. Then all such lines ....

25. The centre of a circle passing through the points (0, 0), (1, 0) and touching the circle $x^2 + y^2 = 9$ is ....

26. If $x + y = k$ is normal to $y^2 = 12x$, then k is ....

27. If α, β are the roots of the equation $x^2 + x + 1 = 0$, then the equation whose roots are α^{19} and β^7 is ....

28. If the quadratic equation $x^2 + ax + b = 0$ and $x^2 + bx + a = 0$ $(a \neq b)$ have a common root, then the numerical value of a + b is ....

29. The sum of rational terms in the expansion of $\left(\sqrt{2}+3^{\frac{1}{15}}\right)^{10}$ is ....

30. In the binomial expansion of $(a-b)^n$, $n \geq 5$, the sum of the 5th and 6th terms is zero. Then $\frac{a}{b}$ is equal to ....

31. If the three digit numbers A28, 3B9 and 62C, where A, B and C are integers between 0 and 9, are divisible by a fixed integer k, then the determinant $\begin{vmatrix} A & 3 & 6 \\ 8 & 9 & C \\ 2 & B & 2 \end{vmatrix}$ is ....

32. If a, b, c are positive and not all equal, then the value of the determinant $\begin{vmatrix} a & b & c \\ b & c & a \\ c & a & b \end{vmatrix}$ is ....

33. The three roots of the equation $\begin{vmatrix} x & 3 & 7 \\ 2 & x & 2 \\ 7 & 6 & x \end{vmatrix} = 0$ are ....

34. The value of k, for which the system of equations $x + ky + 3z = 0$, $3x + ky - 2z = 0$, $2x + 3y - 4z = 0$ posses a non-trival solution over the set of rationals.

35. Find the value of

$$\sqrt{3}\ \text{cosec}\ 20^\circ - \sec 20^\circ.$$

36. Find the value of $\frac{\cot 54^\circ}{\tan 36^\circ} + \frac{\tan 20^\circ}{\cot 70^\circ}$

37. Find the value of

$$\cos\frac{2\pi}{15}\cos\frac{4\pi}{15}\cos\frac{8\pi}{15}\cos\frac{14\pi}{15}.$$

38. Let $f(x) = \begin{vmatrix} x^3 & \sin x & \cos x \\ 6 & -1 & 0 \\ p & p^2 & p^3 \end{vmatrix}$, where p is a constant. Then find $\frac{d^3}{dx^3}[f(x)]$ at $x = 0$.

39. If α is a repeated root of a quadratic equation $f(x) = 0$ and A(x), B(x), C(x) are polynomials of degree > 2, then the determinant $\begin{vmatrix} A(x) & B(x) & C(x) \\ A(\alpha) & B(\alpha) & C(\alpha) \\ A'(\alpha) & B'(\alpha) & C'(\alpha) \end{vmatrix}$ is divisible by ....

40. If $f(x) = \log_x (\ln x)$ then $f'(x)$ at $x = e$ is ....

SYSTEMS OF UNITS

2.1 INTRODUCTION

Physics describes the laws of nature. This description is quantitative and involves the measurement and comparison of physical quantities. To measure a physical quantity we need some standard unit of that quantity. Mr. A is taller than Mrs. A but exactly how many times ? This question could be easily answered if we have a standard length call it a unit length. If Mr. A is 6 times the unit length and Mrs. A is 4 times the unit length then we could say that Mr. A is 1.5 times taller than Mrs. A. If we have a knowledge of unit mass and some one says that a stone is 20 times the unit mass then we might forecast that 10 years old Ricky may lift the stone or he will fail to lift it. Thus, the physical quantities are quantitatively expressed in terms of a unit of that quantity. The measurement of the quantity is mentioned in two parts, the first part gives how many times of the standard unit and the second part gives the name of the unit. Numerical part says that, it is how many times of the unit and second part says what unit is chosen for this quantity or name of unit.

2.2 WHO DECIDES THE UNITS ?

How is a standard unit chosen for a physical quantity ? The first thing is that it should have international acceptance. Otherwise, everyone will choose his or her own unit for the quantity and it will be difficult to communicate freely among the persons distributed over the world. A body named conference Generale des poids et mesures or CGPM also known as General Conference on Weight and Measures in English has been given the authority to decide the units by international agreement. It holds its meetings and any changes in standard units are communicated through the publications of the conference.

2.3 FUNDAMENTAL AND DERIVED QUANTITIES

We can define a set of fundamental quantities as follows:

(a) The fundamental quantities should be independent of each other.

(b) All other quantities might be expressed in terms of the fundamental quantities. It turns out that there are only seven fundamental quantities. The rest may be derived from these quantities by multiplication or division, and they are called derived quantities. The units defined for fundamental quantities are called fundamental units and those derived from these units are called derived units. Fundamental quantities are also called base quantities. Besides the seven fundamental units two supplementary units are defined. They are for plane angle and solid angle. The unit for plane angle is radian and denoted by the symbol **rad** and the unit for the solid angle is steradian and is denoted by the symbol **sr**.

Table 2.1: Fundamental quantities and their units

Quantity	*Name of the unit*	*Symbol*
Length	metre	m
Mass	kilogram	kg
Time	second	s
Electric current	ampere	A
Thermodynamic temperature	kelvin	K
Amount of substance	mole	mol
Luminous Intensity	candela	cd
Supplementary quantities and units		
Plane angle	radian	rad
Solid angle	steradian	sr

2.4 SYSTEMS OF UNITS

Four different systems of units are commonly used in science and engineering. The difference among them arises from the difference in the units used to represent the three fundamental quantities, namely length (l), mass (m) and time (t). They are given below.

Table 2.2: Centimeter-Gramme-Second (C.G.S) system of units

Quantities	*Unit*	*Symbol*
Length	centimetre	cm
Mass	gram	gr
Time	second	s
Derived unit		
Force	dyne	dyne

Note: *Dyne is that amount of force which causes one gram mass to move with an acceleration of 1 cm/s².*

Table 2.3: Foot-Pound-Second (F.P.S.) system of units

Quantities	*Unit*	*Symbol*
Length	foot	ft
Time	second	s
Force	pound	lb
Derived unit		
Mass	slag	slag

Note: *1 slag is the mass which is given an acceleration of 1 ft/s² when acted by a force of one pound.*

Table 2.4: Metre-Kilogramme-Second (M.K.S.) system of units

Quantities	*Unit*	*Symbol*
Length	metre	m
Mass	kilogram	kg
Time	second	s
Derived unit		
Force	kilogramme-weight	kg-wt

Note: *Kilogramme-weight (kg-wt) is the force required to move a mass of one kilogramme with an acceleration equal to the gravitational acceleration (g = 9.81 m/s²).*

i.e., 1 kg-wt = 9.81 N.

In 1971, CGPM held its meeting and decided a system of units which is known as the International System of Units. It is abbreviated as SI from the French name ***Le Systeme International d' Unites***. This system is widely used throughout the world and is also used in this book.

Table 2.5: SI units

Quantities	*Unit*	*Symbol*
Length	metre	m
Mass	kilogram	kg
Time	second	s
Derived unit		
Force	Newton	N

Notes:

1. *1 N is the force which gives an acceleration of 1 m/s² to a mass of 1 kg.*
2. *The difference between MKS system and SI system arises from the difference in mainly selecting unit of force. The unit for force is Newton (N) in SI units while in*

MKS system, it is kilogramme-weight (kg wt).

2.5 SI PREFIXES

The CGPM recommended standard prefixes for certain powers of 10, when quantities are too big or too small.

Table 2.6: SI prefixes

Power of 10	*Prefix*	*Symbol*
18	exa	E
15	peta	p
12	tera	T
9	giga	G
6	mega	M
3	kilo	k
2	hecto	h
1	deka	da
–1	deci	d
–2	centi	c
–3	milli	m
–6	micro	μ
–9	nano	n
–12	pico	p
–15	femto	f
–18	atto	a

2.6 DEFINITIONS OF BASE UNITS

Metre: It is the unit of length. The distance travelled by light in vacuum in $\frac{1}{299792458}$ second is called 1 m.

Kilogram: The mass of a cylinder made of platinum-iridium alloy kept at International Bureau of Weights and Measures, is defined as 1 kg.

Second: Cesium-133 atom emits electromagnetic radiation of several wavelengths. A particular radiation is selected which corresponds to the transition between the two hyperfine levels of the ground state of Cs_{133}. Each radiation has a time period of repetition of certain characteristics. The time duration in 9,192,631,770 time periods of the selected transition is defined as 1 second (1 s).

Kelvin: The fraction $\frac{1}{273.16}$ of the thermodynamic temperature of triple point of water is called 1 K.

Ampere: Suppose two long straight wires with negligible cross-section are placed parallel to each other in vacuum at a distance of 1m and electric currents are established in the two in the same direction. The wires attract each other. If equal currents are maintained in the two wires so that the force between them is 2×10^{-7} N/m, the current in any of the wires is called 1A. Here, Newton is the SI unit of force.

Mole: The amount of a substance that contains as many elementary entities (molecules or atoms if the substance is mono atomic) as there are as the number of atoms in 0.012 kg of carbon-12 is called a mole. This number (number of atoms in 0.012 kg of C_{12}) is called Avogadro's constant and its best value available is 6.022045×10^{23} with an uncertainty of about 0.000031×10^{23}.

Candela: The SI unit of luminous intensity is 1 cd, which is the luminous intensity of a black body of surface area $\frac{1}{600000}$ m^2, placed at the temperature of freezing platinum and at a pressure of 101,325 N/m^2, in the direction perpendicular to its surface.

2.7 SOME DERIVED SI; UNITS USED IN MECHANICS

Table 2.7

Quantity	*Unit*	*SI symbol*
Acceleration (linear)	metre / second^2	m/s^2
Acceleration (angular)	radian / second^2	rad/s^2
Area	metre^2	m^2
Force	Newton	N (= kg m/s^2)
Frequency	hertz (per second)	Hz (= s^{-1})
Impulse (linear)	newton-second	N-s
Impulse (angular)	newton-metre-second	Nms
Moment of a force	newton-metre	Nm
Moment of inertia (area)	metre^4	m^4
Moment of inertia (mass)	kilogram-metre^2	kg-m^2
Momentum (linear)	kilogram-metre/second	kg-m/s
Momentum (angular)	kilogram-metre^2/second	kg m^{2}s^{-1} (= N m/s)
Power	Watt	W (Nm/s)
Spring constant	newton/metre	N/m
Velocity (linear)	metre/second	m/s
Velocity (angular)	radian/second	rad/s
Volume	metre^3	m^3
Work, Energy	Joule	J (= N-m)

2.8 DIMENSION

All the physical quantities of our interest could be derived from the base quantities. When a quantity is expressed in terms of the base quantities, it is expressed as a product of different powers of the base quantities. The exponent of a base quantity that comes into the expression, is called the dimension of that quantity in base. For example, consider the physical quantity **force**. As we know that, force is equal to mass × acceleration.

Thus, Force = mass × acceleration

$$= \text{mass} \times \frac{\text{velocity}}{\text{time}}$$

$$= \text{mass} \times \frac{\text{length/time}}{\text{time}}$$

$$= \text{mass} \times \text{length} \times \text{time}^{-2}$$

Thus, the dimension of force is 1 in mass, 1 in length and –2 in time. The dimension in all other base quantities are zero.

Note that dimension of each physical quantity is fixed.

Force = $(\text{mass})^1 \times (\text{length})^1 \times (\text{time})^{-2} \times (\text{ampere})^0 \times (\text{kelvin})^0 \times (\text{mole})^0 \times (\text{candela})^0$

Therefore, every derived quantity could be expressed in the following form:

Derived quantity = $(\text{mass})^a \times (\text{length})^b \times (\text{time})^c \times (\text{ampere})^d \times (\text{kelvin})^e \times (\text{mole})^f \times (\text{candela})^g$

For convenience, the base quantity is represented by one letter symbol. Generally mass is denoted by M, length by L, time by T and electric current by I. The thermodynamics i.e., temperature, the amount of substance and the luminous intensity are denoted by the symbols of their units K, mol and cd respectively. The physical quantity that is expressed in terms of the base quantities, is enclosed within square bracket to remind that it is a dimension. Thus, the dimensional formula for force is [force] = $M^1L^1T^{-2}$.

Notes:

1. *To find the dimensional formula, constant terms could not be considered.*
2. *It is equality of the type of quantity that enters. Example, for finding the dimensional formula for energy* $\left(E = \frac{1}{2}mv^2\right)$ *only* $E = mv^2$ *is considered since,* $\frac{1}{2}$ *is a number and has no dimension.*

$$\textit{Therefore, } [E] = M^1 \left(\frac{L}{T}\right)^2$$

$$= M^1\, L^2\, T^{-2}$$

2.9 USES OF DIMENSIONS

1. Conversion of units: We could convert one unit into other system of units, out of four systems as discussed in section 2.4. Dimensions could be useful in finding the conversion factor for the unit of a derived physical quantity from one system to other. Consider an example when SI unit is used, the unit of force is 1 Newton (1 N) and when CGS system is used, the unit of force is dyne. Conversion factor between these two units could be calculated as follows:

We know that

$$\text{force} = \text{mass} \times \text{acceleration}$$

Thus,

$$[\text{force}] = [\text{mass}] \times [\text{acceleration}]$$

$$= [\text{mass}] \times \frac{[\text{velocity}]}{[\text{time}]}$$

$$= [\text{mass}] \times \frac{[\text{length}]/[\text{time}]}{[\text{time}]}$$

$$= [\text{mass}]^1 \times [\text{length}]^1 \times [\text{time}]^{-2}$$

$$= M^1L^1T^{-2}$$

So, $1 \text{ N} = (1 \text{ kg}) \times (1 \text{ m}) \times (1 \text{ s})^{-2}$

and $1 \text{ dyne} = (1 \text{ gr}) \times (1 \text{ cm}) \times (1 \text{ s})^{-2}$

Thus,

$$\frac{1\text{N}}{1\text{dyne}} = \left(\frac{1\text{ kg}}{1\text{ gr}}\right) \times \left(\frac{1\text{ m}}{1\text{ cm}}\right) \times \left(\frac{1\text{ s}}{1\text{ s}}\right)^{-2}$$

$$= (10^3) \times (10^2)$$

$$= 10^5$$

Therefore, $1 \text{ N} = 10^5$ dyne.

Thus, knowing the dimensional formula of derived quantity we could be found conversion factor for one out of the four systems of units.

2. Deducing relation among the physical quantities: If one knows the quantities on which a particular physical quantity depends and if one guesses that this dependence is of product type, method of dimension might be helpful in derivation of the relation. For example, suppose we have to derive the expression for the centripetal force. This force depends on mass (m) of body, and velocity (v) of body revolving in a circle of known radius (R), assuming product dependence, we could write,

$$\text{Centripetal force} \propto (\text{mass})^a \times (\text{velocity})^b \times (\text{radius})^c$$

or

$$F_c = k\, m^a v^b r^c$$

where, k is a dimensionless constant and a, b and c are exponents. We have to find the values of a, b and c.

$$[F_c] = [\text{mass}]^a \times [\text{velocity}]^b \times [\text{radius}]^c$$

$$= [\text{mass}]^a \times \frac{[\text{length}]^b}{[\text{time}]^b} \times [\text{radius}]^c$$

or

$$MLT^2 = M^a L^{b+c} T^{-b}$$

$$[\because \ [\text{Force}] = MLT^2]$$

Since, dimension on both sides should be equal and after equating powers, we get

$$a = 1$$

$$b + c = 1$$

and

$$b = 2$$

Thus, we get $a = 1, b = 2, c = -1$

Therefore,

$$F_c = k m^1 v^2 r^{-1}$$

$$= k \frac{mv^2}{r}$$

We know that, $F_c = \frac{mv^2}{r}$ thus, here we are getting the dimensionally equal expression.

Limitations

Although dimensional analysis is very useful in deducing certain relations, it can't lead us too far. First of all we have to know the quantities on which it depends. We are able to deduce those relations that are in product (division) form only. Secondly, numerical value having no dimensions could not be deduced by the method of dimensional analysis as in above example, we could not say that k = 1 using dimensional analysis.

Thirdly, the method works only if there are as many equations available as there are unknowns. In mechanical quantities, only three base quantities length, mass, and time comes into picture so, dimensions of these three might be equated in the guessed relation giving at most three equations in the exponents. If a particular mechanical quantity depends on more than these three quantities, in that case we shall have more unknown and less equations by equating exponent. So, exponents can't be determined uniquely.

3. In checking a equation: If an equation contains several terms which are related by the symbols of plus and/or minus. The principle of homogeneity of dimensions says that, dimension of each term should be equal because we can add or subtract only similar quantities, as we could not add velocity to acceleration, so dimension method help us whether a given relation is correct or not. If dimensions of all the forms are not same, the equation must be wrong. Let us take a equation,

$$\bar{X} = \frac{A_1x_1 + A_2x_2}{A_1 + A_2}$$

or $$\bar{x} = \left(\frac{A_1}{A_1 + A_2}\right)x_1 + \left(\frac{A_2}{A_1 + A_2}\right)x_2$$

or $$\text{Length} = \left(\frac{\text{Area}}{\text{Area} + \text{Area}}\right) \times \text{Length} + \left(\frac{\text{Area}}{\text{Area} + \text{Area}}\right) \times \text{Length}$$

$$= \left(\frac{\text{Area}}{\text{Area}}\right) \times \text{Length} + \left(\frac{\text{Area}}{\text{Area}}\right) \times \text{Length}$$

$$= (\text{Unit less}) \times \text{Length} + (\text{Unit less}) \times \text{Length}$$

$$= \text{Length} + \text{Length}$$

or $$[\text{Length}]_L = [\text{Length}]_R = L$$

Thus the equation is dimensionally correct.

$$\text{Since, } \left[\frac{A_1}{A_1 + A_2}x_1\right] = \left[\frac{A_2}{A_1 + A_2}x_2\right] = K_1\left[\frac{A_1}{A_1 + A_2}x_1\right] = K_2\left[\frac{A_2}{A_1 + A_2}x_2\right]$$

Therefore, dimensionally

$$\bar{x} = \frac{A_1}{A_1 + A_2}x_1 + \frac{A_2}{A_1 + A_2}x_2 = K_1\frac{A_1}{A_1 + A_2}x_1 + K_2\frac{A_2}{A_1 + A_2}x_2$$

$$\text{But, } \bar{x} \neq \frac{K_1A_1x_1 + K_2A_2x_2}{A_1 + A_2}$$

Thus, a dimensionally correct equation need not be actually correct but a dimensionally wrong equation must be wrong.

WORKED OUT EXAMPLES

1. *The SI and CGS units of energy are Joule and erg respectively. How many ergs are equal to one Joule?*

Solution: We have,

$$[E] = M^1L^2T^{-2}$$

Thus, I Joule $= (1\text{ kg})^1\,(1\text{m})^2\,(1\text{s})^{-2}$

and 1 erg $= (1\text{ gr})^1\,(1\text{cm})^2\,(1\text{s})^{-2}$

$$\text{Now, } \frac{1\text{ Joule}}{1\text{ erg}} = \left(\frac{1\text{kg}}{1\text{gr}}\right)^1\left(\frac{1\text{m}}{1\text{cm}}\right)^2\left(\frac{1\text{s}}{1\text{s}}\right)^{-2}$$

$$= (10^3)^1\,(10^2)^2$$

$$= 10^7$$

Therefore, 1 Joule = 10^7 erg.

2. *The distance covered by a particle in t second is given by $x = a \sin bt + cte^{d/t}$, find the dimensions of a, b, c, d.*

Solution:

Since, [x] = L, and **Principle of Homogeneity** of dimensions states that, each term should have same dimension.

Therefore,

$$[a \sin bt] = L$$

or $[a] = L$

where, sin bt is constant.

then $[b] = T^{-1}$

Now $\left[cte^{\frac{d}{t}}\right] = L$

or $[c] = LT^{-1}$ and $[d] = T$

3. *If $\int \frac{dx}{\sqrt{x^2+a^2}} = a^n \ln b\left(x+\sqrt{x^2+a^2}\right)$ then, find the value of n.*

Solution:

Let $[x] = [a] = L$

then, $[x^2 + a^2] = L^2$

then, $\left[\sqrt{x^2+a^2}\right] = L$

$$\text{Now, } \left[\int \frac{dx}{\sqrt{x^2+a^2}}\right] = \int\left[\frac{dx}{\sqrt{x^2+a^2}}\right]$$

$$= \int \frac{[dx]}{\left[\sqrt{x^2+a^2}\right]}$$

$$= \int \frac{L}{L}$$

$$= \int \text{constant}$$

$$= \text{constant} \Rightarrow L^0$$

Since right side should also have the same dimension, therefore, n = 0.

4. *Show that Varignon's theorem is dimensionally correct.*

Solution:

Varignon's theorem i.e., $Rd = \Sigma P_1d_1$ where, R is the resultant force and d is the arm of the force.

Thus, $[Rd] = [R]\,[d]$

$$= M^1L^1T^{-2}\,L^1$$

$$= M^1L^2T^{-2}$$

and, ΣP_1d_1 is sum of P_1d_1 type terms.

Therefore,

$$[\Sigma P_1d_1] = [P_1d_1]$$

$$= [P_1]\,[d_1]$$

$$= [\text{Force}]\,[\text{Length}]$$

$$= M^1L^1T^{-2} \times L^1$$

$$= M^1L^2T^{-2}$$

Thus, theorem is dimensionally correct.

5. *Find the SI units of the followings:*

(a) *moment of inertia of area*

(b) *moment of a force*

(c) *power*

(d) *momentum.*

Solution:

(a) M.I. of area = x^2A, where x is the length and A is the area.

M.I. of area = $(\text{Length})^2 \times (\text{Area})$

$= (\text{Length})^2 \times (\text{Length})^2$

$= m^2 \times m^2$

$= m^4$.

(b) Moment of a force = xF

$= (\text{Length}) \times (\text{Force})$

$= mN$.

(c) Power = Rate of doing work

$= \dfrac{\text{Work}}{\text{Time}}$

$= Js^{-1}$

$= N\,ms^{-1}$

[∵ Work = Force × Displacement = Nm]

(d) Momentum = mv

= Mass × Velocity

$= kgms^{-1}$.

EXERCISE

Model 1: Objective Questions

1. If two quantities have same dimensions then which of the followings is (are) most appropriate answer(s):

(a) they are same physical quantities
(b) must be same physical quantities
(c) may be same physical quantities
(d) none

2. A dimensionless quantity

(a) never has a unit
(b) does not exit
(c) may have a unit
(d) always has a unit

3. A constant quantity

(a) never has a unit
(b) may have a unit
(c) constant with dimension does not exit
(d) is only a number

4. If $\int \dfrac{dx}{\sqrt{x^2+a^2}} = a^n \ln \dfrac{\left(x+\sqrt{x^2+a^2}\right)}{b}$, where x, a, and b stands for length (m), then

(a) n = 0, b = 1
(b) n ≠ 0, a = 1 = b
(c) n may be positive or negative
(d) all are correct

5. A physical quantity is measured and the result is expressed as AU, where U is the unit used and A is the numerical value. If the result is expressed in various unit, then

(a) A ∝ size of u
(b) $A \propto u^n, n \neq -1$
(c) $n \propto \dfrac{1}{\sqrt{u}}$
(d) $n \propto \dfrac{1}{u}$
(e) all are correct

6. A unitless quantity

(a) never has a non-zero dimension
(b) always has a non-zero dimension
(c) may have a non-zero dimension
(d) does not exit

7. Which of the following sets cannot enter into the list of fundamental quantities in any system of unit?

(a) length, mass, acceleration
(b) mass, time, velocity
(c) length, time, mass
(d) mass, length, density

8. The dimension of $M^1L^1T^{-2}$ represents
(a) weight (b) force
(c) newton (d) dyne
(e) all are correct

9. Choose the correct statement(s):
(a) A dimensionally correct equation may be correct.
(b) A dimensionally correct equation must be correct.
(c) A dimensionally incorrect equation must be incorrect.
(d) A dimensionally incorrect may be correct.

10. If centripetal force is equal to $m^a v^b r^c$ then find the values of a, b and c. Take $F_c = \frac{mv^2}{r}$,
(a) a = 1, b = 2, c = – 2
(b) a = 1, b = 2, c = –1
(c) a = –1, b = 2, c = 2
(d) none of these

11. Power of a 100 watt bulb in CGS unit is,
(a) 10^{10} erg/s (b) 10^9 erg/s
(c) 10^7 erg/s (d) 10^5 erg/s

Model 2: Long Questions

12. How many square metres make area of 6.0 km² ?

13. Calculate the number of kilometers in 20.0 mi using only the following conversion factors; 1 mi = 5280 ft, 1 ft = 12 in, 1 in = 2.54 cm, 1m = 100 cm and 1 km = 1000 m.

14. A typical sugar cube has an edge of 1 cm. If you had a cubical box that contained a mole of sugar cubes, what would its edge length be?
(1 mole = 6.02×10^{23} units)

15. How many seconds are there in 1 year (= 365.25 days).

16. (a) Assuming that the density (mass/volume) of water is exactly 1g/cm³, express the density of water in kilogram per cubic meter (kg/cm³).
(b) Suppose that it takes 10 hours to drain a container of 5700 m³ of water. What is the "mass flow rate", in kilogram per second of water from the container?

17. The density of iron is 7.87 g/cm³ and the mass of an iron atoms is 9.27×10^{-26} kg. If the atoms are spherical and tightly packed,
(a) What is the volume of an iron atom and
(b) What is the distance between the centres of adjacent atoms?

18. Suppose the acceleration due to gravity at a place is 10 ms⁻². Find its value in cm min⁻².

19. Test if the following equations are dimensionally correct:

(a) $\bar{x} = \frac{A_1x_1 + A_2x_2 + A_3x_3}{A_1 + A_2 + A_3}$

(b) $x_{cm} = \frac{m_1x_1 + m_2x_2 + m_3x_3}{m_1 + m_2 + m_3}$

(c) $F = \frac{d}{dt}(mv)$ (d) $F = \frac{Gm_1m_2}{r^2}$

where, x_{cm} represent centre of mass of a descrete group of bodies.

20. Taking force, length and time to be the fundamental quantities find the dimensions of
(a) density (b) pressure
(c) momentum (d) energy

21. Theory of relativity reveals that mass could be converted into energy. The energy E so obtained is proportional to certain powers of mass m and the speed of light. Find a relation among the quantities E, m and C using the method of dimensions.

3 VECTOR

3.1 DEFINITIONS

Vector: The physical quantities which have magnitude with specified unit and direction and could be manipulated by vector algebra are called **vector** quantities. Examples are displacement, force, velocity, acceleration, momentum, moment of a force etc.

A vector is geometrically represented by a straight line having an arrow at the end point is called **head**, and the initial point is called **tail**; length of the straight line is proportional to the magnitude of a vector.

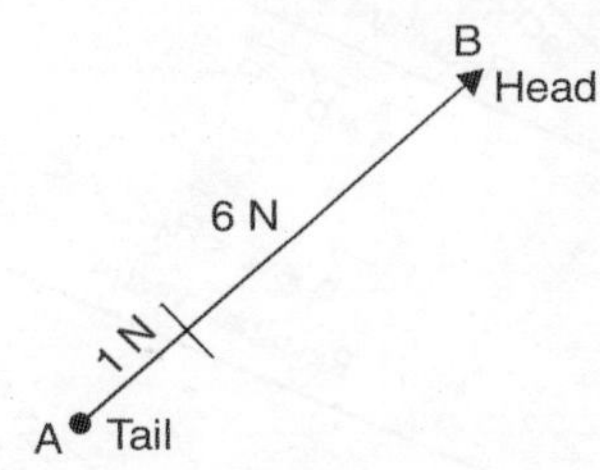

Fig. 3.1

A vector is denoted by putting an arrow over the symbol. For example, the above shown vector is represented by $\overrightarrow{AB}$. Vectors can also be represented by single letter such as $\vec{v}$, $\vec{a}$ and $\vec{F}$. The Arrow denotes sense of direction while the magnitude of the vector is denoted by a letter without an arrow, for example

$$|\vec{a}|, |\overrightarrow{AB}|, a, b, F$$

Scalar: Physical quantities are completely described by magnitude having specified unit and can be manipulated by ordinary algebra are called **scalar** quantities.

Note: *Physical quantity that has magnitude as well as direction but cannot be added by the triangle rule, is not a vector quantity. For example, electric current in a wire has both magnitude and direction but it does not follow the triangle rule, therefore electric current is not a vector quantity. Time is another such example.*

Scalars obeys ordinary algebraic laws of addition and multiplication. These are as followings:

(i) ***Commutative law***

$$a + b = b + a$$
$$a \times b = b \times a$$

(ii) ***Associative law***

$$a + b + c = (a + b) + c = a + (b + c) = (c + a) + b$$
$$a \times (b \times c) = (a \times b) \times c = (c \times a) \times b$$

(iii) ***Distributive law***

$$a \times (b + c) = (a \times b) + (a \times c)$$
$$(a + b) \times c = (a \times c) + (b \times c)$$

3.2 EQUALITY OF TWO VECTORS

Two similar vectors are said to be equal, if their magnitudes and directions are same. Thus, a parallel translation of a vector does not bring any change in a vector.

$$\text{Therefore,} \quad \vec{a} = \vec{b}$$

$$\text{if and only if,} \quad |\vec{a}| = |\vec{b}|$$

$$\text{and} \quad \tan\theta_1 = \tan\theta_2 = \frac{1}{2}$$

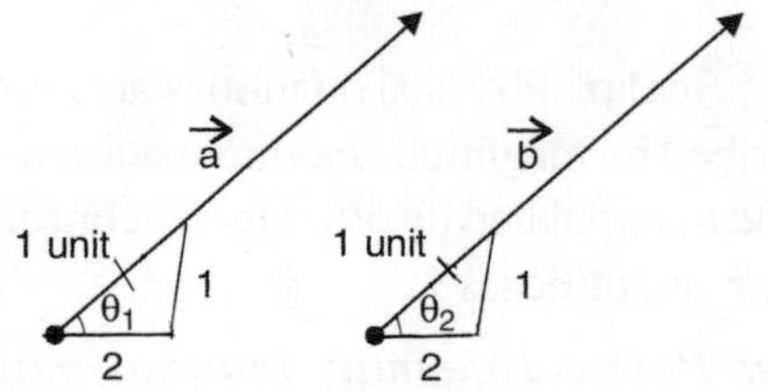

Fig. 3.2

3.3 ANGLE BETWEEN TWO VECTORS

Let $\vec{a}$ and $\vec{b}$ are two vectors and if both vectors are brought together such that the tails or heads of both coincide then out of two angles say θ and $2\pi - \theta$, the smaller one is said to be the angle between them.

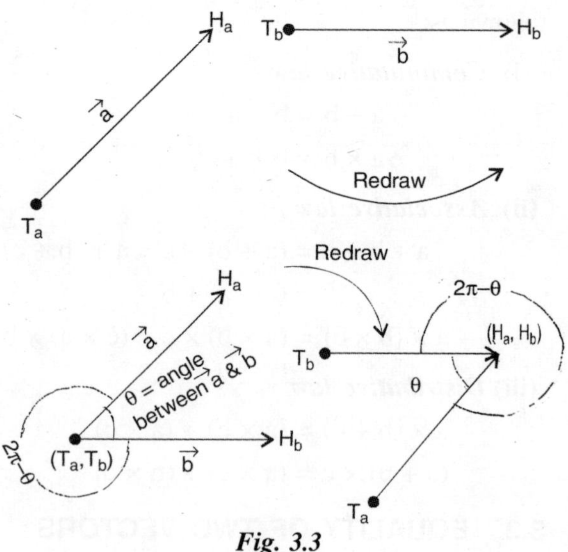

Fig. 3.3

Note: *If $\theta < 2\pi - \theta$ then θ is angle between $\vec{a}$ & $\vec{b}$ otherwise it is $2\pi - \theta$.*

3.4 ADDITION OF TWO VECTORS

3.4.1 Triangle law of vector addition

If we have to add two vectors $\vec{a}$ and $\vec{b}$, then we need to redraw these two vectors such that the head of the first (take any one as first) vector coincides with the tail of second vector and finally join the tail of the first to the head of the second vector and then, put an arrow on the finished point on the line. This arrow is the direction (sense) of the resultant or the sum of these two vectors and the proportional length of the line give magnitude of the resultant vector.

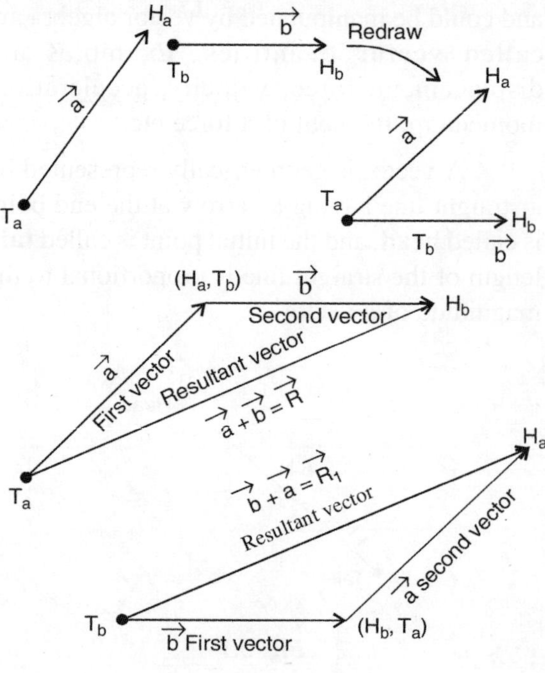

Fig. 3.4

Notes:

1. *Sum of two vectors $\vec{a}$ and $\vec{b}$, is given mathematically by $\vec{a} + \vec{b}$. The plus (+) sign does not have same meaning as that in simple algebra. Because we can see that the plus sign may also function like difference reduce (–). Such type of properties are not found in simple algebra.*

2. *We can add two vectors, if and only if, they represent the same physical quantities and the sum obtained is the same kind of physical quantity say as $\vec{R}$ or $\vec{c}$ etc.*

3. *Further, we see that addition of two vectors obey commutative law of simple algebra i.e.,*

$$\vec{a} + \vec{b} = \vec{b} + \vec{a} = \vec{R} = \vec{R}_1$$

as shown in Fig. 3.4.

3.4.2 Parallelogram law of vector addition

We can state the triangle rule of vector addition in a slightly different way as, let the vectors $\vec{a}$ and $\vec{b}$ be drawn such that both tails should coincide, as shown in Fig. (3.5) and then take these two as the adjacent sides to complete the parallelogram. The diagonal through the common tails gives the sum or resultant of the two vectors $\vec{a}$ and $\vec{b}$. This is said to be the parallelogram law of vector addition.

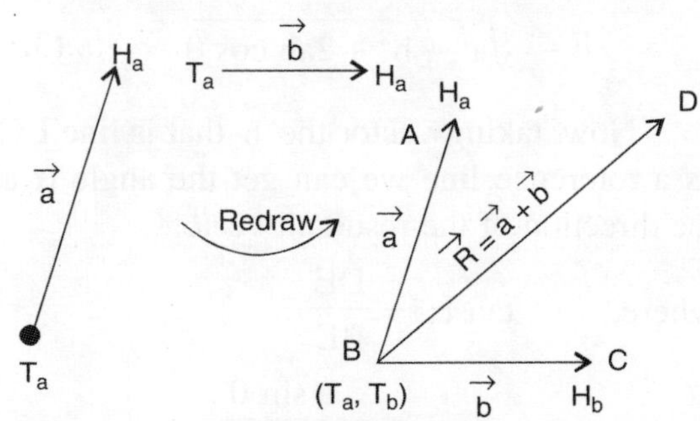

Fig. 3.5

Notes: **1.** *Here, we observe that if we shift $\vec{BC} = \vec{b}$ on line AD, then this gives the same result $\vec{R}$ which is same according to the triangle law.*

2. *Triangle rule or parallelogram rule gives the sum geometrically. On a diagram drawn with scale layout, i.e., draw the vector $\vec{a}$ then draw the vector $\vec{b}$ with its tail at the head of $\vec{a}$ and draw a line from the tail of $\vec{a}$ to the head of $\vec{b}$ to construct a triangle then, third side of the triangle is vector sum $\vec{R}$. This side is equivalent to the magnitude and direction of the resultant vector of $\vec{a}$ and $\vec{b}$. This procedure can be generalized to obtain the sum of a number of vectors because the symbol "+" used here has a different meaning from arithmetic or ordinary algebra. It carries out similar sets of operations. Suppose, we have to add $\vec{a}$, $\vec{b}$, $\vec{c}$ and $\vec{d}$, as shown in the figure 3.6; then,*

Let $\vec{a} + \vec{b} + \vec{c} + \vec{d} = \vec{R}$

or $(\vec{a} + \vec{b}) + \vec{c} + \vec{d} = \vec{R}$

or $\vec{R}_1 + \vec{c} + \vec{d} = \vec{R}$

or $(\vec{R}_1 + \vec{c}) + \vec{d} = \vec{R}$

or $\vec{R}_2 + \vec{d} = \vec{R}$

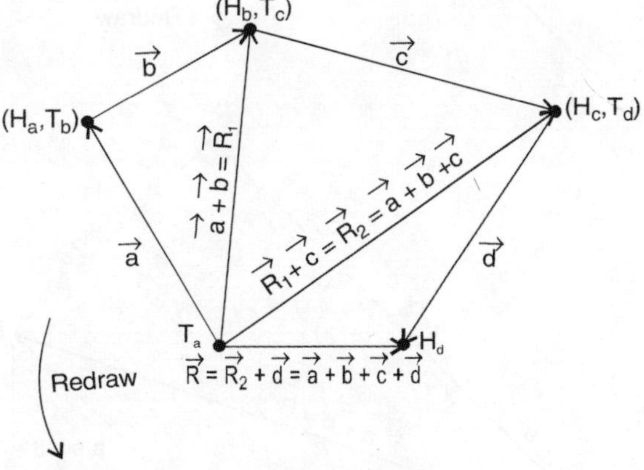

Fig. 3.6

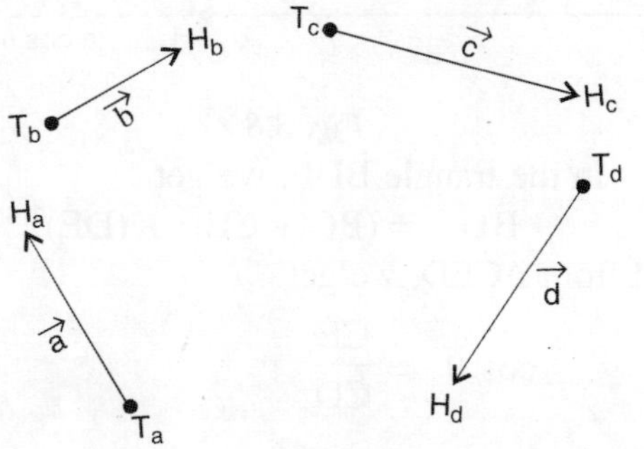

Fig. 3.7

So, in this way we can add a number of vectors by using triangle law of vectors addition.

Now, we have the ability to add two (any number of) vectors geometrically. Since this method is time consuming and requires neat diagram using a proper scale, mathematical analysis (an alternative method) could be employed to find the sum of two vectors.

Suppose, two vectors $\vec{a}$ and $\vec{b}$ are given. Such that magnitude of $\vec{a}$ is a or $|\vec{a}|$ and that of $\vec{b}$ is **b** or $|\vec{b}|$ and let the angle between them be θ. Then, in order to find the resultant vector of $\vec{a}$ and $\vec{b}$ we have to find the magnitude and direction of the resultant, construct the Fig. (3.8).

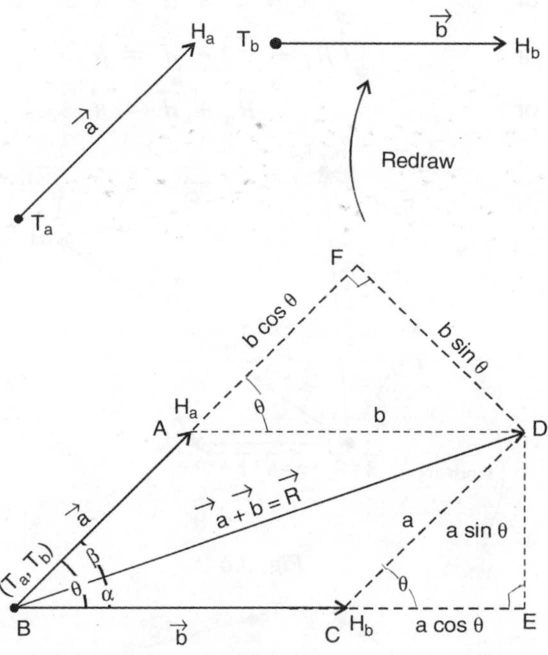

Fig. 3.8

In the trangle BDE, we get

$$BD^2 = (BC + CE)^2 + (DE)^2 \quad ....(i)$$

and from ΔCED, we get

$$\cos\theta = \frac{CE}{CD}$$

$$= \frac{CE}{a}$$

(because ABCD is a parallelogram so, AB $= CD = |\vec{a}| = \mathbf{a}$ and $BC = AD = |\vec{b}| = \mathbf{b}$)

or $$CE = a\cos\theta \quad ....(ii)$$

and $$\sin\theta = \frac{DE}{CD}$$

or $$DE = CD\sin\theta$$

or $$DE = a\sin\theta \quad ....(iii)$$

Substituting CE, DE from equations (ii) and (iii) to equation (i) we get

$$BD^2 = (b + a\cos\theta)^2 + (a\sin\theta)^2$$

$$= b^2 + a^2\cos^2\theta + 2\,ba\cos\theta + a^2\sin^2\theta$$

$$= b^2 + a^2(\cos^2\theta + \sin^2\theta) + 2ab\cos\theta$$

$$= b^2 + a^2 + 2ab\cos\theta \quad [\because \cos^2\theta + \sin^2\theta = 1]$$

$$= a^2 + b^2 + 2ab\cos\theta$$

Thus, the magnitude of resultant (i.e., $|\vec{R}| = BD$) is

$$R = \sqrt{a^2 + b^2 + 2ab\cos\theta} \quad ....(3.1)$$

Now, taking vector the $\vec{b}$ that is line BC) as a reference line we can get the angle α as the direction of the resultant vector.

where, $$\tan\alpha = \frac{DE}{BE}$$

$$= \frac{a\sin\theta}{b + a\cos\theta}$$

Therefore, $$\alpha = \tan^{-1}\frac{a\sin\theta}{b + a\cos\theta} \quad ....(3.2)$$

Now, if we take the vector $\vec{a}$ that is line BA as reference line then the resultant vector $\vec{R}$ makes angle β with the vector $\vec{a}$. We could have angle β to give the direction of resultant vector ($\vec{R}$).

where, $\tan\beta = \dfrac{DF}{FB}$

$$= \frac{b \sin\theta}{a + b\cos\theta}$$

Since from triangle ADF, $\cos\theta = \dfrac{AF}{AD}$

or $AF = b\cos\theta$ and $\dfrac{FD}{AD} = \sin\theta$

or $FD = AD\sin\theta$;

$$BC = |\vec{b}|$$
$$= AD \text{ because } AD \parallel BC.$$

Therefore, $\beta = \tan^{-1}\dfrac{b\sin\theta}{a + b\cos\theta}$ ....(3.3)

Remark 1: $R = (a^2 + b^2 + 2ab\cos\theta)^{1/2}$

$$\tan\alpha = \frac{b\sin\theta}{a + b\cos\theta}$$
$$= \frac{Numerator}{Denominator}$$

and $\tan\beta = \dfrac{a\sin\theta}{b + a\cos\theta}$

$$= \frac{Numerator}{Denominator}$$

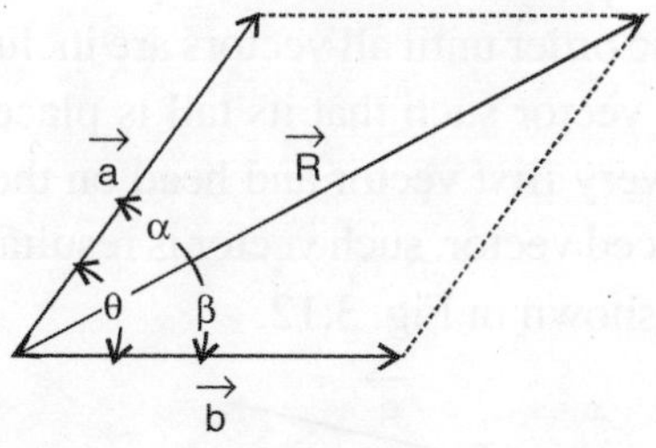

Fig. 3.9

Note: *If we assume an angle which is made by resultant vector with $\vec{a}$ then while calculating angle (α) take b sinθ in numerator and assume a sinθ while we calculate angle made by resultant vector with vector $\vec{b}$.*

3.5 SOME SPECIAL CASES

As seen the resultant of two vectors depends on three parameters. S, magnitudes of two vectors (say $|a| = a$ or $|b| = b$) and angle between them be θ. We get direction (θ) and magnitude of resultant (R), mathematically. It could be represented by $R = f(a, b, \theta)$. Let there be two vectors $\vec{a}$ and $\vec{b}$ we have to add, for further use and we need discussion of some special cases.

Case 1: Let $\vec{a}$ and $\vec{b}$ be parallel i.e., $\theta = 0^o$.

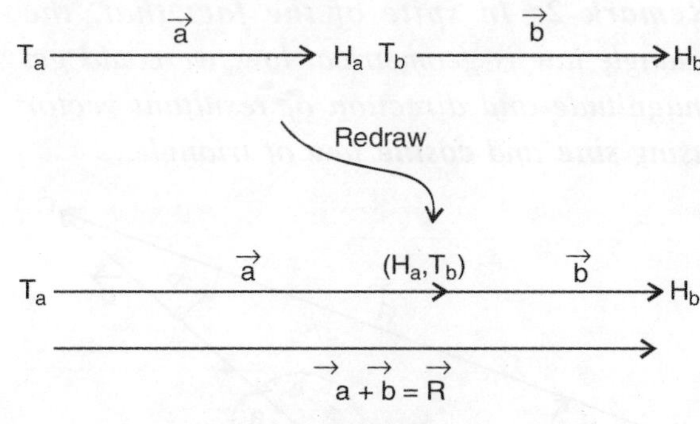

Fig. 3.10

Then as we know that,

$$R = (a^2 + b^2 + 2ab\cos\theta)^{1/2}$$
$$= (a^2 + b^2 + 2ab\cos 0)^{1/2}$$
$$= (a^2 + b^2 + 2ab)^{1/2} \quad [\because \cos 0 = 1]$$
$$= [(a + b)^2]^{\frac{1}{2}}$$
$$= a + b$$

$$\Rightarrow |\vec{R}| = |\vec{a}| + |\vec{b}|$$

Now, $\tan\alpha = \dfrac{b\sin 0}{a + b\cos 0}$; where, α is angle made by $\vec{R}$ with vector $\vec{a}$.

$$= \frac{b \times 0}{a + b}$$
$$[\because \sin 0 = 0 \text{ and } \cos 0 = 1]$$
$$= \frac{0}{a + b}$$
$$= 0$$
$$\Rightarrow \alpha = 0 \quad [\because \tan 0 = 0]$$

Similarly, we can get

$$\beta = 0$$

Thus, $R = a + b$, $\alpha = 0$ and $\beta = 0$, it could be seen in Fig. 3.10.

Note: *We can see that, when parallel vectors are added then the resultant is along the same direction and we can add the magnitude using simple algebra, thus vector algebra follows theory of simple algebra when* $R = f(\vec{a}, \vec{b}, \theta = 0°)$.

Remark 2: *In spite of the fact that, the triangle law is geometrical law, we could get magnitude and direction of resultant vector using* ***sine*** *and* ***cosine*** *law of triangle.*

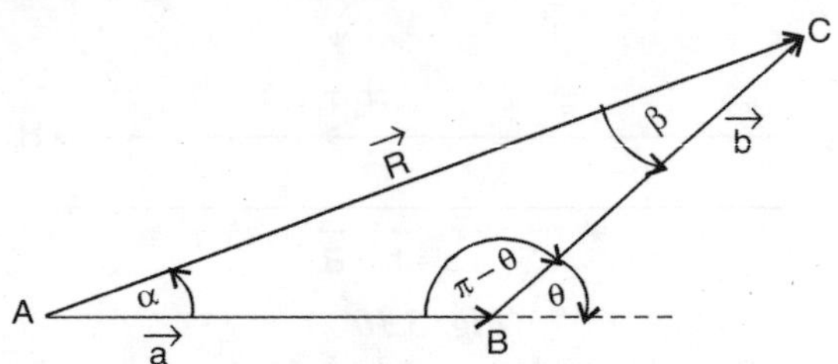

Fig. 3.11

As we know that, in Δ ABC

$$\cos(\pi - \theta) = \frac{AB^2 + BC^2 - AC^2}{2AB \cdot BC}$$

$$= \frac{a^2 + b^2 - R^2}{2ab}$$

or $\quad -2ab\cos\theta = a^2 + b^2 - R^2$

$$[\because \cos(\pi - \theta) = -\cos\theta]$$

or $\quad R^2 = a^2 + b^2 + 2ab\cos\theta$

Therefore,

$$R = \sqrt{a^2 + b^2 + 2ab\cos\theta}$$

Suppose, the resultant vector makes an angle α with $\vec{a}$ and β with $\vec{b}$, then we could write,

$$\frac{b}{\sin\alpha} = \frac{a}{\sin\beta} = \frac{R}{\sin(\pi - \theta)}$$

[sine rule]

or $$\frac{b}{\sin\alpha} = \frac{\sqrt{a^2 + b^2 + 2ab\cos\theta}}{\sin\theta}$$

$$[\because \sin(\pi - \theta) = -\sin\theta]$$

or $$\sin\alpha = \frac{b\sin\theta}{\sqrt{a^2 + b^2 + 2ab\cos\theta}}$$

or $$\alpha = \sin^{-1}\frac{b\sin\theta}{\sqrt{a^2 + b^2 + 2ab\cos\theta}}$$

Similarly, $$\beta = \sin^{-1}\frac{a\sin\theta}{\sqrt{a^2 + b^2 + 2ab\cos\theta}}$$

Regarding vector ($\vec{R}$), it should be noted that,

$R_{min} = |a - b|$, when $\theta = 180°$

$R_{max} = |a + b|$, when $\theta = 0°$

and $R = \sqrt{a^2 + b^2}$, when $\theta = 90°$

keeping magnitudes constant for various orientation, we could get

$$|a - b| \le R \le |a + b|;\ 0 \le \theta \le 180°.$$

3.5.1 Polygon law of vector addition

The law states that when the tail of one vector is placed on the head of preceding vector in cyclic order until all vectors are included then draw a vector such that its tail is placed on the tail of very first vector and head on the head of last placed vector, such vector is resultant vector as it is shown in Fig. 3.12.

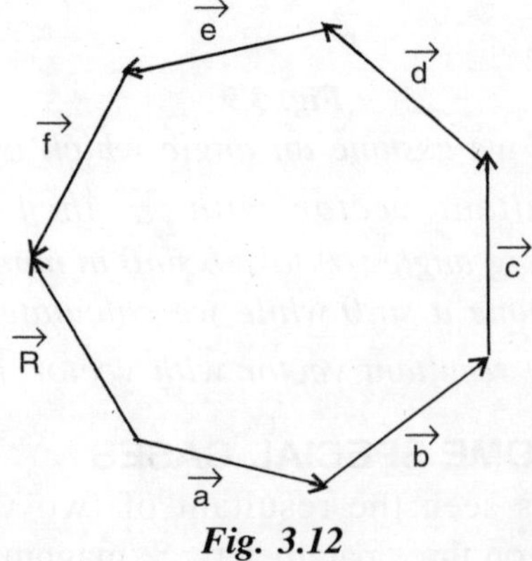

Fig. 3.12

Case 2: If $\vec{a}$ and $\vec{b}$ are antiparallel i.e., $\theta = 180^o$.

Subcase a: $|\vec{a}| > |\vec{b}|$

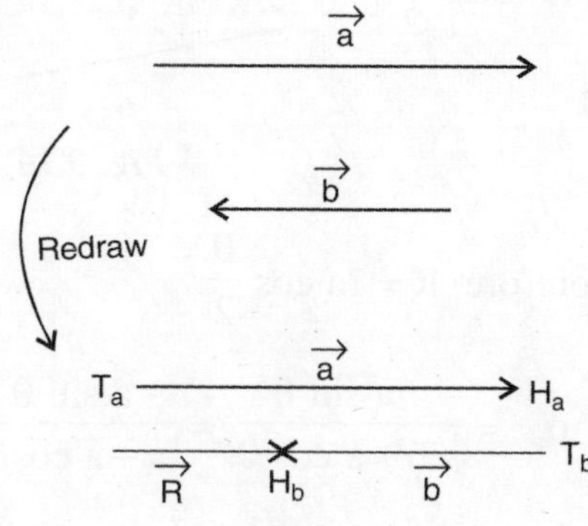

Fig. 3.13

Now magnitude of resultant ($\vec{R}$) is

$$R = (a^2 + b^2 + 2ab \cos 180^o)^{1/2}$$

$$= (a^2 + b^2 - 2ab)^{1/2} \qquad [\because \cos 180^o = -1]$$

$$= \{(a - b)^2\}^{1/2}$$

$$= | a - b |$$

$$= a \sim b$$

$$= a - b \qquad [\because a > b]$$

$$\alpha = \tan^{-1} \frac{a \sin \theta}{b + a \cos \theta}$$

$$= \tan^{-1} \frac{a \sin 180^o}{b + a \cos 180^o}$$

$$= \tan^{-1} \frac{0}{b - a} = \tan^{-1} 0$$

= 0, it is correct because resultant is parallel to vector $\vec{a}$.

and $$\beta = \tan^{-1} \frac{b \sin 180^o}{a + b \cos 180^o}$$

$$= \tan^{-1} \frac{b \times 0}{a - b} = \tan^{-1} 0$$

= 180°, it is also correct because resultant is antiparallel to vector $\vec{b}$.

Subcase b: $| \vec{a} | < | \vec{b} |$

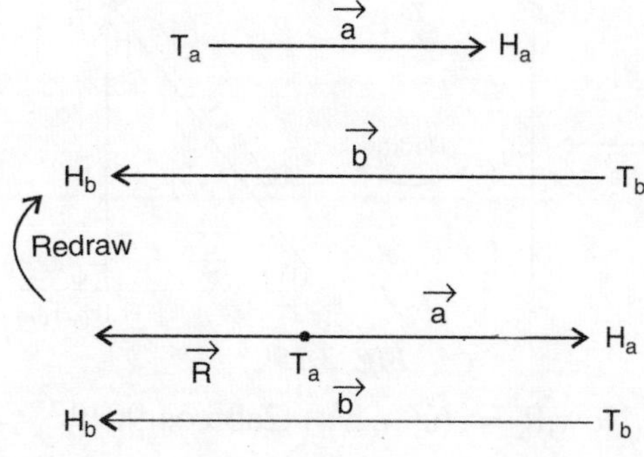

Fig. 3.14

Now $$R = (a^2 + b^2 + 2ab \cos 180^o)^{1/2}$$

$$= (a^2 + b^2 - 2ab)^{1/2}$$

$$= [(a - b)^2]^{1/2}$$

$$= | a - b |$$

$$= a \sim b$$

$$= b - a \qquad [\because b > a]$$

Angle made by resultant with vector $\vec{b}$,

$$\alpha = \tan^{-1} \frac{a \times \sin 180^o}{b + a \cos 180^o}$$

$$= \tan^{-1} 0$$

$$= 0^o$$

and angle made by resultant with vector $\vec{a}$,

$$\beta = \tan^{-1} \frac{b \times \sin 180^o}{a + b \cos 180^o}$$

$$= \tan^{-1} 0$$

$$= 180^o \qquad \text{[according to Fig. 3.14]}$$

Note: *Here we see that in vector algebra plus sign also gives difference when*

$$R = f(\vec{a}, \vec{b}, 180^o).$$

Case 3: When $\vec{a}$ and $\vec{b}$ are perpendicular that is $\theta = 90^o$.

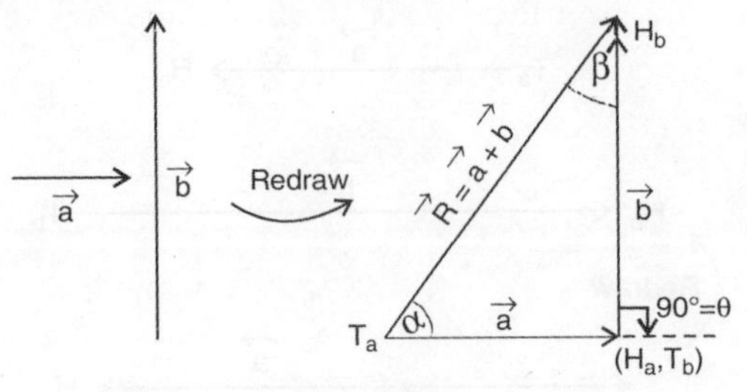

Fig. 3.15

$$\text{Now,} R = (a^2 + b^2 + 2ab \cos 90^o)^{1/2}$$

$$= (a^2 + b^2)^{1/2} \quad [\because \cos 90° = 0]$$

$$= \sqrt{a^2 + b^2}$$

$\tan \alpha$ = angle made by resultant with vector $\vec{a}$

$$= \frac{b \sin \theta}{a + b \cos \theta} = \frac{b \times \sin 90^o}{a + b \cos 90^o}$$

$$= \frac{b \times 1}{a + 0} = \frac{b}{a}$$

and $\tan \beta$ = angle made by resultant with vector $\vec{b}$

$$= \frac{a \sin \theta}{b + a \cos \theta}$$

$$= \frac{a \times \sin 90^o}{b + a \cos 90^o}$$

$$= \frac{a}{b}.$$

Case 4: When two vectors having equal magnitude and angle between them is θ.

$$\text{Now,} \quad R = \sqrt{a^2 + b^2 + 2ab \cos \theta}$$

$$= \sqrt{a^2 + a^2 + 2aa \cos \theta} \quad [\because a = b]$$

$$= \sqrt{2a^2 + 2a^2 \cos \theta}$$

$$= \sqrt{2a^2 (1 + \cos \theta)}$$

$$= \sqrt{2a^2 . 2 \cos^2 \frac{\theta}{2}}$$

$$= \sqrt{4a^2 \cos^2 \frac{\theta}{2}}$$

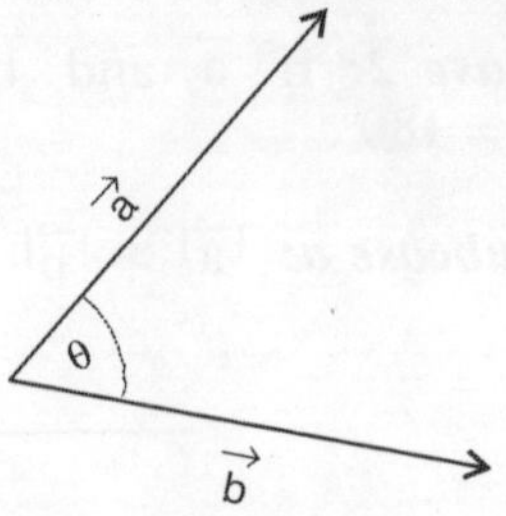

Fig. 3.16

$$\text{Therefore, } R = 2a \cos \frac{\theta}{2} \qquad \text{....(3.4)}$$

$$\text{and } \tan \alpha = \frac{a \sin \theta}{b + a \cos \theta} = \frac{a \sin \theta}{a + a \cos \theta}$$

$$= \frac{a \sin \theta}{a (1 + \cos \theta)}$$

$$= \frac{2a \sin \frac{\theta}{2} \cos \frac{\theta}{2}}{a \times 2 \cos^2 \frac{\theta}{2}}$$

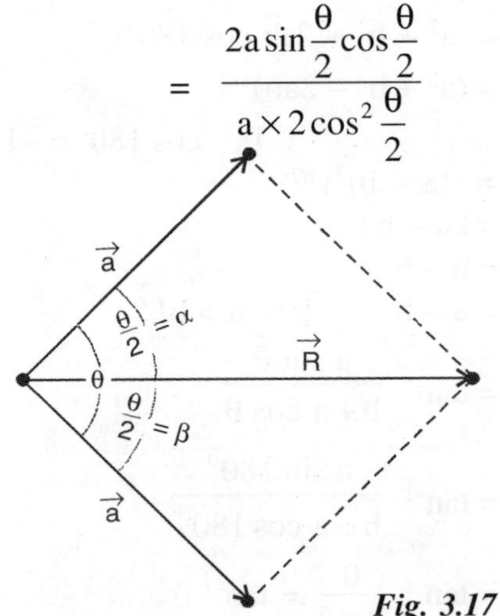

Fig. 3.17

$$= \frac{\sin \frac{\theta}{2}}{\cos \frac{\theta}{2}} = \tan \frac{\theta}{2}$$

$$\Rightarrow \quad \left.\begin{aligned} \alpha &= \frac{\theta}{2} \\ \text{Similarly, we could get, } \beta &= \frac{\theta}{2} \end{aligned}\right\} \qquad \text{...(3.5)}$$

Thus, the resultant of two equal vectors bisects the angle between them as shown in Fig. 3.17.

Example 1: *The magnitudes of displacements* $\vec{a}$ *and* $\vec{b}$ *are 3 mm and 4 mm respectively and* $\vec{c} = \vec{a} + \vec{b}$. *Considering various orientations of* $\vec{a}$ *and* $\vec{b}$, *what is*

(a) the maximum possible magnitude of $\vec{c}$ *and*

(b) the minimum possible magnitude of $\vec{c}$?

Solution: $|\vec{c}|$ = magnitude of $\vec{a} + \vec{b}$

$$= \sqrt{a^2 + b^2 + 2ab\cos\theta}$$

Now, $c_{max} = \sqrt{a^2 + b^2 + 2ab(\cos\theta)_{max}}$

$$= \sqrt{a^2 + b^2 + 2ab} \qquad [\because\ (\cos\theta)_{max} = 1]$$

$$= \sqrt{(a+b)^2}$$

$$= a + b$$

$$= 3 + 4$$

$$= 7 \text{ m}$$

and $c_{min} = \sqrt{a^2 + b^2 + 2ab(\cos\theta)_{min}}$

$$= \sqrt{a^2 + b^2 - 2ab} \qquad [\because\ (\cos\theta)_{min} = -1]$$

$$= \sqrt{(a-b)^2}$$

$$= |a - b|$$

$$= |3 - 4|$$

$$= 1$$

Therefore, $1 \le c \le 7$.

3.6 MULTIPLICATION OF A VECTOR BY A NUMBER

The multiplication of a vector by a scalar has a simple meaning. The product of a scalar **k** by a vector $\vec{a}$ is written as **k**$\vec{a}$, is defined a new vector whose magnitude is **k** times the magnitude of $\vec{a}$. The new vector has the same direction as of $\vec{a}$ if **k** is positive and has the opposite direction if **k** is negative. To divide a vector by a scalar we simply multiply it by the reciprocal (1/**k**) of that scalar.

In particular, multiplication by –1 just inverts the direction of a vector. The vector $\vec{a}$ and $-\vec{a}$ have equal magnitudes but opposite directions.

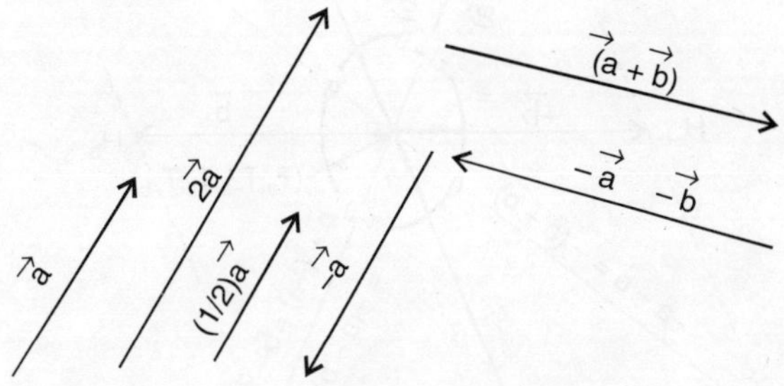

Fig. 3.18

3.7 SUBTRACTION OF TWO VECTORS

Let $\vec{a}$ and $\vec{b}$ be two vectors. We define $\vec{a} - \vec{b}$ as the sum of vectors $\vec{a}$ and $-\vec{b}$. To subtract $\vec{b}$ from $\vec{a}$ invert the direction of $\vec{b}$ and add it to $\vec{a}$. Similarly, to subtract $\vec{a}$ from $\vec{b}$ invert the direction of $\vec{a}$ and add it to $\vec{b}$. Mathematically,

$$\vec{a} - \vec{b} = \vec{a} + (-\vec{b}) = [\vec{a} + (-1)\,\vec{b}]$$

and

$$\vec{b} - \vec{a} = \vec{b} + (-\vec{a}) = [\vec{b} + (-1)\,\vec{a}]$$

Graphically,

$\vec{a}$ $-\vec{a}$ $\vec{b}$ $-\vec{b}$

H_a, H_{-a}, H_b, H_{-b}, $\vec{a}$, $-\vec{a}$, $\vec{b}$, $-\vec{b}$

$\pi - \theta$ = angle between $\vec{a}$ and $-\vec{b}$

θ = angle between $\vec{a}$ and $\vec{b}$

$(T_a, T_b, T_{-a}, T_{-b})$

θ = angle between $-\vec{a}$ and $-\vec{b}$

$\pi - \theta$ = angle between $-\vec{a}$ and $\vec{b}$

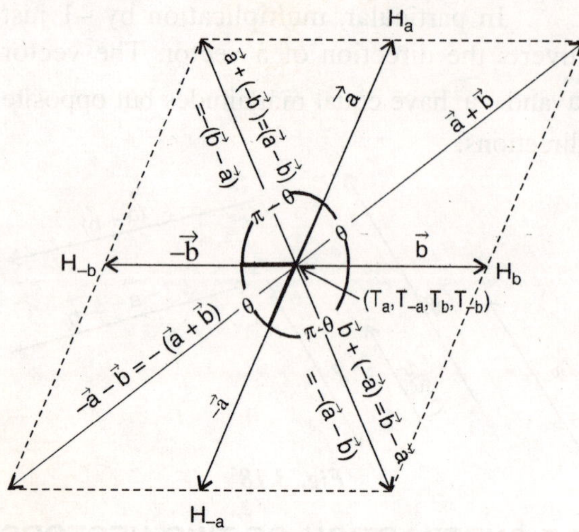

Fig. 3.19

Let $\vec{a}$ and $\vec{b}$ be two given vectors and angle between them be θ, then we can see from the diagram that the angle between $\vec{a}$ and $-\vec{b}$ is $\pi-\theta$, between $-\vec{b}$ and $-\vec{a}$ is θ, and between $-\vec{a}$ and $\vec{b}$ is $\pi-\theta$. From geometry, it is also clear that $|\vec{a}| = |-\vec{a}| = a$ and $|\vec{b}| = |-\vec{b}| = b$.

Now,

$$\left|\vec{a}-\vec{b}\right| = \left|\vec{a}+(-\vec{b})\right|$$

$$= \sqrt{|\vec{a}|^2+|-\vec{b}|^2+2|\vec{a}||-\vec{b}|\cos\angle \text{between } \vec{a} \text{ and } \vec{b}}$$

$$= \sqrt{a^2+b^2+2ab\cos(\pi-\theta)}$$

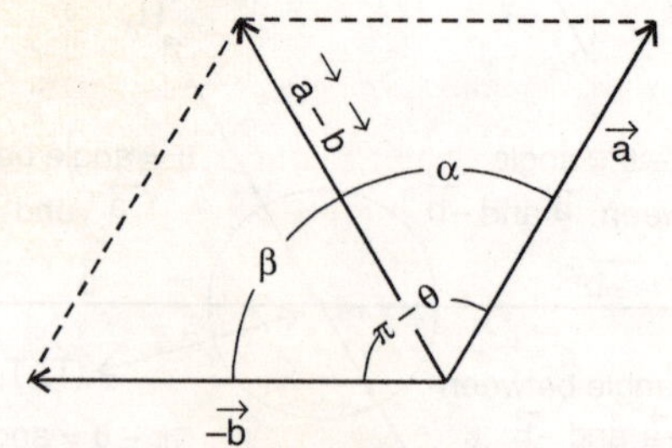

Fig. 3.20

$$= \sqrt{a^2+b^2-2ab\cos\theta} \quad [\because \cos(\pi-\theta) = -1]$$

$$= \sqrt{a^2+b^2-2ab\cos\angle \text{ between } \vec{a} \text{ and } \vec{b}} \text{ and}$$

$$\tan\alpha = \tan(\text{angle between resultant and } \vec{a})$$

$$= \frac{|-\vec{b}|\sin(\pi-\theta)}{a+|-\vec{b}|\cos(\pi-\theta)}$$

$$\left[\because \sin(\pi-\theta)=\theta; \cos(\pi-\theta) = -\cos\theta \text{ and } |-\vec{b}| = b\right]$$

$$= \frac{b\sin\theta}{a-b\cos\theta}$$

where θ is the angle between $\vec{a}$ and $\vec{b}$, not between $\vec{a}$ and $-\vec{b}$ and $\tan\beta = \tan$ (angle between resultant and $-\vec{b}$).

$$= \frac{|\vec{a}|\sin(\pi-\theta)}{|-\vec{b}|+|\vec{a}|\cos(\pi-\theta)}$$

$$= \frac{a\sin\theta}{b-a\cos\theta} \quad [\because |-\vec{b}| = |\vec{b}| = b]$$

where θ is the angle between $\vec{a}$ and $\vec{b}$, not between $\vec{a}$ and $-\vec{b}$.

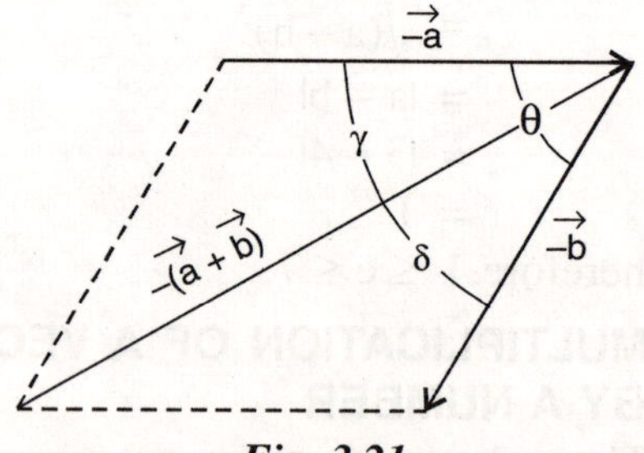

Fig. 3.21

Now, $|-(\vec{a}+\vec{b})|$ = magnitude of $-(\vec{a}+\vec{b})$

$$= |-\vec{a}-\vec{b}|$$

$$= \sqrt{|-\vec{a}|^2+|-\vec{b}|^2+2|-\vec{a}|+|-\vec{b}|\cos \angle\text{angle between } -a \text{ and } -b}$$

$$= \sqrt{a^2+b^2+2ab\cos\theta}$$

$$\tan \gamma = \frac{|-\vec{b}| \sin \theta}{|-\vec{a}| + |-\vec{b}| \cos \theta}$$

$$= \frac{b \sin \theta}{a + b \cos \theta}$$

$$\tan \delta = \frac{|-\vec{a}| \sin \theta}{|-\vec{b}| + |-\vec{a}| \cos \theta}$$

$$= \frac{a \sin \theta}{b + a \cos \theta}$$

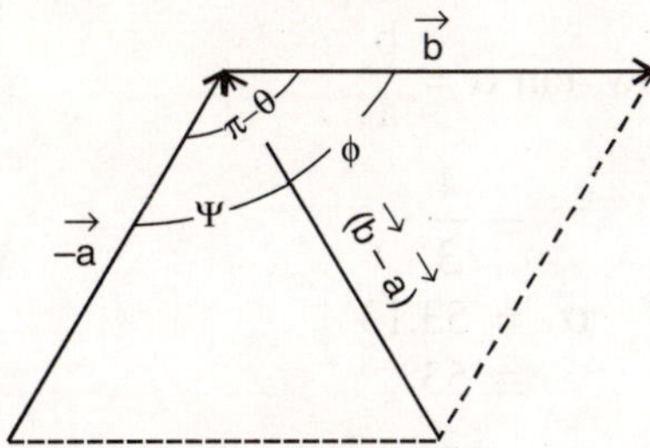

Fig. 3.22

Now, $|\vec{b} - \vec{a}|$

$$= \sqrt{|\vec{b}|^2 + |-\vec{a}|^2 + 2|\vec{b}||-\vec{a}| \cos \text{ angle between } \vec{a} \text{ and} -\vec{b}}$$

= angle between b and –a

$$= \sqrt{b^2 + a^2 + 2ab \cos(\pi - \theta)}$$

$$= \sqrt{a^2 + b^2 + 2ab \cos(\pi - \theta)}$$

$$= \sqrt{a^2 + b^2 - 2ab \cos \theta}$$

$$\tan \phi = \frac{|-\vec{a}| \sin(\pi - \theta)}{|\vec{b}| + |-\vec{a}| \cos(\pi - \theta)}$$

$$= \frac{a \sin \theta}{b - a \cos \theta}$$

and
$$\tan \psi = \frac{|\vec{b}| \sin(\pi - \theta)}{|-\vec{a}| + |\vec{b}| \cos(\pi - \theta)}$$

$$= \frac{b \sin \theta}{a - b \cos \theta}$$

Remark 3: *It is tedious to remember all the above formulae. We remember equations 3.1, 3.2 and 3.3 and accordingly find the angle (θ) between two given vectors and then use equations 3.1, 3.2 and 3.3 straight way for finding the magnitude (R) and directions (α) and (β) of the resultant.*

Example 2: *Two vectors of equal magnitude each of 5 unit have an angle 60° between them. Find the magnitude of (a) the sum of the vectors and (b) the difference of the vectors.*

Solution:

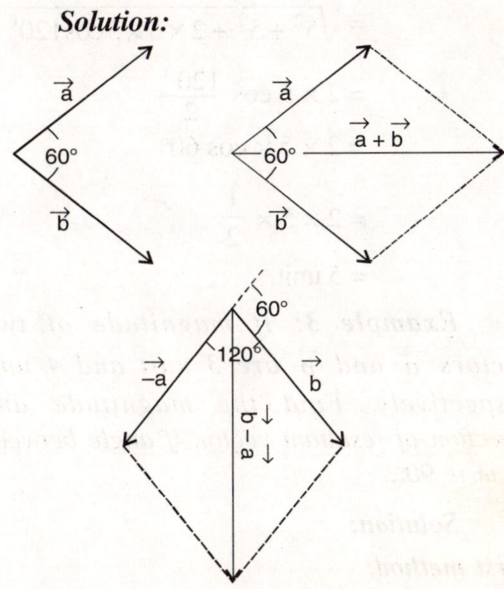

Fig. 3.23

Fig. 3.23 shows the construction of the sum $\vec{a} + \vec{b}$ and the difference $\vec{b} - \vec{a}$.

(a) $\vec{a} + \vec{b}$ is the sum of $\vec{a}$ and $\vec{b}$. Both have magnitude of 5 unit and angle between them is 60°. Thus the magnitude of sum is,

$$|\vec{a} + \vec{b}| = \sqrt{5^2 + 5^2 + 2 \times 5 \times 5 \cos 60^\circ}$$

$$= 2 \times 5 \times \cos \frac{60^\circ}{2}$$

[when equal magnitude then $|\vec{a} + \vec{b}|$

$$= 2a \cos \frac{\theta}{2} = 2\,b \cos \frac{\theta}{2}] \; [\because |\vec{a}| = |\vec{b}|]$$

$$= 10 \times \frac{\sqrt{3}}{2}$$

$$= 5\sqrt{3} \text{ unit.}$$

(b) $\vec{b} - \vec{a}$ is the sum of $\vec{b}$ and $-\vec{a}$. As shown in Fig. 3.23, the angle between $-\vec{a}$ and $\vec{b}$ is 120°. The magnitudes of both $-\vec{a}$ and $\vec{b}$ is 5 unit.

Therefore, $|\vec{a} - \vec{b}| = |\vec{a} - \vec{b}|$

$$= \sqrt{5^2 + 5^2 + 2 \times 5 \times 5 \cos 120^\circ}$$

$$= 2 \times 5 \cos \frac{120^\circ}{2}$$

$$= 2 \times 5 \times \cos 60^\circ$$

$$= 2 \times 5 \times \frac{1}{2}$$

$$= 5 \text{ unit.}$$

Example 3: *If magnitude of two vectors $\vec{a}$ and $\vec{b}$ are 3 unit and 4 unit respectively. Find the magnitude and direction of resultant vector, if angle between them is 90°.*

Solution:

First method:

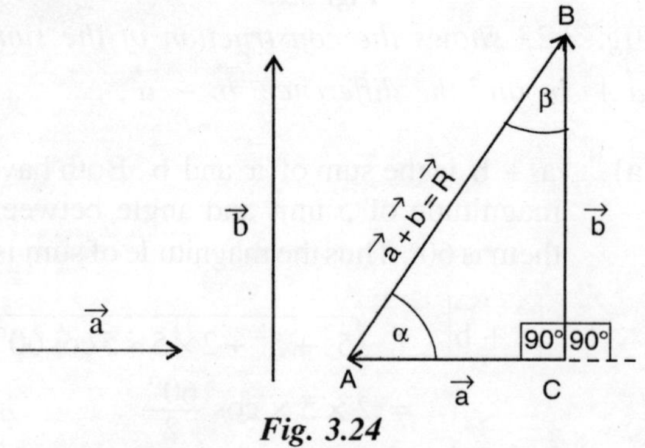

Fig. 3.24

Magnitude of

$$\vec{a} + = |\vec{R}|$$

$$R = \sqrt{|\vec{a}|^2 + |\vec{b}|^2 + 2|\vec{a}|\,|\vec{b}|\cos\theta}$$

$$= \sqrt{a^2 + b^2 + 2ab \cos 90^\circ}$$

$$= \sqrt{a^2 + b^2} \quad [\because \cos 90^\circ = 0]$$

$$= \sqrt{3^2 + 4^2}$$

$$= 5 \text{ unit.}$$

Now, $\tan \alpha = \frac{b}{a}$

$$= \frac{4}{3}$$

$$\therefore \quad \alpha = 53.13$$

$$\cong 53^\circ$$

and $\tan \beta = \frac{a}{b}$

$$= \frac{3}{4}$$

$$= 36.86^\circ$$

$$\cong 37^\circ.$$

Second method:

In ΔABC, we can directly apply **sine** rule.

i.e., $\frac{AC}{\sin\beta} = \frac{BC}{\sin\alpha} = \frac{AB}{\sin 90^\circ}$

or $\frac{a}{\sin\beta} = \frac{b}{\sin\alpha} = \frac{R}{1}$

or $\frac{a}{\sin\beta} = \frac{b}{\sin\alpha} = R$

or $\frac{3}{\sin\beta} = \frac{4}{\sin\alpha} = R$ ....(i)

Now in ΔABC, we have

$$\angle A + \angle B + \angle C = 180^\circ$$

or $\alpha + \beta + 90^\circ = 180^\circ$

or $\alpha + \beta = 90^\circ$ ....(ii)

From (i), $\dfrac{\sin\alpha}{\sin\beta} = \dfrac{4}{3}$

or $\dfrac{\sin\alpha + \sin\beta}{\sin\alpha - \sin\beta} = \dfrac{4+3}{4-3}$

or $\dfrac{2\sin\left(\dfrac{\alpha+\beta}{2}\right)\cos\left(\dfrac{\alpha+\beta}{2}\right)}{2\cos\left(\dfrac{\alpha+\beta}{2}\right)\sin\left(\dfrac{\alpha-\beta}{2}\right)} = \dfrac{7}{1} = 7$

or $\dfrac{\sin\left(\dfrac{\alpha+\beta}{2}\right)}{\sin\left(\dfrac{\alpha-\beta}{2}\right)} = 7$

or $\dfrac{\sin 45^\circ}{\sin\left(\dfrac{\alpha-\beta}{2}\right)} = 7$

[$\because$ from equation (ii), $\alpha + \beta = 90^\circ$]

or $\sin\left(\dfrac{\alpha-\beta}{2}\right) = \dfrac{\sin 45^\circ}{7}$

or $\alpha - \beta = 11.59^\circ \cong 12^\circ$ ....(iii)

From (ii) and (iii),

$$\alpha + \beta = 90^\circ$$
$$\alpha - \beta = 12^\circ$$
$$2\alpha = 102^\circ$$
$$\therefore \quad \alpha = 51^\circ$$

and $\beta = 39^\circ$

From equation (i),

$$\frac{3}{\sin\beta} = R$$

or $\dfrac{3}{\sin 39^\circ} = R$

Therefore, $R_2 = \dfrac{3}{\sin 39^\circ}$

$= 4.76$

$= 5$ unit.

Remark 4: *We can use above result for quick calculation. Fig.3.25. shows 3 – 4 – 5 technique i.e., if AC = 3 m, BC = 4 m and AB = 5 m, then triangle is right angled and acute angles are 37° and 53°, respectively.*

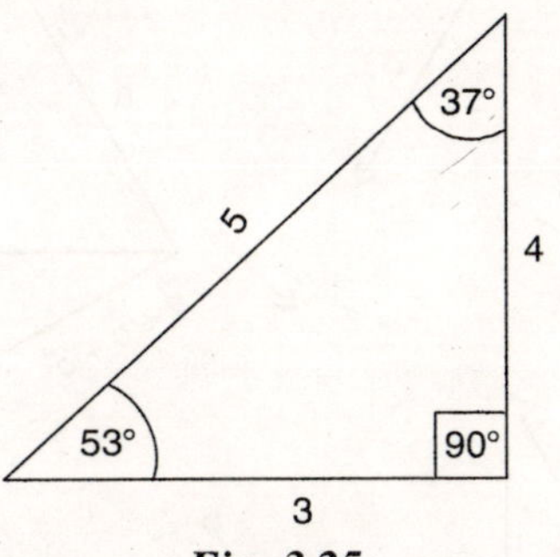

Fig. 3.25

3.8 ALGEBRA OF VECTOR ADDITION

3.8.1 Commutative law

Vector addition, defined in such a way that, it has two important properties. First, the order of addition does not matter. Adding $\vec{a}$ to $\vec{b}$ derives the same result as adding $\vec{b}$ to $\vec{a}$, that is,

$$\vec{a} + \vec{b} = \vec{b} + \vec{a}$$

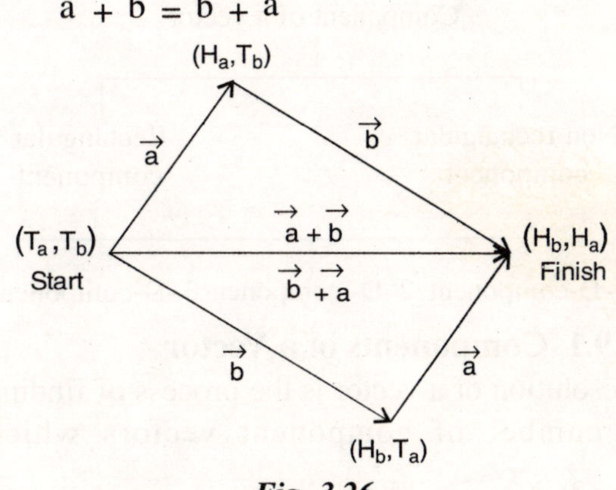

Fig. 3.26

3.8.2 Associative law

Second, when there are more than two vectors, we can group them in any order as we add them. Thus, if want to add vector $\vec{a}$, $\vec{b}$ and $\vec{c}$, we can add $\vec{a}$ and $\vec{b}$ first and then add their vector sum to $\vec{c}$.

We can also add $\vec{b}$ and $\vec{c}$ first and then add their vector sum to $\vec{a}$. That is,

$$(\vec{a} + \vec{b}) + \vec{c} = \vec{a} + (\vec{b} + \vec{c})$$

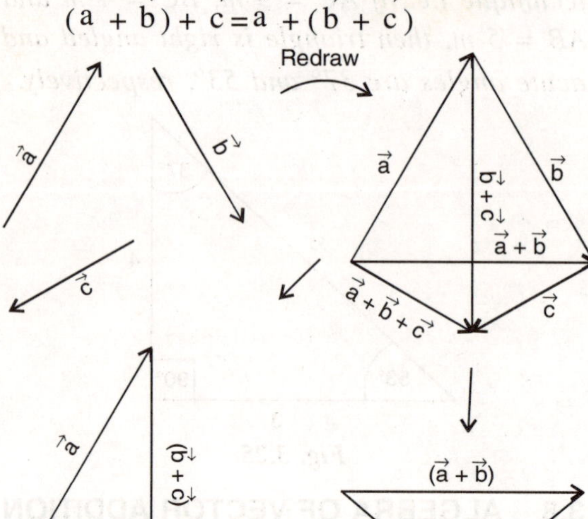

Fig. 3.27

***Note:** Difference of vectors does not obey these two properties.*

3.9 RESOLUTION OF A VECTOR

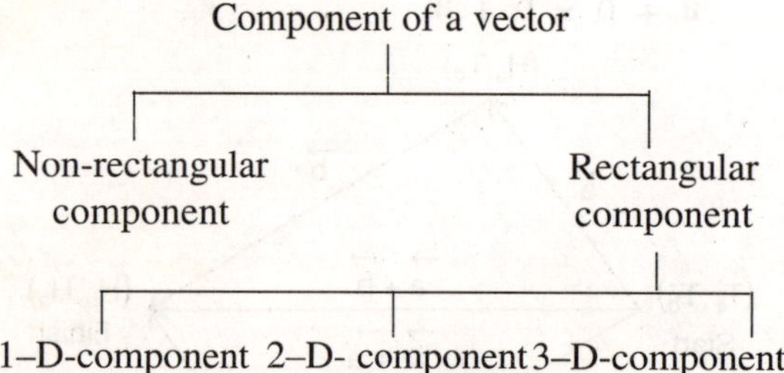

3.9.1 Components of a Vector

Resolution of a vector is the process of finding a number of component vectors which are when added have the same effect as a given vector. With the help of components we can add or subtract two or more vectors analytically instead of geometrically.

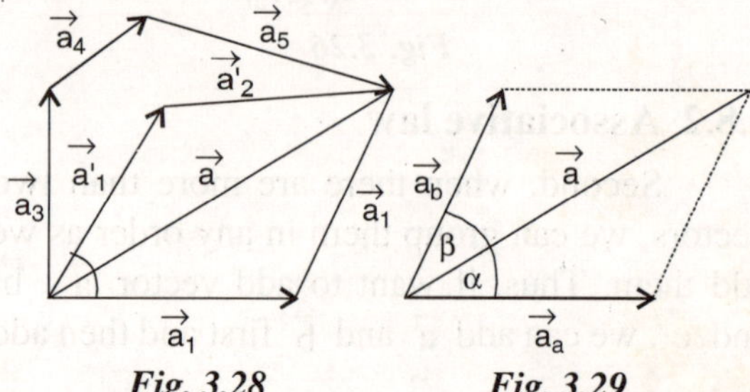

Fig. 3.28 *Fig. 3.29*

Let $\vec{a}$ be a given vector then from triangle law or polygon law we can write,

$$\vec{a} = \begin{cases} \vec{a}_1 + \vec{a}_2 \\ \vec{a}_1' + \vec{a}_2' \\ \vec{a}_3 + \vec{a}_4 + \vec{a}_5 \\ \vec{a}_\alpha + \vec{a}_\beta \end{cases}$$

$\{\vec{a}_3, \vec{a}_4, \vec{a}_5\}$, $\{\vec{a}_1, \vec{a}_2\}$, $\{\vec{a}_1', \vec{a}_2'\}$, $\{\vec{a}_\alpha, \vec{a}_\beta\}$ are components of vector $\vec{a}$ because their sum is equal to $\vec{a}$, these sets are called component vectors and since the angle between these vectors is not 90°, therefore, they are called non-rectangular components.

3.9.2 Rectangular components of a vector

If the angle between two components is equal to 90° then it is called the rectangular component. Thus we see, that for a given vector $\vec{a}$, there are infinite number of sets of componens like $\{\vec{a}_1, \vec{a}_2\}$, $\{\vec{a}_1', \vec{a}_2'\}$, $\{\vec{a}_3, \vec{a}_4, \vec{a}_5\}$, $\{\vec{a}_\alpha, \vec{a}_\beta\}$ and so on.

3.9.2.1 1D-Component

Component of vector $\vec{a}$ along a given direction (η) is called one-dimensional component.

According to the given Fig. 3.30, **a cos θ** is component of $\vec{a}$ along the direction $\vec{\eta}$ and **a cos θ** $\hat{\eta}$ is the component vector along the direction $\vec{\eta}$.

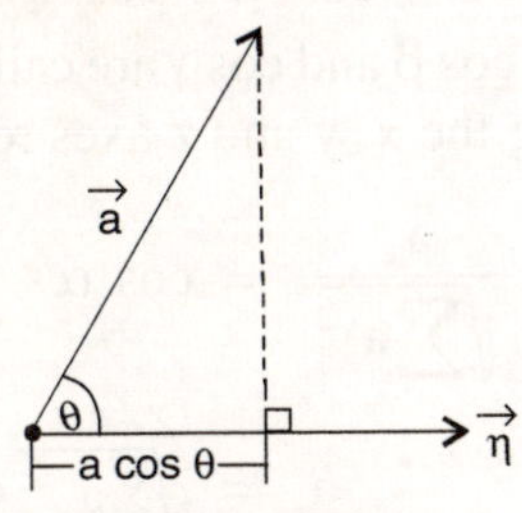

Fig. 3.30

3.9.2.2 2D-Components

Fig. 3.31 shows a vector $\vec{a}$ in the x-y plane drawn from the origin. The vector $\vec{a}$ makes angle α with the x-axis and β with y-axis respectively. Draw the perpendicular AB and AC from A to the x-axis and y-axis respectively. The length OB and OC are projections of $\vec{a}$ on the x-axis and y-axis respectively. They are called components and $a_x \vec{i}$, $a_y \vec{j}$ are component (vector) of vector $\vec{a}$. Now, from triangle rule we can write,

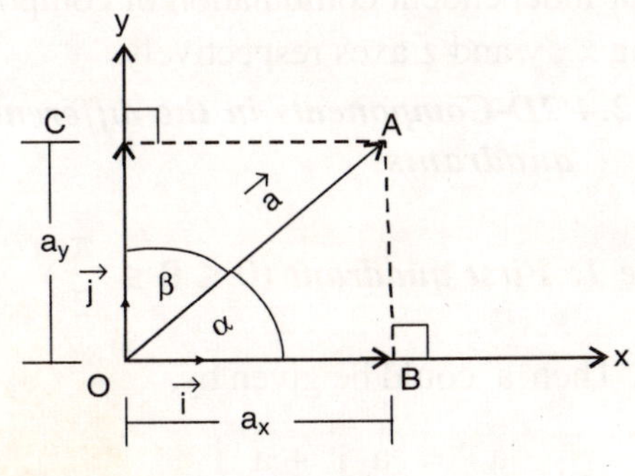

Fig. 3.31

$$\vec{a} = a_x \vec{i} + a_y \vec{j}$$

$$= a \cos\alpha\, \vec{i} + a \cos\beta\, \vec{j}$$

$$= a \cos\alpha\, \vec{i} + a \sin\alpha\, \vec{j}$$

$$= a \sin\beta\, \vec{i} + a \cos\beta\, \vec{j}$$

$$= a \sin\beta\, \vec{i} + a \sin\alpha\, \vec{j}$$

3.9.2.3 3D-Components

Let us suppose that $\vec{a}$ is a vector in space which makes an angles α, β and γ with the co-ordinate axes x, y and z respectively.

Then, from the shown Fig. 3.32, we get

$$\vec{a} = \vec{OB} + \vec{BA}$$

$$= \left(\vec{OD} + \vec{OC}\right) + \vec{BA}$$

$$= \left(\vec{OD} + \vec{OC}\right) + \vec{OE}$$

$$\left[\because \vec{OC} = \vec{DB} \text{ and } \vec{BA} = \vec{OE}\right]$$

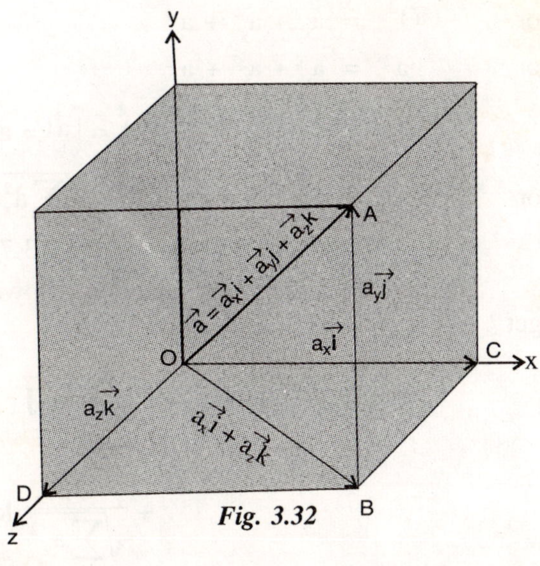

Fig. 3.32

$$= \vec{OC}+\vec{OD}+\vec{OE}$$

$$= \mathbf{a_x}\,\vec{\mathbf{i}}+\mathbf{a_y}\,\vec{\mathbf{j}}+\mathbf{a_z}\,\vec{\mathbf{k}}$$

Therefore,

$$\vec{\mathbf{a}} = \mathbf{a_x}\,\vec{\mathbf{i}}+\mathbf{a_y}\,\vec{\mathbf{j}}+\mathbf{a_z}\,\vec{\mathbf{k}} \quad ....(3.6)$$

where a_x, a_y and a_z are components along the rectangular axes x, y and z and $a_x\vec{\mathbf{i}}$, $a_y\vec{\mathbf{j}}$ and $a_z\vec{\mathbf{k}}$ are component vectors along the axes respectively.

Now from the Fig. 3.32, we can write

$$OA^2 = OB^2 + BA^2$$

(from right angle ΔOBA, OA is the hypotenuse)

$$= (OD^2 + DB^2) + BA^2$$

$$= OD^2 + OB^2 + BA^2$$

(from right angle ΔODB, OB is the hypotenuse)

$$= a_z^2 + a_x^2 + a_y^2$$

$$[\because\ OD = a_z,\ DB = a_x \text{ and } BA = a_y]$$

or $$(\vec{a})^2 = a_x^2 + a_y^2 + a_z^2$$

or $$a^2 = a_x^2 + a_y^2 + a_z^2$$

$$[\because OA = |\vec{a}| = a]$$

or $$a = \sqrt{a_x^2+a_y^2+a_z^2} = \sqrt{\sum a_x^2} \quad ....(3.7)$$

Dividing equation (i) by equation (ii), we get

$$\frac{\vec{a}}{a} = \frac{a_x}{\sqrt{\sum a_x^2}}\vec{\mathbf{i}} + \frac{a_y}{\sqrt{\sum a_y^2}}\vec{\mathbf{j}} + \frac{a_z}{\sqrt{\sum a_z^2}}\vec{\mathbf{k}}$$

or $$\hat{a} = \cos\alpha\ \vec{\mathbf{i}} + \cos\beta\ \vec{\mathbf{j}} + \cos\gamma\ \vec{\mathbf{k}}$$

or $$|\hat{a}| = \sqrt{\cos^2\alpha+\cos^2\beta+\cos^2\gamma}$$

or $$\sqrt{\cos^2\alpha+\cos^2\beta+\cos^2\gamma} = 1$$

$$[\because |\hat{a}| = 1]$$

$$\therefore\ \cos^2\alpha + \cos^2\beta + \cos^2\gamma = 1 \quad ....(3.8)$$

where cos α, cos β and cos γ are called direction cosines along the x, y and z axes respectively.

Since, $$\frac{a_x}{\sqrt{\sum a_x^2}} = \cos\alpha$$

or $$a_x = \sqrt{\sum a_x^2}\cos\alpha$$

$$= a\cos\alpha$$

Similarly, $$a_y = a\cos\beta$$

and $$a_z = a\cos\gamma$$

Therefore,

$$\vec{a} = a_x\vec{\mathbf{i}} + a_y\vec{\mathbf{j}} + a_z\vec{\mathbf{k}}$$

or $$\vec{a} = a\cos\vec{\mathbf{i}}\,\alpha + a\cos\beta\,\vec{\mathbf{j}} + a\cos\gamma\,\vec{\mathbf{k}}$$

Therefore, $$\mathbf{a} = (\mathbf{a}\cos\boldsymbol{\alpha})\ \vec{\mathbf{i}} + (\mathbf{a}\cos\boldsymbol{\beta})\ \vec{\mathbf{j}} + (\mathbf{a}\cos\boldsymbol{\gamma})\ \vec{\mathbf{k}} \quad ....(3.9)$$

So, a vector in space could be given by the linear independent combination of components along x , y and z axes respectively.

3.9.2.4 2D-Components in the different quadrants

Case 1: First quadrant $(0 \le \theta \le \frac{\pi}{2})$

Then $\vec{a}$ could be given by,

$$\vec{a} = a_x\vec{\mathbf{i}} + a_y\vec{\mathbf{j}}$$

$$= a\cos\theta\,\vec{\mathbf{i}} + a\sin\theta\,\vec{\mathbf{j}}$$

where θ is the angle made by $\vec{a}$ with positive direction of x-axis and $$a = \sqrt{a_x^2 + a_y^2}$$

$$\tan\theta = \frac{a_y}{a_x}$$

$$\therefore \quad \theta = \tan^{-1}\frac{a_y}{a_x}$$

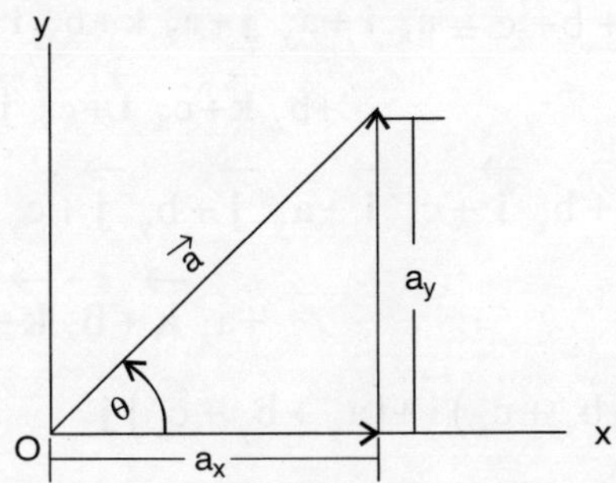

Fig. 3.33

Note: *Here a_x and a_y both are positive. Since, $0 \le \theta \le \frac{\pi}{2}$ therefore $\cos\theta$, $\sin\theta > 0$.*

Case 2: Second quadrant ($\frac{\pi}{2} < \theta \le \pi$)

$$\text{Then, } \vec{a} = a_x\vec{\mathbf{i}} + a_y\vec{\mathbf{j}}$$

$$= a\cos\theta\,\vec{\mathbf{i}} + a\sin\theta\,\vec{\mathbf{j}}$$

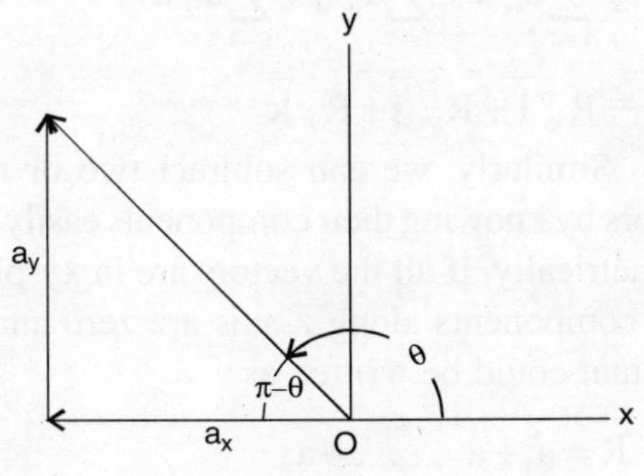

Fig. 3.34

where, θ is the angle made by $\vec{a}$ with positive direction of x-axis.

and $$a = \sqrt{a_x^{\,2} + a_y^{\,2}}$$

$$\tan\theta = \frac{a_y}{a_x}$$

Therefore,

$$\theta = \tan^{-1}\frac{a_y}{a_x}$$

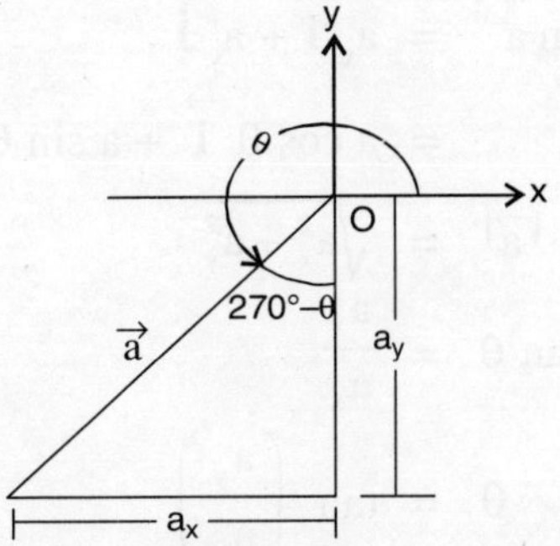

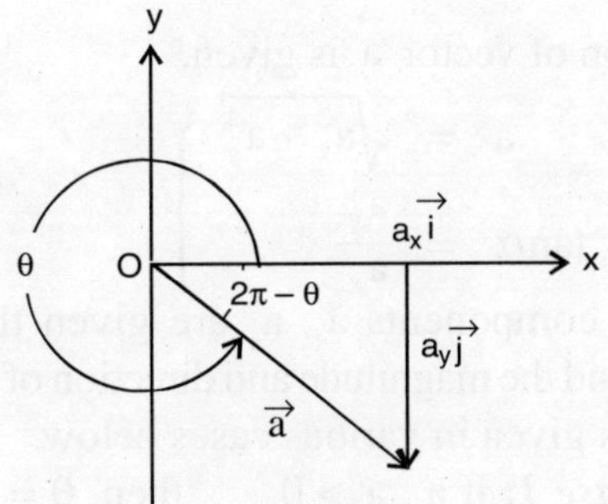

Fig. 3.35

Note: *Here a_x is negative and a_y is positive since, $\frac{\pi}{2} < \theta \le \pi$.*

$\therefore$ $\cos\theta < 0$ and $\sin\theta > 0$.

Case 3: Third quadrant ($\pi < \theta \le \frac{3\pi}{2}$)

$$\text{Then } \vec{a} = a_x\vec{\mathbf{i}} + a_y\vec{\mathbf{j}}$$

$$= a\cos\theta\,\vec{\mathbf{i}} + a\sin\theta\,\vec{\mathbf{j}}$$

Fig. 3.36

and $$|\vec{a}| = \sqrt{a_x^{\,2} + a_y^{\,2}}$$

$$\tan\theta = \frac{a_y}{a_x}$$

$$\therefore \quad \theta = \tan^{-1}\frac{a_y}{a_x}$$

Note: *Here, a_x, a_y both are negative. Since $\pi \le \theta \le \frac{3\pi}{2}$. ∴ sin θ, cos θ < 0.*

Case 4: Fourth quadrant ($\frac{3\pi}{2} \le \theta \le 2\pi$)

$$\text{Then } \vec{a} = a_x \vec{\mathbf{i}} + a_y \vec{\mathbf{j}}$$

$$= a\cos\theta\, \vec{\mathbf{i}} + a\sin\theta\, \vec{\mathbf{j}}$$

$$\text{and} \quad |\vec{a}| = \sqrt{a_x^2 + a_y^2}$$

$$\text{and} \quad \tan\theta = \frac{a_y}{a_x}$$

$$\therefore \quad \theta = \tan^{-1}\left(\frac{a_y}{a_x}\right)$$

Note: *Here, a_x is positive and a_y is negative. Since, $\frac{3\pi}{2} \le \theta \le 2\pi$. ∴ cos θ >0 and sin θ < 0.*

Therefore, we can write generally that,

$$a_x = a\cos\theta$$

$$\text{and} \quad a_y = a\sin\theta$$

$$\text{then,} \quad \vec{a} = a_x \vec{\mathbf{i}} + a_y \vec{\mathbf{j}}$$

where, θ is the angle made by $\vec{a}$ with positive direction of x-axis. While magnitude and direction of vector $\vec{a}$ is given,

$$\left.\begin{aligned} a &= \sqrt{a_x^2 + a_y^2} \\ \text{and} \quad \tan\alpha &= \frac{a_y}{a_x} \end{aligned}\right\} \quad \text{....(3.10)}$$

If components a_x, a_y are given then one could find the magnitude and direction of a vector and it is given in various cases below.

Case 1: if $a_x, a_y > 0$ then, $\theta = \alpha$

Case 2: if $a_x < 0$, $a_y > 0$ then, $\theta = \pi - \alpha$

Case 3: if $a_x < 0$, $a_y < 0$ then, $\theta = \pi + \alpha$

Case 4: if $a_x > 0$, $a_y < 0$ then, $\theta = 2\pi - \alpha$

3.10 ADDITION (SUBTRACTION) OF VECTORS

We can easily add two or more vectors if we know their components along the rectangular coordinate axes.

Let us assume that,

$$\vec{a} = a_x \vec{\mathbf{i}} + a_y \vec{\mathbf{j}} + a_z \vec{\mathbf{k}}$$

$$\vec{b} = b_x \vec{\mathbf{i}} + b_y \vec{\mathbf{j}} + b_z \vec{\mathbf{k}}$$

$$\text{and} \quad \vec{c} = c_x \vec{\mathbf{i}} + c_y \vec{\mathbf{j}} + c_z \vec{\mathbf{k}}$$

then, their resultant is given by

$$\vec{a}+\vec{b}+\vec{c} = a_x \vec{\mathbf{i}} + a_y \vec{\mathbf{j}} + a_z \vec{\mathbf{k}} + b_x \vec{\mathbf{i}} + b_y \vec{\mathbf{j}} + b_z \vec{\mathbf{k}} + c_x \vec{\mathbf{i}} + c_y \vec{\mathbf{j}} + c_z \vec{\mathbf{k}}$$

$$= a_x \vec{\mathbf{i}} + b_x \vec{\mathbf{i}} + c_x \vec{\mathbf{i}} + a_y \vec{\mathbf{j}} + b_y \vec{\mathbf{j}} + c_y \vec{\mathbf{j}} + a_z \vec{\mathbf{k}} + b_z \vec{\mathbf{k}} + c_z \vec{\mathbf{k}}$$

$$= (a_x + b_x + c_x)\vec{\mathbf{i}} + (a_y + b_y + c_y)\vec{\mathbf{j}} + (a_z + b_z + c_z)\vec{\mathbf{k}}$$

$$= \Sigma a_x \vec{\mathbf{i}} + \Sigma a_y \vec{\mathbf{j}} + \Sigma a_z \vec{\mathbf{k}}$$

$$\text{In general } \vec{a}_1 + \vec{a}_2 + \vec{a}_3 + \ldots\ldots + \vec{a}_n$$

$$= (a_{1x} + a_{2x} + a_{3x} + \ldots. + a_{nx})\vec{\mathbf{i}} + (a_{1y} + a_{2y} + a_{3y} + \ldots. + a_{ny})\vec{\mathbf{j}} + (a_{1z} + a_{2z} + a_{3z} + \ldots. + a_{nz})\vec{\mathbf{k}}$$

$$= \sum_{i=1}^{n} a_{ix} \vec{\mathbf{i}} + \sum_{i=1}^{n} a_{iy} \vec{\mathbf{j}} + \sum_{i=1}^{n} a_{iz} \vec{\mathbf{k}}$$

$$= R_x \vec{\mathbf{i}} + R_y \vec{\mathbf{j}} + R_z \vec{\mathbf{k}}$$

Similarly, we can subtract two or more vectors by knowing their components easily than geometrically. If all the vectors are in xy-plane, then components along z-axis are zero and the resultant could be written as

$$\vec{R} = \vec{a}_1 + \vec{a}_2 \ldots\ldots\ldots + \vec{a}_n$$

$$= (a_{1x} + a_{2x} + a_{3x} + \ldots\ldots + a_{nx})\vec{\mathbf{i}} + (a_{1y} + a_{2y} + a_{3y} \ldots\ldots\ldots + a_{ny})\vec{\mathbf{j}}$$

$$= \sum_{i=1}^{n} a_{ix} \vec{\mathbf{i}} + \sum_{i=1}^{n} a_{iy} \vec{\mathbf{j}}$$

$$= R_x \vec{i} + R_y \vec{j}$$

then, $$|\vec{R}| = \sqrt{\left(\sum_{i=1}^{n} a_{ix}\right)^2 + \left(\sum_{i=1}^{n} a_{iy}\right)^2}$$

and $$\tan\theta = \frac{\sum_{i=1}^{n} a_{iy}}{\sum_{i=1}^{n} a_{ix}}$$

where θ is the angle made by resultant with positive x-axis.

Example 4: *Find the resultant of the three vectors as shown in Fig.(3.37).*

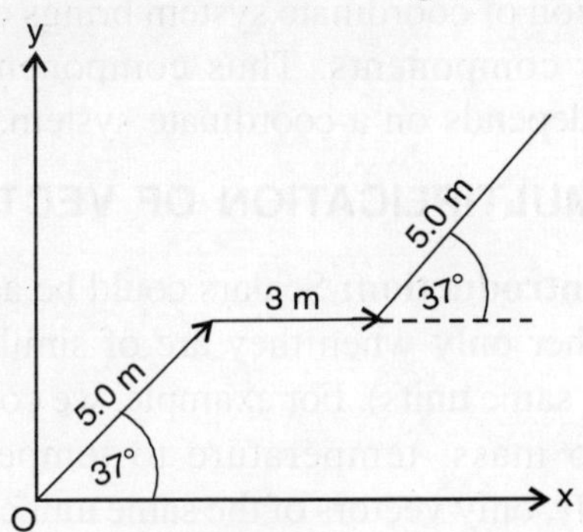

Fig. 3.37

Solution:

Vector	x-component	y-component
5.0 m	5.0 cos 37°	5.0 sin 37°
3.0 m	3.0 cos 0°	3.0 sin 0°
5.0 m	5.0 cos 37°	5.0 sin 37°

Now,

Σ x-components

$= 5.0 \cos 37° + 3.0 \cos 0° + 5.0 \cos 37°$

$= 10.98$ m

$\cong 11.0$ m

and Σ y-components

$= 5.0 \sin 37° + 3.0 \sin 0° + 5.0 \sin 37°$

$= 6.01$ m

$\cong 6.0$ m

$$\therefore |\vec{R}| = \sqrt{(\sum x\text{-components})^2 + (\sum y\text{-components})^2}$$

$$= \sqrt{11^2 + 6^2} = 12.52 \text{ m}$$

and $$\tan\theta = \frac{\Sigma\, y\text{-component}}{\Sigma\, x\text{-component}} = \frac{6}{11} = 0.54$$

∴ θ = angle made by resultant withpositive x-axis

$= \tan^{-1}(0.54) = 28.36°$

Resultant has been graphically represented in Fig. 3.38.

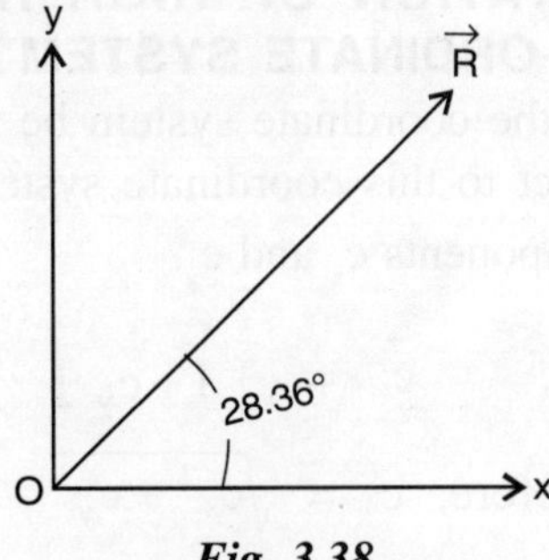

Fig. 3.38

3.11 TRANSLATION OF RIGHT HANDED COORDINATE SYSTEM

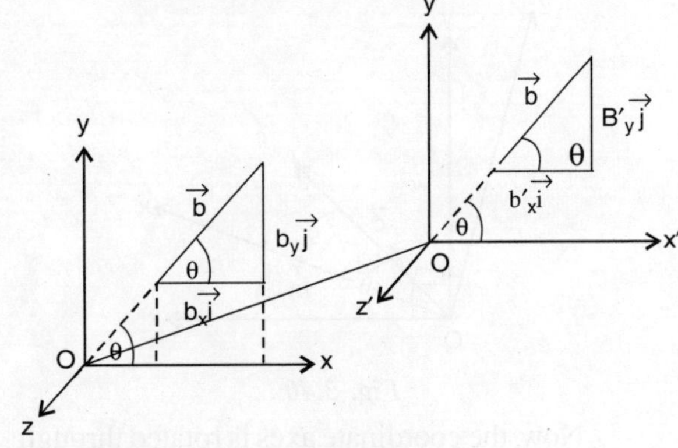

Fig. 3.39

Let us suppose that there are two co-ordinate systems viz. xyz and x′ y′ z′. Let the components for vector $\vec{b}$ be $(b_x, b_y, 0)$ and $(b_x', b_y', 0)$, with respect to two coordinate systems respectively.

Then, $\vec{b} = b_x \vec{\mathbf{i}} + b_y \vec{\mathbf{j}}$ [in xyz-axis]

$= b'_x \vec{i} + b'_y \vec{j}$ [in x′y′z′-axis]

Thus, $|\vec{b}| = \sqrt{b_x^2 + b_y^2} = \sqrt{b_x'^2 + b_y'^2}$

and $\tan\theta = \frac{b_y}{b_x} = \frac{b'_y}{b'_x}$

Therefore, whatever be the coordinate systems the magnitude and direction of a vector remain constant.

3.12 ROTATION OF RIGH HANDED CO-ORDINATE SYSTEM

Let the coordinate system be XOY and with respect to this coordinate system, vector $\vec{c}$ has components c_x and c_y.

Then, $\vec{c} = c_x \vec{\mathbf{i}} + c_y \vec{\mathbf{j}}$

Therefore, $c = \sqrt{c_x{}^2 + c_y{}^2}$

and $\tan\theta = \frac{c_y}{c_x}$

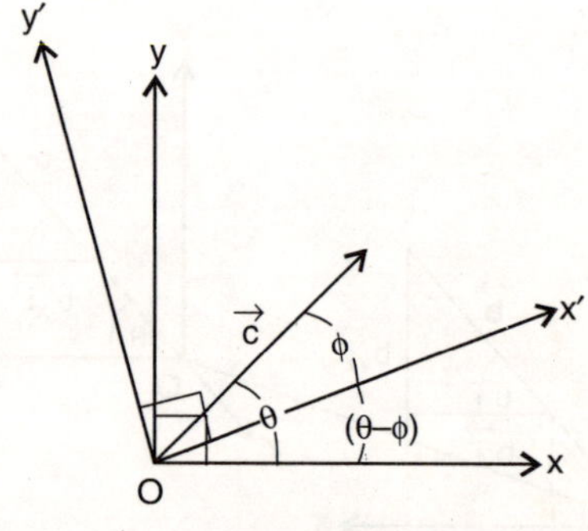

Fig. 3.40

Now, the coordinate axes is rotated through an angle (θ – ϕ), as shown in Fig. 3.40. Let us name this axes as X′OY′ and with respect to this frame of reference suppose c_x' and c_y' are components.

Then, $\vec{c} = c_x' \vec{\mathbf{i}} + c_y' \vec{\mathbf{j}}$

Therefore, $c = \sqrt{(c_x')^2 + (c_y')^2}$

and $\tan\phi = \frac{c_y'}{c_x'}$

Since, magnitude of $\vec{c}$ should be same in both the systems of coordinates.

$\therefore$ $c = \sqrt{c_x^2 + c_y^2}$

$= \sqrt{(c_x')^2 + (c_y')^2}$

Therefore, we see that rotation or translation of coordinate system brings changes to their components. Thus components of a vector depends on a coordinate system.

3.13 MULTIPLICATION OF VECTORS

Introduction: Scalars could be added to each other only when they are of similar kind (having same units). For example, we could add mass to mass, temperature to temperature. Similarly, only vectors of the same unit could be added to each other, *i.e.*, we could add velocity to velocity, torque to torque. But, we could not add velocity to torque. However, as for scalars, vectors of different units could be multiplied to each other to get another physical quantity. For example, force, is a result of mass × acceleration. Two vectors could be multiplied to get a scalar product or a vector product. We will study separately about vector product and scalar product. Multiplication of a vector by a scalar has simple meaning that product of a scalar **k** and a vector $\vec{a}$ written as $\mathbf{k}\vec{a}$, is defined to be a new vector whose magnitude is **k** times the magnitude of $\vec{a}$. The new vector has the same direction as $\vec{a}$ if **k** is positive and opposite direction if **k** is negative. To divide a vector by a scalar we simply multiply the vector by the reciprocal of the scalar.

3.14 SCALAR (DOT) PRODUCT

The scalar product of two vectors $\vec{a}$ and $\vec{b}$, written as $\vec{a} \cdot \vec{b}$, is defined as,

$$\vec{a} \cdot \vec{b} = |\vec{a}||\vec{b}| \cos \phi = ab \cos \phi \quad ....(3.11)$$

where $|\vec{a}| = a$ and $|\vec{b}| = b$, are magnitudes of vectors $\vec{a}$ and $\vec{b}$ respectively, and cos ϕ is the cosine of the angle ϕ between the two vectors that is cosine of vectorial angle. Since, $|\vec{a}| = a$ and $|\vec{b}| = b$ are scalars and cos ϕ is a pure number, the scalar product of the two vectors is scalar (scalar ⊙ scalar pure number = another scalar). Because of the notation $\vec{a} \cdot \vec{b}$, is also called the dot product of $\vec{a}$ and $\vec{b}$ and is spoken as $\vec{a}$ dot $\vec{b}$. As per the definition a number of physical quantities could be described as the scalar product of two vectors. For example, mechanical work, gravitational potential energy etc.

Mathematically

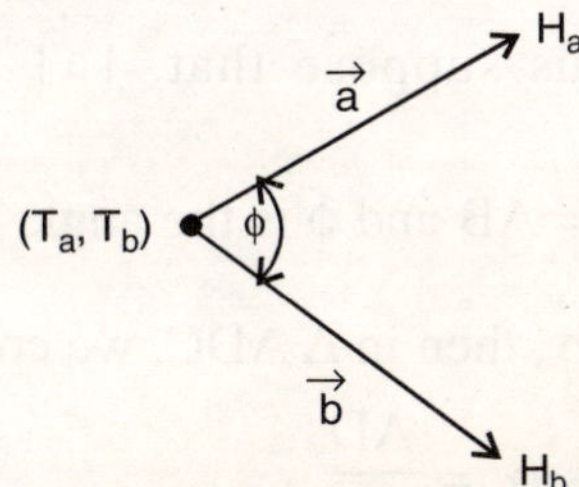

Fig. 3.41

$$\vec{a} \cdot \vec{b} = ab \cos \phi \text{ where, } 0° \leq \phi \leq \pi.$$

3.15 SOME IMPORTANT RESULTS

(i) When $\phi = 0$, i.e., two vectors are parallel then, $\vec{a} \cdot \vec{b} = ab \cos 0° = ab > 0$.

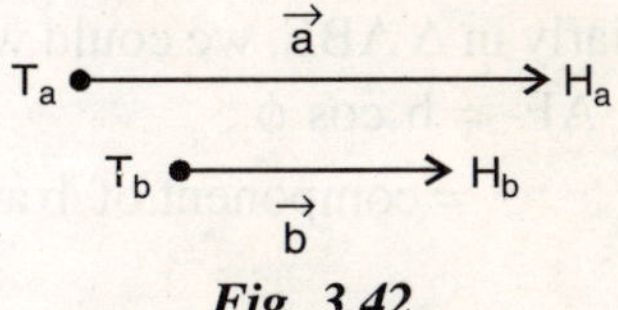

Fig. 3.42

(ii) When $\phi = \pi$, i.e., two vectors are antiparallel.

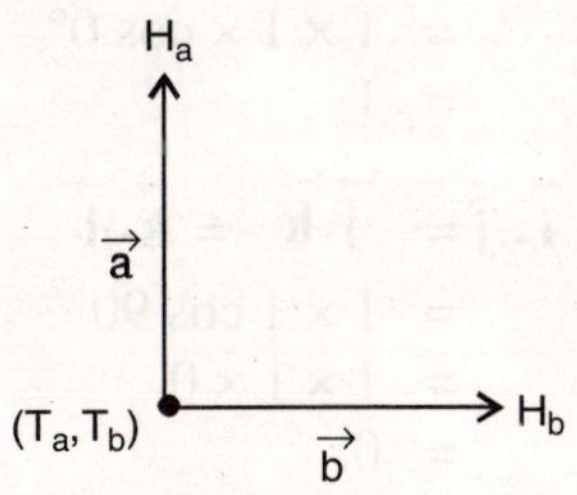

Fig. 3.43

Then, $\vec{a} \cdot \vec{b} = ab \cos \pi$

$= ab\,(-1)$

$= -\,ab < 0$

(iii) When $\phi = \dfrac{\pi}{2}$ i.e., two vectors are perpendicular.

Then, $\vec{a} \cdot \vec{b} = ab \cos \dfrac{\pi}{2}$

$= ab \times 0$

$= 0$

Therefore two vectors are perpendicular if $\vec{a} \cdot \vec{b} = 0$ where, $|\vec{a}| \neq 0$ and $|\vec{b}| \neq 0$.

From the above discussions, we conclude that the dot product of two vectors may be positive, negative or equal to zero.

Hence, we can say that $\vec{a} \cdot \vec{b} > 0$, $\vec{a} \cdot \vec{b} < 0$, $\vec{a} \cdot \vec{b} = 0$, when ϕ is acute, obtuse and equal to 90° respectively.

(iv) Dot product is commutative and distributive,

i.e., $\vec{a} \cdot \vec{b} = ab \cos \phi$

$= ba \cos \phi$

$= \vec{b} \cdot \vec{a}$

and $\vec{a} \cdot \left(\vec{b} + \vec{c}\right) = \vec{a} \cdot \vec{b} + \vec{a} \cdot \vec{c}$

(v) $\vec{i}\cdot\vec{i} = \vec{j}\cdot\vec{j} = \vec{k}\cdot\vec{k}$

$= 1 \times 1 \times \cos 0°$

$= 1$

and $\vec{i}\cdot\vec{j} = \vec{j}\cdot\vec{k} = \vec{k}\cdot\vec{i}$

$= 1 \times 1 \cos 90°$

$= 1 \times 1 \times 0$

$= 0.$

(vi) Consider two vectors $\vec{a}$ and $\vec{b}$, represented in terms of the unit vectors $\vec{i}$, $\vec{j}$ and $\vec{k}$ along coordinate axes then,

$$\vec{a} = a_x\,\vec{i} + a_y\,\vec{j} + a_z\,\vec{k}$$

and

$$\vec{b} = b_x\,\vec{i} + b_y\,\vec{j} + b_z\,\vec{k}$$

Therefore,

$$\vec{a}\cdot\vec{b} = \left(a_x\,\vec{i} + a_y\,\vec{j} + a_z\,\vec{k}\right)\cdot\left(b_x\,\vec{i} + b_y\,\vec{j} + b_z\,\vec{k}\right)$$

$$= a_x\,\vec{i}\cdot\left(b_x\,\vec{i} + b_y\,\vec{j} + b_z\,\vec{k}\right) + \left(b_x\,\vec{i} + b_y\,\vec{j} + b_z\,\vec{k}\right)$$

$$\cdot a_y\,\vec{j} + a_z\,\vec{k}\bullet(b_x\,\vec{i} + b_y\,\vec{j} + b_z\,\vec{k})$$

$$= a_x\,\mathbf{i}\cdot b_x\,\mathbf{i} + a_x\,\mathbf{i}\cdot b_y\,\mathbf{j} + a_x\,\mathbf{i}\cdot b_z\,\mathbf{k}$$

$$+ a_y\,\mathbf{j}\cdot b_x\,\mathbf{i} + a_y\,\mathbf{j}\cdot b_y\,\mathbf{j} + a_y\,\mathbf{j}\cdot b_z\,\mathbf{k}$$

$$+ a_z\,\mathbf{k}\cdot b_x\,\mathbf{i} + a_z\,\mathbf{k}\cdot b_y\,\mathbf{j} + a_z\,\mathbf{k}\cdot b_z\,\mathbf{k}$$

$$= a_x b_x\,\vec{i}\cdot\vec{i} + a_x b_y\,\vec{i}\cdot\vec{j} + a_x b_z\,\vec{i}\cdot\vec{k}$$

$$+ a_y b_x\,\vec{j}\cdot\vec{i} + a_y b_y\,\vec{j}\cdot\vec{j} + a_y b_z\,\vec{j}\cdot\vec{k}$$

$$+ a_z b_x\,\vec{k}\cdot\vec{i} + a_z b_y\,\vec{k}\cdot\vec{j} + a_z b_z\,\vec{k}\cdot\vec{k}$$

$$= a_x b_x + 0 + 0 + 0 + a_y b_y + 0 + 0 + 0 + a_z b_z$$

$$\left[\because\ \vec{i}\cdot\vec{i} = \vec{j}\cdot\vec{j} = \vec{k}\cdot\vec{k} = 1 \text{ and } \vec{i}\cdot\vec{j} = \vec{j}\cdot\vec{k} = \vec{k}\cdot\vec{i} = 0\right]$$

$$= a_x b_x + a_y b_y + a_z b_z$$

$$\therefore\quad \vec{\mathbf{a}}\cdot\vec{\mathbf{b}} = \mathbf{a_x b_x + a_y b_y + a_z b_z} \quad \text{....(3.12)}$$

or $|\vec{a}|\,|\vec{b}| \cos\theta = a_x b_x + a_y b_y + a_z b_z$

or $$\cos\theta = \frac{a_x b_x + a_y b_y + a_z b_z}{\sqrt{a_x^2 + a_y^2 + a_z^2}\,\sqrt{b_x^2 + b_y^2 + b_z^2}}$$

$\because$ $|\vec{a}| = \sqrt{a_x^2 + a_y^2 + a_z^2}$

and $|\vec{b}| = \sqrt{b_x^2 + b_y^2 + b_z^2}$

Therefore,

$$\mathbf{\cos\theta = \frac{\Sigma a_x b_x}{\sqrt{\Sigma a_x^2}\,\sqrt{\Sigma b_x^2}}} \quad \text{....(3.13)}$$

(vii) Geometrical meaning of dot product.

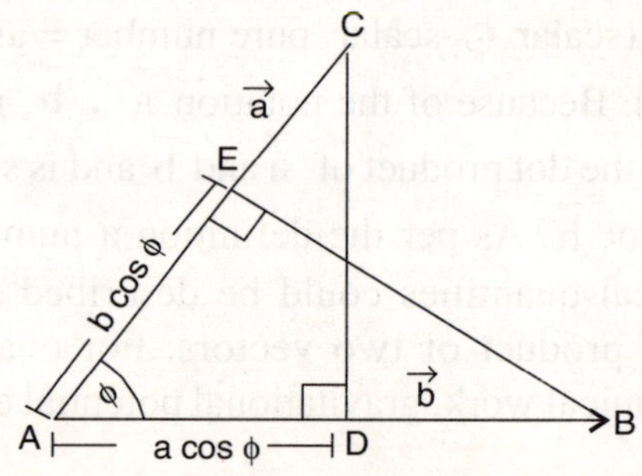

Fig. 3.44

Let us suppose that, $|\vec{a}|$ = AC and $|\vec{b}|$ = AB and ϕ is the angle between $\vec{a}$ and $\vec{b}$, then in Δ ADC, we could write

$$\cos\phi = \frac{AD}{AC}$$

or AD = AC cos ϕ

$= |\vec{a}| \cos\phi$

$\therefore$ AD = a cos ϕ

= component of $\vec{a}$ along $\vec{b}$

Similarly in Δ ABE, we could write

AE = b cos ϕ

= component of $\vec{b}$ along $\vec{a}$.

Now, $\vec{a} \cdot \vec{b} = ab \cos \phi$

$= a (b \cos \phi)$

$= (\text{magnitude of } \vec{a})$

$\times (\text{component of } \vec{b} \text{ along } \vec{a})$

$= ba \cos \theta$

$= (\text{magnitude of } \vec{b})$

$\times (\text{component of } \vec{a} \text{ along } \vec{b})$

So, we came to know that the scalar product of two vectors could be given by the product of magnitude of one vector and the component of the other vector in the direction of the first vector or projection of other vector in the direction of the first vector.

(viii) $\vec{a}^2 = \vec{a} \cdot \vec{a}$

$= |\vec{a}||\vec{a}| \cos 0°$

$= a \times a \cos 0°$

$= a^2 \times 1$

$= a^2$

$= |\vec{a}|^2$

That is square of a vector is equal to square of magnitude of the vector.

(ix) $\left(\vec{a}+\vec{b}\right)\cdot\left(\vec{a}-\vec{b}\right) = \vec{a}\cdot\vec{a}-\vec{a}\cdot\vec{b}+\vec{b}\cdot\vec{a}-\vec{b}\cdot\vec{b}$

$= a^2 - \vec{a}\cdot\vec{b}+\vec{a}\cdot\vec{b}-b^2$

$\left[\because \vec{a}\cdot\vec{b}=\vec{b}\cdot\vec{a}\right]$

$= a^2 - b^2$

(x) $\left\{\vec{a}+\vec{b}\right\}^2 = \left(\vec{a}+\vec{b}\right)\cdot\left(\vec{a}+\vec{b}\right)$

$= \vec{a}\cdot\vec{a}+\vec{a}\cdot\vec{b}+\vec{b}\cdot\vec{a}+\vec{b}\cdot\vec{b}$

$= a^2+\vec{a}\cdot\vec{b}+\vec{a}\cdot\vec{b}+b^2$

$= a^2 + 2\vec{a}\cdot\vec{b} + b^2$

$= \vec{a}^2+2\vec{a}\cdot\vec{b}+\vec{b}^2$

$\left[\because a^2 = \vec{a}^2, b^2 = \vec{b}^2\right]$

(xi) $\left\{\vec{a}-\vec{b}\right\}^2 = \left(\vec{a}-\vec{b}\right)\cdot\left(\vec{a}-\vec{b}\right)$

$= \vec{a}\cdot\vec{a}-\vec{a}\cdot\vec{b}-\vec{b}\cdot\vec{a}+\vec{b}\cdot\vec{b}$

$= a^2-2\vec{a}\cdot\vec{b}+\vec{b}^2$

(xii) $\vec{a}\cdot\vec{b} = ab \cos\theta$

or $\dfrac{\vec{a}\cdot\vec{b}}{ab} = \cos\theta$

or $\left(\dfrac{\vec{a}}{a}\right)\cdot\left(\dfrac{\vec{b}}{b}\right) = \cos\theta$

or $\hat{a}\cdot\hat{b} = \cos\theta$

where $\hat{a}$ and $\hat{b}$ are unit vectors in direction of $\vec{a}$ and $\vec{b}$ respectively and θ is the angle between $\vec{a}$ and $\vec{b}$.

(xiii)

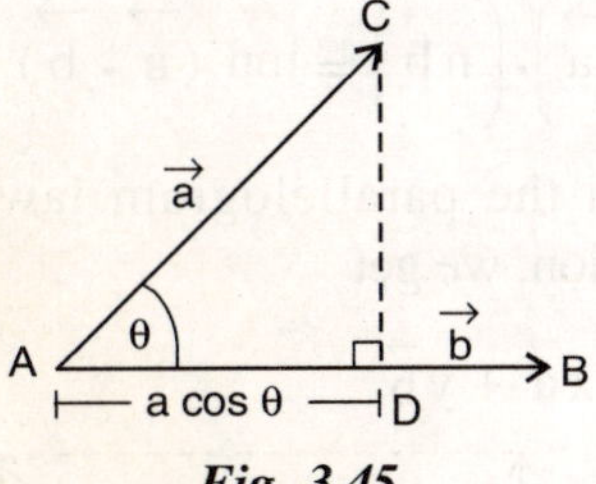

Fig. 3.45

As we know that, $\vec{a}\cdot\vec{b} = ab\cos\theta$

$= b\,(a \cos\theta)$

$= b\,(\text{projection of } \vec{a} \text{ in the direction of } \vec{b})$

$= b\,(\text{component of } \vec{a} \text{ in the direction of } \vec{b})$

$= b \times AD$

Therefore component (projection) of $\vec{a}$ in the direction of $\vec{b} = \dfrac{\vec{a}\cdot\vec{b}}{b}$

$$= \vec{a}\cdot\hat{b}$$

where $\hat{b}$ is the unit vector in the direction of $\vec{b}$.

Component vector of $\vec{a}$ in the direction of

$$\vec{b} = \left(\vec{a}\cdot\hat{b}\right)\hat{b}$$

$$= \left(\frac{\vec{a}\cdot\vec{b}}{b}\right)\hat{b}$$

$$= \left(\frac{(\vec{a}\cdot\vec{b})}{b}\right)\frac{\vec{b}}{b} = \frac{(\vec{a}\cdot\vec{b})\vec{b}}{b^2}$$

$$\text{Therefore, } \vec{AD} = \left(\frac{\vec{a}\cdot\vec{b}}{b^2}\right)\vec{b}$$

Similarly, component vector of $\vec{b}$ in the direction of $\vec{a}$ is given by $\left(\dfrac{\vec{a}\cdot\vec{b}}{a^2}\right)\vec{a}$

(xiv) $\left(m\vec{a}\right)\cdot\left(n\vec{b}\right) = mn\ (\vec{a}\cdot\vec{b})$

(xv) From the parallelogram law of vector addition, we get

$$\vec{r}_1 = x\vec{a} + y\vec{b}$$

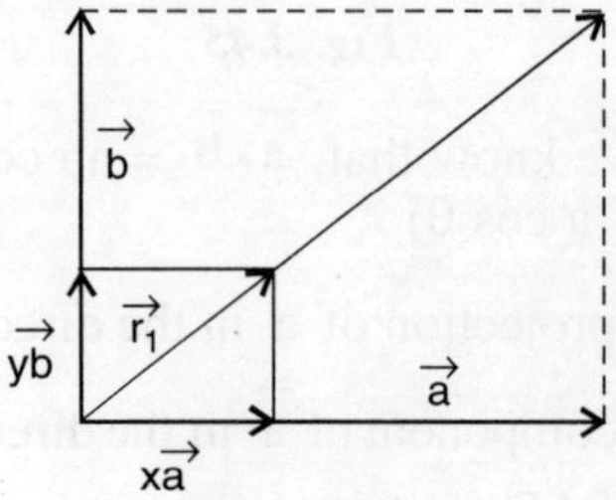

Fig. 3.46

where, $0 < x < 1$ and $0 < y < 1$ ....(i)

$$\text{or} \quad \vec{r}_1\cdot\vec{a} = x\,\vec{a}\cdot\vec{a} + y\,\vec{b}\cdot\vec{a}$$

$$= xa^2 + y\,\vec{a}\cdot\vec{b}$$

$$\left[\because \vec{a}\cdot\vec{a} = \vec{a}^2 = a^2 \text{ and } \vec{a}\cdot\vec{b} = \vec{b}\cdot\vec{a}\right]$$

$$= x\,a^2 + y \times 0$$

$$\left[\because \vec{a}\perp\vec{b} \therefore \vec{a}\cdot\vec{b} = 0\right]$$

$$= xa^2$$

$$\therefore \quad x = \frac{\vec{r}_1\cdot\vec{a}}{a^2}$$

$$\therefore \quad x\hat{a} = \left(\frac{\vec{r}_1\cdot\vec{a}}{a^3}\right)\vec{a} \quad \text{....(ii)}$$

From equation (i), we can write

$$y\vec{b} = \vec{x}_1 - x\vec{a}$$

$$\text{or} \quad y\vec{b} = \vec{r}_1 - \left(\frac{\vec{r}_1\cdot\vec{a}}{a^2}\right)\vec{a}$$

Component of vector $\vec{r}_1$ in the perpendicular direction to $\vec{a}$, where a is in the plane of $\vec{r}$ and $\vec{a}$ given vector $\vec{a}$ in the plane of $\vec{r}$ and a and $x\vec{a}$ = $\left(\dfrac{\vec{r}_1\cdot\vec{a}}{a^2}\right)\vec{a}$ = component of a rector $\vec{r}_1$ in the direction of a given vector.

i.e., $\vec{r}_1 = \vec{r}_1$, parallel to $\vec{a}$ + $\vec{r}_1$, perpendicular to $\vec{a}$

$$= \vec{r}_1 \parallel \text{to } \vec{a} + \vec{r}_1 \perp \vec{a}$$

where, $\vec{r}_1 \parallel \vec{a} = x\,\vec{a}$

$$= \left(\frac{\vec{r}_1 \cdot \vec{a}}{a^2}\right)\vec{a}$$

and $\vec{x}_1 \perp \vec{a} = y\,\vec{b}$

$$= \vec{r}_1 - \left(\frac{\vec{r}_1 \cdot \vec{a}}{a^2}\right)\vec{a}$$

(xvi)

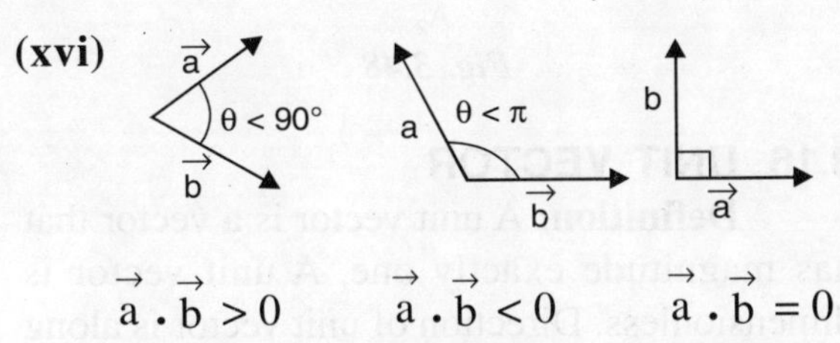

$\vec{a} \cdot \vec{b} > 0$ $\quad \vec{a} \cdot \vec{b} < 0$ $\quad \vec{a} \cdot \vec{b} = 0$

Fig. 3.47

(xvii) We can represent the orthogonal unit scalar product as given in table (3.1).

Table 3.1

•	$\vec{i}$	$\vec{j}$	$\vec{k}$
$\vec{i}$	1	0	0
$\vec{j}$	0	1	0
$\vec{k}$	0	0	1

(xviii) Let $\vec{r} = r_x\,\vec{\mathbf{i}} + r_y\,\vec{\mathbf{j}} + r_z\,\vec{\mathbf{k}}$

Then $\vec{r} = \left(\vec{r} \cdot \vec{\mathbf{i}}\right)\vec{\mathbf{i}} + \left(\vec{r} \cdot \vec{\mathbf{j}}\right)\vec{\mathbf{j}} + \left(\vec{r} \cdot \vec{\mathbf{k}}\right)\vec{\mathbf{k}}$

As we know that, component of $\vec{a}$ in the direction of

$$\vec{b} = \vec{a} \cdot \hat{b}$$

Therefore, $r_x = \vec{r} \cdot \vec{\mathbf{i}},\ x_y = \vec{r} \cdot \vec{\mathbf{j}},\ r_z = \vec{r} \cdot \vec{\mathbf{k}}$

Example 5: *If the angle between two vectors* $\vec{a} = a_x\,\vec{\mathbf{i}} + 3\,\vec{\mathbf{j}} + 4\,\vec{\mathbf{k}}$ *and* $\vec{b} = \vec{\mathbf{i}} - 2\,\vec{\mathbf{j}} + 3\,\vec{\mathbf{k}}$ *is 66.6°. Find* a_x.

Solution: $\vec{a} = a_x\,\vec{\mathbf{i}} + 3\,\vec{\mathbf{j}} + 4\,\vec{\mathbf{k}}$

and $\vec{b} = \vec{\mathbf{i}} - 2\,\vec{\mathbf{j}} + 3\,\vec{\mathbf{k}}$

$\therefore\ a_x = a_x,\ a_y = 3,\ a_z = 4$

and $b_x = 1,\ b_y = -2,\ b_z = 3$

As we know that,

$$\vec{a} \cdot \vec{b} = a_x b_x + a_y b_y + a_z b_z$$
$$= a_x 1 + 3 - 2 + 4\ 3$$
$$= a_x - 6 + 12$$
$$= a_x + 6$$

or $|\vec{a}|\,|\vec{b}| \cos\theta = a_x + 6$

or $\left(\sqrt{a_x^2 + 3^2 + 4^2} \times \sqrt{1^2 + (-2)^2 + 3^2}\right)\cos\theta$

$$= a_x + 6$$

or $\left(\sqrt{a_x^2 + 25} \times \sqrt{14}\right)\cos 66.6° = a_x + 6$

Squaring both sides, we get

or $(a_x^2 + 25) \times (14) \cos^2 66.6°$

$$= (a_x + 6)^2$$
$$= a_x^2 + 12a_x + 36$$

or $(14\,a_x^2 + 350) \times \cos^2 66.6°$

$$= a_x^2 + 12a_x + 36$$

or $14 \times \cos^2 66.6\ a_x^2 + 350 \times \cos^2 66.6°$

$$= a_x^2 + 12a_x + 36$$

or $14 \times 0.397 \times 0.397\ a_x^2 + 350$

$$\times 0.397 \times 0.397$$
$$= a_x^2 + 12a_x + 36$$

$[\because \cos 66.6° = 0.397]$

or $2.2\,a_x^2 + 55.16 = a_x^2 + 12a_x + 36$

or $1.2\,a_x^2 - 12\,a_x + 19.16 = 0$

$$\therefore a_x = \frac{-(-12) \pm \sqrt{12^2 - 4 \times 1.2 \times 19.16}}{1.2 \times 2}$$

$$= \frac{12 \pm \sqrt{144 - 91.96}}{2.4} = \frac{12 \pm 7.21}{2.4}$$

$$= \frac{12 + 7.21}{2.4}, \frac{12 - 7.21}{2.4}$$

$$= \frac{19.21}{2.4}, \frac{4.79}{2.4}$$

$$= 8, 2$$

$$\therefore a_x = 2$$

Since, $a_x = 8$ does not satisfy that the condition that the angle between them is 66.6°. Note that the extra result comes by squaring.

Example 6: *Let N be an integer greater than one and even number; then prove that,*

$$\cos 0 + \cos\frac{2\pi}{N} + \cos\frac{4\pi}{N} + ... + \cos (N-1)\frac{2\pi}{N} = 0 \text{ this is } \sum_{n=0}^{n=N-1} \cos\frac{2\pi n}{N} = 0$$

Solution:

Let origin be the centre of a regular polygon of sides n.

[∵ Orign is the centre of polygon]

$$\therefore \angle A_1OA_2 = \angle A_2OA_3 = ... = A_nOA_1 = \frac{2\pi}{N}$$

From the symmetry of the circle, we can get

$$\vec{OA}_1 + \vec{OA}_2 + \vec{OA}_3 + ... + \vec{OA}_n = 0$$

or $(\vec{OA}_1)\cdot(\vec{OA}_1 + \vec{OA}_2 + \vec{OA}_3 + ... + \vec{OA}_n) = 0$

or $\vec{OA}_1 \cdot \vec{OA}_1 + \vec{OA}_1 \cdot \vec{OA}_2 + ... + \vec{OA}_1 \cdot \vec{OA}_n = 0$

or $\cos 0 + \cos\frac{2\pi}{N} + \cos\frac{4\pi}{N} + ... + \cos(N-1)\frac{2\pi}{N} = 0$

$$\left[\because \left|\vec{OA_1}\right| = \left|\vec{OA_2}\right| = ... = \left|\vec{OA_n}\right|\right]$$

$$\therefore \cos 0 + \cos\frac{2\pi}{N} + \cos\frac{4\pi}{N} + ... + \cos (N-1)\frac{2\pi}{N} = 0$$

or, $$\sum_{n=0}^{n=N-1} \cos\frac{2n\pi}{N} = 0$$

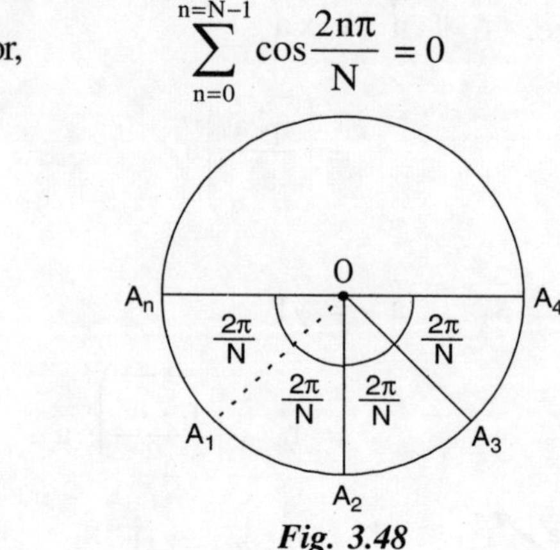

Fig. 3.48

3.16 UNIT VECTOR

Definition: A unit vector is a vector that has magnitude exactly one. A unit vector is dimensionless. Direction of unit vector is along the given vector.

Let $\vec{a}$ be a unit vector then unit vector along

$$\vec{a} = \frac{\vec{a}}{|\vec{a}|} \text{ that is } \hat{U}_a = \frac{\vec{a}}{|\vec{a}|}$$

$$\therefore \quad \vec{a} = |\vec{a}|\,\hat{U}_a \qquad ....(3.14)$$

where, $\hat{U}_a$ is a unit vector in the direction of $\vec{a}$. In the rectangular coordinate system the special symbols $\vec{i}$, $\vec{j}$ and $\vec{k}$ are usually used for unit vectors in the positive x, y, and z axes respectively.

A vector is represented in terms of unit vectors $\vec{i}$, $\vec{j}$ and $\vec{k}$.

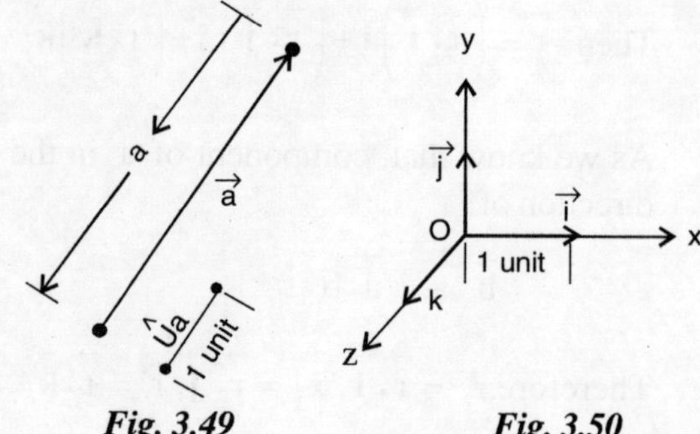

Fig. 3.49 *Fig. 3.50*

$\vec{a} = a_x\vec{i} + a_y\vec{j} + a_z\vec{k}$ in space

$\vec{b} = b_x\vec{i} + b_y\vec{j} + 0\vec{k}$ in xy-plane

$\vec{c} = c_x\vec{i} + c_z\vec{k} + 0\vec{j}$ in xz-plane

$\vec{d} = d_y\vec{j} + d_z\vec{k} + 0\vec{i}$ in yz-plane

Example 7: *Let $\vec{b} = \vec{i} + \vec{j} - \vec{k}$ then, using the definition of unit vector, find the unit vector along $\vec{b}$.*

Solution:

Here, $|\vec{b}| = \sqrt{3}$

$$\therefore \quad \hat{U}_b = \frac{\vec{b}}{|\vec{b}|}$$

$$= \frac{\vec{i} + \vec{j} - \vec{k}}{\sqrt{3}}$$

3.17 ZERO VECTOR

If we add two vectors which have equal magnitude but are in opposite directions then we get the zero or null vector. The direction of a zero vector is indeterminant. In other words, vector in which, tail and head points coincide or vector product of two equal or parallel vectors is nothing but zero vector. Zero vector is denoted by $\vec{0}$.

Some properties of zero vector

(*i*) $\vec{a} + \vec{0} = \vec{a}$

(*ii*) $\vec{a} - \vec{0} = \vec{a}$

(*iii*) $\vec{a} \bullet \vec{0} = 0$

(*iv*) $\vec{a} \times \vec{0} = \vec{0}$

(*v*) $\lambda\vec{0} = \vec{0}$ where, λ is scalar.

3.18 VECTOR (CROSS) PRODUCT

The vector product of two vectors, $\vec{a}$ and $\vec{b}$ written as $\vec{a} \times \vec{b}$, produces a third vector $\vec{c}$ given by $\vec{c} = (ab \sin \theta)\,\hat{n}$ ....(3.15)

where θ is smaller of the two angles between $\vec{a}$ and $\vec{b}$ and $\hat{n}$ is the unit vector perpendicular to plane containing vectors $\vec{a}$ and $\vec{b}$, the sense of $\hat{n}$ is given by right handed rule. Because of the notation, $\vec{a} \times \vec{b}$ is knows as cross product and in speech it is "$\vec{a}$ cross $\vec{b}$". From the above definition, it is clear that $|\vec{c}| = ab \sin \theta$.

3.19 RIGHT HAND RULE

Draw the two vectors $\vec{a}$ and $\vec{b}$ such that the tails coincide without altering their orientation and imagine a line perpendicular to their plane passes through the coinciding point of the tails. Now, place your stretched right hand palm perpendicular to the plane of $\vec{a}$ and $\vec{b}$ in such a way that the fingers are along the vector $\vec{a}$ and when the fingers are closed they go towards $\vec{b}$. Then, direction of the thumb gives the direction of the vector $\vec{c}$ as shown in the following figures.

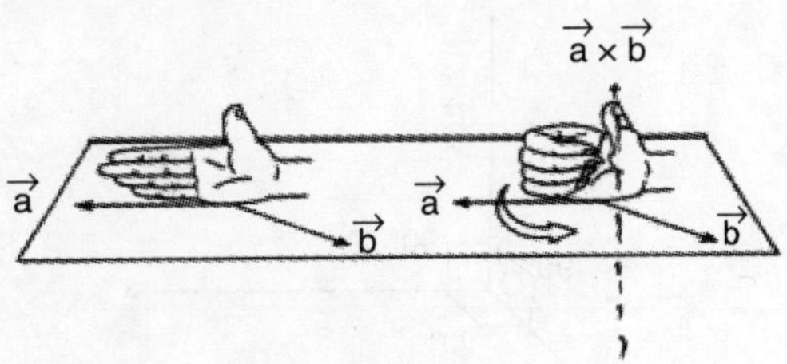

Fig. 3.51

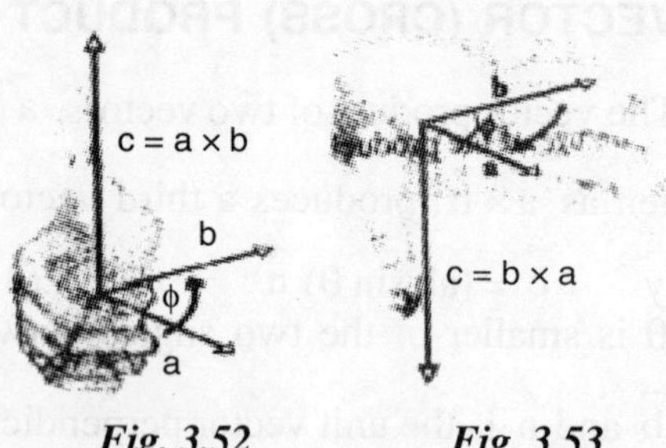

Fig. 3.52 *Fig. 3.53*

From the figures we conclude that $\vec{a} \times \vec{b} = -\vec{b} \times \vec{a}$, i.e., cross product is non commutative

Therefore $\vec{a} \times \vec{b} \neq \vec{a} \times \vec{c}$. Some examples of vector product are torque, angular momentum etc.

3.20 IMPORTANT RESULTS ON CROSS PRODUCT

(i) Cross product follows the distributive law

i.e., $\vec{a} \times \left(\vec{b} + \vec{c}\right) = \vec{a} \times \vec{b} + \vec{a} \times \vec{c}$

(ii) Cross product does not follow associative law.

i.e., $\vec{a} \times \left(\vec{b} \times \vec{c}\right) \neq \left(\vec{a} \times \vec{b}\right) \times \vec{c}$

(iii) If $\theta = 90°$ is the angle between vector $\vec{a}$, $\vec{b}$ and $\vec{c} = \vec{a} \times \vec{b}$. Then all are at right angles to one another and form a three dimensional right handed co-ordinate system.

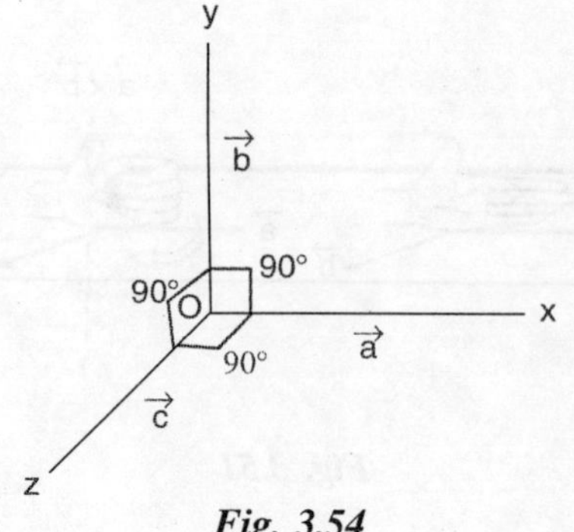

Fig. 3.54

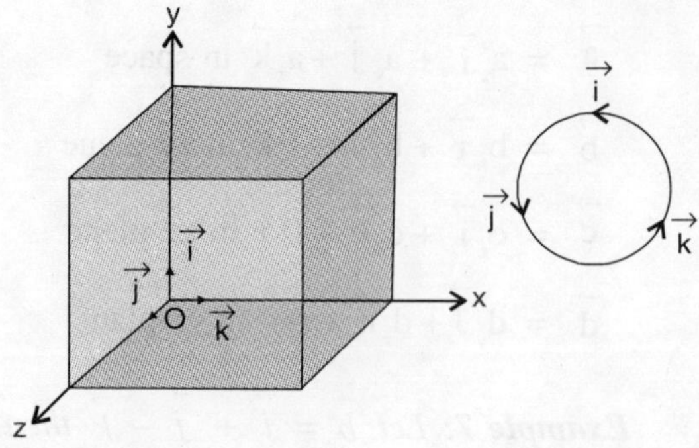

Fig. 3.55 *Fig. 3.56*

(iv) In right handed coordinate system

$\vec{i} \times \vec{j} = \vec{k}$, $\vec{j} \times \vec{k} = \vec{i}$ and $\vec{k} \times \vec{i} = \vec{j}$

or $\vec{j} \times \vec{i} = -\vec{k}$, $\vec{k} \times \vec{j} = -\vec{i}$ and $\vec{i} \times \vec{k} = -\vec{j}$

(v) $\vec{i} \times \vec{i} = \vec{j} \times \vec{j} = \vec{k} \times \vec{k} = (1 \times 1 \times \sin 0)\, \vec{n}$

$= (1 \times 1 \times 0)\, \vec{n} = \vec{0}$

(vi) When $\phi = 0°$, i.e., when $\vec{a}$ and $\vec{b}$ are parallel and non-zero vectors.

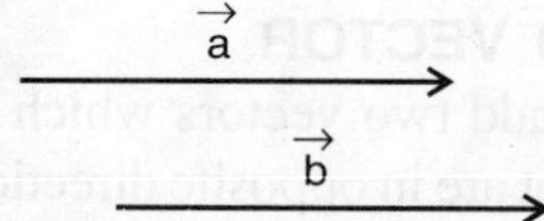

then, $|\vec{a} \times \vec{b}| = ab \sin \phi = ab \sin 0 = ab.0 = 0$. Therefore two vectors are parallel or collinear if $|\vec{a} \times \vec{b}| = 0$.

(vii) When $\phi = 90°$ i.e., when $\vec{a}$ and $\vec{b}$ are perpendicular.

then, $|\vec{a} \times \vec{b}| = ab \sin \dfrac{\pi}{2}$

$\vec{a}$

$\vec{b}$

$= ab \times 1$

$= ab$ (maximum)

(viii) When $\phi = 180°$ i.e., when $\vec{a}$ and $\vec{b}$ are antiparallel.

then $|\vec{a}\times\vec{b}| = ab \sin \pi$
$= ab \times 0$
$= 0$

From above discussion we came to know that

$$\textbf{(min) } 0 \le |\vec{a}\times\vec{b}| \le ab \textbf{ (max).}$$

(ix) $$\vec{a}\times\vec{b} = \begin{vmatrix} \vec{i} & \vec{j} & \vec{k} \\ a_x & a_y & a_z \\ b_x & b_y & b_z \end{vmatrix} \quad \text{..... (3.16)}$$

where $\vec{a} = a_x \vec{i} + a_y \vec{i} + a_z \vec{k}$ and

$\vec{b} = b_x \vec{i} + b_y \vec{j} + b_z \vec{k}$.

(x) Geometrical property: If $\vec{a}$ and $\vec{b}$ are two vectors such that angle between them is θ then,

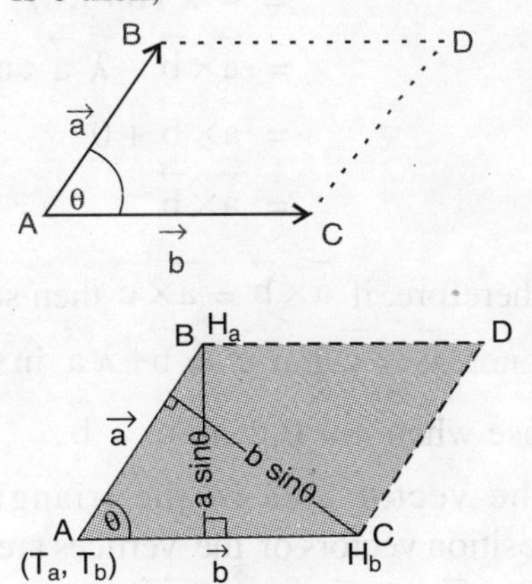

Fig. 3.57

$|\vec{a}\times\vec{b}| = ab \sin\theta$
$= a\,(b \sin\theta)$
$= b\,(a \sin\theta)$
= base × height
= area of parallelogram ABCD

Therefore magnitude of the vector product of two vectors represents the area of the parallelogram formed by taking $\vec{a}$ and $\vec{b}$ as adjacent sides. Vector area of parallelogram is given by $\vec{a}\times\vec{b}$.

(xi) Cross product of two vectors in terms of the components along the right handed co-ordinate axes.

Let two vectors $\vec{a}$ and $\vec{b}$ are given by

$$\vec{a} = a_x \vec{i} + a_y \vec{j} + a_z \vec{k}$$

and $\vec{b} = b_x \vec{i} + b_y \vec{j} + b_z \vec{k}$ then,

$(\vec{a}\times\vec{b})$

$$= \left(a_x \vec{i} + a_y \vec{j} + a_z \vec{k}\right)\times\left(b_x \vec{i} + b_y \vec{j} + b_z \vec{k}\right)$$

As per distributive law $\vec{a}\times\vec{b}$

$$= a_x b_x\, \vec{i}\times\vec{i} + a_x b_y\, \vec{i}\times\vec{j} + a_x b_z\, \vec{i}\times\vec{k}$$
$$+ a_y b_x\, \vec{j}\times\vec{j} + a_y b_y\, \vec{j}\times\vec{i} + a_y b_z\, \vec{j}\times\vec{k}$$
$$+ a_z b_x\, \vec{k}\times\vec{i} + a_z b_y\, \vec{k}\times\vec{j} + a_z b_z\, \vec{k}\times\vec{k}$$
$$= a_x b_x (0) + a_x b_y \vec{k} + a_x b_z (-\vec{j})$$
$$+ a_y b_x (-\vec{k}) + a_y b_y \times (0) + a_y b_z (\vec{i})$$
$$+ a_y b_x (\vec{j}) + a_z b_y (-i) + a_z b_z \times 0$$
$$= \left(a_y b_z - a_z b_y\right)\vec{i} + (a_z b_x - a_x b_z)\,\vec{j}$$
$$+ (a_x b_y - a_y b_x)\,\vec{k}$$

(xii) $\vec{a}\times\vec{0} = \vec{0}$

$$\because\ \vec{a}\times\vec{0} = |a| \times 0 \times \sin\theta\ \hat{n} = \vec{0}$$

(xiii) The vector product of any vector with itself is zero.

Thus $\vec{a}\times\vec{a} = |a||a| \sin 0°\ \hat{n}$

$$= a \times a \times 0\ \hat{n}$$

$$= 0\ \hat{n}$$

$$= 0$$

(xiv) $|\vec{a} \times \vec{b}| = ab \sin\theta$

or $\dfrac{|\vec{a} \times \vec{b}|}{ab} = \sin\theta$

or $\left|\left(\dfrac{\vec{a}}{a}\right)\left(\dfrac{\vec{b}}{b}\right)\right| = \sin\theta$

or $|\hat{a} \times \hat{b}| = \sin\theta$

$\therefore\ \theta = \sin^{-1} |\hat{a} \times \hat{b}|$ where $\hat{a}$ and $\hat{b}$ are two unit vectors along $\vec{a}$ and $\vec{b}$ respectively.

(xv) $(m\vec{a}) \times \vec{b} = m(\vec{a} \times \vec{b}) = \vec{a} \times (m\vec{b})$ that is vector product is associative with respect to a scalar.

(xvi) Expression for $\sin\theta$ in terms of components of the two vectors $\vec{a}$ and $\vec{b}$.

Let $\vec{a} = a_x \vec{i} + a_y \vec{j} + a_z \vec{k}$ and $\vec{b} = b_x \vec{i} + b_y \vec{j} + b_z \vec{k}$ then

$$|\vec{a}| = \sqrt{a_x^2 + a_y^2 + a_z^2}$$

and $|\vec{b}| = \sqrt{b_x^2 + b_y^2 + b_z^2}$

Now as we know that, $\vec{a} \times \vec{b} = ab \sin\theta\ \hat{n}$

where, $\hat{n}$ is the unit vector perpendicular to the plane confining $\vec{a}$ and $\vec{b}$ then

$$(\vec{a} \times \vec{b}) \cdot (\vec{a} \times \vec{b}) = (ab \sin\theta\ \hat{n}) \cdot (ab \sin\theta\ \hat{n})$$

or $\left\{(a_y b_z - a_z b_y)\vec{i} + (a_z b_x - a_x b_z)\vec{j} + (a_x b_y - a_y b_x)\vec{k}\right\} \cdot \left\{(a_y b_z - a_z b_y)\vec{i} + (a_z b_x - a_x b_z)\vec{j} + (a_x b_y - a_y b_x)\vec{k}\right\}$

$$= a^2 b^2 \sin^2\theta \qquad [\because\ \hat{n} \cdot \hat{n} = 1]$$

or $(a_y b_z - a_z b_y)^2 + (a_z b_x - a_x b_z)^2 + (a_x b_y - a_y b_x)^2$

$$= (a_x^2 + a_y^2 + a_x^2)(b_x^2 + b_y^2 + b_z^2)\sin^2\theta$$

or $\sin^2\theta$

$$= \frac{(a_y b_z - a_z b_y)^2 + (a_z b_x - a_x b_z)^2 + (a_x b_y - a_y b_x)^2}{(a_x^2 + a_y^2 + a_z^2)(b_x^2 + b_y^2 + b_z^2)}$$

$\therefore\ \sin\theta$

$$= \left[\frac{(a_x b_z - a_z b_y)^2 + (a_z b_x - a_x b_z)^2 + (a_x b_y - a_y b_x)^2}{(a_x^2 + a_y^2 + a_z^2)(b_x^2 + b_y^2 + b_z^2)}\right]^{\frac{1}{2}}$$

(xvii) if $\vec{a} \times \vec{b} = \vec{a} \times \vec{c}$ then $\vec{c} = \vec{b} + \lambda\vec{a}$, where λ is a scalar may or may not be zero.

***Proof*:** $\vec{a} \times \vec{b} = \vec{a} \times \vec{c}$

$$= \vec{a} \times (\vec{b} + \lambda\vec{a})$$

$$= \vec{a} \times \vec{b} + \lambda\, \vec{a} \times \vec{a}$$

$$= \vec{a} \times \vec{b} + 0$$

$$= \vec{a} \times \vec{b}$$

Therefore, if $\vec{a} \times \vec{b} = \vec{a} \times \vec{c}$ then solution is not $\vec{b} = \vec{c}$ but $\vec{c} = \vec{b} + \lambda\vec{a}$ in special case when $\lambda = 0$ then $\vec{c} = \vec{b}$.

(xviii) The vector area of the triangle, the position vectors of the vertices are given as $A(\vec{a})$, $B(\vec{b})$ and $C(\vec{c})$ with respect to the origin.

Proof:

Let ABC be the given triangle where,

$\vec{OA} = \vec{a}$, $\vec{OB} = \vec{b}$ and $\vec{OC} = \vec{c}$ then,

$\vec{BA} = \vec{a} - \vec{b}$ and $\vec{BC} = \vec{c} - \vec{b}$.

Therefore, vector area of triangle ABC is,

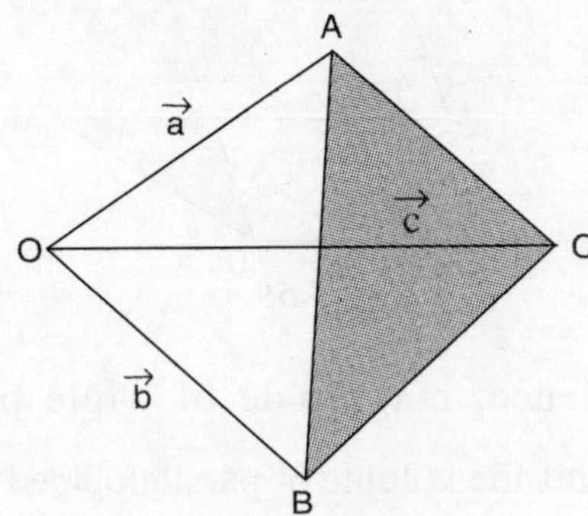

Fig. 3.58

$$= \frac{1}{2}\left(\vec{BA} \times \vec{BC}\right)$$

$$= \frac{1}{2}\left[\left(\vec{a} - \vec{b}\right) \times \left(\vec{c} - \vec{b}\right)\right]$$

$$= \frac{1}{2}\left[\vec{a} \times \vec{c} - \vec{a} \times \vec{b} - \vec{b} \times \vec{c} + \vec{b} \times \vec{b}\right]$$

$$= \frac{1}{2}\left[-\vec{c} \times \vec{a} - \vec{a} \times \vec{b} - \vec{b} \times \vec{c} + 0\right]$$

$$= -\frac{1}{2}\left[\vec{a} \times \vec{b} + \vec{b} \times \vec{c} + \vec{c} \times \vec{a}\right]$$

$$= \frac{1}{2}\left[\vec{b} \times \vec{a} + \vec{c} \times \vec{b} + \vec{a} \times \vec{c}\right]$$

$$\therefore \quad \text{area } (\Delta ABC) = \frac{\vec{b} \times \vec{a} + \vec{c} \times \vec{b} + \vec{a} \times \vec{c}}{2}$$

If point A $\left(\vec{a}\right)$, B $\left(\vec{b}\right)$ and C $\left(\vec{c}\right)$ are collinear (on a line) then vector area of $\Delta ABC = 0$.

$$\text{i.e.,} \quad \frac{\vec{b} \times \vec{a} + \vec{c} \times \vec{b} + \vec{a} \times \vec{c}}{2} = 0$$

$$\therefore \quad \vec{a} \times \vec{b} + \vec{b} \times \vec{c} + \vec{c} \times \vec{a} = 0$$

(xix) Unit vector perpendicular to plane containing $\vec{a}$ and $\vec{b}$. Since

$$\vec{a} \times \vec{b} = ab \sin\theta\, \hat{n} = |\vec{a} \times \vec{b}|\hat{n},$$

where $\hat{n}$ is the unit vector perpendicular to the plane containing $\vec{a}$ and $\vec{b}$.

$$\therefore \quad \hat{n} = \pm \frac{\vec{a} \times \vec{b}}{|\vec{a} \times \vec{b}|}$$ where, $\pm$ shows either up the plane or down the plane.

Example 8: *The vector $\vec{a}$ has magnitude of 5 unit and $\vec{b}$ has magnitude of 6 unit and cross product of $\vec{a}$ and $\vec{b}$ has magnitude of 15 unit. Find the angle between $\vec{a}$ and $\vec{b}$.*

Solution:

If θ is the angle between $\vec{a}$ and $\vec{b}$.

Then, $|\vec{a} \times \vec{b}| = |\vec{a}||\vec{b}| \sin\theta$

or $\quad 15 = 5 \times 6 \times \sin\theta$

$$\sin\theta = \frac{15}{30}$$

$$= \frac{1}{2}$$

$$\therefore \quad \theta = 36° \text{ or } 150°$$

Example 9: *Prove that $\vec{a} \cdot (\vec{a} \times \vec{b}) = 0$.*

Solution:

As we know that, $\vec{a} \times \vec{b}$ is perpendicular to the plane containing $\vec{a}$ and $\vec{b}$ i.e., $\vec{a} \times \vec{b}$ is perpendicular to $\vec{a}$ and $\vec{b}$ both.

Therefore, $\vec{a} \cdot (\vec{a} \times \vec{b}) = |\vec{a}||\vec{a} \times \vec{b}| \cos 90°$

$$= 0$$

3.21 SCALAR TRIPLE PRODUCT

Definition: If $\vec{a}$, $\vec{b}$ and $\vec{c}$ are three vectors then the scalar product of $\vec{a}\times\vec{b}$ with $\vec{c}$ is called scalar triple product of $\vec{a}$, $\vec{b}$ and $\vec{c}$. It is denoted by $[\vec{a}\,\vec{b}\,\vec{c}]$ thus $(\vec{a}\times\vec{b})\cdot\vec{c} = [\vec{a}\,\vec{b}\,\vec{c}]$. In this type of product we finally get a quantity scalar so it is called as the scalar triple product. It contains word triple because three vectors are used. It is also called mixed product because the dot and the cross products are used in the scalar product.

Geometrical meaning of scalar triple product

Consider a parallelopiped whose coterminous edges are OA, OO' and OC and has the length and direction of the vectors $\vec{a}$, $\vec{b}$ and $\vec{c}$ respectively. Let v be its volume then $\vec{c}\times\vec{a}$ represent the vector area of the parallelogram OABC whose two adjacent sides are represented by $\vec{a}$ and $\vec{c}$.

Therefore area (OABC) = $|\vec{c}\times\vec{a}|$

and vector area (OABC) = $|\vec{c}\times\vec{a}|\hat{n}$

where, $\hat{n}$ is the unit vector perpendicular to base of parallelogram OABC. Now $\vec{b}\cdot\hat{n}$ = Projection of OO′ along $\hat{n}$ = perpendicular length to base OABC of parallelopped. Therefore, v = (area OABC) × perpendicular length,

$$= (|\vec{c}\times\vec{a}|).(\vec{b}\cdot\hat{n})$$

$$= (|\vec{c}\times\vec{a}|)\hat{n}\cdot\vec{b}$$

$$= (\vec{c}\times\vec{a})\cdot\vec{b} \qquad [\because |\vec{c}\times\vec{a}|\hat{n} = \vec{c}\times\vec{a}]$$

$$= [\vec{c}\,\vec{a}\,\vec{b}]$$

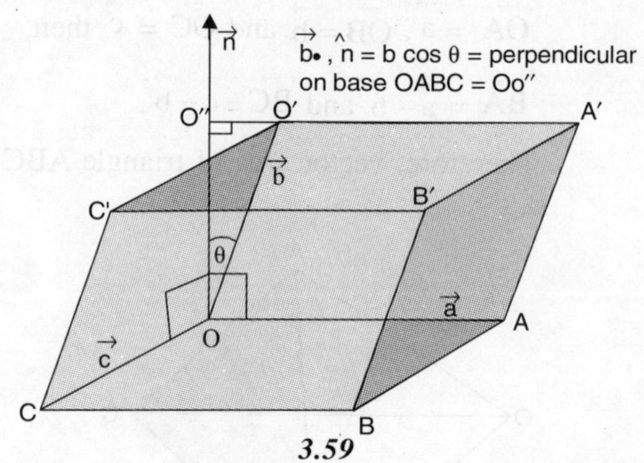

3.59

Hence, magnitude of triple product represents the volume of parallelopiped whose coterminous edges are $\vec{a}$, $\vec{b}$ and c. Note that $[\vec{a}\,\vec{b}\,\vec{c}] = [\vec{b}\,\vec{c}\,\vec{a}] = [\vec{c}\,\vec{a}\,\vec{b}]$ represents the volume of the same parallelopiped.

Properties of scalar triple product

(i) If any two vectors are identical or parallel to each other then volume of parallelopiped is zero.

i.e., $[\vec{a}\,\vec{a}\,\vec{c}] = 0$ where, $\vec{b} = \vec{a}$

Analytical proof:

$$= [\vec{a}\,\vec{a}\,\vec{c}]$$

$$= (\vec{a}\times\vec{a})\cdot\vec{c}$$

$$= \vec{0}\cdot\vec{c}$$

$$= 0$$

Similarly, $[\vec{a}\,\vec{b}\,\vec{b}] = [\vec{a}\,\vec{b}\,\vec{a}] = 0$

(ii) Scalar triple product of orthogonal unit vectors $\vec{i}$, $\vec{j}$ and $\vec{k}$ of right handed coordinate system is 1.

i.e., $[\vec{i}\,\vec{j}\,\vec{k}] = [\vec{j}\,\vec{k}\,\vec{i}] = [\vec{k}\,\vec{i}\,\vec{j}] = 1$

(iii) The necessary and sufficient condition that three non-parallel and non-zero vectors $\vec{a}, \vec{b}$ and $\vec{c}$ are co-plannar is $[\vec{a}\,\vec{b}\,\vec{c}] = 0$.

(iv) Let

$$\vec{a} = a_x \vec{i} + a_y \vec{j} + a_z \vec{k},$$

$$\vec{b} = b_x \vec{i} + b_y \vec{j} + b_z \vec{k}$$

and

$$\vec{c} = c_x \vec{i} + c_y \vec{j} + c_z \vec{k}$$

Then

$$\vec{b} \times \vec{c} = \begin{vmatrix} \vec{i} & \vec{j} & \vec{k} \\ b_x & b_y & b_z \\ c_x & c_y & c_z \end{vmatrix}$$

$$= (b_y c_z - c_y b_z)\vec{i} + (c_x b_z - b_x c_z)\vec{j} + (b_x c_y - c_x b_y)\vec{k}$$

$$= \{a_x \vec{i} + a_y \vec{j} + a_z \vec{k}\} \cdot \{(b_y c_z - c_y b_z)\vec{i} + (c_x b_z - b_x c_z)\vec{j} + (b_x c_y - c_x b_y)\vec{k}$$

$$= (b_y c_z - c_y b_z) a_x + (c_x b_z - b_x c_z) a_y + (b_x c_y - c_x b_y) a_z$$

i.e.,

$$\vec{a} \cdot (\vec{b} \times \vec{c}) = \begin{vmatrix} a_x & a_y & a_z \\ b_x & b_y & b_z \\ c_x & c_y & c_z \end{vmatrix}$$

$$= [\vec{a}\,\vec{b}\,\vec{c}] \quad \text{....(3.17)}$$

(v) Volume of triangular prism

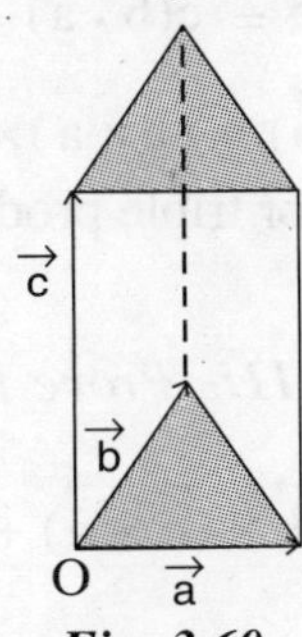

Fig. 3.60

$$V = \frac{1}{2}[\vec{a}\,\vec{b}\,\vec{c}]$$

= half of the volume of parallelopiped

(vi) Volume of tetrahedron

$$V = \frac{1}{6}[\vec{a}\,\vec{b}\,\vec{c}]$$

Examples 10: *Find the volume of the parallelopiped whose edges are represented by* $\vec{a} = 2\vec{i} - 3\vec{j} + 4\vec{k}$, $\vec{i} - 2\vec{j} - \vec{k}$ *and* $\vec{c} = 3\vec{i} - \vec{j} + 2\vec{k}$.

Solution: The required volume of parallelopiped

$$= \vec{a} \cdot (\vec{b} \times \vec{c})$$

$$= \begin{vmatrix} a_x & a_y & a_z \\ b_x & b_y & b_z \\ c_x & c_y & c_z \end{vmatrix}$$

$$= \begin{vmatrix} 2 & -3 & 4 \\ 1 & 2 & -1 \\ 3 & -1 & 2 \end{vmatrix}$$

$$= 2(4-1) + 3(2+3) + 4(-1-6)$$

$$= 6 + 15 - 28$$

$$= 7$$

$= 7$ (numerically)

(vii) Volume of tetrahedron

Let OACB be a tetrahedral whose three co-terminous edges are $\vec{OA}$, $\vec{OB}$ and $\vec{OC}$ and are represented by the vectors $\vec{a}$, $\vec{b}$ and $\vec{c}$ respectively.

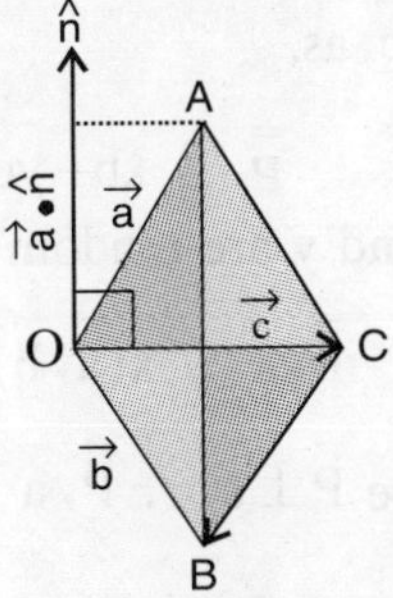

Fig. 3.61

We know that, volume of tetrahedral

$= \frac{1}{3}$ (area of the base) × height

$= \frac{1}{3}[\frac{1}{2} \times |\vec{b} \times \vec{c}|] \times (\vec{a} \cdot \hat{n})$

$= \frac{1}{6}|\vec{a} \cdot (\vec{b} \times \vec{c})\hat{n}|$

$= \frac{1}{6}\vec{a} \cdot (\vec{b} \times \vec{c})$

$= \frac{1}{6}[\vec{a}\,\vec{b}\,\vec{c}]$.

3.22 VECTOR TRIPLE PRODUCT

If $\vec{a}$, $\vec{b}$ and $\vec{c}$ are three vectors then the vector product of $\vec{a}$ with $\vec{b} \times \vec{c}$ is called the vector triple product and is denoted by $\vec{P} = \vec{a} \times (\vec{b} \times \vec{c})$. Vector triple product of three vectors gives a vector as a result.

Geometrical meaning of $\vec{a} \times (\vec{b} \times \vec{c})$

Let $\vec{P} = \vec{a} \times (\vec{b} \times \vec{c})$ then $\vec{P}$ should be perpendicular to the vector $\vec{a}$ as well as vector $\vec{b} \times \vec{c}$. But $\vec{b} \times \vec{c}$ is a vector perpendicular to plane containing $\vec{b}$ and $\vec{c}$. Therefore, $\vec{P}$ is a vector perpendicular to $\vec{a}$ and lying in the plane of $\vec{b}$ and $\vec{c}$.

Hence, $\vec{P}$ can be expressed in the terms of of $\vec{b}$ and $\vec{c}$ as,

$$\vec{P} = x\vec{b} + y\vec{c} \qquad \text{....(i)}$$

where, x and y are random scalars.

or $\vec{P} \cdot \vec{a} = x(\vec{a} \cdot \vec{b}) + y(\vec{a} \cdot \vec{c})$

Since $\vec{P} \perp \vec{a}$ ∴ $\vec{P} \cdot \vec{a} = 0$

∴ $x(\vec{a} \cdot \vec{b}) + y(\vec{a} \cdot \vec{c}) = 0$

or $x(\vec{b} \cdot \vec{a}) = -y(\vec{c} \cdot \vec{a})$

or $\frac{x}{\vec{c} \cdot \vec{a}} = -\frac{y}{\vec{a} \cdot \vec{b}} = k$ (say)

∴ $x = k\,\vec{c} \cdot \vec{a}$ and $y = -k\,\vec{a} \cdot \vec{b}$

Putting the values of x and y in (i), we could get

$$\vec{P} = k(\vec{c} \cdot \vec{a})\vec{b} - k(\vec{a} \cdot \vec{b})\vec{c}$$

or $\vec{a} \times (\vec{b} \times \vec{c})$

$$= k[(\vec{a} \cdot \vec{c})\vec{b} - (\vec{a} \cdot \vec{b})\vec{c}]$$

For right handed coordinate system we could get k = 1.

∴ $\vec{P} = \vec{a} \times (\vec{b} \times \vec{c})$

$$= (\vec{a} \cdot \vec{c})\vec{b} - (\vec{a} \cdot \vec{b})\vec{c} \qquad \text{....(3.18)}$$

That is linear combination of vectors $\vec{b}$ and $\vec{c}$ and lies in the plane of $\vec{b}$ and $\vec{c}$. Similarly, we could get

$$\vec{b} \times (\vec{c} \times \vec{a}) = (\vec{b} \cdot \vec{a})\vec{c} - (\vec{b} \cdot \vec{c})\vec{a}$$

$$\vec{c} \times (\vec{a} \times \vec{b}) = (\vec{c} \cdot \vec{b})\vec{a} - (\vec{c} \cdot \vec{a})\vec{b}$$

and $(\vec{c} \times \vec{a}) \times \vec{b} = \vec{c}(\vec{b} \cdot \vec{a}) - (\vec{b} \cdot \vec{c})\vec{a}$

⇒ $\vec{c} \times (\vec{a} \times \vec{b}) \neq (\vec{c} \times \vec{a}) \times \vec{b}$

That is vector triple product does not obey associative law.

Example 11: *Prove that*

$$\vec{a} \times (\vec{b} \times \vec{c}) + \vec{b} \times (\vec{c} \times \vec{a}) + \vec{c} \times (\vec{a} \times \vec{b}) = 0.$$

Solution:

As we know that,

$$\vec{a} \times (\vec{b} \times \vec{c}) = (\vec{a} \cdot \vec{c})\vec{b} - (\vec{a} \cdot \vec{b})\vec{c}$$

$$\vec{b}\times(\vec{c}\times\vec{a}) = (\vec{b}\cdot\vec{a})\vec{c}-(\vec{b}\cdot\vec{c})\vec{a}$$

and $$\vec{c}\times(\vec{a}\times\vec{b}) = (\vec{c}\cdot\vec{b})\vec{a}-(\vec{c}\cdot\vec{a})\vec{b}$$

Therefore $\vec{a}\times(\vec{b}\times\vec{c})+\vec{b}\times(\vec{c}\times\vec{a})+\vec{c}\times(\vec{a}\times\vec{b})$

$$=(\vec{a}\cdot\vec{c})\vec{b}-(\vec{a}\cdot\vec{b})\vec{c}+(\vec{b}\cdot\vec{a})\vec{c}-(\vec{b}\cdot\vec{c})\vec{a}$$

$$+(\vec{c}\cdot\vec{b})\vec{a}-(\vec{c}\cdot\vec{a})\vec{b}$$

$$=0+0+0=0$$

Example 12: If $\vec{A}=2\vec{i}+\vec{j}-3\vec{k}$,

$\vec{B}=\vec{i}+2\vec{j}+\vec{k}$ *and* $\vec{C}=\vec{i}+\vec{j}-4\vec{k}$

then, find $\vec{A}\times(\vec{B}\times\vec{C})$.

Solution:

Now $$\vec{B}\times\vec{C} = \begin{vmatrix} \vec{i} & \vec{j} & \vec{k} \\ 1 & -2 & 1 \\ -1 & 1 & -4 \end{vmatrix}$$

$$= (8-1)\vec{i}+(-1+4)\vec{j}+(1-2)\vec{k}$$

$$= 7\vec{i}+3\vec{j}-\vec{k}$$

Then, $$\vec{A}\times\left(\vec{B}\times\vec{C}\right) = \begin{vmatrix} \vec{i} & \vec{j} & \vec{k} \\ 2 & 1 & -3 \\ 7 & 3 & -1 \end{vmatrix}$$

$$= (-1+9)\vec{i}+(-21+2)\vec{j}+(6-7)\vec{k}$$

$$= 8\vec{i}-19\vec{j}-\vec{k}$$

3.23 SCALAR PRODUCT OF FOUR VECTORS

Definition: If $\vec{a}$, $\vec{b}$, $\vec{c}$ and $\vec{d}$ are any four vectors then the product of the form $(\vec{a}\times\vec{b})\bullet(\vec{c}\times\vec{d})$ is called the scalar product of four vectors.

Lagrange's identity

If $\vec{a}$, $\vec{b}$, $\vec{c}$ and $\vec{d}$ are any four vectors.

Then, $$(\vec{a}\times\vec{b})\cdot(\vec{c}\times\vec{d}) = \begin{vmatrix} \vec{b}\cdot\vec{c} & \vec{a}\cdot\vec{c} \\ \vec{b}\cdot\vec{d} & \vec{a}\cdot\vec{d} \end{vmatrix} \quad ....(3.19)$$

Proof : Let $\vec{a}\times\vec{b} = \vec{m}$

then, $$= \vec{m}\cdot\left(\vec{c}\times\vec{d}\right) = \left(\vec{m}\times\vec{c}\right)\cdot\vec{d}$$

(As dot and cross are interchangeable in case of scalar triple product.)

$$= \{(\vec{a}\times\vec{b})\times\vec{c}\}\cdot\vec{d}$$

$$= \{(\vec{c}\cdot\vec{b})\vec{a}-(\vec{c}\cdot\vec{a})\vec{b}\}\cdot\vec{d}$$

$$= \{(\vec{c}\cdot\vec{b})(\vec{a}\cdot\vec{d})-(\vec{c}\cdot\vec{a})(\vec{b}\cdot\vec{d})\}$$

$$= \begin{vmatrix} \vec{b}\cdot\vec{c} & \vec{a}\cdot\vec{c} \\ \vec{b}\cdot\vec{d} & \vec{a}\cdot\vec{d} \end{vmatrix}$$

3.24 VECTOR PRODUCT OF FOUR VECTORS

Definition: $(\vec{a}\times\vec{b})\times(\vec{c}\times\vec{d})$ is called the vector product of four vectors $\vec{a}, \vec{b}, \vec{c}$ and $\vec{d}$.

Geometrical meaning

Let $\vec{P} = (\vec{a}\times\vec{b})\times(\vec{c}\times\vec{d})$

then $\vec{P}\perp(\vec{a}\times\vec{b})$ and $\vec{P}\perp(\vec{c}\times\vec{d})$

$\Rightarrow$ $\vec{P}$ is co-planar with $\vec{a}$ and $\vec{b}$ as well as $\vec{P}$ is co-planar with $\vec{c}$ and $\vec{d}$.

$\Rightarrow$ $\vec{P}$ represents a vector parallel to the line of intersection of two planes one of this is parallel to the plane containing $\vec{a}$ and $\vec{b}$ and other is parallel to the plane containing $\vec{c}$ and $\vec{d}$.

Theorem (i)

$$(\vec{a}\times\vec{b})\times(\vec{c}\times\vec{d}) = [\vec{a}\,\vec{b}\,\vec{d}]\vec{c}-[\vec{a}\,\vec{b}\,\vec{c}]\vec{d}$$

That is linear combination of $\vec{c}$ and $\vec{d}$.

Theorem (ii)

$$(\vec{a}\times\vec{b})\times(\vec{c}\times\vec{d}) = [\vec{a}\,\vec{c}\,\vec{d}]\vec{b}-[\vec{b}\,\vec{c}\,\vec{d}]\vec{a}$$

That is linear combination of $\vec{a}$ and $\vec{b}$.

Proof of theorem (i).

$$(\vec{a}\times\vec{b})\times(\vec{c}\times\vec{d}) = [\vec{a}\,\vec{b}\,\vec{d}]\vec{c}-[\vec{a}\,\vec{b}\,\vec{c}]\vec{d}$$

Let $\vec{a}\times\vec{b}=\vec{m}$

$\therefore \quad \vec{m}\times(\vec{c}\times\vec{d})$

$= (\vec{m}\cdot\vec{d})\vec{c} - (\vec{m}\cdot\vec{c})\vec{d}$

$= \{(\vec{a}\times\vec{b})\cdot\vec{d}\}\vec{c}-\{(\vec{a}\times\vec{b})\cdot\vec{c}\}\vec{d}$

$= [\vec{a}\,\vec{b}\,\vec{d}]\vec{c}-[\vec{a}\,\vec{b}\,\vec{c}]\vec{d}$ ....(i)

Proof of theorem (ii)

Let $\vec{c}\times\vec{d} = \vec{n}$

Then, $(\vec{a}\times\vec{b})\times(\vec{c}\times\vec{d})$

$= (\vec{a}\times\vec{b})\times \vec{n}$

$= -\{\vec{n}(\vec{a}\times\vec{b})\}$

$= \{(\vec{n}\cdot\vec{b})\vec{a}-(\vec{n}\cdot\vec{a})\vec{b}\}$

$= (\vec{n}\cdot\vec{a})\vec{b}-(\vec{n}\cdot\vec{b})\vec{a}$

$= \{(\vec{c}\times\vec{d})\cdot\vec{a}\}\vec{b}-\{(\vec{c}\times\vec{d})\cdot\vec{b}\}\vec{a}$

$= [\vec{c}\;\vec{d}\;\vec{a}]\;\vec{b} -[\vec{c}\;\vec{d}\;\vec{b}]\;\vec{a}$ ....(ii)

From equations (i) and (ii), we get

$= [\vec{a}\,\vec{b}\,\vec{d}]\vec{c}-[\vec{a}\,\vec{b}\,\vec{c}]\vec{d}$

$= [\vec{c}\;\vec{d}\;\vec{a}]\;\vec{b} -[\vec{c}\;\vec{d}\;\vec{b}]\;\vec{a}$

$[\vec{a}\,\vec{b}\,\vec{d}]\vec{c}+[\vec{c}\,\vec{d}\,\vec{b}]\vec{a}-[\vec{c}\,\vec{d}\,\vec{a}]\vec{b}$

$= [\vec{a}\,\vec{b}\,\vec{c}]\vec{d}$

or $[\vec{a}\,\vec{b}\,\vec{d}]\vec{c}+[\vec{c}\,\vec{d}\,\vec{b}]\vec{a}+[\vec{c}\,\vec{a}\,\vec{d}]\vec{b}$

$= [\vec{a}\,\vec{b}\,\vec{c}]\vec{d}$

$$\therefore\ \vec{d} = \frac{[\vec{a}\,\vec{b}\,\vec{d}]\vec{c}+[\vec{c}\,\vec{d}\,\vec{b}]\vec{a}+[\vec{c}\,\vec{a}\,\vec{d}]\vec{b}}{[\vec{a}\,\vec{b}\,\vec{c}]} \quad ...(3.20)$$

where, $[\vec{a}\,\vec{b}\,\vec{c}] \neq 0$ i.e., $\vec{a}$, $\vec{b}$, and $\vec{c}$ are not co-planar.

WORKED OUT EXAMPLES

1. *If* $|\vec{a}+\vec{b}| = |\vec{a}-\vec{b}|$ *then, prove that* $|\vec{a}\times\vec{b}|= ab$ *where* $|\vec{a}|$, $|\vec{b}| \neq 0$.

Solution:

First method: As we know that,

$|\vec{a}+\vec{b}| = (a^2+b^2+2ab\cos\theta)^{1/2}$

and $|\vec{a}-\vec{b}| = (a^2+b^2-2ab\cos\theta)^{1/2}$

where θ is the angle between $\vec{a}$ and $\vec{b}$.

Now it is given that $|\vec{a}+\vec{b}| = |\vec{a}-\vec{b}|$ then

$(a^2+b^2+2ab\cos\theta)^{1/2}$

$= (a^2+b^2-2ab\cos\theta)^{1/2}$

Squaring both sides, we get

$a^2 + b^2 + 2ab\cos\theta = a^2 + b^2 - 2ab\cos\theta$

or $\quad 4ab\cos\theta = 0$

or $\quad \cos\theta = 0 \quad \because \quad [\,|a|, |\vec{b}| \neq 0\,]$

Therefore, θ = 90° i.e., $\vec{a}$ and $\vec{b}$ are perpendicular to each other.

$\therefore \quad |\vec{a}\times\vec{b}| = ab\sin\theta$

$= ab\sin 90°$

$= ab$

Second method: As we know that,

$\therefore \quad |\vec{a}\times\vec{b}|^2 = (\vec{a}+\vec{b})\cdot(\vec{a}+\vec{b})$

$= a^2 + b^2 + 2\,\vec{a}\cdot\vec{b} \quad [\because a^2 = \vec{a}\cdot\vec{a}]$

and $|\vec{a}-\vec{b}|^2 = (\vec{a}-\vec{b})\cdot(\vec{a}-\vec{b})$

$= a^2 + b^2 - 2\,\vec{a}.\vec{b}$

According to the question $|\vec{a}+\vec{b}| = |\vec{a}-\vec{b}|$ is given.

or $|\vec{a}\times\vec{b}|^2 = |\vec{a}-\vec{b}|^2$

after squaring both sides, we have

or $a^2 + b^2 + 2\,\vec{a}\cdot\vec{b} = a^2 + b^2 - 2\,\vec{a}\cdot\vec{b}$

or $4\,\vec{a}\cdot\vec{b} = 0$

or $\vec{a}\cdot\vec{b} = 0$

$\Rightarrow$ $\vec{a}$ and $\vec{b}$ are perpendicular to each other i.e., $\theta = 90°$.

$\therefore\ |\vec{a}\times\vec{b}| = ab\sin\theta$

$= ab\sin 90°$

$= ab$

2. *If $\vec{A} = 2\vec{i} - 3\vec{j} + 7\vec{k}$, $\vec{B} = \vec{i} + 2\vec{k}$ and $\vec{C} = \vec{j} - \vec{k}$ then prove that $\vec{A}$, $\vec{B}$ and $\vec{C}$ are co-planar.*

Solution:

First method: As we know that,

$$(\vec{A}\times\vec{B})\cdot\vec{C} = \begin{vmatrix} x_1 & y_1 & z_1 \\ x_2 & y_2 & z_2 \\ x_3 & y_3 & z_3 \end{vmatrix} = \begin{vmatrix} 2 & -3 & 7 \\ 1 & 0 & 2 \\ 0 & 1 & -1 \end{vmatrix}$$

$= 2(0 - 2) + 3(-1 - 0) + 7(1 - 0)$

$= -4 - 3 + 7 = 0$

Since $(\vec{A}\times\vec{B})\cdot\vec{C} = 0$ i.e., volume of parallelopiped formed by co-terminus edges $\vec{A}$, $\vec{B}$ and $\vec{C}$ is zero. Therefore, $\vec{A}$, $\vec{B}$ and $\vec{C}$ are co-planar vectors.

Second method: Suppose that $\vec{A}$, $\vec{B}$ and $\vec{C}$ are co-planar vectors then we can write $\vec{A}$ as linear combination of $\vec{B}$ and $\vec{C}$.

That is, $\vec{A} = x\vec{B} + y\vec{C}$

or $2\vec{i} - 3\vec{j} + 7\vec{k}$

$= x(\vec{i} + 2\vec{k}) + y(\vec{j} - \vec{k})$

$= x\vec{i} + 2x\vec{k} + y\vec{j} - y\vec{k}$

$= x\vec{i} + y\vec{j} + (2x - y)\vec{k}$

or $(2 - x)\vec{i} + (-3 - y)\vec{j} + (7 - 2x + y)\vec{k} = 0$

$\Rightarrow$ $2 - x = 0$, $-3 - y = 0$,

and $7 - 2x - y = 0$

or $x = 2$ ....(i)

$y + 3 = 0$ ....(ii)

$2x - y = 7$ ....(iii)

From equation (ii) $y = -3$

Putting $y = -3$ in equation (iii), we get

$2x - (-3) = 7$

or $2x + 3 = 7$

or $2x = 4$

$\therefore$ $x = 2$

Then $x = 2$, $y = -3$. Since here we get unique solution, therefore our assumption was correct.

3. *Give the solution for $\vec{a}\cdot\vec{b} = \vec{c}\cdot\vec{b}$ where $\vec{a} \neq \vec{c}$.*

Solution :

$\because$ $\vec{a}\cdot\vec{b} = \vec{c}\cdot\vec{b}$

or $\vec{a}\cdot\vec{b} - \vec{c}\cdot\vec{b} = 0$

or $(\vec{a} - \vec{c})\cdot\vec{b} = 0$

$\Rightarrow \vec{b}$ and $\vec{a}-\vec{c}$ are perpendicular to each other that means $\vec{a}$ and $\vec{c}$ are collinear vectors.

$\therefore \quad \vec{c} = k\vec{a}$, $k \neq 1$ is the general solution.

4. *The force on a charged particle due to electric and magnetic fields is given by* $\vec{F} = q\vec{E} + q(\vec{v}\times\vec{B})$, *considering* $\vec{E}$ *is along the x-axis and* $\vec{B}$ *along the y-axis. In what direction and with what minimum speed* v_{min}, *should a positively charged particle be thrown so that the net force on it is zero?*

Solution:

Since $\vec{E}$ is along x-axis then we can write $\vec{E} = E\vec{i}$ and as $\vec{B}$ is along y-axis then $\vec{B} = B\vec{j}$.

Now since net force is zero on charged particle therefore, $\Sigma\vec{F} = 0$

or $\quad q\vec{E} + q(\vec{v}\times\vec{B}) = 0$

or $\quad q\{\vec{E} + \vec{v}\times\vec{B}\} = 0$

or $\quad \vec{E} = -\vec{v}\times\vec{B}$

or $\quad \vec{E} = \vec{B}\times\vec{v}$

$[\because \vec{a}\times\vec{b} = -\vec{b}\times\vec{a}]$

or $\quad |\vec{E}| = |\vec{B}\times\vec{v}|$

or $\quad E = Bv\sin\theta$

or $\quad v = \dfrac{E}{B\sin\theta}$

Since for v_{min} $\sin\theta$ should be maximum that is $\sin\theta = 1$.

$$\therefore \quad v_{min} = \frac{E}{B}$$

Now since θ is the angle between $\vec{v}$ and $\vec{B}$ therefore $\vec{v}$ is either along z-axis or x-axis.

If $\vec{v}$ is along x-axis then, $\vec{B}\times\vec{v}$ has to be in negative z-axis, so we can't equate $\vec{E}$ and $\vec{B}\times\vec{v}$. Therefore $\vec{v}$ must be along z-axis to satisfy equation $\vec{E} = \vec{B}\times\vec{v}$.

5. *Suppose* $\vec{a}$, $\vec{b}$ *and* $\vec{c}$ *are three non-co-planar vectors. They are not necessarily mutually at right angles, show that* $\vec{a}\cdot(\vec{b}\times\vec{c}) = \vec{b}\cdot(\vec{c}\times\vec{a}) = \vec{c}\cdot(\vec{a}\times\vec{b})$.

Solution:

Let $\vec{a} = a_x\vec{i} + a_y\vec{j} + a_z\vec{k}$, $\vec{b} = b_x\vec{i} + b_y\vec{j} + b_z\vec{k}$ and $\vec{c} = c_x\vec{i} + c_y\vec{j} + c_z\vec{k}$ be three vectors then as we know that, $\vec{a}\cdot(\vec{b}\times\vec{c}) = (\vec{b}\times\vec{c})\cdot\vec{a}$

$$= \begin{vmatrix} b_x & b_y & b_z \\ c_x & c_y & c_z \\ a_x & a_y & a_z \end{vmatrix} \quad [\because \vec{a}\cdot\vec{b} = \vec{b}\cdot\vec{a}]$$

$$= \begin{vmatrix} b_x & b_y & b_z \\ a_x & a_y & a_z \\ c_x & c_y & c_z \end{vmatrix}$$

Using property of determinant; row interchange.

$$= \begin{vmatrix} a_x & a_y & a_z \\ b_x & b_y & b_z \\ c_x & c_y & c_z \end{vmatrix} \quad [R_1 \to R_2]$$

Now

$\vec{b}\cdot(\vec{c}\times\vec{a}) = (\vec{c}\times\vec{a})\cdot\vec{b}$

$$= \begin{vmatrix} c_x & c_y & c_z \\ a_x & a_y & a_z \\ b_x & b_y & b_z \end{vmatrix}$$

$$= -\begin{vmatrix} a_x & a_y & a_z \\ c_x & c_y & c_z \\ b_x & b_y & b_z \end{vmatrix} \qquad [R_1 \to R_2]$$

$$= \begin{vmatrix} a_x & a_y & a_z \\ b_x & b_y & b_z \\ c_x & c_y & c_z \end{vmatrix} \qquad [R_2 \to R_3]$$

and $\vec{c}\cdot(\vec{a}\times\vec{b}) = (\vec{a}\times\vec{b})\cdot\vec{c}$

$$= \begin{vmatrix} a_x & a_y & a_z \\ b_x & b_y & b_z \\ c_x & c_y & c_z \end{vmatrix}$$

Therefore,

$$\vec{a}\cdot(\vec{b}\times\vec{c}) = \vec{b}\cdot(\vec{c}\times\vec{a}) = \vec{c}\cdot(\vec{a}\times\vec{b}) = [\vec{a}\,\vec{b}\,\vec{c}]$$

$$= \begin{vmatrix} a_x & a_y & a_z \\ b_x & b_y & b_z \\ c_x & c_y & c_z \end{vmatrix}$$

6. *Prove that* $(\vec{b}\times\vec{c})\times(\vec{c}\times\vec{a}) = [\vec{a}\,\vec{b}\,\vec{c}]\vec{c}$.

Solution:

Let $(\vec{b}\times\vec{c}) = \vec{A}$

then $\vec{A}\times(\vec{c}\times\vec{a})$

$$= \left(\vec{A}\cdot\vec{a}\right)\vec{c} - \left(\vec{A}\cdot\vec{c}\right)\vec{a}$$

$$= \{(\vec{b}\times\vec{c})\cdot\vec{a}\}\vec{c} - \{(\vec{b}\times\vec{c})\cdot\vec{c}\}\,\vec{a}$$

$$= [\vec{b}\,\vec{c}\,\vec{a}]\vec{c} - [\vec{b}\,\vec{c}\,\vec{c}\,]\vec{a}$$

$$= [\vec{b}\,\vec{c}\,\vec{a}]\vec{c} \qquad [\because [\vec{b}\,\vec{c}\,\vec{c}] = 0]$$

$$= [\vec{a}\,\vec{b}\,\vec{c}]\vec{c}$$

7. *If the vectors* $\vec{a}$, $\vec{b}$, $\vec{c}$ *and* $\vec{d}$ *are co-planar then show that* $(\vec{a}\times\vec{b})\times(\vec{c}\times\vec{d}) = 0$.

Solution:

Let $\vec{a}\times\vec{b} = \vec{A}$ then $\vec{A}\times(\vec{c}\times\vec{d})$

$$= (\vec{A}\cdot\vec{d})\vec{c} - (\vec{A}\cdot\vec{c})\vec{d}$$

$$= \{(\vec{a}\times\vec{b})\cdot\vec{d}\}\vec{c} - \{(\vec{a}\times\vec{b})\cdot\vec{c}\}\vec{d}$$

$$= [\vec{a}\,\vec{b}\,\vec{d}]\vec{c} - [\vec{a}\,\vec{b}\,\vec{c}]\,\vec{d}$$

$= 0\vec{c} - 0\vec{d}$ [$\because$ $\vec{a}, \vec{b}, \vec{c}$ and $\vec{d}$ are co-planar therefore $[\vec{a}\,\vec{b}\,\vec{d}] = [\vec{a}\,\vec{b}\,\vec{c}] = 0$.] $= 0$

8. *If* $\vec{a} = \vec{i} + \vec{j} + \vec{k}$ *and* $\vec{b} = -\vec{i} + \vec{j} - \vec{k}$; *find the unit vector perpendicular to both vectors.*

Solution:

As vectors $\vec{a}$ and vector $\vec{b}$ are given therefore,

$$\vec{a}\times\vec{b} = \begin{vmatrix} \vec{i} & \vec{j} & \vec{k} \\ 1 & 1 & 1 \\ -1 & 1 & -1 \end{vmatrix}$$

$$= \vec{i}\,(-1-1) - \vec{j}\,(-1+1) + \vec{k}\,(1+1)$$

$$= -2\vec{i} + 2\vec{k}$$

Since $\vec{a}\times\vec{b}$ is perpendicular to the plane containing $\vec{a}$ and $\vec{b}$, therefore unit vector perpendicular to both vector is unit vector along $\vec{a}\times\vec{b}$.

Now, $|\vec{a}\times\vec{b}| = \sqrt{(-2)^2 + 2^2}$

$$= \sqrt{4+4}$$

$= \sqrt{8}$

$= 2\sqrt{2}$

Therefore unit vector perpendicular to both vector is,

$$\hat{U}_{ab} = \frac{-2\vec{i}+2\vec{k}}{2\sqrt{2}}$$

$$= \frac{-\vec{i}+\vec{k}}{\sqrt{2}}.$$

9. *Find the resultant of the vectors shown in Fig.3-W1.*

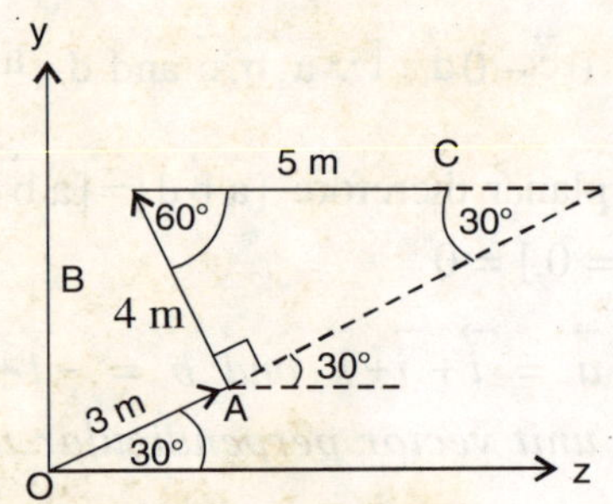

Fig. 3-W1

Solution:

First method: Geometrical method

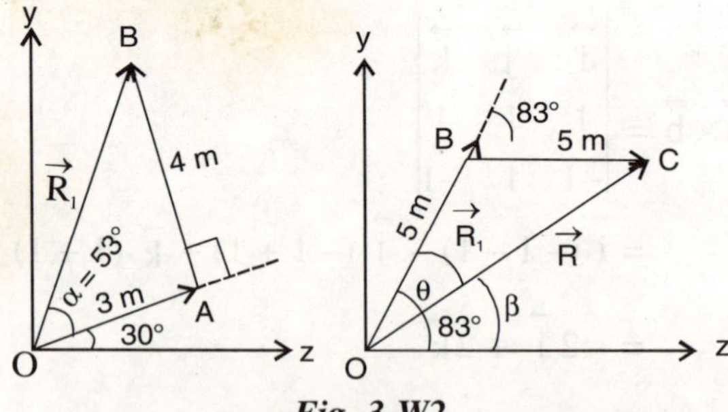

Fig. 3-W2

First find the resultant of $\vec{OA}$ and $\vec{AB}$ using the triangle rule. Since $\vec{OA}$ and $\vec{AB}$ are perpendicular to each other.

$$\therefore \quad |\vec{R}| = \sqrt{3^2+4^2}$$

$$= \sqrt{5^2} = 5 \text{ unit}$$

and $\tan\alpha = \dfrac{4}{3}$

$\therefore \quad \alpha = 53°$

Since angle between $\vec{R_1}$ and $\vec{BC}$ is 83°.

$\therefore \quad |\vec{R}|$ = magnitude of resultant of $\vec{R_1}$

$$= \sqrt{5^2+5^2+2\times5\times5\times\cos 83°}$$

$= 7.48$ m

and $\tan\theta = \tan$ (angle with vector $\vec{R_1}$)

$$= \frac{5\sin 83°}{5+5\cos 83°}$$

$= 0.88$

$\therefore \quad \theta = 41.5°$

$\therefore \quad \beta = 83° - 41.5°$

$= 41.5°$

$\therefore \quad \vec{R_1}$ is 7.48 m in magnitude and 41.5° with z-axis.

Second method : Component method

Table 3-W1

Vector	x-component	y-component
3 m	3 cos 30°	3 sin 30°
4 m	4 cos 120°	4 sin 120°
5 m	5 cos 0°	5 sin 0°

Σ x-component

$= 3\cos 30° + 4\cos 120° + 5\cos 0°$

$= 5.59$ m

Σy-component

$= 3\sin 30° + 4\sin 120° + 5\sin 0°$

$= 4.96$

$$\therefore \quad \vec{R_1} = 5.59\,\vec{k} + 4.96\,\vec{j}$$

where $\quad |\vec{R}| = \sqrt{(5.59)^2+(4.96)^2}$

$= 7.48$ m

and $\tan\beta = \frac{4.96}{5.59}$

$= 0.88$

$\therefore \beta = 41.5°$

10. *Find the value of*

$\vec{i}\times(\vec{j}\times\vec{k})+\vec{j}(\vec{k}\times\vec{i})+\vec{k}\times(\vec{i}\times\vec{j})$

Solution: As we know that,

$$\vec{i}\times(\vec{j}\times\vec{k}) = (\vec{i}\cdot\vec{k})\vec{j}-(\vec{i}\cdot\vec{j})\vec{k}$$
$$= 0\vec{j}-0\vec{k}$$
$$= \vec{0}$$
$$\vec{j}\times(\vec{k}\times\vec{i}) = (\vec{j}\cdot\vec{i})\vec{k}-(\vec{j}\cdot\vec{k})\vec{i}$$
$$= 0\vec{j}-0\vec{k}$$
$$= \vec{0}$$

and
$$\vec{k}\times(\vec{i}\times\vec{j}) = (\vec{k}\cdot\vec{j})\vec{i}-(\vec{k}\cdot\vec{i})\vec{j}$$
$$= 0\vec{j}-0\vec{k}$$
$$= \vec{0}$$

$$\therefore \vec{i}\times(\vec{j}\times\vec{k})+\vec{j}(\vec{k}\times\vec{i})+\vec{k}\times(\vec{i}\times\vec{j})$$
$$= \vec{0}+\vec{0}+\vec{0}$$
$$= \vec{0}$$

EXERCISE

1. An airplane travels 130 miles on a straight course making an angle of 22.5° east of due north. How far in the north and how far in the east did the plane travel from its starting point?

2. Three co-planar vectors are expressed with respect to a certain rectangular right handed co-ordinate system as

$\vec{a} = 4\vec{i}-\vec{j}, \vec{b} = -3\vec{i}+2\vec{j}$ and $\vec{c} = -3\vec{j}$

and units of components are given in meter. Find the resultant vector.

3. A man flies from Washington to Manila. Describe the displacement vector. What is its magnitude, if the latitude and longitude of the two cities are 39° N, 77° W , 15° N and 121° E?

4. Fig. (3-E1) shows two vectors $\vec{a}$ and $\vec{b}$ and two system of coordinates which differ in that the x and x′ axes and the y and y' axes each makes an angle ϕ with each other. Prove analytically that, $\vec{a}+\vec{b}$ has same magnitude and direction no matter which system is used to carry out the analysis.

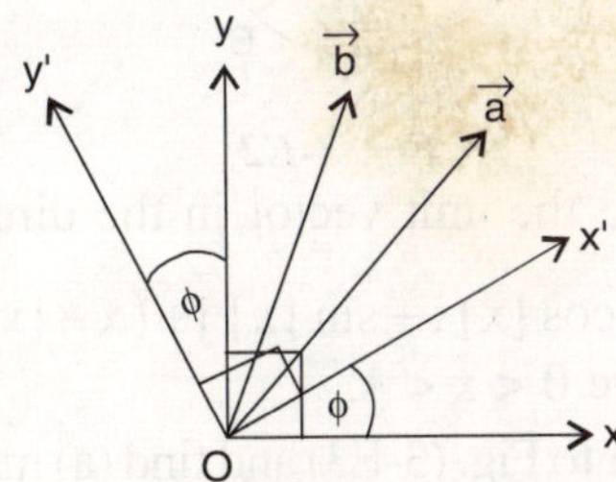

Fig. 3-E1

5. What is the angle θ between vectors $\vec{a} = 3\vec{i}-4\vec{j}$ and $\vec{b} = -2\vec{i}+3\vec{k}$?

6. Find the angles between the body diagonal of a cube having edge length a.

7. Find (a) north cross west (b) down dot south (c) east cross up (d) west dot west and (e) south cross south. Let each vector has unit magnitude.

8. (a) Show that $\vec{a} \cdot \left(\vec{b} \times \vec{a}\right)$ is zero for all vectors $\vec{a}$ and $\vec{b}$. (b) What is the value of $\vec{a} \times \left(\vec{b} \times \vec{a}\right)$, if there is an angle ϕ between the directions of $\vec{a}$ and $\vec{b}$?

9. The work done by a force $\vec{F}$ during a displacement $\vec{r}$ is given by $\vec{F} \cdot \vec{r}$. Suppose a force of 12 kN acts on a particle in vertically downward direction and the particle is displaced through 2.0 m in vertically upward direction. Find the work done by the force during the displacement.

10. Find the magnitude of the resultant of the three vectors $\vec{OA}, \vec{OB}$ and $\vec{OC}$ shown in Fig. (3-E2). Radius of circle is 5 m.

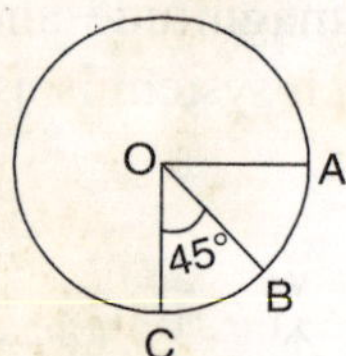

Fig. 3-E2

11. Write the unit vector in the direction of $\vec{A} = \cos[x]\,\vec{i} + \sin[x]\,\vec{j} + (x - [x])\,\vec{k}$ where $0 < x < 1$.

12. Refer to Fig. (3-E3) and find (a) magnitude of resultant (b) x and y components of resultant and (c) the angle made by resultant with the forces $\vec{OA}, \vec{BC}$ and $\vec{DE}$.

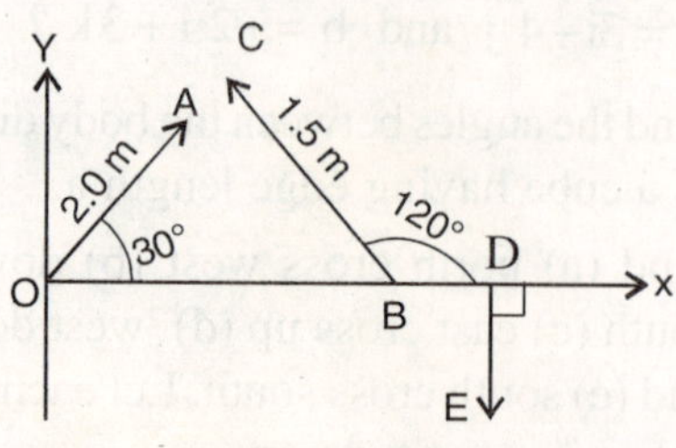

Fig. 3-E3

13. A vector $\vec{\epsilon}$ has magnitude of 6 unit and is in the direction of the x-axis; vector $\vec{\psi}$ has magnitude of 4 unit and lies in the xy-plane; making an angle of 30° with x-axis and an angle of 60° with the y-axis. Find the vector product of $\vec{\epsilon} \times \vec{\psi}$.

14. Given two vectors $\vec{A} = (2, 3, 4)$ and $\vec{B} = (1, -2, 3)$ obtain the following :

(a) The magnitude of each vector.

(b) An expression for the vector sum using unit vectors.

(c) The magnitude of vector sum.

(d) An expression for the vector difference using unit vectors.

(e) The magnitude of the vector $\vec{A} - \vec{B}$. Is this the same as the magnitude of $\vec{B} - \vec{A}$? Explain.

15. Two points P_1 and P_2 are described by their x and y coordinates (x_1, y_1) and (x_2, y_2) respectively. Show that the components of the displacement A from P_1 to P_2 are $A_x = \Delta x$ and $A_y = \Delta y$. Also derive expression for the magnitude and direction of the displacement.

16. Find the vector $\vec{r}$ which satisfies

$$\vec{r} \times \vec{b} = \vec{c} \times \vec{b},\ \vec{r} \cdot \vec{a} = 0 \text{ and } \vec{a} \cdot \vec{b} \neq 0.$$

17. If $\vec{a} = \vec{i} + 2\vec{j} - \vec{k}, \vec{b} = 2\vec{i} + \vec{j} + 3\vec{k}, \vec{c} = \vec{i} - \vec{j} + \vec{k}$ and $\vec{d} = 3\vec{i} + \vec{j} + 2\vec{k}$ then evaluate,

(i) $\left(\vec{a} \times \vec{b}\right) \cdot \left(\vec{c} \times \vec{d}\right)$

(ii) $\left(\vec{a} \times \vec{b}\right) \times \left(\vec{c} \times \vec{d}\right)$.

18. Prove that $\vec{a}\times\left(\vec{b}\times\vec{c}\right), \vec{b}\times\left(\vec{c}\times\vec{a}\right), \vec{c}\times\left(\vec{a}\times\vec{b}\right)$ are co-planar.

19. For what value(s) of k, the points A (1, 0, 3), B (–1, 3, 4), C (1, 2, 1) and D (k, 2, 5) are co-planar ?

20. Find the value of $\left[\vec{a}-\vec{b}\;\; \vec{b}-\vec{c}\;\; \vec{c}-\vec{a}\right]$.

21. Determine the resultant and direction of the four forces act on the body shown in Fig. (3-E4).

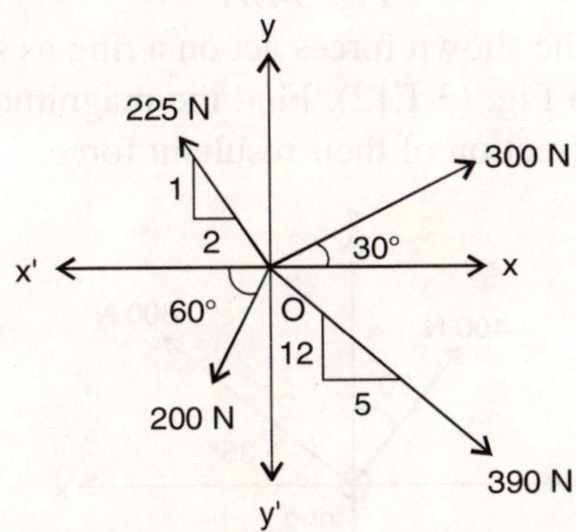

Fig. 3-E4

22. Determine the resultant of the co-planar system of forces shown in Fig. (3-E5), using vector approach.

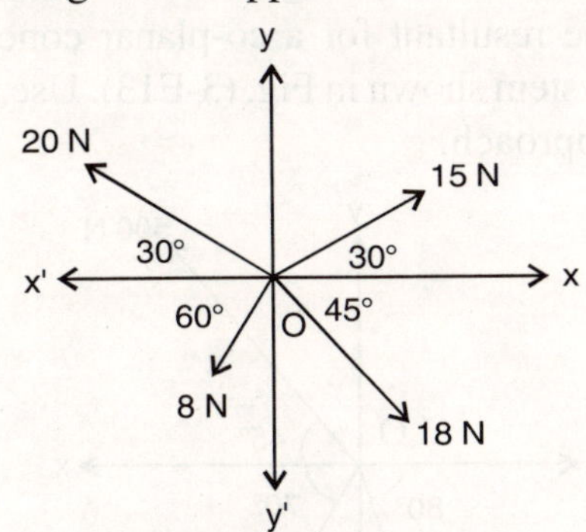

Fig. 3-E5

23. Distinguish between dot product and cross product of two vectors.

24. Determine the magnitude and direction of the force F. If F_R = 270 N, acts along y-direction as shown in Fig. (3-E6). Use vector approach only.

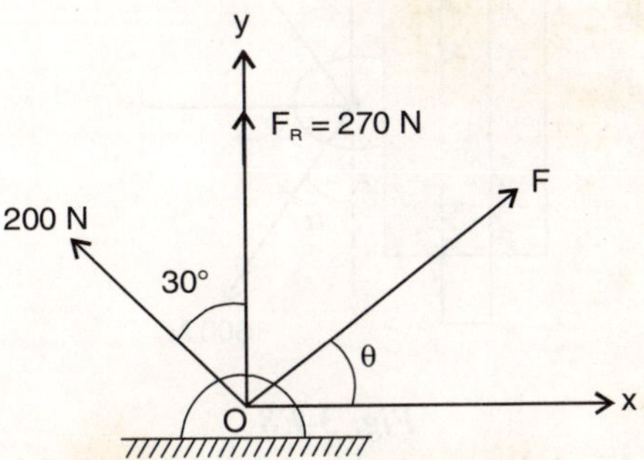

Fig. 3-E6

25. ABCD is a square of side 2 m long. On the sides AB, BC, CD and DA, forces acting are I N, 2 N, 8 N and 5 N respectively. Also, along diagonals AC and DB the forces acting are $5\sqrt{2}$ N and $2\sqrt{2}$ N respectively. Find the resultant of these forces. Use vector approach only.

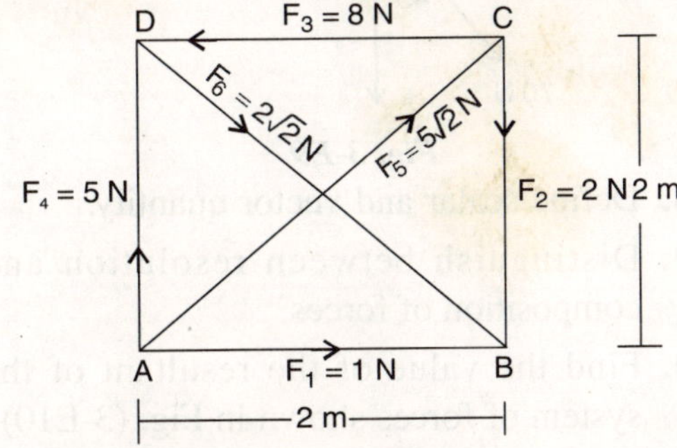

Fig. 3-E7

26. A collar which may slide on a vertical rod is subjected to three forces as shown in Fig. (3-E8). Determine (i) the value of α for which the resultant of these three forces is horizontal (ii) the corresponding magnitude of resultant.

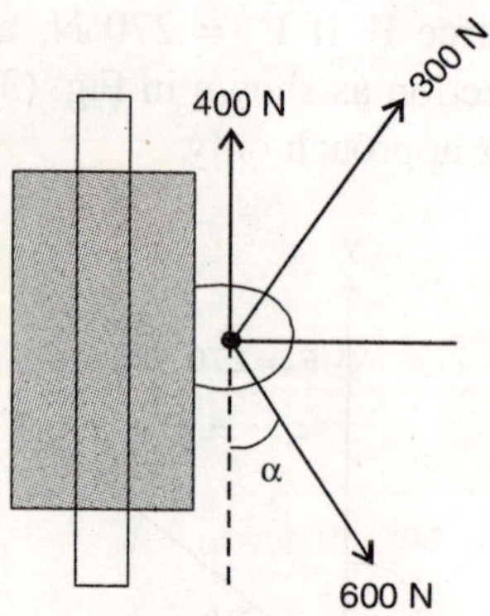

Fig. 3-E8

27. Find the resultant of the system of forces shown in Fig. (3-E9) and locate their resultant.

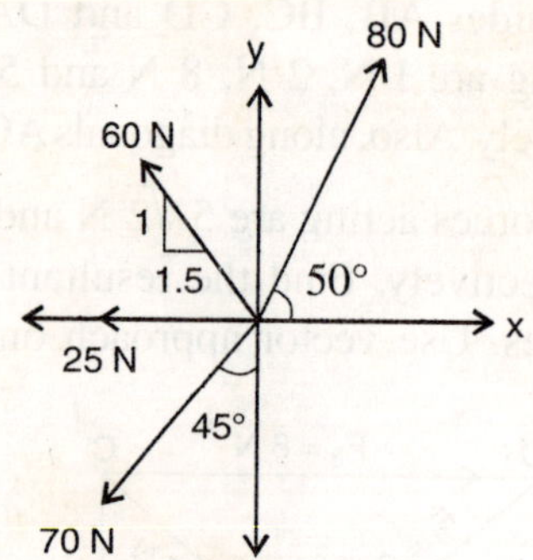

Fig. 3-E9

28. Define scalar and vector quantity.

29. Distinguish between resolution and composition of forces.

30. Find the value of the resultant of the system of forces shown in Fig. (3-E10)

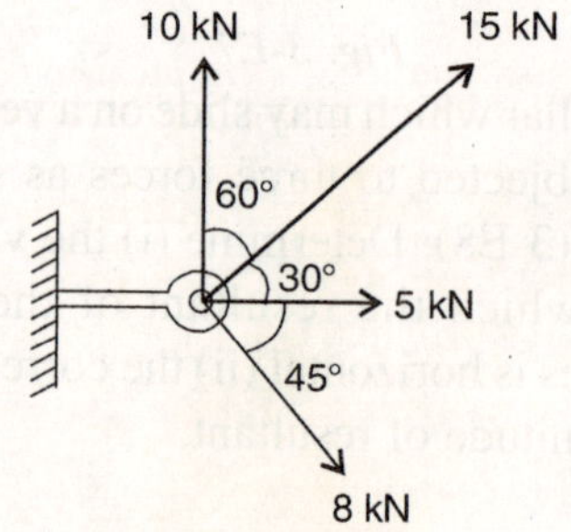

Fig. 3-E10

31. Determine the resultant of the forces act at a point shown in Fig. (3-E11). Use the vector approach.

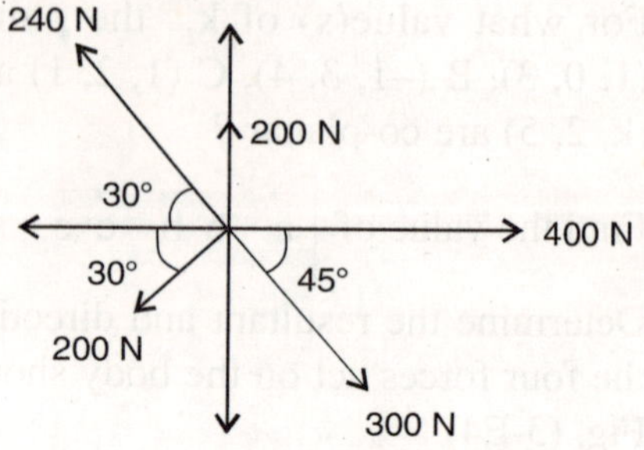

Fig. 3-E11

32. The shown forces act on a ring as shown in Fig. (3-E12). Find the magnitude and direction of their resultant force.

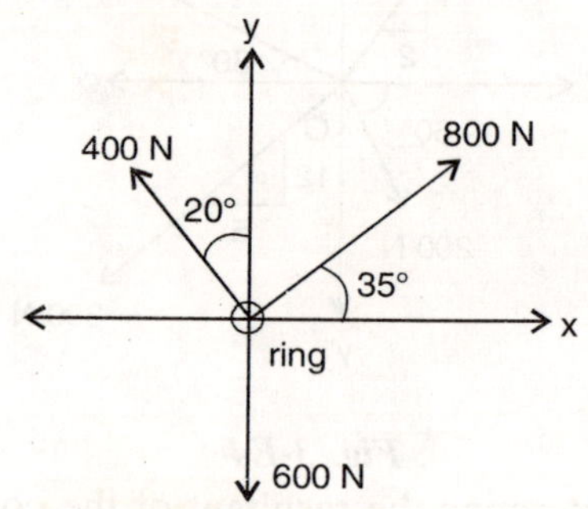

Fig. 3-E12

33. Determine the magnitude and direction of the resultant for a co-planar concurrent system shown in Fig. (3-E13). Use vector approach.

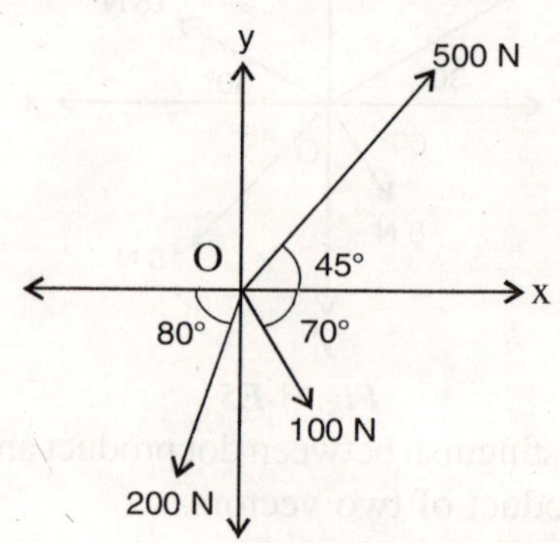

Fig. 3-E13

34. The co-planar forces act a at point as shown in Fig. (3-E14). The resultant is 500 N and acts along x-axis. Determine the force P and its inclination θ with x-axis.

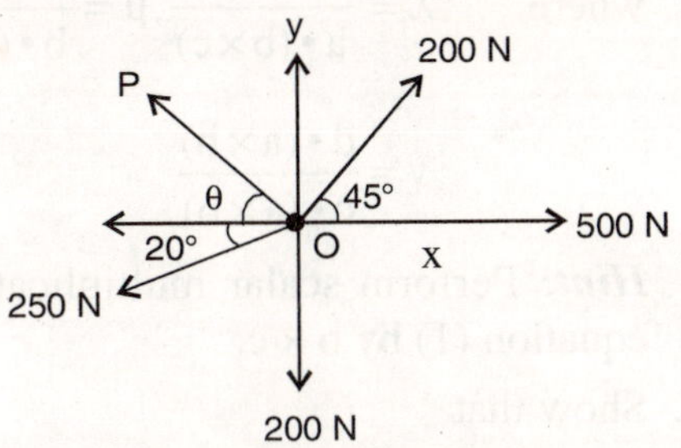

Fig. 3-E14

35. The two forces $\vec{P}$ and $\vec{Q}$ act at the angle α and are applied to a body at point A, find formulae for calculating the magnitude of their resultant R and angles β and γ which its line of action makes with those of the given forces.

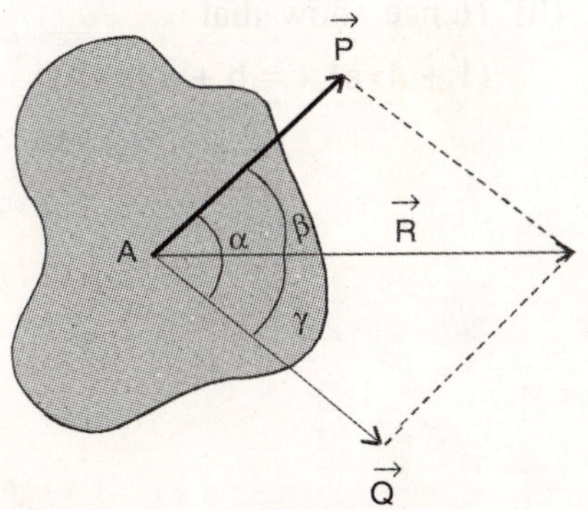

Fig. 3-E15

36. A man of weight W holds one end of a rope that passes over a pulley vertically above his head and the other end of the rope is attached with a weight Q. Find the force with which the man's feet presses against the floor.

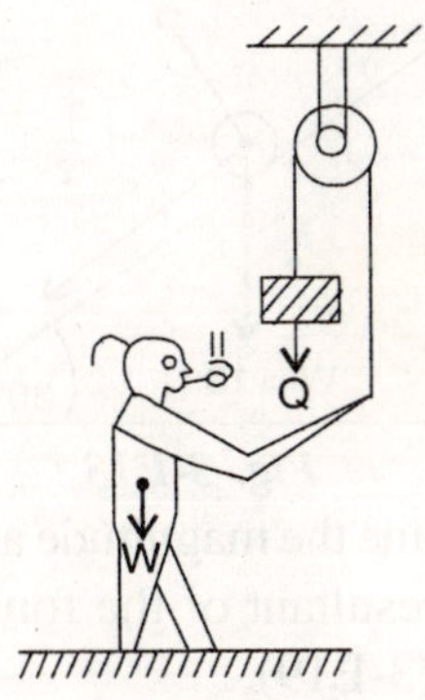

Fig. 3-E16

37. A boat moves uniformly along a canal by two horses pulling with forces P = 200 N and Q = 240 N acting at an angle α = 60°. Determine the magnitude of the resultant pull on the boat and angles β and γ as shown in figure.

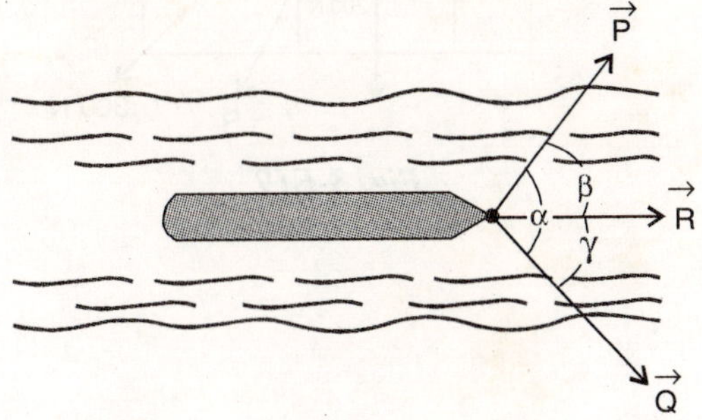

Fig. 3-E17

38. A force F combined with P = 6 N is directed vertically upwards and gives a horizontal resultant R = 40 N. Find magnitude and the inclination of F.

39. A small block of weight Q = 10 N is placed on an inclined plane which makes an angle α = 30° with the horizontal. Resolve the weight into two rectangular components Q_p and Q_n acting parallel and normal with respect, to the inclined plane.

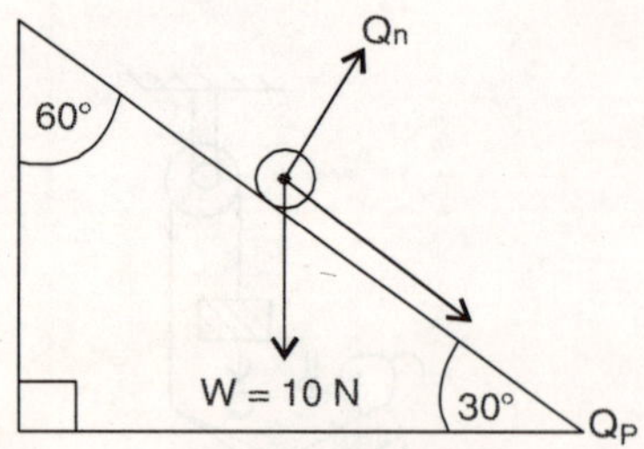

Fig. 3-E18

40. Determine the magnitude and the direction of the resultant of the four forces shown in Fig. (3-E19).

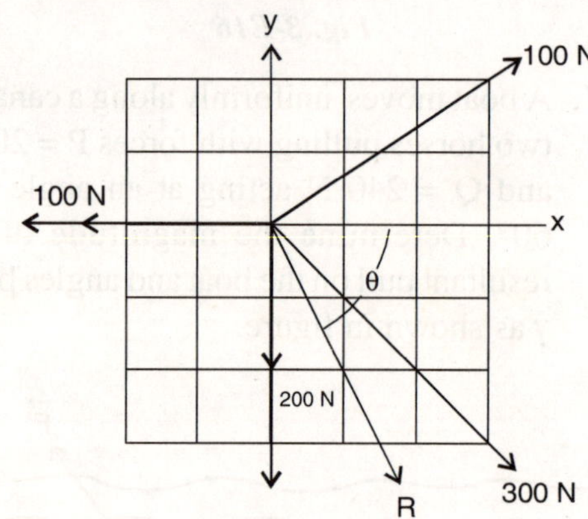

Fig. 3-E19

41. Show that if a, b and c are non-coplanar vectors then any other vector d can be written as the linear combination as

$$d = \lambda a + \mu b + \nu c \quad ....(1)$$

where, $\lambda = \dfrac{d \bullet (b \times c)}{a \bullet (b \times c)}, \mu = \dfrac{d \bullet (c \times a)}{b \bullet (c \times a)},$

$$\nu = \frac{d \bullet (a \times b)}{c \bullet (a \times b)} \quad ....(2)$$

Hint: Perform scalar multiplication of equation (1) by $b \times c$.

42. Show that

(a) $a \times (b \times c) + b \times (c \times a) + c \times (a \times b) = 0$

(b) $(a + b) \bullet [(b + c) \times (c + a)] = 2a \bullet (b \times c)$

(c) $(a \times b) \bullet [(b \times c) \times (c \times a)] = [a \bullet (b \times c)]^2$

43. The vector x satisfies the vector equation $a \times (x \times a) + x = b$

(a) Show that $a \bullet x = a \bullet b$

(b) Hence show that

$(1 + a \bullet a)\, x = b + a\, (a \bullet b).$

NEWTON'S LAWS OF MOTION

4.1 INTRODUCTION TO MECHANICS

Mechanics is the branch of science which describes and predicts the condition of rest and motion of a body under the action of forces. It is divided into three parts viz. Mechanics of rigid bodies, mechanics of deformable bodies and mechanics of fluids.

The mechanics of rigid body is sub-divided into statics and dynamics, statics deals about bodies at rest (equilibrium) and dynamics about bodies in motion when subjected to external forces. Dynamics may be classified into kinematics and kinetics. Kinematics deals with geometry of motion of bodies without reference to the agents causing the motion. Kinetics deals with the motion of bodies as a consequence of the application of forces and couples on bodies.

Mechanics of deformable bodies is studied under the title, 'Strength of Material'. Third division of mechanics is mechanics of fluids is further divided into the study of incompressible fluids and of compressible fluids.

Mechanics of incompressible fluids is hydromechanics or hydraulics. Hydromechanics deals with the condition under which fluid could remains at rest or in motion. The subject of hydromechanics can further be divided into hydrostatics and hydrodynamics. Hydrostatics deals with fluid at rest and hydrodynamics deals with fluids in motion.

Throughout in this textbook authors have concentrated only on mechanics of rigid bodies.

4.2 IDEALIZATION OF MECHANICS

4.2.1 Space

Space is the geometrical region occupied by bodies whose positions are described by linear and angular measurements relative to coordinate system. For three-dimensional bodies three independent coordinates are needed. For two dimensional (plane) bodies only two coordinates are required.

4.2.2 Time

Time is the measure of a succession of events and is a basic quantity in dynamics. Time is not directly involved in the analysis of problems in statics.

4.2.3 Mass

Mass is a measure of the inertia of a body, and is its resistance to change in velocity. Mass could also be defined as the quantity of matter in body. Mass is the property of every body by which it experiences mutual attraction to other bodies. Units of mass are kilogram, gram, tonne, etc.

4.2.4 Particle

Particle may be defined as an object which has only mass and no size. By particle we mean a very small amount of matter which could be assumed to occupy a single point. In other words, particle could be defined as a body of infinitely small volume and is considered to be concentrated at a point. Mathematically, a body is therefore represented as a particle, if its dimensions are small compared to the co-ordinates describing its motion. Body C is a particle with respect to x-o-y because r >>> l (size of the body) where as body C is still a body with respect to x_1-o_1-y_1 because r_1 > l.

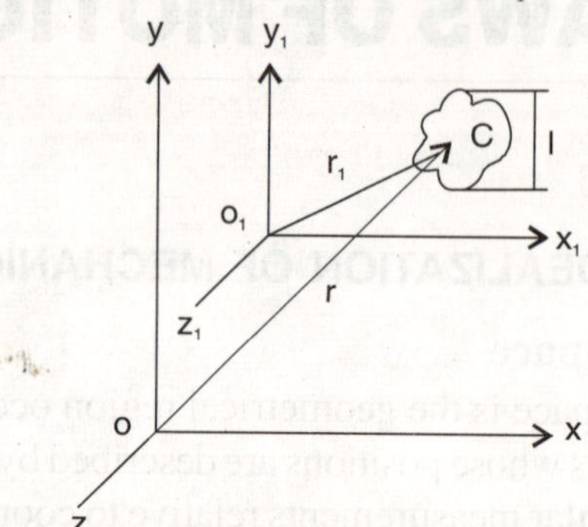

Fig. 4.1

Example 1: *A planet as seen from another planet, star as seen from earth.*

4.2.5 Continuum

It is a well known fact that a body consists of several particles like electrons, protons and neutrons, etc. It is not feasible to solve any engineering problem by treating a body as a conglomeration of such discrete particles. To solve such type of problems we assume that a body is made up of continuum. In other words, the body is treated as a continuum. The concept of continuum of bodies enables us to treat the bodies as rigid bodies and this simplifies the problems in engineering mechanics.

4.2.6 Point force

Point force or concentrated force is yet another idealization, very commonly used in engineering mechanics. Actually, the weight of a book on a table can't pass through a point CG of shown block. There are certain areas of contact between the book and the table which are small compared to the other dimensions involved in the problem. Therefore, much accuracy is not lost by treating it as a point force and thereby, simplifying the problem.

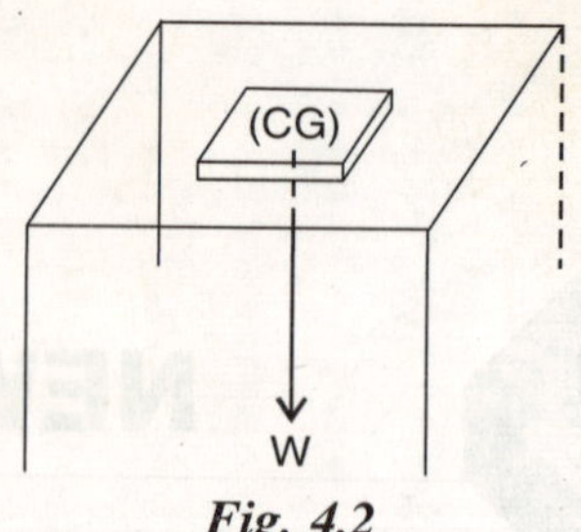

Fig. 4.2

4.2.7 Rigid body

A rigid body subjected to an external system of forces does not undergo any deformation. In other words, the relative position of two particles does not change under the action of forces. The study of Newtonian Mechanics of particles is obviously a prerequisite to that of rigid bodies.

Actually any kind of body is not a perfectly rigid body but deforms slightly under the action of the load it carries. But we may safely ignore this deformation and assume that body is a rigid body.

Example 2:*Consider two particles A and B a body C with two particles A and B, 2 mm apart. After the application of forces, the distance between the two particles A and B is still 2 mm.*

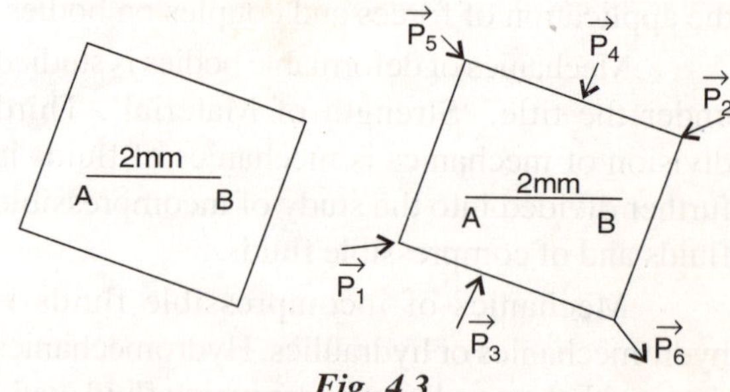

Fig. 4.3

4.2.8 Deformable body

A body is said to be deformable when the distance between its particles change under the action of forces.

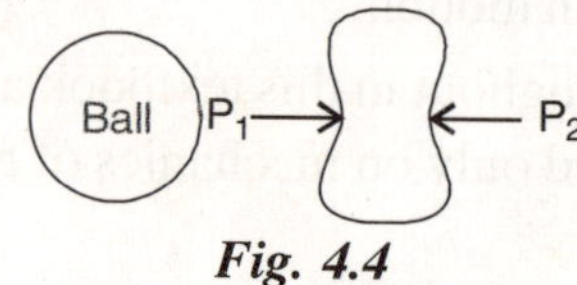

Fig. 4.4

Example 3: *When a soft ball is subjected to a pair of opposite forces, its shape changes i.e., the distance between its particles is varied due to the applied forces.*

4.3 FUNDAMENTAL LAWS

4.3.1 Newton's first law

Every particle continues in its state of rest or of uniform motion in a straight line unless it is compelled to change its state by forces acting on it. If the resultant or net force acting on a particle is zero, then it remains at rest (if originally at rest) or it moves with constant speed in a straight line (if originally in motion).

If the vector sum or algebraic sum of all the components of the forces acting on a particle is zero then the particle is said to be remained unaccelerated that is remains at rest when initially at rest and moves with constant velocity when initially in motion.

One can reverse the statement by saying that, when the particle is in a state of rest under the action of a concurrent co-planar system of forces then the algebraic sum of components of the forces along any direction is zero.

Mathematically,

$$\Sigma \vec{F} = 0 \text{ if and only if } \vec{a} = 0$$

or

$$\vec{a} = 0 \text{ if and only if } \sum \vec{F} = 0$$

where, Force (N) = Mass (kg) × acceleration (ms^{-2}). In other words, we say Newton's second law as "the acceleration of a particle as measured from an inertial frame is given by the vector sum per unit mass of all the forces act on the particle."

Mathematically, $\Sigma \vec{F} = m\vec{a}$

Thus three laws of Newton's are concluded as,

(i) $\Sigma F_x = ma_x$, $\Sigma F_y = ma_y$, $\Sigma F_z = ma_z$ in general $\Sigma F_{\text{any direction}} = m \times a_{\text{in that direction}}$

(ii) If $\vec{a} = 0$, i.e., body is at rest or moving with constant velocity then, $\Sigma \vec{F} = 0$

(iii) (a) If $\vec{a} = 0$ then $\Sigma \vec{F} = 0$

(b) If $\Sigma \vec{F} = 0$ then $\vec{a} = 0$.

(iv)

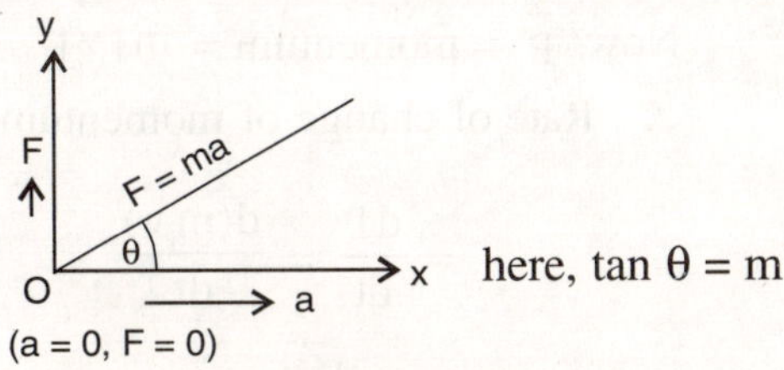

here, $\tan\theta = m$

Fig. 4.5

(v) We can note that the first law of motion is contained in the second law as a special case. For $\vec{F} = 0$, we get $\vec{a} = 0$. Therefore in the absence of applied force a body moves with a constant velocity or continues to be at rest (zero velocity) which is first law. Therefore, out of Newton's three laws of motion only two are independent laws.

(vi) If $\vec{F} = 0$, condition of motion is said to be static.

(vii) $\vec{F} = m\vec{a}$

or $(F_x \vec{i} + F_y \vec{j} + F_z \vec{k})$

$$= m(a_x \vec{i} + a_y \vec{j} + a_z \vec{k})$$

or $F_x \vec{i} + F_y \vec{j} + F_z \vec{k}$

$$= ma_x \vec{i} + ma_y \vec{j} + ma_z \vec{k}$$

$$\Rightarrow F_x = ma_x$$

$$F_y = ma_y$$

$$F_z = ma_z$$

or in general,

$$F_{\text{in any direction}} = ma_{\text{in that direction}}$$

4.3.2 Newton's second law

Statement: The rate of change of momentum of a body (particle) is directly proportional to the impressed force and it takes place in the same direction in which the force acts.

i.e., $F \propto$ rate of change of momentum

Now, $\vec{P}$ = momentum = $m|\vec{v}|$

$\therefore$ Rate of change of momentum

$$= \frac{d\vec{p}}{dt} = \frac{d(m\vec{v})}{dt}$$

$$= m\frac{d\vec{v}}{dt}$$

$$= m\vec{a}$$

Since, $\vec{F} \propto m\vec{a}$

or $|\vec{F}| = km|\vec{a}|$

or $F = kma$

where, F is the magnitude force, m is mass, a is magnitude of acceleration and k is constant of proportionality.

One Newton force could be defined as force which is required to move an object having 1 kg mass with 1 ms^{-2} acceleration.

i.e., $1\text{ N} = k\,(1\text{ kg}) \times (1\text{ ms}^{-2})$

or $1\text{ N} = k \times (1\text{ kg ms}^{-2})$

Let $N = \text{kg ms}^{-2}$

Therefore, $k = 1$

So with this assumption the most fundamental relation between force, mass and acceleration is obtained. Where, $|\vec{F}|$ itself represents vector sum of forces acting on an object.

$$|\vec{F}| = m|\vec{a}| \qquad \text{....(4.1)}$$

4.3.3 Newton's third law

Forces acting on a body originate due to other bodies that make up its environment. A single force is due to the mutual interaction between two bodies. Experimentally it has been found that when one body exerts a force on the second body, the second body also exerts a force on the first. Further more these forces are equal in magnitude but opposite in direction. A single isolated force is therefore an impossibility. One of the two forces involved in the interaction between two bodies is called **action** and other is called **reaction**. These property of forces was first stated by Newton in his third law of motion. "For every action there is always an equal reaction." In simple words, "for every action there is an equal and opposite reaction." We can say that, if body A exerts a force on body B, body B also exerts an equal but opposite force on body A and further more the forces lie along the same line joining the bodies. Note that, the action and reaction always occur as a pair, act on different bodies. If they were to act on the same body, we could never have accelerated motion because the resultant force on every body would have been zero. In other word, if a body A exerts a force $\vec{F}$ on a body B, then B exerts a force $-\vec{F}$ on A. We can note that the forces exerted by two bodies, connected by the third law act on two different bodies and hence never appear together in the list of forces acting on one body to apply either the first or second law. We can understand this law, in the shown figures.

Example 4:

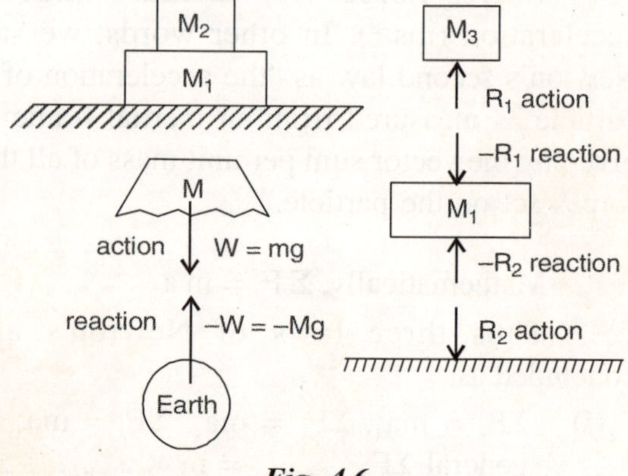

Fig. 4.6

4.3.4 Law of gravitation

Any two particles are attracted towards each other along a line connecting their centres with a mutual force whose magnitude is directly proportional to the product of their masses and inversely proportional to the square of the distance between them.

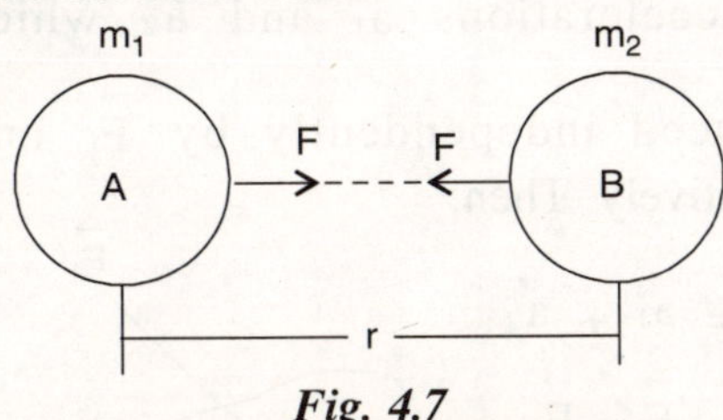

Fig. 4.7

Let A and B be two particles of masses m_1 and m_2 separated by the distance r therefore,

$$F = \frac{G m_1 m_2}{r^2}$$, where G is the universal constant of gravitation its value is 6.67×10^{-11} Nm^2/kg^2 or $m^3/kg\ s^2$. Consider $m_1 = m_2 = 1$ kg then,

$$F = \frac{6.67 \times 10^{-11} \times 1 \times 1}{1^2}$$

$$= 6.67 \times 10^{-11} \text{ N.}$$

The law of gravitation helps us to define the weight of a body. The weight of a body is nothing but the force exerted by the planet (Earth) on the body under consideration. For an Earth bound object of mass m, its weight in gravitational field is given by,

$$W = \frac{GM_{earth}\, m}{R^2}$$

$$= \left(\frac{GM_{earth}}{R^2}\right) m$$

$$= mg \quad \text{where,} \quad g = \frac{GM_{earth}}{R^2};$$

$$M_{earth} = 5.976 \times 10^{24} \text{ and } R = 6.371 \times 10^6 \text{ m}$$

$$\therefore \quad g = \frac{6.67 \times 10^{-11} \times 5.976 \times 10^{24}}{6.371 \times 6.371 \times 10^{12}} \text{ ms}^{-2}$$

$$= 9.82 \text{ ms}^{-2}$$

g is also known as gravitational field intensity or gravitational acceleration which is constant near the surface of the earth. It is also acceleration of body when it falls freely in the gravitational field without any resistance thus,

$$W = mg; \quad g = 9.82 \text{ ms}^{-2}$$

If m is given in kg then the unit of the weight of body is N.

4.3.5 Principle of transmissibility of forces

The state of rest or of motion of a rigid body is unaltered if a force acting on the body is replaced by another force of the same magnitude and direction but, acts anywhere on the body along the line of action of the replaced force. Law of transmissibility of forces hold true for rigid bodies only. By the transmission of the force only the state of the body is unaitered, but not the internal stresses which may develop within the body.

In the shown diagram $\vec{P}$ is the force acting on rigid body at point A. According to the principle of transmissibility of forces this force has the same effect on the body as force $\vec{P}$ when applied at point B.

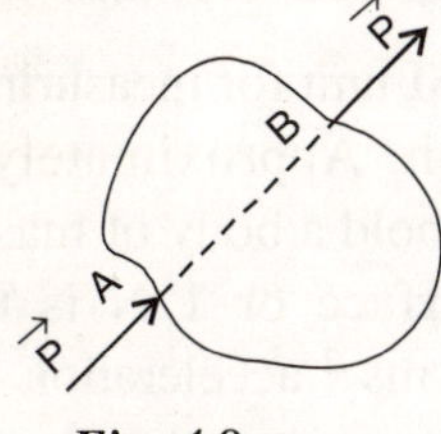

Fig. 4.8

4.4 FORCE

Force could be defined as an action that tends to change the state of rest or motion of a body to which it is applied for the investigation of problems in statics. In our day-to-day language force is associated with a push or a pull, perhaps exerted by our muscles. If one pushes a body, he is exerting a force away from his body. Similarly, when he pulls a body, he exerts a force towards himself. When you hold a heavy body in your hand, you exert a large force; when you hold a light block you exert a small force. The force can also be applied by a non-living body. For example, as we feel that wind exerts force on us. Further more force is a vector quantity that changes the acceleration of a body.

There are three quantities which completely define a force are called its specifications. They are (1) its magnitude (2) its point of application and (3) its direction, the magnitude of a force is obtained by comparing it with a certain standard unit of force. The point of application of a force acting upon a body is that point on the body at which the force can be assumed to be concentrated. Physically it is impossible to concentrate a force at a single point i.e., every force must have some finite area or volume over which its action is distributed. Force obeys principle of the superposition that is when several forces act on a body simultaneously, each produces its own acceleration independently. The resulting acceleration is the vector sum of the several independent accelerations.

The SI unit for measuring a force is given by Newton. Approximately, it is the force needed to hold a body of mass 102 gr, near the earth's surface or 1 N is that force which produce 1 ms^{-2} acceleration in 1 kg body. All forces in nature can be classified under three headings, each with a different relative strength. (1) gravitational forces, which are relatively very weak, (2) electromagnetic forces, which are of intermediate strength and (3) nuclear forces. Nuclear forces are of two types, those which bind neutrons and protons in the nucleus (very strong) and those responsible for beta decay (weak). These forces are real in the sense, that we can associate them with specific objects. Since force is a vector, their sum (differences) can be found by triangle law or parallelogram law of vector addition. Let m be the mass of a body and two forces $\vec{F_1}$ and $\vec{F_2}$ are acting simultaneously then the net acceleration of the body is given by vector sum of accelerations $\vec{a_1}$ and $\vec{a_2}$ which are produced independently by $\vec{F_1}$ and $\vec{F_2}$ respectively. Then,

$$\vec{a} = \vec{a_1} + \vec{a_2}$$

$$= \frac{\vec{F_1}}{m} + \frac{\vec{F_2}}{m}$$

$$= \frac{\vec{F_1} + \vec{F_2}}{m}$$

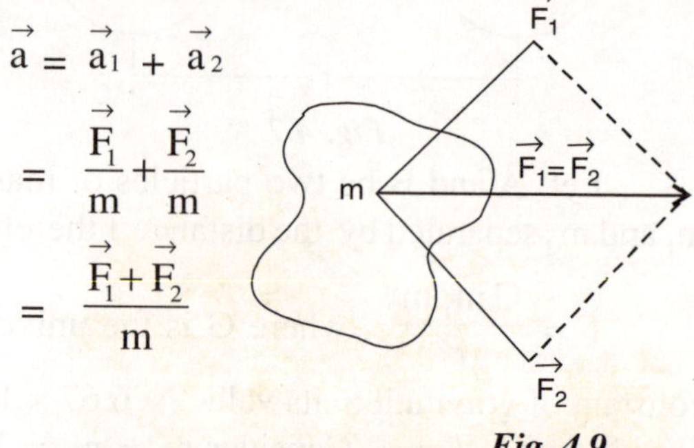

Fig. 4.9

4.4.1 Frame of reference

There are certain frame of references in which Newton's laws are valid. Measure acceleration from such a frame and you are allowed to say that $\Sigma \vec{a} = 0$ iff $\Sigma \vec{F} = 0$. But there are frame of references in which Newton's laws are not valid. We may find that even if the sum of forces $(\Sigma \vec{F})$ is not zero, even the net acceleration is equal to zero (i.e., $\Sigma \vec{a} \neq 0$). Sometimes we may find that sum of forces is not zero, though the net acceleration is equal to zero. So the validity of Newton's laws depends on the chosen frame of reference. A frame of reference in which Newton's laws are valid is called an inertial frame of reference. This is a frame of reference that is either at rest or is moving at constant velocity with respect to the fixed stars.Nevertheless we can, if we find it convenient, apply Newton's laws in a non-inertial frame, such a frame might be one that is accelerating with respect to the fixed stars or inertial frame of reference. But directly we can't apply, rather we have to suppose one extra force apart from the real forces induced by environment. Later we call this as pseudo force. Thus a frame of reference in which Newton's laws are not valid is called a non-inertial frame of reference.

Newton's first law, thus, reduces to a definition of an inertial frame why do we call it a law then? Suppose after going through this chapter you keep the book on your table fixed rigidly with the Earth. The book is at rest with respect to the earth's frame. The acceleration of the book with respect to the earth is to be zero. The forces on the book are (a) the gravitational force $\vec{W}$ exerted by the Earth and (b) the contact force $\vec{N}$ by the table. Is the sum of $\vec{W}$ and $\vec{N}$ zero? A very accurate measurement will give the answer no. The sum of forces is not zero although the book is at rest. Therefore, Earth is not an inertial frame of reference, However, the sum of $-\vec{W}$ and $\vec{N}$ is not too different from zero and we can say that the Earth is an inertial frame of reference to a good approximation. Thus, for routine affairs, $\vec{a} = 0$, iff $\vec{F} = 0$ is true in the Earth frame of reference or a frame of reference which is moving with constant velocity with respect to Earth. This fact was identified and formulated by Newton and is said to be Newton's first law. If we restrict that all measurements are made from the Earth frame, indeed it becomes frames, it becomes a definition. We shall assume that unless stated otherwise, we work from an inertial frame of reference.

4.4.2 Inertial forces

If we work with classical mechanics in non-inertial frames, we have to introduce a force(s) is/are called inertial force(s) or pseudo force(s). They are so called because unlike the forces those are acted by its environment on a body but we can't associate pseudo force with a particular body in the environment by which they act rather this is a non real force. Finally, if we view the body (particle) from the inertial frame, the pseudo force disappears. These forces are, then, simply a technique that permits us to apply classical or Newtonian mechanics in the non-inertial frame.

Consider the situation shown in Fig. (4.10).

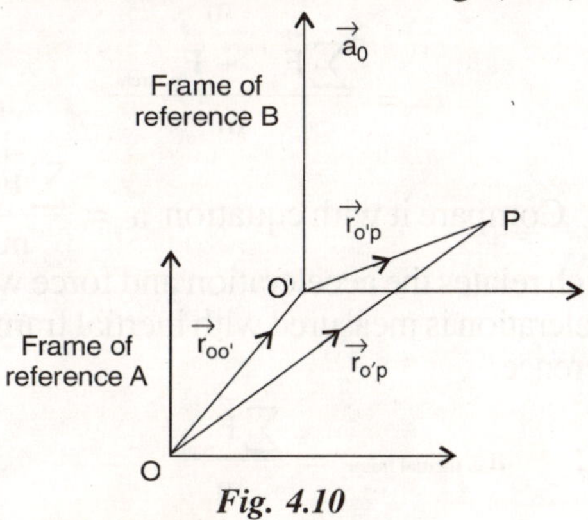

Fig. 4.10

Suppose the frame of reference B moves with a constant acceleration $\vec{a}_0$ with respect to an inertial frame of reference A.

Let $\vec{r}_{OO'}$, be a position vector of frame (B) with respect to inertial frame of reference (A), then frame of reference (B) is moving with an acceleration, $\vec{a}_0$ therefore, it becomes a non-inertial frame of reference. Suppose, $\vec{r}_{o'p}$ and $\vec{r}_{op}$ are position vectors of particle p with respect to frame of reference, B and frame of reference, A respectively, then from triangle law of vector addition we get,

$$\vec{r}_{op} = \vec{r}_{oo'} + \vec{r}_{o'p}$$

or $$\ddot{\vec{r}}_{op} = \ddot{\vec{r}}_{oo'} + \ddot{\vec{r}}_{o'p}$$

or $\quad \vec{a}_{op} = \vec{a}_{oo'} + \vec{a}_{o'p}$

or $\quad \vec{a}_{op} = \vec{a}_o + \vec{a}_{o'p}$

or $\quad \vec{a}_{o'p} = \vec{a}_{op} - \vec{a}_o$

or $\quad m\vec{a}_{o'p} = m\vec{a}_{op} - m\vec{a}_o$

or $\quad \vec{a}_{o'p} = \dfrac{m\vec{a}_{op} - m\vec{a}_o}{m}$

$\therefore \quad \vec{a}_{o'p} = \dfrac{\sum \vec{F}_{real} - m\vec{a}_o}{m}$

$$= \frac{\sum \vec{F}_{real} + m(-\vec{a}_o)}{m}$$

$$= \frac{\sum \vec{F}_{real} + \vec{F}_{nonreal}}{m} \qquad \text{....(i)}$$

Compare it with equation $\vec{a} = \dfrac{\sum \vec{F}_{real}}{m}$, which relates the acceleration and force when acceleration is measured with inertial frame of reference.

i.e., $\quad \vec{a}_{\text{in inertial frame}} = \dfrac{\sum \vec{F}_{real}}{m} \qquad \text{...(ii)}$

and $\quad \vec{a}_{\text{in non-inertial frame}} = \dfrac{\sum \vec{F}_{real} + (-m\vec{a}_o)}{m} \qquad \text{...(iii)}$

Therefore equation (ii) is not valid in a non-inertial frame of reference. An extra term $-m\vec{a}_0$ has to be added to sum of all the forces act on the particle of mass m. Note that in this extra term m is the mass of the particle under consideration and $\vec{a}_0$ is the acceleration of non-inertial frame of references. Thus acceleration of the non-inertial frame of reference comes into the equation of motion of particle. Such correction term $-m\vec{a}_0$ in the list of forces is called pseudo force.

Therefore, while solving the problems in mechanics we have two choices:

1. Select an inertial frame as a reference frame and consider only real forces that is, forces those could be associated with definite bodies in the environment.
2. Select a non-inertial frame as a reference frame and consider not only the real forces but suitably defined pseudo forces. Although we usually choose the first alternative, we sometimes select the second; both are completely equivalent and choice is a matter of convenience.

Example 5: *A pendulum (5 gr) is hanging from the ceiling of a car moving with an acceleration 5 ms^{-2} with respect to the road. Find the angle made by the string with horizontal.*

Solution:

The situation is shown in Fig. (4.11).

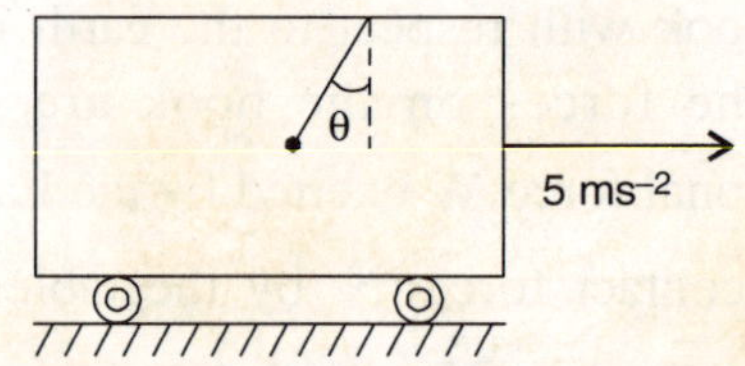

Fig. 4.11

Suppose the mass of the bob is m = 5 gr and string makes an angle θ with the vertical let a_0 = 5 ms^{-2} be the acceleration of car with respect to the road.

Consider inertial frame

Let us consider our frame of reference on the road, since road is not accelerating with respect to inertial frame (stars frame) therefore directly we can apply Newton's second law. Real forces on bob are :

(a) tension in string T,

(b) mg download, by Earth.

F.B.D. of bob is shown below.

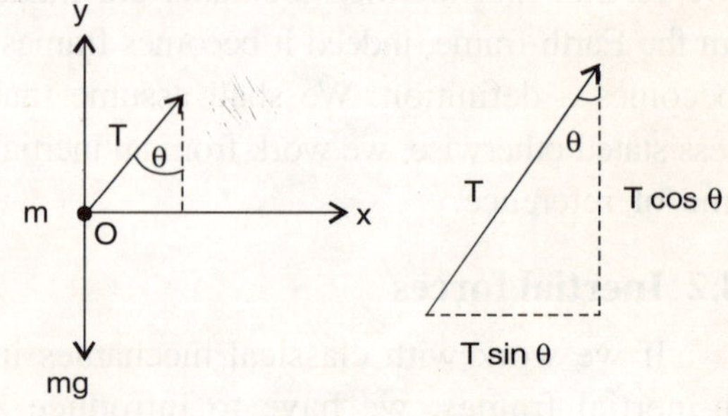

Fig. 4.12 *Fig. 4.13*

Because body is rigidly attached to car, therefore, acceleration of car along horizontal = acceleration of bob in horizontal direction. The net component of forces along the x-axis is,

$$F_x = ma_x$$

or $$T \sin\theta = ma_0 \quad \text{....(i)}$$

and the net component of forces along y-axis is,

$$F_y = ma_y$$

or $$T\cos\theta - mg = m \times 0$$
$$= 0$$

as body is not moving vertically, so $a_y = 0$

or $$T\cos\theta = mg \quad \text{....(ii)}$$

Dividing equation (i) by (ii), we get

$$\frac{T\sin\theta}{T\cos\theta} = \frac{ma_0}{mg}$$

or $$\tan\theta = \frac{a_0}{g}$$
$$= \frac{5}{9.8}$$
$$= 0.51$$

$\therefore$ $$\theta = \tan^{-1}(0.51)$$
$$= 27°.$$

Therefore, angle made by string with horizontal is,

$$\theta_h = 90° - \theta$$
$$= 90° - 27°$$
$$= 63°.$$

Consider non-inertial frame

Now let us work with the car frame. This frame is non-inertial as it has an acceleration 5 ms^{-2} with respect to an inertial frame (the road). Hence, directly we can't use Newton's second law rather we have to include a pseudo force.

Taking bob as the system the forces on the bob are:

(a) T along the string, by string.

(b) mg downward, by the Earth.

(c) ma_0 towards left because $\vec{a}$ is towards right. As we know that pseudo force is $-m\vec{a}_0$ i.e., in opposite direction to $\vec{a}_0$.

F.B.D of bob is shown in Fig. (4.14).

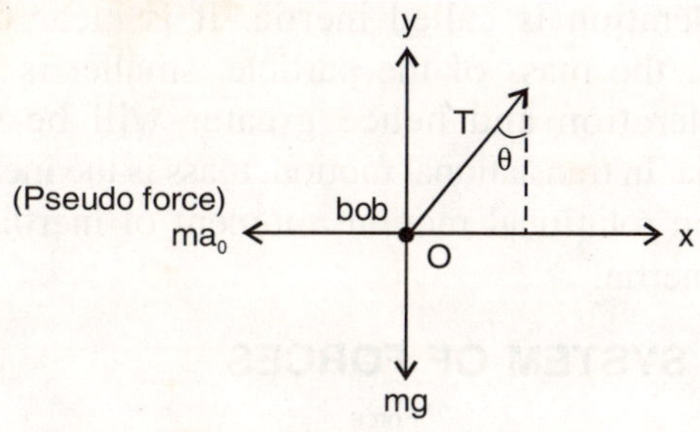

Fig. 4.14

Since bob is at rest in the car as seen by the people on the back seat so, acceleration of bob in car frame is zero in both x and y directions.

Mathematically,

$$\Sigma F_x = 0$$

or $$T\sin\theta - ma_0 = 0$$

or $$T\sin\theta = ma_0 \quad \text{....(iii)}$$

and $$\Sigma F_y = 0$$

or $$T\cos\theta - mg = 0$$

or $$T\cos\theta = mg \quad \text{....(iv)}$$

Dividing (iii) by (iv), we get

$$\tan\theta = \frac{a_0}{g} = \frac{5}{9.8} = 0.51$$

$\therefore$ $$\theta = 27°$$

$\therefore$ $$\theta_h = 90° - 27°$$
$$= 63°.$$

4.5 INERTIA

A particle is accelerated (in an inertial frame) if and only if, a resultant force acts on it. In other words, one can say that a particle does not change its state of rest or of uniform motion along a straight line unless it is forced to do this. This unwillingness of a particle to change its state of rest or of uniform motion along a straight line is called inertia. We can understand the property of inertia in more precise terms as follows. If equal forces are applied on two particles, in general, the acceleration of the particles are different. The property of a particle to allow a smaller acceleration is called inertia. It is clear that larger the mass of the particle, smaller is the acceleration and hence greater will be the inertia. In translational motion, mass is the inertia and in rotational motion, moment of inertia is the inertia.

4.6 SYSTEM OF FORCES

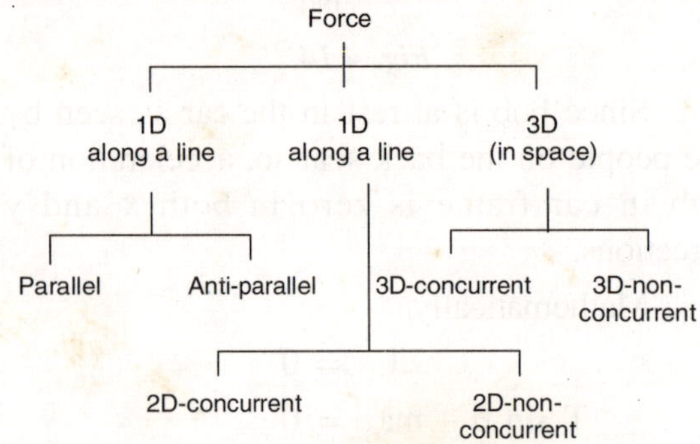

4.6.1 1D-Parallel (anti parallel) forces

If the lines of action of all the forces lie along a single line then it is a 1-D (co-linear) system of forces. If their sense of direction is same then it is co-linear parallel and if their sense of direction is opposite then, it is antiparallel colinear system of forces.

$\vec{P_1}$ $\vec{P_2}$ Parallel

$\vec{P_1}$ $\vec{P_2}$ Anti parallel

4.6.2 2D-concurrent forces

If all the forces lie on a single plane (xy-plane or yz-plane or zx-plane), then it is known as co-planar system of forces. It is known as 2D-concurrent, if all these forces pass through a single point.

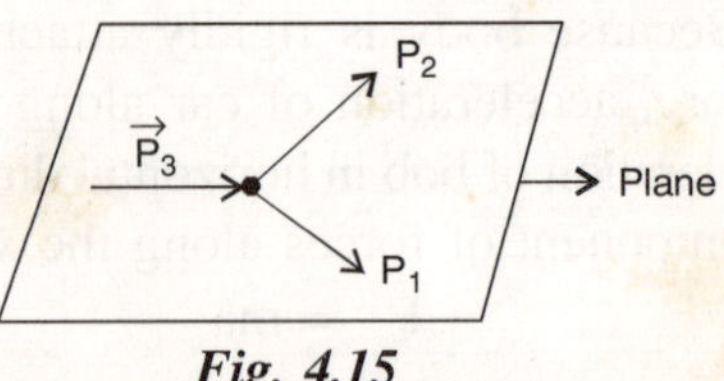

Fig. 4.15

4.6.3 2D-non-concurrent forces

If all the forces lie on a single plane, but their line of actions do not pass through a single point, then this system of forces is known as co-planar non-concurrent.

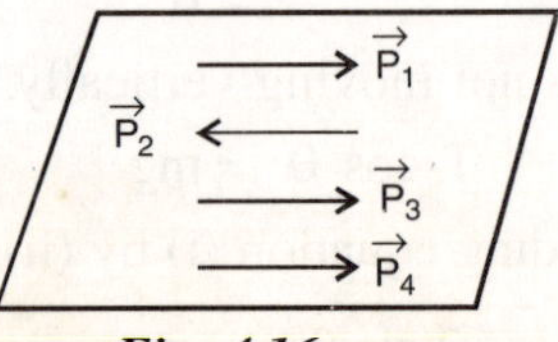

Fig. 4.16

4.6.4 3D-concurrent forces

If forces lie in the space and their line of actions pass through a single point, then it is known as 3D-concurrent system of forces.

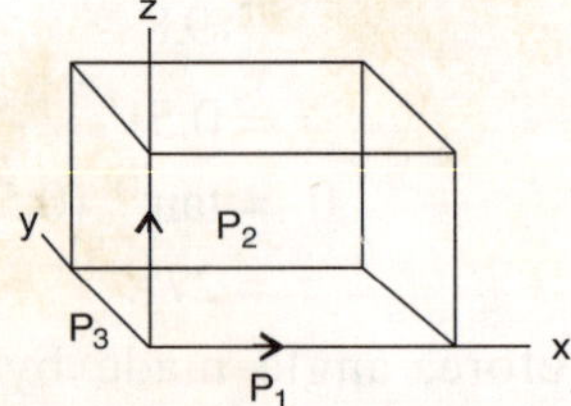

Fig. 4.17

4.6.5 3D-non-concurrent forces

If forces lie in the space but their lines of action do not pass through a single point, then it is known as 3D-non-concurrent system of forces.

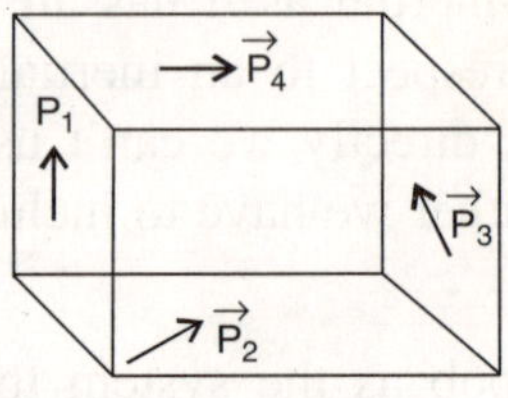

Fig. 4.18

4.7 SYSTEM

For applying Newton's law, first step is to decide the system. The system may be a single particle, a block, a combination of blocks, a string, combination of blocks and strings. The only restriction is, that all parts of the system should have same acceleration. Consider the situation shown in Fig. (4.19).

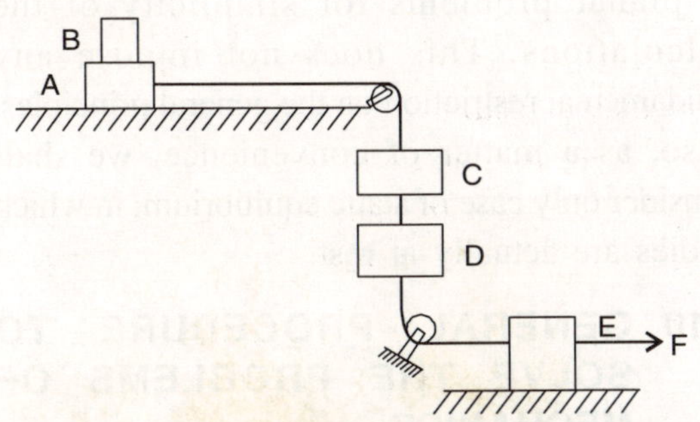

Fig. 4.19

If block B does not slip over A and all parts of the string are tight, then any combination of A, B and E could be taken as a system because they moves equal distances and in the same direction for a given time of interval. Similarly, C and D could be taken as a system but combinations from combination (A, B, E) and combination (C, D) can't be taken as a system, because they moves equal distance but in different directions.

4.8 FREE BODY DIAGRAM

We can make a separate vector diagram of the body alone, showing the frame of reference and all the forces acting on the body. This is said to be free body diagram (FBD) of that body. In free body diagram, optionally a body (particle) could be represented by a point and all the forces which acts on the body, directly or indirectly should pass through this point; this point will represent the mass of that body. The forces may lie along a line or may be distributed over the space (non-planar). Tails of all the forces act on the body and head of the forces away from the body in its specified direction. Note, that at the time when we apply Newton's third law, out of action-reaction pair, only action on the body is considered for the free body diagram, reaction by body can't be considered because if we will consider action-reaction pair on the body for which we are going to draw the free body diagram, then net force on this body is always zero. In this way, the universe will be unaccelerated and this is not possible practically. Let us consider the situation shown in Fig. (4.20) and we have to draw free body diagram of the block.

Considering, the environment surrounding the block(m), there is (1) a man applying a force T through a string (2) earth, which is applying gravitational attraction (3) a reaction by surface and (4) a opposing force by the surface. Drawing these forces on the block we get the free body diagram of the block.

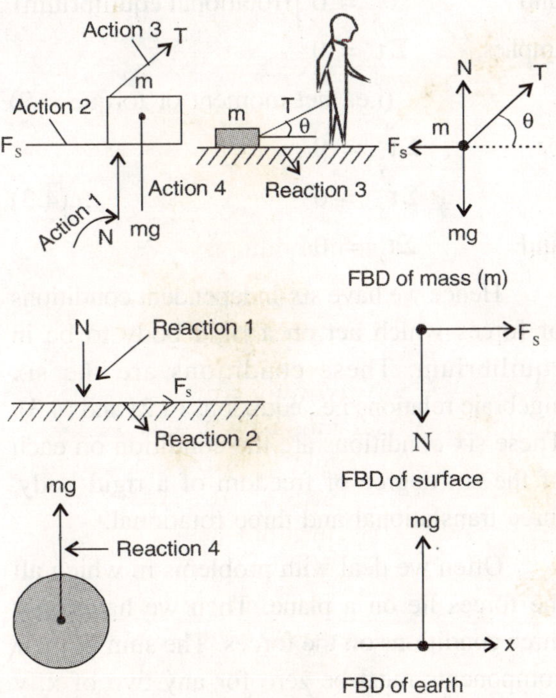

Fig. 4.20

4.9 EQUILIBRIUM OF RIGID BODY

A rigid body is said to in mechanical equilibrium when it is viewed from an inertial reference frame if and only if (i) the linear acceleration ($\vec{a}$) of the body is zero and (ii) its angular acceleration ($\vec{\alpha}$) about any axis fixed in this reference frame is zero. This definition does not require body to be at rest with respect to the observer but only to be unaccelerated. This state of body is said to be non static equilibrium and if body is at rest then it is said to be in static equilibrium.

Now, $\vec{a} = 0$ (translational equilibrium)

implies, $\Sigma \vec{F} = 0$ (i.e., net force = 0)

or $\Sigma F_x = 0$

$$\left.\begin{aligned} \Sigma F_y &= 0 \\ \text{and} \quad \Sigma F_z &= 0 \end{aligned}\right\} \quad \text{....(4.2)}$$

and $\vec{\alpha} = 0$ (rotational equilibrium)

implies, $\Sigma \tau = 0$

(i.e., net moment or torque = 0)

or

$$\begin{aligned} \Sigma \tau_x &= 0 \\ \Sigma \tau_y &= 0 \quad \text{....(4.3)} \\ \text{and} \quad \Sigma \tau_z &= 0 \end{aligned}$$

Hence we have six-independent conditions of forces which act on a rigid body to be in equilibrium. These conditions are the six algebraic relations i.e., equations (4.2) and (4.3). These six conditions are the condition on each of the six degree of freedom of a rigid body, three translational and three rotational.

Often we deal with problems in which all the forces lie on a plane. Then we have only three conditions on the forces. The sum of their components must be zero for any two of x, y and z directions in a plane, because sum along third direction automatically will be zero and sum of their torques about one axis perpendicular to the plane, must be zero (because sum of torques about the axes which lie in the plane of the body automatically will be zero). These corresponds to the three degree of freedom for the equilibrium of rigid body, two of translational and one of rotational.

We shall limit ourselves henceforth mostly to planar problems for simplicity of the calculations. This does not impose any fundamental restriction on the general principles. Also, as a matter of convenience, we shall consider only case of static equilibrium, in which bodies are actually at rest.

4.10 GENERAL PROCEDURE TO SOLVE THE PROBLEMS OF MECHANICS

It will be helpful to write down some procedures for solving problems in classical mechanics. They are: (1) identifying the system (body) to which the problem is referred; (2) after selecting the system (the body), we next turn our attention to the objects in the environment because these objects (planes, strings, springs, cords, the Earth etc.) exert forces on the body; (3) the next step is to select a suitable (inertial or non-inertial) reference frame. We should position the origin and orient the coordinate axes so as to simplify the problem; (4) then make a free body diagram of system chosen and (5) finally apply Newton's second law of motion in form of equation as follows:

$$\left.\begin{aligned} \Sigma F_x &= ma_x \\ \Sigma F_y &= ma_y \\ \Sigma F_z &= ma_z \end{aligned}\right\} \text{for translational motion}$$

and

$$\left.\begin{aligned} \Sigma \tau_x &= I\alpha_x \\ \Sigma \tau_y &= I\alpha_y \\ \Sigma \tau_z &= I\alpha_z \end{aligned}\right\} \text{for rotational motion}$$

WORKED OUT EXAMPLES

1. *A rod of length L is acted upon by two forces F_1 and F_2 applied to its ends and directed oppositely. (Fig. 4-W1). The mass per unit length of the rod decreases linearly from left to right and is equal to half of initial value at the right end. What is the tension in the rod in cross-section at a distance x from left end?*

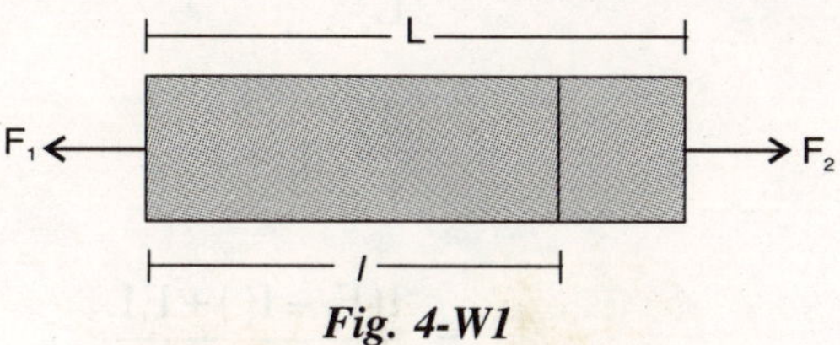

Fig. 4-W1

Solution:

As the mass per unit length λ of the rod decreases linearly, let it be

$$\lambda = \lambda_0 - ax \qquad \text{....(i)}$$

where, λ_0 is the mass per unit length at the left end and a is some constant.

But, according to question at

$$x = L; \lambda = \frac{\lambda_0}{2}$$

$$\therefore \quad \frac{\lambda_0}{2} = \lambda_0 - aL$$

or $$aL = \frac{\lambda_0}{2}$$

or $$a = \frac{\lambda_0}{2L}$$

So, $$\lambda = \lambda_0 - \frac{\lambda_0}{2L}x$$

$$= \lambda_0\left(1 - \frac{x}{2L}\right) \qquad \text{....(ii)}$$

Thus mass of the rod is given by,

$$M = \int_0^L \lambda dx$$

$$= \int_0^L \lambda_0\left(1 - \frac{x}{2L}\right)dx$$

$$= \lambda_0\left[x - \frac{x^2}{4L}\right]_0^L$$

$$= \lambda_0\left[L - \frac{L}{4}\right]$$

$$= \frac{3\lambda_0 L}{4} \qquad \text{....(iii)}$$

Let m be the mass of rod of one of the section of length l then,

$$m = \int_0^{\ell} \lambda\, dx$$

$$= \int_0^l \lambda_0\left(1 - \frac{x}{2L}\right)dx$$

$$= \lambda_0 \int_0^l \left(1 - \frac{x}{2L}\right)dx$$

$$= \lambda_0\left[x - \frac{x^2}{4L}\right]_0^l$$

$$= \lambda_0\left[l - \frac{l}{4}\right]$$

$$= \lambda_0\left[\frac{3l}{4}\right]$$

$$= \frac{3\lambda_0 l}{4} \qquad \text{....(iv)}$$

Note: *In equation (iii), directly we can put 'l' instead of L to get mass of section of length l.*

Therefore, mass of the other section

$$= M - m$$

$$= \left[\frac{3\lambda_0 L}{4}\right] - \frac{3\lambda_0 l}{4}$$

$$= \frac{3\lambda_0 (L - l)}{4} \quad ....(v)$$

To determine the tension at the specified position vertually, divide the rod into two parts as shown below.

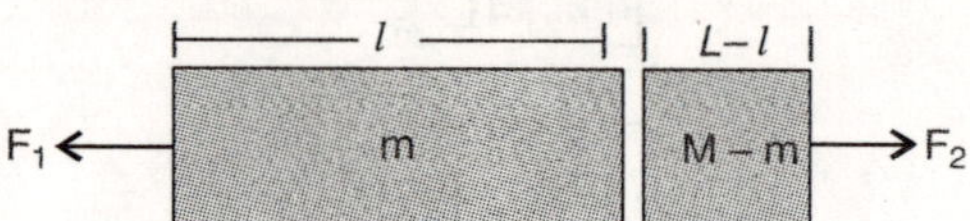

Fig. 4-W2

Let tension at the cross-section be T, then free body diagrams of each part is as shown below.

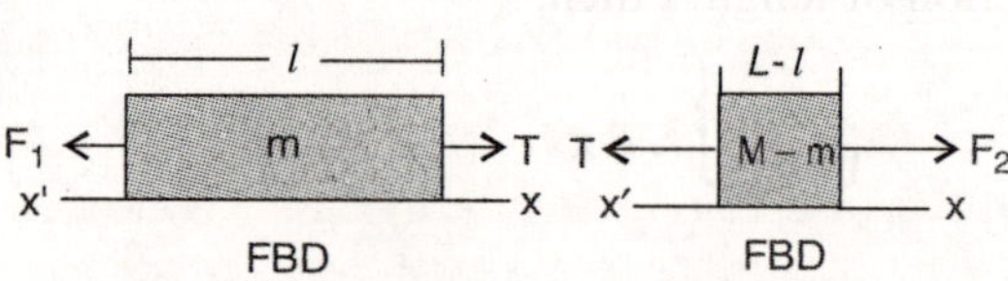

Fig. 4-W3

Let acceleration of rod be a towards right, since acceleration of both the blocks are same as we have divided vertually. So, equations of motion for them are as follows:

$$\sum F_{x,m} = 0$$

or $$T - F_1 = m\,a$$

or $$T - F_1 = \frac{3\lambda_0 l}{4} a \quad ....(vi)$$

and $$\sum F_{x,M-m} = 0$$

or $$F_2 - T = (M - m)\,a$$

or $$F_2 - T = \frac{3\lambda_0 (L - l)}{4} a \quad ....(vii)$$

Adding equations (vi) and (vii), we get

$$F_2 - F_1 = \frac{3\lambda_0 L}{4} a$$

[It is obvious if we suppose entire rod as one system.]

or $$a = \frac{4\Delta F}{3\lambda_0 La}$$

where, $\Delta F = F_2 - F_1$

Putting, a in equation (vi), we have

$$T = \frac{3\lambda_0 l}{4} \times \frac{4\Delta F}{3\lambda_0 L_1} + F_1$$

$$= \frac{l}{L}\,\Delta F + F_1$$

$$= \frac{l}{L}\,\Delta F + F_1$$

$$= \frac{l(F_2 - F_1) + F_1 L}{L}$$

$$= \frac{lF_2 + (L - l)F_1}{L}$$

$$= F_1 \frac{L - l}{L} + F_2 \frac{l}{L}$$

$$\therefore \quad T = F_1 \frac{L - l}{L} + F_2 \frac{l}{L}$$

Note: **1.** *From this result, we see that tensions at various points are functions of L, F_1, F_2 and l. It is independent of mass distribution for a given length of rod and mass.*

2. *The mass per unit length of a ladder increases linearly from the top to the bottom of the ladder and is n times (n > 1) as great at the top. The ladder stands on a horizontal plane and rests against a vertical wall and makes an angle θ with the horizontal, as shown in Fig. (4-W4). The coefficients of friction at top and at bottom are μ_1 and μ_2 respectively. Find the angle θ for which the ladder is just about to slip.*

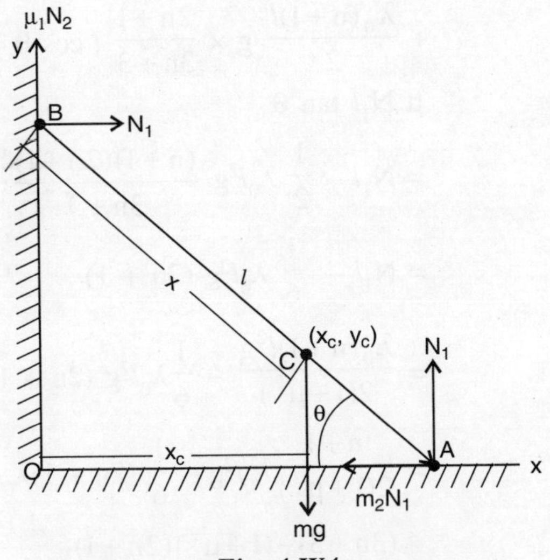

Fig. 4-W4

Solution:

Let the ladder be AB and centre of gravity or centre of mass passes through C, rest on the wall and be in the position such that it is about to slip. Fig. (4-W4) shows the forces acting on rod including the limiting forces of friction ($f_{s, max} = \mu N$) whenever required. Let the mass per unit length of the ladder be λ where,

$$\lambda = \lambda_0 + ax$$

where, λ_0 is the mass per unit length at the upper end and a is some constant.

But, when

$$x = l, \lambda = n\lambda_0$$

Therefore,

$$n\lambda_0 = \lambda_0 + al$$

or $$a = \frac{(n-1)\lambda_0}{l}$$

$$\therefore \lambda = \lambda_0 + \frac{(n-1)\lambda_0}{l}x$$

$$= \lambda_0 \left(1 + \frac{(n-1)}{l}x\right)$$

Mass m of the ladder is given by,

$$m = \int_0^l \lambda dx$$

$$= \int_0^l \lambda_0\left(1+\frac{n-1}{l}x\right)dx$$

$$= \lambda_0\left[x+\frac{n-1}{l}\frac{x^2}{2}\right]_0^l$$

$$= \lambda_0\left[l+\frac{n-1}{l}\cdot\frac{l^2}{2}\right]$$

$$= \lambda_0\left[l+\frac{n-1}{2}l\right]$$

$$= \lambda_0\left[\frac{2l+nl-l}{2}\right]$$

$$= \lambda_0\left[\frac{nl+l}{2}\right]$$

$$= \frac{\lambda_0(n+1)l}{2}$$

$$\therefore m = \frac{\lambda_0(n+1)l}{2}$$

The distance x of the centre of mass of the ladder from the end B is $\dot{x}_c$.

Therefore, $$m\dot{x}_c = \int_0^l \lambda x dx$$

$$= \int_0^l \lambda_0\left(1+\frac{n-1}{l}x\right)x\,dx$$

$$= \lambda_0\int_0^l\left(x+\frac{n-1}{l}x^2\right)dx$$

$$= \lambda_0\left[\frac{x^2}{2}+\frac{n-1}{l}\frac{x^3}{3}\right]_0^l$$

$$= \lambda_0\left[\frac{l^2}{2}+\frac{n-1}{l}\times\frac{l^3}{3}\right]$$

$$= \lambda_0\left[\frac{l^2}{2}+\frac{n-1}{3}l^2\right]$$

$$= \lambda_0\left[\frac{3+2n-2}{6}\right]l^2$$

$$= \lambda_0\left[\frac{2n+1}{6}\right]l^2$$

$$\therefore \quad \dot{x}_c = \frac{\lambda_0(2n+1)l^2}{6m}$$

$$= \frac{\lambda_0(2n+1)l^2}{6} \times \frac{2}{\lambda_0(n+1)l}$$

$$= \frac{2n+1}{3n+3}l$$

$$= \frac{2n+1}{3n+3}l$$

From the translational equilibrium of the ladder $N_2 = \mu_2 N_1$ in horizontal direction.

$$\text{And } N_1 + \mu_1 N_2 = mg$$

$$= \frac{\lambda_0(n+1)l}{2}g;$$

[vertical equilibrium] ....(i)

So, from equation (i)

$$N_2 = \mu_2 N_1$$

$$N_1 + \mu_1\mu_2 N_1 = \frac{\lambda_0(n+1)lg}{2}$$

or

$$N_1 = \frac{\lambda_0(n+1)lg}{2(1+\mu^2)}$$

where

$$\mu = \sqrt{\mu_1\mu_2}$$

and

$$N_2 = \frac{\lambda_0(n+1)lg\mu_2}{2(1+\mu^2)}$$

Since the ladder is in equilibrium, therefore

$$\tau_0 = 0$$

That is net moment about point O should be zero. Note we can assume any point as moment centre which is in the plane of the ladder for convenience assume O.

or $N_2(OB) + N_1(OA) + (mg\, x_c) = 0$

or $N_2\, l \sin\theta - N_1 l \cos\theta + mg\,\dot{x}_c \cos\theta = 0$

or $\mu_2 N_1 l \sin\theta - N_1 l \cos\theta + mg\,\dot{x}_c \cos\theta = 0$

or

$$\mu_2 N_1 l \sin\theta - N_1 l \cos\theta + \frac{\lambda_0(n+1)l}{2} g \times \frac{2n+1}{3n+3} l \cos\theta = 0$$

or

$$\mu_2 N_1 l \tan\theta$$

$$= N_1 l - \frac{1}{2}\lambda_0 l^2 g \frac{(n+1)(2n+1)}{3n+3}$$

$$= N_1 l - \frac{1}{6}\lambda_0 l^2 g\,(2n+1)$$

$$= \frac{\lambda_0(n+1)l^2 g}{2(1+\mu^2)} - \frac{1}{6}\lambda_0 l^2 g\,(2n+1)$$

$$= \frac{n+1}{2(1+\mu^2)}\lambda_0 l^2 g - \frac{1}{6}(2n+1)\lambda_0 l^2 g$$

$$= \frac{(3n+3)-(1+\mu^2)(2n+1)}{6(1+\mu^2)}\lambda_0 l^2 g$$

$$= \frac{3n+3-2n-2n\mu^2-1-\mu^2}{6(1+\mu^2)}\lambda_0 l^2 g$$

$$= \frac{(n+2)-(2n+1)\mu^2}{6(1+\mu^2)}\lambda_0 l^2 g$$

$$= \frac{(n+2)-(2n+1)\mu^2}{6(1+\mu^2)}\lambda_0 l^2 g \times \frac{1}{\mu_2 l} \times \frac{2(1+\mu^2)}{\lambda_0(n+1)lg}$$

or,

$$\tan\theta = \frac{(n+2)-(2n+1)\mu^2}{3.\mu_2(n+1)}$$

$$\therefore \quad \theta = \tan^{-1}\frac{\overline{n+2}-\overline{2n+1}\mu^2}{3\mu_2\,\overline{n+1}}$$

3. *At the moment t = 0, the force F = a × t is applied to a small block of mass m resting on a smooth horizontal plane (a is a constant). The fixed direction of the force forms an angle α with horizontal (Fig. 4-W5). Find,*

(a) the velocity of the block at the moment of its breaking off the plane

(b) the distance traversed by the block up to this moment.

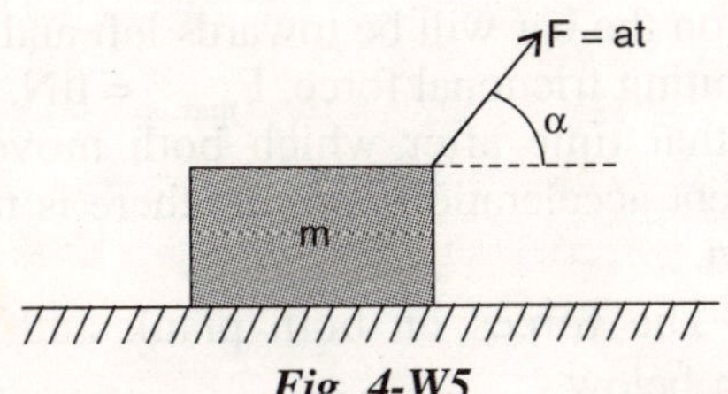

Fig. 4-W5

Solution:

Let a_h and a_n be accelerations of block in horizontal and normal directions respectively.

Forces on block are:

1. Applied force, F.
2. Weight of block, mg, downward.
3. Normal reaction by plane.

Free body diagram of block is shown in Fig. (4-W6).

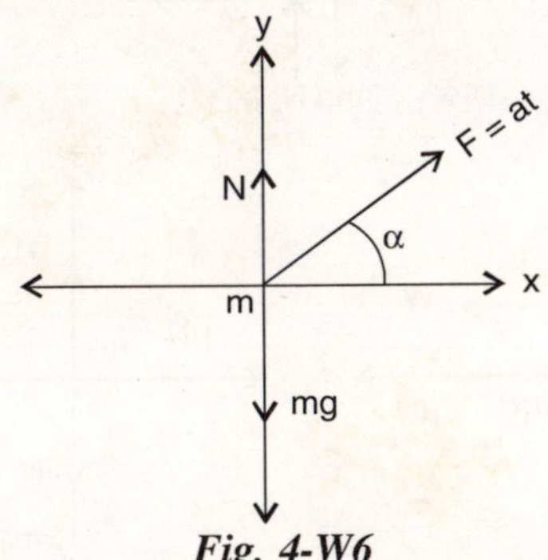

Fig. 4-W6

Now translational equation of motion for block is,

$$\Sigma F_x = ma_x$$

or $$\Sigma F_h = ma_h$$

or $$F \cos \alpha = ma_h$$

or $$at \cos \alpha = m\frac{dv_h}{dt} \qquad \left[\because a = \frac{dv}{dt}\right]$$

or $$\frac{m}{a\cos\alpha} dv_h = t\,dt$$

or $$KV_h = \frac{t^2}{2};$$

where, $$K = \frac{m}{a\cos\alpha}$$

or $$V_h = \frac{t^2}{2K}$$

and since there is no motion along normal direction.

Therefore $$\Sigma F_y = 0$$

or $$N - mg + F \sin \alpha = 0$$

or $$N - mg + at \sin \alpha = 0$$

at the breaking off the plane, there will be no contact with plane. In this case normal reaction (N) should be zero.

$$\therefore \quad t_b = \frac{mg}{a\sin\alpha}$$

(a) $$V_h = V_b$$

$$= \text{velocity at this moment}$$

$$= \frac{t^2{}_b}{2K} = \frac{m^2g^2}{a^2 \sin^2 \alpha \times 2K}$$

$$= \frac{m^2g^2}{a^2 \sin^2 \alpha} \times \frac{1}{2\times\dfrac{m}{a\cos\alpha}}$$

$$= \frac{mg^2 \cos\alpha}{2a\sin^2 \alpha}$$

(b) $$\because \quad V_h = \frac{ds}{dt} = \frac{t^2}{2K}$$

or, $$S = \frac{t^3}{6K} + C$$

$$\therefore \quad S_b = \frac{t_b^3}{6K}$$

$$= \frac{m^3g^3}{a^3 \sin^3 \alpha} \times \frac{1}{6} \times \frac{a\cos\alpha}{m}$$

$$= \frac{m^2g^3 \cos\alpha}{6a^2 \sin^3 \alpha}.$$

4. *A plank of mass m_1 with a bar of mass m_2 placed on it lies on a smooth horizontal plane. A horizontal force changing with time t as F = a × t (a is constant) is applied to bar. Find how the acceleration of the plank w_1*

and of the bar w_2 depends on t, if the coefficient of friction between the plank and the bar is equal to μ. Draw the approximate plots of these dependences.

Solution:

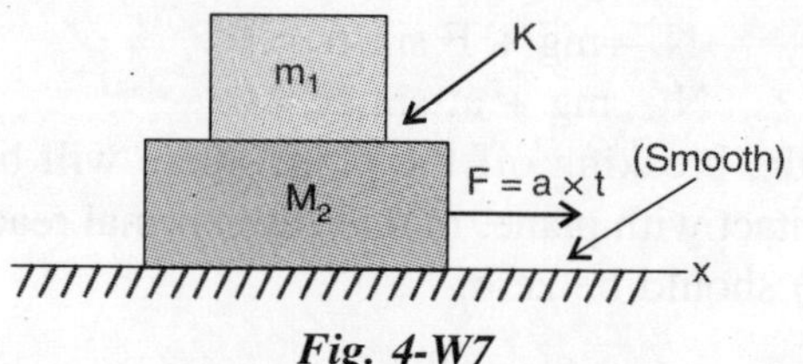

Fig. 4-W7

Let us suppose that up to a certain time t_0 both planks and the bar move together with same acceleration, $w_1 = w_2$. In this situation, we can assume both plank and bar as a system and F = at, is the only external force acts on the system.

Therefore, applying Newton's second law on the system, we get

$$\vec{F} = M_{system}\ \vec{a}_{system}$$

or $$F_x = M_{system}\ a_{system,\ x}$$

or $$at_0 = (m_1 + m_2)w$$

or $$at_0 = (m_1 + m_2)w$$

or $$w = \frac{at_0}{m + m_2}$$

$$\therefore \quad w = w_1 = w_2 = \left(\frac{a}{m_1 + m_2}\right)t_0 = kt_0$$

Therefore, up to $t = t_0$ both plank and bar move together with acceleration kt_0 where $k = \dfrac{a}{m_1 + m_2}$, is a constant. Now for finding the time t_0 consider one out of the plank and the bar as a single system. Let the bars be separate systems. Now since on plank (m_1) the force is only frictional force along right because plank is moving along right. Therefore, frictional force on the bar will be towards left and it will be limiting frictional force, $F_{max,\ s} = \mu N$. Since t_0 is that time after which both move with different accelerations, so that there is relative motion.

The forces on both plank and bar as shown below.

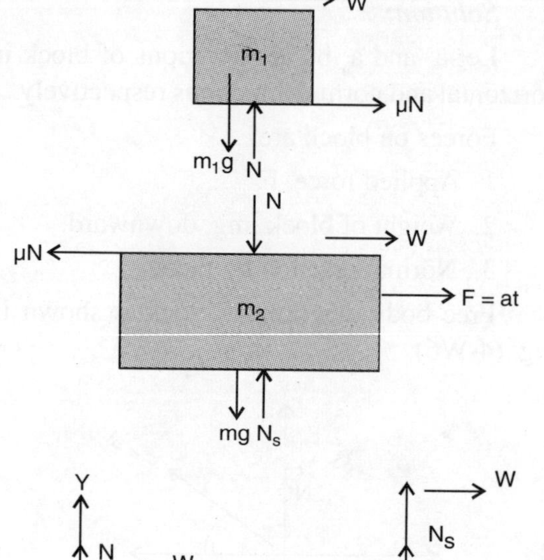

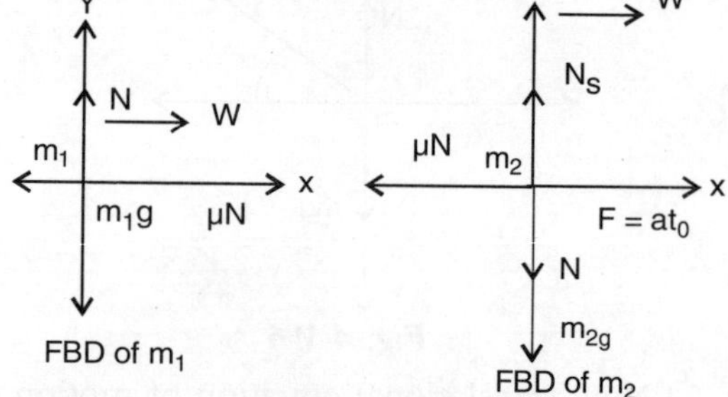

Fig. 4-W8

Equation of motion for plank gives

$$\Sigma F_{Plank,\ y} = 0;$$

because there is no motion along y-axis

or $$N - m_1 g = 0$$

or $$N = m_1 g$$

and $$\mu N = m_1 w$$

or $$\mu m_1 g = m_1 w$$

or $$w = \mu g$$

Equation of motion for bar gives

$\Sigma F_{bar,\ y} = 0$; because there is no motion along y-axis.

or $N_S - N - m_2g = 0$

or $N_s = N + m_2g$

$= m_1g + m_2g$

$\therefore$ $N_s = (m_1 + m_2)\, g$

and $\Sigma F_{bar,\, x} = wm_2$

or $at_0 - \mu m_1 g = m_2\, Kg$

or $at_0 = \mu m_2 g + \mu m_1 g$

or $t_0 = \dfrac{\mu g\,(m_1 + m_2)}{a}$

$\therefore$ $t_0 = \dfrac{\mu(m_1 + m_2)g}{a}$

$= \dfrac{\mu M_{system}\, g}{a}$

Now for $t \geq t_0$ plank moves with acceleration w_1 and bar with w_2, with respect to inertial frame of reference.

Therefore, $\mu N = m_1 w_1$

or $w_1 = \dfrac{\mu N}{m_1}$

$= \dfrac{\mu m_1 g}{m_1}$

$= \mu g$

and $at - \mu N = m_2 w_2$

or $at - m_1 gK = m_2 w_2$

or $w_2 = \dfrac{(at - m_1 gK)}{m_2}$

$\therefore$ $w_2 = \dfrac{at - m_1 gK}{m_2}$

$$\therefore a(t) = \begin{cases} w_1 = w_2 \\ = \dfrac{at}{m_1 + m_2}; & t \leq t_0 \\ = W_1 \\ = \mu g;\ t \geq t_0 & \text{....(i)} \\ W_2 = \dfrac{at - m_1 g\mu}{m_2}; & t \geq t_0 \end{cases}$$

Drawing function(i) on xy-plane, we get Fig. (4-W9).

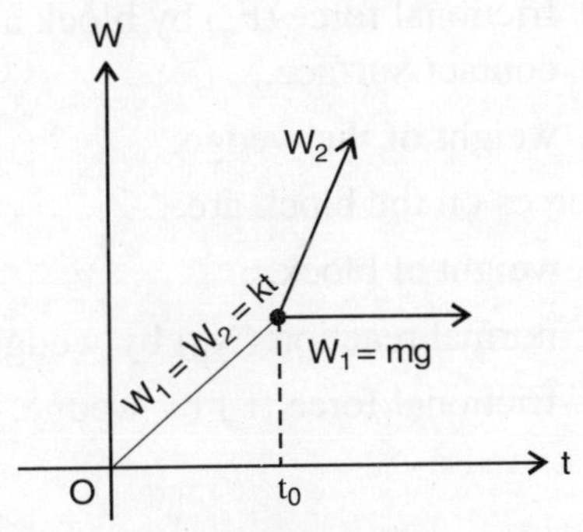

Fig. 4-W9

Note :

1. *In this problem we consider* $\mu_s = \mu_k = \mu$
2. *With acceleration* $w_2 - w_1$, *plank moves back with respect to bar frame.*

5. *Block-2 of mass m is placed on the inclined surface of wedge 1. The block gets a horizontal acceleration (a) directed to the right as shown in Fig. (4-W10). At what inclination will the block start its motion? Find the acceleration of block for inclination for which* $\cot\theta < \dfrac{g - a\mu}{a + g\mu}$, *Mass of wedge is M and wedge is on horizontal smooth surface and coefficient of friction between them is m.*

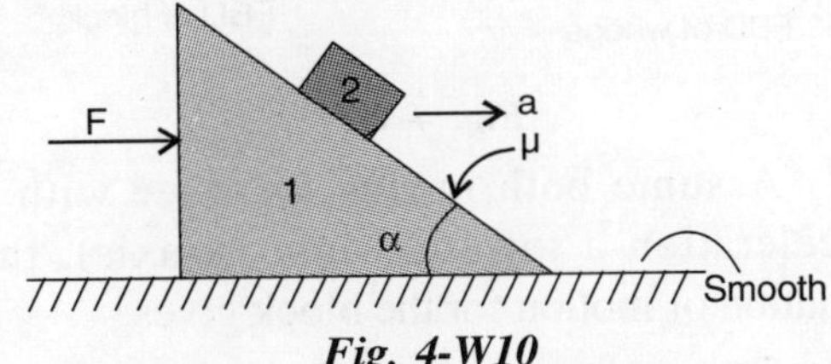

Fig. 4-W10

Solution: Forces and free body diagrams of wedge and block are shown below.

Forces on wedge are:

1. applied force F,
2. normal reaction (N_s) by horizontal smooth surface,
3. normal reaction (N_B) by block,

4. frictional force (F_S) by block along the contact surface,
5. weight of the wedge.

Forces on the block are:

1. weight of block,
2. normal reaction (N_B) by wedge,
3. frictional force (f_s) by wedge.

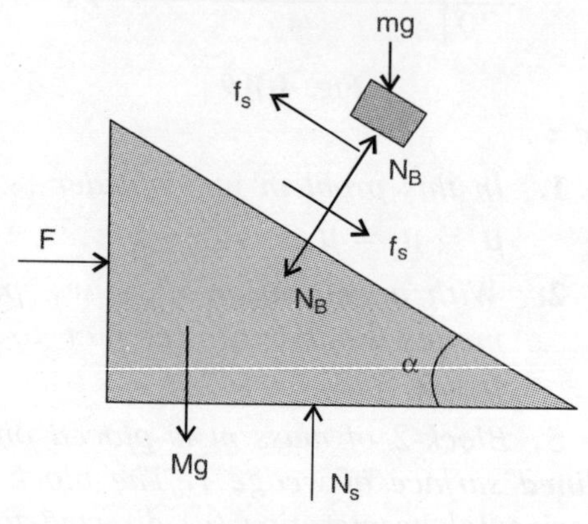

Fig. 4-W11

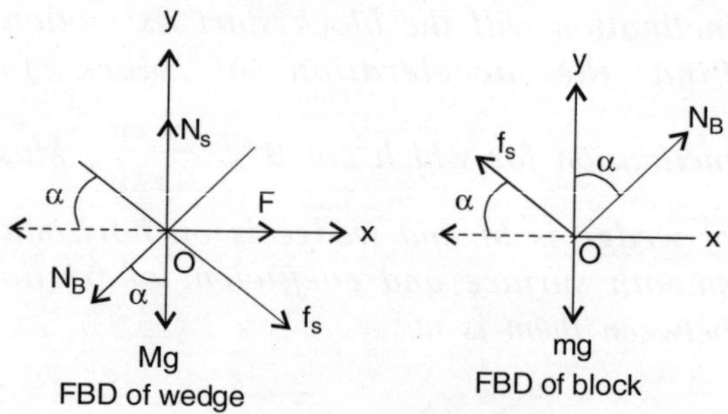

Fig. 4-W12

Assume both moves together with an acceleration a towards right (x-axis), then equation of motion for the block gives.

$\Sigma F_{m,y} = 0$ (As there is no motion along vertical direction.)

or $f_s \sin\alpha + N_B \cos\alpha - mg = 0$

or $f_s \sin\alpha + N_B \cos\alpha - mg = 0$

or $f_s \sin\alpha + N_B \cos\alpha = mg$ .... (i)

and $\Sigma F_{m,x} = ma$

or $N_B \sin\alpha - f_s \cos\alpha = ma$ ....(ii)

Multiplying (i) by cos α and (ii) by sin α and after adding (i) and (ii), we get

$$f_s \sin\alpha \cos\alpha + N_B \cos^2\alpha = mg \cos\alpha$$

$$N_B \sin^2\alpha - f_s \sin\alpha \cos\alpha = ma \sin\alpha$$

$$\therefore \quad N_B = m(a \sin\alpha + g\cos\alpha) \quad ....(iii)$$

Similarly, we can get

$$f_s = m(-a\cos\alpha + g\sin\alpha) \quad ....(iv)$$

Now time when block will start motion, let corresponding inclination be α_m.

$$f_s = f_{limiting}$$
$$= N_B\mu$$

Therefore, dividing equation (iv) by (iii), we have

$$\mu = \frac{-a\cos\alpha_m + g\sin\alpha_m}{a\sin\alpha_m + g\cos\alpha_m}$$

$$= \frac{-a + g\tan\alpha_m}{a\tan\alpha_m + g}$$

or $\mu a \tan\alpha_m + \mu g = -a + g\tan\alpha_m$

or $\tan\alpha_m(\mu a - g) = -a - \mu g$

or $\tan\alpha_m = \dfrac{a+\mu g}{g - \mu a}$;

where, $a = \dfrac{F}{m+M}$

$\therefore \alpha_m = \tan^{-1}\dfrac{a+\mu g}{g-\mu a}$ is the critical angle beyond that block will start moving.

Now let a_{rel} be the acceleration of block relative to wedge for inclination beyond α_m i.e., θ. Consider frame of reference attached to wedge therefore acceleration of block in this frame is a_{rel} along inclined plane. Since this frame is non-inertial frame, so we need to consider pseudo force.

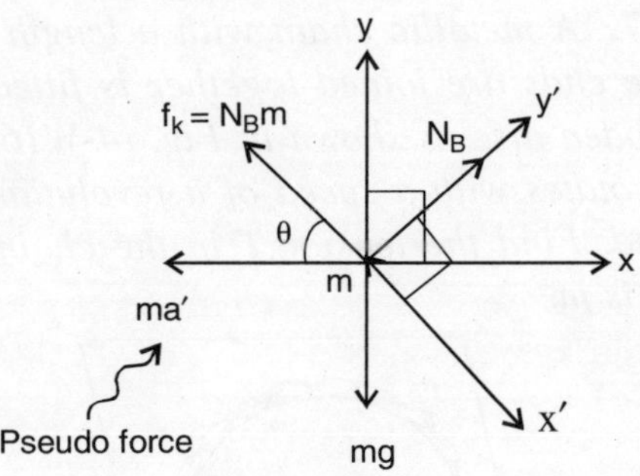

Fig. 4-W13

Since acceleration along y′ (⊥ to plane) is zero,

$$\therefore \quad \Sigma F_{m, y'} = 0$$

or $$N_B - mg\cos\theta - ma'\sin\theta = 0$$

$$N_B = mg\cos\theta + ma'\sin\theta$$

where, a′ is acceleration of wedge in inertial frame.

Now, $\Sigma F_{m, x'} = a_{rel}\, m$

or $mg\sin\theta - N_B\mu - ma'\cos\theta = ma_{rel}$

or $mg\sin\theta - \mu(mg\cos\theta + ma'\sin\theta) - ma'\cos\theta = ma_{rel}$

or $a_{rel} = g\sin\theta - \mu g\cos\theta - \mu a'\sin\theta - a'\cos\theta$

$\therefore \quad a_{rel} = (\sin\theta - \mu\cos\theta)\, g - (\mu\sin\theta + \cos\theta)a'$ ....(v)

Now consider the equation of motion for wedge.

Then, $\Sigma F_{M, x'} = Ma'$

$F + f_k\cos\theta - N_B\sin\theta = Ma'$

or $F + \mu N_B\cos\theta - N_B\sin\theta = Ma'$

or $F + \mu(mg\cos\theta + ma'\sin\theta) - (mg\cos\theta + ma'\sin\theta)\sin\theta = Ma'$

or $F + \mu mg\cos\theta - mg\cos\theta\sin\theta - a' \times (M + m\sin^2\theta - \mu m\sin\theta)$

or $$a' = \frac{F + (\mu - \sin\theta)mg\cos\theta}{M + m\sin\theta(-\mu + \sin\theta)}$$

$$\therefore \quad a' = \frac{F + (\mu - \sin\theta)mg\cos\theta}{M - (\mu - \sin\theta)m\sin\theta}$$

Therefore from equation (v)

$$a_{rel} = (\sin\theta - \mu\cos\theta)\, g - (\mu\sin\theta + \cos\theta) \times \left[\frac{F + (\mu - \sin\theta)mg\cos\theta}{M - (\mu - \sin\theta)m\sin\theta}\right]$$

Do it by yourself:

1. *Find a_{rel} for $\theta = 43°$, $\mu = 0.2$, $F = 30$ N, $M = 4$ kg, $m = 1$ kg, take $g = 10\ ms^{-2}$.*
2. *Using these datas find acceleration of block in inertial frame attached to plane surface.*

6. *Two particles each of mass m is connected by a light string of length 2l as shown in Fig. (4-W14). A constant force F is being applied at the mid-point of the string (x = 0) at right angle to the initial position of the string. Show that acceleration of m in the direction at right angle to F is given by $a_x = \frac{F}{2m}\frac{x}{\sqrt{l^2 - x^2}}$. Here x is the perpendicular distance of one of the particle from the line of action of F. Discuss the situation when x = l.*

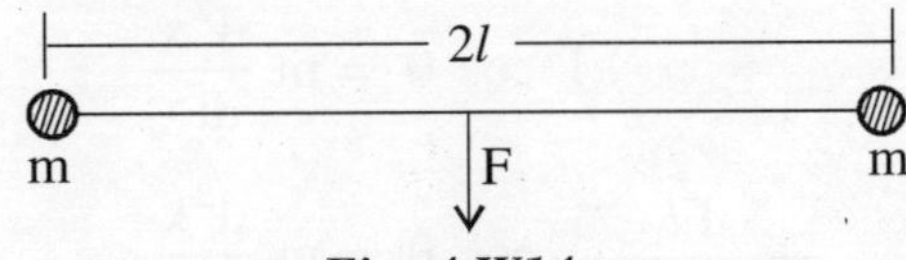

Fig. 4-W14

Solution:

Let us consider at some instant of time situation would be as shown below.

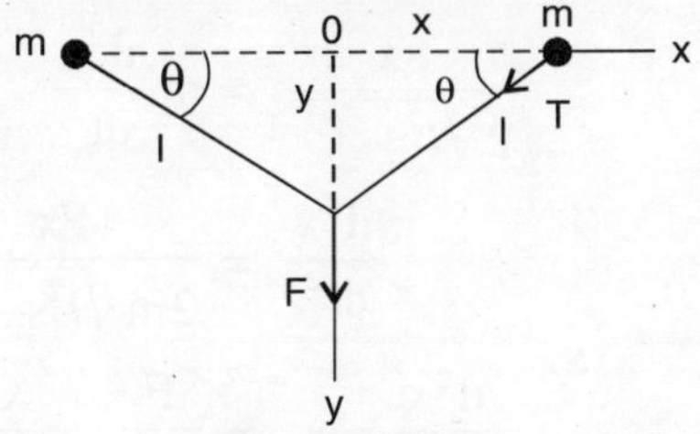

Fig. 4-W15

Suppose that T is the tension in the string.

Therefore,

$$T \cos\theta = m\frac{d^2x}{dt^2} \quad \text{....(i)}$$

and $$T \sin\theta = m\frac{d^2y}{dt^2} \quad \text{....(ii)}$$

or $$2T \sin\theta = 2m\frac{d^2y}{dt^2}$$

But $F = 2T \sin\theta$

$\therefore$ $F = 2m\, a_y$

or $$a_y = \frac{F}{2m} \quad \text{....(iii)}$$

or $$T = \frac{F}{2\sin\theta}$$

$$= \frac{F}{2\frac{y}{l}}$$

$$= \frac{Fl}{2y}$$

$$= \frac{Fl}{2\sqrt{l^2 - x^2}}$$

$\because$ $y^2 = l^2 - x^2$

Therefore from equation (i),

$$T \cos\theta = m\frac{d^2x}{dt^2}$$

or $$\frac{Fl}{2\sqrt{l^2 - x^2}} \cos\theta = m\frac{d^2x}{dt^2}$$

or $$\frac{Fl}{2\sqrt{l^2 - x^2}} \times \frac{x}{l} = m\frac{d^2x}{dt^2}$$

or $$\frac{Fx}{2\sqrt{l^2 - x^2}} = m\frac{d^2x}{dt^2}$$

or $$\frac{d^2x}{dt^2} = \frac{Fx}{2m\sqrt{l^2 - x^2}}$$

$\therefore$ $$\frac{d^2x}{dt^2} = a_x = \frac{F}{2m}\frac{x}{\sqrt{l^2 - x^2}}$$

When $x = l$ then there is no force along x-axis and so $a_x = 0$.

7. *A metallic chain with a length l and whose ends are joined together is fitted onto a wooden disc as shown in Fig. (4-W16). The disc rotates with a speed of n revolutions per second. Find the tension, T in the chain if its mass is m.*

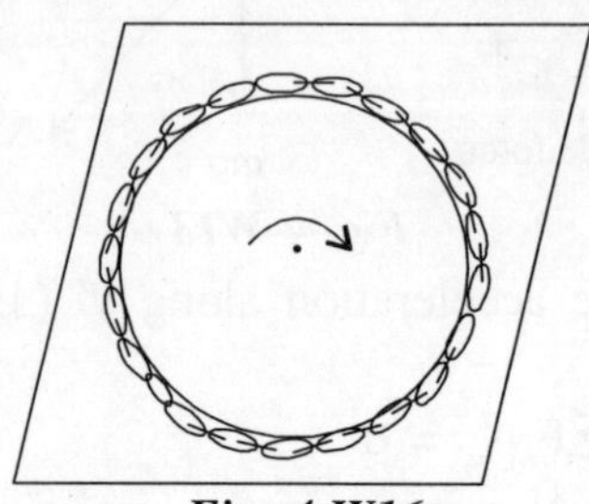

Fig. 4-W16

Solution:

If R be the radius of the chain then,

$$2\pi R = l$$

$\therefore$ $$R = \frac{l}{2\pi}$$

Let λ is the mass per unit of the chain.

Then, $$\lambda = \frac{m}{l}$$

Consider an elementary part of the chain which makes an angle $\Delta\theta$ at the centre, then mass of this elementary part is given by,

$$\Delta m = \lambda \Delta l$$

$$= \lambda \frac{l}{2\pi} \Delta\theta$$

$\therefore$ $$\Delta m = \frac{m}{2\pi}\Delta\theta$$

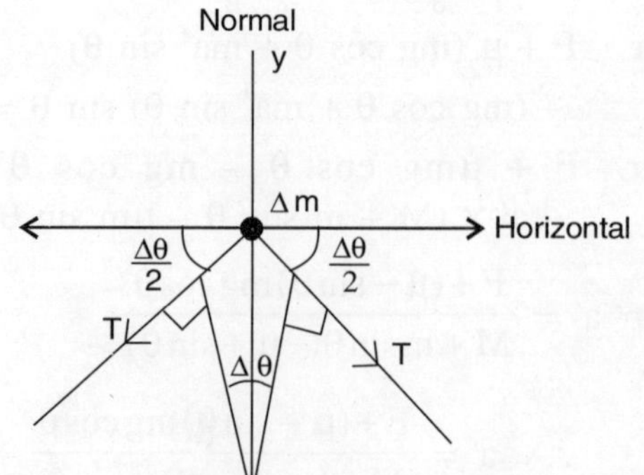

Fig. 4-W17

Let T be the tension in the chain then net force perpendicular to normal is,

$$\Sigma F\perp = T\cos\frac{\Delta\theta}{2} - T\cos\frac{\Delta T}{2}$$

$$= 0$$

and net force along perpendicular is,

$$2T\sin\frac{\Delta\theta}{2} = \Delta m w^2 R$$

or $$2T\sin\frac{\Delta\theta}{2} = \Delta m w^2 \times \frac{l}{2\pi}$$

or $$2T\frac{\Delta\theta}{2} = \frac{\Delta m w^2 l}{2\pi}$$

[∵ $\sin\alpha \cong \alpha$ for small angle α]

or $$T\,\Delta\theta = \frac{w^2 l}{2\pi} \times \frac{m\Delta\theta}{2\pi}$$

or $$T = \frac{w^2 l m}{4\pi^2}$$

Since, $$w = \frac{\theta}{t}$$

$$= 2\pi n$$

where n = rev/sec.

∴ $$T = \frac{m}{n^2}.$$

8. *A massless string thrown over a stationary pulley is passed through a slit. As the string moves, it is acted upon by a constant friction force F on the side of the slit. The ends of the string carry the masses m_1 and m_2 ($m_1 > m_2$). Find the acceleration of each block.*

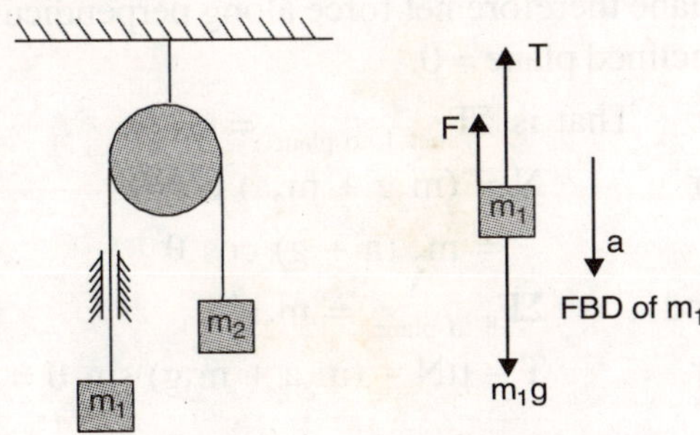

Fig. 4-W18 *Fig. 4-W19*

Solution:

Let m_1 (since $m_1 > m_2$) go down with acceleration a. Therefore, as string is unstretchable, m_2 will go up with the same acceleration. Let T be the tension in the string. The free body diagrams of the block shown below, forces on block m_1 are,

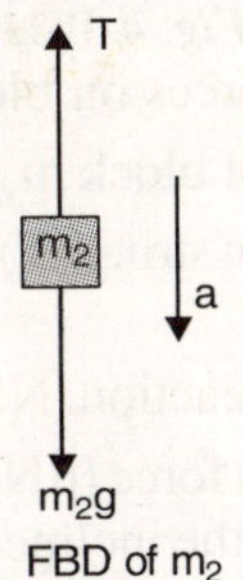

Fig. 4-W20

1. weight of block, $m_1 g$ downward,
2. tension T in the string,
3. friction F by the string, and forces on block m_2 are:

(a) weight of block m_2 i.e., $m_2 g$ downward,

(b) tension T by the string.

Equations of motion for both block gives,

$$m_1 g - T - F = m_1 a \quad \text{....(i)}$$

$$T - m_2 g = m_2 a \quad \text{.....(ii)}$$

and after adding equations (i) and (ii), we get

$$m_1 g - F - m_2 g = (m_1 + m_2)a$$

$$\therefore\ a = \frac{(m_1 - m_2)g - F}{m_1 + m_2}.$$

9. *Two blocks on a rough incline are connected by a light string that passes over a frictionless pulley as shown in the Fig. (4-W21). Assuming $m_1 > m_2$ and taking the coefficient of kinetic friction for each block is μ, find the acceleration of block m_1 with respect to incline, if it (incline) ascends with an acceleration a.*

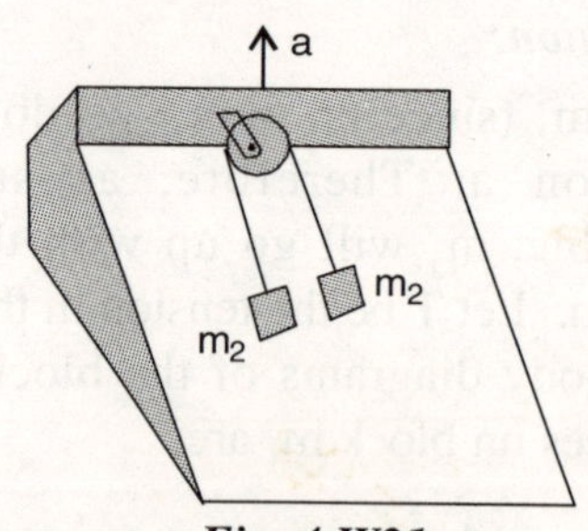

Fig. 4-W21

Solution: Forces on block, m_1 are:

1. weight of block m_1 along downward,
2. tension in string (T) along the inclined plane,
3. normal reaction, N by inclined plane,
4. frictional force (μN) by inclined plane along to the inclined plane,
5. pseudo force m_1a downward.

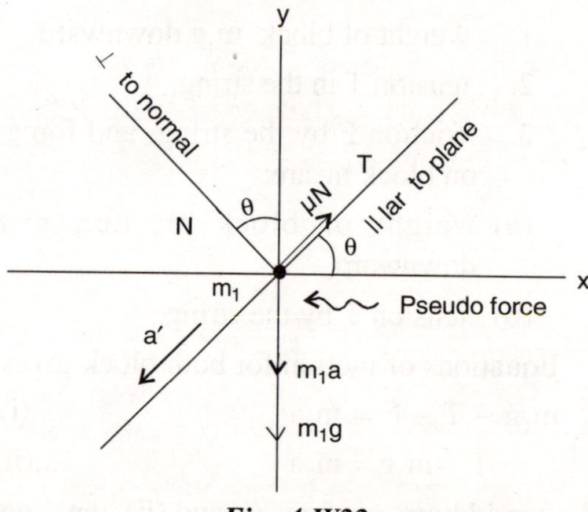

Fig. 4-W22

Let m_1 go down the inclined plane with an acceleration a′ with respect to inclined plane. Now fixe a frame of reference on the inclined plane and write down equation of motion for m_1 along the inclined plane and perpendicular to inclined plane respectively.

Since there is no motion perpendicular to plane.

Therefore, $\Sigma F_{\perp,\text{ inclined plane}} = 0$

or $m_1a \cos\theta + m_1g \cos\theta - N = 0$

∴ $N = m_1 (a + g) \cos\theta$ ....(i)

and $\Sigma F||_{,\text{ inclined plane}} = m_1a'$

or $m_1g \sin\theta + m_1a \sin\theta - T - N\mu = m_1a'$

or $m_1g \sin\theta + m_1a \sin\theta - T - \mu m_1 (a + g) \cos\theta = m_1a'$ [From (i)]

or $m_1 (a + g) \sin\theta - T - \mu m_1 (a + g) \cos\theta = m_1a'$

or $m_1 (\sin\theta - \mu \cos\theta)(a + g) - T = m_1a'$ ....(ii)

Forces on block m_2 are:

1. weight of block, m_2g, downward,
2. tension in string, T along the inclined plane,
3. normal reaction (N) perpendicular to the inclined plane,
4. frictional force (μN) by inclined plane along to inclined plane,
5. pseudo force m_2a downward.

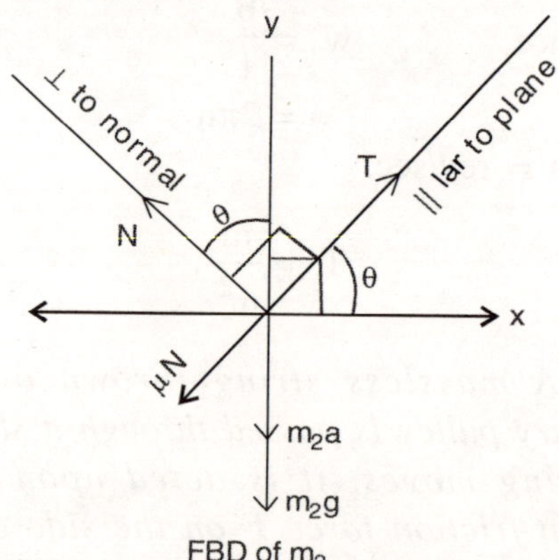

Fig. 4-W23

Now there is no motion perpendicular to plane therefore net force along perpendicular to inclined plane = 0.

That is $\Sigma F_{\text{net} \perp \text{ to plane}} = 0$

or $N = (m_2g + m_2a) \cos\theta$

$= m_2 (a + g) \cos\theta$ ....(iii)

and $\Sigma F_{\parallel \text{ to plane}} = m_2a'$

or $T - \mu N - (m_2a + m_2g) \sin\theta = m_2a'$

or $\quad T - m_2\mu (a + g) \cos\theta - m_2 (a + g) \sin\theta = m_2 a'$

or $\quad T - m_2 (\sin\theta + \mu \cos\theta) (a + g) = m_2 a' \quad$ ....(iv)

Adding equations (ii) and (iv), we get

$m_1 (\sin\theta - \mu \cos\theta) (a + g) - m_2 (\sin\theta + \mu \cos\theta) (a + g) = (m_2 + m_1) a'$

or $\quad [m_1(\sin\theta - \mu \cos\theta) - m_2 (\sin\theta + \mu \cos\theta)] (a + g) = (m_1 + m_2) a'$

or $\quad [(m_1 - m_2) \sin\theta - (m_1 + m_2) \mu \cos\theta] (a + g) = (m_1 + m_2) a'$

$$\therefore a' = \frac{[(m_1 - m_2)\sin\theta - (m_1 + m_2)\mu\cos\theta](a+g)}{m_1 + m_2}.$$

10. *In the arrangement shown in Fig.(4-W24) the bodies have masses m_0, m_1 and m_2. The friction is absent, the masses of the pulleys and the threads are negligible. Find the acceleration of the body m_1. Look into possible all possible cases.*

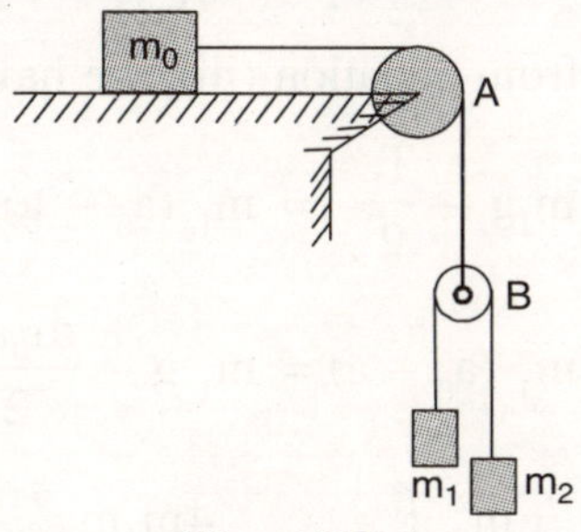

Fig. 4-W24

Solution:

Suppose acceleration of m_0 is a_0 towards right. This will also be the downward acceleration of the pulley B because the string connecting m_0 and pulley B is constant in length. Also the string connecting m_1 and m_2 has a constant length. This implies that the decrease in separation between m_1 and pulley B equals the increase in the separation between m_2 and pulley B. So, the upward acceleration of m_1 with respect to pulley B equals the downward acceleration of m_2 with respect to same. Let this acceleration be a.

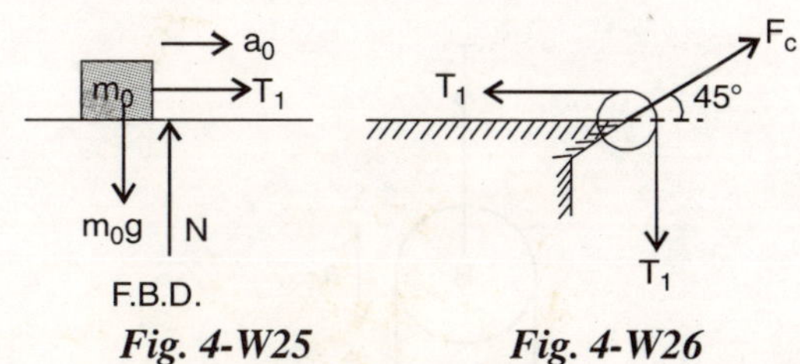

Fig. 4-W25 *Fig. 4-W26*

The acceleration of m_1 with respect to the ground (inrtial frame) = $a - a_1$, upward and the acceleration of m_2 with respect to ground = $a + a_0$, downward. Since these acceleration a_0, $a - a_0$, $a + a_0$ are measured from ground therefore directly we can apply Newton's law.

Let the tension be T_1 in the upper string and T_2 in the lower string. Forces on all parts and FBDs are shown below.

Forces on m_0 are:

(a) T_1 by the string, horizontally,

(b) $m_0 g$ by the earth, downward and,

(c) N by the table, upward.

In horizontal direction, the equation of motion for m_0 is,

$$T_1 = m_0 a_0 \qquad .....(i)$$

Forces on pulley A are:

(a) T_1 by the string horizontally away from it,

(b) T_1 by the string downward,

(c) contact reaction by surface, F_c.

Since pulley A is light and is fixed, then $F_c = \sqrt{2}\ T_1$.

The forces on pulley B are:

(a) T_1 by upper string, upward,

(b) $2T_2$ by lower string, downward.

As the mass of the pulley B is negligible.

$\therefore \quad -T_1 + 2T_2 = m_B a_0$

or $\quad 2T_2 - T_1 = 0 \times a_0 = 0$

or $2T_2 = T_1$

or $T_2 = \frac{T_1}{2}$ ....(ii)

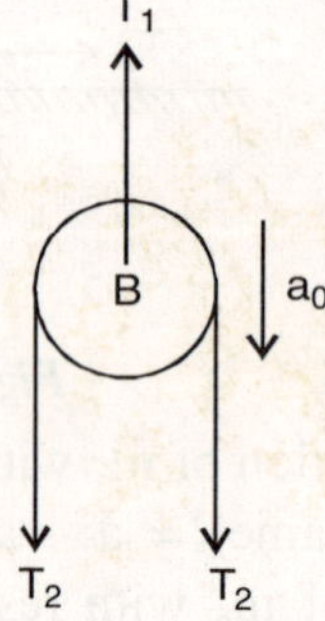

Fig. 4-W27

The forces on m_1 are:

(a) T_2 by the string, upward,

(b) $m_1 g$ downward, by Earth.

So, by equation of motion for m_1 in vertical direction gives,

$$T_2 - m_1 g = m_1 (a - a_0) \quad ....(iii)$$

or $m_1 g - T_2 = m_1 (a_0 - a)$

or $m_1 g - \frac{T_1}{2} = m_1 (a_0 - a)$

$\because \quad 2T_2 = T_1$

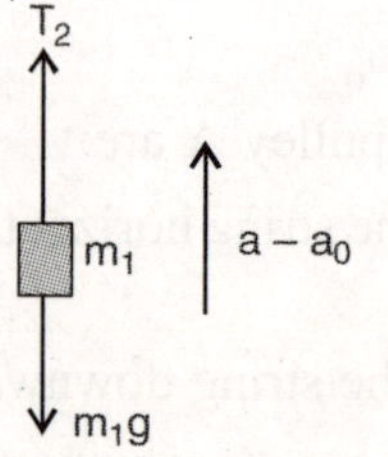

Fig. 4-W28

or $m_1 g - \frac{m_0 a_0}{2} = m_1 (a_0 - a)$ [From (i)]

or $a = \frac{(m_0 + 2m_1)a_0 - 2m_1 g}{2}$ ....(iv)

Forces on m_2 are:

(a) $m_2 g$ downward by Earth,

(b) T_2 upward by the string.

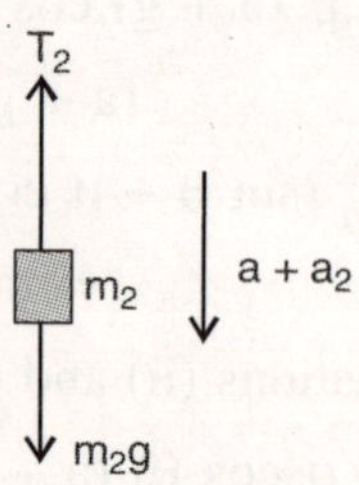

Fig. 4-W29

$\therefore$ Equation of motion along vertical direction gives,

$$m_2 g - T_2 = m_2 (a + a_0) \quad ....(v)$$

or $m_2 g - \frac{T_1}{2} = m_2 (a + a_0)$

or $m_2 g - \frac{m_0 a_0}{2} = m_2 (a + a_0)$

or $a = \frac{2m_2 g - (m_0 + 2m_2)a_0}{2m_2}$ ....(vi)

Equating equations (iv) and (vi), we get

$$a_0 = \frac{4m_1 m_2}{4m_1 m_2 + m_0(m_1 + m_2)} g$$

Now, from equation (iii), we have

$$m_1 g - \frac{T_1}{2} = m_1 (a_0 - a)$$

or $m_1 (a_0 - a) = m_1 g - \frac{m_0 a_0}{2}$

$$= m_1 g - \frac{m_0}{2} \left(\frac{4m_1 m_2}{4m_1 m_2 + m_0(m_1 + m_2)} g \right)$$

or $a_0 - a = g - \frac{2m_0 m_2 g}{4m_1 m_2 + m_0(m_1 + m_2)}$

$$= \frac{4m_1 m_2 + m_0(m_1 + m_2) - 2m_0 m_2}{4m_1 m_2 + m_0(m_1 + m_2)} g$$

$$= \frac{4m_1 m_2 + m_0(m_1 + m_2 - 2m_2)}{4m_1 m_2 + m_0(m_1 + m_2)} g$$

$$= \frac{4m_1m_2 + m_0(m_1 - m_2)}{4m_1m_2 + m_0(m_1 + m_2)} g$$

$\therefore$ $a - a_0$= acceleration of m_1 with respect to inertial frame, i.e., with respect to ground $= \frac{(m_2 - m_1)m_0 - 4m_1m_2}{(m_2 + m_1)m_0 + 4m_1m_2} g$; $m_2 > m_1$

Try yourself: All the possible special cases.

11. *Two forces act at an angle 120°. The bigger force is 40 N and the resultant is perpendicular to the smaller one. Find the smaller force.*

Solution:

First method:

Let us suppose that magnitude of smaller force is $|\vec{a}|$, therefore, $|\vec{a}| < 40$ N and $\vec{R}$ is the resultant force.

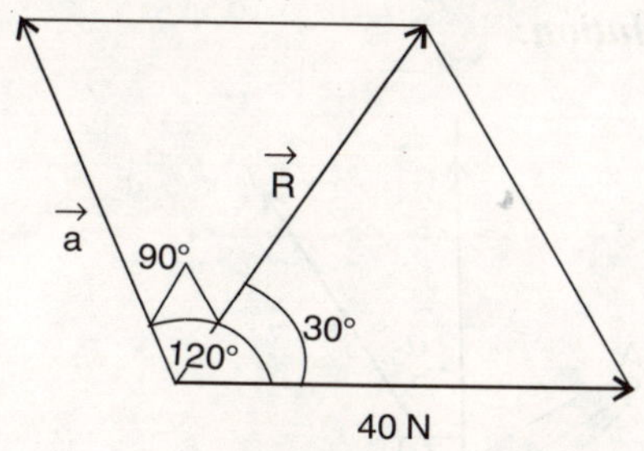

Fig. 4-W30

Let θ be the angle between forces then θ = 120° and α is angle made by resultant force with smaller force ($|\vec{a}|$) and β is the angle made by resultant with 40 N, therefore α = 90° as given in equation and then β = 120° – 90° = 30°.

Now as we know that,

$$\tan\alpha = \frac{40\sin 120°}{a + 40\cos 120°}$$

or $$\tan 90° = \frac{40\sin 120°}{a + 40\cos 120°}$$

$\because$ $\tan 90° = \infty$

$\therefore$ $a + 40\cos 120° = 0$

or $a = -40\cos 120°$

$= -40 \times -\frac{1}{2}$

$= 20$ N.

Second method: As we know that,

$$\tan\beta = \frac{a\sin\theta}{40 + a\cos\theta}$$

or $$\tan 30° = \frac{a\sin 120°}{40 + a\cos 120°}$$

or $$\frac{1}{\sqrt{3}} = \frac{\frac{\sqrt{3}}{2}a}{40 - \frac{1}{2}a}$$

or $$40 - \frac{1}{2}a = \frac{3}{2}a$$

or $$40 = \frac{3}{2}a + \frac{1}{2}a$$

$$= 2a$$

or $$a = \frac{40}{2}$$

$$= 20$$

$\therefore$ $a = 20$ N.

12. *To move a boat uniformly along a canal at a given speed requires a resultant force R = 500 N. This is accomplished by two horses pulling with forces P and Q on two ropes as shown in Fig. (4-W31). If the angles that the tow ropes make with the axis of the canal are β = 39° and γ = 21°. What are the corresponding tensions in the ropes?*

Solution:

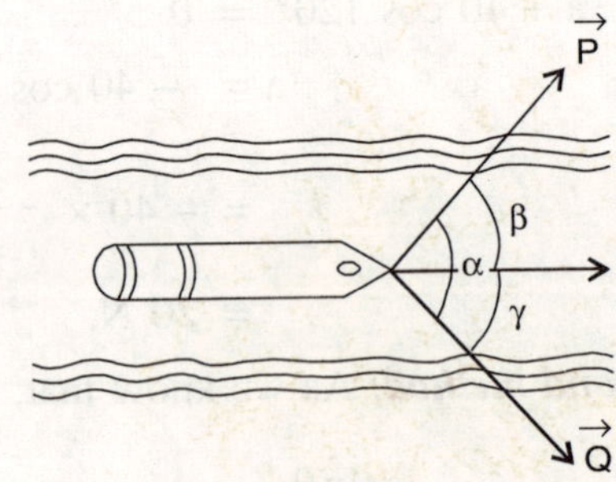

Fig. 4-W31

Resultant of two forces $\vec{P}$ and $\vec{Q}$ is given by

$R^2 = P^2 + 2PQ \cos \theta$; where θ is the angle between forces $\vec{P}$ and $\vec{Q}$.

Here, $\alpha = \beta + \gamma = 39° + 21° = 60°$

or $\quad 500^2 = P^2 + Q^2 + 2PQ \cos 60°$

or $\quad 500^2 = P^2 + Q^2 + PQ \qquad$ ....(i)

Since β is the angle made by resultant force $(\vec{R})$ with force $\vec{P}$.

Then, $\tan \beta = \dfrac{Q \sin \alpha}{P + Q \cos \alpha}$

or $\quad \tan 39° = \dfrac{Q \sin 60°}{P + Q \cos 60°}$

or $\quad P + Q \cos 60° = \dfrac{Q \sin 60°}{\tan 39°}$

$= 1.0694Q$

or $\quad P = 1.0694\ Q - 0.5\ Q$

or $\quad P = 0.5694\ Q$

or $\quad Q = 1.756\ P \qquad$ ....(ii)

Putting value of Q in terms of P in equation (i), we have

$$500^2 = P^2 + Q^2 + PQ$$
$$= P^2 + (1.756P)^2 + 1.756P^2$$
$$= P^2\ (1 + 1.756^2 + 1.756)$$
$$= P^2\ (5.8395)$$

or $\quad P^2 = \dfrac{500^2}{5.8395}$

or $\quad P = \sqrt{\dfrac{500^2}{5.8395}}$

$\therefore \quad P = 206.91$ N

$\therefore \quad Q = 1.756$ P

$= 1.756 \times 206.91$

$= 363.33$ N.

Therefore tensions in ropes are P = 206.91 N and Q = 363.33 N respectively.

13. *In level flight, the chord AB of an aeroplane wing makes an angle α = 7° with the horizontal Fig.(4-W32). The resultant wind pressure on the wing for such conditions is defined by its lift and drag components L = 1600 N and D = 250 N, which are vertical and horizontal respectively as shown. Resolve this force into rectangular components x and y coinciding with chord AB and normal respectively.*

Solution:

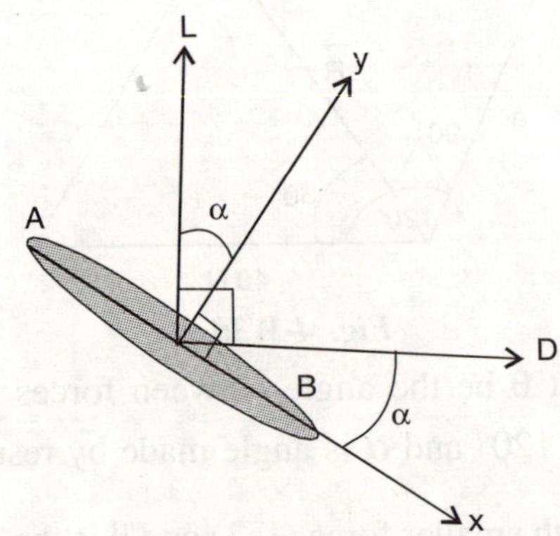

Fig. 4-W32

x-component $= D \cos \alpha + L \cos (90 + \alpha)$

$= D \cos \alpha + L\ (- \sin \alpha)$

$= D \cos \alpha - L \sin \alpha$

$= 250 \cos 7° - 1600 \sin 7°$

$= 53.14$ N

y-component $= L \cos \alpha + D \cos (90 - \alpha)$

$= L \cos \alpha + D \sin \alpha$

$= 1600 \cos \alpha + 250 \sin \alpha$

$= 1600 \cos 7° + 250 \sin 7°$

$= 1612.54$ N.

14. *For a particular position as shown in Fig. (4-W33), the connecting rod BA of an engine exerts a force P = 400 N on the crank pin at A. Resolve this force into two rectangular components P_h and P_v acting horizontally and vertically respectively, at A.*

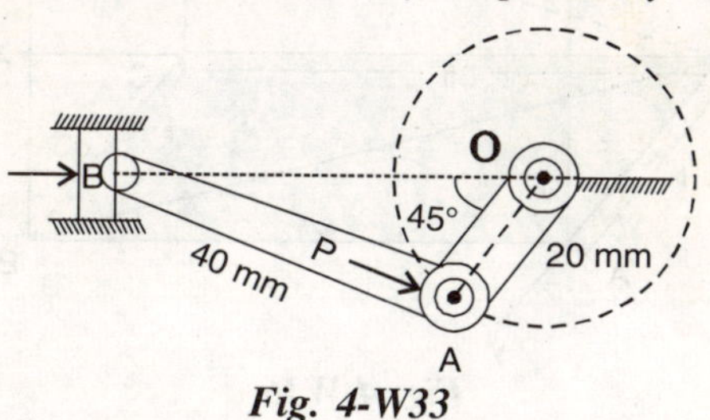

Fig. 4-W33

Solution:

In triangle BAO, applying sine rule we get,

$$\frac{40}{\sin 45°} = \frac{20}{\sin\theta}; \quad \text{Let } \angle ABO = \theta$$

or $$\sin\theta = \frac{20 \sin 45°}{40}$$

$$= \frac{\sin 45°}{2}$$

or $$\theta = \sin^{-1}\left(\frac{\sin 45°}{2}\right)$$

$\therefore$ $$\theta = 20.7°$$

Therefore, $\angle BAO = 180° - 45° - 20.7°$

$= 114.3°$

Let us redraw the Fig. (4-W33).

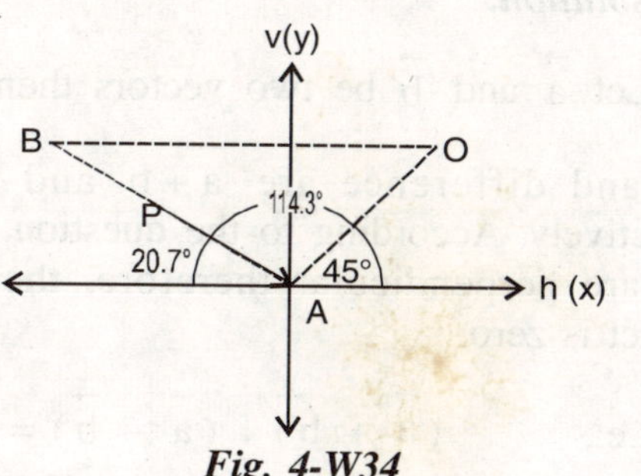

Fig. 4-W34

Therefore, from the drawn figure, we have

Horizontal components, $P_h = P\cos 20.7°$

$= 400 \cos 20.7°$

$= 374.17$ N

Vertical components, $P_v = P \sin 20.7$

$= 400 \sin 20.7$

$= 141.38$ N.

15. *A carrom board (6ft × 6ft) has the queen at the centre. The queen is hit by the striker moves to the front edge, rebounds and goes in the hole behind the striking line. Find the magnitude of displacement of the queen (a) from the centre to the front front edge (b) from front edge to the hole and (c) from the centre to the hole.*

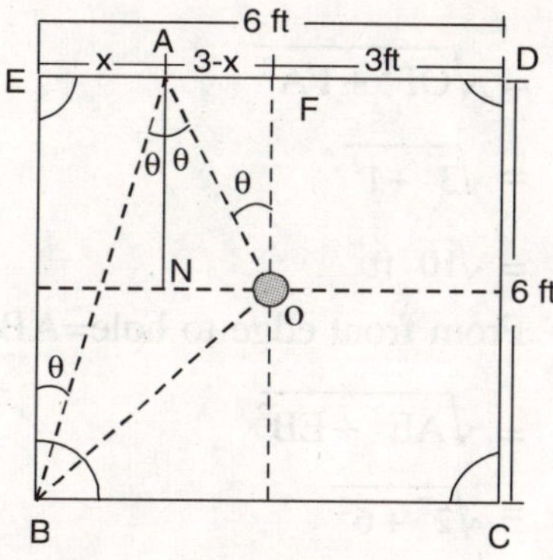

Fig. 4-W35

Solution:

Let us suppose that queen strikes at point A at a distance x from the hole E. Therefore EA = x and FA = 3 – x. Let us suppose that striking of queen from front edge obeys the rules of reflection of a lighter elastic rigid body from heavier elastic rigid body. Let AN be normal at point A

$\therefore$ $\angle NAB = \angle NAO = \theta$

Now, $$\tan\theta = \frac{3 - x}{3}$$

....(i) in Δ OAF

[∵ AF = NO = 3 – x] and [AN = OF = 3]

and $\tan\theta = \dfrac{x}{6}$ [$\because$ AE = X and BF = 6]

....(ii) in Δ BAE

Comparing equations (i) and (ii), we get

$$\frac{x}{6} = \frac{3-x}{3}$$

or $3x = 18 - 6x$

or $9x = 18$

$\therefore$ $x = 2$ ft

= AE

$\therefore$ $3 - x = 1$ ft

= FA

Therefore magnitude of displacement from the centre to the front edge

= OA

$= \sqrt{OF^2 + FA^2}$

$= \sqrt{3^2 + 1^2}$

$= \sqrt{10}$ ft

From front edge to hole=AB

$= \sqrt{AE^2 + EB^2}$

$= \sqrt{2^2 + 6^2}$

$= \sqrt{40}$

From centre to hole, OB $= \sqrt{3^2 + 3^2}$

$= 3\sqrt{2}$ ft.

16. *A mosquito net over a 6 ft × 4 ft bed is 3 ft high. The net has a hole at one corner of the bed through which a mosquito enters the net. It flies and sits at the diagonally opposite upper corner of the net. (a) Find the magnitude of the displacement of the mosquito. (b) Taking the hole as the origin, the length of the bed as x-axis its width as the y-axis, and vertically up as the z-axis, write components of the displacement vector.*

Solution:

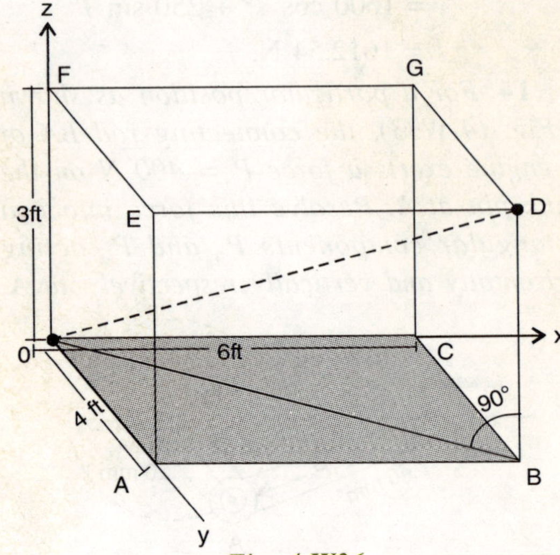

Fig. 4-W36

If initially mosquito was at origin and finally at D then (a) magnitude of displacement = OD.

$= \sqrt{OB^2 + BD^2}$

$= \sqrt{OA^2 + AB^2 + BD^2}$

$= \sqrt{6^2 + 4^2 + 3^2}$

$= \sqrt{61}$

= 7.81 ft

(b) $\vec{d} = 6\,\vec{i} + 4\,\vec{j} + 3\,\vec{k}$

$\therefore$ $d_x = 6$, $d_y = 4$ and $d_z = 3$.

17. *If sum and difference of two vectors is perpendicular to each other, then show that their magnitude must be equal.*

Solution:

Let $\vec{a}$ and $\vec{b}$ be two vectors then their sum and difference are $\vec{a} + \vec{b}$ and $\vec{a} - \vec{b}$ respectively. According to the question, since they are perpendicular therefore, their dot product is zero.

i.e., $(\vec{a} + \vec{b}) \cdot (\vec{a} - \vec{b}) = 0$

or $\vec{a} \cdot \vec{a} + \vec{b} \cdot \vec{a} - \vec{a} \cdot \vec{b} - \vec{b} \cdot \vec{b} = 0$

or $a^2 + \vec{a} \cdot \vec{b} - \vec{a} \cdot \vec{b} - b^2 = 0$

or $a^2 - b^2 = 0$

or $(a + b)(a - b) = 0$

$\Rightarrow$ $a - b = 0$

or $a = b$

$[\because a + b \neq 0]$

That means magnitude of $\vec{a}$ = magnitude of $\vec{b}$ and $a + b = 0$. But it is not possible because both magnitudes are positive therefore their sum cannot be zero.

18. *The resultant of vectors $\vec{a}$ and $\vec{b}$ is perpendicular to $\vec{a}$ as shown in Fig. (4-W37). Find the angle between them.*

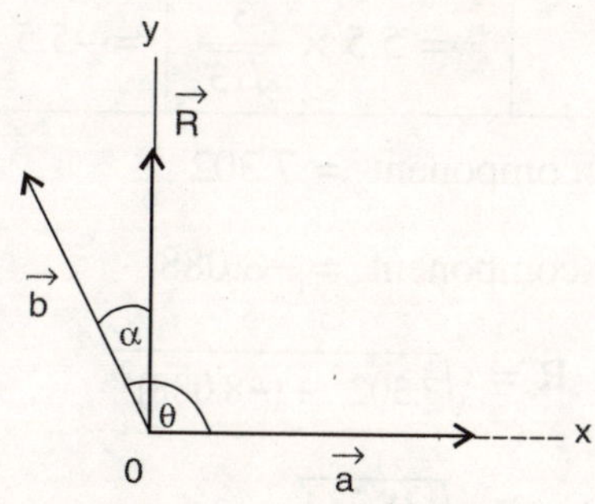

Fig. 4-W37

Solution:

First method

x-component of resultant

$= a \cos 0° + b \cos \theta$

$= a + b \cos \theta$

Since resultant is perpendicular to $\vec{a}$, therefore, its projection along $\vec{a}$ must be zero. that is *x*-component is equal to zero.

or $a + b \cos \theta = 0$

or $\cos \theta = \frac{-a}{b}$

$\therefore$ $\theta = \cos^{-1}\left(\frac{-a}{b}\right).$

Second method

Let α be the angle made by resultant with $\vec{b}$ vector.

Then, $\tan \alpha = \frac{a \sin \theta}{b + a \cos \theta}$

or $\tan (\theta - 90°) = \frac{a \sin \theta}{b + a \cos \theta}$

$[\because \alpha = \theta - 90°]$

or $-\cot \theta = \frac{a \sin \theta}{b + a \cos \theta}$

or $-\frac{\cos \theta}{\sin \theta} = \frac{a \sin \theta}{b + a \cos \theta}$

or $-(b \cos \theta + a \cos^2 \theta)$

$= a \sin^2 \theta$

or $a \sin^2 \theta + a \cos^2 \theta = -b \cos \theta$

or $a = -b \cos \theta$

or $\cos \theta = -\frac{a}{b}$

$\Rightarrow$ $\theta = 90° + \cos^{-1}\left(\frac{a}{b}\right).$

Third method

If β is the angle made by resultant with $\vec{a}$ then it is equal to 90° given in question.

Therefore, $\tan 90° = \frac{b \sin \theta}{a + b \cos \theta}$

or $\infty = \frac{b \sin \theta}{a + b \cos \theta}$

or $\frac{1}{0} = \frac{b \sin \theta}{a + b \cos \theta}$

or $a + b \cos \theta = 0$

or $$\cos\theta = \left(\frac{-a}{b}\right)$$

$$\therefore \quad \theta = 90° + \cos^{-1}\left(\frac{a}{b}\right).$$

19. *Prove that* $\vec{b}\cdot(\vec{a}\times\vec{b}) = 0$.

Solution:

From the definition of cross-product of vectors $\vec{a}$ and $\vec{b}$ we know that $\vec{a}\times\vec{b}$ is perpendicular to plane made by vectors $\vec{a}$ and $\vec{b}$ i.e., $\vec{a}\times\vec{b}$ is perpendicular to $\vec{b}$. Now we know that dot product of perpendicular vectors be zero.

Therefore, $\vec{b}\cdot(\vec{a}\times\vec{b}) = 0$

also $\vec{a}\cdot(\vec{a}\times\vec{b}) = 0$

20. *A system of five forces acting on a body is as shown in Fig. (4-W38). Determine the resultant.*

Solution:

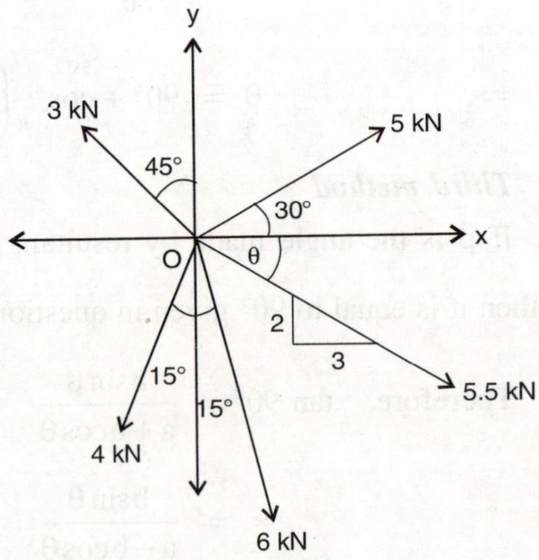

Fig. 4-W38

If θ be the inclination of 5.5 kN force then,

$$\tan\theta = \frac{2}{3}$$

$$\therefore \quad \sin\theta = \frac{2}{\sqrt{13}}$$

$$\cos\theta = \frac{3}{\sqrt{13}}$$

Now $R_x = \Sigma$x-component

and $R_y = \Sigma$y-component.

Table 4-W1: Components of forces

Force	x-component	y-component
5 kN	5 sin 60°	5 cos 60°
3 kN	–3 cos 45°	3 sin 45°
4 kN	–4 sin 15°	–4 cos 15°
6 kN	6 sin 15°	–6 cos 15°
5.5 kN	5.5 cos θ	–5.5 sin θ
	$= 5.5 \times \frac{3}{\sqrt{13}}$	$= -5.5 \times \frac{2}{\sqrt{13}}$

Σx-component = 7.302

Σy-component = –8.088

$$\therefore \quad R = \sqrt{7.302^2 + (-8.088)^2}$$

$$= \sqrt{118.734}$$

$$= 10.89 \text{ kN}$$

$$\tan\alpha = \frac{\Sigma y}{\Sigma x}$$

$$= \frac{-8.080}{7.302}$$

$$= -1.107$$

$$\therefore \quad \alpha = -\tan^{-1} 1.107$$

$$= -47.92°$$

$$\cong -48°.$$

EXERCISE

1. In the arrangement of Fig. (4-E1) the masses m_0, m_1 and m_2 of bodies are equal, the masses of the pulley and the threads are negligible, and there is no friction in the pulley. Find the acceleration 'w' with which the body m_0 comes down, and the tension of the thread binding together the bodies m_1 and m_2, if the coefficient of friction between these bodies and the horizontal surface is equal to k. Consider possible cases.

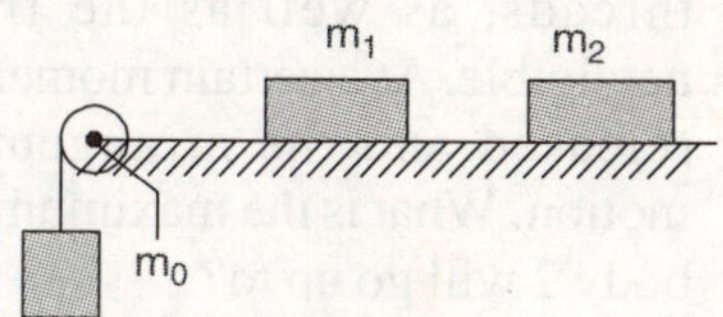

Fig. 4-E1

2. A small body was launched up an inclined plane set at an angle $\alpha = 15°$ against the horizontal. Find the coefficient of friction, if the time of the ascent of the body is $\eta = 2.0$ times less than the time of its descent.

3. The following parameters of the arrangement of Fig. (4-E2) are available; the angle α which inclined plane forms with the horizontal, and the coefficient of friction k between the body m_1 and the inclined plane. The masses of the pulley and the threads as well as the friction in the pulley, are negligible. Assuming both bodies to be motionless at the initial moment, find the mass ratio $\frac{m_2}{m_1}$ at which the body m_2 (a) starts coming down (b) starts going up (c) is at rest.

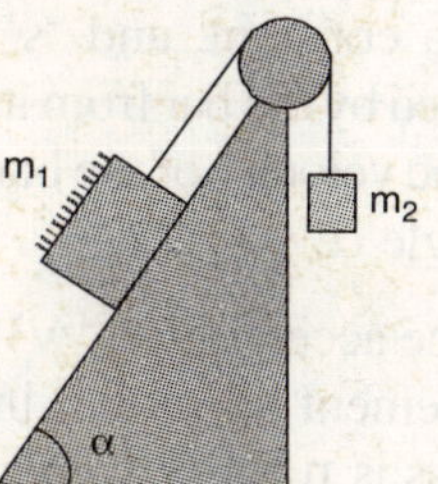

Fig. 4-E2

4. A small body A starts sliding down from the top of a wedge (Fig. 4-E3) whose base is equal to $l = 2.10$ cm. The coefficient of friction between the body and the wedge surface is $k = 0.140$. At what value of the angle α will the time of sliding be the least? What will it be equal to?

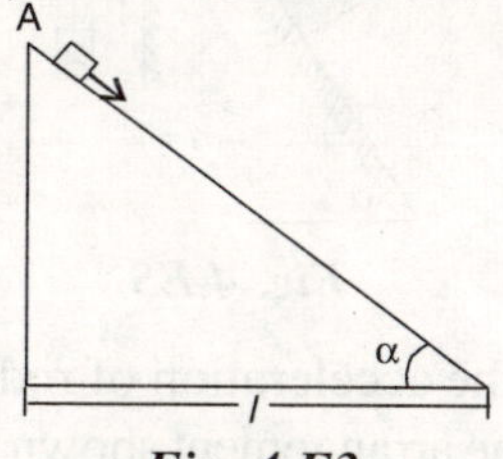

Fig. 4-E3

5. A bar of mass m is pulled by means of a thread up an inclined plane forming an angle α with the horizontal (Fig. 4-E4). The coefficient of friction is equal to k. Find the angle β which the thread must form with the inclined plane for the tension of the thread to be minimum. What is it equal to?

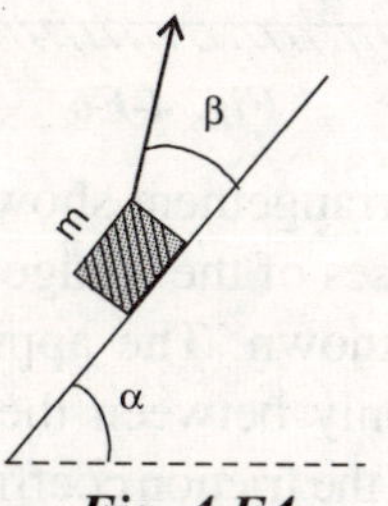

Fig. 4-E4

6. A bar of mass m resting on a smooth horizontal plane starts moving due to the force F = mg/3 of constant magnitude. In the process of its rectilinear motion the angle α between the direction of this force and the horizontal varies as $\alpha = as$, where 'a' is a constant, and 's' is the distance traversed by the bar from its initial position. Find the velocity of the bar as a function of the angle α.

7. Find the acceleration 'w' of body 2 in the arrangement as shown in Fig. (4-E5), if its mass is η time as great as the mass of bar 1 and the angle that the inclined plane forms with the horizontal is equal to α. The masses of the pulleys and the threads, as well as the friction, are assumed to the negligible. Look into possible cases.

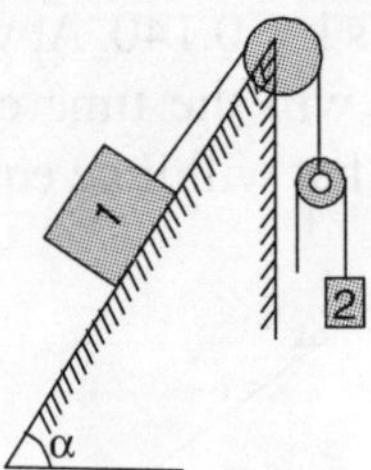

Fig. 4-E5

8. Find the acceleration of rod A and wedge B in the arrangement shown in Fig. (4-E6). If the ratio of the masses of the wedge to that of the rod equals η, and the friction between all contact surfaces is negligible.

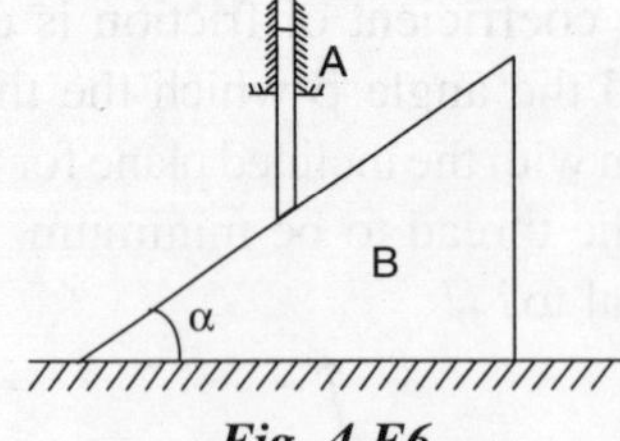

Fig. 4-E6

9. In the arrangement shown in Fig. (4-E7) the masses of the wedge M and the body m are known. The appreciable friction exists only between the wedge and the body m, the friction coefficient being equal to k. The masses of the pulley and the thread are negligible. Find the acceleration of the body m relative to horizontal surface on which the wedge slides.

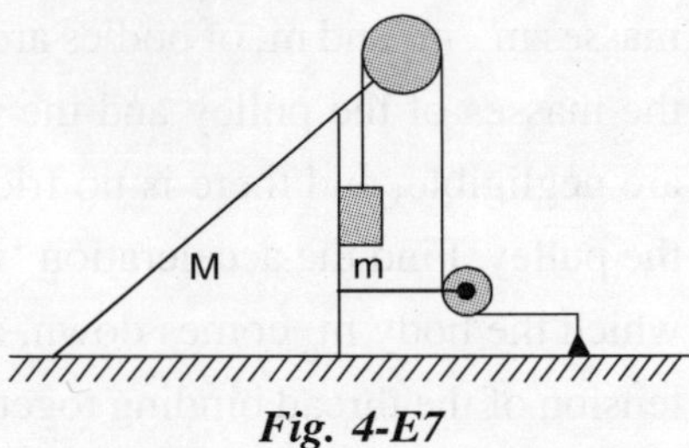

Fig. 4-E7

10. In the arrangement shown in Fig. (4-E8), the mass of body 1 is $\eta = 4.0$ times as great as that of body 2. The height h = 20 cm. The masses of the pulleys and the threads, as well as the friction are negligible. At a certain moment body 2 is released and the arrangement set in motion. What is the maximum height that body 2 will go up to?

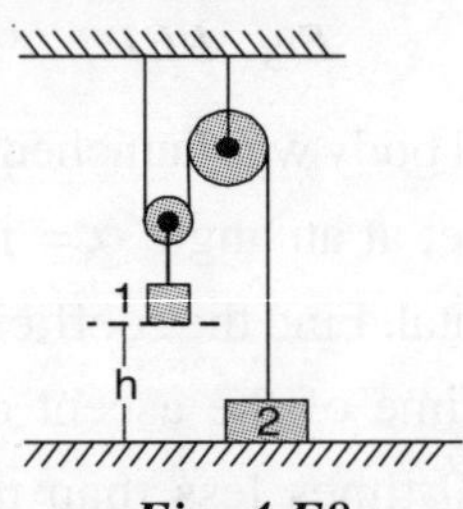

Fig. 4-E8

11. What is the minimum acceleration with which bar A (Fig. 4-E9) should be shifted horizontally to keep bodies 1 and 2 stationary relative the bar? The masses of the bodies are equal, and the coefficient of friction between the bar and the bodies is equal to k. The masses of the pulley and the threads are negligible, the friction in the pulley is absent.

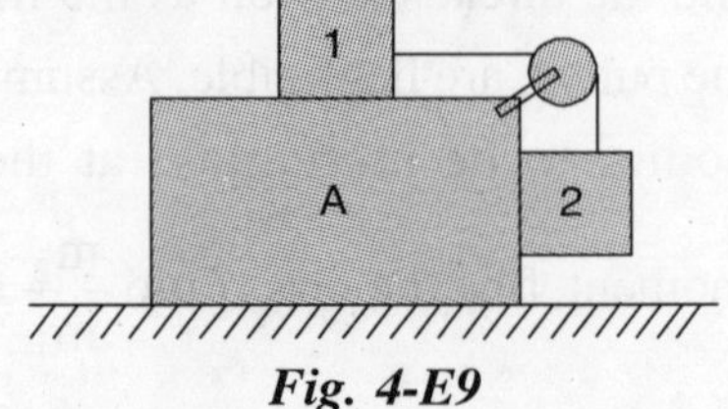

Fig. 4-E9

12. Prism 1 with bar 2 of mass m placed on it gets a horizontal acceleration w directed to the left Fig. (4-E10). At what maximum value of this acceleration will the bar be still stationary relative to the prism, if the coefficient of friction between them, $k < \cot \alpha$?

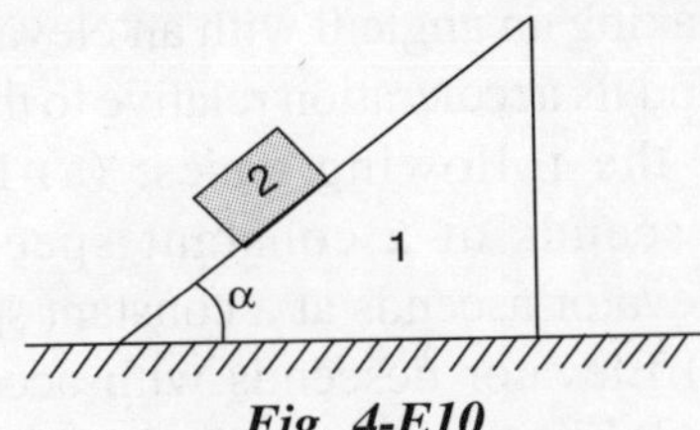

Fig. 4-E10

13. In the arrangement shown in Fig. (4-E11) the masses m of the bar and M of the wedge, as well as the wedge angle α are known. The masses of the pulley and the thread are negligible. The friction is absent. Find the acceleration of the wedge M.

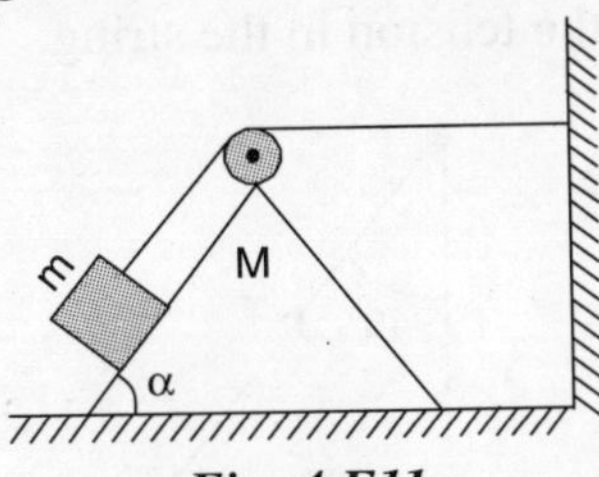

Fig. 4-E11

14. A particle of mass m moves along a circle of radius R. Find the modulus of the average vector of the force acting on the particle over the distance equal to a quarter of the circle, if the particle moves:

(a) uniformly with speed v;

(b) With constant tangential acceleration w; the initial velocity being equal to zero.

15. A small sphere of mass m suspended by a thread is first taken a side so that the thread forms the right angle with the vertical and then released. Find,

(a) the total acceleration of the sphere and the thread tension as a function of θ, the angle of deflection of the thread from the vertical;

(b) the thread tension at the moment when the vertical component of the velocity of sphere is maximum;

(c) the angle θ between the thread and the vertical at the moment when the total acceleration vector of the sphere is directed horizontally.

16. A fixed pulley carries a weightless thread with masses m_1 and m_2 at its ends. There is friction between the thread and the pulley. It is such that the thread starts slipping when the ratio $\frac{m_2}{m_1} = \eta_0$. Find,

(a) the friction coefficient;

(b) the acceleration of the masses, when

$$\frac{m_2}{m_1} = \eta > \eta_0.$$

17. A chain of mass m forming a circle of radius R is slipped on a smooth round cone with half-angle θ. Find the tension of the chain, if it rotates with a constant angular velocity (ω) about a vertical axis coinciding with the symmetry axis of the cone.

18. A small bar starts sliding down on an inclined plane forming an angle α with the horizontal. The friction coefficient depends on the distance x covered as $k = ax$, where a is a constant. Find the distance covered by the bar till it stops, and its maximum velocity over this distance.

19. A body of mass m is suspended by two strings making angles 90° and θ with the horizontal, as shown in Fig. (4-E12). Find the tension in the string, T_1.

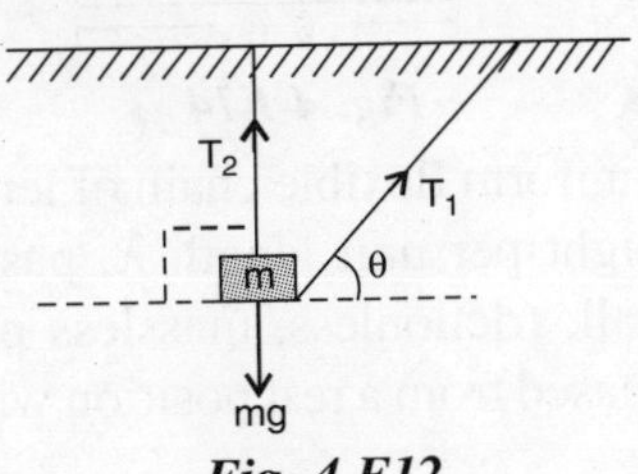

Fig. 4-E12

20. A block of mass 5 kg is set on a box of mass 10 kg. They descend together with an acceleration 8 ms^{-2}. Find the force exerted by the block on the floor of box. Take g = 10 ms^{-2}.

21. A smooth ring A of mass m can slide on a fixed horizontal rod. A string tied to ring passes over a fixed pulley B and carries a block C of mass M as shown in Fig. (4-E13). At an instant the string between ring and pulley makes an angle θ with the horizontal surface of fixed block. With what acceleration will the ring start if the system is released from rest.

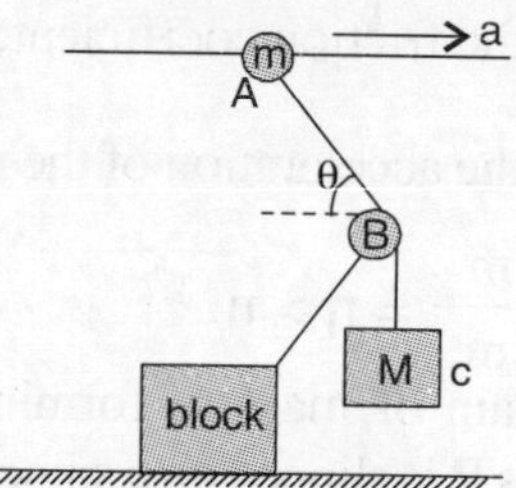

Fig. 4-E13

22. A particle of mass, m slides down a smooth inclined plane of elevation θ, fixed in an elevator going up with an acceleration 10 ms^{-2}. Fig. (4-E14). The base of the inclined plane has a length 5 m. Find the minimum time taken by the particle to reach the bottom. Take g = 10 ms^{-2}.

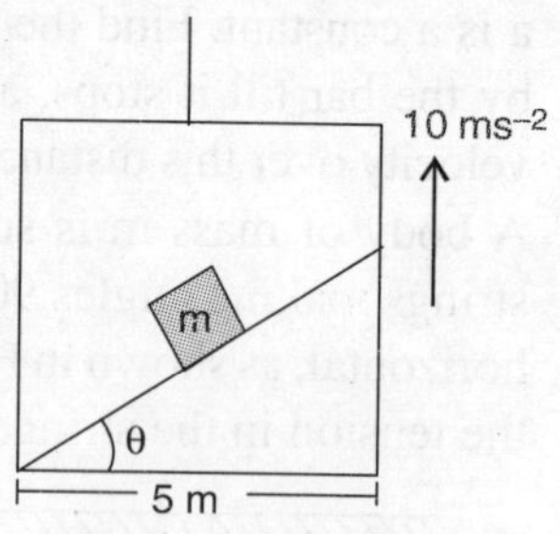

Fig. 4-E14

23. A uniform flexible chain of length l, with weight per unit length λ, passes over a small, frictionless, massless pulley. It is released from a rest position with a length of chain x hanging from one side and a length $l–x$ from the other side. (a) Under what circumstances will it accelerate? (b) (Assuming that these circumstances are not there), find the acceleration a as function of x.

24. A block slides down a frictionless incline making an angle θ with an elevator floor. Find its acceleration relative to the incline in the following cases. (a) Elevator descends at a constant speed v. (b) Elevator ascends at a constant speed 'v'. (c) Elevator descends with acceleration a. (d) Elevator descends with deceleration 'a'. (e) Elevator cable breaks.

25. A charged sphere of mass 4.0 × 10^{-4} kg is suspended from a string. An electric force acts horizontally on the sphere so that the string makes an angle of 40° with the vertical when at rest, Fig. (4-E15). Find (a) the magnitude of the electric force (b) the tension in the string.

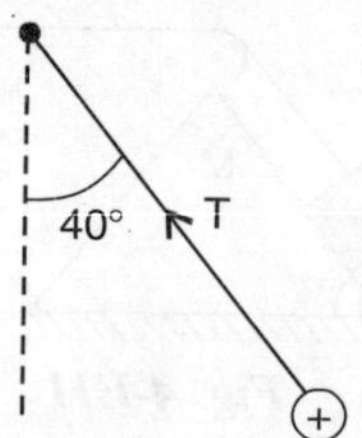

Fig. 4-E15

26. Three blocks are connected, as shown in Fig. (4-E16) on a horizontal frictionless table and are being pulled to the right with a force T_3 = 60 N. If m_1 = 10 kg, m_2 = 20 kg and m_3 = 30 kg. Find the tensions T_1 and T_2.

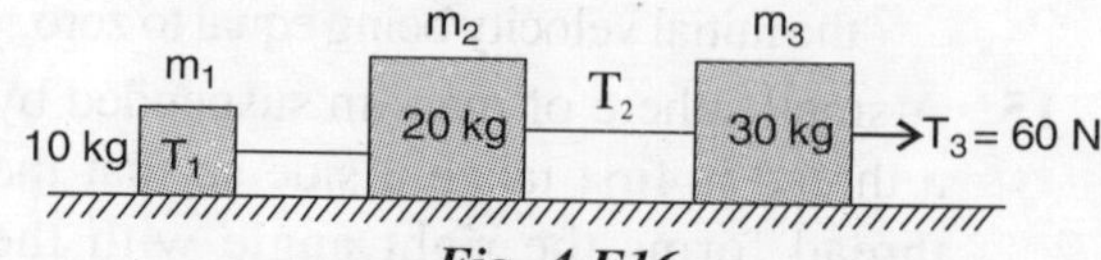

Fig. 4-E16

27. A particle of mass m, originally at rest, is subjected to a force whose direction is constant but whose magnitude varies with the time according to the relation,

$$F = F_0\left[1-\left(\frac{t-T}{T}\right)^2\right];$$ where F_0 and T are constants. The force acts only for the interval 2T. Prove that the speed v of the particle after time 2T elapsed is equal to $\frac{4F_0T}{3m}$.

28. A particle of mass m is moving under the action of an external force $\vec{F} = km\,\vec{r}$ where $\vec{r}$ is the particle's radius vector. Determine the path of the particle if its initial position vector is $\vec{r}_0$ and its initial velocity $\vec{v}_0$ is perpendicular to $\vec{r}_0$.

29. At t = 0, both masses m_1 = 1 kg and m_2 = 2 kg touch the ground and the string is taut which passes over a movable pulley Fig. (4-E17). A vertically upward, time dependent force F = 2t (F in N, t in s) is applied to the pulley. Calculate (i) the time when m_1 is lifted off the ground and (ii) m_2 is lifted off the ground. (iii) when t = 25 s, what is the acceleration of the masses as seen by an observer outside.

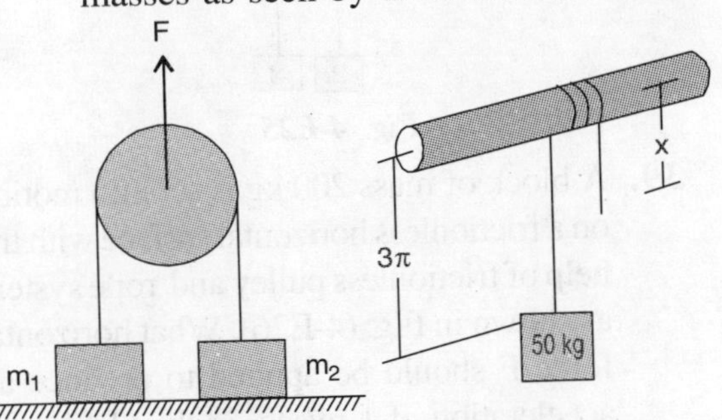

Fig. 4-E17 *Fig. 4-E18*

30. A rope having a mass per unit length of 0.8 kg/m is wound $2^1/_2$ times around a horizontal rod. What length x of rope should be left hanging if a 50 kg load is to be supported? The coefficient of friction between the rope and the rod is 0.25. Refer Fig. (4-E18).

31. A block of mass M rests on a smooth horizontal surface. Two blocks A and B of mass m and m′ respectively are connected by a string passing over a smooth pulley fixed to M. B rests on the smooth horizontal top surface of M. A can move along a frictionless vertical slot in block M. Find the acceleration of A.

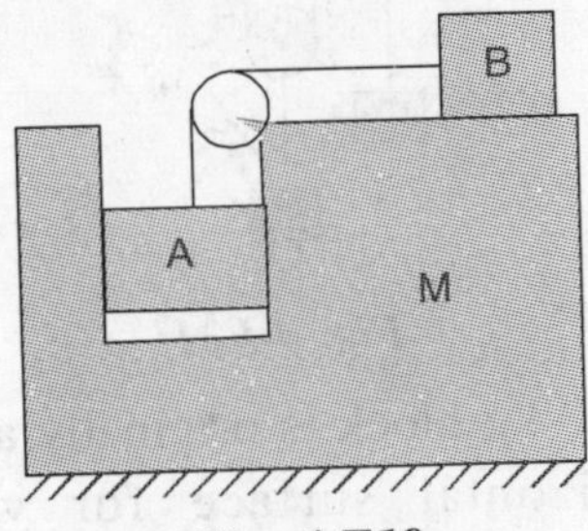

Fig. 4-E19

32. If body of mass m moves through a liquid at low velocity, the force of resistance due to viscosity is given by,

$$F = kv$$

where k is the resistance force at unit velocity, show that the velocity would decrease exponentially with time and linearly with displacement.

33. In the system shown in Fig. (4-E20) the masses are 1, 2, 3 and 4 kg respectively. Determine the accelerations of masses 1 and 2.

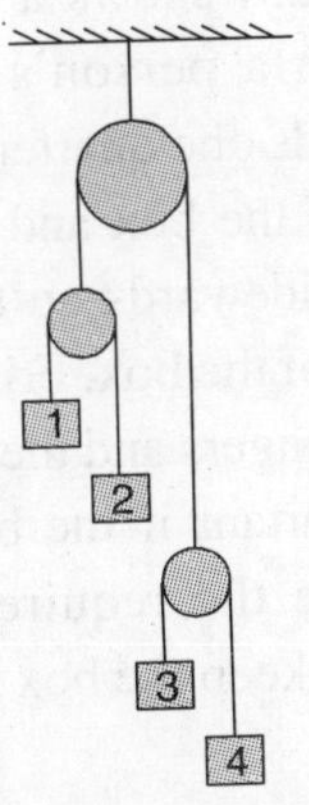

Fig. 4-E20

34. Determine the acceleration of the blocks in the pulley system as shown in Fig. (4-E21). Neglect the masses of pulleys and also friction.

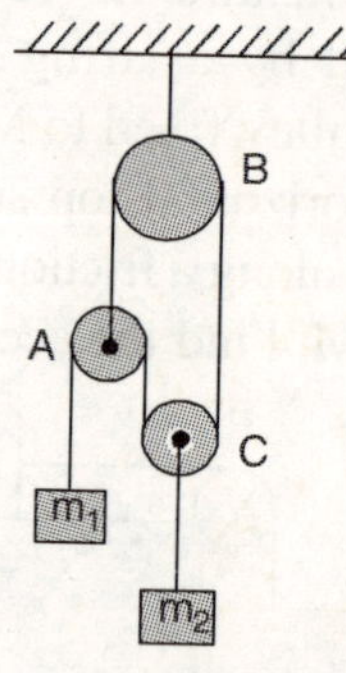

Fig. 4-E21

35. A 20 kg block is originally at rest on a horizontal surface for which the coefficient of friction is 0.6. If a horizontal force F is applied such that it varies with time, as shown in Fig. (4-E22). Determine the speed of the block in 10 s.

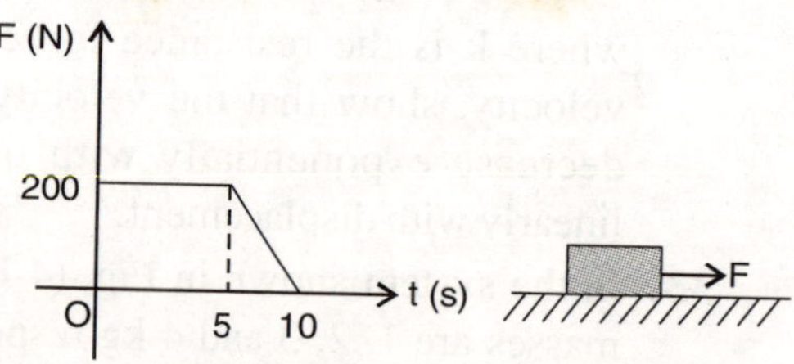

Fig. 4-E22

36. Fig. (4-E23) shows a small square box held with a person's fingers pushing downwards one quarter of the way along the top of the box and with their thumb pushing sidewards on the bottom of the side face of the box. Friction between the person's fingers and the box is obviously, very important if the box is not to fall. Determine the required coefficient of friction to keep the box from falling.

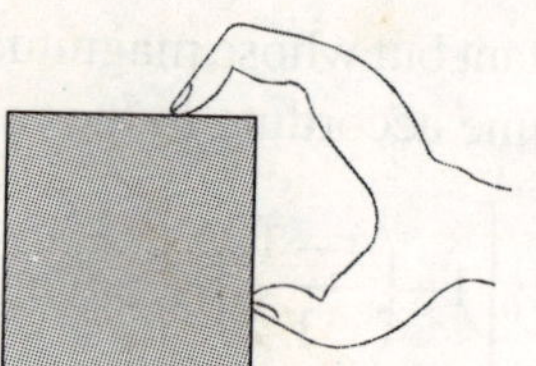

Fig. 4-E23

37. As shown in Fig. (4-E24) the wedge is fixed to the ground the prism B of mass M and the block C of mass m is placed as shown. Find the acceleration of the block C with respect to B when system is set free. Neglect any friction.

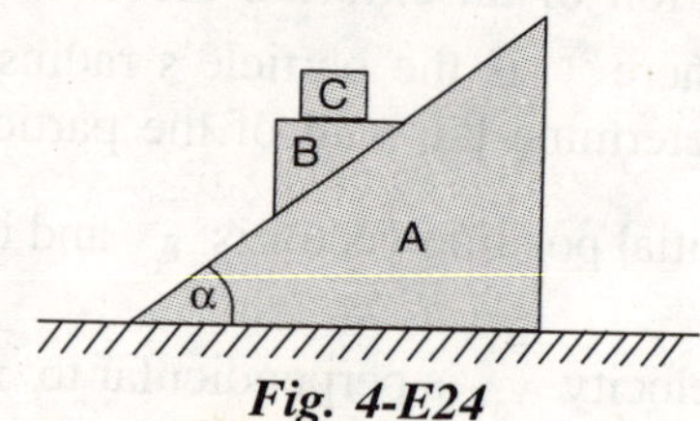

Fig. 4-E24

38. Find the constraint relation between a_1, a_2 and a_3.

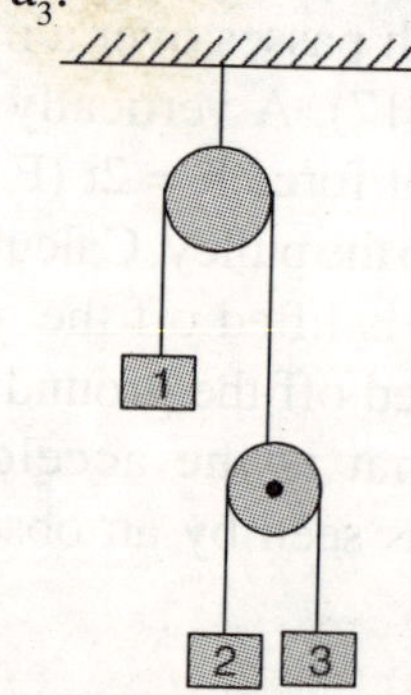

Fig. 4-E25

39. A block of mass 200 kg is set into motion on a frictionless horizontal surface with the help of frictionless pulley and rope system as shown in Fig. (4-E26). What horizontal force F should be applied to produce an acceleration of 1 m/sec² in the block.

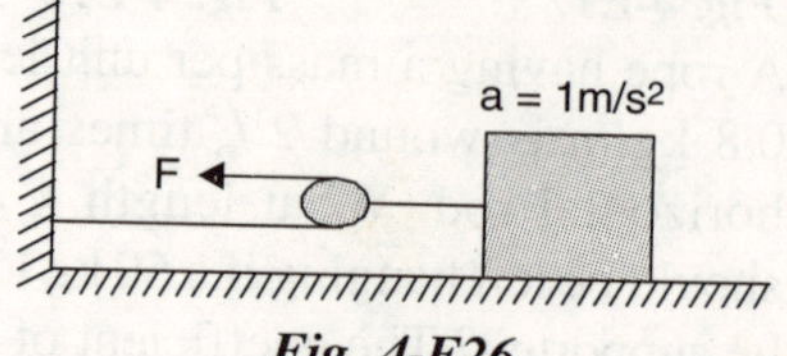

Fig. 4-E26

40. In Fig. (4-E27) the coefficient of friction of all the surfaces is μ. The string and pulley are light. The blocks are moving with constant speed. Then find the values of F and T.

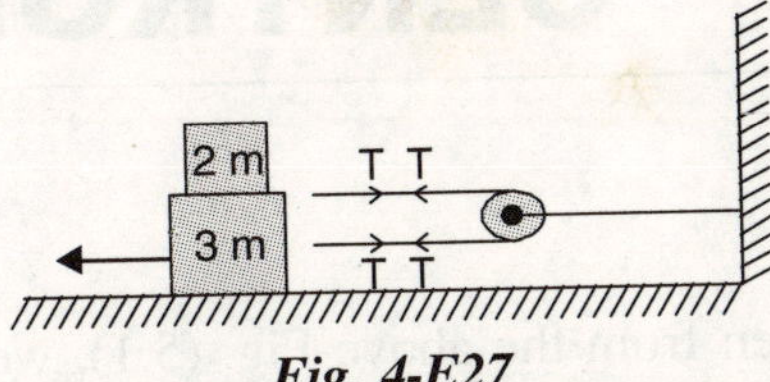

Fig. 4-E27

41. Find the acceleration w of body 2 as shown in Fig. (4-E28). If its mass is η times as great as the mass of bar 1 and the angle that the inclined plane form with horizontal is equal to α. The masses of the pulley and threads as well as friction are assumed to be negligible (look into all possible cases).

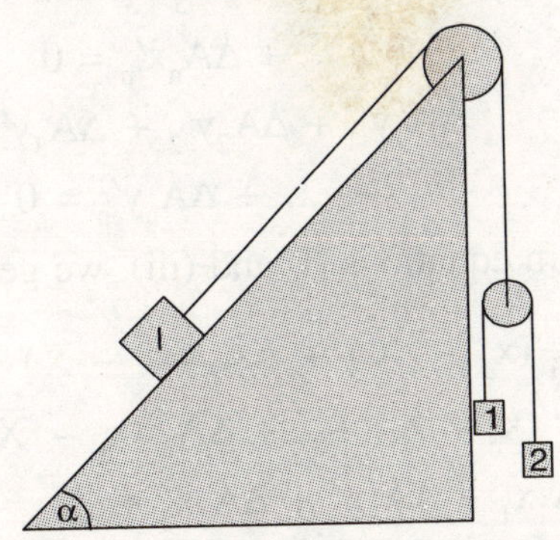

Fig. 4-E28

42. Using constraint equations find the relation between a_1 and a_2.

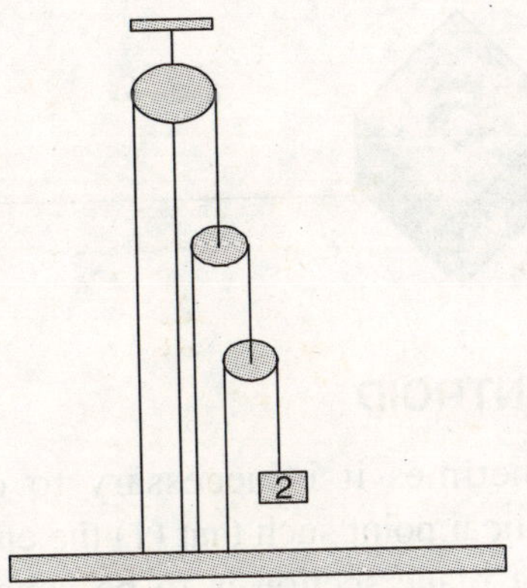

Fig. 4-E29

43. On a smooth plane surface two blocks A and B each of mass m are in contact and accelerated by applying a force F as shown in Fig. (4-E29). What is the force of contact between the two bodies?

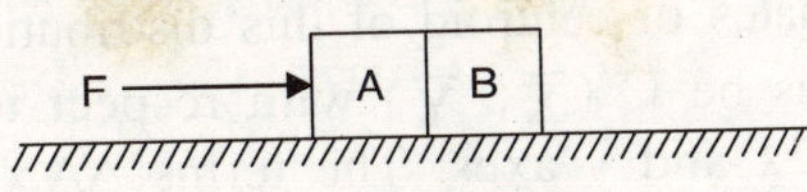

Fig. 4-E30

(a) F (b) $\frac{F}{2}$ (c) $\frac{F}{4}$ (d) 2Fs.

5

CENTROID

5.1 CENTROID

Sometimes it is necessary to define a mathematical point such that (1) the entire area of the the plane section is to be concentrated at that very point and (2) summation of the first moment of the areas about a axis passing through it should be zero.

Let us consider a collection of n-particles (Fig. 5.1). Let the area of the *ith* particle be ΔA_i and its coordinates with respect to the centroidal axes x′ and y′ be (x'_i, y'_i) and coordinates of centroid of this distribution of particles be C $(\overline{X}, \overline{Y})$ with respect to the chosen x and y axes. The terms $\Delta A_i x_i$ and $\Delta A_i y_i$ are known as the first moments about y-axis and x-axis respectively. Therefore, $x_i = \overline{X} + x'_i$ and $y_i = \overline{Y} + y'_i$ are coordinates of ith particle with respect to x and y axes. Let area of point C be ΔA.

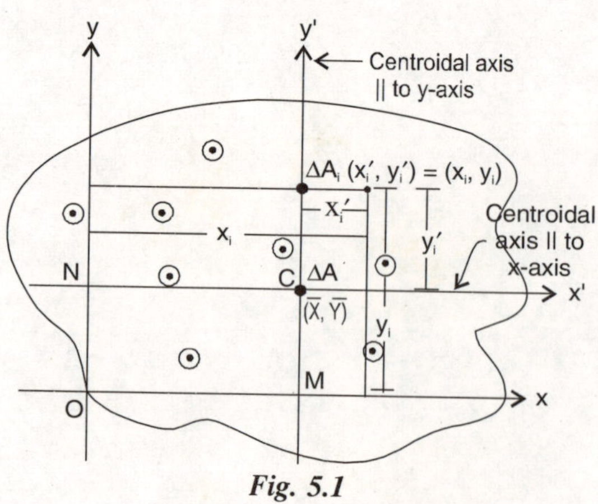

Fig. 5.1

Then from the above Fig. (5.1), we have

$$\left.\begin{aligned} x_i &= \overline{X} + x'_i \\ \text{and} \quad y_i &= \overline{Y} + y'_i \end{aligned}\right\} \quad \text{....(i)}$$

Now using condition (1), we get

$$\Delta A = \Delta A_1 + \Delta A_2 + \Delta A_3 + ... + \Delta A_n$$

or

$$\Delta A = \sum \Delta A_i \quad \text{....(ii)}$$

and using condition (2), we get

$$\Delta A_1 x'_1 + \Delta A_2 x'_2 + \Delta A_3 x'_3 + ... + \Delta A_n x'_n = 0 \quad \text{....(iii)}$$

and

$$\Delta A_1 y'_1 + \Delta A_2 y'_2 + \Delta A_3 y'_3 + ... + \Delta A_n y'_n = 0 \quad \text{....(iv)}$$

From equations (i) and (iii), we get

$$\Delta A_1 (x_1 - \overline{X}) + \Delta A_2 (x_2 - \overline{X}) + \Delta A_3 \times (x_3 - \overline{X}) + ... + \Delta A_n (x_n - \overline{X}) = 0$$

or

$$(\Delta A_1 x_1 + \Delta A_2 x_2 + \Delta A_3 x_3 + ... + \Delta A_n x_n) - (\Delta A_1 + \Delta A_2 + \Delta A_3 + ... + \Delta A_n) \times \overline{X} = 0$$

or

$$\Delta A_1 x_1 + \Delta A_2 x_2 + \Delta A_3 x_3 + ... + \Delta A_n x_n = \Delta A \overline{X} \quad \text{[from equation (ii)]}$$

That is the summation of moments about the y-axis = the moment of centroid about y-axis therefore

$$\overline{X} = \frac{\Delta A_1 x_1 + \Delta A_2 x_2 + \Delta A_3 x_3 + ... + \Delta A_n x_n}{\Delta A} \quad \text{....(5.1)}$$

Similarly from equations (iv), (ii) and (i), we can get

$$\overline{Y} = \frac{\Delta A_1 y_1 + \Delta y_2 y_2 + \Delta A_3 y_3 + \ldots\ldots\ldots + \Delta A_n y_n}{\Delta A} \quad \ldots(5.2)$$

For continuous plane area the number of points tends to be infinity therefore $\Delta A_i \rightarrow 0$. Thus,

$$\overline{X} = \lim_{\Delta A_i \to 0} \frac{\sum \Delta A_i x_i}{\sum \Delta A_i}$$

$$= \frac{\int_A x_i dA}{\int_A dA}$$

$$\therefore \quad \overline{X} \int_A dA = \int_A x_c dA \quad \ldots(5.3)$$

and
$$\overline{y} \int_A dA = \int_A y_c dA \quad \ldots(5.4)$$

Notes:

1. *Since a plane is two dimensional, therefore $\overline{Z} = 0$ if it lies in the xy-plane.*
2. *Centroid of a section is a completely mathematical point. Physically we can't concentrate the whole area at a single point.*
3. *Centroid may lie out-side the given plane area. For example, a centroid of a centrally punch circular plate is the geometrical centre of the circle and there is no area at all as shown in Fig. (5.2).*
4. *A centroid is the property of a surface. It depends on the shape and the size of the surface.*

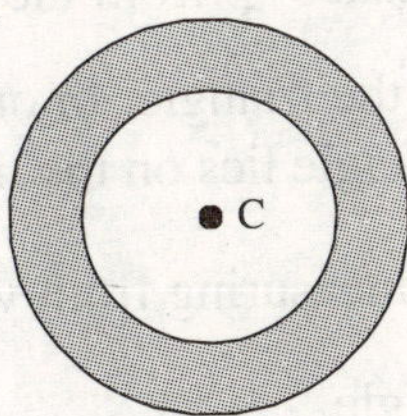

Fig. 5.2

5.2 DETERMINATION OF CENTROID OF SIMPLE SECTIONS FROM FIRST PRINCIPLE

For any types of sections we have general expressions for calculating centroid $G(\overline{X}, \overline{Y})$ of lying in xy-plane by,

$$\overline{Y} \int_A dA = \int_A x_c dA$$

and
$$\overline{X} \int_A dA = \int_A y_c dA \quad \ldots(5.5)$$

where dA is the elementary area, x_c and y_c are the coordinates of centroid of elementary area from y and x axes respectively.

The location of a centroid using equations (5.3) and (5.4) are known as the first principles of finding the centroid.

Here, let us find centroids of some standard sections like triangle, semi-circle and quarter circle etc.

5.2.1 Centroid of triangle

Let us consider a triangle ABC of base b and height h as shown in Fig. (5.3).

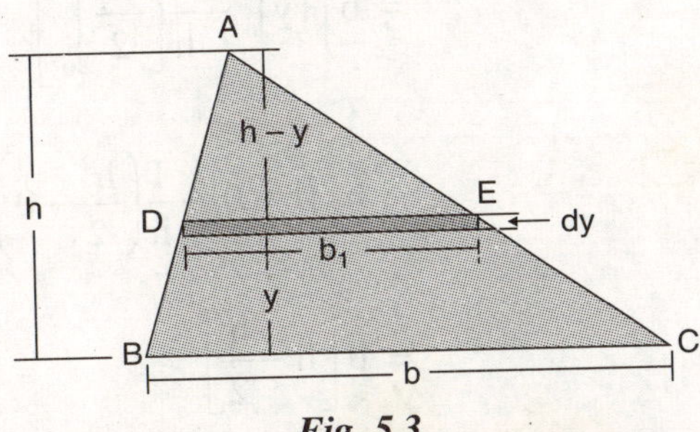

Fig. 5.3

Now let us consider a rectangular strip of length b_1 and width dy at a distance y from the base as shown in Fig. (5.3).

Since geometrically, Δ ABC and Δ ADE are similar triangles then,

$$\frac{b}{b_1} = \frac{h}{h-y}$$

[Ratio of similar sides should be equal.]

or $$\frac{h-y}{h} = \frac{b_1}{b}$$

or $$b_1 = \left(1-\frac{y}{h}\right)b \quad ....(i)$$

Now the elementary area of the rectangular strip is,

$$dA = \text{base} \times \text{height}$$
$$= b_1 dy$$
$$= \left(1-\frac{y}{h}\right) b\, dy \quad ....(ii)$$

Integrating equation (ii) over appropriate limits, we get the area (A) of the triangle.

That is $\int_0^A dA = \int_0^h \left(1-\frac{y}{h}\right) b\, dy$

or $$A = b\int_0^h \left(1-\frac{y}{h}\right)dy$$

$$= b\left[\int_0^h dy - \int_0^h \frac{y}{h}dy\right]$$

$$= b\left[[y]_0^h - \frac{1}{h}\left[\frac{y^2}{2}\right]_0^h\right]$$

$$= b\left[[h-0] - \frac{1}{h}\left(\frac{h^2}{2}-0\right)\right]$$

$$= b\left[h - \frac{h}{2}\right]$$

$$= b\left[\frac{h}{2}\right]$$

$$= \frac{1}{2}bh$$

Now taking the base of the triangle as a reference to locate the centroid of triangle and using equation (5.4) i.e., $\overline{Y}\int_A dA = \int_A y\, dA$ where, y is the distance of centroid of the elementary area of rectangular strip from base axis.

or $$\overline{Y}\left(\frac{1}{2}bh\right) = \int_A y\, dA$$

or $$\overline{Y}\left(\frac{1}{2}bh\right) = \int_0^h y\left(1-\frac{y}{h}\right) b\, dy$$

$$\left[A_{Triangle} = \frac{1}{2}bh\right]$$

$$= \int_0^h \left(y - \frac{y^2}{h}\right) b\, dy$$

$$= b\left[\frac{y^2}{2} - \frac{1}{h}\frac{y^3}{3}\right]_0^h$$

$$= b\left[\frac{h^2}{2} - \frac{1}{h}\frac{h^3}{3}\right]$$

$$= \left[\frac{h^2}{2} - \frac{h^2}{3}\right]b$$

$$= \frac{bh^2}{6}$$

or $$\overline{Y} = \frac{bh^2}{6} \times \frac{2}{bh}$$

$$= \frac{1}{3}h$$

$$\therefore \quad \overline{Y} = \frac{h}{3}$$

Thus the centroid of the triangle lies on the line at a distance $\frac{h}{3}$ from the base where, h is the height of the triangle. In other words the centroid of a triangle lies on the line which is at distance of $\frac{2h}{3}$ measuring from vertex where h is height of triangle.

Example 5.1: *Find the centroid of the triangle as shown in Fig. (5.4). All the dimensions are in cm.*

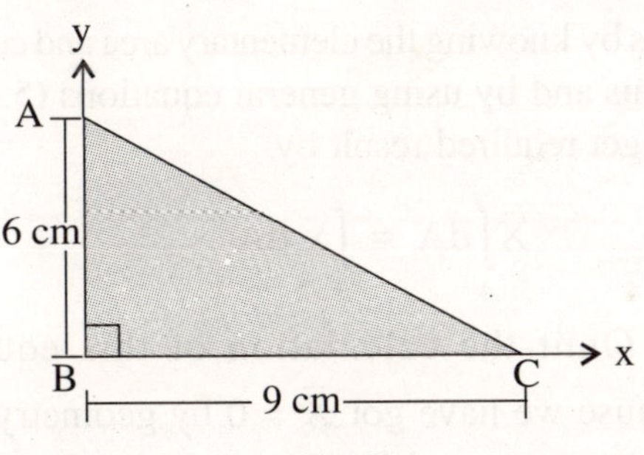

Fig. 5.4

Solution:

We have already seen that the centroid lies on the line which is at a distance of one third of the height measured from the base of the triangle.

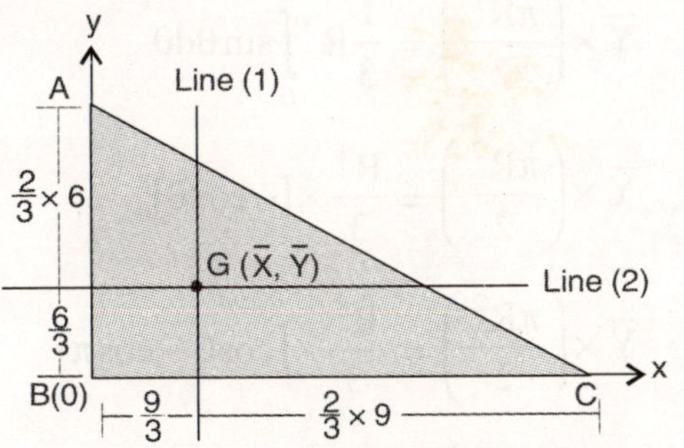

Fig. 5.5

Now take BC as the base and BA the as height then the centroid lies on line (2) and similarly AB as the base and BC as the height, then the centroid lies on lime (1). As there is only one centroid for one figure, therefore common point is centroid G $(\overline{X},\overline{Y}) \equiv (3, 2)$.

5.2.2 Centroid of semicircle

First method: Consider a semicircle of radius R as shown in Fig. (5.6). Since the semi-circle is symmetrical about the y-axis and we know that the centroid always lies on the symmetrical line hence the centroid of the shown semi-circle must lie on the y-axis i.e., x-coordinate of centroid $(\overline{X}) = 0$. Now we want to find $\overline{Y}$ i.e., the distance of the centroid from the x-axis measured along the y-axis.

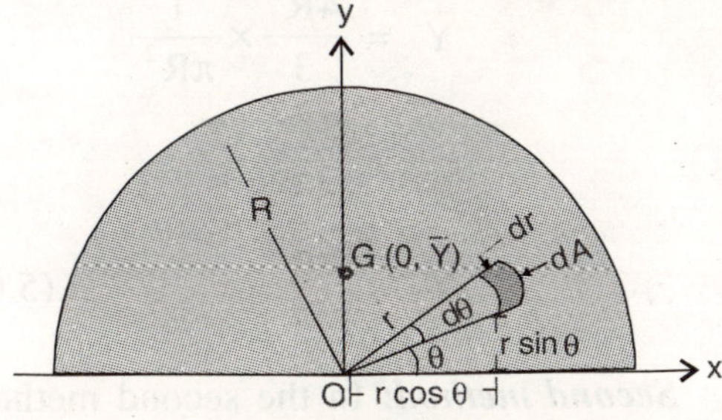

Fig. 5.6

For finding $\overline{Y}$, let us consider an elementary area dA as shown in the Fig. (5.6). Now we can treat the elementary area as a rectangular strip of length (r dθ) and width dr (increment in r).

Therefore, $dA = (r\,d\theta) \times (dr)$

$$= r\,dr\,d\theta \quad \text{....(i)}$$

Since this elementary area (dA) is at a distance (r sin θ) from x-axis then we can use the general formula (5.4) for calculating $\overline{Y}$ as,

$$\overline{Y}\int_A dA = \int_A y_c\,dA$$

where y_c is y-coordinate of the centroid of the elementary area dA.

$$\text{or } \overline{Y}\int_{r=0}^{r=R}\int_{\theta=0}^{\theta=\pi} r\,dr\,d\theta = \int_{r=0}^{r=R}\int_{\theta=0}^{\theta=\pi}(r\sin\theta)(r\,dr\,d\theta)$$

$$\text{or } \overline{Y}\left(\int_{r=0}^{r=R} r\,dr\right)\left(\int_{\theta=0}^{\theta=\pi} d\theta\right) = \left(\int_{r=0}^{r=R} r^2 dr\right)\left(\int_{\theta=0}^{\theta=\pi}\sin\theta\,d\theta\right)$$

$$\text{or } \overline{Y}\left[\frac{r^2}{2}\right]_0^R \times [\theta]_0^\pi = \left[\frac{r^3}{3}\right]_0^R \times [-\cos\theta]_0^\pi$$

$$\text{or } \overline{Y}\left[\frac{R^2}{2} - 0\right] \times [\pi - 0]$$

$$= \left[\frac{R^3}{3} - 0\right] \times [\cos 0 - \cos\pi]$$

$$\text{or } \overline{Y}\left[\frac{R^2}{2}\right] \times \pi = \frac{R^3}{3} \times \{1 - (-1)\}$$

or $$\overline{Y}\frac{1}{2}\pi R^2 = \frac{2}{3}R^3$$

or $$\overline{Y} = \frac{4R^3}{3} \times \frac{1}{\pi R^2}$$

$$= \frac{4R}{3\pi}$$

$$\therefore \quad \overline{Y} = \frac{4R}{3\pi} \qquad \text{....(5.6)}$$

***Second method*:** In the second method we make use of a single variable instead of two variables as used in the first method. Here we use the result obtained for the centroid of the triangle that centroid lies at a distance $\frac{2h}{3}$ from the vertex measured along the height of the triangle. Consider the Fig. (5.7). Since shown semi-circle is symmetrical about y-axis therefore $\overline{X} = 0$. We have to find $\overline{Y}$. Consider radial line OP = R at an angle θ measured in anticlockwise direction as shown in Fig. (5.7).

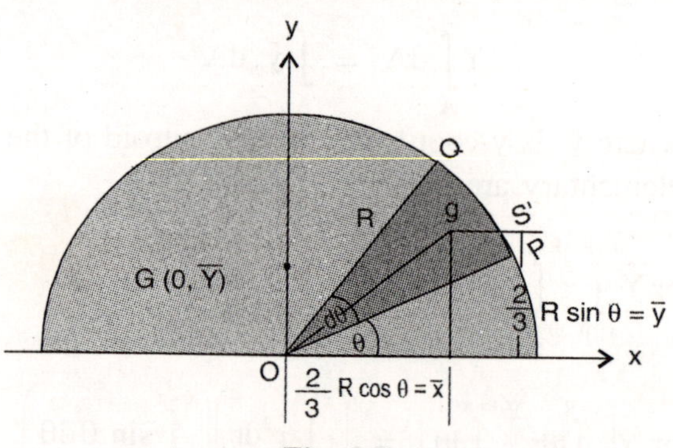

Fig. 5.7

Now increasing θ to θ + dθ draw radial line OQ. Since dθ is an infinitesimally small angle i.e., point Q is very close to P therefore we can treat arc (PS′Q) as a straight line thus OPQ is triangle of area dA = area of sector (OPQ) = $\frac{1}{2}$ $R^2 d\theta$, where θ measured in radian

$$\therefore \quad dA = \frac{1}{2}R^2\, d\theta \qquad \text{....(i)}$$

Now the centroid of this triangle of elementary area is $g \equiv \left(\frac{2}{3}R\cos\theta, \frac{2}{3}R\sin\theta\right)$.

Thus by knowing the elementary area and centroid of this and by using general equations (5.3), we can get required result by.

$$\overline{X}\int_A dA = \int_A x_c\, dA \qquad \text{.....(ii)}$$

Omit the calculation of this equation because we have got $\overline{X} = 0$ by geometry.

and $$\overline{Y}\int dA = \int y_c\, dA$$

or $$\overline{Y}\,A = \int_0^{\pi}\left(\frac{2}{3}R\sin\theta\right)\frac{1}{2}R^2 d\theta$$

or $$\overline{Y}\times\left(\frac{\pi R^2}{2}\right) = \frac{1}{3}R^3\int_0^{\pi}\sin\theta\, d\theta$$

or $$\overline{Y}\times\left(\frac{\pi R^2}{2}\right) = \frac{R^3}{3}\times[-\cos\theta]_0^{\pi}$$

or $$\overline{Y}\times\left(\frac{\pi R^2}{2}\right) = \frac{R^3}{3}\times[\cos 0 - \cos\pi]$$

or $$\overline{Y}\times\left(\frac{\pi R^2}{2}\right) = \frac{R^3}{3}\times 2$$

$$\therefore \quad \overline{Y} = \frac{4R}{3\pi}$$

Therefore the centroid of the semicircle is $G\left(\overline{X} = 0, \overline{Y} = \frac{4R}{3\pi}\right)$.

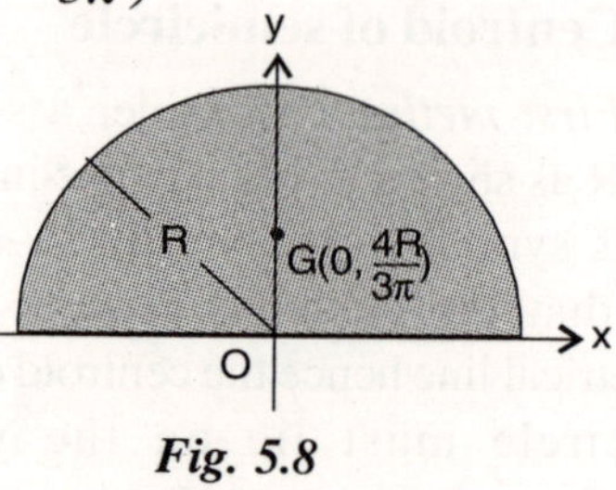

Fig. 5.8

Example 5.2: *Find out the centroid of the semicircle shown in Fig. (5.9).*

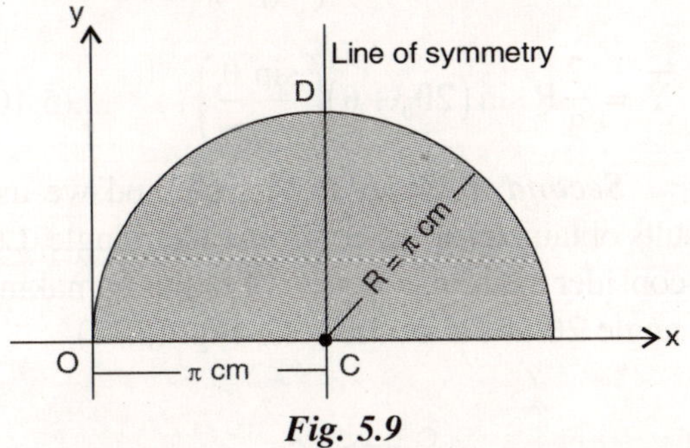

Fig. 5.9

Solution:

In the given coordinate system the co-ordinate of the centre of the semi-circle is at C (π, 0). Since CD is a symmetrical line for the semi-circle then the centroid of the semi-circle must lie on CD. Now since the x-coordinate of CD is fixed i.e., x = π. Using the result obtain above centroid of the semi-circle is,

$$G = \left(\overline{X}, \overline{Y}\right) = \left(\pi, \frac{4}{3}\right).$$

5.2.3 Centroid of circular section

First method: Consider a circular section of the radius (R) making an angle $2\theta_c$ at the centre as shown in the Fig. (5.10).

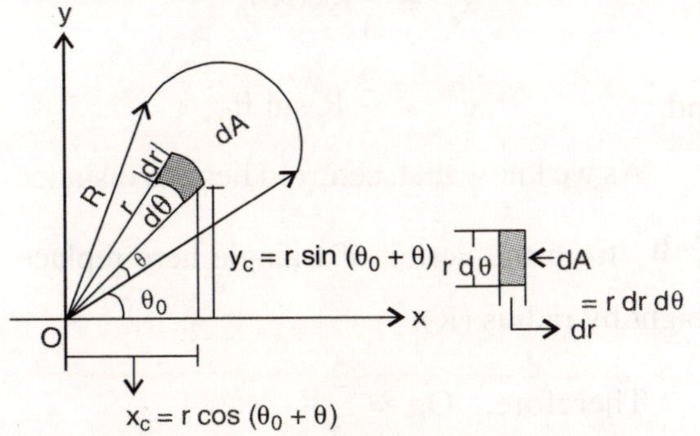

Fig. 5.10

Now refering to Fig (5.11) consider an intermediate stage of the differential area dA as shown. Since dθ is an infinitesimally small angle therefore one could treat r dθ as a straight line. Now dA is the area of the rectangle having r dθ and dr as its sides.

$$\begin{aligned} \text{Then, } dA &= \text{product of sides} \\ &= r\,d\theta \times dr \\ &= r\,dr\,d\theta \qquad \text{....(i)} \end{aligned}$$

Now the centroid of the elementary area dA is $g \equiv (x_c, y_c) = (r\cos(\theta_0 + \theta), r\sin(\theta_0 + \theta))$and by knowing the centroid of the elementary area one could find the centroid of circular sector with the help of equation (5.3) and (5.4) as discussed in section (5.1).

That is,

$$\overline{X}A_T = \int_A x_c\,dA \qquad \text{....(5.7)}$$

and

$$\overline{Y}A_T = \int_A y_c\,dA \qquad \text{....(5.8)}$$

where, $A_T = \int_A dA$ = total area of the circular section.

$$= \left(\frac{\pi R^2}{2\pi}\right) \times 2\theta_c \qquad [\theta_c \text{ in radian only}]$$

$$= R^2\theta_c$$

Now let proceed further on equation (5.7).

$$\text{i.e., } \overline{X}\left(R^2\theta\right) = \int_{\theta_0}^{\theta_0+2\theta}\int_{r=0}^{r=R} r\cos(\theta_0+\theta)\,r\,d\theta\,dr$$

$$= \int_{\theta_0}^{\theta_0+2\theta}\int_{\theta}^{R} \cos(\theta_0+\theta)\,r^2\,dr\,d\theta$$

$$= \int_{\theta_0}^{\theta_0+2\theta} \cos(\theta_0+\theta)\,d\theta \int_0^R r^2 dr$$

$$= \int_{\theta_0}^{\theta_0+2\theta} \cos(\theta_0+\theta)\,d\theta \left[\frac{r^3}{3}\right]_0^R$$

$$= \frac{R^3}{3}\int_{\theta_0}^{\theta_0+2\theta} \cos(\theta_0+\theta)\, d\theta$$

$$= \frac{R^3}{3}\left[\sin(\theta_0+\theta)\right]_{\theta_0}^{\theta_0+2\theta}$$

$$= \frac{R^3}{3}\left[\sin(\theta_0+\theta_0+2\theta)-\sin(\theta_0+\theta_0)\right]$$

$$= \frac{R^3}{3}\left[\sin(2\theta_0+2\theta)-\sin(2\theta_0)\right]$$

$$= \frac{R^3}{3}\left[2\cos\left(\frac{2\theta_0+2\theta+2\theta_0}{2}\right) \times\sin\left(\frac{2\theta_0+2\theta-2\theta_0}{2}\right)\right]$$

$$= \frac{R^3}{3}\left[2\cos(2\theta_0+\theta)\sin\theta\right]$$

$$= \frac{2}{3}R^3\cos(2\theta_0+\theta)\sin\theta$$

$$= \frac{\frac{2}{3}R^3\cos(2\theta_0+\theta)\sin\theta}{R^2\theta}$$

$$\therefore\ \overline{X} = \frac{2}{3}R\cos(2\theta_0+\theta)\left(\frac{\sin\theta}{\theta}\right) \quad \text{....(5.9)}$$

Now let proceed further on equation (5.8)

i.e., $$\overline{Y}(R^2\theta) = \int_0^R\int_{\theta_0}^{\theta_0+2\theta} r\sin(\theta_0+\theta)\, r\, dr\, d\theta$$

$$= \frac{R^3}{3}\left[-\cos(\theta_0+\theta)\right]_{\theta_0}^{\theta_0+2\theta}$$

$$= \frac{R^3}{3}\left[\cos 2\theta_0-\cos(2\theta_0+2\theta)\right]$$

$$= \frac{R^3}{3}\times 2\sin\left(\frac{2\theta_0+2\theta_0+2\theta}{2}\right)\sin\left(\frac{2\theta_0+2\theta-2\theta_0}{2}\right)$$

$$= \frac{R^3}{3}\times 2\times\sin(2\theta_0+\theta)\sin\theta$$

$$= \frac{\frac{2}{3}\times R^3\times\sin(2\theta_0+\theta)\sin\theta}{R^2\theta}$$

$$= \frac{2}{3}R\sin(2\theta_0+\theta)\left(\frac{\sin\theta}{\theta}\right)$$

$$\therefore\ \overline{Y} = \frac{2}{3}R\sin(2\theta_0+\theta)\left(\frac{\sin\theta}{\theta}\right) \quad \text{....(5.10)}$$

Second method: In this method we use results obtained for the centroid of a triangle. Let us consider a circular section of radius R making an angle 2θ at the centre refer Fig. (5.11).

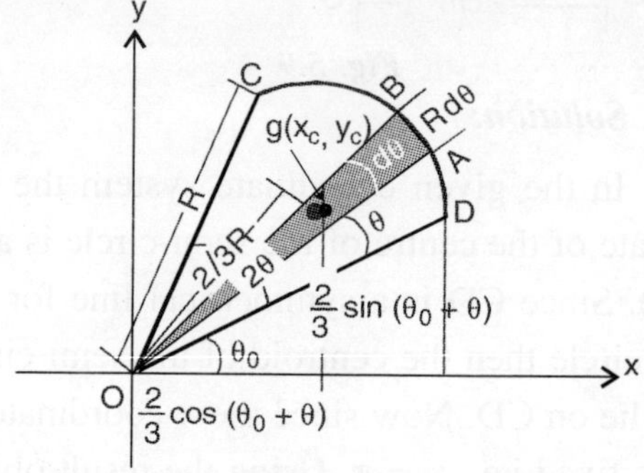

Fig. 5.11

Let us consider the elementary circular section of an infinitesimally small area dA as shown in the figure. Now we approximate the circular section (OABO) by a triangle of area, $dA = \frac{\pi R^2}{2\pi}\times d\theta = \frac{1}{2}R^2\, d\theta$ where, θ is in radian.

Let g (x_c, y_c) be the centroid of the elementary circular section. Then, we can get

$$x_c = \frac{2}{3}R\cos\theta_0$$

and $$y_c = \frac{2}{3}R\sin\theta_0.$$

As we know that, centroid lies at a distance $\frac{2}{3}h$ from the vertex of triangle here replace height by radius (R).

Therefore, $$Og = \frac{2}{3}R$$

and using right (angle) triangle OgM, we get x_c and y_c in terms of R and θ_0. Now using general equations, we can get

$$\overline{X}A_T = \oint_A x_c dA \qquad ....(5.11)$$

$$\text{or } \overline{X}(R^2\theta) = \int_{\theta_0}^{\theta_0+2\theta} \left(\frac{2}{3}R\cos(\theta_0+\theta)\right)\left(\frac{1}{2}R^2 d\theta\right)$$

where, $A_T = R^2\theta$ = Total area of circular section $\overline{OACO}$.

$$= \int_{\theta_0}^{\theta_0+2\theta} \frac{1}{3}R^3\cos(\theta_0+\theta)d\theta$$

$$= \frac{1}{3}R^3\left[\sin(\theta_0+\theta)\right]_{\theta_0}^{\theta_0+2\theta}$$

$$= \frac{1}{3}R^3\left[\sin(\theta_0+\theta_0+2\theta)-\sin(\theta_0+\theta_0)\right]$$

$$= \frac{1}{3}R^3\left[2\cos\left(\frac{2\theta_0+2\theta+2\theta_0}{2}\right)\sin\left(\frac{2\theta_0+2\theta-2\theta_0}{2}\right)\right]$$

$$= \frac{1}{3}R^3\left[2\cos(2\theta_0+\theta)\sin\theta\right]$$

$$= \frac{2}{3}R^3\cos(2\theta_0+\theta)\sin\theta$$

$$= \frac{\frac{2}{3}\times R^3\times\cos(2\theta_0+\theta)\sin\theta}{R^2\theta}$$

$$\therefore \overline{X} = \frac{2}{3}R\cos(2\theta_0+\theta)\left(\frac{\sin\theta}{\theta}\right)$$

$$\text{and} \quad \overline{Y}A_T = \int_{\theta_0}^{\theta_0+2\theta}\frac{2}{3}R\sin(\theta+\theta)\frac{1}{2}R^2 d\theta$$

$$= \frac{1}{3}R^3\int_{\theta_0}^{\theta_0+2\theta}\sin(\theta_0+\theta)d\theta$$

$$= \frac{1}{3}R^3\left[-\cos(\theta_0+\theta)\right]_{\theta_0}^{\theta_0+2\theta}$$

$$= \frac{1}{3}R^3\left[\cos(\theta_0+\theta_0)-\cos(\theta_0+\theta_0+2\theta)\right]$$

$$= \frac{1}{3}R^3\left[\cos 2\theta_0-\cos(2\theta_0+2\theta)\right]$$

$$= \frac{1}{3}R^3\left[2\sin\left(\frac{2\theta_0+2\theta_0+2\theta}{2}\right)\sin\left(\frac{2\theta_0+2\theta-2\theta}{2}\right)\right]$$

$$= \frac{1}{3}R^3\left[2\sin(2\theta_0+\theta)\sin\theta\right]$$

$$\text{or} \quad \overline{Y}(R^2\theta) = \frac{1}{3}\times R^3\times 2\times\sin(2\theta_0+\theta)\sin\theta$$

$$\text{or} \quad \overline{Y} = \frac{2R^3\sin(2\theta_0+\theta)\sin\theta}{3\times R^2\theta}$$

$$= \frac{2R\sin(2\theta_0+\theta)\sin\theta}{3\theta}$$

$$= \frac{2}{3}R\sin(2\theta_0+\theta)\frac{\sin\theta}{\theta}$$

$$\therefore \quad \overline{Y} = \frac{2}{3}R\sin(2\theta_0+\theta)\left(\frac{\sin\theta}{\theta}\right)$$

Therefore the centroid of the circular section is,

$$G \equiv (\overline{X}, \overline{Y})$$

$$\equiv \left\{\frac{2}{3}R\cos(2\theta_0+\theta)\left(\frac{\sin\theta}{\theta}\right), \frac{2}{3}R\sin(2\theta_0+\theta)\left(\frac{\sin\theta}{\theta}\right)\right\}$$

***Third method*:** In this method we use results obtained for the centroid of the circular arc i.e., equations (5.11) and (5.12) as discussed in section (5.2.3).

Let us consider a circular sector of radius R of an area $R^2\theta$, (θ in radian measurement) making an angle of 2θ at the centre refer Fig. 5.12.

Consider the intermediate arc $\overline{AB}$ of length $2r\theta$, here the arc $\overline{AB}$ could be expressed in the form of the ring of an area $dA = 2r\,\theta\,dr$.

Now for the circular arc $\overline{AB}$ we know the centroid is,

$$x_c = \frac{r\cos(2\theta_0+\theta)\sin\theta}{\theta} \qquad ....(5.12)$$

and $$y_c = \frac{r\sin(2\theta_0+\theta)\sin\theta}{\theta} \quad ....(5.13)$$

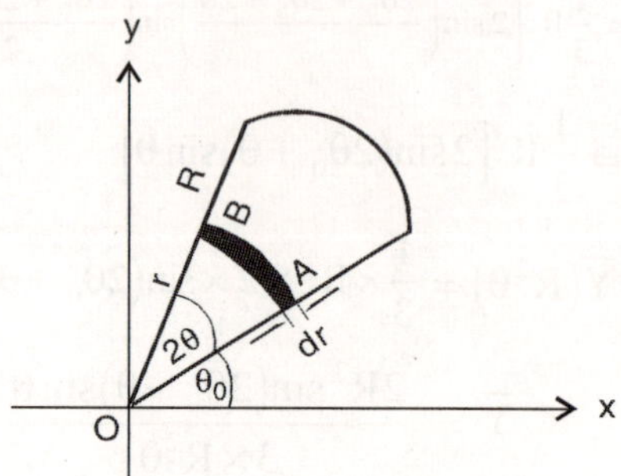

Fig. 5.12

Now knowing dA, x_c and y_c we can find centroid G $(\overline{X},\overline{Y})$ for the circular section using general equations.

i.e., $$\overline{X}\,A_T = \int_A x_c\,dA$$

or $$\overline{X}(R^2\theta) = \int_A \frac{r\cos(2\theta_0+\theta)\sin\theta}{\theta}\,dA$$

or $$\overline{X}(R^2\theta) = \int_0^R \frac{r\cos(2\theta_0+\theta)\sin\theta}{\theta}\,2r\theta\,dr$$

$$= \int_0^R \frac{2r^2\theta\cos(2\theta_0+\theta)\sin\theta}{\theta}\,dr$$

$$= 2\cos(2\theta_0+\theta)\sin\theta\int_0^R r^2\,dr$$

$$= 2\cos(2\theta_0+\theta)\sin\theta\times\left[\frac{r^3}{3}\right]_0^R$$

$$= \frac{2}{3}\cos(2\theta_0+\theta)\sin\theta\times\left[\frac{R^3}{3}-0\right]$$

$$= \frac{2}{3}R^3\cos(2\theta_0+\theta)\sin\theta$$

$$\therefore\quad \overline{X} = \frac{2R\cos(2\theta_0+\theta)\sin\theta}{3\theta}$$

and $$\overline{Y}\,A_T = \int_A y_c\,dA$$

or $$\overline{Y}(R^2\theta) = \int_0^R \frac{r\sin(2\theta_0+\theta)\sin\theta}{\theta}\,2r\theta\,dr$$

$$\therefore\quad \overline{Y} = \frac{2R\sin(2\theta_0+\theta)\sin\theta}{3\theta}$$

Therefore G = centroid of circular section

$$\equiv (\overline{X},\overline{Y})$$

$$\equiv\left\{\frac{2R\cos(2\theta_0+\theta)\sin\theta}{3\theta},\frac{2R\sin(2\theta_0+\theta)\sin\theta}{3\theta}\right\}$$

Finally we see that in the first method the radius (r) and the angle (θ) both are varying over the required limit i.e., $0 \le r \le R$ and $\theta_0 \le \theta \le \theta_0 + 2\theta$. In the second method, the radius (r = R) is constant and angle (θ) is varying in the interval $\theta_0 \le \theta \le \theta_0 + 2\theta$, and in the third method angle (θ) is kept constant and radius (r) is varying from $0 \le r \le R$.

From all the above discussed methods, we get the same result for the circular section shown in (5.14),

$$\overline{X} = \frac{2R\cos(2\theta_0+\theta)\sin\theta}{3\theta}$$

and $$\overline{Y} = \frac{2R\sin(2\theta_0+\theta)\sin\theta}{3\theta}$$

In different cases we use this general result to find the centroid for semicircles, quarter circle and circle.

Case 1**:** When $\theta_0 = 0$ and $2\theta = \frac{\pi}{2}$ i.e., $\theta = \frac{\pi}{4}$, quarter circle.

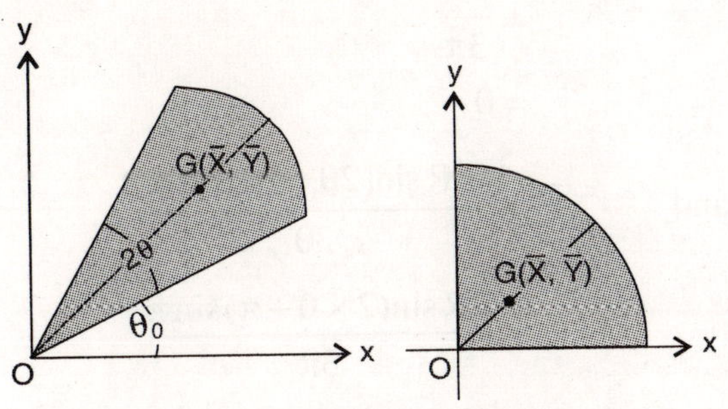

Fig. 5.13 *Fig. 5.14*

$$\overline{X} = \frac{2R\cos(2\theta_0 + \theta)\sin\theta}{3\theta}$$

$$= \frac{2R\cos\left(2\times 0 + \frac{\pi}{4}\right)\sin\frac{\pi}{4}}{3\times\frac{\pi}{4}}$$

$$= \frac{2R\cos\left(\frac{\pi}{4}\right)\times\sin\frac{\pi}{4}}{3\times\frac{\pi}{4}}$$

$$= \frac{2R\times\frac{1}{\sqrt{2}}\times\frac{1}{\sqrt{2}}}{3\times\frac{\pi}{4}}$$

$$= \frac{R}{3\times\frac{\pi}{4}}$$

$$= \frac{4R}{3\pi}$$

and $$\overline{Y} = \frac{2R\sin(2\theta_0 + \theta)\sin\theta}{3\theta}$$

$$= \frac{2R\times\sin\left(2\times 0 + \frac{\pi}{4}\right)\sin\left(\frac{\pi}{4}\right)}{3\times\frac{\pi}{4}}$$

$$= \frac{2R\times\sin\frac{\pi}{4}\times\sin\frac{\pi}{4}}{3\times\frac{\pi}{4}}$$

$$= \frac{4R}{3\pi}$$

$$\therefore \quad G \equiv \left(\overline{X}, \overline{Y}\right)$$

$$\equiv \left(\frac{4R}{3\pi}, \frac{4R}{3\pi}\right).$$

Case 2: When $\theta_0 = 0$ and $2\theta = \pi$ i.e.,

$\theta = \frac{\pi}{2}$; semicircle.

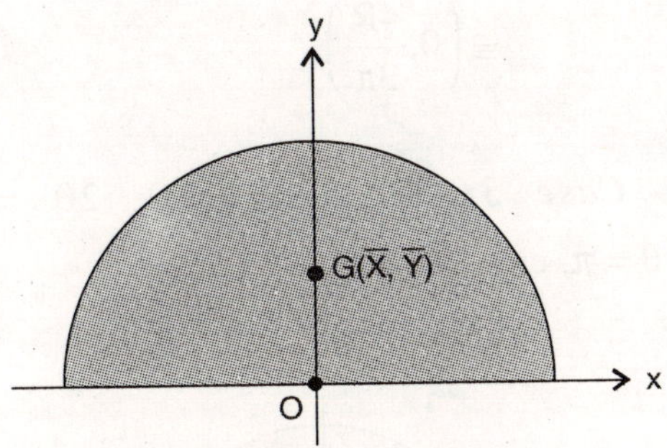

Fig. 5.15

$$\overline{X} = \frac{2R\cos(2\theta_0 + \theta)\sin\theta}{3\theta}$$

$$= \frac{2R\cos\left(2\times 0 + \frac{\pi}{2}\right)\sin\left(\frac{\pi}{2}\right)}{3\times\frac{\pi}{2}}$$

$$= \frac{2R\cos\frac{\pi}{2}\times\sin\frac{\pi}{2}}{3\times\frac{\pi}{2}}$$

$$= 0$$

and $$\overline{Y} = \frac{2R\sin(2\theta_0 + \theta)\sin\theta}{3\theta}$$

$$= \frac{2R\sin\left(2\times 0+\frac{\pi}{2}\right)\times \sin\frac{\pi}{2}}{3\times\frac{\pi}{2}}$$

$$= \frac{2R\sin\frac{\pi}{2}\times \sin\frac{\pi}{2}}{3\times\frac{\pi}{2}}$$

$$= \frac{2R}{3\frac{\pi}{2}}$$

$$= \frac{4R}{3\pi}$$

$$\therefore\ G \equiv \left(\overline{X}, \overline{Y}\right)$$

$$\equiv \left(0, \frac{4R}{3\pi}\right)$$

***Case 3*:** When $\theta_0 = 0$, $2\theta = 2\pi$ or $\theta = \pi$, circle.

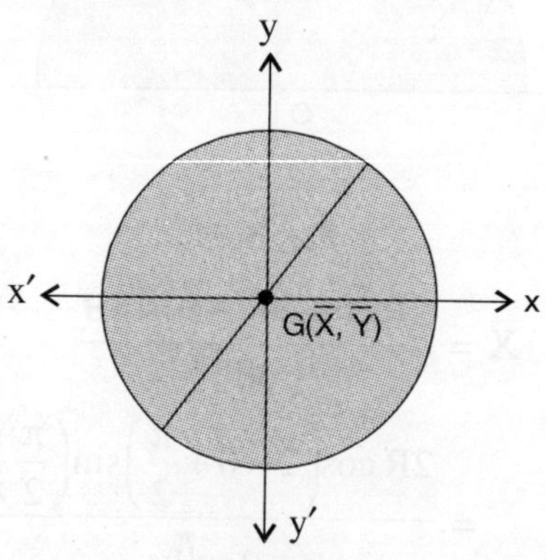

Fig. 5.16

$$\overline{X} = \frac{2R\cos(2\theta_0+\theta)\sin\theta}{3\theta}$$

$$= \frac{2R\cos(2\times 0+\pi)\sin\pi}{3\times\pi}$$

$$= \frac{2R\cos\pi.\sin\pi}{3\pi}$$

$$= \frac{2Rx(-1)\times 0}{3\pi}$$

$$= \frac{0}{3\pi}$$

$$= 0$$

and $$\overline{Y} = \frac{2R\sin(2\theta_0+\theta)\sin\theta}{3\theta}$$

$$= \frac{2R\sin(2\times 0+\pi)\sin\pi}{3\pi}$$

$$= \frac{2R\sin\pi\times\sin\pi}{3\pi}$$

$$= \frac{2R\times 0\times 0}{3\pi}$$

$$= \frac{0}{3\pi}$$

$$= 0$$

$$\therefore\ G \equiv \left(\overline{X}, \overline{Y}\right)$$

$$\equiv (0, 0).$$

***Case 4*:** When $\theta_0 = 0$, circular section as shown in Fig. (5.18).

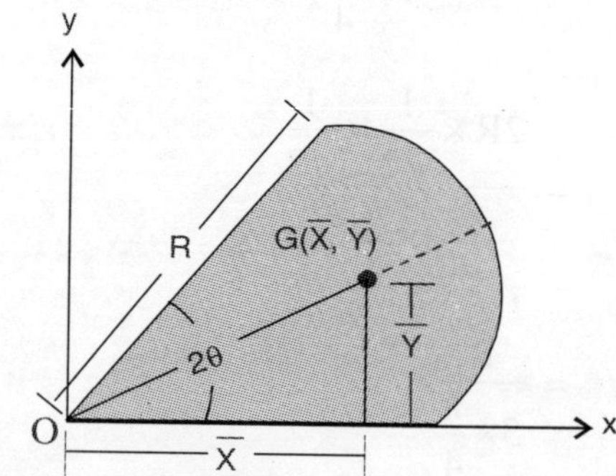

Fig. 5.17

Now we know,

$$\overline{X} = \frac{2R\cos(2\theta_0+\theta)\sin\theta}{3\theta}$$

$$= \frac{2R\{\cos(2\times 0+\theta)\}\sin\theta}{3\theta}$$

$$= \frac{2R\cos\theta\sin\theta}{3\theta}$$

and $$\overline{Y} = \frac{2R\sin(2\theta_0+\theta)\sin\theta}{3\theta}$$

$$= \frac{2R\sin(2\times 0+\theta)\sin\theta}{3\theta}$$

$$= \frac{2R\sin\theta\sin\theta}{3\theta}$$

$$\therefore OG^2 = \overline{X}^2 + \overline{Y}^2$$

$$= \left(\frac{2R\cos\theta\sin\theta}{3\theta}\right)^2 + \left(\frac{2R\sin\theta\sin\theta}{3\theta}\right)^2$$

$$= \frac{4R^2}{9\theta^2}\left[\cos^2\theta\sin^2\theta + \sin^2\theta\sin^2\theta\right]$$

$$= \frac{4R^2}{9\theta^2}\left[\sin^2\theta(\cos^2\theta + \sin^2\theta)\right]$$

$$= \frac{4R^2}{9\theta^2}\times\left[\sin^2\theta\right]$$

or $$OG = \left[\frac{4R^2\sin^2\theta}{9\theta^2}\right]^{\frac{1}{2}}$$

$$\therefore OG = \frac{2R\sin\theta}{3\theta}$$

Now rotate the circular section such that the x-axis bisects the section. Since the centroid always lies on the line of symmetry, therefore $\overline{Y} = 0$ and G′ becomes to G.

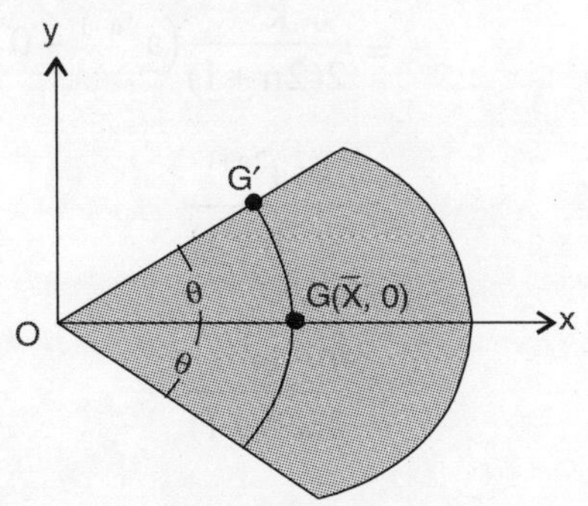

Fig. 5.18

Thus, $$\overline{X} = OG' = OG$$

$$= \frac{2R\sin\theta}{3\theta}$$

5.2.4 Centroid of parabolic spandrel of nth Order

Consider the parabolic spandrel of order n shown in Fig. 5.19. Height of the elementary strip at a distance x along x-axis from origin is $y = kx^n$.

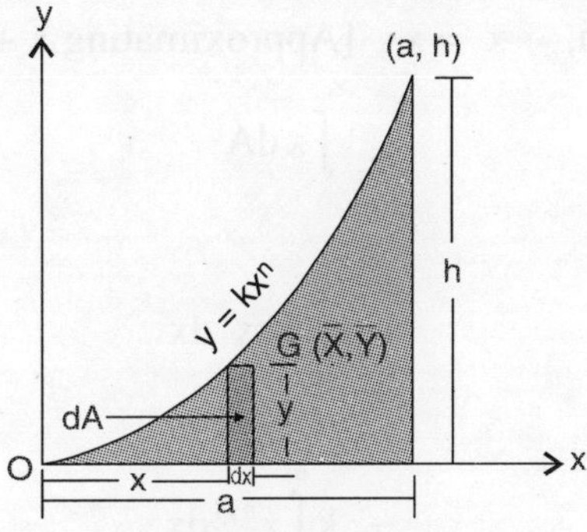

Fig. 5.19

Therefore area of the elementary strip,

$$dA = ydx$$
$$= kx^n dx$$

Therefore total area (A_T) bounded by the curve $y = kx^n$ is,

$$A_T = \int_A dA$$

$$= \int_0^a kx^n\,dx$$

$$= k\left[\frac{x^{n+1}}{n+1}\right]_0^a$$

$$= \frac{k}{n+1}\left[a^{n+1} - 0\right]$$

$$= \frac{ka^{n+1}}{n+1}$$

$$= \frac{ka\times a^n}{n+1}$$

$$= \frac{a(ka^n)}{n+1}$$

$$= \frac{ah}{n+1} \qquad [h = ka^n]$$

Now as we know that,

$$\overline{X}\,A_T = \int_A x_c dA;$$

where, x_c is the centroid of the elementary rectangular strip.

Then, $x_c = x$ [Approximating x + dx by x]

$$= \int_A x\,dA$$

$$= \int_0^a x\,kx^n dx$$

$$= k\int_0^a x^{n+1} dx$$

$$= k\left[\frac{x^{n+1+1}}{n+1+1}\right]_0^a$$

$$= \left[\frac{kx^{n+2}}{n+2}\right]_0^a$$

$$= \frac{ka^{n+2}}{n+2}$$

$$= \frac{a^2(ka^n)}{n+2}$$

$$= \frac{a^2h}{n+2}$$

or $\overline{X}\times\left(\frac{ah}{n+1}\right) = \frac{a^2h}{n+2}$

$$\therefore \quad \overline{X} = \frac{a^2h}{n+2}\times\frac{n+1}{ah}$$

$$= \frac{a(n+1)}{n+2}$$

$$= \left(\frac{n+1}{n+2}\right)a$$

$$\overline{X} = \left(\frac{n+1}{n+2}\right)a \qquad \text{....(5.14)}$$

Similarly, as we know that $\overline{Y}\,A_T = \int_A y_c dA;$

where, y_c is the y coordinate of the centroid of the elementary rectangular strip.

Then, $y_c = \frac{y}{2}$

$$= \int_A \frac{y}{2} dA$$

$$= \frac{1}{2}\int_0^a kx^n \times kx^n dx$$

$$= \frac{k^2}{2}\int_0^a x^{2n} dx$$

$$= \frac{k^2}{2}\left[\frac{x^{2n+1}}{2n+1}\right]_0^a$$

$$= \frac{k^2}{2(2n+1)}(a^{2n+1} - 0)$$

$$= \frac{k^2a^{2n+1}}{2(2n+1)}$$

$$= \frac{a.(ka^n)^2}{2(2n+1)}$$

$$= \frac{ah^2}{2(2n+1)}$$

or $\overline{Y}\times\frac{ah}{n+1} = \frac{ah^2}{2(2n+1)}$

$$\therefore \quad \overline{Y} = \frac{ah^2}{2(2n+1)} \times \frac{n+1}{ah}$$

$$= \frac{h}{2}\left(\frac{n+1}{2n+1}\right) \qquad .....(5.15)$$

Thus the centroid (G) = $(\overline{X}, \overline{Y})$

$$= \left[\frac{n+1}{n+2}a, \frac{n+1}{(2n+1)}\frac{h}{2}\right]$$

Table 5.1

Order (n) of spandrel	Cartesian equation y = f(x)	$\overline{X}$	$\overline{Y}$	$G(\overline{X}, \overline{Y})$
First (n = 1)	y = kx	$\frac{2a}{3}$	$\left(\frac{2h}{3}\right)\frac{1}{2}$	$\left(\frac{2a}{3}, \frac{h}{3}\right)$
Second (n = 2)	$y = kx^2$	$\frac{3a}{4}$	$\left(\frac{3h}{5}\right)\frac{1}{2}$	$\left(\frac{3a}{4}, \frac{3h}{10}\right)$
Third (n = 3)	$y = kx^3$	$\frac{4a}{5}$	$\left(\frac{4h}{7}\right)\frac{1}{2}$	$\left(\frac{4a}{5}, \frac{2h}{7}\right)$

5.2.5 General procedure to find the centroid of an area bounded by a curve f(x, y, c) = 0 and ordinates at abscissa $x = x_1$ and $x = x_2$ where, $x_1 < x_2$

Let us consider for the curve shown in Fig. (5.20) for the certesian function f (x, y, c) =0. Now there are two methods in the first one consider the elementary rectangular strip of elementary area, dA perpendicular to x-axis at an intermediate position i.e., $x_1 < x < x_2$.

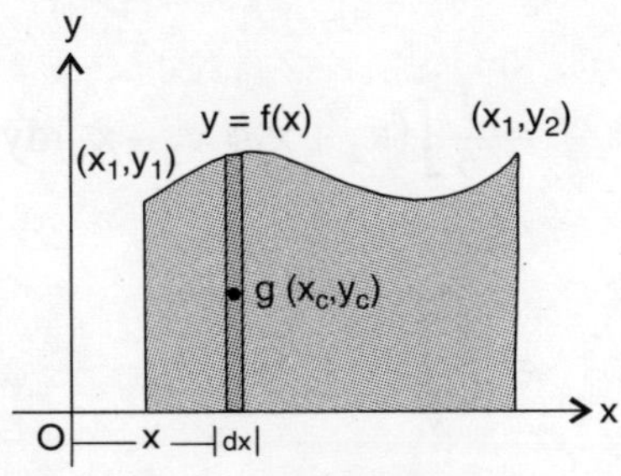

Fig. 5.20

We get dA = f(x) dx. Now x_c = x-coordinate of the centroid of the elementary rectangular area = x and y_c = y-coordinate of the centroid of the elementary rectangular area $= \frac{f(x)}{2}$. Now knowing x_c, y_c and dA we could use the general formula for finding $\overline{X}$ and $\overline{Y}$

that is, $$\overline{X}\int_A dA = \int_A x_c\, dA$$

or $$\overline{X}\int_{x_1}^{x_2} f(x)dx = \int_{x_1}^{x_2} x f(x)dx$$

Similarly, $$\overline{Y}\int_A dA = \int_A y_c dA$$

or $$\overline{Y}\int_{x_1}^{x_2} f(x)dx = \int_{x_1}^{x_2} \frac{f(x)}{2} f(x)dx$$

$$= \frac{1}{2}\int_{x_1}^{x_2} f^2(x)dx$$

$$\text{or} \qquad \overline{Y} = \frac{\int_{x_1}^{x_2} f^2(x)\,dx}{\int_{x_1}^{x_2} f(x)\,dx}$$

Therefore coordinates of centroid that is $\overline{X}$ and $\overline{Y}$ are given by the equation (5.16).

$$\left.\begin{aligned} \overline{X} &= \frac{\int_{x_1}^{x_2} x f(x)\,dx}{\int_{x_1}^{x_2} f(x)\,dx} \\ \text{and} \quad \overline{Y} &= \frac{\frac{1}{2}\int_{x_1}^{x_2} \{f(x)\}^2\,dx}{\int_{x_1}^{x_2} f(x)\,dx} \end{aligned}\right\} \quad \text{....(5.16)}$$

When a rectangular strip is parallel to x-axis.

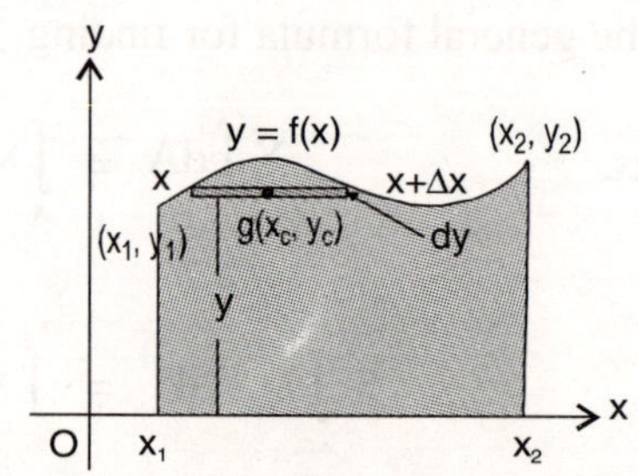

Fig. 5.21

Now, $dA = \Delta x\, dy$ = the area of the elementary strip which is parallel to x-axis as shown in Fig. (5.21). Let g $x_c y_c$ be the centroid of the elementary strip.

$$\text{Then,} \qquad x_c = x + \frac{\Delta x}{2}$$

$$\text{and} \qquad y_c = y$$

$$\text{Therefore,} \quad \overline{X} = \frac{\int_A x_c\, dA}{\int_A dA}$$

$$= \frac{\int_{y_1}^{y_2} \left(x + \frac{\Delta x}{2}\right)\Delta x\, dy}{A_T}$$

$$\text{or} \qquad \overline{X} A_T = \frac{1}{2}\int_{y_1}^{y_2} (2x + \Delta x)\Delta x\, dy$$

$$\text{Similarly,} \quad \overline{Y} A_T = \int_{y_1}^{y_2} y \Delta x dy$$

$$\left.\begin{aligned} \therefore \quad \overline{X} &= \frac{\int_{y_1}^{y_2} \left(x + \frac{\Delta x}{2}\right)\Delta x\, dy}{A_T} \\ \text{and} \quad \overline{Y} &= \frac{\int_{y_1}^{y_2} y \Delta x\, dy}{A_T} \end{aligned}\right\} \quad \text{....(5.17)}$$

where symbols have their own meaning.

$$A_T \overline{Y} = \int_{y_1}^{y_2} \left(x_1 + \frac{x_2 - x_1}{2}\right)\left(\frac{x_2 - x_1}{2}\right) dy$$

$$= \frac{1}{2}\int_{y_1}^{y_2} (x_2 + x_1)(x_2 - x_1)\, dy$$

$$= \frac{1}{2}\int_{y_1}^{y_2} (x_2^2 - x_1^2)\, dy \qquad \text{....(5.18)}$$

$$\text{Similarly,} \quad A_T \overline{X} = \int_{y_1}^{y_2} y \Delta x\, dy$$

$$= \int_{y_1}^{y_2} y(x_2 - x_1)\,dy. \qquad ...(5.19)$$

Example 5.3: *Find the centroid of the area bounded by the curve y = f(x) = x, x-axis and ordinates at x = 0 and x = 2.*

Solution:

By drawing the curve y = x, we get the following diagram:

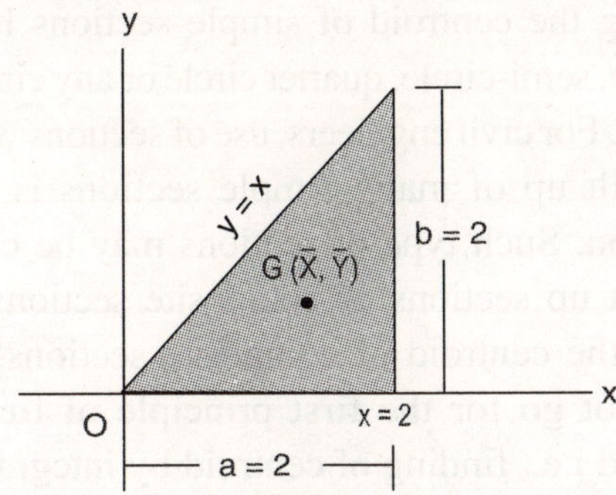

Fig. 5.22

First method: Let the elementary strip be parallel to the y-axis.

Then, $$\overline{X}\,A_T = \int_{x_1}^{x_2} x f(x)\,dx$$

$$= \int_0^2 xy\,dx$$

$$= \int_0^2 x^2\,dx$$

$$[\because\ f(x) = y = x]$$

$$= \left[\frac{x^3}{3}\right]_0^2$$

$$= \frac{1}{3}\left[2^3 - 0\right]$$

$$= \frac{2^3}{3}$$

or $$2\overline{X} = \frac{2^3}{3}$$

or $$\overline{X} = \frac{4}{3}$$

$$= \frac{2a}{3}$$

and $$\overline{Y}\,A_T = \frac{1}{2}\int_{x_1}^{x_2}\left[f(x)\right]^2 dx$$

$$= \frac{1}{2}\int_0^2 y^2\,dx$$

$$= \frac{1}{2}\int_0^2 x^2\,dx$$

$$= \frac{1}{2}\times\left[\frac{x^3}{3}\right]_0^2$$

$$= \frac{4}{3}$$

$$\therefore \quad \overline{Y} = \frac{2}{3}$$

$$= \frac{b}{3}$$

Therefore $G \equiv (\overline{X}, \overline{Y}) = \left(\frac{4}{3}, \frac{2}{3}\right)$ we see that the above result is also matches with the result obtain for a triangle, i.e., equation (5.5) in section (5.2.1).

Second method: Let the elementary strip be parallel to x-axis.

Then, $$\overline{Y}\,A_T = \frac{1}{2}\int_{y_1}^{y_2}\left(x_2^2 - x_1^2\right)dy$$

$$= \frac{1}{2}\int_0^2 (x_2^2 - 0)\,dy$$

[Put $x_2 = x$ and $x_1 = 0$]

$$= \frac{1}{2}\int_0^2 x^2 dy$$

$$= \frac{1}{2}\int_0^2 y^2 dy \qquad [\because y = x]$$

$$= \frac{1}{3}[y]_0^2 \times \frac{1}{2}$$

$$= \frac{1}{2} \times \frac{1}{3}[2^3 - 0]$$

$$= \frac{4}{3}$$

or $\quad \overline{Y} = \frac{2}{3}$

$$= \frac{b}{3}$$

Similarly, $\overline{X}\, A_T = \int_{y_1}^{y_2} y(x_2 - x_1)\, dy$

$$= \int_0^2 y(x_2 - 0)\, dy$$

$$= \int_0^2 y x_2\, dy$$

$$= \int_0^2 y x\, dy$$

$$2\overline{X} = \int_0^2 y^2\, dy$$

$$= \left[\frac{y^3}{3}\right]_0^2$$

$$= \frac{8}{3}$$

$$\therefore \quad \overline{X} = \frac{4}{3}$$

$$= \frac{2}{3} \times 2$$

$$= \frac{2}{3} \times h$$

$$= \frac{2}{3} \times a.$$

5.3 CENTROID OF COMPOSITE SECTIONS

So far, the discussion was confined to locating the centroid of simple sections like a triangle, semi-circle, quarter circle or any circular section. For civil engineers, use of sections which are built up of many simple sections is very common. Such type of sections may be called as built up sections or composite sections. To locate the centroid of composite sections, one need not go for the first principle of finding centroid i.e., finding of centroid by integration. For finding the centroid of given composite section one can split the given composite section into suitable simple sections and centroid of the composite section can be found by using the standard formula for them listed in table (5.2). As per definition of centroid we can assume the total area of a simple section is concentrated at its centroid. Now its moment about an axis (moment = area × distance of point of area from axis) can be found. After determining the moment of each area about the axes, add them. Then the centroid of the composite section could be found by dividing the moments by the total area of the composite section.

Let us consider a composite section C. This section could be split into the sections C_1, C_2, C_3, C_4 having their centroids g_1 (x_1, y_1), g_2 (x_2, y_2), g_3 (x_3, y_3) and g_4 (x_4, y_4) as shown in Fig. (5.23), where G $\left(\overline{X}, \overline{Y}\right)$ is the centroid of the composite section.

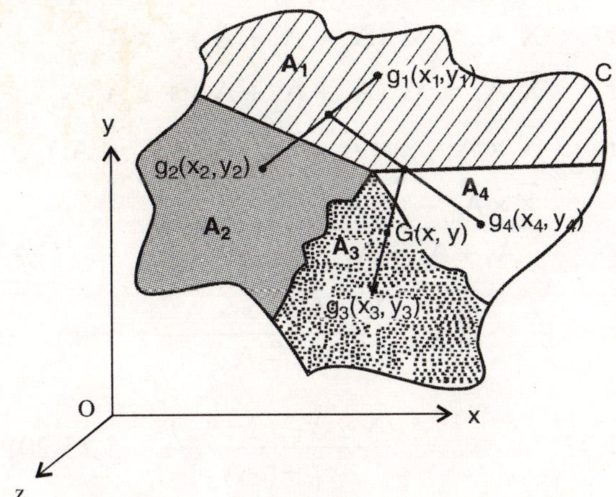

Fig. 5.23

Now the areas of sections $C_1 : A_1$, $C_2 : A_2$, $C_3 : A_3$ and $C_4 : A_4$ could be assumed to be concentrated at their centroids i.e., assume A_1, A_2, A_3 and A_y are concentrated at g_1, g_2, g_3 and g_4 respectively.

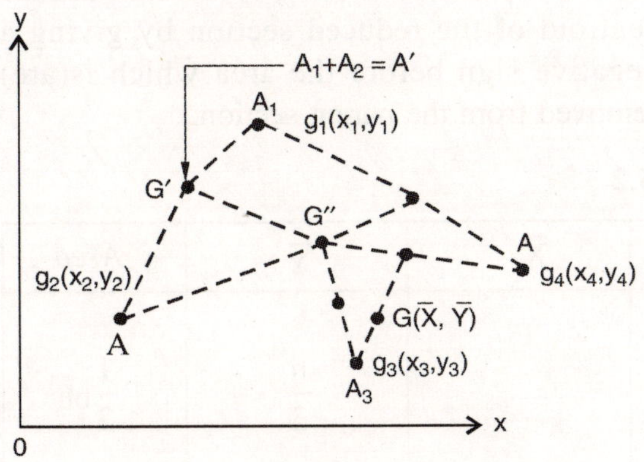

Fig. 5.24

Now A_1 and A_2 could be assumed to be concentrated at G′ where,

$$G'_x = \frac{A_1x_1 + A_2x_2}{A_1 + A_2} = x'$$

and $$G'_y = \frac{A_1y_1 + A_2y_2}{A_1 + A_2} = y'$$

Now $A_1 + A_2 = A'$ and A_4 can be assumed to be concentrated at G″ where,

$$G''_x = \frac{x'A' + x_4A_4}{A' + A_4}$$

$$= \frac{A_1x_1 + A_2x_2 + A_4x_4}{A_1 + A_2 + A_4}$$

Similarly $$G''_y = \frac{A_1y_1 + A_2y_2 + A_4y_4}{A_1 + A_2 + A_4}$$

Now at $G''_x(x'', y'')$ one could suppose that areas A_1, A_2, and A_4 are concentrated i.e., $A_1 + A_2 + A_4 = A''$ (say).

Similarly $A_1 + A_2 + A_4 = A''$ and A_3 could assume to be concentrated at $G\left(\overline{X}, \overline{Y}\right)$ where,

$$\overline{X} = \frac{x''A'' + A_3x_3}{A'' + A_3}$$

$$= \frac{\left(\dfrac{A_1x_1 + A_2x_2 + A_4x_4}{A_1 + A_2 + A_3}\right)A'' + x_3A_3}{A_1 + A_2 + A_4 + A_3}$$

$$\therefore \quad \overline{X} = \frac{A_1x_1 + A_2x_2 + A_3x_3 + A_4x_4}{A_1 + A_2 + A_3 + A_4} \quad ....(i)$$

and $$\overline{Y} = \frac{A_1y_1 + A_2y_2 + A_3y_3 + A_yy_4}{A_1 + A_2 + A_3 + A_4} \quad ....(ii)$$

Therefore the entire area $A = A_1 + A_2 + A_3 + A_4$ is assumed to be concentrated at $G\left(\overline{X}, \overline{Y}\right)$. Where $\overline{X}$ and $\overline{Y}$ are given by equation (i) and equation (ii) respectively.

For n-sections equation (i) and equation (ii) could be extended to n-terms.

$$\overline{X} = \frac{A_1x_1 + A_2x_2 + A_3x_3 + ... + A_nx_n}{A_1 + A_2 + A_3 .... + A_n}$$

$$= \frac{\Sigma x_1A_1}{\Sigma A_1}$$

and $$\overline{Y} = \frac{A_1y_1+A_2y_2+A_3y_3+....+A_ny_n}{A_1+A_2+A_3+....+A_n}$$

$$= \frac{\Sigma y_1 A_1}{\Sigma A_1}.$$

Now by knowing the centroid of the composite section C we can find the centroid of a sub-section. Say $C_1 = A_1 + A_2 + A_3$. Consider that G $\left(\overline{X}, \overline{Y}\right)$ is the centroid of section C and centroid of $C_1 : A_1 + A_2 + A_3$ is G_1 (x', y') and that of section A_4 is g(x, y). Now the composition of C_1 and A_4 is section C itself. Then,

$$\overline{X} = \frac{x_1'(A_1 + A_2 + A_3) + xA_4}{A_1 + A_2 + A_3 + A_4}$$

or $$\overline{X} = \frac{x_1'(A_1 + A_2 + A_3) + xA_4}{A_1 + A_2 + A_3 + A_4}$$

or $$\overline{X}(A_1 + A_2 + A_3 + A_4) = x_1' \ (A_1 + A_2 + A_3) + x\,A_4$$

or $$\overline{X}\,A_T = x_1' \ (A_1 + A_2 + A_3) + x\,A_4$$

or $$\overline{X}\,A_T = x_1' \ (A_1 + A_2 + A_3) + x\,A_4$$

or $$\overline{X} \times A_T - x \times A_4 = x_1' \ (A_1 + A_2 + A_3)$$

$$\frac{\overline{X} \times A_T - x \times A_4}{A_1 + A_2 + A_3} = x_1'$$

or $$x_1' = \frac{\overline{X} \times A_T - x \times A_4}{A_T - A_4}$$

$\therefore$ $$x_1' = \frac{\overline{X} \times A_T + (-A_4)x}{A_T + (-A_4)} \quad ....(5.20)$$

Similarly, $$y_1' = \frac{\overline{Y} \times A_T + (-A_4)y}{A_T + (-A_4)} \quad ....(5.21)$$

where, $A_T = \Sigma A_1$.

Since we can take section C_1 as reduced section of C by A_4, therefore knowing the centroid of whole section (C) we can find the centroid of the reduced section by giving a negative sign before the area which is(are) removed from the parent section.

Table 5.2

Section	*Figure*	$\overline{X}$	$\overline{Y}$	*Area*
Triangle	h; $\frac{h}{3}$; x'; x	—	$\frac{h}{3}$	$\frac{1}{2}bh$
Semicircle	y; G; $\frac{4R}{3\pi}$; R; x; O	0	$\frac{4R}{3\pi}$	$\frac{\pi R^2}{2}$
Quartercircle	y; G; $\frac{4R}{3\pi}$; O; $\frac{4R}{3\pi}$; x	$\frac{4R}{3\pi}$	$\frac{4R}{3\pi}$	$\frac{\pi R^2}{4}$

Section	*Figure*	$\overline{X}$	$\overline{Y}$	*Area*
Sector of circle	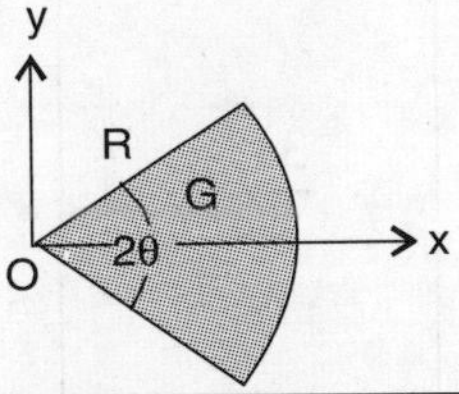 y, R, G, O, 2θ, x	$\frac{2R\sin\theta}{3\theta}$	0	θR^2
Parabolic spandrel	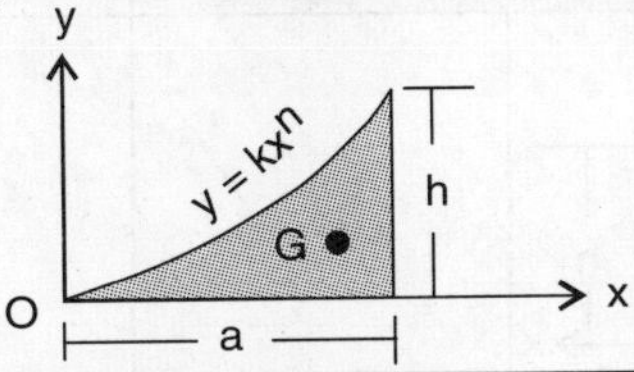 y, $y = kx^n$, h, G, O, x, a	$\left(\frac{n+1}{n+2}\right)a$	$\left(\frac{n+1}{4n+2}\right)h$	$\frac{ah}{n+1}$
Parabola	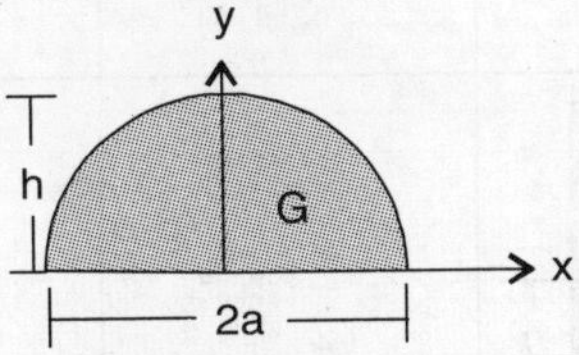 y, h, G, x, 2a	0	$\frac{3h}{5}$	$\frac{4ah}{3}$
Semiparabola	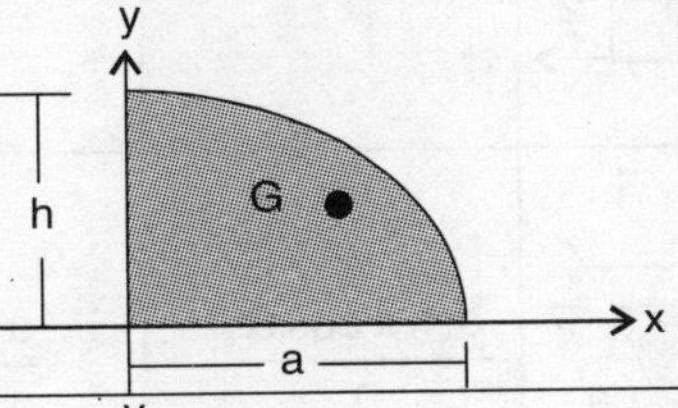 y, h, G, x, a	$\frac{3}{8}a$	$\frac{3h}{5}$	$\frac{2ah}{3}$
Arc of circle	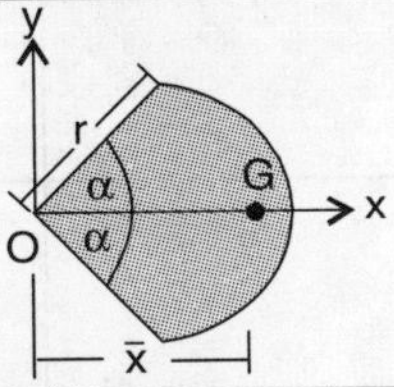 y, r, α, α, G, O, x, $\bar{x}$	$\frac{r\sin\alpha}{\alpha}$	0	$2r\alpha$
Quarter circular arc	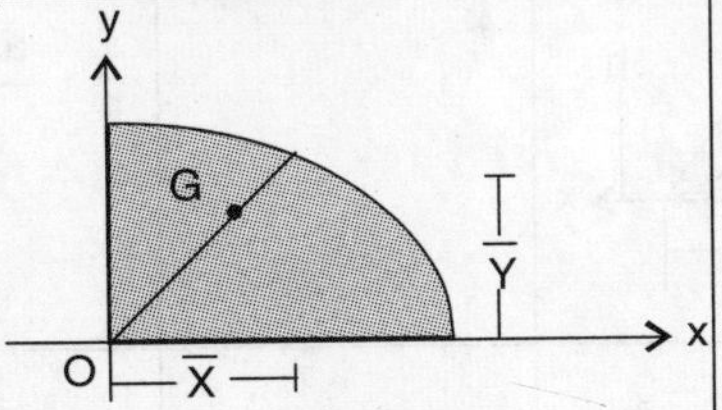 y, G, Y, O, X, x	$\frac{2r}{\pi}$	$\frac{2r}{\pi}$	$\frac{\pi r}{2}$
Semicircular arc	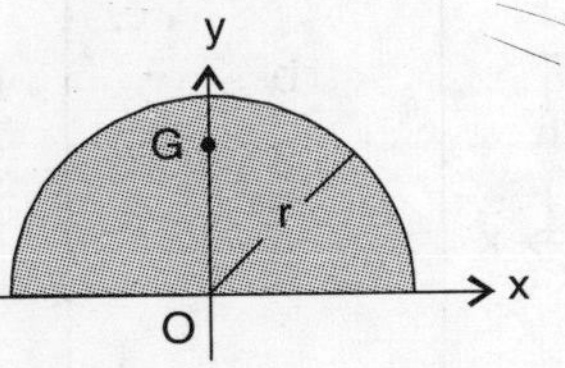 y, G, r, O, x	0	$\frac{2r}{\pi}$	πr

Section	*Figure*	$\overline{X}$	$\overline{Y}$	*Area*
Parabolic area		$\frac{3a}{8}$	$\frac{3h}{5}$	$\frac{2ah}{3}$
Parabolic area		0	$\frac{3h}{5}$	$\frac{4ah}{3}$
Rectangle		$\frac{a}{2}$	$\frac{h}{2}$	bh
Parallelogram		$\frac{b+a\cos\alpha}{2}$	$\frac{a\sin\alpha}{2}$	Gb sin α
Triangle			$\frac{a+b}{3}$	$\frac{h}{3}$
Trapezium		$\frac{b}{2}$	$\frac{h}{3}\left(\frac{b+2a}{b+a}\right)$	$\left(\frac{a+b}{2}\right)h$

Example 5.4: *Locate the centroid of the T-section shown in Fig. (5.25). All the dimensions are in mm.*

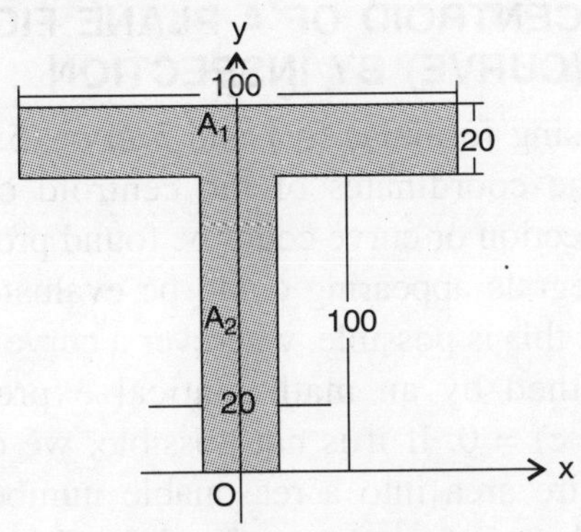

Fig. 5.25

Solution:

Split the given T-section into two rectangular sections having areas $A_1 = 100 \times 20$ mm² and $A_2 = 20 \times 100$ mm² respectively. The centroids of sections A_1 and A_2 could be found by dividing them along two symmetrical lines as shown below.

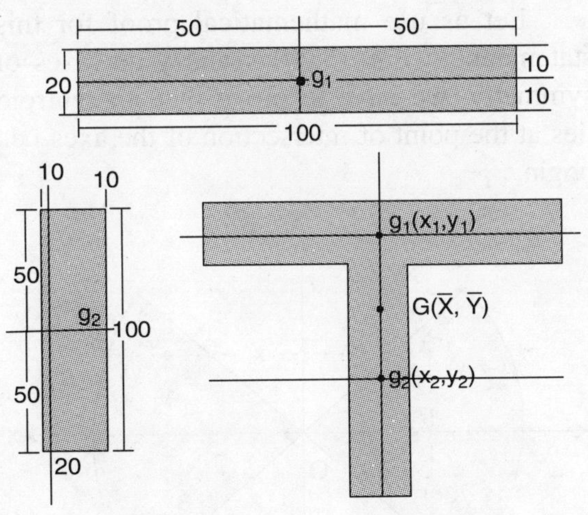

Fig. 5.26

where g_1 and g_2 are centroids of rectangles of areas A_1 and A_2 respectively. Now in the shown coordinate system, we can get

$$g_1(x_1, y_1) \equiv g_1\left(0, 100 + \frac{20}{2}\right)$$

$$\equiv g_1(0, 110)$$

and

$$g_2(x_2, y_2) \equiv g_2\left(0, \frac{100}{2}\right)$$

$$\equiv g_2(0, 50)$$

since the given T-section is the composition of A_1 and A_2 therefore, we get $\overline{X}$ and $\overline{Y}$ by

$$\overline{X} = \frac{x_1A_1 + A_2x_2}{A_1 + A_2}$$

$$= \frac{0 \times 2000 + 0 \times 2000}{2000 + 2000}$$

$$= 0$$

$$\overline{Y} = \frac{y_1A_1 + y_2A_2}{A_1 + A_2}$$

$$= \frac{2000 \times 110 + 2000 \times 50}{2000 + 2000}$$

$$= 80 \text{ mm.}$$

Example 5.5: *With respect to the given co-ordinate axes x and y locate the centroid of the shaded area shown in Fig. (5.27). All the dimensions are in mm.*

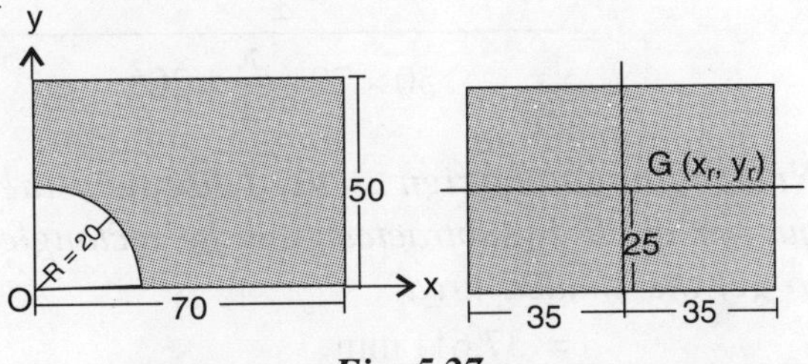

Fig. 5.27

Solution:

Since the shaded area is the reduced part of the rectangle by means of the quarter circle.

Therefore,

$$A_{shaded} = A_{rectangle} - A_{quarter}$$

$$= 50 \times 70 - \frac{\pi}{4} \times (20)^2$$

Let $G(x_r, y_r)$ and $g(x_q, y_q)$ be centroids of the rectangle and the quarter circle then,

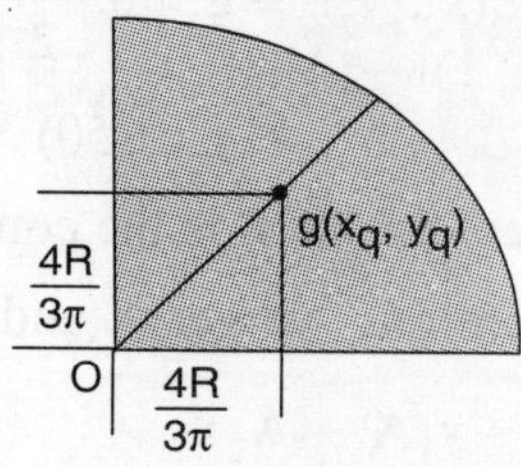

Fig. 5.28

$$G(x_r, y_r) \equiv (35, 25)$$

and $$g(x_q, y_q) \equiv \left(\frac{4\times 20}{3\times\pi}, \frac{4\times 20}{3\times\pi}\right)$$

$$\equiv (8.5, 8.5).$$

Now since the rectangle is being reduced by removing quarter circle, therefore knowing the centroids of the rectangle and quarter circle we could get the centroid of the shaded area by using the following formulae.

$$\overline{X}_s = \frac{A_r \times x_r - A_q \times r_q}{A_r - A_q}$$

$$= \frac{50\times 70\times 35 - \frac{\pi}{4}\times 20\times 20\times 8.5}{50\times 70 - \frac{\pi}{4}\times 20^2}$$

Note: *A negative sign is used because the quarter circle is subtracted from he rectangle to get the shaded area.*

$$= 37.611 \text{ mm}$$

and $$\overline{Y}_s = \frac{A_r y_r - A_q y_q}{A_r - A_q}$$

$$= \frac{70\times 50\times 25 - \frac{\pi}{4}\times 20\times 20\times 8.5}{70\times 50 - \frac{\pi}{4}\times 20\times 20}$$

$$= 26.626 \text{ mm}$$

therefore centroid of shaded area is,

$$G(\overline{X}, \overline{Y}) \equiv (37.611 \text{ mm}, 26.62 \text{ mm}).$$

5.4 CENTROID OF A PLANE FIGURE (CURVE) BY INSPECTION

Using formulae (5.1 and 5.2) or (5.3 and 5.4), the coordinates of the centroid of any plane section or curve could be found provided the integrals appearing could be evaluated. In general this is possible, wherever a curve could be defined by an mathematical expression f (x, y, c) = 0. If it is not possible, we divide the entire area into a reasonable numbers of small but finite elements, then find their individual centroids and then for finding the centroid of the complete section we use the expressions (5.1) and (5.2).

Sometimes the position of the centroid of a plane figure could be found by inspection. For example, if a plane figure has two axes of symmetry then its centroid lies at their intersection.

Let us use mathematical proof for this statement. In Fig. (5.29) x and y are axes of symmetry, we have to prove that its centroid lies at the point of intersection of the axes i.e., origin.

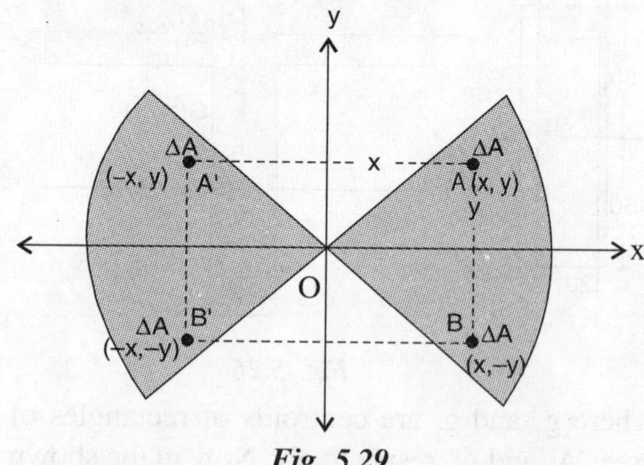

Fig. 5.29

Take four points A, A′, B and B′ such that they have coordinates A(x, y), A′ (–x, y), B(x, –y) and B′ (–x, –y) respectively and each having an elementary area ΔA. This distribution is known as symmetrical distribution (for four points). Now from expression (5.1) and (5.2) we have,

$$\overline{X} = \frac{A_1x_1 + A_2x_2 + A_3x_3 + A_4x_4}{A_1 + A_2 + A_3 + A_4}$$

$$= \frac{\Delta Ax_A + \Delta Ax_{A'} + \Delta Ax_B + \Delta Ax_{B'}}{\Delta A + \Delta A + \Delta A + \Delta A}$$

$$= \frac{\Delta Ax + \Delta A(-x) + \Delta Ax + \Delta A(-x)}{4\Delta A}$$

$$= \frac{(\Delta Ax - \Delta Ax) + (\Delta Ax - \Delta Ax)}{4\Delta A}$$

$$= \frac{0+0}{4\Delta A}$$

$$= 0.$$

Similarly $\overline{Y} = 0$. Now we can increase the strength of symmetrical distribution to a level, so that we can cover the whole figure (the number of such type of points should be even, 2 m). Then also we have,

$$\overline{X} = \frac{\Delta Ax_1 + \Delta A_1' + \Delta Ax_2 + \Delta A_2' + \ldots + \Delta Ax_m + \Delta Ax_m'}{\Delta A + \Delta A + \Delta A + \ldots \text{to } 2m \text{ terms}}$$

$$= \frac{(\Delta Ax_1 + \Delta Ax_1') + (\Delta Ax_2 + \Delta Ax_2') + \ldots + (\Delta Ax_m + \Delta Ax_m')}{A_{Total}}$$

$$= \frac{(\Delta Ax_1 - \Delta Ax_1) + (\Delta Ax_2 - \Delta Ax_2) + \ldots + (\Delta Ax_m - \Delta Ax_m)}{A_{Total}}$$

$$\because \quad (x'_i = -x_i) = \frac{0 + 0 + \ldots\ldots\ldots + 0}{A_{total}} = 0$$

Similarly $\overline{Y} = 0$.

Therefore the centroid lies at the origin of the coordinates system, i.e., at the intersection of the axes of symmetry.

Some plane section, while not having any axes of symmetry, they might be symmetrical about a point. That is, there is one point C in the figure which is the mid-point of every conceivable diameters. In this case also we can make use of expressions (5.1 and 5.2) and show that this point C is the centroid of the Fig. (5.30).

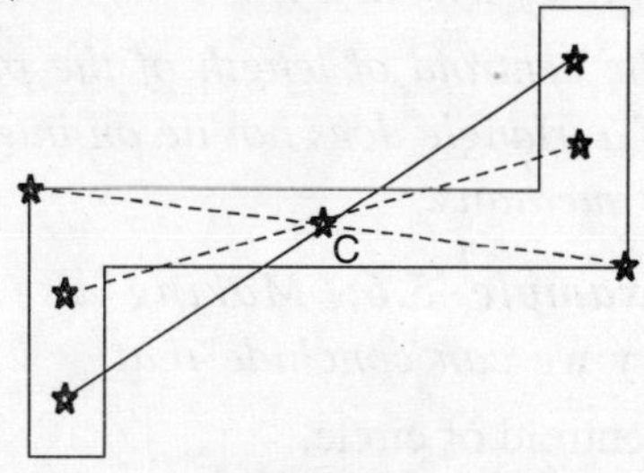

Fig. 5.30

The centroid of the area of a triangle lies at the intersection of its medians. This could be proved as follows: Consider ΔABC, Fig. (5.31), let its area is divided into infinitesimal strips parallel to the sides BC, CA and BA respectively as shown in Figs. (5.32a), (532b) and (5.32c). Now it is an evident that the centroid of each strip lies at the mid-point of its length. The locus of such points are the medians Aa, Bb and Cc.

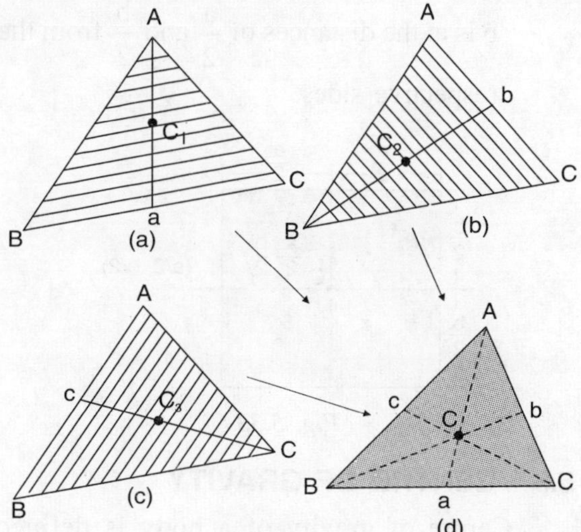

Fig. 5.31

Suppose on Aa, C_1, on Bb, C_2 and on Cc, C_3 are the centroids. But for a triangle there should be only one centroid and that is at their intersection point C.

Notes:

1. *The centroid of curves can also be analysed by the line of symmetry or the point of symmetry.*
2. *The centroid of length of the perimeter of a triangle does not lie on intersection of medians.*

Example 5.6: *Making use of the symmetry we can conclude that*

1. Centroid of circle.

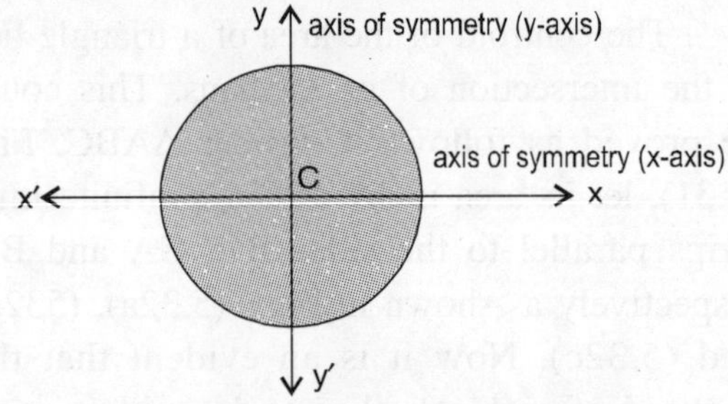

Fig. 5.32

2. The centroid of a rectangle of sides a and b is at the distances of $\frac{a}{2}$ and $\frac{b}{2}$ from the respective sides.

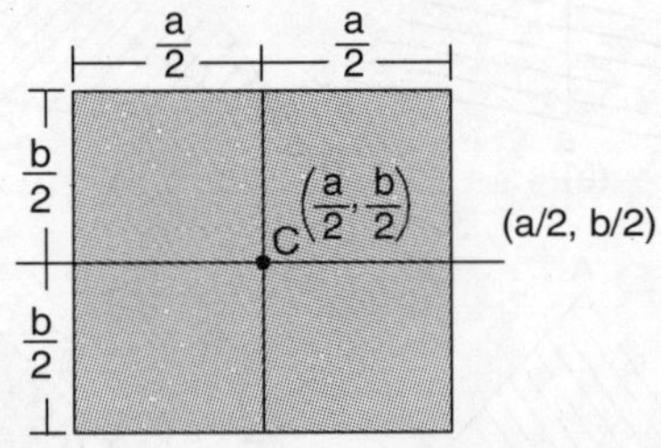

Fig. 5.33

5.5 CENTRE OF GRAVITY

Centre of gravity of a body is defined mathematically as follows:

1. The centre of gravity is a point on the body or outside the body through which the net weight of the body passes and
2. The net torque of individual elements (the atoms) about an axis which passes through centre of gravity is zero, where the net weight is the vector sum of gravitational forces acting vertically downward on individual elements (the atoms).

Fig. (5.34) shows an extended body of mass M and masses of its elements are given by m_i; i = 1, 2, 3, ...n. Each of such element has weight $m_i g$, where g is the acceleration due to the gravity at the location of the element.

Let (x_i, y_i, z_i) be the coordinates of ith element with respect to x-o-y axes and (x'_c, y'_c, z'_c) be the coordinates of same with respect to x′-cg-y axes.

Let (x_{cg}, y_{cg}, z_{cg}) be the coordinates of the centre of gravity with respect to x-o-y axes.

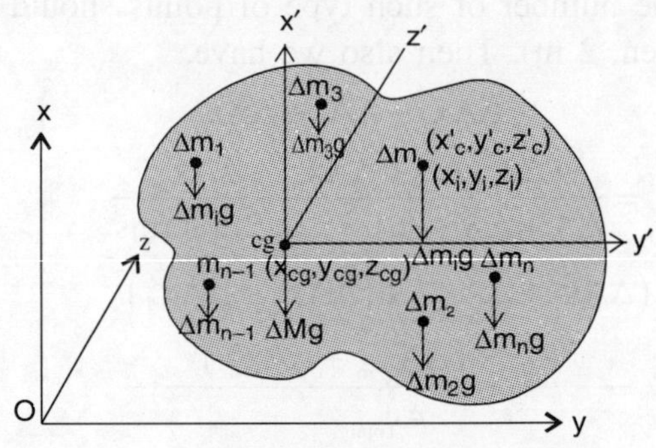

Fig. 5.34

From the above Fig. (5.35), we have

$$\left.\begin{aligned} x_i &= x'_c + x_{cg} \\ y_i &= y'_i + y_{cg} \\ \text{and} \quad z_i &= z_i + z_{cg} \end{aligned}\right\} \quad \text{....(i)}$$

Now by using the first property of cg, we get

$$m_1 \vec{g} + m_2 \vec{g} + ...+ m_n \vec{g} = M \vec{g}$$

$$\text{or} \quad \vec{w}_1 + \vec{w}_2 + \vec{w}_3 + ... + \vec{w}_n = \vec{w} \quad \text{....(ii)}$$

Now the net moment (torque) about the centre of gravity is,

$$\tau_{net} = \sum x'_i \Delta m_i g$$
$$= x'_1\Delta m_1 g + x'_2\Delta m_2 g + ... + x'_n \Delta mg$$
$$= x'_1\Delta m_1 g + x'_2\Delta m_2 g + ... + x'_n\Delta m_n g$$
$$[\because \quad \tau = r_\perp F]$$

or $x'_1\Delta m_1 g + x'_2\Delta m_2 g + ... + x'_n\Delta m_n g = 0$

[using second property]

or $(x_1 - x_{cg})\ \Delta m_1 g + (x_2 - x_{cg})\ \Delta m_2 g + ... + (x_n - x_{cg})\ \Delta m_n g = 0$

or $x_1\Delta m_1 g + x_2\Delta m_2 g + ... + x_n\Delta m_n g = x_{cg}(\Delta m_1 + \Delta m_2 + ... + \Delta m_n)g$

or
$$x_{cg} = \frac{x_1\Delta m_1 g + x_2\Delta m_2 g + x_3\Delta m_3 g + ... + x_n\Delta m_n g}{\begin{pmatrix}\Delta m_1 + \Delta m_2 + \\ .... + \Delta m_n\end{pmatrix} g}$$
$$= \frac{x_1\Delta w_1 + x_2\Delta w_2 + ... + x_n\Delta w_n}{\Delta w_1 + \Delta w_2 + ... + \Delta w_n}$$

or $\Delta w_T\, x_{cg} = \sum x_i\Delta w_i\ ;\ [\Delta w_T = \Sigma\Delta w_i]$

or $x_{cg}\sum_{i=1}^{n}\Delta w_i = \sum_{i=1}^{n} x_i\Delta w_i$

[for discrete distribution] ...(5.22)

and for a continuous distribution we take

$$\Delta w_i \to dw_i$$

Therefore $x_{cg}\int_w dw = \int_w x dw$ ...(5.23)

Similarly $y_{cg}\sum_{i=1}^{n}\Delta w_i = \sum_{i=1}^{n} y_i\Delta w_i$

[for discrete distribution]

$$y_{cg}\int_w dw = \int_w y dw$$

[for continuous distribution]

and $z_{cg}\sum_{i=1}^{n}\Delta w_i = \sum_{i=1}^{n} z_i\Delta w_i$

[for discrete distribution]

$$z_{cg}\int_w dw = \int_w z\, dw$$

[for continuous distribution].

Notes:

1. *For constant gravitational acceleration the position of the centre of gravity and the centre of the mass is same. i.e., $x_{cg} = x_{cm}$.*
2. *The centre of gravity is the property of the body and it depends on the shape and the size of the body and it is independent of the choices of co-ordinate axes.*
3. *The centre of gravity is independent of the orientation of the body and it does not rotate at all when it is suspended from the centre of gravity.*

5.6 DIFFERENCE BETWEEN CENTRE OF GRAVITY AND CENTROID

From the above discussion we can differentiate between centre of gravity and centroid.

1. A centroid refers to a lamina i.e., a body with negligible thickness but with considerable area. This is the mathematical point where the complete area of the lamina seems to be concentrated.
2. Centre of gravity refers to a lamina with considerable mass and dimensions. It is the mathematical point where the complete weight of the body seems to be concentrated.
3. The centre of gravity of a body is a point through which the resultant gravitational force (weight) acts regardless of the orientation of the body in space. Whereas a centroid is a point in a plane area such that the moment of the area about any axis through that point is zero.

WORKED OUT EXAMPLES

1. *Locate the centroid of the T-section shown in the Fig. (5-W1). All the dimensions are in mm.*

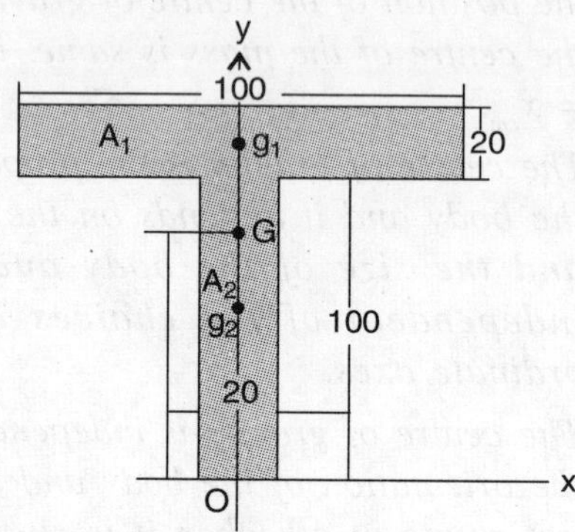

Fig. 5-W1

Solution:

Since the given T-section is symmetrical about the y-axis therefore supposing it as the centroidal axis, we can get $\overline{X} = 0$.

Now the given T-section could be divided into two rectangles A_1 and A_2 each of size 100 × 20 and 20 × 100.

The centroids of A_1 and A_2 are g_1 (0, 110) and g_2(0, 50) respectively.

∴ The centroid $(0, \overline{Y})$ of the composite section is,

$$\overline{Y} = \frac{100\times 20\times 110 + 100\times 20\times 50}{100\times 20 + 100\times 20}$$

$$= 80 \text{ mm}$$

Hence, the centroid of the T-section is on the symmetrical y-axis at a distance of 80 mm from the bottom.

2. *Find the centroid of the L-section shown in the Fig. (5-W2). All the dimensions are in mm.*

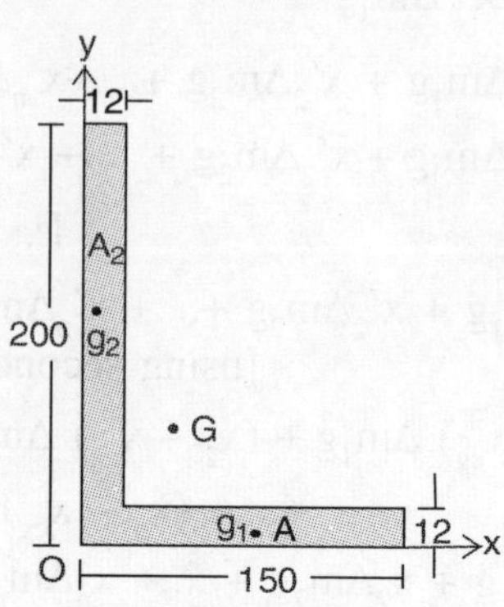

Fig. 5-W2

Solution:

Now the given composite section could be divided into two rectangles A_1 and A_2.

$$A_1 = (150 - 12) \times 12$$
$$= 1656 \text{ mm}^2$$

and $$A_2 = 200 \times 12$$
$$= 2400 \text{ mm}^2$$

Total area, $$A_T = A_1 + A_2$$
$$= 1656 \text{ mm}^2 + 2400 \text{ mm}^2$$
$$= 4056 \text{ mm}^2$$

Selecting the reference axis as shown, the centroid of A_1, $g_1 \left[12 + \frac{138}{2}, \frac{12}{6}\right]$ and that of A_2, $g_2 \left[\frac{12}{2}, \frac{200}{2}\right]$.

$$\therefore \overline{X} = \frac{\text{Moment of } A_1 \text{ and } A_2 \text{ about y-axis}}{\text{Total area}}$$

$$= \frac{A_1 x_1 + A_2 x_2}{A_1 + A_2}$$

$$= \frac{1656\times 81 + 2400\times 6}{4056}$$

$= 36.62$ mm.

and $\overline{Y} = \dfrac{\text{Moment of } A_1 \text{ and } A_2 \text{ about x-axis}}{\text{Total area}}$

$= \dfrac{A_1 y_1 + A_2 y_2}{A_1 + A_2}$

$= \dfrac{1656 \times 6 + 2400 \times 100}{4056}$

$= 61.62$ mm.

3. *Locate the centroid of I section shown in Fig. (5-W3). All the dimensions are in mm.*

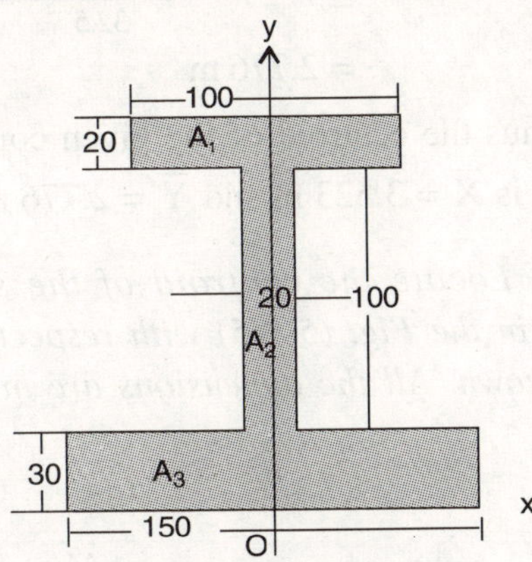

Fig. 5-W3

Solution:

Since the given section is symmetrical about the y-axis, therefore $\overline{X} = 0$.

Now the composite sections could be split into three rectangles of areas A_1, A_2 and A_3 where,

$$A_1 = 100 \times 20 = 2000 \text{ mm}^2$$

The centroid of A_1 from the origin is,

$$y_1 = 30 + 100 + \frac{20}{2} \text{ mm} = 140 \text{ mm}$$

Similarly $A_2 = 20 \times 100 \text{ mm}^2 = 2000 \text{ mm}^2$

$$y_2 = 30 + \frac{100}{2} = 30 + 50 = 80 \text{ mm}$$

and $A_3 = 30 \times 150 = 4500 \text{ mm}^2$

$$y_3 = \frac{30}{2} = 15 \text{ mm}$$

Therefore the centroid of the given composite section is,

$$\overline{Y} = \frac{A_1 Y_1 + A_2 Y_2 + A_3 Y_3}{A_1 + A_2 + A_3}$$

$$= \frac{2000 \times 140 + 2000 \times 80 + 4500 \times 15}{2000 + 2000 + 4500}$$

$= 59.7$ mm

Thus, the centroid is on the symmetrical y-axis at a distance of 59.7 mm from the bottom section.

4. *Determine the centroid of the section of the concrete dam shown in Fig. (5-W4). All the dimensions are in metre.*

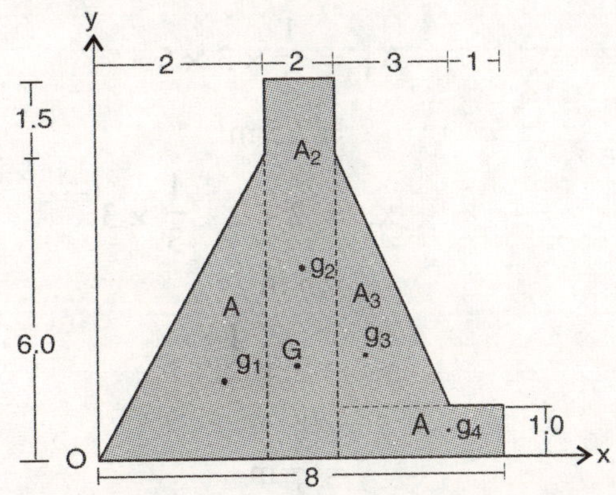

Fig. 5-W4

Solution:

Let the axes are as chosen in the Fig. (5-W4). The composite section could be conveniently divided into two triangles and two rectangles as shown in the Fig. (5-W4).

Then $A_1 = \dfrac{1}{2} \times 2 \times 6 = 6 \text{ m}^2$

Now the centroid of A_1 from the origin is,

$$x_1 = 2\frac{h}{3}$$

$$= 2 \times \frac{2}{3}$$

$$= \frac{4}{3} \text{ m}$$

$$y_1 = \frac{h}{3}$$

$$= \frac{6}{3}$$

$$= 2 \text{ m}$$

Similarly, $A_2 = 7.5 \times 2$

$$= 15 \text{ m}^2$$

$$x_2 = 2 + \frac{2}{2}$$

$$= 3 \text{ m}$$

$$y_2 = \frac{7.5}{2}$$

$$= 3.75 \text{ m}$$

$$A_3 = \frac{1}{2} \times 3 \times 5$$

$$= 7.5 \text{ m}^2$$

$$x_3 = 2 + 2 + \frac{1}{3} \times 3$$

$$= 5 \text{ m}$$

$$y_3 = 1 + \frac{1}{3} \times 5$$

$$= \frac{8}{3} \text{ m}$$

and $A_4 = 1 \times 4$

$$= 4 \text{ m}^2$$

$$x_4 = 2 + 2 + \frac{4}{2}$$

$$= 6 \text{ m}$$

$$y_4 = \frac{1}{2}$$

$$= 0.5 \text{ m}$$

$$\therefore \quad \overline{X} = \frac{A_1x_1 + A_2x_2 + A_3x_3 + A_4x_4}{A_T}$$

$$= \frac{6 \times \frac{4}{3} + 15 \times 3 + 7.5 \times 5 + 4 \times 6}{6 + 15 + 7.5 + 4}$$

$$= 3.523 \text{ m}$$

and

$$\overline{Y} = \frac{A_1y_1 + A_2y_2 + A_3y_3 + A_4y_4}{A_T}$$

$$= \frac{6 \times 2 + 15 \times 3.75 + 7.5 \times \frac{8}{3} + 4 \times 0.5}{32.5}$$

$$= 2.776 \text{ m}$$

Thus the centroid of the given composite section is $\overline{X} = 3.523$ m and $\overline{Y} = 2.776$ m.

5. *Locate the centroid of the section shown in the Fig. (5-W5) with respect to the axes shown. All the dimensions are in metre.*

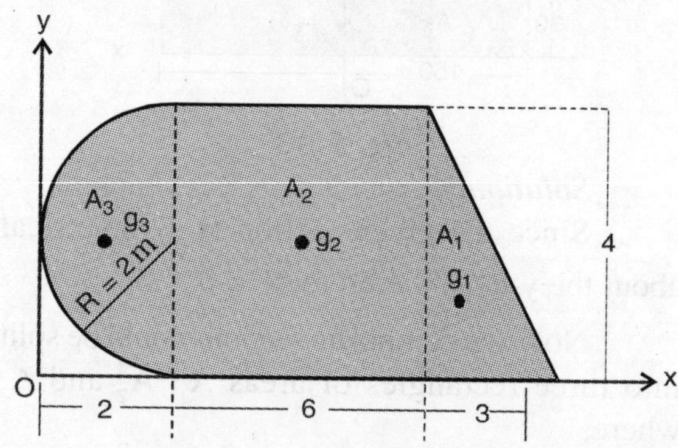

Fig. 5-W5

Solution:

Divide the composite section into three known sections viz a triangle, a rectangle and a semicircle.

Now the areas and the centroids of the three simple (sections) are,

$$A_1 = \frac{1}{2}bh$$

$$= \frac{1}{2} \times 3 \times 4$$

$$= 6 \text{ m}^2$$

$$x_1 = 2 + 6 + \frac{h}{3}$$

$$= 8 + \frac{3}{3}$$

$$= 9 \text{ m}$$

$$y_1 = \frac{h}{3}$$

$$= \frac{4}{3} \text{ m}$$

$$A_2 = b \times h$$

$$= 6 \times 4$$

$$= 24 \text{ m}^2$$

$$x_2 = 2 + \frac{6}{2}$$

$$= 5 \text{ m}$$

$$y_2 = \frac{h}{2}$$

$$= \frac{4}{2}$$

$$= 2 \text{ m}$$

and $$A_3 = \frac{\pi R^2}{2}$$

$$= \frac{\pi \times 4}{2}$$

$$= 2\pi$$

$$= 6.2857 \text{ m}^2$$

$$x_3 = R - \frac{4R}{3\pi}$$

$$= 2 - \frac{4 \times 2}{3\pi}$$

$$= 1.1515 \text{ m}$$

$$y_3 = 2 \text{ m}$$

$$\therefore \quad \overline{X} = \frac{A_1x_1 + A_2x_2 + A_3x_3}{A_1 + A_2 + A_3}$$

$$= \frac{6 \times 9 + 24 \times 5 + 6.2857 \times 1.1515}{6 + 24 + 6.2857}$$

$$= 4.994 \text{ m}$$

and $$\overline{Y} = \frac{6 \times \frac{4}{3} + 24 \times 2 + 6.2857 \times 2}{36.2857}$$

$$= 1.889 \text{ m}$$

Therefore the centroid of the composite section is $\overline{X} = 4.994$ m and $\overline{Y} = 1.889$ m.

6. *Determine the coordinates x_c and y_c of the centre of a 100 mm diameter circular hole cut in a thin plate so that this point will be the centroid of the remaining shaded area shown in the Fig. (5-W6). All the dimensions are in mm.*

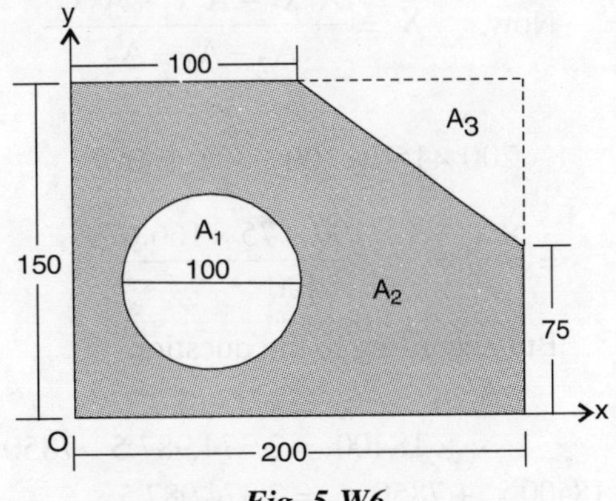

Fig. 5-W6

Solution:

If x_c and y_c are the coordinates of the centre of the circle then according to the question this is the centroid of the shaded area too. The shaded area may be considered as a rectangle of size 200 mm × 150 mm minus a triangle of sides 100 mm and 75 mm and a circle of diameter 100 mm.

$\therefore$ The total shaded area

$$= A_2 - A_1 - A_3$$

$$= 200 \times 150 - \frac{1}{2} \times 100 \times 75 - \frac{\pi}{4} \times (100)^2$$

$$= 18400 \text{ mm}^2 \text{ (take } \pi = 3.14)$$

If the centroid of the rectangle is g_2 (x_2, y_2) then,

$$x_2 = \frac{200}{2}$$
$$= 100 \text{ mm}$$
$$y_2 = \frac{150}{2}$$
$$= 75 \text{ mm}$$

Similarly for triangle g_3 (x_3, y_3) then,

$$y_3 = 150 - \frac{75}{3}$$
$$= 125 \text{ mm}$$
$$x_3 = 100 + 2 \times \frac{100}{3}$$
$$= 166.67 \text{ mm}$$

Now, $$\overline{X} = \frac{A_2x_2 - A_1x_1 - A_3x_3}{A_2 - A_1 - A_3}$$

$$= \frac{200 \times 150 \times 100 - \frac{\pi}{4} \times 100^2 \times x_c - \frac{1}{2} \times 100 \times 75 \times 166.67}{18400}$$

But according to the question

$$\overline{X} = x_c$$
$$\therefore \quad x_c \times 18400 = 2{,}374{,}987.5 - 7850\, x_c$$
or $$18400\, x_c + 7850\, x_c = 2{,}374{,}987.5$$
or $$26250\, x_c = 2{,}374{,}987.5$$
or $$x_c = \frac{2{,}374{,}987.5}{26250}$$
or $$x_c = 90.47$$
$$\therefore \quad x_c = 90.47 \text{ mm}$$

Similarly,

$$18400 \times y_c = 200 \times 150 \times 75 - \frac{1}{2} \times 100 \times 75 \times 125 - \frac{\pi}{4} \times 100^2 \times y_c$$
or $$26{,}250\, y_c = 1{,}781{,}250$$
or $$y_c = \frac{1{,}781{,}250}{26{,}250} = 67.85 \text{ mm}$$

Thus the centre of the circle should be located at (90.47, 67.85) so that this point will be the centroid of remaining shaded area too.

7. *Find the centroid of the end shield of a bulldozer blade shown in Fig. (5-W7). All the dimensions are in mm.*

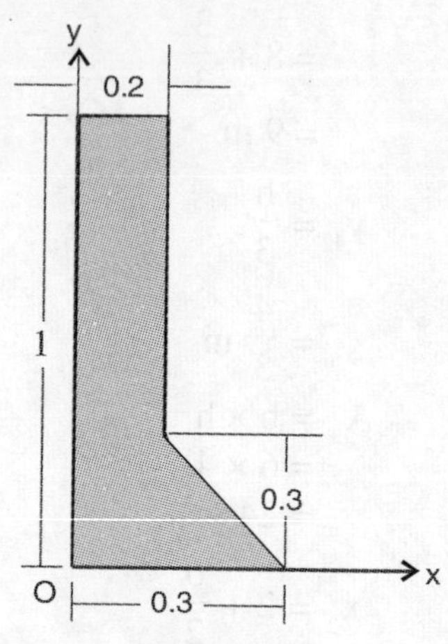

Fig. 5-W7

Solution:

Consider the blade in two parts (i) a rectangle (A_1) of sides 0.2 m and 1 m and (ii) triangle (A_2) of sides 0.3 m and 0.1 m.

Let (x_1, y_1) and (x_2, y_2) be the centroids of the areas A_1 and A_2 respectively.

Then, $$A_1 = 0.2 \times 1$$
$$= 0.2 \text{ m}^2$$
$$x_1 = \frac{0.2}{2}$$
$$= 0.1 \text{ m}$$
$$y_1 = \frac{1}{2}$$
$$= 0.5 \text{ m}$$

and $$A_2 = \frac{1}{2} \times 0.1 \times 0.3$$
$$= 0.015 \text{ m}^2$$
$$x_2 = 0.2 + \frac{1}{3} \times 0.1$$
$$= 0.233 \text{ m}$$

$$y_2 = 0.3 \times \frac{1}{3}$$

$$= 0.1 \text{ m}$$

Therefore, $$\overline{X} = \frac{A_1x_1 + A_2x_2}{A_1 + A_2}$$

$$= \frac{0.2 \times 0.1 + 0.015 \times 0.233}{0.2 + 0.015}$$

$$= 0.1093 \text{ m}$$

$$\overline{Y} = \frac{A_1y_1 + A_2y_2}{A_1 + A_2}$$

$$= \frac{0.2 \times 0.5 + 0.015 \times 0.1}{0.215}$$

$$= 0.4720 \text{ m}$$

Thus the centroid of the bulldozer blade is $\overline{X} = 0.1093$ m and $\overline{Y} = 0.4720$ m.

8. *Find the coordinates of the centroid of the shaded area with respect to the axes shown in the Fig. (5-W8). All the dimensions are in mm.*

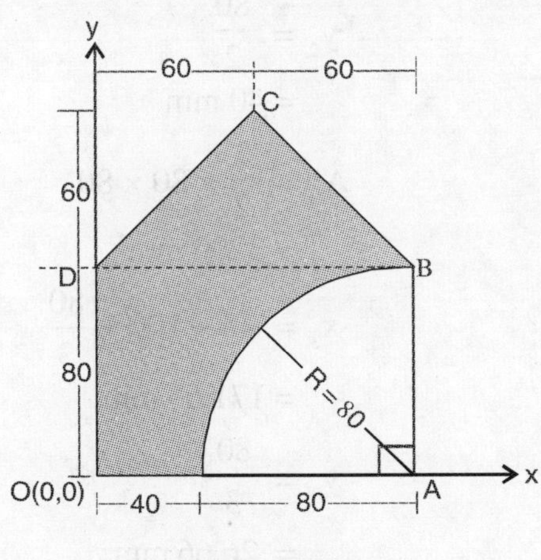

Fig. 5-W8

Solution:

Consider the shaded area as a triangle of sides 120 mm × 60 mm, rectangle of size 120 mm × 80 mm, minus a quarter circle of radius (R = 80 mm).

Let A_1, A_2 and A_3 are the areas of the rectangle, triangle and quarter circle respectively.

Then the total shaded area, $A_T = A_1 + A_2 - A_3$

$$= \frac{1}{2} \times 120 \times 60 + 120 \times 80 - \frac{\pi}{4} \times 80^2$$

$$= 60 \times 60 + 120 \times 80 - 20 \times 80 \times \pi$$

$$= 8{,}171.42 \text{ mm}^2 \qquad \left[\pi = \frac{22}{7}\right]$$

Let (x_1, y_1), (x_2, y_2) and (x_3, y_3) be the individual centroids of the rectangle, triangle and quarter circle respectively.

Then, $$x_1 = \frac{b}{2}$$

$$= \frac{120}{2}$$

$$= 60 \text{ mm}$$

$$y_1 = \frac{h}{2}$$

$$= \frac{80}{2}$$

$$= 40 \text{ mm}$$

$$x_2 = \frac{120}{2}$$

$$= 60 \text{ mm.}$$

Since the triangle is symmetrical about the axis which passes through the vertex (C) of the triangle and perpendicular to x-axis.

$$y_2 = 80 + \frac{h}{3}$$

$$= 80 + \frac{60}{3}$$

$$= 80 + 20$$

$$= 100 \text{ m.}$$

$$x_3 = 120 - \frac{4R}{3\pi}$$

$$= 120 - \frac{4 \times 80}{3 \times \pi}$$

$= 86.07 \text{ mm}$

$$y_3 = \frac{4R}{3\pi}$$

$$= \frac{4 \times 80}{3 \times \pi}$$

$= 33.93 \text{ mm}$

Therefore, $$\overline{X} = \frac{A_1x_1 + A_2x_2 - A_3x_3}{A_1 + A_2 - A_3}$$

$$= \frac{3600 \times 60 + 9600 \times 60 - 5028.57 \times 86.07}{3600 + 9600 - 5028.57}$$

$= 43.96 \text{ mm}$

$$\overline{Y} = \frac{A_1y_1 + A_2y_2 - A_3y_3}{A_1 + A_2 - A_3}$$

$$= \frac{3600 \times 40 + 9600 \times 100 - 5028.57 \times 33.93}{3600 + 9600 - 5028.57}$$

$= 114.22 \text{ mm.}$

9. *With respect to the shown system of the coordinate axes locate the centroid of the shaded area shown in the Fig. (5-W9). All the dimensions are in mm.*

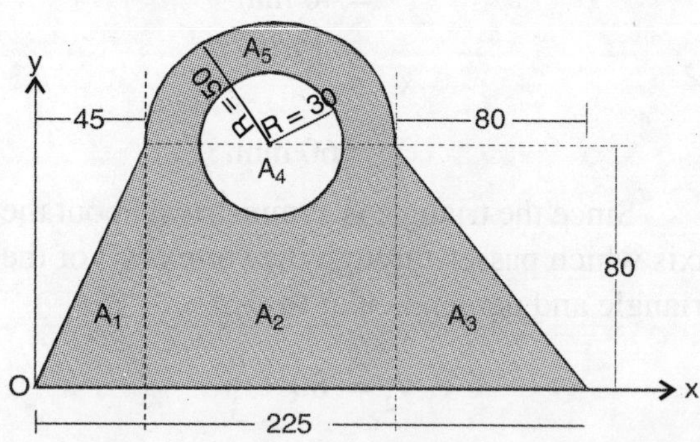

Fig. 5-W9

Solution:

Let divide the shaded area as shown in Fig. (5-W9), so that one could assume that the shaded area is a combination of triangle (A_1), rectangle (A_2), triangle (A_3), semicircle (A_5) and circle (A_4). Let (x_i , y_i) be the centroids of the areas, A_i where, i = 1, 2, 3, 4, 5 respectively.

Then, $$A_1 = \frac{1}{2} \times 45 \times 80$$

$= 1800 \text{ mm}^2$

$$x_1 = \frac{2h}{3}$$

$$= \frac{2}{3} \times 45$$

$= 30 \text{ mm}$

$$y_1 = \frac{h}{3}$$

$$= \frac{80}{3}$$

$= 26.66 \text{ mm}$

$A_2 = (225 - 45 - 80) \times 80$

$= 8000 \text{ mm}^2$

$$x_2 = 45 + \frac{100}{2}$$

$= 45 + 50$

$= 95 \text{ mm}$

$$y_2 = \frac{80}{2}$$

$= 40 \text{ mm}$

$$A_3 = \frac{1}{2} \times 80 \times 80$$

$= 3200 \text{ mm}^2$

$$x_3 = 45 + 100 + \frac{80}{3}$$

$= 171.66 \text{ mm}$

$$y_3 = \frac{80}{3}$$

$= 26.66 \text{ mm}$

$A_4 = -\pi \times 30^2$

$= -2826 \text{ mm}^2$

[$\pi = 3.14$]

$x_4 = 95 \text{ mm}$

$y_4 = 80 \text{ mm}$

and $$A_5 = \frac{\pi}{2} \times (50)^2$$
$$= 3925 \text{ mm}^2$$
$$x_5 = 95 \text{ mm}$$
$$y_5 = 80 + \frac{4 \times 50}{3\pi}$$
$$= 101.23 \text{ mm}$$

Therefore, the centroid of the composite section is,

$$\overline{X} = \frac{A_1x_1 + A_2x_2 + A_3x_3 - A_4x_4 + A_5x_5}{A_1 + A_2 + A_3 - A_4 + A_5}$$

$$= \frac{1800 \times 30 + 8000 \times 95 + 3200 \times 171.66 - 2826 \times 95 + 3925 \times 95}{1800 + 8000 + 3200 - 2826 + 3925}$$

$$= 104.10 \text{ mm}$$

$$= \frac{A_1y_1 + A_2y_2 + A_3y_3 - A_4y_4 + A_5y_5}{A_1 + A_2 + A_3 - A_4 + A_5}$$

and

$$\overline{Y} = \frac{1800 \times 26.66 + 8000 \times 40 + 3200 \times 26.66 - 2826 \times 80 + 3925 \times 101.23}{1800 + 800 + 3200 - 2826 + 3925}$$

$$= 44.29 \text{ mm}$$

10. *Determine the coordinates of the centroid of the area shown in Fig. (5-W10) with reference to the axes shown. Take x = 40 mm.*

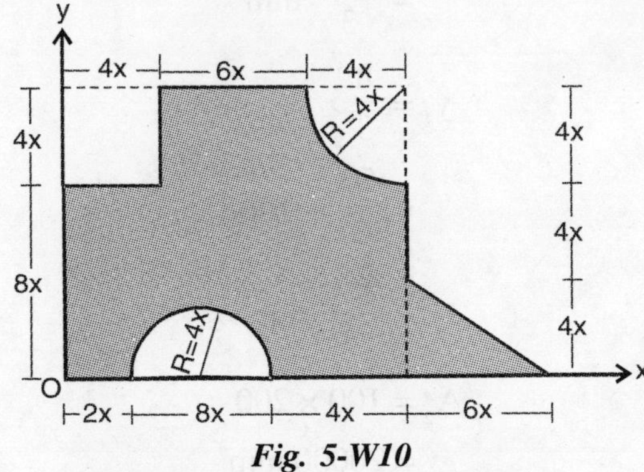

Fig. 5-W10

Solution:

The composite section is divided into the following known sections.

A rectangle, $A_1 = 14x \times 12x$

$$= 168x^2 \text{ mm}^2$$
$$x_1 = \frac{14x}{2}$$
$$= 7x \text{ mm}$$
$$y_1 = \frac{12x}{2}$$
$$= 6x \text{ mm}$$

A triangle,

$$A_2 = \frac{1}{2} \times 6x \times 4x$$
$$= 12x^2 \text{ mm}^2$$
$$x_2 = 2x + 8x + 4x + \frac{6x}{3}$$
$$= 16x \text{ mm}$$
$$y_2 = \frac{4x}{3}$$
$$= 1.3333x \text{ mm}$$

Since square of side 4x is being cut out from rectangle of size 14x × 12x therefore its area has to be negative.

Hence, $A_3 = -4x \times 4x$
$$= -16x^2 \text{ mm}^2$$
$$x_3 = \frac{4x}{2}$$
$$= 2x \text{ mm}$$
$$y_3 = 8x + \frac{4x}{2}$$
$$= 10x \text{ mm.}$$

Similarly semicircle is being subtracted.

Hence $A_4 = -\frac{1}{2}\pi (4x)^2$
$$= -8\pi x^2 \text{ mm}^2$$
$$= -25.1428x^2 \text{ mm}^2;$$
$$\pi = \frac{22}{7}$$
$$x_4 = 2x + \frac{8x}{2}$$
$$= 6x \text{ mm}$$

$$y_4 = \frac{4R}{3\pi}$$

$$= 4 \times \frac{4x}{3\pi}$$

$$= \frac{16x}{3\pi} \text{ mm}$$

and $$A_5 = -\frac{1}{4} \times \pi \times (4x)^2$$

$$= -4\pi x^2$$

$$= -12.5714x^2 \text{ mm}^2$$

$$X_5 = 14x - \frac{4R}{3\pi}$$

$$= 14x - \frac{16x}{3\pi}$$

$$= 12.3030x \text{ mm}$$

Total area, $$A_T = A_1 + A_2 - A_3 - A_4 - A_5$$

$$= 168x^2 + 12x^2 - 16x^2 - 8\pi x^2 - 4\pi x^2$$

$$= 126.5542x^2 \text{ mm}^2$$

Now the centroid of the composite section

$$\overline{X} = \frac{A_1x_1 + A_2x_2 - A_3x_3 - A_4x_4 - A_5x_5}{A_1 + A_2 - A_3 - A_4 - A_5}$$

$$= \frac{168x^2 \times 7x + 12x^2 \times 16x^2 - 16x^2 \times 2x - 25.142x^2 \times 6x - 12.5714x^2 \times 12.3030x}{126.5542x^2}$$

$$= 325.70 \text{ mm } (x = 40 \text{ mm})$$

$$\overline{Y} = \frac{A_1y_1 + A_2y_2 - A_3y_3 - A_4y_4 - A_5y_5}{A_1 + A_2 - A_3 - A_4 - A_5}$$

$$= \frac{168x^2 \times 6x + 12x^2 \times 1.3333x - 16x^2 \times 2x - 25.142x^2 \times 6x - 12.5714x^2 \times 12.3030x}{126.5542x^2}$$

$$= 216.97 \text{ mm.}$$

Therefore the centroid of the composite section is at (325.70, 216.97).

11. *With respect to the coordinate axes x and y locate the centroid of the shaded area shown in Fig. (5-W11). All the dimensions are in mm.*

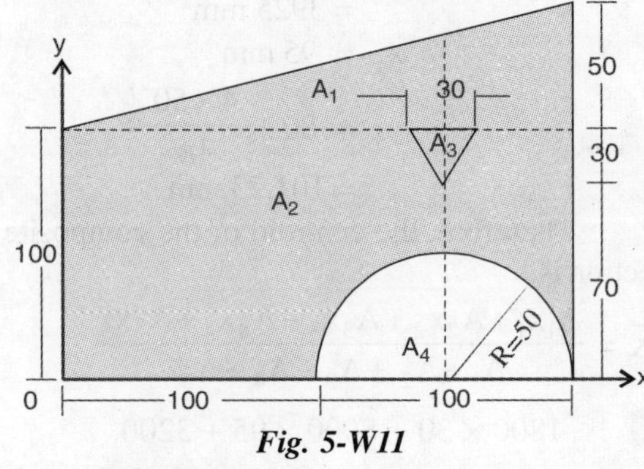

Fig. 5-W11

Solution:

Divide the given composite section in known sections like triangle (A_1), rectangle (A_2), triangle (A_3) and semicircle (A_4) having centroids (x_1, y_1), (x_2, y_2), (x_3, y_3), and (x_4, y_4) respectively.

Then, $$A_1 = \frac{1}{2} \times 200 \times 50$$

$$= 5{,}000 \text{ mm}^2$$

$$x_1 = \frac{2h}{3}$$

$$= 2 \times \frac{200}{3}$$

$$= \frac{400}{3} \text{ mm}$$

$$y_1 = \frac{h}{3}$$

$$= \frac{50}{3} + 100$$

$$= \frac{350}{3} \text{ mm}$$

$$A_2 = 100 \times 200$$

$$= 20000 \text{ mm}^2$$

$$x_2 = \frac{200}{2}$$

$$= 100 \text{ mm}$$

$$y_2 = \frac{100}{2}$$
$$= 50 \text{ mm}$$
$$A_3 = \frac{1}{2} \times 30 \times 30$$
$$= 450 \text{ mm}^2$$
$$x_3 = 100 + 50$$
$$= 150 \text{ mm}$$
$$y_3 = 100 - 30 \times 1/3$$
$$= 100 - 10$$
$$= 90 \text{ mm}$$

and $$A_4 = \frac{\pi}{2} \times 50^2$$
$$= 3925 \text{ mm}^2$$
$$\pi = 3.14$$
$$x_4 = 100 + 50$$
$$= 150 \text{ mm}$$
$$y_4 = \frac{4R}{3\pi}$$
$$= \frac{4 \times 50}{3 \times \pi}$$
$$= 21.2314 \text{ mm}$$

Therefore, the centroid of the composite section is given by,

$$\overline{X} = \frac{A_1x_1 + A_2x_2 - A_3x_3 - A_4x_4}{A_1 + A_2 - A_3 - A_4}$$

$$= \frac{5000 \times \frac{400}{3} + 20000 \times 100 - 450 \times 150 - 3925 \times 150}{5000 + 20000 - 450 - 3925}$$

$$= 97.47 \text{ mm.}$$

$$\overline{Y} = \frac{A_1y_1 + A_2y_2 - A_3y_3 - A_4y_4}{A_1 + A_2 - A_3 - A_4}$$

$$\overline{Y} = \frac{5000 \times \frac{350}{3} + 20000 \times 50 - 450 \times 90 - 3925 \times 21.2314}{5000 + 20000 - 450 - 3925}$$

$$= 70.76 \text{ mm.}$$

12. *Find the position of the centroid of the T-section shown in Fig. (5-W12). All the dimensions are in mm.*

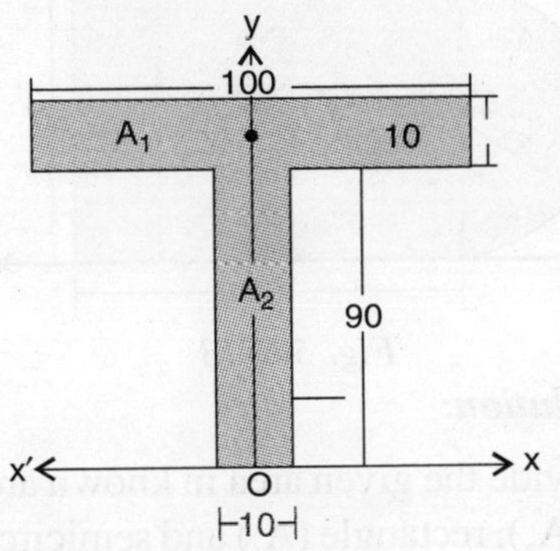

Fig. 5-W12

Solution:

Divide the given T-section into two rectangles of sizes A_1 (10 mm × 100 mm) and A_2 (90 mm × 10 mm) respectively. Since the T-section is symmetrical about the y-axis therefore, $\overline{X} = 0$.

Now $$A_1 = 10 \times 100$$
$$= 1000 \text{ mm}^2$$
$$y_1 = 90 + \frac{10}{2}$$
$$= 90 + 5$$
$$= 95 \text{ mm}$$
$$A_2 = 90 \times 10$$
$$= 900 \text{ mm}^2$$
$$y_2 = \frac{90}{2}$$
$$= 45 \text{ mm}$$
$$\therefore \quad \overline{Y} = \frac{A_1y_1 + A_2y_2}{A_1 + A_2}$$
$$= \frac{1000 \times 95 + 900 \times 45}{1000 + 900}$$
$$= 71.31 \text{ mm}$$

Therefore the centroid is (0,71.31).

13. *Find the centroid of the given area shown in Fig. (5-W13) with resect to shown axes. All the dimensions are in mm.*

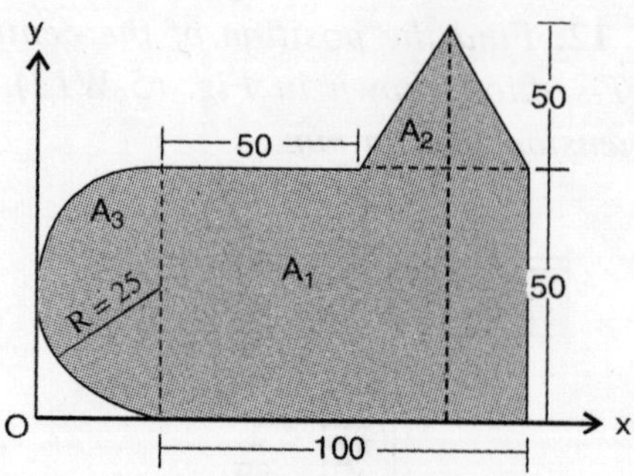

Fig. 5-W13

Solution:

Divide the given area in known areas like triangle (A_2), rectangle (A_1) and semicircle (A_3). Let the centroids of the rectangle, triangle and semicircle be g_1 (x_1, y_1), g_2 (x_2, y_2) and g_3 (x_3, y_3) respectively.

Then, $A_1 = 100 \times 50$

$= 5000 \text{ mm}^2$

$x_1 = 25 + \frac{100}{2}$

$= 25 + 50$

$= 75 \text{ mm}$

$y_1 = \frac{50}{2}$

$= 25 \text{ mm}$

$A_2 = \frac{1}{2} \times 50 \times 50$

$= 1250 \text{ mm}^2$

$y_2 = 50 + \frac{50}{3}$

$= 66.66 \text{ mm}$

$x_2 = 25 + 50 + 25$

$= 100 \text{ mm}$

and $A_3 = \frac{\pi}{2} \times 25^2$

$= 981.25 \text{ mm}^2$

$\pi = 3.14$

$x_3 = 25 - \frac{4 \times 25}{3 \times \pi}$

$= 14.3842 \text{ mm}$

$y_3 = 25 \text{ mm}$

Therefore x coordinate of the cenroid is given by,

$$\overline{X} = \frac{A_1 x_1 + A_2 x_2 + A_3 x_3}{A_1 + A_2 + A_3}$$

$$= \frac{5000 \times 75 + 1250 \times 100 + 981.25 \times 14.3842}{5000 + 1250 + 981.25}$$

$= 71.09$ mm.

and y-coordinate of the centroid is givin by,

$$\overline{Y} = \frac{A_1 y_1 + A_2 y_2 + A_3 y_3}{A_1 + A_2 + A_3}$$

$$= \frac{5000 \times 25 + 1250 \times 66.66 + 981.25 \times 25}{5000 + 1250 + 981.25}$$

$= 32.20$ mm.

Therefore the centroid of the composite area is (70.09, 32.20).

14. *Locate the centroid of the shaded area obtained by cutting a semicircle of diameter 2a from a quarter circle of radius 2a as shown in the Fig. (5-W14).*

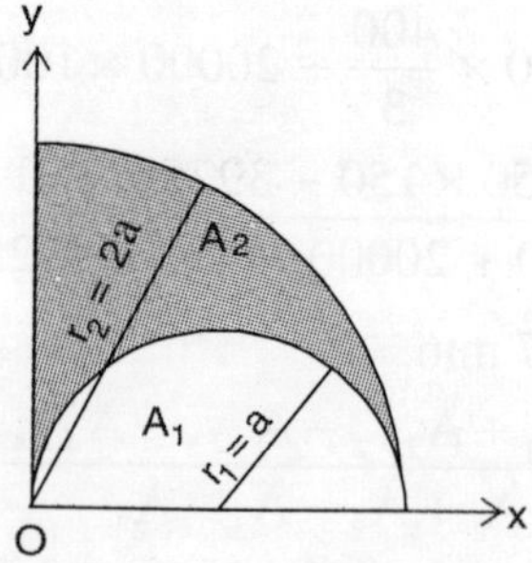

Fig. 5-W14

Solution:

The given shaded area is the area of the quarter circle of radius r_2 = 2a minus the area of the semicircle of radius r_1 = a.

Let the area of the quarter circle be A_2 and the centroid $g_2(x_2, y_2)$ and that of semicircle be A_1 and centroid $g_1(x_1, y_1)$ respectively.

Then, $A_2 = \frac{\pi}{4} r_2^2$

$= \frac{\pi}{4}(2a)^2 \text{ unit}^2$

$x_2 = \frac{4r_2}{3\pi}$

$= \frac{4(2a)}{3\pi}$ unit

$y_2 = \frac{4r_2}{3\pi}$

$= \frac{4(2a)}{3\pi}$ unit

and $A_1 = \frac{\pi}{2} r_1^2$

$= \frac{\pi}{2} a^2 \text{ unit}^2$

$x_1 = r_1$

$= a$ unit

$y_1 = \frac{4r_1}{3\pi}$

$= \frac{4a}{3\pi}$ unit

Therefore, $\overline{X} = \frac{A_2 x_2 - A_1 x_1}{A_2 - A_1}$

$= \frac{\frac{\pi}{4}(4a^2)\times\frac{4(2a)}{3\pi} - \frac{\pi}{2}a^2 \times a}{\frac{\pi}{4}(4a^2) - \frac{\pi}{2}a^2}$

$= \frac{(16-3\pi)a}{3}$ unit

$\overline{Y} = \frac{A_2 y_2 - A_1 y_1}{A_2 - A_1}$

$= \frac{\pi a^2 \times \frac{8a}{3\pi} - \frac{\pi}{2}a^2 \times \frac{4a}{3\pi}}{\frac{\pi}{2}\times a^2}$

$= \frac{4a}{\pi}$ unit.

15. *Determine the abscissa of the centroid of the trapezoid shown in Fig (5-W15) in terms of h_1, h_2 and a. All the dimensions are in mm.*

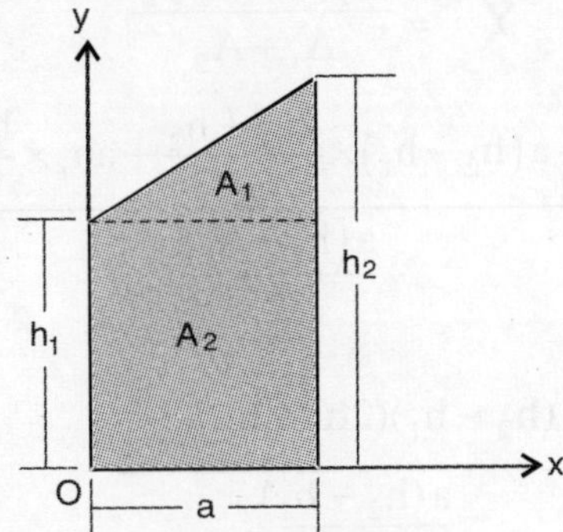

Fig. 5-W15

Solution:

One can consider the given trapezoid is combination of two sections viz,

1. Triangle of area, A_1 and centroid, $g_1(x_1, y_1)$ and

2. Rectangle of area, A_2 and centroid, $g_2(x_2, y_2)$.

Then, $A_1 = \frac{1}{2} a (h_2 - h_1) \text{ mm}^2$

$x_1 = \frac{2a}{3}$ mm

$y_1 = \frac{h_2 + 2h_1}{3}$ mm

and $A_2 = a \times h_1 \text{ mm}^2$

$x_2 = \frac{a}{2}$ mm

$y_2 = \frac{h_1}{2}$ mm.

Therefore the centroid of the trapezoid is,

$$\overline{X} = \frac{A_1 x_1 + A_2 x_2}{A_1 + A_2}$$

$$= \frac{\frac{1}{2}a(h_2 - h_1)\times\frac{2a}{3} + ah_1 \times \frac{a}{2}}{\frac{1}{2}a(h_2 - h_1) + ah_1}$$

$$= \frac{2h_2 + h_1}{3h_2 + 3h_1} a \text{ mm}$$

and $$\overline{Y} = \frac{A_1 y_1 + A_2 y_2}{A_1 + A_2}$$

$$= \frac{\frac{1}{2}a(h_2 - h_1)\times\frac{2h_1 + h_2}{3} + ah_1 \times \frac{h_1}{2}}{\frac{a(h_2 + h_1)}{2}}$$

$$= \frac{\frac{a}{6}(h_2 - h_1)(2h_1 + h_2)\frac{ah_1^2}{2}}{\frac{a(h_2 + h_1)}{2}}$$

$$= \frac{h_2^2 + h_1 h_2 - h_1^2}{3(h_2 + h_1)} \text{ mm.}$$

16. *Locate the centroid of the shaded three quarters of the area of a square of dimension a in x and y axes as shown in the Fig.(5-W16). All the dimensions are in mm.*

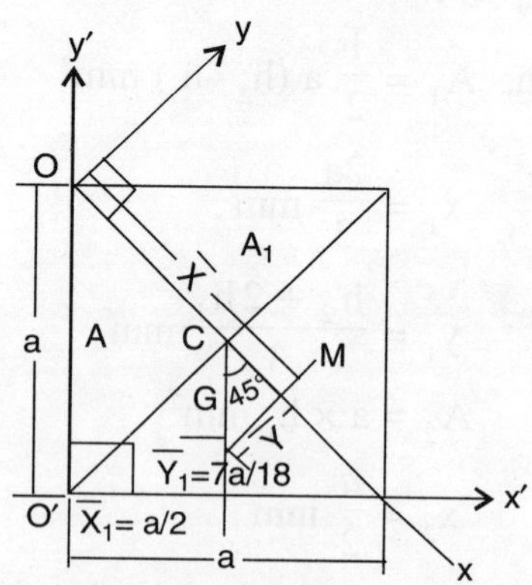

Fig. 5-W16

Solution:

The given shaded area is considered as the area (A) of the square of side a minus the area of the triangle (A_1). If the centroids of the square and triangle are g (x , y) and g_1 (x_1, y_1) respectively, with respect to reference frame x′ and y′.

Then,

$$A = a^2 \text{ mm}^2$$

$$x = \frac{a}{2} \text{ mm}$$

$$y = \frac{a}{2} \text{ mm}$$

and $$A_1 = \frac{1}{2} \times \frac{a}{2} \times a$$

$$= \frac{a^2}{4} \text{ mm}^2$$

$$x_1 = \frac{a}{2} \text{ mm}$$

$$y_1 = a - \frac{1}{3}\left(\frac{a}{2}\right)$$

$$= a - \frac{a}{6}$$

$$= \frac{5a}{6} \text{ mm.}$$

Let the centroid of the shaded area be $(\overline{X}_1, \overline{Y}_1)$ with respect to x′ and y′ then,

$$\overline{X}_1 = \frac{Ax - A_1 x_1}{A - A_1}$$

$$= \frac{a^2 \times \frac{a}{2} - \frac{a^2}{4} \times \frac{a}{2}}{a^2 - \frac{a^2}{4}}$$

$$= \frac{\frac{a^3}{2} - \frac{a^3}{8}}{\frac{3a^2}{4}}$$

$$= \frac{4a^3 - a^3}{8} \times \frac{4}{3a^2}$$

$$= \frac{a}{2}$$

and $$\overline{Y}_1 = \frac{Ay - A_1y_1}{A - A_1}$$

$$= \frac{a^2 \times \frac{a}{2} - \frac{a^2}{4} \times \frac{5a}{6}}{a^2 - \frac{a^2}{4}}$$

$$= \frac{12a^3 - 5a^3}{24} \times \frac{4}{3a^2}$$

$$= \frac{7a^3 \times 4}{24 \times 3a^2}$$

$$= \frac{7a}{18}$$

Now in ΔCGM, $\angle CMG = 90°$ and $\angle GCM = 45°$.

$\therefore$ $$\sin 45° = \frac{GM}{GC}$$

or $$GM = CG \sin 45°$$

$$= \frac{2a}{18} \sin 45°$$

$$= \frac{a}{9\sqrt{2}}$$

and $$\cos 45° = \frac{CM}{CG}$$

or $$CM = CG$$

$$\cos 45° = \frac{a}{9\sqrt{2}}$$

Now $$\overline{X} = OC + CM$$

$$= \frac{a}{\sqrt{2}} + \frac{a}{9\sqrt{2}}$$

$$= \frac{10a}{9\sqrt{2}}$$

and $$\overline{Y} = -MG$$

$$= -\frac{a}{9\sqrt{2}}$$

Therefore, centroid with respect to x and y axes is $\left(\frac{10a}{9\sqrt{2}}, -\frac{a}{9\sqrt{2}}\right)$.

17. *Determine the length of a thin homogeneous wire which is bent into a semicircular arc of radius R together with an extension on either ends as shown in the Fig. (5-W17) such that the centroid of the region is located at the centre of the semicircle.*

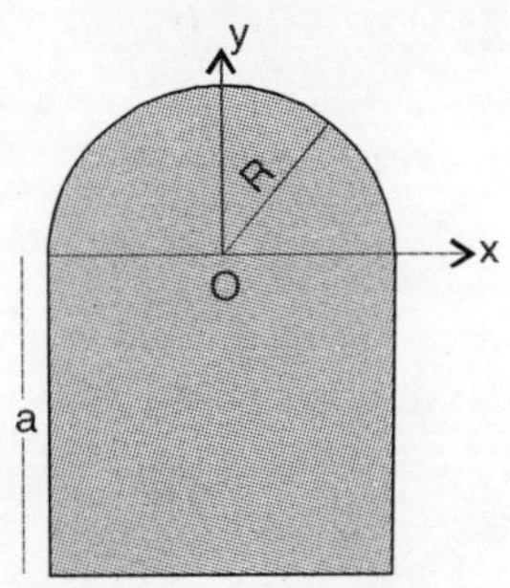

Fig. 5-W17

Solution:

Divide the given region into a rectangle (A_1) and a semicircle (A_2). Let the centroids of these two sections be g_1 (x_1, y_1) and g (x_2, y_2) respectively then,

$$A_1 = 2aR \text{ unit}^2$$

$$x_1 = 0 \text{ unit}$$

$$y_1 = -\frac{a}{2} \text{ unit}$$

$$A_2 = \frac{\pi R^2}{2} \text{ unit}^2$$

$$x_2 = 0 \text{ unit}$$

$$y_2 = \frac{4R}{3\pi} \text{ unit}$$

According to the question $\overline{X} = 0$ and $\overline{Y} = 0$.

Now centroid of the region is given by,

$$\overline{Y} = \frac{2aR \times \left(\frac{-a}{2}\right) + \frac{\pi R^2}{2} \times \frac{4R}{3\pi}}{2aR + \frac{\pi R^2}{2}}$$

or $$0 = 2aR\left(\frac{-a}{2}\right) + \frac{\pi R^2}{2} \times \frac{4R}{3\pi}$$

or $$a^2R = \frac{4R^3\pi}{6}$$

or $$a^2 = \frac{4R^2\pi}{6}$$

or $$a^2 = \frac{2R^2\pi}{3}$$

$\therefore$ $$a = \sqrt{\frac{2\pi}{3}}R$$

$\therefore$ The length of the wire, L

$$= \pi R + 2a$$

$$= \pi R + 2 \times \sqrt{\frac{2\pi}{3}}R$$

$$= \pi R + \sqrt{\frac{8\pi}{3}}R$$

$$= R\left(\pi + \sqrt{\frac{8\pi}{3}}\right)$$

$$= 6.03 \text{ R unit.}$$

18. *Find the centroids of the shaded areas shown in the following figures.*

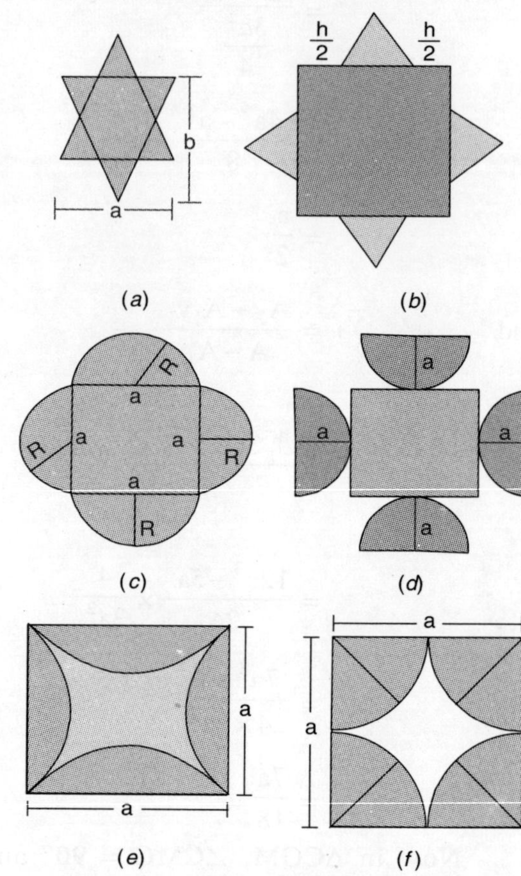

Fig. 5-W18

Solution:

All the figures are symmetrical about the centroidal axes, taking these axes as reference axis, we could get $\overline{X} = 0$ and $\overline{Y} = 0$.

19. *Find the centroid of the area bounded by the curve $|x||y| = ke$.*

Solution:

Given, $|x|\,|y| = ke$

$\Rightarrow$ $|x| = k$

and $|y| = e$

or $|x| = e$

and $|y| = k$

$\Rightarrow$ $x = \pm k$

and $\quad y = \pm e$

or $\quad x = \pm e$

and $\quad y = \pm k$

Since $x = \pm k$ and $y = \pm e$ or $x = \pm e$ and $y = \pm k$ represents two rectangles of sides $2k \times 2e$ and $2e \times 2k$ respectively.

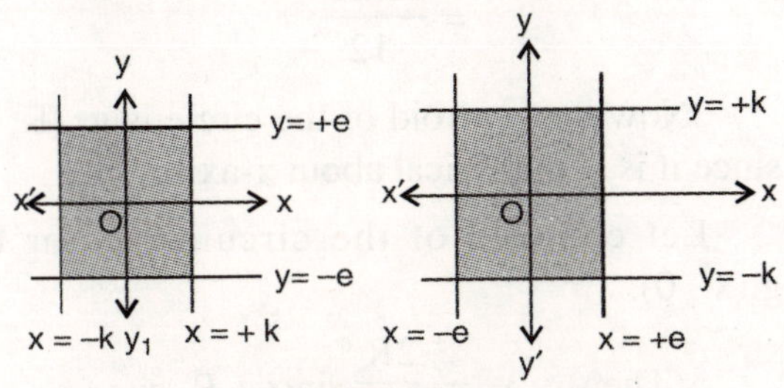

Fig. 5-W19

Since these rectangles are symmetrical about the x and the y-axes therefore the centroid lies on the orgin so $\overline{X} = 0$ and $\overline{Y} = 0$.

20. *A semicircle of 90 mm radius is cut out from a trapezium as shown in the Fig. (5-W20). Determine the centroid of the rest of the area. All the dimensions are given in mm.*

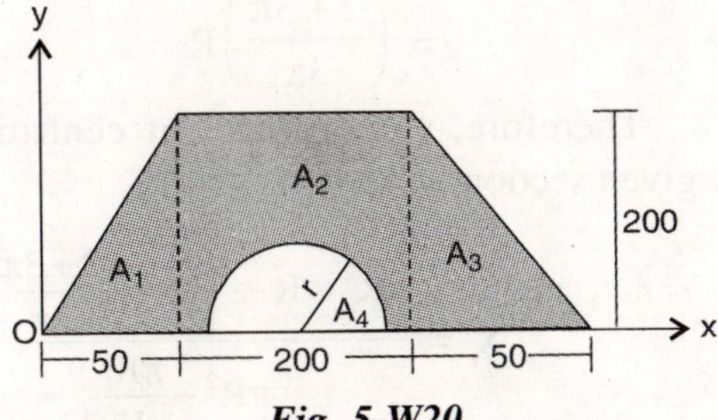

Fig. 5-W20

Solution:

Divide the given area in known areas such as triangle (A_1), square (A_2), triangle (A_3) and a semicircle (A_4) of radius, r = 90 mm. If their centroids with respect to the given axes are $g_1(x_1, y_1)$, $g_2(x_2, y_2)$, $g_3(x_3, y_3)$ and $g_4(x_4, y_4)$ respectively.

Then the total shaded area,

$$A = A_1 + A_2 + A_3 - A_4$$

$$= \frac{1}{2} \times 50 \times 200 + 200 \times 200$$

$$+ \frac{1}{2} \times 50 \times 200 - \frac{\pi}{2} \times 90^2 \text{ mm}^2$$

$$= 37283 \text{ mm}^2$$

$$x_1 = 2\,\frac{h}{3}$$

$$= 2 \times \frac{50}{3}$$

$$= \frac{100}{3} \text{ mm}$$

$$y_1 = \frac{h}{3}$$

$$= \frac{200}{3} \text{ mm}$$

$$x_2 = 50 + \frac{200}{2}$$

$$= 150 \text{ mm}$$

$$y_2 = \frac{200}{2}$$

$$= 100 \text{ mm}$$

$$x_3 = 50 + 200 + \frac{h}{3}$$

$$= 250 + \frac{50}{3}$$

$$= \frac{800}{3} \text{ mm}$$

$$y_3 = \frac{h}{3}$$

$$= \frac{200}{3} \text{ mm}$$

and $\quad x_4 = 50 + 100$

$$= 150 \text{ mm}$$

$$y_4 = \frac{4R}{3\pi}$$

$$= \frac{4 \times 90}{3\pi}$$

$$= \frac{120}{\pi} \text{ mm.}$$

Let x_c and y_c be the coordinates of the centroid of the rest of the area then,

$$x_c = \frac{A_1x_1 + A_2x_2 + A_3x_3 - A_4x_4}{A_1 + A_2 + A_3 - A_4}$$

$$= \frac{5000 \times \frac{100}{3} + 40000 \times 150 + 5000 \times \frac{800}{3} - 12717 \times 150}{37283}$$

$$= \frac{5592450}{37283}$$

$= 150$ mm

$$y_c = \frac{5000 \times \frac{200}{3} + 40000 \times 100 + 5000 \times \frac{200}{3} - 12717 \times \frac{120}{\pi}}{37283}$$

$= 112.13$ mm.

Thus the centroid of the rest of the area is (150, 112.13).

21. *Find the first moment of the area of the given section in Fig. (5-W21).*

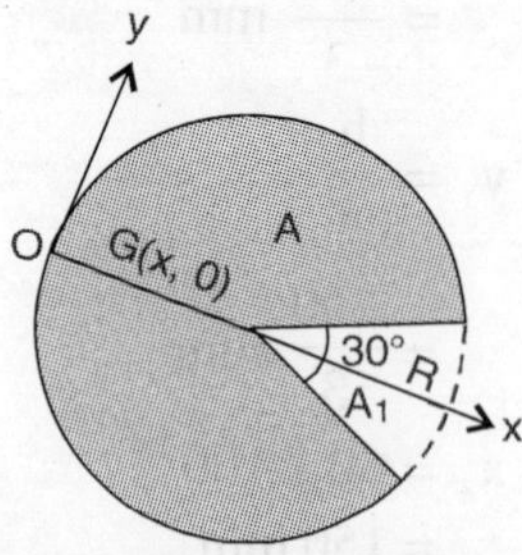

Fig. 5-W21

Solution:

Since given section is symmetrical about x-axis thus $y_c = 0$. Now the given shaded area is equal to the area of a circle of radius, R minus area of circular sector of angle $\alpha = 30°$.

$$A_s = A - A_1$$

$$= \pi R^2 - \frac{\pi R^2}{12}$$

$$= \frac{11\pi R^2}{12}$$

Now the centroid of the circle is g_1 (R, 0) since it is symmetrical about x-axis.

Let centroid of the circular sector be $g_2(x, 0)$.

Then $$x = \frac{2R}{3\alpha} \sin\alpha + R$$

$$= \frac{2 \times R}{3 \times \frac{\pi}{6}} \times \sin\left(\frac{\pi}{6}\right) + R$$

$$= \frac{4R}{3\pi} \times \frac{1}{2} + R$$

$$= \frac{2R}{3\pi} + R$$

$$= \left(\frac{2 + 3\pi}{3\pi}\right) R$$

Therefore, x-coordinate of centroid of the given section,

$$x_c = \frac{\pi R^2 \times R - \frac{\pi R^2}{12} \times \left(\frac{2+3\pi}{3\pi}\right) R}{\pi R^2 - \frac{\pi R^2}{12}}$$

$$= \frac{\pi R^3 - \frac{\pi(2+3\pi)R^3}{36\pi}}{\frac{11\pi R^2}{12}}$$

$$= \frac{(36\pi^2 - 2 - 3\pi)\pi R^3}{36\pi} \times \frac{12}{11\pi R^2}$$

$= 3.316$ R

Thus the centroid of the section is (3.316 R, 0).

22. *Locate the centroid of the given section with respect to x and y axes.*

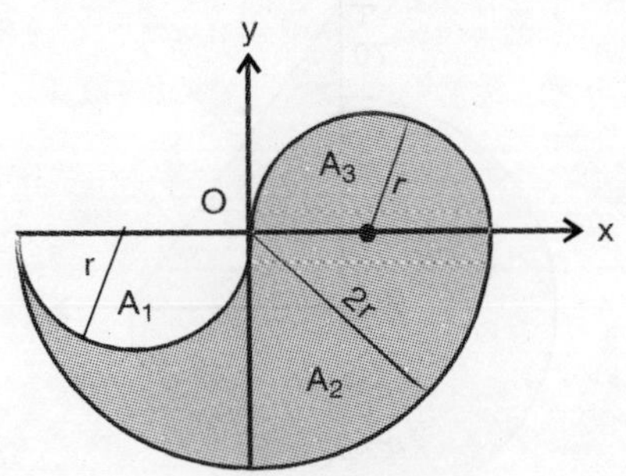

Fig. 5-W22

Solution:

Divide the given section into semicircles of radii r, 2r and r respectively.

Now the total area of the given section is,

$$A_T = A_2 - A_1 + A_3$$

$$= \frac{\pi}{2}(2r)^2 - \frac{\pi}{2}r^2 + \frac{\pi}{2}r^2$$

$$= 2\pi r^2 \text{ unit}^2.$$

Let the centroids of the semicircles be (x_1, y_1), (x_2, y_2) and (x_3, y_3) respectively.

Then, $(x_1, y_1) = \left(\frac{-r}{2}, \frac{-4r}{3\pi}\right)$

$(x_2, y_2) = \left(0, -\frac{4(2r)}{3\pi}\right)$

and $(x_3, y_3) = \left(\frac{r}{2}, \frac{4r}{3\pi}\right)$

Assume that the centroid of the given section is $\overline{X}$ and $\overline{Y}$.

Then, $\overline{X} = \frac{-A_1x_1 + A_2x_2 + A_3x_3}{-A_1 + A_2 + A_3}$

$$= \frac{-\frac{\pi}{2}r^2 \times \frac{-r}{2} + \frac{\pi}{2}(2r)^2 \times 0 + \frac{\pi}{2}r^2 \times \frac{r}{2}}{2\pi r^2}$$

$$= \frac{\frac{\pi r^3}{4} + \frac{\pi r^3}{4}}{2\pi r^2}$$

$$= \frac{\pi r^3}{2} \times \frac{1}{2\pi r^2}$$

$$= \frac{r}{4} \text{ unit.}$$

and $\overline{Y} = \frac{-A_1y_1 + A_2x_2 + A_3x_3}{-A_1 + A_2 + A_3}$

$$= \frac{-\frac{\pi}{2}r^2 \times \frac{-4r}{3\pi} + \frac{\pi}{2}(2r)^2 \times \left(-\frac{4(2r)}{3\pi}\right) + \frac{\pi}{2}r^2 \times \frac{4r}{3\pi}}{-\frac{\pi}{2}r^2 + \frac{\pi}{2}(2r)^2 + \frac{\pi}{2}r^2}$$

$$= \frac{\frac{2r^3}{3} - \frac{16r^3}{3} + \frac{2r^3}{3}}{2\pi r^2}$$

$$= \frac{-12r^3}{3 \times 2\pi r^2} = \frac{-4r}{2\pi}$$

$$= \frac{-2r}{\pi}.$$

23. *Determine the centroid of the section shown in the Fig. (5-W23). All the dimensions are in mm.*

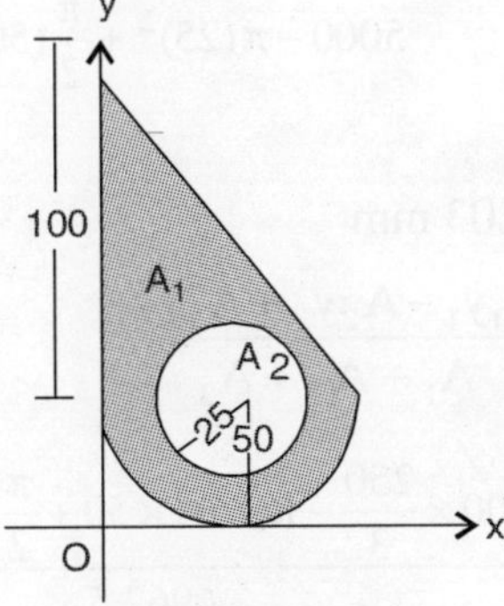

Fig. 5-W23

Solution:

Divide the given sections into:

1. Triangle (A_1) of size 100 mm×100 mm,

2. Circle (A_2) of radius 25 mm,

3. Semicircle of radius 50 mm.

The total shaded area

$$A_T = A_1 - A_2 + A_3$$

$$= \frac{1}{2} \times 100 \times 100 - \pi 25^2$$

$$+ \frac{\pi}{2} \times 50^2$$

$$= 6962.5 \text{ mm}^2$$

The centroids of the respective sections are,

$$g_1(x_1, y_1) = \left(\frac{100}{3}, \frac{100}{3} + 50\right)$$

$$= \left(\frac{100}{3}, \frac{250}{3}\right)$$

$$g_2(x_2, y_2) = (50, 50)$$

and $$g_3(x_3, y_3) = \left(50, 50 - \frac{4 \times 50}{3\pi}\right)$$

$$= (50, 28.76)$$

Then, x-coordinate of the shaded area is,

$$\overline{X} = \frac{A_1 x_1 - A_2 x_2 + A_3 x_3}{A_1 - A_2 + A_3}$$

$$= \frac{5000 \times \frac{100}{3} - \pi(25)^2 \times 50 + \frac{\pi}{2}(50)^2 \times 50}{5000 - \pi(25)^2 + \frac{\pi}{2}(50)^2}$$

$$= 38.03 \text{ mm}$$

$$\overline{Y} = \frac{A_1 y_1 - A_2 y_2 + A_3 y_3}{A_1 - A_2 + A_3}$$

$$= \frac{5000 \times \frac{250}{3} - \pi(25)^2 \times 50 + \frac{\pi}{2}(50)^2 \times 28.76}{6962.5}$$

$$= 61.96 \text{ mm.}$$

Therefore the centroid of the given section is at a point (38.03 mm, 61.96 mm).

24. *Find the centroid of the section shown in Fig. (5-W24). All the dimensions are in mm.*

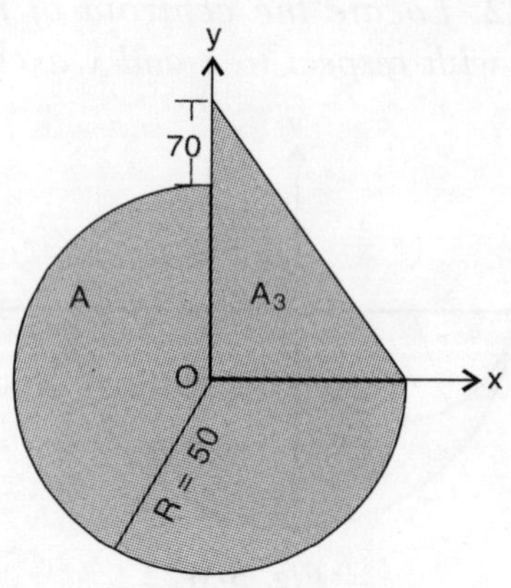

Fig. 5-W24

Solution:

Divide the given sections into:

1. A circle (A) of radius, R = 50 mm,

2. A quarter circle (A_2) of radius, R = 50 mm, with a negative sign because it has to be deducted.

3. Triangle (A_3) of size 120 × 50 mm².

The total area of the section,

$$A_T = A_1 - A_2 + A_3$$

$$= \pi(50)^2 - \frac{\pi}{4}(50)^2 + \frac{1}{2} \times 120 \times 50$$

$$= 7850 - 1962.5 + 3000$$

$$= 8887.5 \text{ mm}^2$$

Now with respect to the given coordinate system, the centroids of the circle, quarter circle and triangle are,

$$g_1(x_1, y_1) = (0, 0)$$

[Because circle is symmetrical about both the axes.]

$$g_2(x_2, y_2) = \left(\frac{4 \times 50}{3 \times \pi}, \frac{4 \times 50}{3 \times \pi}\right)$$

$$= (21.23, 21.23)$$

$$g_3(x_3, y_3) = \left(\frac{b}{3}, \frac{h}{3}\right)$$

$$= \left(\frac{50}{3}, \frac{120}{3}\right)$$

$$= (16.66, 40)$$

Therefore, the centroid of the composite section is,

$$\overline{X} = \frac{A_1x_1 - A_2x_2 + A_3x_3}{A_1 - A_2 + A_3}$$

$$= \frac{7850 \times 0 - 1962.5 \times 21.23 + 3000 \times 16.66}{8887.5}$$

$= 0.93$ mm

and $$\overline{Y} = \frac{A_1y_1 - A_2y_2 + A_3y_3}{A_1 - A_2 + A_3}$$

$$= \frac{7850 \times 0 - 1962.5 \times 21.23 + 3000 \times 40}{8887.5}$$

$= 8.81$ mm.

Thus the centroid of the composite given section is point G (0.93, 8.81).

25. *Find the centroid of the area bounded by the curve,* $|x| + |y| = 1$ *with the given axes α-B-β shown in Fig. (5-W25).*

Solution:

Case: (a) $x > 0, y > 0$

Then, $|x| = x$ and $|y| = y$

$\therefore \quad x + y = 1$ .....(i)

Case: (b) $x < 0, y > 0$

Then, $|x| = -x$ and $|y| = y$

$\therefore \quad -x + y = 1$ .....(ii)

Case: (c) $x > 0, y < 0$

Then, $|x| = x$ and $|y| = -y$

$\therefore \quad x - y = 1$ .....(iii)

Case: (d) $x < 0, y < 0$

Then, $|x| = -x$ and $|y| = y$

$\therefore \quad -x - y = 1$ .....(iv)

Now if we draw the equations (i), (ii), (iii) and (iv) on the coordinate axes, we get the area bounded by lines as shown in Fig.(5-W25).

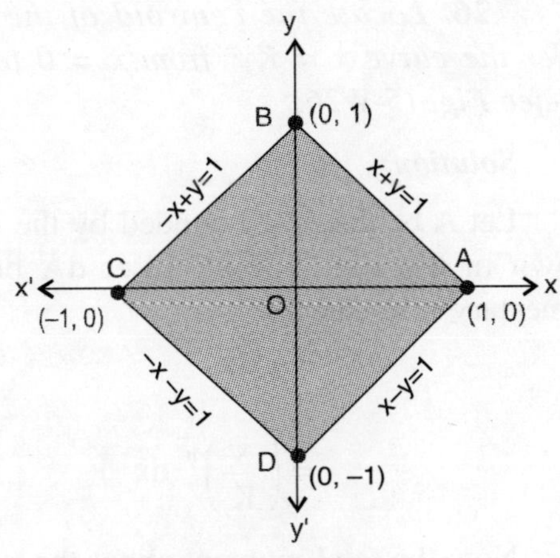

Fig. 5-W25

Now, $x + y = 1$ .....(i)

Putting $x = 0$, we get

$y = 1$

$\Rightarrow$ B (0, 1)

Putting $y = 0$, we get

$x = 1$

$\Rightarrow$ A (1 , 0); $-x + y = 1$ .....(ii)

Putting $x = 0$, we get

$y = 1$

$\Rightarrow$ B(0, 1)

Putting $y = 0$, we get

$x = -1$

$\Rightarrow$ C(–1 , 0); $x - y = 1$ .....(iii)

Putting $x = 0$, we get

$y = -1$

$\Rightarrow$ D (0, –1)

Putting $y = 0$, we get

$x = 1$

$\Rightarrow$ A (1, 0)

Since the area bounded by lines is symmetrical about the x and y-axes, therefore the centroid is O (0 , 0) with respect to x and y axes. But it is $\left(\frac{1}{\sqrt{2}}, \frac{1}{\sqrt{2}}\right)$ with respect to axis α-B-β since AB = BC = CA = DA = $\sqrt{2}$.

26. *Locate the centroid of the area under the curve $x = Ky^n$ from $x = 0$ to $x = a$. Refer Fig. (5-W26).*

Solution:

Let A be the area bounded by the curve shown in the Fig.(5-W26). Let dA be the elementary area then,

$$dA = y\,dx$$

$$= \left(\frac{x}{K}\right)^{\frac{1}{n}} dx$$

Now the total moment about the y-axis, M_y is,

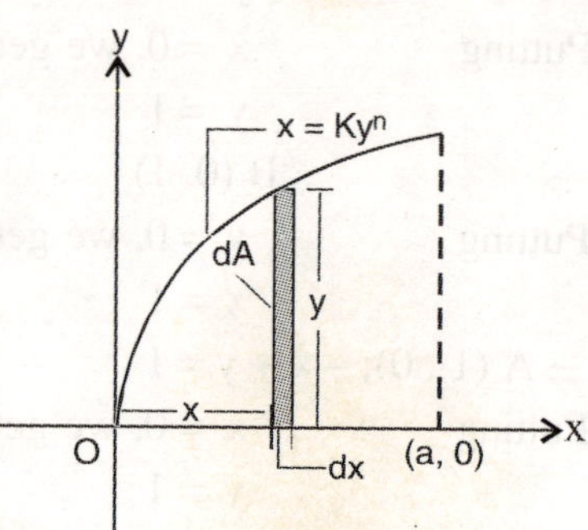

Fig. 5-W26

$$M_y = \int_A x\,dA$$

$$= \int_{x=0}^{x=a} xy\,dx$$

$$= \int_{x=0}^{x=a} x\left(\frac{x}{K}\right)^{\frac{1}{n}} dx$$

$$= \left(\frac{1}{K}\right)^{\frac{1}{n}} \int_{x=0}^{x=a} x^{1+\frac{1}{n}} dx$$

$$= \left(\frac{1}{K}\right)^{\frac{1}{n}} \left[\frac{x^{2+\frac{1}{n}}}{2+\frac{1}{n}}\right]_0^a$$

$$= \frac{\left(\frac{1}{K}\right)^{\frac{1}{n}}}{\left(2+\frac{1}{n}\right)}\left[a^{2+\frac{1}{n}} - 0^{2+\frac{1}{n}}\right]$$

$$= \frac{n}{2n+1}\left(\frac{1}{K}\right)^{\frac{1}{n}} a^{\frac{2n+1}{n}}$$

The moment of area, M_x about x-axis is,

$$M_x = \int_A \frac{y}{2} y\,dx$$

$$= \frac{1}{2}\int_{x=0}^{x=a} \left(\frac{x}{K}\right)^{\frac{2}{n}} dx$$

$$= \frac{1}{2}\left(\frac{1}{K}\right)^{\frac{2}{n}} \int_{x=0}^{x=a} x^{\frac{2}{n}} dx$$

$$= \frac{1}{2}\left(\frac{1}{K}\right)^{\frac{2}{n}} \left[\frac{x^{\frac{2}{n}+1}}{\frac{2}{n}+1}\right]_{x=0}^{x=a}$$

$$= \frac{1}{2}\left(\frac{1}{K}\right)^{\frac{2}{n}} \frac{n}{n+2}\left[a^{\frac{n+2}{n}} - 0\right]$$

$$= \frac{1}{2}\left(\frac{1}{K}\right)^{\frac{2}{n}} \frac{n}{n+2} a^{\frac{n+2}{n}}$$

$$= \frac{1}{2}\left(\frac{1}{K}\right)^{\frac{2}{n}} \left(\frac{n}{n+2}\right) a^{\frac{n+2}{n}}$$

Now the total area,

$$A = \int_A dA$$

$$= \int_A y\,dx$$

$$= \int_{x=0}^{x=a} \left(\frac{x}{K}\right)^{\frac{1}{n}} dx$$

$$= \left(\frac{1}{K}\right)^{\frac{1}{n}} \left[\frac{x^{1+\frac{1}{n}}}{1+\frac{1}{n}} \right]_0^a$$

$$= \left(\frac{1}{K}\right)^{\frac{1}{n}} \frac{n}{n+1} \left[a^{1+\frac{1}{n}} - 0 \right]$$

$$= \left(\frac{1}{K}\right)^{\frac{1}{n}} \frac{n}{n+1} a^{\frac{n+1}{n}}$$

∴ The x-coordinate of the centroid,

$$\overline{X} = \frac{M_y}{A}$$

$$= \frac{n}{2n+1} \times \frac{n+1}{n} a^{\frac{2n+1}{n} - \frac{n+1}{n}}$$

$$= \frac{n+1}{2n+1} a$$

and the y-coordinate of the centroid,

$$\overline{Y} = \frac{M_x}{A}$$

$$= \frac{\frac{1}{2}\left(\frac{1}{K}\right)^{\frac{2}{n}}\left(\frac{n}{n+2}\right) a^{\frac{n+2}{n}}}{\left(\frac{1}{K}\right)^{\frac{1}{n}}\left(\frac{n}{n+1}\right) a^{\frac{n+1}{n}}}$$

$$= \frac{1}{2}\left(\frac{1}{K}\right)^{\frac{1}{n}} \frac{n+1}{n+2} a^{\frac{1}{n}}$$

$$\therefore \quad G = \left\{ \frac{n+1}{2n+1} a, \frac{1}{2}\left(\frac{1}{K}\right)^{\frac{1}{n}} \frac{n+1}{n+2} a^{\frac{1}{n}} \right\}$$

27. *Determine the coordinates of the centroid of the shaded area shown in the Fig. (5-W27).*

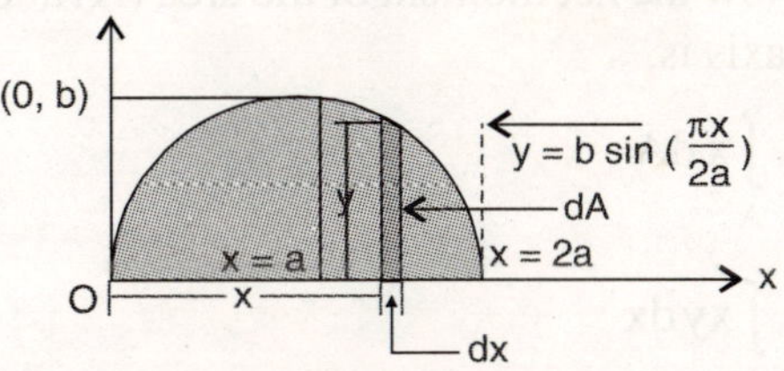

Fig. 5-W27

Solution:

Let dA be the elementary area of the rectangular strip.

$$\therefore \quad dA = \text{area of rectangular strip}$$
$$= \text{width} \times \text{height}$$
$$= dx \times y$$
$$= y\, dx$$

Integrating over the proper limits, we get the area of bounded region,

$$A = \int_A dA$$

$$= \int_{x=a}^{x=2a} y\, dx$$

$$= \int_{x=a}^{x=2a} b \sin\frac{\pi x}{2a}\, dx$$

$$= b\left[-\cos\frac{\pi x}{2a}\right]_{x=a}^{x=2a} \times \frac{1}{\frac{\pi}{2a}}$$

$$= \frac{2ab}{\pi}\left[-\cos\frac{\pi x}{2a}\right]_a^{2a}$$

$$= -\frac{2ab}{\pi}\left[\cos\frac{\pi}{2a}2a - \cos\frac{\pi}{2a}a\right]$$

$$= -\frac{2ab}{\pi}\left[\cos\pi - \cos\frac{\pi}{2}\right]$$

$$= -\frac{2ab}{\pi}(-1-0)$$

$$= \frac{2ab}{\pi}$$

Now the net moment of the area (A) about the y-axis is,

$$M_y = \int_A x\,dA$$

$$= \int_A xy\,dx$$

$$= \int_{x=a}^{x=2a} x\,b\sin\frac{\pi x}{2a}dx$$

$$= b\int_{x=a}^{x=2a} x\sin\left[\frac{\pi x}{2a}\right]dx$$

$$= b\left[x\int\sin\frac{\pi x}{2a}dx - \int\left(\frac{dx}{dx}\int\sin\frac{\pi x}{2a}dx\right)dx\right]_{x=a}^{x=2a}$$

$$= b\left[x\left(-\cos\frac{\pi x}{2a}\right)\frac{2a}{\pi} - \int -\frac{2a}{\pi}\cos\frac{\pi x}{2a}dx\right]_{x=a}^{x=2a}$$

$$= b\left[-\frac{2a}{\pi}x\cos\frac{\pi x}{2a} + \left(\frac{2a}{\pi}\right)^2\sin\frac{\pi x}{2a}\right]_{x=a}^{x=2a}$$

$$= b\left[-\frac{2a}{\pi}\times 2a\cos\frac{\pi}{2a}\times 2a + \left(\frac{2a}{\pi}\right)^2\sin\frac{\pi 2a}{2a}\right.$$

$$\left. + \frac{2a}{\pi}\times a\times\cos\frac{\pi a}{2a} - \left(\frac{2a}{\pi}\right)^2\sin\frac{\pi\times a}{2a}\right]$$

$$= b\left[\frac{4a^2}{\pi} - \frac{4a^2}{\pi^2}\right]$$

$$= 4a^2b\left[\frac{1}{\pi} - \frac{1}{\pi^2}\right]$$

The total moment about x-axis is,

$$M_y = \int_A x\,dA$$

$$= \int_A \frac{y}{2}dA$$

$$= \int_{x=a}^{x=2a} \frac{y}{2}y\,dx$$

$$= \frac{1}{2}\times\int_{x=a}^{x=2a} y^2dx$$

$$= \frac{1}{2}\times\int_{x=a}^{x=2a}\left(b\sin\frac{\pi x}{2a}\right)^2dx$$

$$= \frac{1}{2}b^2\int_{x=a}^{x=2a}\frac{1-\cos\frac{2\pi x}{2a}}{2}dx$$

$$= \frac{1}{4}b^2\int_{x=a}^{x=2a} 1-\cos\frac{\pi x}{a}dx$$

$$= \frac{b^2}{4}\left[x - \frac{a}{\pi}\sin\frac{\pi x}{a}\right]_{x=a}^{x=2a}$$

$$= \frac{b^2}{4}\left[2a - \frac{a}{\pi}\sin\frac{2a\pi}{a} - a + \frac{a}{\pi}\sin\frac{a\pi}{a}\right]$$

$$= \frac{ab^2}{4}$$

$$\overline{X} = \frac{M_y}{A}$$

$$= \frac{4a^2b\left[\frac{1}{\pi} - \frac{1}{\pi^2}\right]}{\frac{2ab}{\pi}}$$

$$= 4a^2b\left[\frac{\pi-1}{\pi^2}\right]\times\frac{\pi}{2ab}$$

$$= 2a\left[\frac{\pi-1}{\pi}\right]$$

and $$\overline{Y} = \frac{M_x}{A}$$

$$= \frac{\frac{ab^2}{4}}{\frac{2ab}{\pi}}$$

$$= \frac{ab^2}{4} \times \frac{\pi}{2ab}$$

$$= \frac{b}{8}\pi$$

$$= \frac{\pi b}{8}$$

The centroid of the area bounded by the curve is $G = \left[\frac{2a}{\pi}(\pi - 1), \frac{\pi b}{8}\right]$.

28. *Find the centroid of area bounded by the straight lines $y = m_1x$ and $y = m_2x + K$, $(m_2>m_1)$ and the ordinates at $x = 0$ and $x = a$.*

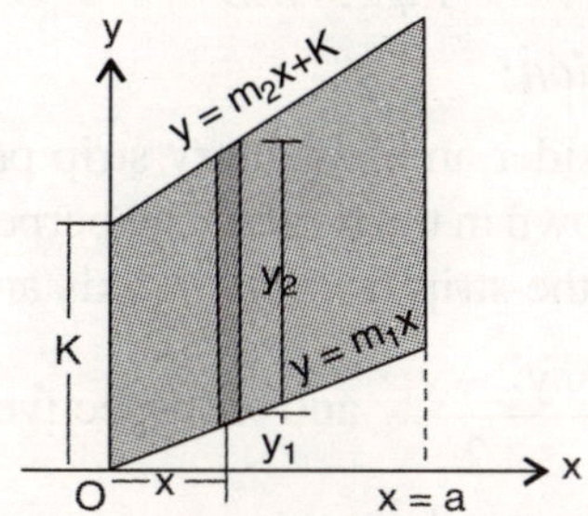

Fig. 5-W28

Solution:

The area bounded by the curves and the ordinates is shown in the Fig. (5-W28). Let us consider a rectangular elementary strip of area, therefore,

$$dA = \text{width} \times \text{height}$$

$$= dx\ y_2 - y_1$$

$$= (y_2 - y_1)\ dx$$

where, $y_2 = m_2x + K$

and $y_1 = m_1x$

$$\therefore \quad dA = (m_2 - m_1)\ x + K\ dx$$

Now we can integrate dA over proper limits to get the total area bounded.

$$\therefore \quad A = \int_{x=0}^{x=a} [(m_2 - m_1)x + K]dx$$

$$= \left[\Delta m \frac{x^2}{2} + Kx\right]_{x=0}^{x=a}$$

$$[\Delta m = m_2 - m_1]$$

$$= \frac{\Delta m}{2}a^2 + Ka$$

Now suppose that the moment of dA about the y-axis is dM_y then,

$$dM_y = x\ dA$$

$$= x\ (\Delta mx + K)\ dx$$

The moment of the area bounded about the y-axis is,

$$M_y = \int dM_y$$

$$= \int_{x=0}^{x=a} x(\Delta mx + K)dx$$

$$= \int_{x=0}^{x=a} (\Delta mx^2 + Kx)dx$$

$$= \left[\Delta m \frac{x^3}{3} + K\frac{x^2}{2}\right]_{x=0}^{x=a}$$

$$= \frac{\Delta ma^3}{3} + \frac{Ka^2}{2}$$

Similarly, suppose that moment of dA about x-axis, $dM_x = Y_c dA$ where, Y_c = y-coordinate of the centroid of the elementary rectangular strip.

$$= y_1 + \frac{y_2 - y_1}{2}$$

$$= \frac{1}{2}(y_1 + y_2)$$

$$\therefore M_x = \int_{x=0}^{x=a} \frac{1}{2}(y_1 + y_2)(\Delta mx + K)\ dx$$

$$= \frac{1}{2}\int_{x=0}^{x=a}\left(\Delta m^{1}x+K\right)\left(\Delta mx+K\right)dx$$

[where $\Delta m' = m_1 + m_2$]

$$= \frac{1}{2}\int_{x=0}^{x=a}\left(\Delta m'\Delta mx^{2}+K(\Delta m'+\Delta m)x+K^{2}\right)dx$$

$$= \frac{1}{2}\left[\Delta m'\Delta m\frac{x^{3}}{3}+K(\Delta m'+\Delta m)\frac{x^{2}}{2}+K^{2}x\right]_{x=0}^{x=a}$$

$$= \frac{1}{2}\left[\frac{\Delta m'\Delta ma^{3}}{3}+\frac{K(\Delta m'+\Delta m)a^{2}}{2}+K^{2}a\right]$$

$$\therefore\quad \overline{X} = \frac{M_y}{A}$$

$$= \frac{\dfrac{\Delta ma^{3}}{3}+\dfrac{Ka^{2}}{2}}{\dfrac{\Delta m}{2}a^{2}+Ka}$$

$$= \frac{2\Delta ma^{3}+3Ka^{2}}{6}\times\frac{2}{\Delta ma^{2}+2Ka}$$

$$= \frac{2\Delta ma^{3}+3Ka^{2}}{3(\Delta ma^{2}+2Ka)}$$

$$= \frac{2\Delta ma^{2}+3Ka}{3(\Delta ma+2K)}$$

and

$$\overline{Y} = \frac{\dfrac{2\Delta m'\Delta ma^{3}+3Ka^{2}(\Delta m'+\Delta m)+6K^{2}a}{12}}{\dfrac{2\Delta ma^{3}+3Ka^{2}}{6}}$$

$$= \frac{2\Delta m'\Delta ma^{3}+3Ka^{2}(\Delta m'+\Delta m)+6K^{2}a}{12}\times\frac{6}{2\Delta ma^{3}+Ka^{2}}$$

$$= \frac{2\Delta m'\Delta ma^{3}+3Ka^{2}(\Delta m'+\Delta m)+6K^{2}a}{(4\Delta ma^{3}+2Ka^{2})}$$

29. *Find the centroid of the figure bounded by the ellipse $4x^2 + 9y^2 = 36$ and thecircle $x^2 + y^2 = 9$ and which is situated in the first quadrant as shown in Fig. (5-W29).*

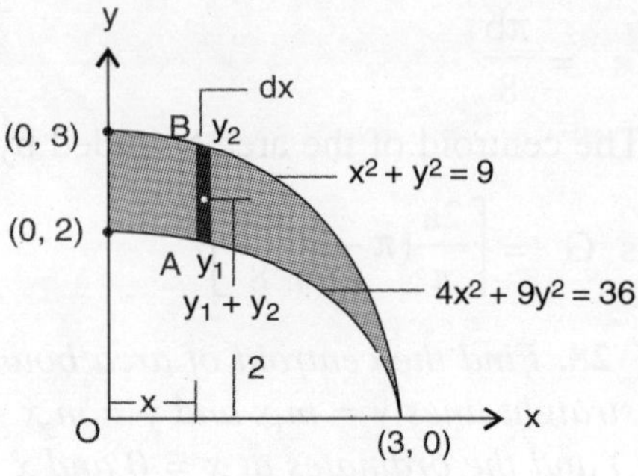

Fig. 5-W29

Solution:

Consider an elementary strip parallel to y-axis as shown in the figure. The perpendicular distance of the strip from the x-axis and y-axis $\pi/2$ are $y_1 + \frac{y_2 - y_1}{2}$ and x respectively. Now the total moment about the y-axis is,

$$M_y = \int_A x\,dA$$

$$= \int_A x(y_2 - y_1)\,dx$$

$$= \int_0^3 x\left\{\sqrt{9-x^2}-\frac{2}{3}\sqrt{9-x^2}\right\}dx$$

$$= \int_0^3 \frac{1}{3}x\sqrt{9-x^2}\,dx$$

Substituting $x = 3\sin\theta$

Then, $dx = 3\cos\theta\, d\theta$

As x varies from 0 to 3 then θ will vary from 0 to $\frac{\pi}{2}$.

So, $$M_y = \int_0^{\frac{\pi}{2}} \sin 3 \sin\theta \times 3\cos\theta \times 3 \times \cos\theta \, d\theta$$

$$= 9\int_0^{\frac{\pi}{2}} \sin\theta \cos^2\theta \, d\theta$$

Now substitute $z = \cos\theta$. As θ varies from 0 to 1 therefore z will vary from 1 to 0.

$$\therefore \quad M_y = -9\left[\frac{z^3}{3}\right]_1^0 = 3$$

Similarly the moment of the area about the x-axis is,

$$M_x = \int_A y\,dA$$

$$= \int_A \left(y_1 + \frac{y_2 - y_1}{2}\right)(y_2 - y_1)dx$$

$$= \frac{1}{2}\int_A \left(y_2^2 - y_1^2\right)dx$$

$$= \frac{1}{2}\int_0^3 \left(\sqrt{9-x^2}\right)^2 - \left(\frac{2}{3}\sqrt{9-x^2}\right)^2 dx$$

$$= \frac{1}{2}\int_0^3 (9-x^2) - \frac{4}{9}(9-x^2)dx$$

$$= \frac{5}{18}\int_0^3 (9-x^2)dx$$

$$= \frac{5}{18}\left[9x - \frac{x^3}{3}\right]_0^3$$

$$= \frac{5}{18}(27-9)$$

$$= 5$$

Now area of the bounded region,

$$A = \int_A dA$$

$$= \int_0^3 (y_2 - y_1)\,dx$$

$$= \int_0^3 \left(\sqrt{9-x^2} - \frac{2}{3}\sqrt{9-x^2}\right)dx$$

$$= \frac{1}{3}\int_0^3 \sqrt{9-x^2}\,dx = \frac{3\pi}{4}$$

$$\therefore \quad \bar{X} = \frac{M_y}{A} = \frac{3}{\left(3\frac{\pi}{4}\right)} = \frac{4}{\pi}$$

and $$\bar{Y} = \frac{M_x}{A} = \frac{5}{3\frac{\pi}{4}} = \frac{20}{3\pi}$$

30. *Find the centroid of the area bounded by the arc of the parabola $y^2 = 2x$ between $x = 0$ and $x = 2$ $(y > 0)$.*

Solution:

Area bounded by the given curve is shown in Fig. (5-W30). Let an elementary strip as shown in figure having elementary area, $dA = y\,dx$.

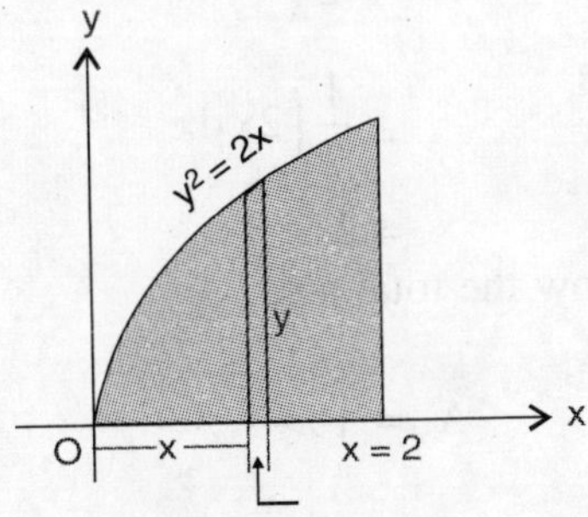

Fig. 5-W30

Now the moment of the area about the y-axis

$$M_y = \int_A x\,dA$$

$$= \int_A xy\,dx$$

$$= \int_0^2 x\sqrt{2x}\,dx$$

$$= \sqrt{2}\int_0^2 x^{3/2}dx$$

$$= \sqrt{2}\left(\frac{x^{\frac{3}{2}+1}}{\frac{3}{2}+1}\right)_0^2$$

$$= \sqrt{2}\times\frac{2}{5}\left[x^{\frac{5}{2}}\right]_0^2$$

$$= \frac{2\sqrt{2}}{5}2^{\frac{5}{2}}$$

$$= \frac{2\sqrt{2}}{5}\times 4\sqrt{2} = \frac{16}{5}$$

Similarly the moment of the area about the x-axis

$$M_x = \frac{1}{2}\int_A y\,dA$$

$$= \frac{1}{2}\int_A y^2\,dx$$

$$= \frac{1}{2}\int_0^2 2x\,dx$$

$$= 2$$

Now the total area,

$$A = \int_0^2 y\,dx$$

$$= \sqrt{2}\int_0^2 x^{1/2}dx$$

$$= \sqrt{2}\left(\frac{x^{\frac{3}{2}}}{\frac{3}{2}}\right)$$

$$\frac{2\sqrt{2}}{3}2^{3/2} = \frac{2\sqrt{2}\times 2\sqrt{2}}{3} = \frac{8}{3}$$

$$\therefore \quad \overline{X} = \frac{M_y}{A} = \frac{16}{5}\times\frac{3}{8} = \frac{6}{5}$$

and $$\overline{Y} = \frac{M_x}{A} = 2\times\frac{3}{8} = \frac{3}{4}$$

31. *Find the centroid of the arc of curve $y^2 = 2x$ between $x \in [0, 2]$ as shown in Fig. (5-W31).*

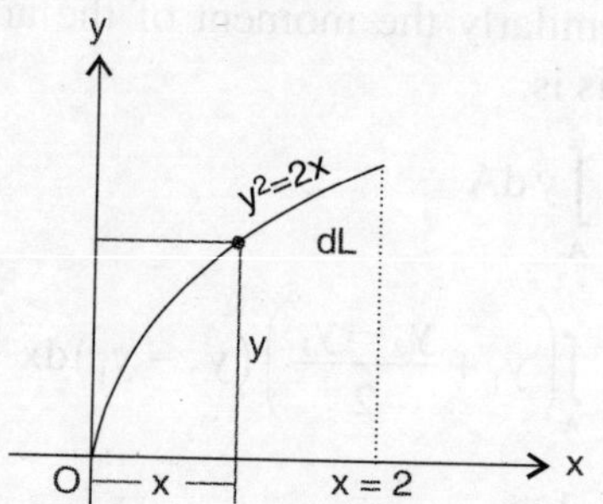

Fig. 5-W31

Solution:

Let the length of the elementary arc be dL, then by the definition we know that

$$L\overline{X} = \int_L x\,dL$$

$$= \int_0^2 x\left(\frac{dL}{dx}\right)dx$$

$$= \int_0^2 x\sqrt{1+\left(\frac{dy}{dx}\right)^2}\,dx$$

$$= \int_0^2 x\sqrt{1+\left\{\frac{d}{dx}\left(\sqrt{2x}\right)\right\}^2}\,dx$$

$$= \int_0^2 x\sqrt{1+2\left(\frac{d\left(\sqrt{x}\right)}{dx}\right)^2}\,dx$$

$$= \int_0^2 x\sqrt{1+2\left(\frac{1}{2\sqrt{x}}\right)^2}\,dx$$

$$= \int_0^2 x\sqrt{1+\frac{1}{2x}}\,dx$$

$$= \quad x \in [0, 2]$$

$$= \int_0^2 \sqrt{x^2+\frac{x}{2}}\,dx$$

$$= \int_0^2 \sqrt{x^2+2.x.\frac{1}{4}+\frac{1}{16}-\frac{1}{16}}\,dx$$

$$= \int_0^2 \sqrt{\left(x+\frac{1}{4}\right)^2-\left(\frac{1}{4}\right)^2}\,dx$$

Let us supose that $x+\frac{1}{4}=\frac{1}{4}\sec\theta$

$$\therefore\ dx = \frac{1}{4}\sec\theta\tan\theta\,d\theta$$

$$= \int_{\theta=0}^{\theta=\sec^{-1}9} \sqrt{\frac{1}{16}\sec^2\theta-\frac{1}{16}} \times \frac{1}{4}\sec\theta\tan\theta\,d\theta$$

$$= \frac{1}{16}\int_{\theta=0}^{\theta=\sec^{-1}9} \tan\theta\,.\,\sec\theta\,.\,\tan\theta\,d\theta$$

$$= \frac{1}{16}\int_{\theta=0}^{\theta=\sec^{-1}9} \tan^2\theta\sec\theta\,d\theta$$

$$= \frac{1}{16}\int_{\theta=0}^{\theta=\sec^{-1}9} (\sec^3\theta-\sec\theta)\,d\theta$$

$$= \frac{1}{16}\left[\frac{\sec\theta\tan\theta-\ln(\sec\theta+\tan\theta)}{2}\right]_{\theta=0}^{\theta=\sec^{-1}9}$$

$$= \frac{1}{16}\left[\frac{9\times\sqrt{80}-\ln\left(9+\sqrt{80}\right)}{2}-\frac{1\times 0-\ln(1+0)}{2}\right]$$

$$= \frac{9\sqrt{80}-\ln\left(9+\sqrt{80}\right)}{32}$$

$$= \frac{9\sqrt{20}-\frac{1}{2}\ln\left(9+\sqrt{80}\right)}{16}$$

$$= \frac{18\sqrt{5}}{16}-\frac{1}{32}\ln\left(9+\sqrt{80}\right)$$

$$= \frac{9}{8}\sqrt{5}+\frac{1}{16}\ln\left(\sqrt{9+\sqrt{80}}\right)^{-1}$$

$$= \frac{9}{8}\sqrt{5}+\frac{1}{16}\ln\left(\frac{1}{\sqrt{9+\sqrt{80}}}\right)$$

$$= \frac{9}{8}\sqrt{5}+\frac{1}{16}\ln\sqrt{9-\sqrt{80}}$$

Similarly, we have $L\overline{Y} = \int_L y\,dL$

$$= \int_{x=0}^{x=2} \sqrt{2x}\sqrt{1+\left(\frac{dy}{dx}\right)^2}\,dx$$

$$= \int_{x=0}^{x=2} \left(\sqrt{2x}\right)\sqrt{1+\left(\frac{d}{dx}\left(\sqrt{2x}\right)\right)^2}\,dx$$

$$= \int_{x=0}^{x=2} \sqrt{2x}\sqrt{1+\frac{1}{2x}}\,dx$$

$$= \int_{x=0}^{x=2} \sqrt{1+2x}\,dx$$

$$= \left[\frac{(1+2x)^{1+\frac{1}{2}}}{2\left(1+\frac{1}{2}\right)}\right]_{x=0}^{x=2}$$

$$= \frac{1}{3}\left[(2x+1)^{\frac{3}{2}}\right]_{x=0}^{x=2}$$

$$= \frac{1}{3}\left[5^{\frac{3}{2}}-1^{\frac{3}{2}}\right]$$

$$= \frac{1}{3}\left[5\sqrt{5}-1\right]$$

$$= \frac{1}{3}\left[5\sqrt{5}-1\right]$$

Now as we know that,

$$dL = \sqrt{1+\left(\frac{dx}{dy}\right)^2}\,dy$$

$$= \sqrt{1+y^2}\,dy \quad \left[\because \frac{dx}{dy} = y\right]$$

$$\therefore \quad L = \int_{y=0}^{y=2} \sqrt{1+y^2}\,dy$$

By substituting y = tan ψ, we have

$$\therefore \quad dy = \sec^2\psi\, d\psi$$

$$\therefore \quad L = \int_0^{\tan^{-1}2} \sec^3\psi\, d\psi$$

$$= [\sec\psi.\tan\psi + \ln(\sec\psi + \tan\psi)]_0^{\tan^{-1}2}$$

$$= 2\sqrt{9} + \ln\left(2+\sqrt{9}\right)$$

$$= 6 + \ln 5$$

Therefore, $$\overline{X} = \frac{\left(\frac{9}{8}\sqrt{5} + \frac{1}{16}\ln\sqrt{9-\sqrt{80}}\right)}{6+\ln 5}$$

and $$\overline{Y} = \frac{\frac{5\sqrt{5}-1}{3}}{6+\ln 5}$$

$$= \frac{5\sqrt{5}-1}{18+\ln 125}.$$

32. *Determine the coordinates of the centroid of the shaded area as shown in the Fig (5-W32).*

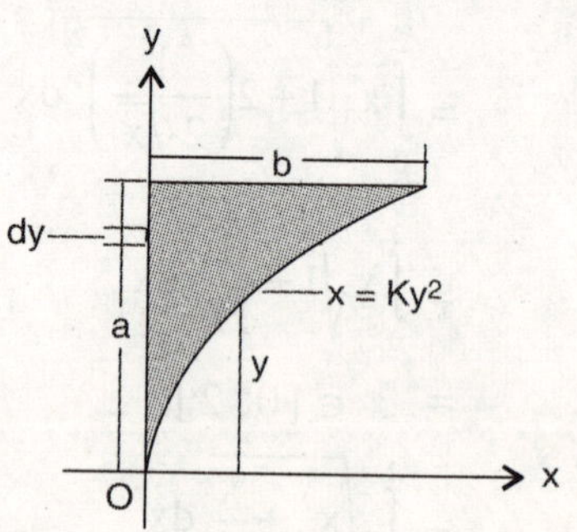

Fig. 5-W32

First method:

Solution:

Consider the rectangular strip parallel to the x-axis of an elementary area dA.

$$\therefore \quad dA = \text{width} \times \text{height}$$

$$= x\,dy$$

Integrating dA over the proper limits, we get

$$A = \int_A dA$$

$$= \int_{y=0}^{y=a} dA$$

$$= \int_{y=0}^{y=a} x\,dy$$

$$= \int_{y=0}^{y=a} Ky^2 dy$$

$$= K\int_{y=0}^{y=a} y^2 dy$$

$$= K\left[\frac{y^3}{3}\right]_{y=0}^{y=a}$$

$$= \frac{K}{3}\left[a^3 - 0^3\right]$$

$$= \frac{Ka^3}{3}$$

$$= \frac{b}{a^2} \times \frac{a^3}{3}$$

$$= \frac{ab}{3} \qquad [\because b = Ka^2]$$

Now the moment, dM_y of an elementary area, dA about the y-axis is,

$$dM_y = x_c dA$$

where, $x_c = \frac{x}{2}$ is the x-coordinate of the the centroid of the elementary area, dA

or $$dM_y = \frac{x}{2} dA$$

$$= \frac{x}{2} x\,dy$$

$$= \frac{x^2}{2} dy$$

$$\therefore \quad M_y = \int_{y=0}^{y=a} \frac{x^2}{2} dy$$

$$= \frac{1}{2} \int_{y=0}^{y=a} (Ky^2)^2 dy$$

$$= \frac{1}{2} \int_{y=0}^{y=a} K^2 y^4 dy$$

$$= \frac{K^2}{2} \left[\frac{y^{4+1}}{4+1} \right]_{y=0}^{y=a}$$

$$= \frac{K^2}{2} \left[\frac{a^5}{5} - \frac{0^5}{5} \right]$$

$$= \frac{K^2 a^5}{10}$$

$$= \left(\frac{b}{a^2} \right)^2 \times \frac{a^5}{10}$$

$$= \frac{ab^2}{10}$$

Similarly, the moment, dM_x of the elementary area dA about the x-axis is,

$$dM_x = y_c\, dA;$$

where $y_c = y$ is y-coordinate of centroid of the elementary area, dA

$$= y\, dA$$

$$= yx\, dy$$

$$= xy\, dy$$

$$\therefore \quad M_x = \int_{M_X} dM_x$$

$$= \int_{y=0}^{y=a} xy\, dy$$

$$= \int_{y=0}^{y=a} Ky^2 \times y\, dy$$

$$= K \int_{y=0}^{y=a} y^3 dy$$

$$= K \left[\frac{y^4}{4} \right]_{y=0}^{y=a}$$

$$= \frac{K}{4} \left[a^4 - 0^4 \right]$$

$$= \frac{Ka^4}{4}$$

$$= \left(\frac{b}{a^2} \right) \times \frac{a^4}{4}$$

$$= \frac{a^2 b}{4}$$

As we know that, x-coordinate of centroid of the shaded area is given by,

$$A\overline{X} = \int_A x_c dA$$

$$\therefore \quad \overline{X} = \frac{\int x_c dA}{\int dA}$$

$$= \frac{M_y}{A}$$

$$= \frac{ab^2}{10} \times \frac{3}{ab}$$

$$= \frac{3b}{10}$$

Similarly, y-coordinate of centroid of the shaded area is given by,

$$A\overline{Y} = \int_A y_c \, dA$$

$$= M_x$$

$$\therefore \quad \overline{Y} = \frac{M_x}{A}$$

$$= \frac{a^2 b}{4} \times \frac{3}{ab}$$

$$= \frac{3a}{4}$$

Therefore the centroid (G) of the area bounded by the given curve is $\left(\frac{3b}{10}, \frac{3a}{4}\right)$.

Second method: Consider an elementary rectangular strip parallel to the y-axis of an elementary area, dA.

Therefore,

$$dA = \text{width} \times \text{height}$$

$$= dx \times (a - y)$$

$$= (a - y)\, dx$$

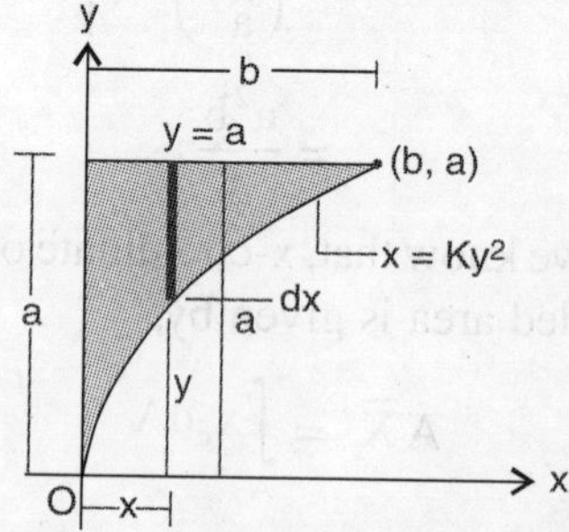

Fig. 5-W33

Integrating dA over the proper limits, we get

$$A = \int_A dA$$

$$= \int_{x=0}^{x=b} (a - y)\, dx$$

$$= \int_{x=0}^{x=b} \left(a - \frac{\sqrt{x}}{\sqrt{K}}\right) dx$$

$$= \left[ax - \frac{1}{\sqrt{K}} \frac{2}{3} x^{\frac{3}{2}}\right]_{x=0}^{x=b}$$

$$= ab - \frac{1}{\sqrt{K}} \frac{2}{3} b^{\frac{3}{2}}$$

$$= ab - \frac{a}{\sqrt{b}} \frac{2}{3} b^{\frac{3}{2}}$$

$$= ab - \frac{2}{3} ab$$

$$= \frac{ab}{3}$$

Now we know that, y-coordinate of centroid of the shaded area is given by,

$$A\overline{Y} = \int_A y_c dA$$

or

$$\frac{ab}{3}\overline{Y} = \int_A \frac{a + y}{2} dA$$

$$\left[\because y_c = y + \frac{a - y}{2} = \frac{a + y}{2}\right]$$

$$= \frac{1}{2}\int_A (a + y) dA$$

$$= \frac{1}{2}\int_{x=0}^{x=b} (a + y)(a - y)\, dx$$

$$= \frac{1}{2}\int_{x=0}^{x=b} (a^2 - y^2)\, dx$$

$$= \frac{1}{2}\left[a^2x - \frac{x^2}{2K}\right]_{x=0}^{x=b}$$

$$= \frac{1}{2}\left[a^2b - \frac{b^2}{2K}\right]$$

$$= \frac{1}{2}\left[a^2b - \frac{b^2}{2} \times \frac{a^2}{b}\right]$$

$$= \frac{1}{2}\left[a^2b - \frac{a^2b}{2}\right]$$

$$= \frac{a^2b}{4}$$

$$\therefore \quad \overline{Y} = \frac{a^2b}{4} \times \frac{3}{ab}$$

$$= \frac{3a}{4}$$

Similarly, x-coordinate of centroid of the shaded area could be given by,

$$A\overline{X} = \int_A x_c \, dA$$

or $$\frac{ab}{3}\overline{X} = \int_A x \, dA \qquad [\because x_c = x]$$

$$= \int_{x=0}^{x=b} x(a-y)\,dx$$

$$= \int_{x=0}^{x=b} (ax - xy)\,dx$$

$$= \int_{x=0}^{x=b} (ax - ax\frac{\sqrt{x}}{\sqrt{K}})\,dx$$

$$= \left[a\frac{x^2}{2} - \frac{a}{\sqrt{K}}\frac{x^{1+\frac{3}{2}}}{1+\frac{3}{2}}\right]_{x=0}^{x=b}$$

$$= \left[\frac{ab^2}{2} - \frac{a}{\sqrt{K}} \times \frac{2}{5} b^{\frac{5}{2}}\right]$$

$$= \left[\frac{ab^2}{2} - \frac{2a}{5} \times \frac{1}{\sqrt{K}} b^{\frac{5}{2}}\right]$$

$$= \frac{ab^2}{2} - \frac{2a}{5} \times \frac{1}{\sqrt{K}} \times b^{\frac{5}{2}}$$

$$= \frac{ab^2}{2} - \frac{2}{5} \times \frac{a}{\sqrt{b}} b^{\frac{5}{2}}$$

$$= \frac{ab^2}{2} - \frac{2ab^2}{5}$$

$$= \frac{ab^2}{10}$$

$$\therefore \quad \overline{X} = \frac{ab^2}{10} \times \frac{3}{ab}$$

$$= \frac{3b}{10}$$

Therefore centroid of the shaded area is,

$$G \equiv \left(\frac{3b}{10}, \frac{3a}{4}\right).$$

33. *Determine the coordinates of the centroid of the circular arc in the coordinate system,as shown in Fig. (5-W34).*

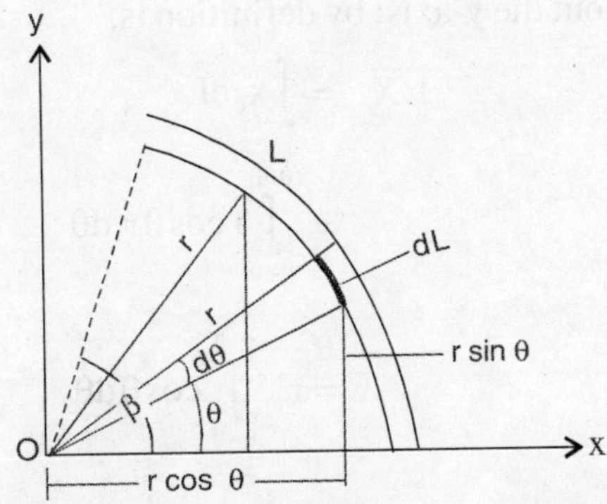

Fig. 5-W34

Solution:

Let dL be the elementary arc length,

$$\therefore \quad dL = r\,d\theta$$

Now the moment of the elementary arc-length about the x-axis; by definition is,

$$L\overline{Y} = \int y_c dL$$

or

$$r\beta\overline{Y} = \int_{\theta=0}^{\theta=\beta} r\sin\theta\, r\, d\theta$$

$$= r^2 \int_{\theta=0}^{\theta=\beta} \sin\theta\, d\theta$$

$$= r^2 \left[-\cos\theta\right]_0^\beta$$

$$= r^2 \left[-\cos\beta + \cos 0\right]$$

$$= r^2 \left[1 - \cos\beta\right]$$

$$= r^2\, 2\sin^2\frac{\beta}{2}$$

$$= 2r^2 \sin^2\frac{\beta}{2}$$

$$\therefore \quad \overline{Y} = \left(\frac{2r\sin^2\left(\frac{\beta}{2}\right)}{\beta}\right)$$

The moment of the elementary arc length dL about the y-axis; by definition is,

$$L\overline{X} = \int x_c dL$$

$$= \int_{\theta=0}^{\theta=\beta} r\cos\theta\, r\, d\theta$$

$$= r^2 \int_{\theta=0}^{\theta=\beta} \cos\theta\, d\theta$$

$$= r^2 \left[\sin\theta\right]_0^\beta$$

$$= r^2 \left[\sin\beta - \sin 0\right]$$

$$= r^2 \sin\beta$$

$$\therefore \quad \overline{X} = \frac{r^2 \sin\beta}{r\beta}$$

$$= \frac{r\sin\beta}{\beta}$$

$$\therefore \text{ The centroid, } G \equiv \left(\frac{r\sin\beta}{\beta}, \frac{2r\sin^2\frac{\beta}{2}}{\beta}\right)$$

34. *Find the centroid of the area between the curves $y^2 = x$ and $x^2 = y$.*

Solution:

The area between the curves shown in the Fig. (5-W35). Point of intersection P is (1,1).

Given that, $\quad y^2 = x \qquad .....(i)$

and $\quad x^2 = y \qquad ....(ii)$

From (i) and (ii), we get

$$\left(x^2\right)^2 = x \quad [\because y = x^2]$$

or $\quad x^4 = x$

or $\quad x^4 - x = 0$

or $\quad x(x^3 - 1) = 0$

$\therefore \quad x = 0$

and $\quad x = 1$

[imaginary roots are omitted]

Now $x = 0 \Rightarrow y = 0$ i.e., point O (0, 0)

and $\quad x = 1 \Rightarrow y = 1$ i.e., point P (1, 1).

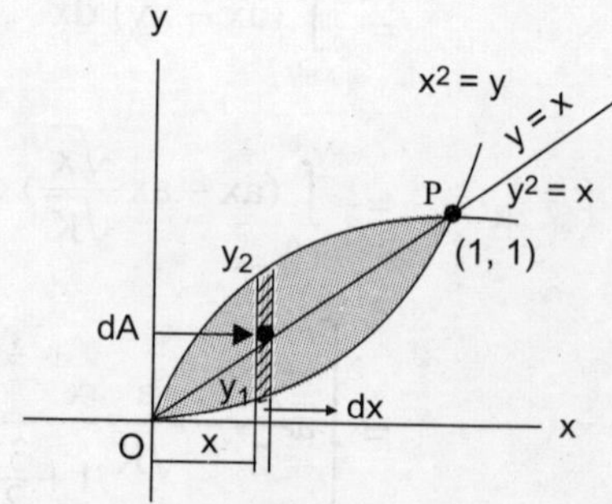

Fig. 5-W35

Consider an elementary area dA parallel to the y-axis.

Hence, $dA = \text{width} \times \text{height}$

$= dx \times (y_2 - y_1)$

$= \left(\sqrt{x} - x^2\right) dx$

Since $y_2^2 = x$ and $y_1 = x^2$

Now the moment of dA about the y-axis,

$$\int_0^{M_y} dM_y = \int_A x_c dA$$

$$\therefore \quad M_y = \int_0^1 x\left(\sqrt{x} - x^2\right)dx$$

$$= \int_0^1 (x^{\frac{3}{2}} - x^3)dx$$

$$= \left[\frac{x^{\frac{3}{2}+1}}{\frac{3}{2}+1} - \frac{x^4}{4}\right]_0^1$$

$$= \left[\frac{2}{5}x^{\frac{5}{2}} - \frac{x^4}{4}\right]_0^1$$

$$= \frac{2}{5} - \frac{1}{4}$$

$$= \frac{8-5}{20}$$

$$= \frac{3}{20}$$

Similarly, $M_x = \int_A y_c \, dA$

$$= \int_0^1 \left(y_1 + \frac{y_2 - y_1}{2}\right)(y_2 - y_1)dx$$

$$= \frac{1}{2}\int_0^1 \left(y_2^2 - y_1^2\right)dx$$

$$= \frac{1}{2}\int_0^1 \left(x - x^4\right)dx$$

$$= \frac{1}{2}\left(\frac{x^2}{2} - \frac{x^5}{5}\right)_0^1$$

$$= \frac{1}{2}\left(\frac{1}{2} - \frac{1}{5}\right)$$

$$= \frac{1}{2}\left(\frac{5-2}{10}\right)$$

$$= \frac{3}{20}$$

Now as, $dA = \left(\sqrt{x} - x^2\right)dx$

$$\therefore \quad A = \int_0^1 \left(\sqrt{x} - x^2\right)dx$$

$$= \left[x^{\frac{3}{2}} - \frac{x^3}{3}\right]_0^1$$

$$= \left[1 - \frac{1}{3}\right]$$

$$= \frac{2}{3}$$

As we know that,

$$A\overline{X} = \int_A x_c \, dA$$

$$\therefore \quad \overline{X} = \frac{\frac{3}{20}}{\frac{2}{3}}$$

$$= \frac{3}{20} \times \frac{3}{2}$$

$$= \frac{9}{40}$$

and $$A\overline{Y} = \int_A y_c \, dA$$

$$\therefore \quad \overline{Y} = \frac{\frac{3}{20}}{\frac{2}{3}}$$

$$= \frac{3}{20} \times \frac{3}{2}$$

$$= \frac{9}{40}$$

$$\therefore \quad G \equiv \left(\frac{9}{40}, \frac{9}{40}\right).$$

***Note**: Since the area enclosed within the two curves $y^2 = x$ and $x = y^2$ is symetrical about the $y = x$ line therefore $\overline{X} = \overline{Y} = \frac{9}{40}$.*

35. *Find the centroid of the area bounded by the cardioid, $r = a\ (1 + \cos\theta)$.*

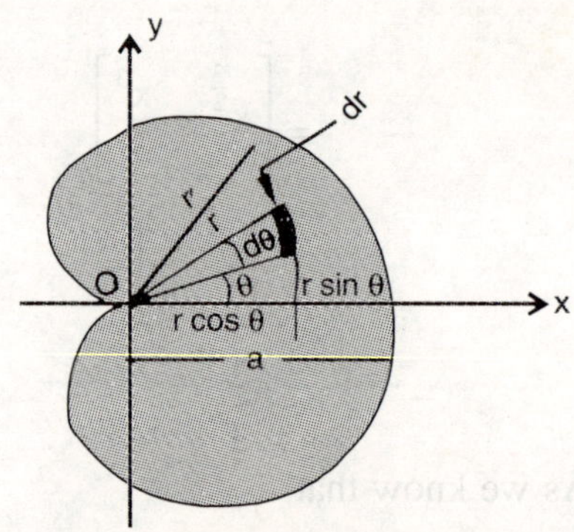

Fig. 5-W36

Solution:

The area bounded by the curve r = a (1+ cos θ) is shown in the Fig. (5-W36). Let us consider an elementary area dA in an intermediate position, then

dA = the area of the rectangular elementary strip
= width × height
= dr × r dθ
= r dθ dr

Now as we know that,

$$A\overline{X} = \int_A x_c \, dA$$

$$\text{or} \quad \left(\int_A dA\right)\overline{X} = \int_{\theta=0}^{\theta=2\pi} \int_{r=0}^{r=a(1+\cos\theta)} r\cos\theta \, r\, d\theta\, dr$$

$$\text{or} \quad \left(\int_{\theta=0}^{\theta=2\pi} \int_{r=0}^{r=a(1+\cos\theta)} r\, d\theta\, dr\right)\overline{X}$$

$$= \int_{\theta=0}^{\theta=2\pi} \int_{r=0}^{r=a(1+\cos\theta)} r^2 \cos\theta\, dr\, d\theta$$

$$\text{or} \quad \left(\int_0^{2\pi} \left(\frac{r^2}{2}\right)_0^{r=a(1+\cos\theta)} d\theta\right)\overline{X}$$

$$= \int_{\theta=0}^{\theta=2\pi} \left[\frac{r^3}{3}\right]_{\theta=0}^{\theta=a(1+\cos\theta)} \cos\theta\, d\theta$$

$$\text{or} \quad \frac{1}{2}\left(\int_0^{2\pi} a^2\, (1+\cos\theta)^2\, d\theta\right)\overline{X}$$

$$= \int_0^{2\pi} \frac{a^3}{3}[1+\cos\theta]^3 \cos\theta\, d\theta$$

$$\text{or} \quad \frac{a^2}{2}\left(\int_0^{2\pi} \left(1+\cos^2\theta + 2\cos\theta\right) d\theta\right)\overline{X}$$

$$= \frac{a^3}{3}\int_0^{2\pi} \left(1+\cos^3\theta + 3\cos^2\theta + 3\cos\theta\right)\cos\theta\, d\theta$$

$$\text{or} \quad \frac{a^2}{2}\left(\int_0^{2\pi} \left(1+\frac{1+\cos 2\theta}{2} + 2\cos\theta\right) d\theta\right)\overline{X}$$

$$= \frac{a^3}{3}\int_0^{2\pi} (\cos\theta + \cos^4\theta + 3\cos^3\theta + 3\cos^2\theta)\, d\theta$$

$$\text{or} \quad \left(\int_0^{2\pi} \left(\frac{3}{2} + \frac{\cos 2\theta}{2} + 2\cos\theta\right) d\theta\right)\overline{X}$$

$$= \frac{2}{3} a \int_0^{2\pi} \left(\cos^1 \theta + 3\cos^2 \theta + 3\cos^3 \theta + \cos^4 \theta \right) d\theta$$

or $\left[\frac{3}{2}\theta + \frac{1}{2}\frac{\sin 2\theta}{2} + 2\sin\theta \right]_0^{2\pi} \overline{X}$

$$= \frac{2a}{3} \int_0^{2\pi} \left(\cos\theta + 3\left(\frac{1+\cos 2\theta}{2}\right) + \left(\frac{1+\cos 2\theta}{2}\right)^2 + 3\left(\frac{\cos 3\theta + 3\cos\theta}{4}\right)\right) d\theta$$

or

$$(3\pi)\overline{X} \int_0^{2\pi} \left(\cos\theta + \frac{3}{2} + \frac{3}{2}\cos 2\theta + \frac{3}{4}(\cos 3\theta + 3\cos\theta) + \frac{1}{4}(1 + \cos^2 2\theta + 2\cos 2\theta) \right) d\theta$$

$$= \frac{2a}{3} \int_0^{2\pi} \left(\cos\theta + \frac{3}{2} + \frac{3}{2}\cos 2\theta + \frac{3}{4}\cos 3\theta + \frac{9}{4}\cos\theta + \frac{1}{4} + \frac{1}{4}\left(\frac{1+\cos 4\theta}{2}\right) + \frac{1}{2}\cos 2\theta \right) d\theta$$

$$= \frac{2a}{3} \int_0^{2\pi} \left(\frac{13}{4}\cos\theta + \frac{15}{8} + 2\cos 2\theta + \frac{3}{4}\cos 3\theta + \frac{1}{8}\cos 4\theta \right) d\theta$$

$$= \frac{2a}{3} \left[\frac{13}{4}\sin\theta + \frac{15}{8} + 2\left(\frac{\sin 2\theta}{2}\right) + \frac{3}{4}\left(\frac{\sin 3\theta}{3}\right) + \frac{1}{8}\left(\frac{\sin 4\theta}{4}\right) \right]_0^{2\pi}$$

$$= \frac{2a}{3}\left[\frac{15}{8} \times 2\pi\right]$$

$$= \frac{2a}{3} \times \frac{15}{8} \times 2\pi$$

$$= \frac{5}{2}\pi$$

$$\because \overline{X}(3\pi) = \frac{5}{2}\pi$$

$$\therefore \quad \overline{X} = \frac{5}{6}$$

Since the given curve is symmetrical about x-axis therefore $\overline{Y} = 0$.

$$\therefore G \equiv \text{centroid} = \left(\frac{5}{6}, 0\right).$$

EXERCISE

1. Find the y-coordinate of the centroid of the area bounded by only the upper part of the ellipse $\frac{x^2}{a^2} + \frac{y^2}{b^2} = 1$, where eccentricity of ellipse is e.
2. Find the y-coordinate of the centroid of the area bounded by the curve $y = \cos x$ between $x_1 = -\frac{\pi}{2}$ and $x_2 = \frac{\pi}{2}$.
3. Find the centroid of the area bounded by the curve $y = \frac{e^x + e^{-x}}{2} = \cos hx$ between A (0, 1) and B (a, cos ha).
4. Find the centroid of the first arc of the cycloid $x = a(t - \sin t)$, $y = a(1 - \cos t)$ where, $0 \le t \le 2\pi$.
5. Find the centroid of the area bounded by the straight line $y = \frac{2}{\pi}x$ and the sinusoidal $y = \sin x$ where, $x \ge 0$.
6. Find the centroid of a trapezium with the base b and the parallel sides h_1 and h_2.
7. Determine the coordinates of the centroid of the shaded area as shown in Fig. (5-E1).

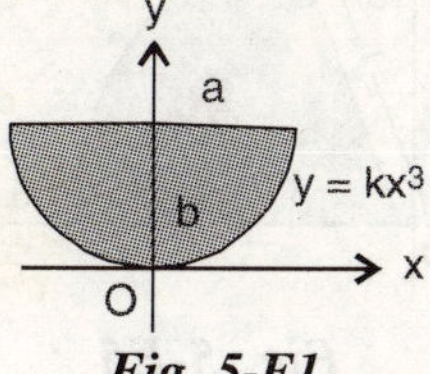

Fig. 5-E1

8. Determine the centroid of the shaded area as shown in Fig. (5-E2).

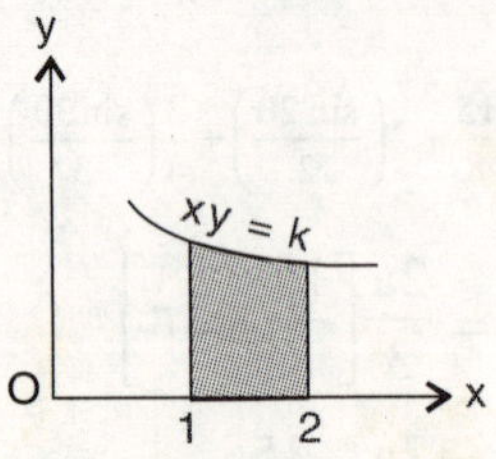

Fig. 5-E2

9. Locate the centroid of the shaded area between the ellipse and the circle.

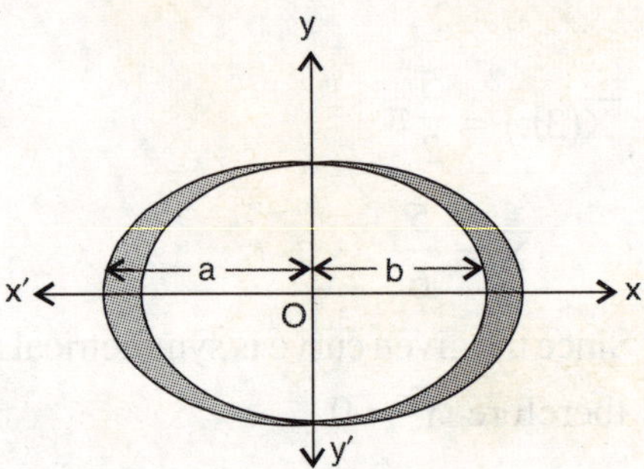

Fig. 5-E3

10. Locate the centroid of the shaded area shown in Fig. (5-E4).

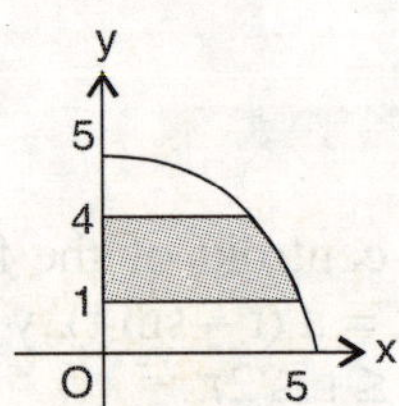

Fig. 5-E4

11. Determine the coordinates of the centroid of the area of the sector in the coordinate system shown in Fig. (5-E5).

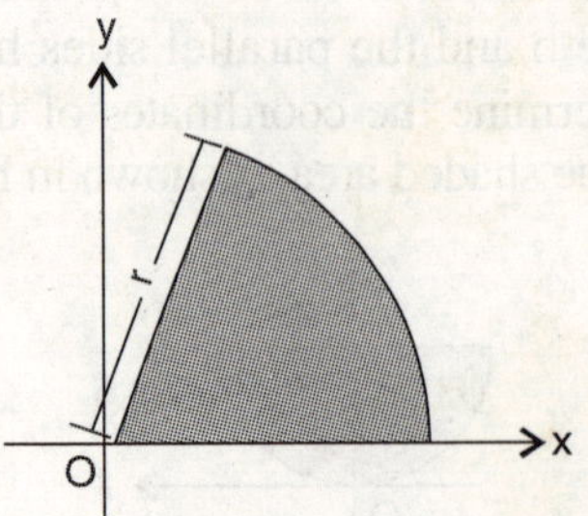

Fig. 5-E5

12. Determine the abscissa of the centroid of the circular segment in terms of r and α.

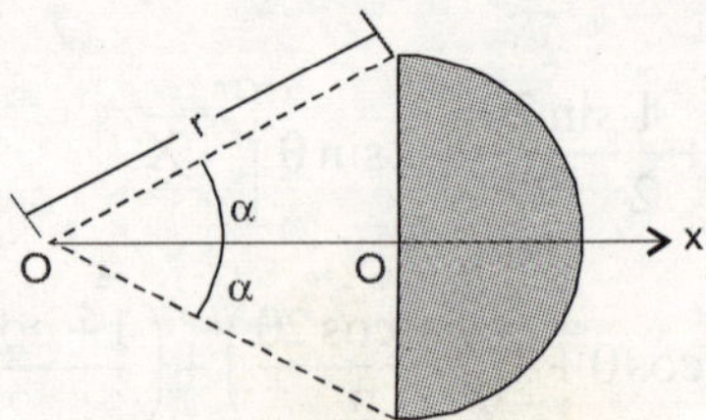

Fig. 5-E6

13. Determine the centroid of the area shown in Fig. (5-E7) by direct integration method.

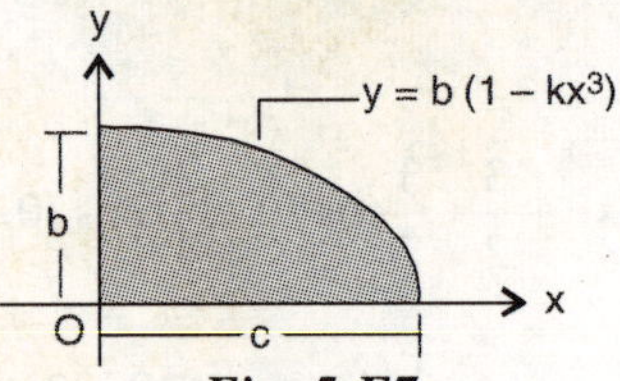

Fig. 5-E7

14. Determine the centroid of the area bounded by curve, $y = xe^{-1/x}$ when (i) x = h (ii) x → ∞.

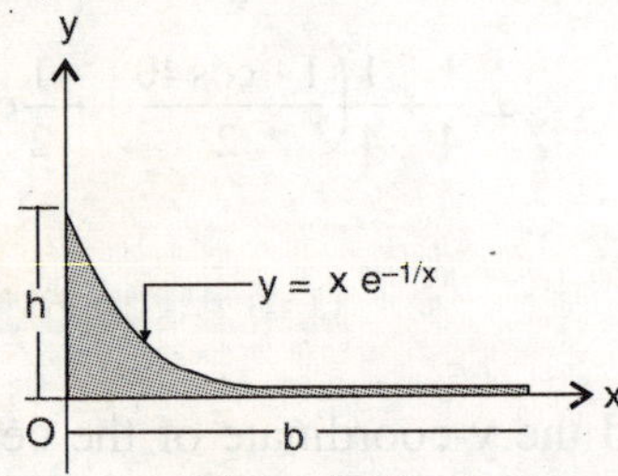

Fig. 5-E8

15. What are the centroidal coordinates for the shaded area ? The curved boundary is that of a parabola. (**Hints:** The general equation for the parabola of the shape shown is $y = ax^2 + b$). All the dimensions are in mm.

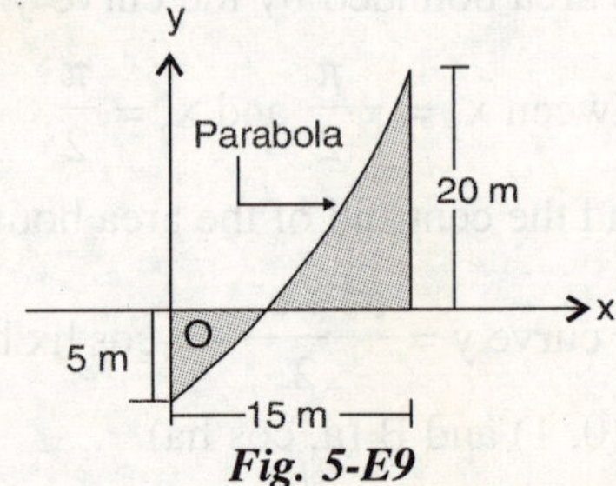

Fig. 5-E9

16. Determine the position of the centroid of the plane section as shown in Fig. (5-E10). All the dimensions are in mm.

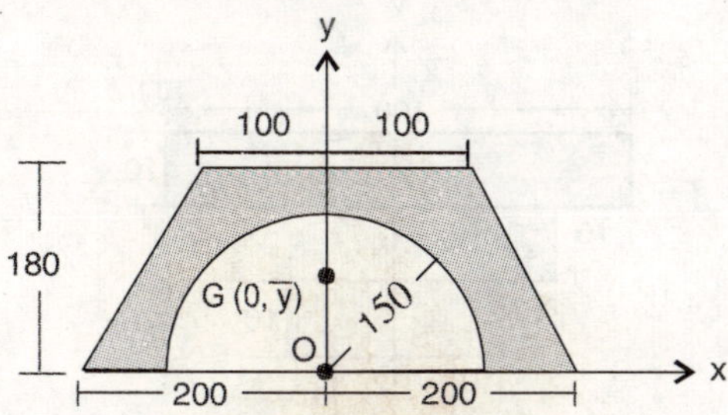

Fig. 5-E10

17. In a circular sheet of metal of radius R = 100 mm, a hole of radius r = 30 mm is made as shown in Fig. (5-E11). Determine the centroid of the remaining sheet. All the dimensions are in mm.

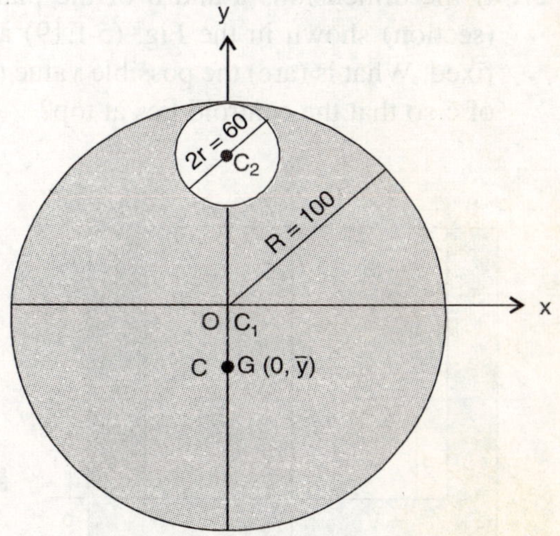

Fig. 5-E11

18. From the first principle find the co-ordinates of the centroid of (a) right angled isosceles triangle of base 60 mm (b) an isosceles triangle of base angle 70° and one of the equal side is 50 mm.

19. Find the position of the centroid of the T section shown in the Fig. (5-E12). All the dimensions are in mm.

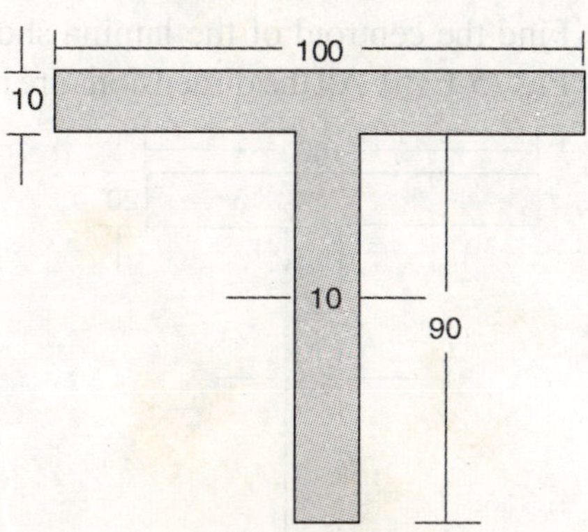

Fig. 5-E12

20. Locate the centroid of a right angled triangle with base b and height h.

21. Determine the centroidal distances $\overline{x}$ and $\overline{y}$ for the section shown in the Fig. (5-E13). All the dimensions are in mm.

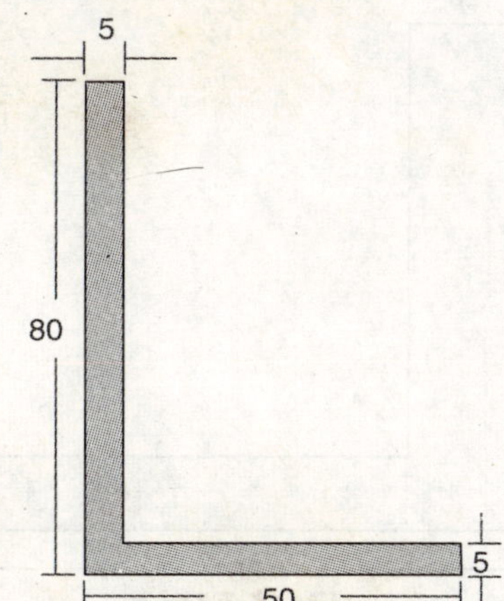

Fig. 5-E13

22. Find the centroid of the lamina shown in Fig. (5-E14). All the dimensions are in mm.

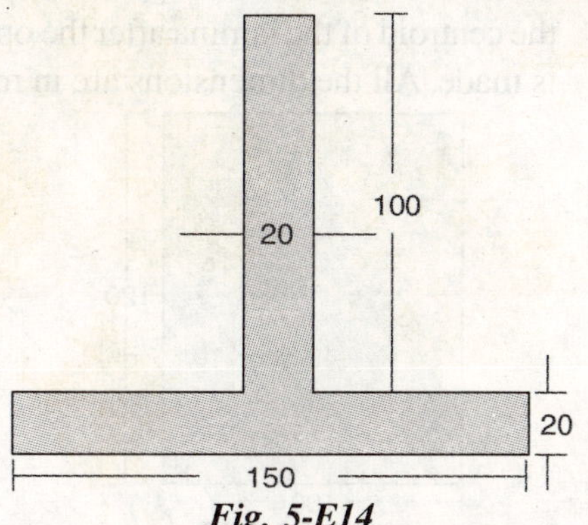

Fig. 5-E14

23. Find the centroid of the lamina shown in Fig. (5-E15). All the dimensions are in mm.

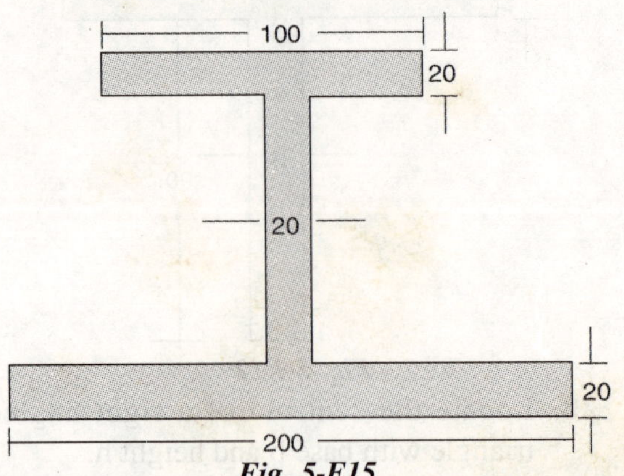

Fig. 5-E15

24. Find the centroid of the L-section shown in Fig. (5-E16). All the dimensions are in mm.

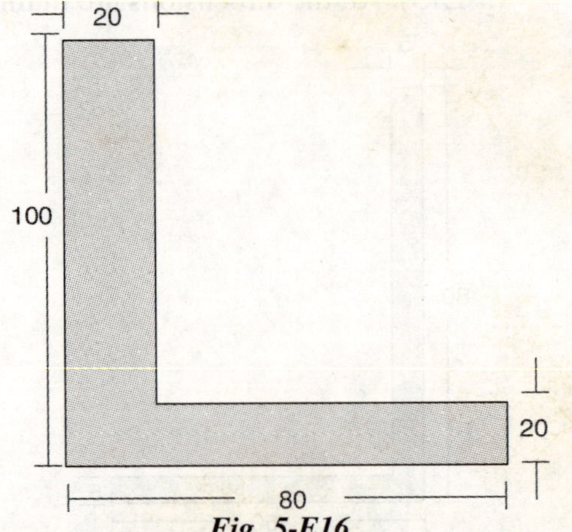

Fig. 5-E16

25. In a rectangular lamina 100 mm × 120 mm a rectangular opening 30 mm × 40 mm is made as shown in the Fig. (5-E17). Find the centroid of the lamina after the opening is made. All the dimensions are in mm.

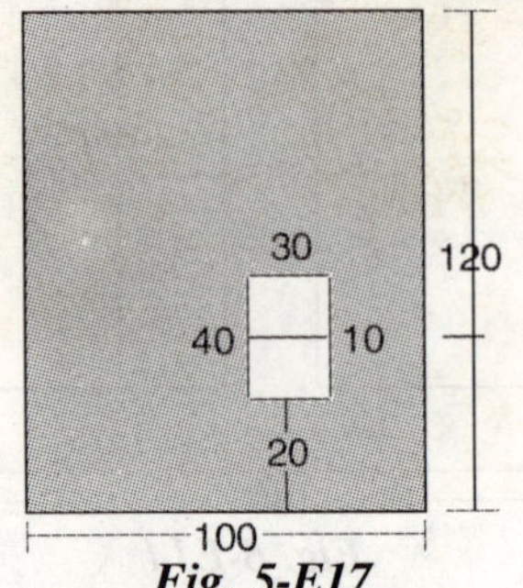

Fig. 5-E17

26. Find the centroid of the section in Fig. (5-E18) with respect to the axes as shown. All the dimensions are in mm.

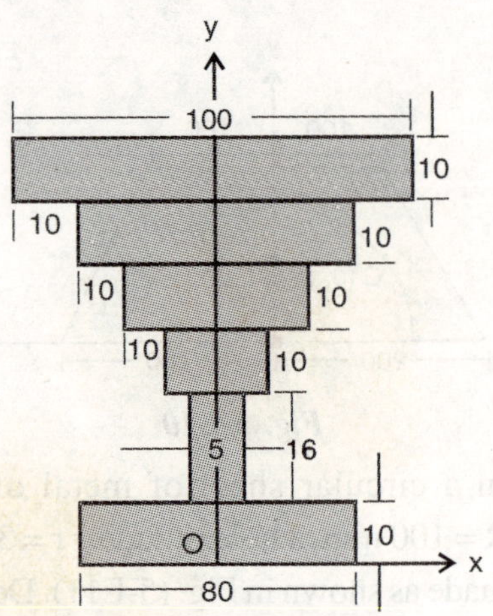

Fig. 5-E18

27. If the dimensions a and b of the plane (section) shown in the Fig. (5-E19) are fixed. What is (are) the possible value (s) of c so that the centroid lies at top?

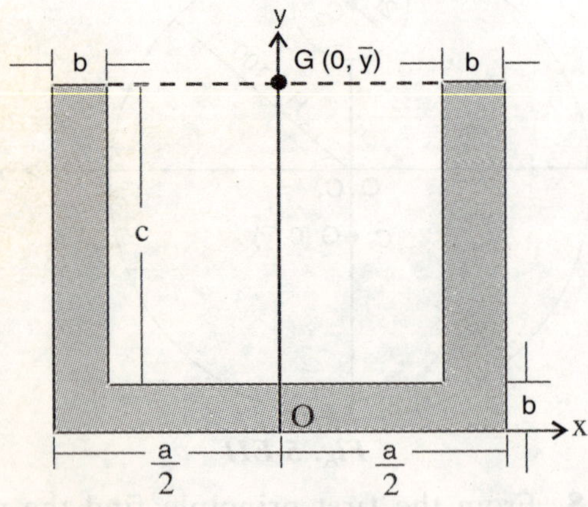

Fig. 5-E19

(a) $\dfrac{\sqrt{a^2 - 2ab} - a}{2}$ (b) $\dfrac{\pm\sqrt{a^2 - 2bc} - a}{2}$

(c) Not physically possible position of centroid.

(d) $\dfrac{-a \pm \sqrt{a^2 - 2bc}}{2}$.

28. Locate the centroid of the shaded three-quarters of the area of a square of dimension a as shown in Fig. (5-E20).

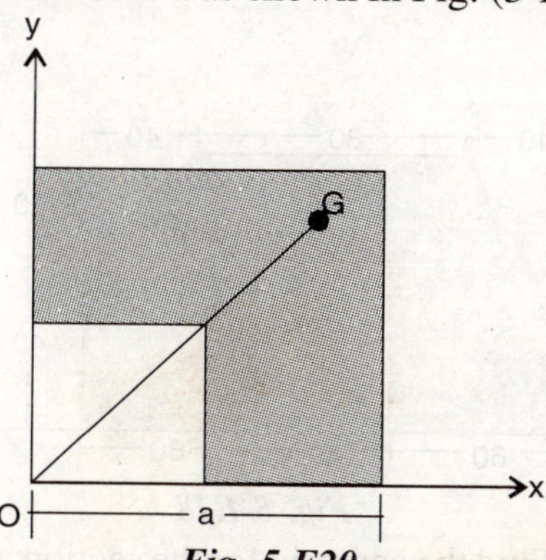

Fig. 5-E20

29. Locate the centroid of the shaded three-quarters of the area of a square of dimension a as shown in Fig. (5-E21).

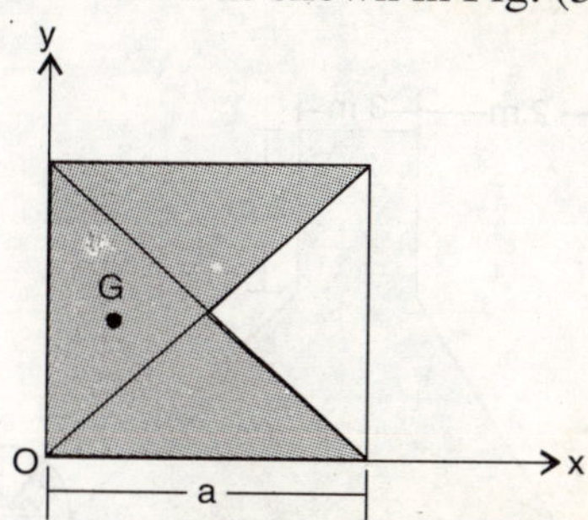

Fig. 5-E21

30. Locate the centroid of the shaded area as shown in Fig. (5-E22). All the dimensions are in mm.

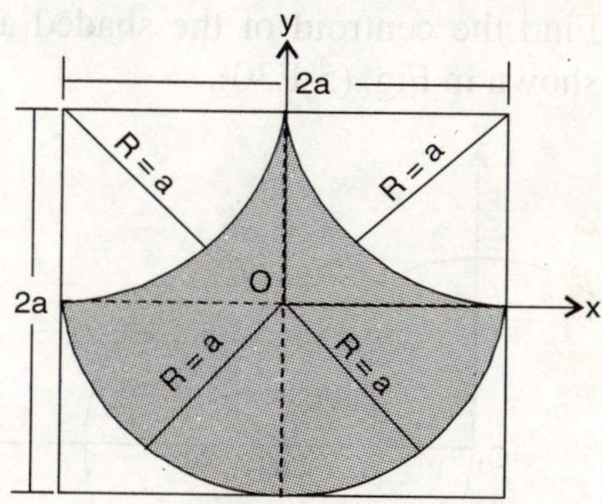

Fig. 5-E22

31. Determine the position of the centroid of the lamina with a circular cut out as shown in Fig. (5-E23). All the dimensions in mm.

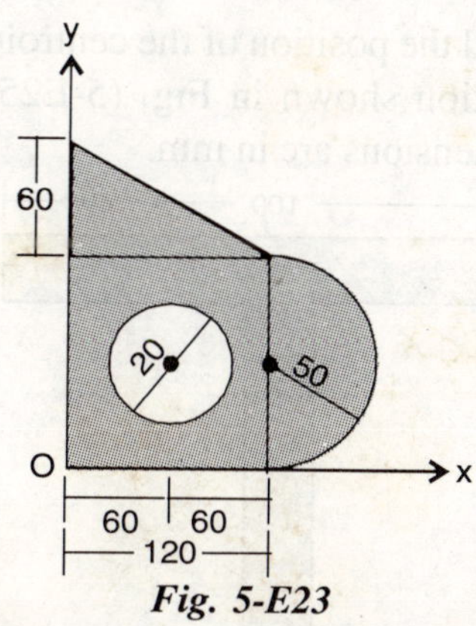

Fig. 5-E23

32. Locate the centroid of an area shown in Fig. (5-E24) with respect to the cartesian coordinate system as shown.

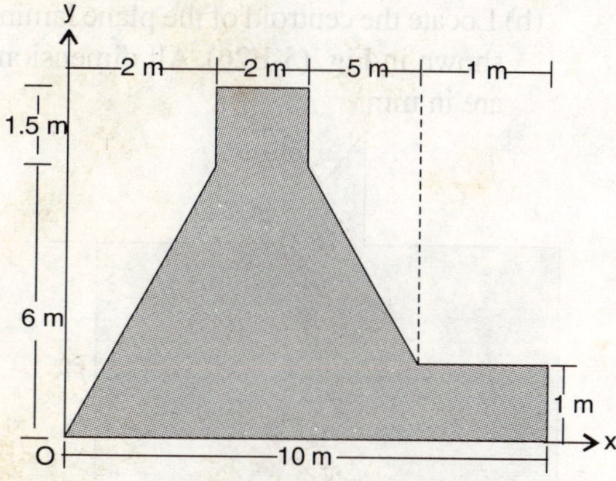

Fig. 5-E24

33. (a) From the first principles, find the coordinates of the centroid of an isosceles triangle.

(b) Locate the centroid of a semicircular lamina.

(c) Prove that the centroid of a triangle of height h is at a distance $\frac{h}{3}$ from its base or from the first principle derive an expression for locating the centroid of a triangle.

(d) Locate the centroid of a right-angled triangle with respect to base axis.

34. Find the position of the centroid of the T-section shown in Fig. (5-E25). All the dimensions are in mm.

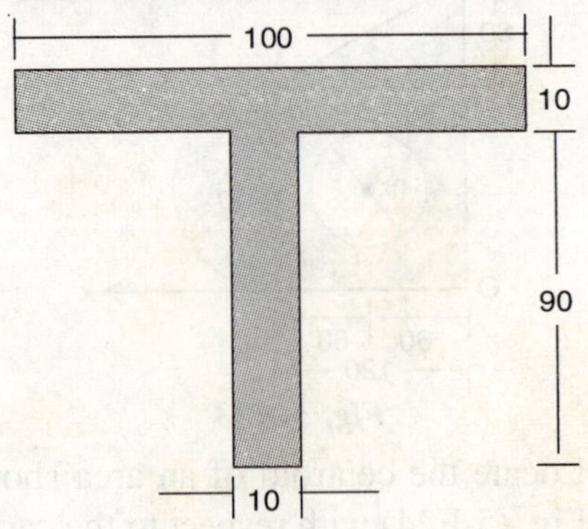

Fig. 5-E25

35. (a) Define centroid.

(b) Locate the centroid of the plane lamina shown in Fig. (5-E26). All dimensions are in mm.

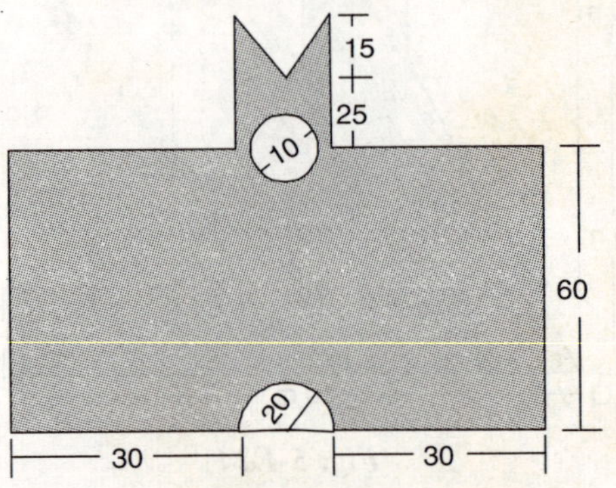

Fig. 5-E26

36. Locate the centroid of the area OABD shown in Fig. (5-E27). All the dimensions are in mm.

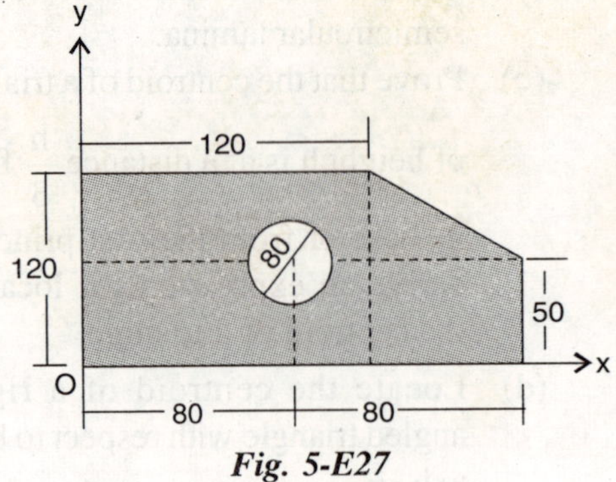

Fig. 5-E27

37. Locate the centroid of the plane lamina shown in Fig. (5-E28). All the dimensions are in mm.

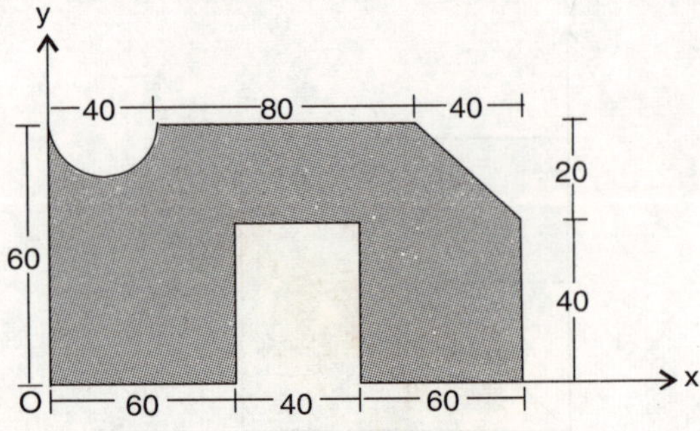

Fig. 5-E28

38. Find the centroid for the section as shown in Fig. (5-E29). All the dimensions are in mm.

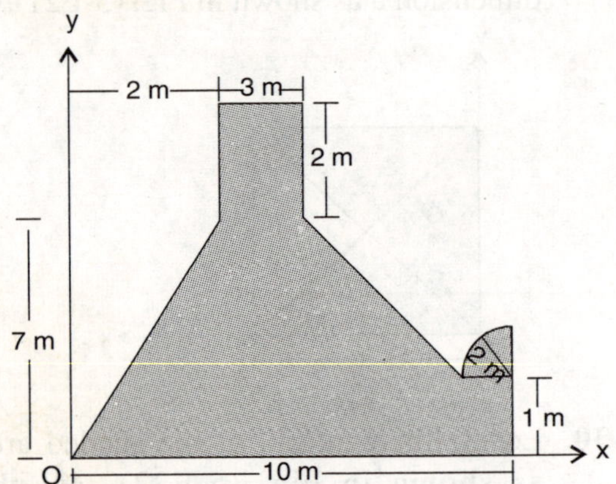

Fig. 5-E29

39. Find the centroid of the shaded area as shown in Fig. (5-E30).

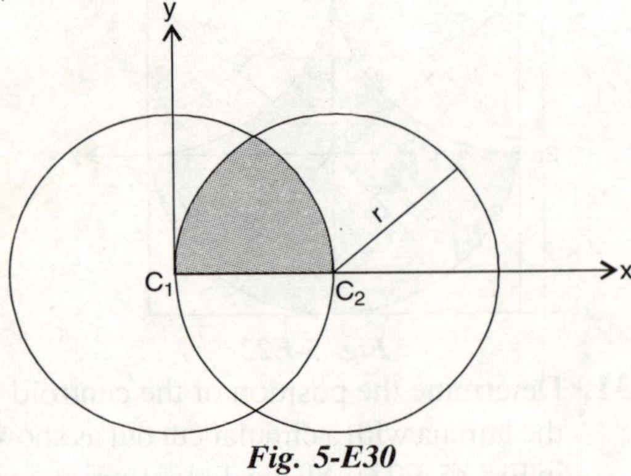

Fig. 5-E30

40. Find the centroid of the section shown in Fig. (5-E31).

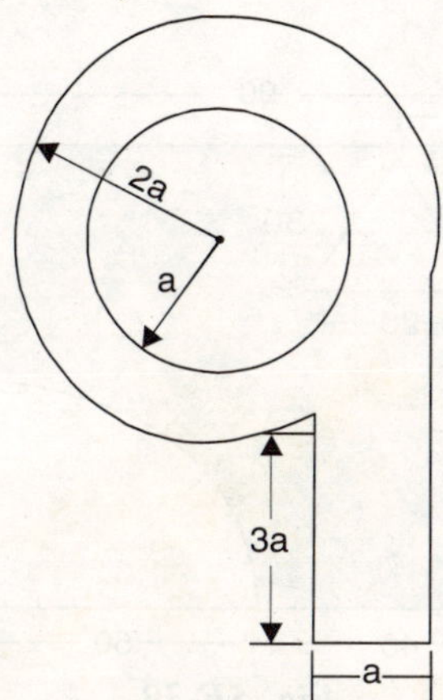

Fig. 5-E31

41. Locate the centroid of the following areas. Refer Fig. (5-E32a) and (5-E32b). All the dimensions are in mm.

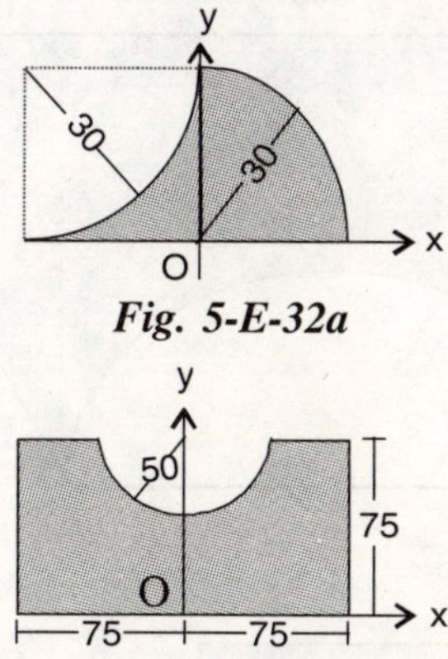

Fig. 5-E-32a

Fig. 5-E-32b

42. A semicircular area of radius 90 mm is cut out from a trapezium section as shown in the Fig. (5-E33). Find its centroid. All the dimensions are in mm.

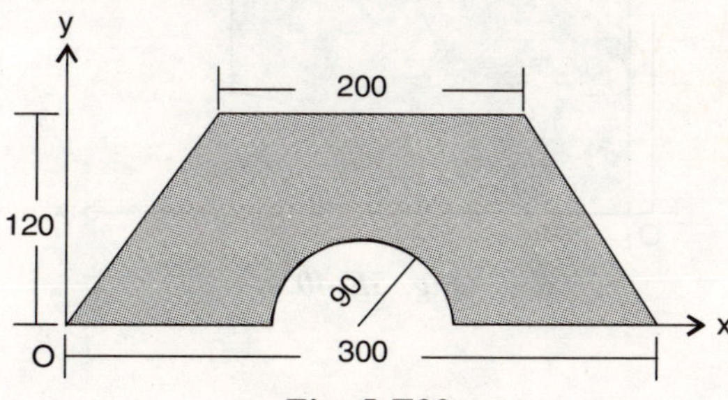

Fig. 5-E33

43. Determine the centroid of the shown (section in) Fig. (5-E34). All dimensions are in meter.

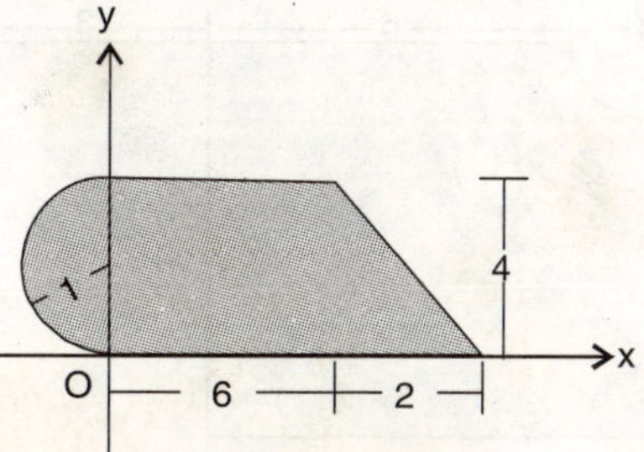

Fig. 5-E34

44. Find the centroid of the area as shown in Fig. (5-E35). All the dimensions are in mm.

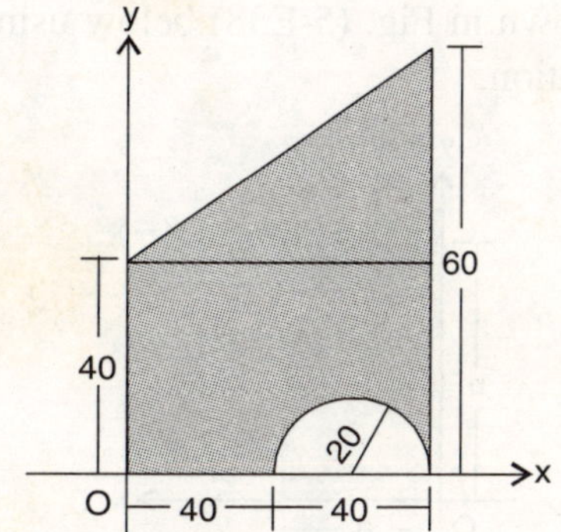

Fig. 5-E35

45. Determine the centroid of the section of the concrete dam as shown in Fig. (5-E36).

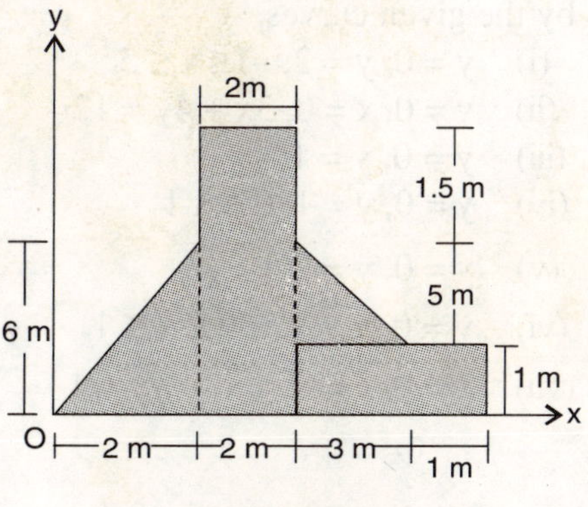

Fig. 5-E36

46. Locate the centroid of the area shown in Fig. (5-E37). All the dimensions are in mm.

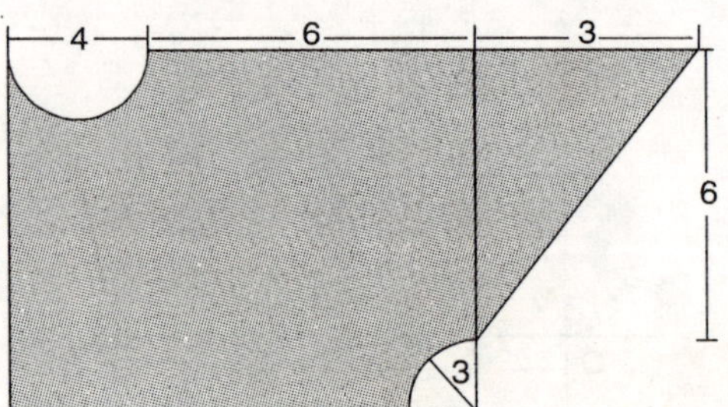

Fig. 5-E37

47. Find the centroid of the area bounded by $y = x^3$, $x = 2$ and the x-axis.

48. Find the centroid of the parabolic area shown in Fig. (5-E38) below using integration.

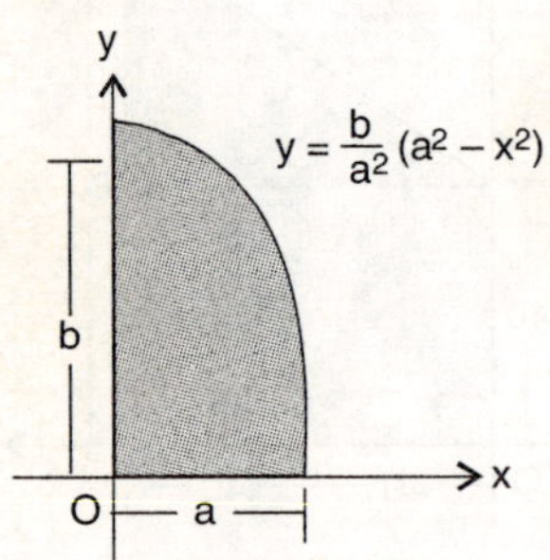

Fig. 5E-38

49. Find the centroid of the triangle with verfices A (1, 7), B (– 3, 2) and C (7,–6).

50. Find the centroids of the regions bounded by the given curves.

(i) $y = 0, y = 2, -1 \le x \le 5$

(ii) $y = 0, x = 0, 3x + 4y = 12$

(iii) $y = 0, y = 1 - x^2$

(iv) $y = 0, y = 1, 0 \le x\ 1$

(v) $y = 0, y = \sqrt{9 - x^2}$

(vi) $y = 0, y = x^{1/3}, 0 \le x \le 1$

(vii) $x = 0, y = 0, \sqrt{x} + \sqrt{y} = 1$

51. Find the centroid of the section as shown in Fig. (5-E39).

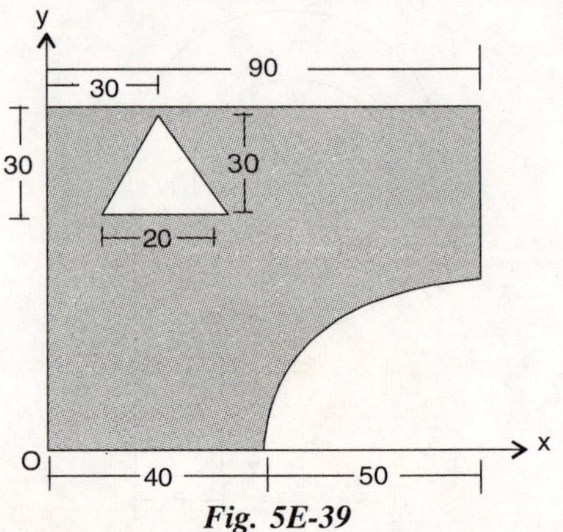

Fig. 5E-39

52. Locate the centroid of the shaded area shown in Fig. (5-E40) with respect to the axes shown. All the dimensions are in mm.

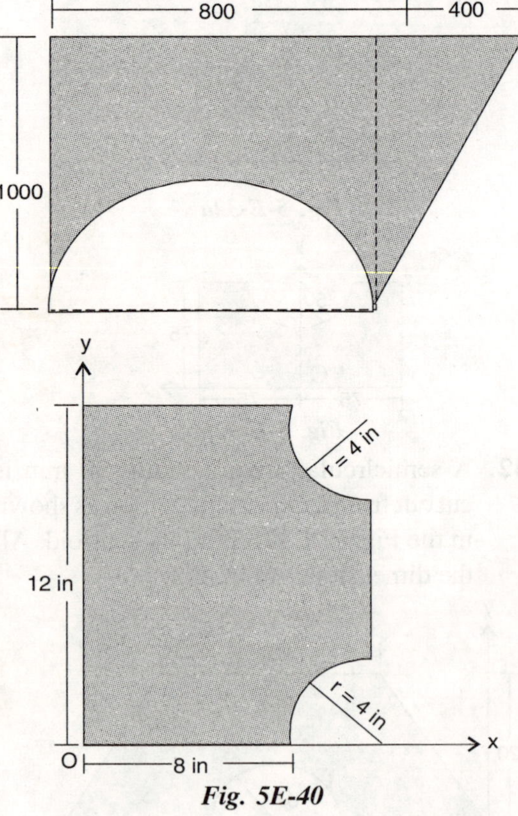

Fig. 5E-40

6 MOMENT OF INERTIA (AREA)

6.1 MOMENT OF INERTIA

The moment of inertia of any type of area distribution with respect to x and y axes in its plane are defined, respectively, by the integrals or summations as given in equation (6.1).

$$\left.\begin{aligned} I_x &= \int y^2 dA = \sum_{i=1}^{\infty} y^2 \Delta A \\ I_y &= \int x^2 dA = \sum_{i=1}^{\infty} x^2 \Delta A \end{aligned}\right\} \quad \text{....(6.1)}$$

In which an element of an area dA or ΔA is multiplied by the square of its distance from the corresponding axis and integration (summation) is carried out over the entire area of the figure.

Consider the distributions shown in the Figs. 6.1(a) and 6.1(b), dA or ΔA is an elementary area having coordinates as x and y for continuous area distribution and (x_i, y_i) for discrete area distribution. Therefore,

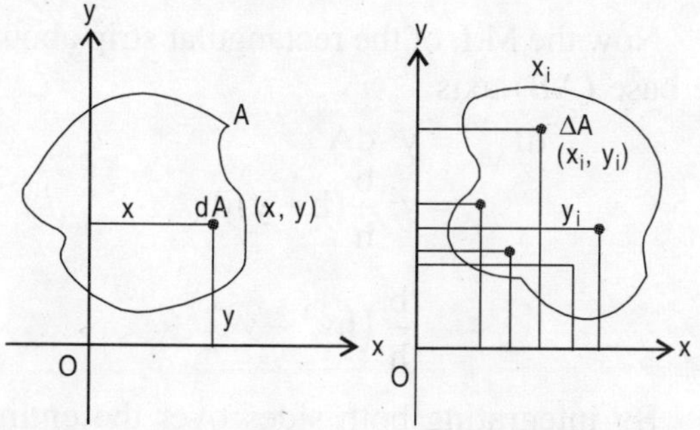

Fig. 6.1 (a) Continuous area distribution

Fig. 6.1 (b) Discrete area distribution

$$I_x \text{ or } I_{xx} = \text{M.I. about the x-axis} = \int y^2 dA$$

for a continuous body.

$$\text{and } I_x \text{ or } I_{xx} = \sum_{i=1}^{n} y^2 \Delta A \text{ for 'n' discrete}$$

distribution of areas.

Similarly, I_y or I_{yy} = MI about y-axis

$$= \int x^2 dA \text{ or } \sum_{i=1}^{n} x^2 \Delta A.$$

Since the term x dA is called as the first moment of the area, hence the moment of inertia i.e., the term x^2 dA = x (x dA) = moment of the moment of the area or the second moment of the area. Thus, the moment of inertia is nothing but the second moment of the area. Infact the term 'second moment of the area' appears to correctly signify the meaning of the expression Σx^2 dA or Σy^2 dA.

Though the moment of inertia is a purely mathematical term, it is one of the important properties of area which depends on the shape and size of the area distribution and frame of reference about which it is carried out.

In strength of material, subject to bending depends on the moment of inertia of its cross sectional area. Students will find this property of area is very useful when they study subjects like SOM, structural design and machine design and rotational motion.

The moment of inertia is a fourth dimensional term. Since it is a term obtained by multiplying area ($unit^2$) by the square of the distance ($unit^2$). Hence, in SI unit m^4 is an unit for moment of inertia. Other units mm^4, cm^4, ft^4 or in^4 are commonly used as units for moment of inertia.

Calculation of the moment of inertia using equation, (6.1) is known as calculation of MI from the first principle.

6.2 SIGNIFICANCE OF MOMENT OF INERTIA (AREA)

The concept of inertia is provided in Newton's first law of motion. The property of matter by virtue of which it resists any changes in its state of rest or of uniform motion is called inertia. The translatory inertia is identified by mass whereas the rotational inertia is termed as moment of inertia. In other words, the moment of inertia is the rotational analogue of mass, we could see it mathematically also as we have,

$F = ma$; for translational motion

$\tau = I\alpha$; for rotational motion.

It plays the role of resisting a change in rotational motion in quite the same sense as mass plays the role of resisting change in translatory motion. For small thickness and homogeneous density, the mass moment of inertia and the area moment of inertia is the same.

6.3 THE MOMENT OF INERTIA OF A TRIANGLE ABOUT ITS BASE AXIS FROM THE FIRST PRINCIPLE

Consider a triangle having base 'b' and height 'h', we have to determine the M.I. of the triangle about its base (AB) axis and the centroidal axis (GG′) shown in Fig. (6.2).

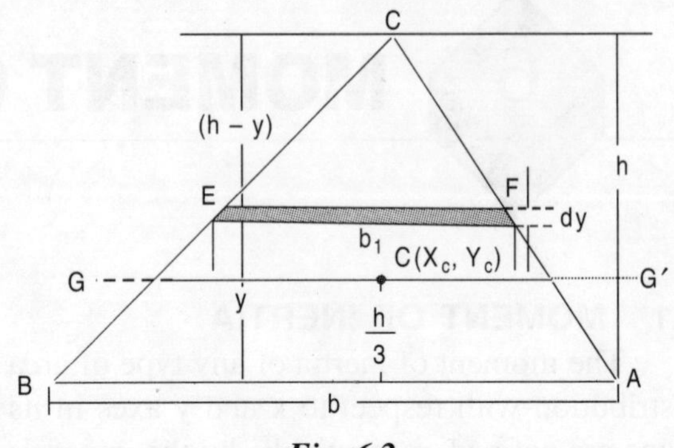

Fig. 6.2

Let us consider an elementary rectangular strip having length b_1 and width dy is at a distance 'y' from the base (AB) of the triangle.

Since the triangles ABC and FEC are similar triangles.

$$\text{Therefore,}\quad \frac{h}{h-y} = \frac{b}{b_1}$$

$$\text{or,}\quad b_1 = \frac{b}{h}(h-y)$$

by multiplying both sides by 'dy', we get

dA = the elementary area of rectangular strip

$$= b_1 dy$$

$$= \frac{b}{h}(h-y)\,dy$$

Now the M.I. of the rectangular strip about the base (AB) axis.

$$dI_{AB} = y^2\, dA$$

$$= y^2 \frac{b}{h}(h-y)\,dy$$

$$= \frac{b}{h}\left(hy^2 - y^3\right) dy$$

By integrating both sides over the entire area ($\frac{1}{2}bh$), we have

$$\int_0^{I_{AB}} dI_{AB} = \frac{b}{h}\int_0^h (hy^2 - y^3)dy$$

or $$I_{AB} = \frac{b}{h}\left[h\left[\frac{y^3}{3}\right] - \left[\frac{y^4}{4}\right]\right]_0^h$$

$$= \frac{b}{h}\left[h\left[\frac{h^3}{3} - 0\right] - \left[\frac{h^4}{4} - 0\right]\right]$$

$$= \frac{b}{h}\left[\frac{h^4}{3} - \frac{h^4}{4}\right]$$

$$= \frac{b}{h}\left[\frac{h^4}{12}\right]$$

$\therefore$ $$I_{AB} = \frac{bh^3}{12} \text{ unit}^4 \quad ...(6.2)$$

The M.I. about the centroidal (GG′) axis can be evaluated by using the parallel axis theorem,

i.e., $$I_{AB} = I_{GG'} + AY_c^2$$

or $$\frac{bh^3}{12} = I_{GG'} + \frac{1}{2}bh \times Y_c^2$$

$$= I_{GG'} + \frac{1}{2}bh \times \left(\frac{h}{3}\right)^2$$

$$= I_{GG'} + \frac{bh^3}{18}$$

or $$I_{GG'} = \frac{bh^3}{12} - \frac{bh^3}{18}$$

$$= bh^3\left[\frac{1}{12} - \frac{1}{18}\right]$$

$$= bh^3\left[\frac{3-2}{36}\right]$$

$$= \frac{bh^3}{36}$$

$\therefore$ $I_{GG'}$ = M.I. about centroidal axis

$$= \frac{\pi D^4}{128} \text{ unit}^4. \quad ...(6.3)$$

Example 1: *A right angled triangle and a rectangle having base (b) and height (h) respectively. If the M.I. of the triangle about the base is $I_{T,\,AB}$ and that of the rectangle is $I_{R,\,AB}$. As A_T is half of A_R then, is it $I_{T,\,AB} = \frac{I_{R,AB}}{2}$? Refer Fig. (6.3).*

Solution:

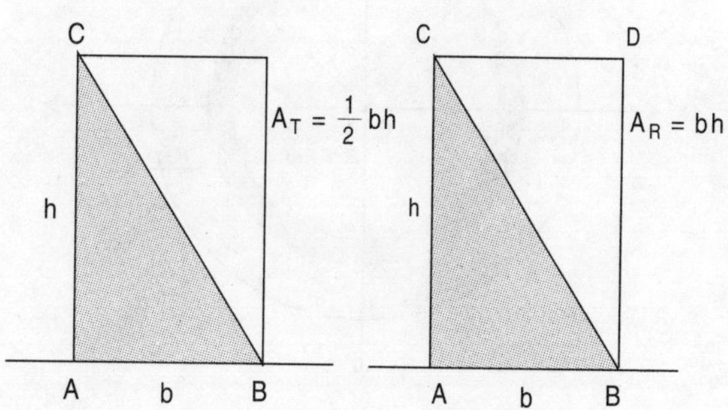

Fig. 6.3

As we know that, $$I_{T,\,AB} = \frac{bh^3}{36} \text{ unit}^4$$

and $$I_{R,\,AB} = \frac{bh^3}{3}$$

Therefore $$\frac{I_{R,AB}}{I_{T,AB}} = \frac{bh^3}{3} \times \frac{36}{bh^3}$$

$$= 12$$

or $$I_{R,\,AB} = 12\, I_{T,\,AB}$$

i.e., the M.I. of the rectangle about the base AB

= 12 time the M.I. of the triangle about base AB.

Thus if the area is double then it is not necessary that the M.I. will be also double.

6.4 MOMENT OF INERTIA OF THE PLANE FIGURES WITH RESPECT TO AN AXIS PERPENDICULAR TO THE PLANES

6.4.1 Moment of inertia of a circle about its diametrical axis

First method

Let us consider a circle of radius 'R', we have to find the moment of inertia of the circle about its diametrical axes xx′ and yy′ as shown in Fig. (6.4).

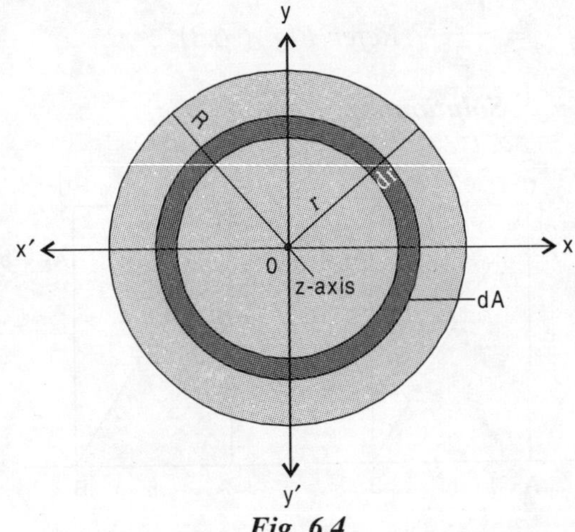

Fig. 6.4

Since xx′ and yy′ are two symmetrical axes of the circle.

Therefore, $I_{xx'} = I_{yy'}$ ...(i)

z-axis is perpendicular to both the x-axis and y-axis and then from the perpendicular axis theorem, we have,

$$I_{zz'} = I_{xx'} + I_{yy'} \quad ...(ii)$$

From (i) and (ii), we get,

$$I_{xx'} = I_{yy'} = \frac{I_{zz'}}{2} \quad ...(iii)$$

Since the z-axis is perpendicular to the plane of the circle.

Therefore, $I_{zz'} = \int_{area} r^2 dA$

where r = the perpendicular distance of an elementary area dA from the z-axis where, dA = an area of the elementary circular strip.

= an area of the folded rectangular strip

= 2πrdr

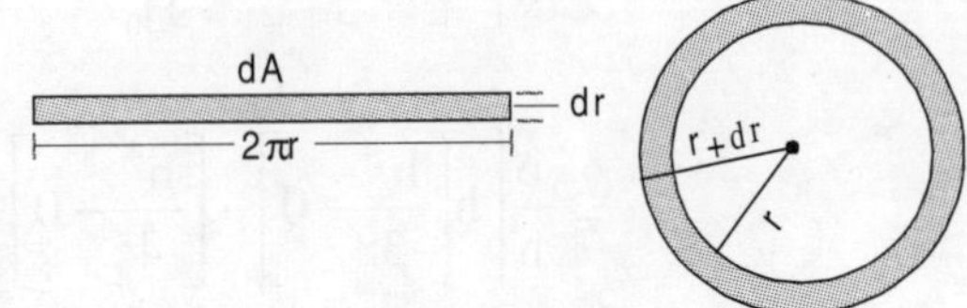

Fig. 6.5

$$\therefore I_{zz'} = \int_{r=0}^{r=R} r^2 2\pi r dr$$

$$= 2\pi \int_0^R r^3 dr$$

$$= 2\pi \left[\frac{r^4}{4}\right]_0^R$$

$$= \frac{2\pi R^4}{4}$$

$$= \frac{\pi R^4}{2}$$

or $$I_{zz'} = \frac{\pi}{2} \times \left[\frac{D}{2}\right]^4$$

$$= \frac{\pi D^4}{32} \text{ unit}^4$$

$\therefore I_{zz'} = \dfrac{\pi D^4}{32}$ unit^4; where 'D' is the diameter of the circle.

Therefore, the polar moment of the inertia of a circle of diameter (D) is,

$$I_{zz'} = I_{polar} = \frac{\pi D^4}{32} \text{ unit}^4$$

Now, from the equation (iii)

$$I_{xx'} = I_{yy'} = \frac{\pi D^4}{64}$$

= the moment of inertia of the circle about the diametrical axes.

$$\therefore I_x = I_y = \frac{\pi D^4}{64} = \frac{I_z}{2} \qquad ...(6.4)$$

Second method

Consider an elementary area as shown in the Fig. (6.6) 'r sin θ' is the distance of an elementary area from the diametrical axis xx′.

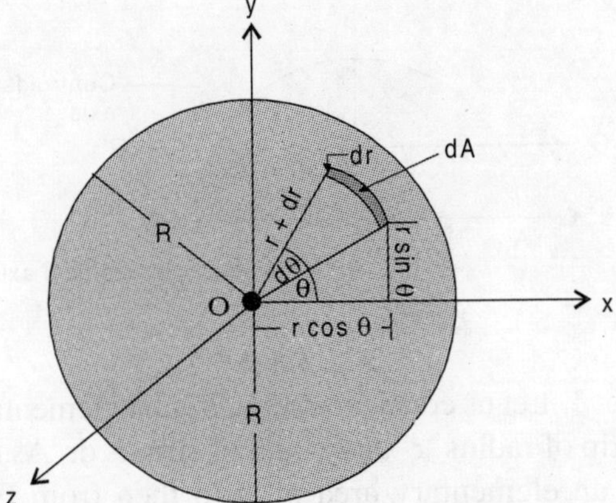

Fig. 6.6

Now since dθ is very small, therefore, elementary area could be treated as an area of rectangle of length r dθ and width dr

$$\therefore \quad dA = r\, d\theta\, dr$$

$$\text{Therefore, } I_{xx'} = \int_{\theta=0}^{\theta=2\pi}\int_{r=0}^{r=R} (r\sin\theta)^2\, dA$$

$$= \int_0^{2\pi}\int_0^{R} r^2 \sin^2\theta \times r\, dr\, d\theta$$

$$= \int_0^{2\pi}\int_0^{R} r^2 dr \sin^2\theta\, dr$$

$$= \int_0^{2\pi}\left[\frac{r^4}{4}\right]_0^R \sin^2\theta\, d\theta$$

$$= \int_0^{2\pi}\left[\frac{R^4}{4}\right] \sin^2\theta\, d\theta$$

$$= \frac{R^4}{4}\int_0^{2\pi} \sin^2\theta\, d\theta$$

$$= \frac{R^4}{4}\int_0^{2\pi}\left(\frac{1-\cos 2\theta}{2}\right) d\theta$$

$$= \frac{R^4}{8}\int_0^{2\pi}(1-\cos 2\theta)\, d\theta$$

$$= \frac{R^4}{8}\left(\theta - \frac{\sin 2\theta}{2}\right)_0^{2\pi}$$

$$= \frac{\pi R^4}{4}$$

$$= \frac{\pi}{4}\times\left[\frac{D}{2}\right]^4$$

$$= \frac{\pi}{4}\times\frac{D^4}{16}$$

$$= \frac{\pi D^4}{64}$$

Similarly $I_{yy'} = \dfrac{\pi D^4}{64}$; because a circle is symmetrical about the xx′ and yy′ axes

As we know that,

$$I_{zz'} = I_{xx'} + I_{yy'} = \frac{\pi D^4}{64} + \frac{\pi D^4}{64} = \frac{\pi D^4}{32}$$

$$= \frac{\pi}{32} \times (2R)^4$$

$$= \frac{\pi}{32} \times 16R^4$$

$$\therefore \quad I_{zz'} = \frac{\pi R^4}{2}$$

$$= I_{Polar}.$$

6.4.2 Moment of inertia of a circular ring

$I_{xx'}$ = M.I. of the circle of radius R about xx′ – M.I. of circle of radius r about zx′.

$$= \frac{\pi R^4}{4} - \frac{\pi r^4}{4}$$

$$= \frac{\pi}{4}\left(R^4 - r^4\right)$$

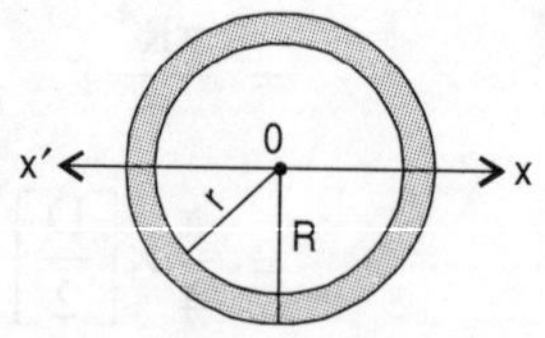

Fig. 6.7

Similary, $I_{zz'}$ = M.I. of the circle of radius R (or diameter D)–M.I. of the circle of radius r (or diameter d)

$$= \frac{\pi D^4}{32} - \frac{\pi d^4}{32}$$

$$= \frac{\pi}{32}\left[D^4 - d^4\right]$$

$$\therefore \quad I_{xx'} = \frac{p}{64}\left(D^4 - d^4\right)$$

$$= I_{yy'} \quad \text{...(6.5)}$$

6.4.3 Moment of inertia of a semicircle

First method

(a) *About a diametrical axis*

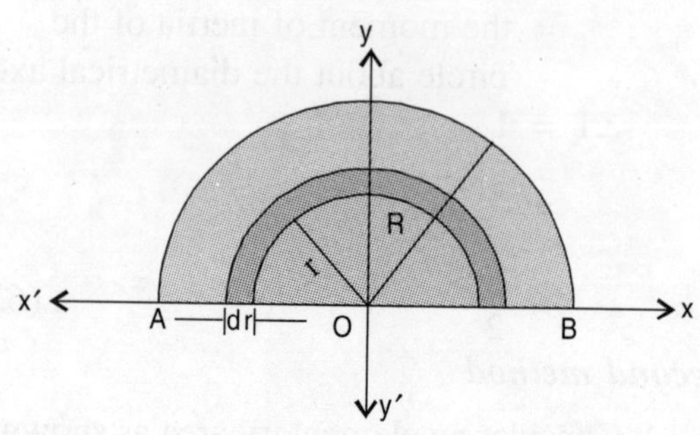

Fig. 6.8 (a)

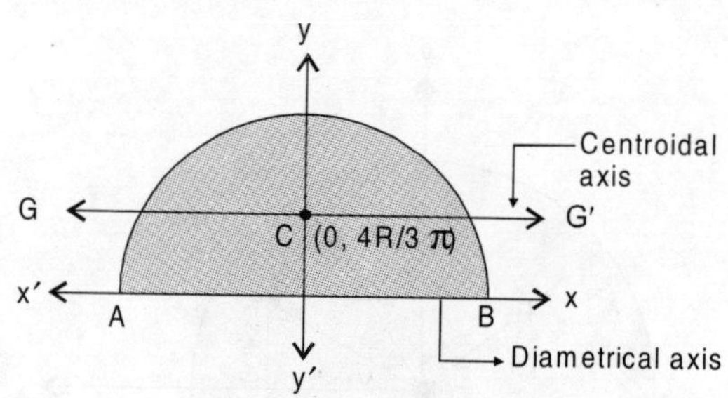

Fig. 6.8 (b)

Let us consider a semicircular elementary strip of radius 'r' and width of strip is dr. As dA is an elementary area of strip then from Fig. (6.8c), we have

$$dA = \left(\frac{2\pi r}{2}\right) dr$$

$$= \pi r \, dr \quad \text{...(i)}$$

Therefore M.I. of semicircular elementary strip about z-axis is

$$dI_z = dI_{Polar}$$

$$= r^2 dA$$

On integrating dI_z over proper limits, we get

$$I_z = \int_0^R (\pi r)\, r^2 dr$$

$$= \pi \int_0^R r^3 dr$$

$$= \pi \left[\frac{r^4}{4} \right]_0^R$$

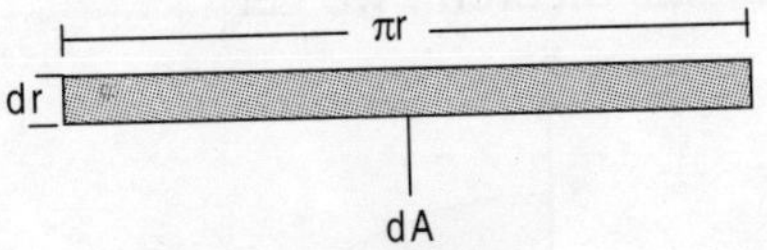

Fig. 6.8 (c)

$$= \frac{\pi R^4}{4}$$

$$= \frac{\pi D^4}{4 \times 16}$$

$$= \frac{\pi D^4}{64} \text{ unit}^4$$

Since given semicircle is symmetrical about the y-axis, therefore,

$$I_x = I_y$$

Now applying perpendicular axis theorem, we get,

$$I_z = I_x + I_x$$

or
$$I_x = \frac{I_z}{2} \text{ unit}^4$$

$$= \frac{\pi D^4}{128} \text{ unit}^4$$

$$\therefore \quad I_x = I_{AB}$$

$$= \frac{\pi D^4}{128} \text{ unit}^4$$

= M.I. about diametrical axis. ...(6.6)

(b) *About a centroidal axis*

Now using the parallel axis theorem, we have

$$I_{AB} = I_{GG'} + AY_C^2$$

or
$$\frac{\pi D^4}{128} = I_{GG'} + \frac{\pi R^2}{2} \times \left(\frac{4R}{3\pi}\right)^2$$

$$= I_{GG'} + \frac{\pi D^2}{8} \times \left(\frac{2D}{3\pi}\right)^2$$

$$= I_{GG'} + \frac{4\pi D^4}{8 \times 9\pi^2}$$

$$\therefore \quad I_{GG'} = \frac{\pi D^4}{128} - \frac{\pi D^4}{18\pi^2}$$

$$= \left(\frac{\pi}{128} - \frac{1}{18\pi}\right) D^4 \quad \text{...(6.7)}$$

Therefore, the moment of inertia of semicircle about the centroidal axis is $I_{GG'} = 0.0068598D^4$ unit^4 and that of about the diametrical axis is,

$$I_{AB} = I_x$$

$$= \frac{\pi}{128} D^4 \quad \text{...(6.8)}$$

$$= 0.024543692 D^4 \text{ unit}^4.$$

Second method

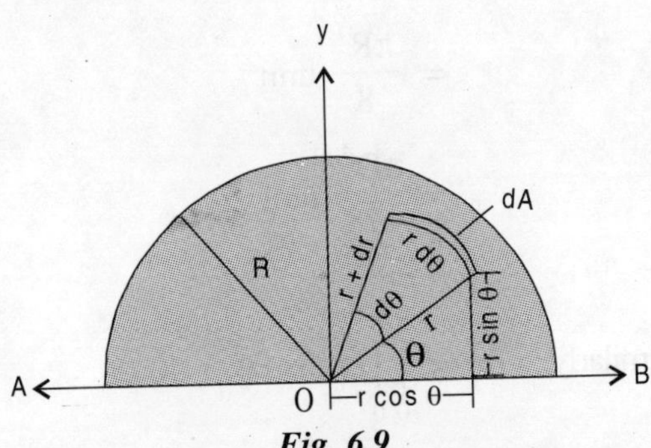

Fig. 6.9

Let us consider a semicircle of radius R, we have to find the moment of inertia of the semicircle about the diametrical axis-xx′. For this, consider a radius 'r' which is rotated through 'dθ' from angle θ, increasing 'r' by dr we have get an elementary arc strip, this arc strip having an area dA, can be treated as a rectangle of sides dr and r dθ respectively.

Therefore, $\quad dA = \text{length} \times \text{width}$

$$= r \, dr \, d\theta$$

Now this, elementary area is at a distance r sin θ from xx′ or the diametrical axis, and at a distance r cos θ from the y-axis.

Therefore, $I_x = \int_0^{\pi}\int_0^{R} (r\sin\theta)^2 r\,dr\,d\theta$

$$= \int_0^{\pi}\int_0^{R} r^3 dr \sin^2\theta\, d\theta$$

$$= \left(\int_0^{\pi} \sin^2\theta\, d\theta\right)\left(\int_0^{R} r^3 dr\right)$$

$$= \left(\int_0^{\pi} \frac{1-\cos 2\theta}{2}\, d\theta\right) \times \left[\frac{r^4}{4}\right]_0^R$$

$$= \frac{1}{2}\left[\theta - \frac{\sin 2\theta}{2}\right]_0^{\pi} \times \left(\frac{R^4}{4} - 0\right)$$

$$= \frac{1}{2}\left[(\pi - 0) - (0-0)\right] \times \frac{R^4}{4}$$

$$= \frac{\pi R^4}{8} \text{ unit}^4$$

$$= \frac{\pi D^4}{128} \text{ unit}^4$$

Similarly, $I_y = \int_0^{R}\int_0^{\pi} (r\cos\theta)^2\, r\, d\theta\, dr$

$$= \frac{\pi D^4}{128} \text{ unit}^4$$

Now, from the perpendicular axis theorem,

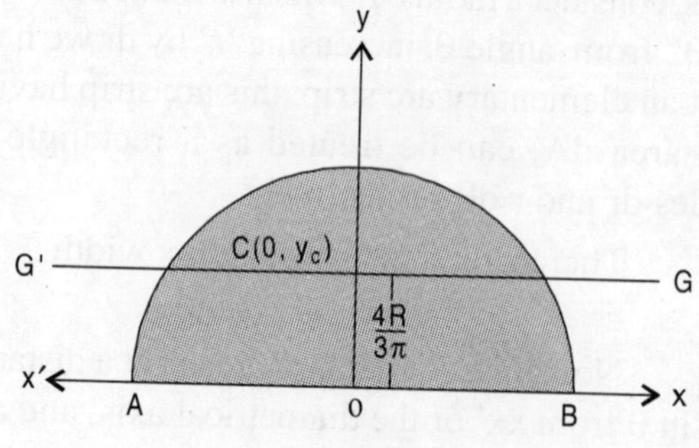

Fig. 6.10

we get $I_z = I_x + I_y$

$$= \frac{\pi D^4}{128} + \frac{\pi D^4}{128} \quad [\because 2R = D]$$

$$= \frac{\pi D^4}{64} \text{ unit}^4$$

Now by referring Fig. (6.10), from the parallel axis theorem, we get

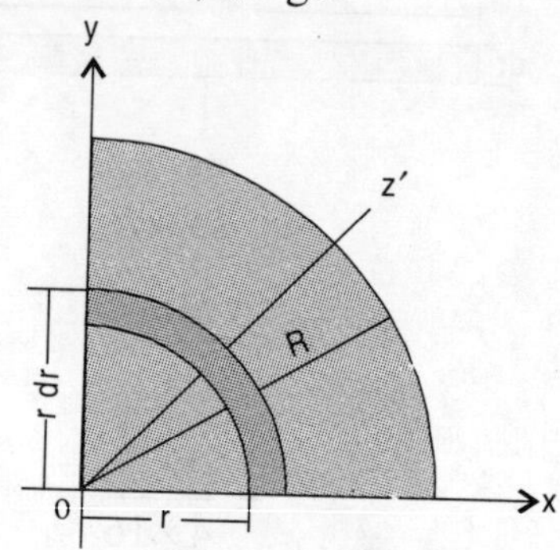

Fig. 6.11

$$I_{AB} = I_{GG'} + AY_c^2$$

or $\frac{\pi D^4}{128} = I_{GG'} + \frac{\pi D^2}{8} \times \left(\frac{2D}{3\pi}\right)^2$

$$\therefore \quad I_{GG'} = \frac{\pi D^4}{128} - \frac{\pi D^4}{2 \times 9\pi^2}$$

$$= \frac{\pi D^4}{128} - \frac{D^4}{18\pi}$$

$$= 0.0068598\, D^4 \text{ unit}^4.$$

6.4.4 Moment of inertia of a quarter of a circle

(a) *About the base axis*

First method

Consider a quarter circle of radius R, we have to find the moment of inertia about x, y and z-axes respectively.

Consider an elementary quarter circular ring of differential area dA then,

$$dA = \frac{\pi r}{2} dr$$

$$= \frac{\pi r\, dr}{2}$$

Now as this elementary ring is at a distance of r from the z-axis then,

$$I_z = I_{polar}$$
$$= \int_A r^2 dA$$
$$= \int_0^R r^2 \frac{\pi r\, dr}{2}$$
$$= \frac{\pi}{2}\int_0^R r^3 dr$$
$$= \frac{\pi}{2} \times \left[\frac{r^4}{4}\right]_0^R$$
$$= \frac{\pi R^4}{8}$$

Hence, $I_z = \frac{\pi R^4}{8}$ unit⁴.

Now, from the perpendicular axis theorem, we get,

$$I_z = I_x + I_y$$

For quarter of circle we observe that I_x and I_y are equal, therefore

$$I_x = I_y$$
$$= \frac{I_z}{2}$$

Hence, $I_x = I_y$

$$= \frac{\pi R^4}{16}$$
$$= \frac{\pi D^4}{256} \text{ unit}^4 \quad ...(6.9)$$

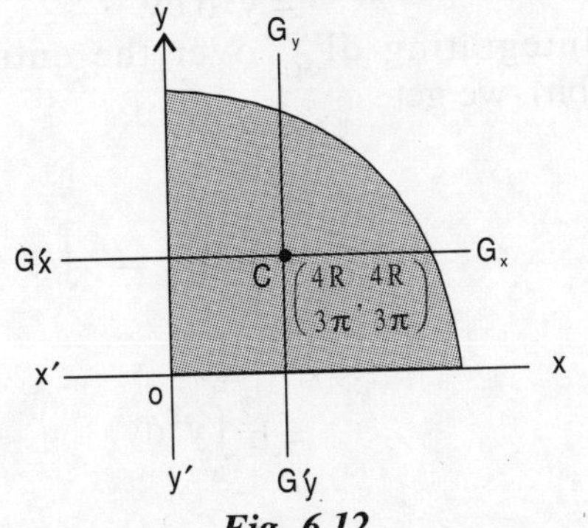

Fig. 6.12

Now from the parallel axis theorem, we get that

$$I_{xx'} = I_{G_xG'_x} + AY_C^2$$

or $$\frac{\pi D^4}{256} = I_{G_xG'_x} + \frac{\pi R^2}{4} \times \left(\frac{4R}{3\pi}\right)^2$$

or $$\frac{\pi D^4}{256} = I_{G_xG'_x} + \frac{\pi D^2}{16} \times \frac{4D^2}{9\pi^2}$$

or $$\frac{\pi D^4}{256} = I_{G_xG'_x} + \frac{4D^4\pi}{9 \times 16\pi^2}$$

$$\therefore \quad I_{G_xG'_x} = \frac{\pi D^4}{256} - \frac{D^4}{36\pi}$$
$$= D^4 \left[\frac{\pi}{256} - \frac{1}{36\pi}\right]$$
$$= 0.00343 D^4 \text{unit}^4 \quad ...(6.10)$$

Second method

Consider an elementary ring as shown in the Fig. (6.13) then differential area.

$$dA = \text{area of the rectangular strip}$$
$$= r\, d\theta\, dr$$

Now this differential area is at distance r from the z-axis.

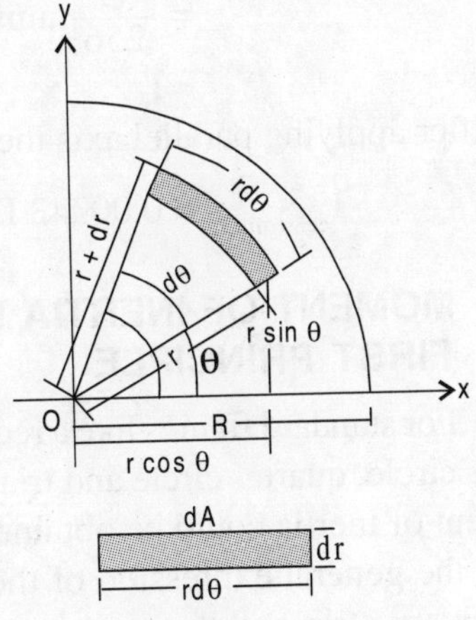

Fig. 6.13

Therefore, $I_z = I_{polar}$

$$= \int_0^R \int_0^{\frac{\pi}{2}} r^2\, dA$$

$$= \int_0^R \int_0^{\frac{\pi}{2}} r^2 \, r \, dr \, d\theta$$

$$= \int_0^R \int_R^{\frac{\pi}{2}} r^3 \, dr \, d\theta$$

$$= \int_0^R r^3 dr \times \int_0^{\frac{\pi}{2}} d\theta$$

$$= \left[\frac{r^4}{4}\right]_0^R \times [\theta]_0^{\frac{\pi}{2}}$$

$$= \frac{\pi}{2} \times \frac{R^4}{4}$$

$$= \frac{\pi R^4}{8} \text{ unit}^4.$$

Hence from the perpendicular axis theorem, we get

$$I_x = I_y$$

$$= \frac{\pi R^4}{16}$$

$$= \frac{\pi D^4}{256} \text{ unit}^4$$

$$= I_y.$$

and after applying parallel axis theorem, we get

$$I_{centroidal} = 0.00343 \, D^4 \text{ unit}^4.$$

6.5 MOMENT OF INERTIA FROM THE FIRST PRINCIPLE

For standard figures like a rectangle, semi-circle, circle, quarter circle and triangle etc., the moment of inertia could be obtained by writing down the general expression of the M.I. for an elementary strip and then carry out integration over the whole area. For this we use following formulae:

I_x = M.I. about the x-axis $= \int y^2 dA$

and I_y = M.I. about the y-axis $= \int x^2 dA$ where dA is an elementary area of strip. This method of finding M.I. is called the first principle.

6.6 MOMENT OF INERTIA OF A RECTANGLE ABOUT ITS CENTROIDAL AXES

Consider a rectangle having width 'b' and height 'h'. We have to find out the moment of inertia about the centroidal axis GG′, which is parallel to one of its side as shown in Fig. (6.14).

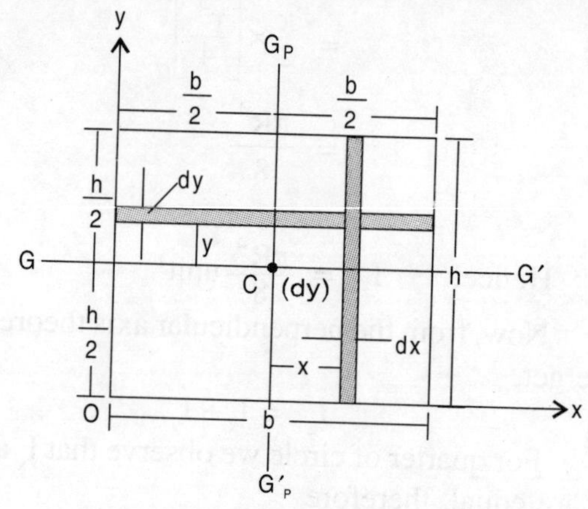

Fig. 6.14

Consider an elementary rectangular strip having width 'b' and elementary height dy, then the elementary area of the strip is,

$$dA = b \, dy$$

Therefore, the moment of inertia of the elementary rectangular strip about the centroidal axis GG′ is,

$$dI_{GG'} = y^2 \, dA$$

$$= y^2 b \, dy$$

Integrating $dI_{GG'}$ over the entire area (A = bh), we get

$$I_{GG'} = \int_A dI_{GG'} = \int_{-\frac{h}{2}}^{\frac{h}{2}} y^2 b \, dy$$

$$= b \int_{-\frac{h}{2}}^{\frac{h}{2}} y^2 dy$$

$$= b\left[\frac{y^3}{3}\right]_{-\frac{h}{2}}^{\frac{h}{2}}$$

$$= \frac{b}{3}\left[\left(\frac{h}{2}\right)^3 - \left(\frac{-h}{2}\right)^3\right]$$

$$= \frac{b}{3}\left[\frac{h^3}{8} + \frac{h^3}{8}\right]$$

Therefore, $I_{GG'} = \frac{bh^3}{12}$ unit^4 ...(6.11)

Similarly, now consider an elementary rectangular strip parallel to the y-axis, therefore

$$dA = \text{elementary area} = h\,dx$$

Now the moment of inertia of the elementary rectangular strip about the centroidal, ($G_P G_P'$) axis which is parallel to the y-axis is,

$$dI'_{G_P G_P'} = x^2\,dA = x^2\,h\,dx$$

Integrating $dI_{yy'}$ over the entire area A = bh will get,

$$I_{G_P G_P'} = \int_A dI_{G_P G_P'}$$

$$= \int_{-\frac{b}{2}}^{\frac{b}{2}} x^2\,h\,dx$$

$$= h\int_{-\frac{b}{2}}^{\frac{b}{2}} x^2 dx$$

$$= h\left[\frac{x^3}{3}\right]_{-\frac{b}{2}}^{\frac{b}{2}}$$

$$= \frac{h}{3}\left[\frac{b^3}{8} - \left(\frac{-b^3}{8}\right)\right]$$

$$= \frac{h}{3}\left[\frac{b^3}{8} + \frac{b^3}{8}\right] = \frac{h}{3}\left[\frac{b^3}{4}\right]$$

$\therefore \quad I_{G_P G_P'} = \frac{hb^3}{12}$ unit^4 ...(6.12)

Now suppose that we have to calculate the M.I. about the base axes of the rectangle then we could have integrated $dI_{G_P G_P'}$ over proper limits $x \to 0 \to b$ or $y \to 0 \to h$.

i.e., $\quad I_{base} = \int_0^h y^2\,b\,dy$ or $\int_0^b x^2 h\,dx$

$$= b\left[\frac{y^3}{3}\right]_0^4 \text{ or } h\left[\frac{x^3}{3}\right]_0^b$$

$$= \frac{bh^3}{3} \text{ or } \frac{hb^3}{3}$$

Thus $I_{base,\ x\text{-axis}} = \frac{bh^3}{3}$ unit^4

and $I_{base,\ y\text{-axis}} = \frac{bh^3}{3}$ unit^4.

By knowing the M.I. about the centroidal axis we could also find the M.I. about the base axis using the parallel axis theorem.

i.e., $\quad I_{bx} = \frac{bh^3}{12} + A\bar{y}^2$ ...(6.13)

$$= \frac{bh^3}{12} + A\left[\frac{h}{2}\right]^2$$

$$= \frac{bh^3}{12} + \frac{bh^3}{4}$$

$$= \frac{bh^3 + 3bh^3}{12}$$

$$= \frac{4bh^3}{12}$$

$$= \frac{bh^3}{3} \text{ unit}^4$$

$$\text{Similarly, } I_{by} = \frac{hb^3}{12} + A\bar{y}^2$$

$$= \frac{hb^3}{12} + bh \times \left[\frac{b}{2}\right]^2$$

$$= \frac{hb^3}{12} + \frac{hb^3}{4} = \frac{hb^3}{3} \text{ unit}^4$$

Now the M.I. about the axis passing through the centroid and perpendicular to the plane of rectangle

i.e., $I_{polar} = I_x + I_y$

[perpendicular axis theorem]

$$= \left[\frac{hb^3}{12}\right] + \left[\frac{bh^3}{12}\right]$$

$$= \frac{bh}{12}[b^2 + h^2] \qquad \text{...(6.14)}$$

$$\therefore \quad I_{Polar} = I_{Z,C} = \frac{A}{12}[b^2 + h^2] \text{ unit}^4$$

= Polar M.I. through centroid

Similarly, $I_{Z,O} = I_x + I_y$ = polar M.I. through the origin

$$= \frac{bh^3}{3} + \frac{hb^3}{3}$$

$$\therefore \quad I_{Z,O} = \frac{1}{3} bh\left[h^2 + b^2\right]$$

$$= \frac{A}{3}\left[h^2 + b^2\right]. \qquad \text{...(6.15)}$$

The results finally obtained are:

1. $I_{GG\|xx'} = \dfrac{bh^3}{12}$

$I_{xx'} = \dfrac{bh^3}{3}$

2. $I_{GG'\|yy'} = \dfrac{hb^3}{12}$

$I_{yy'} = \dfrac{hb^3}{3}$

3. $I_{Z,C} = \dfrac{A}{12}(b^2 + h^2)$

$I_{Z,O} = \dfrac{A}{3}(b^2 + h^2)$

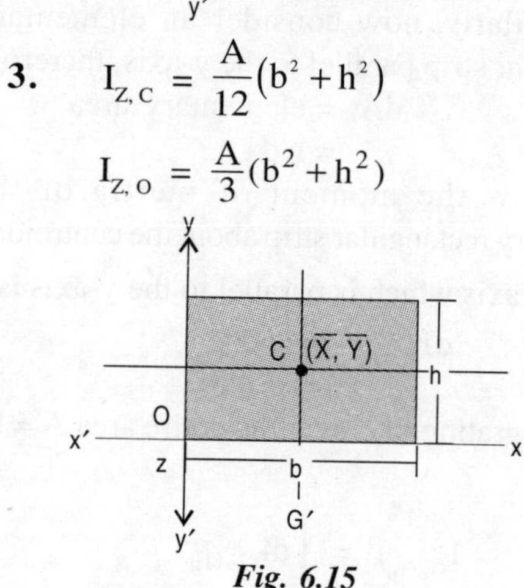

Fig. 6.15

6.6.1 Moment of inertia of rectangular hollow section

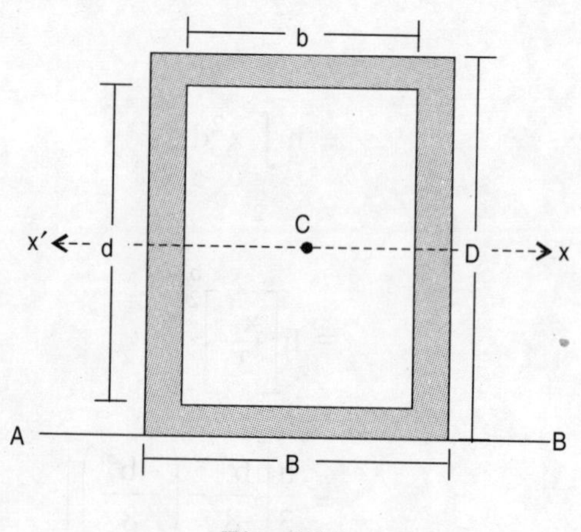

Fig. 6.16

The moment of inertia of a hollow rectangle about the centroidal axis xx′ is

$I_{xx'}$ = he moment of inertia of rectangle (B × D) – the moment of inertia of rectangle (b × d)

$$= \frac{BD^3}{12} - \frac{bd^3}{12}$$

$$= \frac{1}{12}\left[BD^3 - bd^3\right]\text{unit}^4$$

Similarly, about the centroidal axis- yy′and base-axis AB.

$$I_{yy'} = \frac{1}{12}\left[\frac{DB^3}{12} - \frac{db^3}{12}\right]$$

$$= \frac{1}{12}\left[DB^3 - db^3\right]\text{unit}^4$$

and $$I_{AB} = \frac{1}{3}\left[BD^3 - bd^3\right]\text{unit}^4 \qquad ...(6.16)$$

Table 6.1 : The moment of inertia of standard sections

Section	*Shape*	*Axis*	*Moment of inertia*
Rectangle	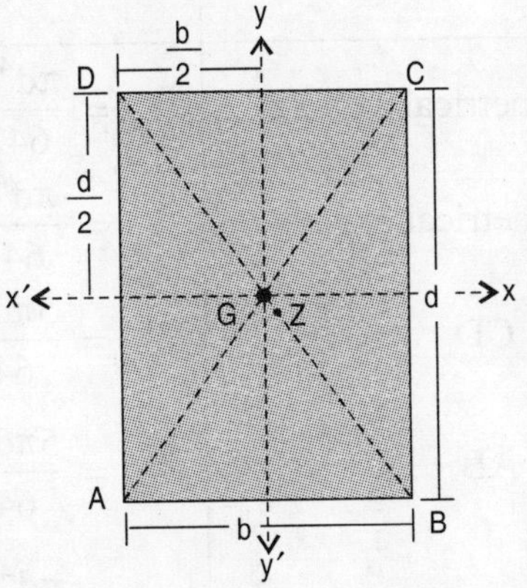	centroidal axis xx′	$I_{xx'} = \frac{bd^3}{12}$
		centroidal axis yy′	$I_{yy'} = \frac{db^3}{12}$
		axis AB	$I_{AB} = \frac{bd^3}{3}$
		axis BC	$I_{BC} = \frac{db^3}{3}$
		axis zz′	$I_{zz'} = \frac{1}{12}bh\left(b^2 + h^2\right)$
Square	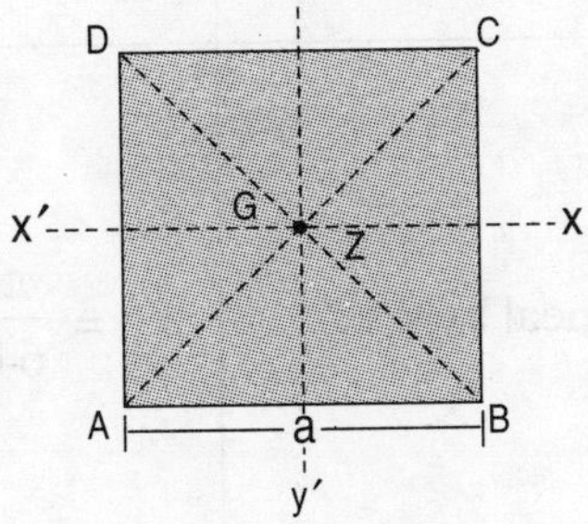	centroidal axis xx′	$I_{xx'} = \frac{a^4}{12}$
		centroidal axis yy′	$I_{yy'} = \frac{a^4}{12}$
		axis AB	$I_{AB} = \frac{a^4}{3}$
		axis BC	$I_{BC} = \frac{a^4}{3}$
		diagonal axis AC	$I_{AC} = \frac{a^4}{12}$
		diagonal axis BC	$I_{BD} = \frac{a^4}{12}$

Section	*Shape*	*Axis*	*Moment of inertia*
Hollow rectangle	y, b, x′, d, D, X, B, y′	centroidal axis xx′ centroidal axis yy′	$\frac{BD^3}{12} - \frac{bd^3}{12}$ $\frac{DB^3}{12} - \frac{db^3}{12}$
Triangle	A, G, x′, x, h, h/3, B, b, C	centroidal axis xx′ base-axis AB	$I_{xx'} = \frac{bh^3}{36}$ $I_{AB} = \frac{bh^3}{12}$
Circle	B, Y, x′, x, r = d/2, Z, C, D, Z′, A, Y′	diametrical axis xx′ diametrical axis yy′ axis CD axis AB axis zz′ axiszz′	$I_{xx'} = \frac{\pi d^4}{64}$ $I_{yy'} = \frac{\pi d^4}{64}$ $I_{CD} = \frac{5\pi d^4}{64}$ $I_{AB} = \frac{5\pi d^4}{64}$ $I_{ZZ'} = \frac{\pi d^4}{32}$ $I_{ZZ'} = \frac{5\pi d^4}{32}$
Hollow circle	x′, d, x, D	diametrical axis xx′	$I_{xx'} = \frac{\pi}{64}\left(D^4 - d^4\right)$

Section	*Shape*	*Axis*	*Moment of inertia*
Rectangle		AC or BD	$I_{AC} = \dfrac{b^3h^3}{6\left(b^2+h^2\right)} = I_{BD}$
		J_C	$I_{J_C} = \dfrac{bh}{12}\left(b^2+h^2\right)$
		J_O	$I_{J_O} = \dfrac{bh}{3}\left(b^2+h^2\right)$
Rectangular ring		xx′	$I_{xx'} = \dfrac{bh^3}{12} - \dfrac{(b-2t)(h-2t)^3}{12}$
		yy′	$I_{yy'} = \dfrac{hb^3}{12} - \dfrac{(h-2t)(b-2t)^3}{12}$
		x	$I_x = \dfrac{ab^3}{21}$
		y	$I_y = \dfrac{a^3b}{5}$

Section	*Shape*	*Axis*	*Moment of inertia*
Ellipse	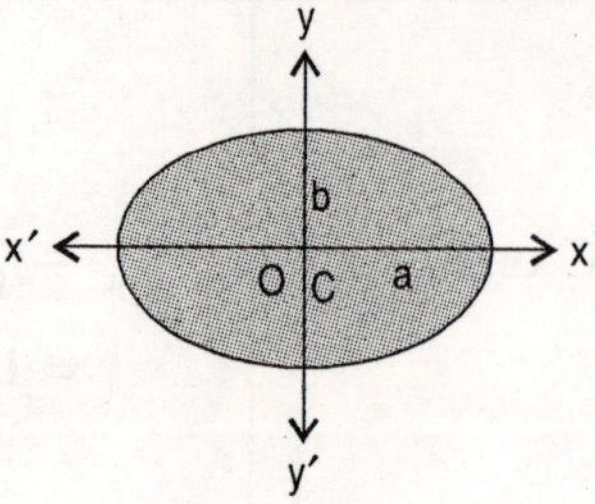 y, y′, x′, x, b, O, C, a	xx′ yy′ J_C	$I_{xx'} = \frac{\pi ab^3}{4}$ $I_{yy'} = \frac{\pi ba^3}{4}$ $I_{J_C} = \frac{\pi ab}{4}\left(a^2 + b^2\right)$
Cylindrical ring	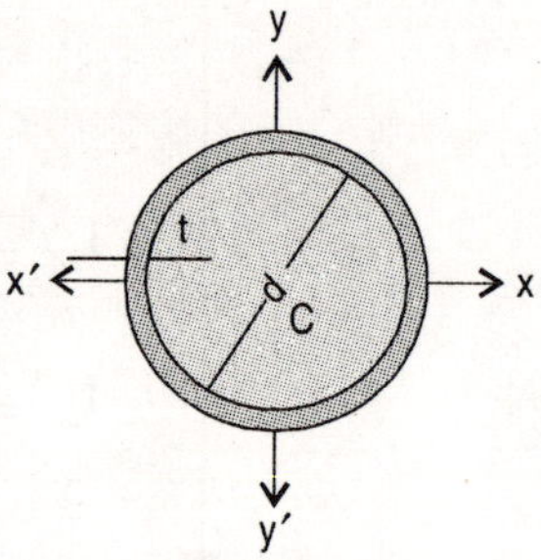 y, y′, x′, x, t, d, C	xx′ yy′ J_C	$I_{xx'} = \frac{\pi td^3}{8}$ $I_{yy'} = \frac{\pi td^3}{8}$ $I_{J_C} = \frac{\pi td^3}{4}$
Hollow cylinder	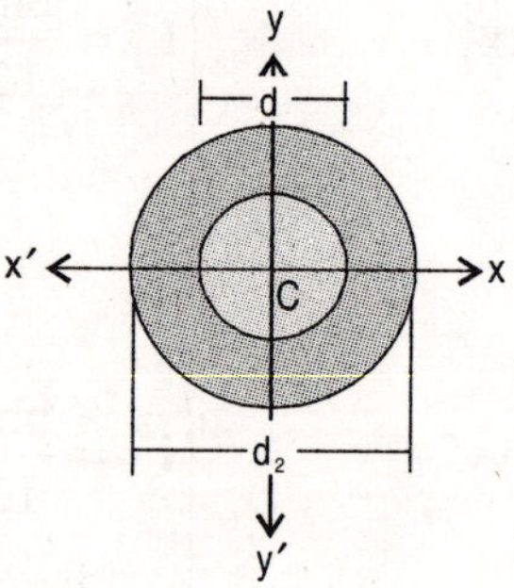 y, y′, x′, x, d, C, d_2	xx′ yy′ $J_{C\ (polar)}$	$I_{xx'} = \frac{\pi\left(d_2^4 - d_1^4\right)}{64}$ $I_{yy'} = \frac{\pi\left(d_2^4 - d_1^4\right)}{64}$ $I_{J_C} = \frac{\pi\left(d_2^4 - d_1^4\right)}{32}$
Semicircle	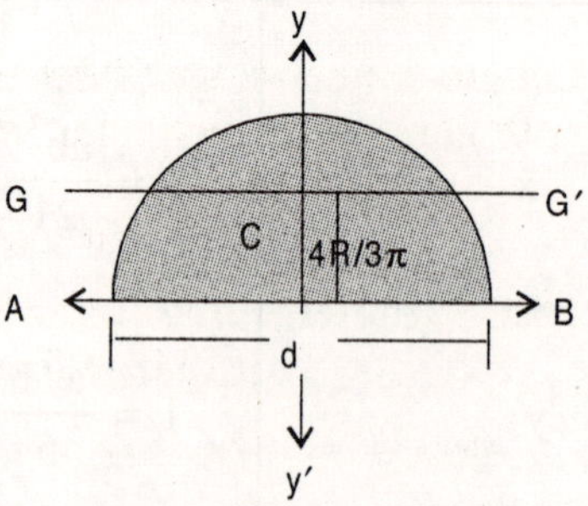 y, y′, G, G′, C, 4R/3π, A, B, d	centroidal axis GG′ diametrical axis axis yy′	$I_{GG'} = 0.0068598d^4$ $I_{AB} = \frac{\pi d^4}{128}$ $I_{GG'} = 0.11R^4$ $I_{yy'} = \frac{\pi d^4}{128}$

Section	*Shape*	*Axis*	*Moment of inertia*
Quarter circle	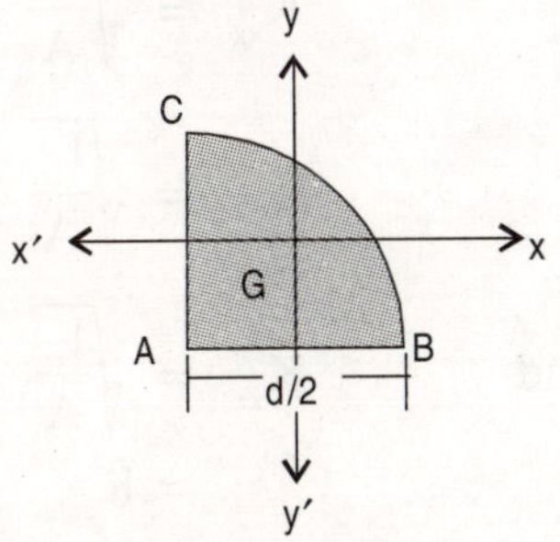	centroidal axis xx′ and yy′ axis AB, AC	$I_{xx'} = I_{yy'}$ $= 0.00343d^4$ $= 0.055R^4$ $I_{AB} = I_{AC} = \frac{\pi d^4}{256}$ $= \frac{\pi R^4}{16}$
Isoceles triangle	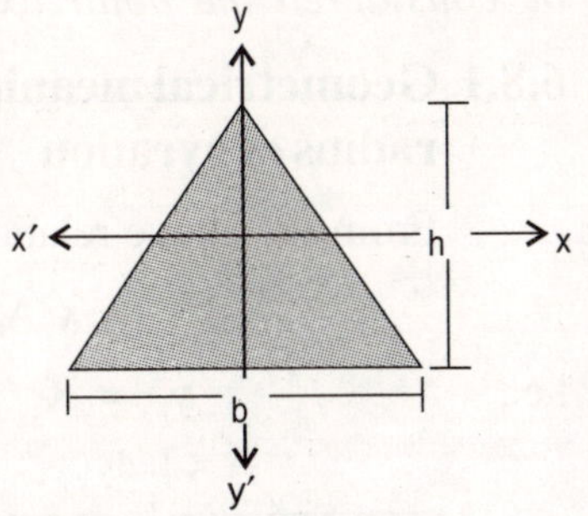	yy′	$I_{yy'} = \frac{bh^3}{48}$

6.7 POLAR MOMENT OF INERTIA

The moment of inertia about an axis perpendicular to the plane of an area is known as the polar moment of inertia. It could be denoted by J or $J_{zz'}$. Thus, the moment of inertia about an axis perpendicular to the plane of the area (A) and passes through origin as shown in Fig. (6.17), it is the z-axis. The moment of inertia about this axis is called the polar moment of inertia and is given by $I_{zz} = \Sigma r^2 \Delta A$; for descrete particles system and $I_{zz} = \int_A r^2 dA$; for continuous body.

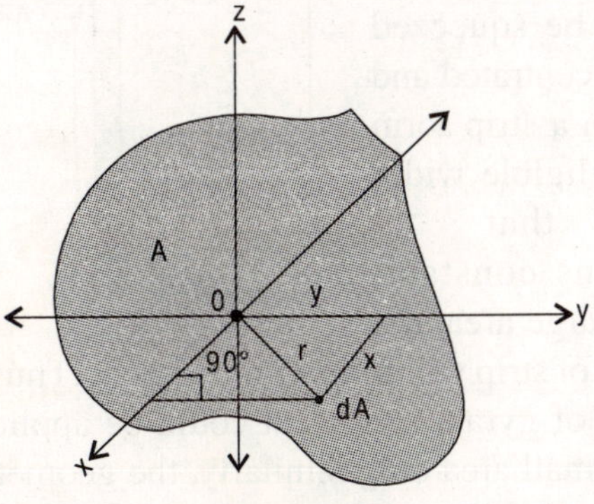

Fig. 6.17

Now in rectangular frame of reference xyz, we have $r^2 = x^2 + y^2$ [Pythagorous theorem.]

Therefore, $I_{polar} = I_{zz}$

$= \Sigma\, r^2\, \Delta A$

or $\int_A r^2 dA = \Sigma\,(x^2 + y^2)\, dA$

or $\int_A \left(x^2 + y^2\right) dA = \Sigma x^2\, dA + \Sigma y^2\, dA$

Therefore, $I_{polar} = I_{zz}$

$= I_{yy} + I_{xx}$ ...(6.17)

where, I_{yy} and I_{xx} are defined by

$$I_{xx} = \int y^2 dA$$
$$= \Sigma\, y^2\, \Delta A$$

and
$$I_{yy} = \int x^2 dA$$
$$= \Sigma\, x^2\, \Delta A$$

***Example 2:** Find the polar moment of inertia of rectangle as shown in Fig. (6.18).*

Solution: As we know that,

$$I_{polar} = I_{zz}$$
$$= I_{xx} + I_{yy}$$
$$= \frac{bd^3}{12} + \frac{db^3}{12}$$
$$= \frac{bd}{12}\left(b^2 + d^2\right)\text{unit}^4$$

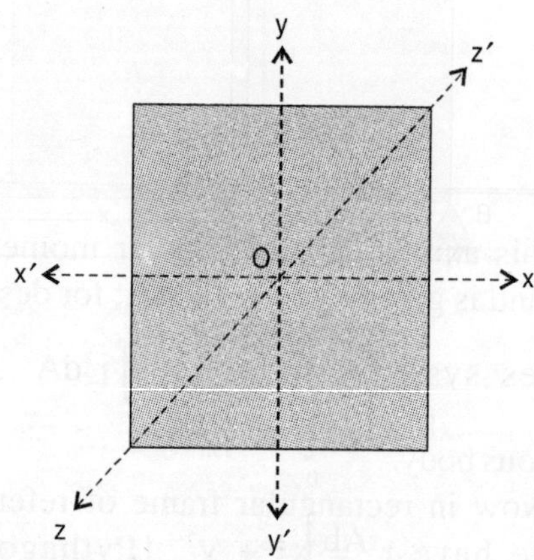

Fig. 6.18

6.8 RADIUS OF GYRATION

The radius of gyration is mathematically defined by the relation

$$K_a^2 A = I_a \quad \text{...(6.18a)}$$

Where 'K_a' is the radius of gyration about the a-axis 'a' and 'I_a' is the moment of inertia about the same axis, 'A' is area of the plane figure of which we have to find the radius of gyration. The unit of the radius of gyration is meter or millimeter etc.

From the above definition, we get,

$$\left.\begin{aligned} K_{xx} &= \sqrt{\frac{I_{xx}}{A}} \\ K_{yy} &= \sqrt{\frac{I_{yy}}{A}} \\ \text{and} \quad K_{zz} &= \sqrt{\frac{I_{zz}}{A}} \\ &= K_{polar} \end{aligned}\right\} \quad \text{....(6.18b)}$$

***Note:** In some cases x-axis or y-axis may also be considered the polar axis.*

6.8.1 Geometrical meaning of the radius of gyration

From the above relation, we have

$$I_{xx} = y^2 A = K_{xx}^2 A$$

i.e.,
$$y = K_{xx'}$$

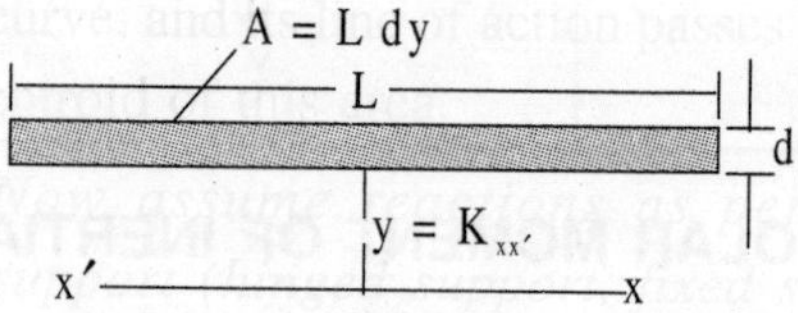

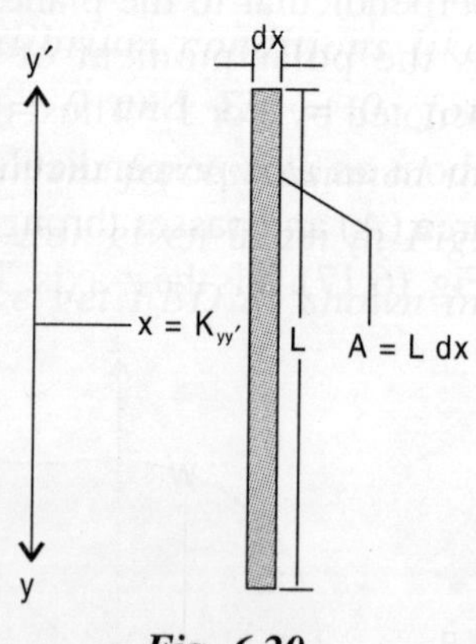

Fig. 6.19

That means the radius of gyration $K_{xx'}$ could be considered as the distance at which the complete area could be squeezed or concentrated and kept in a strip form of negligible width such that 'y' remains constant, for a large area, the length of strip will be of infinite length. Thus, the radius of gyration concept could be applicable for a small area only. Similarly, the geometrical meaning of K_{yy} can be seen in similar manner and it is shown in Fig.(6.20).

Fig. 6.20

Mathematically, $K_{yy'} = \sqrt{\dfrac{I_{yy}}{A}} = x$

or $I_{yy'} = K_{yy'}^2 A = x^2 A$

The geometrical meaning of the polar radius of gyration is given in Fig. (6.21).

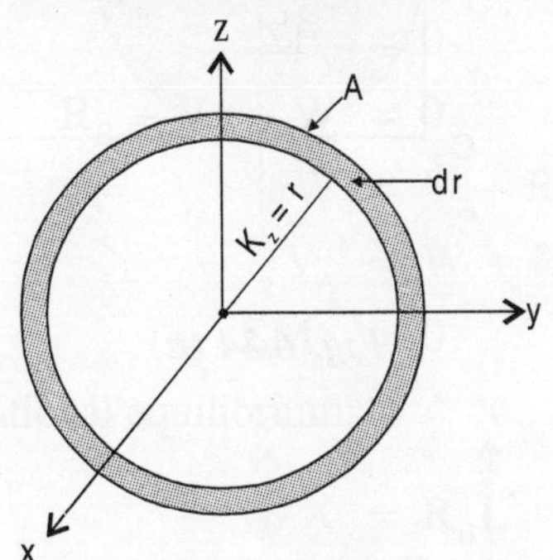

Fig. 6.21

Now $I_{polar} = K_{zz}^2 A = r^2 A$ where, $A = 2\pi$ rdr; and 'r' depends on area (A) of plane figure. In this case we can consider 'K' as the radius of ring of width dr at which the complete area could be squeezed or concentrated.

6.9 USE OF AXIS OF SYMMETRY

Some times the position of the centroid of a plane figure or curve would be determined by inspection. If the given area is assumed to cut along any axis (axes) and each part is a mirror replica of the other corresponding part then it is said to be that area is symmetrical about that axis (axes). Now it is a fact that the centroid of an area lies on the axis of symmetry if it (they) exit(s). Thus this is a useful result to locate the centroid of the area. This result can be proved as follows.

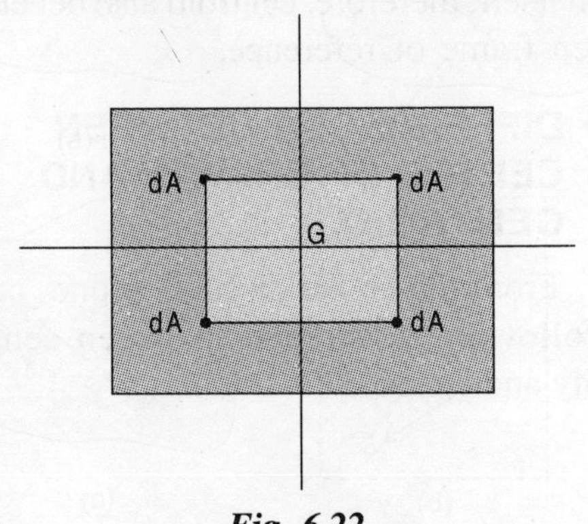

Fig. 6.22

When a plane figure is symmetrical then it has an equal area on either side of the symmetrical axis (axes) as shown in Fig. (6.22) then the sum of the moments of these areas cancel each other and the total moment becomes zero about that axis. Hence, the distance of the centroid from this symmetrical axis is zero. i.e., it lies on the axis of symmetry. If an area has more than one axis of symmetry then the centroid lies on their intersection.

Mathematically, we know that,

$$\overline{X} = \frac{x_1A_1 + x_2A_2 + \ldots.. + A_n x_n}{A_1 + A_2 + \ldots.. + A_n}$$

$$= \frac{\Sigma x_1 A_1}{\Sigma A_1}$$

and

$$\overline{Y} = \frac{y_1A_1 + y_2A_2 + \ldots.. + A_n y_n}{A_1 + A_2 + \ldots.. + A_{n_1}}$$

$$= \frac{\Sigma y_1 A_1}{\Sigma A_1}$$

Now the numerator of $\overline{X}$ (or $\overline{Y}$), there is a minus x dA (or xA) or y dA (or yA) for every plus x dA (or xA) or y dA (or yA), therefore, we get $\overline{X} = 0$ and $\overline{Y} = 0$ for better understanding refer the given Fig. (6.22).

Note that if the symmetrical axis is x-axis then $\overline{Y} = 0, \overline{X} = C_1$. Similarly, if the symmetrical axis is y-axis then $\overline{X} = 0$, $\overline{Y} = C_2$ if both the axes are symmetrical axes then $\overline{X} = 0$, $\overline{Y} = 0$ i.e., the origin itself is the centroid. May be a point other than the origin that depends on the choice of the reference frame.

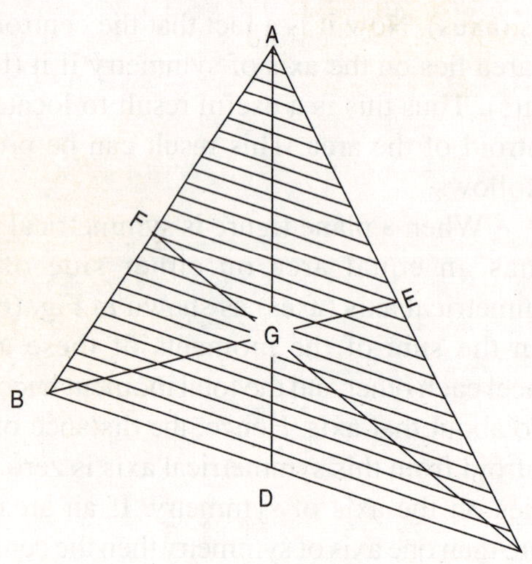

Fig. 6.23

Sometimes a plane figure would not have any axes of symmetry rather it could be symmetrical about a point, that is there is one point which is the mid-point of every conceivable diameter. In the Fig. (6.23) point C_T is the centroid. The centroid of the area of any type of triangle lies at the intersection of its diameters.

Consider a set of parallel strips parallel to the side BC. Now, the centroid of each strip lies on the mid-point of the strip. Therefore, we can say that the centroid must lie on the median AD.

Similarly we can consider set of parallel strips parallel to sides AB and AC then the centroid also lies on the medians BE and CF therefore, the centroid lies at the intersection of the medians.

Now consider the circle and the rectangle shown in Fig. (6.24a) and Fig. (6.24b).

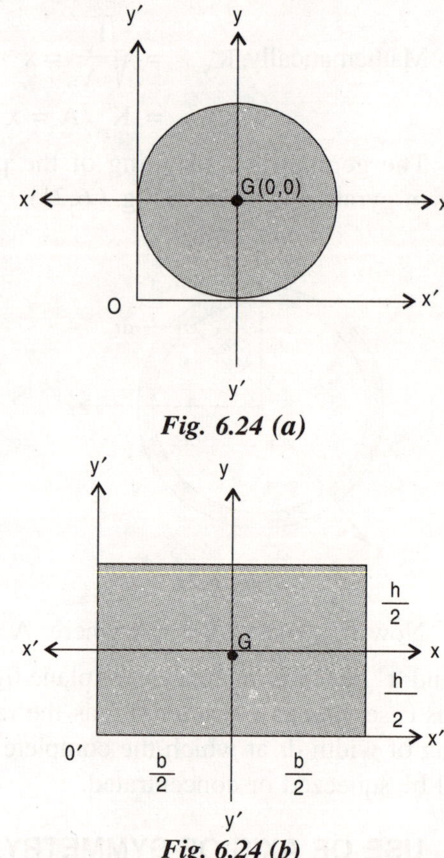

Fig. 6.24 (a)

Fig. 6.24 (b)

There are two axes of symmetry and they are x-axis and y-axis therefore, centroid is origin itself in x-o-y and in x′- o′-y′ frame of references it is G(R, R) and $G_R\left(\frac{b}{2}, \frac{h}{2}\right)$ and in x-O-y it is origin itself, therefore, centroid also depends on chosen frame of reference.

6.10 DIFFERENCES BETWEEN CENTRE OF GRAVITY AND CENTROID

From the above discussion one can get the following differences between centre of gravity and centroid.

Table 6.2

<table>
<tr><th>Centre of gravity</th><th>Centroid</th></tr>
<tr><td>(a) The term centre of gravity is associated with 3D bodies for referring mass and weight.</td><td>(a) The term centroid is associated with 2D-bodies to refer plane area.</td></tr>
<tr><td>(b) The centre of gravity of a body is a point through which the resultant gravitational force (weight) acts for any orientation of the body.</td><td>(b) A centroid is that mathematical point in a plane area where we can assume the whole area is concentrated. That is we could assume the whole area at centroid.</td></tr>
<tr><td>(c) Centre of gravity is a mathematical point given by position vector,
$\vec{r}_{CG} = x_{CG}\vec{i} + y_{CG}\vec{j} + z_{CG}\vec{k}$
$= \dfrac{(\Sigma x_i A_i)\vec{i} + (\Sigma y_i A_i)\vec{j} + (\Sigma z_i A_i)\vec{k}}{\Sigma A_i}$
[For descreate area distribution]
$= \dfrac{\left(\int x\,dw\right)\vec{i} + \left(\int y dw\right)\vec{j} + \left(\int z dw\right)\vec{k}}{\left(\int dw\right)}$
[For continuous mass distribution]
where (x_i, y_i, z_i) is coordinates of ith particle having weight w_i and dw is an elementary change in weight for changes $x \rightarrow x + dx$, $y \rightarrow y + dx$ and $z \rightarrow z + dz$ in intelligent choice of reference axes.</td><td>(c) Position vector of centroid is given by position vector,
$\vec{r}_c = x_c\vec{i} + y_c\vec{J} + z_c\vec{k}$
$= \dfrac{(\Sigma x_i w_i)\vec{i} + (\Sigma y_i w_i)\vec{j} + (\Sigma z_i w_i)\vec{k}}{\Sigma w_i}$
[For descrete mass distribution]
$= \dfrac{\left(\int x\,dA\right)\vec{i} + \left(\int y dA\right)\vec{j} + (z dA)\vec{j}}{\int dA}$
[For continuous area distribution]
Where (x_i, y_i, z_i) is coordinates of ith particle having area A_i and dA is an elementary change in area for changes $x \rightarrow x + dx, y \rightarrow y + dy$, and $lz \rightarrow z + dz$ in intelligent choice of reference axes.</td></tr>
</table>

6.11 MOMENT OF INERTIA OF COMPOSITE SECTION

The moment of inertia of a composite section could be found by the following procedures:

(a) Divide the given composite sections into a number of simple sections for which the moment of inertia is already calculated in the Table (6.1).

(b) Find out the centroid of the divided sections individually, with respect to the given axis (axes) or considered axis (axes) to efficient calculation then find the centroid of composite section.

(c) For some problems the axis about which the moment of inertia has to be calculated might be given or in some places it has to be calculated about the centroidal axis i.e., axis passes through the centroid of the composite section.

(d) Find the moment of inertia of the individual divided sections about their own centroidal axis (refer table 6.1).

(e) Finally using the parallel axis theorem their moment of inertia could be found in the given axis or the centroidal axis.

The procedure given above are illustrated below.

Referring to the Fig. (6.25), it is required to find out the moment of inertia of the composite section about the x-axis and centroidal axis.

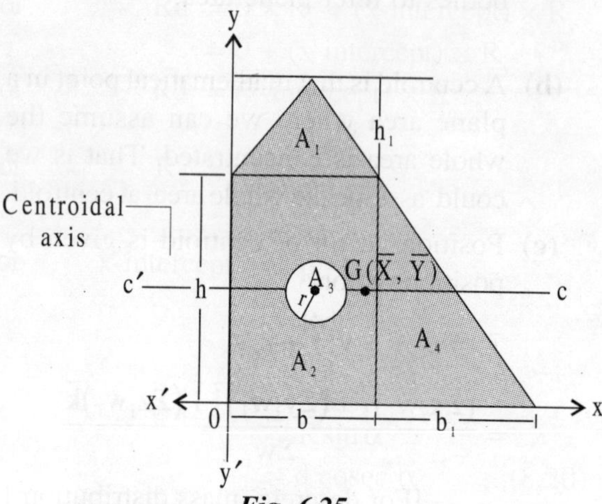

Fig. 6.25

Let us divide the composite section into the following sections:

(a) a rectangle (A_2),

(b) a triangle (A_1),

(c) a triangle (A_4) and

(d) a cut circle of radius r (A_3).

If A is the area of the composite section then,

$$A = A_1 + A_2 + A_4 - A_3$$

Considering x-axis as the reference axis, calculate the centroids of each of the section.

They are, $g_R \equiv \left(\bar{x}_R, \bar{y}_R\right) = \left(\frac{b}{2}, \frac{h}{2}\right)$

$$g_{T_1} \equiv \left(\bar{x}_{T_1}, \bar{y}_{T_1}\right) = \left(\frac{b}{2}, h + \frac{h_1}{3}\right)$$

$$g_{T_2} \equiv \left(\bar{x}_{T_2}, \bar{y}_{T_2}\right) = \left(b + \frac{b_1}{3}, \frac{h}{3}\right)$$

and $g_{cir} = \left(\bar{x}_{cir}, \bar{y}_{cir}\right)$

$$= \left(\frac{-b}{2}, \frac{-h}{2}\right);$$

[Negative sign due to cut-section.]

Therefore, $\bar{X}$ = the x-coordinate of the centroid of the composite section

$$= \frac{\frac{b}{2} \times A_2 + \frac{b}{2} \times A_1 + \left(b + \frac{b_1}{3}\right) A_4 - \frac{b}{2} \times A_3}{A_2 + A_1 + A_4 - A_3}$$

= k (say)

and

$$\bar{Y} = \frac{\frac{h}{2} \times A_2 + \left(h + \frac{h_1}{3}\right) A_1 + \frac{h}{3} \times A_4 - \frac{h}{2} \times A_3}{A_2 + A_1 + A_4 - A_3}$$

= k′ (say)

$\therefore G\left(\bar{X}, \bar{Y}\right) \equiv$ (k, k′) is centroid of composite section.

Now the M.I. of a triangle (A_1) about its centroidal axis,

$$\bar{I}_{T_1} = \frac{1}{36} b h_1^3$$

Similarly, $\bar{I}_{T_2} = \frac{1}{36} b_1 h^3$

$$\bar{I}_R = \frac{1}{12} b h^3$$

and $\bar{I}_{cir} = -\frac{1}{4} \pi r^4$

[Minus sign due to cut-section.]

Therefore, by the parallel axis theorem,

$$I_{x,\ composite\ section} =$$

$$\left\{\bar{I}_{T_1} + A_1\left(\bar{y}_{T_1}\right)^2\right\} + \left\{\bar{I}_{T_2} + A_4 \times \left(\bar{y}_{T_2}\right)^2\right\} + \left\{\bar{I}_R + A_2\left(\bar{y}_R\right)^2\right\} - \left\{\bar{I}_{cir} + A_3\left(\bar{y}_{cir}\right)^2\right\}$$

$$= \left[\bar{I}_{T_1} + A_1\left(h + \frac{h_1}{3}\right)^2\right] + \left[\bar{I}_{T_2} + A_4\left(\frac{h}{3}\right)^2\right]$$

$$+ \left[\bar{I}_R + A_2\left(\frac{h}{2}\right)^2\right] - \left[\bar{I}_{cir} + A_3\left(\frac{h}{2}\right)^2\right]$$

$\therefore\ I_{x,\ composite\ section}$

$$= \left[\frac{1}{36}bh_1^3 + A_1\left(h + \frac{h_1}{3}\right)^2 + \frac{1}{36}b_1h^3 + A_4\left(\frac{h}{3}\right)^2\right.$$

$$\left. + \frac{1}{12}bh^3 + A_2\left(\frac{h}{2}\right)^2 - \frac{1}{4}\pi r^4 - A_3\left(\frac{h}{2}\right)^2\right] \text{unit}^4$$

Similarly, we can find the $I_{y,\ composite\ section}$ by using the parallel axis theorem and the standard formulae given in the table (6.1).

Now, we have to calculate M.I. about the centroidal axis parallel to the x-axis, then from the parallel axis theorem.

$$I_{centroidal\ \| xx'} = \left[\bar{I}_{T_1} + A_1\left(\bar{Y} - \bar{y}_{T_1}\right)^2\right] + \left[\bar{I}_{T_2} + A_4\left(\bar{Y} - \bar{y}_{T_2}\right)^2\right]$$

$$+ \left[\bar{I}_R + A_2\left(\bar{Y} - \bar{y}_R\right)^2\right] - \left[\bar{I}_{cir} + A_3\left(\bar{Y} - \bar{y}_{cir}\right)^2\right]$$

Similarly, we can calculate the $I_{centroidal\ \|\ yy'}$

$$= \left[\frac{1}{36}bh_1^3 + A_1\left(\bar{X} - \frac{b}{2}\right)^2 + \frac{1}{36}hb_1^3 + A_4\right.$$

$$\left(\bar{X} - b + \frac{b_1}{3}\right)^2 + \frac{1}{12}hb^3 + A_2\left(\bar{X} - \frac{b}{2}\right)^2$$

$$\left. - \frac{\pi r^4}{4} - A_3\left(\bar{X} - \frac{b}{2}\right)^2\right] \text{unit}^4$$

Thus finally, for finding the moment of inertia about the given axis if this axis is the x-axis then,

$I_{x\text{-axis}}$

$$= \left[\pm\left(\bar{I}_{1x} + A_1\left(\bar{Y}_1\right)^2\right)\right] + \left[\pm\left(\bar{I}_{2x} + A_2\left(\bar{Y}_2\right)^2\right)\right]$$

+ ... up to number of sections divided and

$$I_{y\text{-axis}} = \left[\pm\left(\bar{I}_{1y} + A_1\left(\bar{X}_1\right)^2\right)\right] + \left[\pm\left(\bar{I}_{2y} + A_2\left(\bar{X}_2\right)^2\right)\right] + \ldots$$

$I_{centroidal\ axis\ parallel\ to\ x\text{-axis}}$

$$= \left[\pm\left(\bar{I}_{1x} + A_1\left(\bar{Y} - \bar{Y}_1\right)^2\right)\right] + \left[\pm\left(\bar{I}_{2x} + A_2\left(\bar{Y} - \bar{Y}_2\right)^2\right)\right] + \ldots$$

and

$I_{centroidal\ axis\ parallel\ to\ y\text{-axis}}$

$$= \left[\pm\left(\bar{I}_{1y} + A_1\left(\bar{x} - \bar{x}_1\right)^2\right)\right] + \left[\pm\left(\bar{I}_{2y} + A_2\left(\bar{x} - \bar{x}_2\right)^2\right)\right] + \ldots$$

where symbols have their own meaning and if a cut section is there then take the negative sign out of ± sign.

6.12 PARALLEL AXIS THEOREM

There is a simple and very useful relation between the moment of inertia of a plane figure (having an area A) about an axis and a axis which passes through the centroid of plane figure such that these axes are parallel to each other.

In another words, the moment of inertia of a plane figure with respect to a axis in its plane (perpendicular to plane) is equal to the moment of inertia of the area with respect to a parallel centroidal axis plus the product of the total area with the square of the distance between these two axes.

Mathematically :

$$I_{about\ any\ axis} = I_{about\ a\ parallel\ axis\ at\ centroid} + Ad^2$$

First method:

Proof : Let $C(\bar{X}, \bar{Y})$ be the centroid of a plane figure of an arbitrary shape. The plane figure is shown in Fig. (6.26). Choose the x-y plane to include 'C' so that $\bar{Z} = 0$. Consider an axis passes through $C(\bar{X}, \bar{Y}, \bar{Z} = 0)$ and it is perpendicular to the plane of the given area and an another axis parallel to it and passes through $P(\overline{X}+a, \overline{Y}+b, \overline{Z}=0)$. The distance between these axes is $h=\sqrt{a^2+b^2}$. Then the square of the distance of the elementary area dA from the axis through 'C' is x^2+y^2 where 'x' and 'y' measure the coordinates of dA relative to the axis through 'C'. The square of its distance from the axis through 'P' is $(x-a)^2+(y-b)^2$, therefore from the Fig. (6.26) we can get

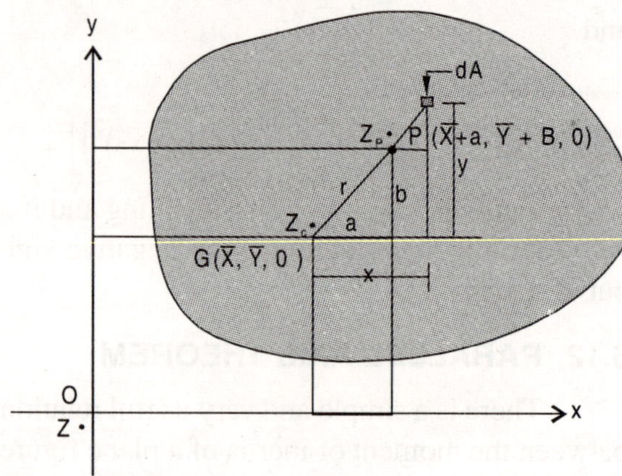

Fig. 6.26

$$dI_{zp} = \left[(y-b)^2+(x-a)^2\right]dA$$

$$= \left[y^2-2yb+b^2+x^2-2xa+a^2\right]dA$$

$$= \left[(x^2+y^2)-2yb-2xa+(a^2+b^2)\right]dA$$

$$= (x^2+y^2)dA - 2yb\,dA - 2xa\,dA + (a^2+b^2)dA$$

Integrating dI_{zp} over the entire area A we get,

$$I_{zp} = \int_A (x^2+y^2)\,dA - 2b\int_A y\,dA - 2a\int_A x\,dA + \int_A (a^2+b^2)\,dA$$

$$= \int_A r^2 dA - 2bA\left(\frac{\int_A y dA}{A}\right) - 2aA\left(\frac{\int_A x dA}{A}\right) + (a^2+b^2)\int_A dA$$

$= I_{ZC} - 2Ab$ [distance of the centroid from the centroidal axis along the y-axis]

$-2aA$ [the distance of the centroid from the centroidal axis along the x-axis]

$+(a^2+b^2)A$

$= I_{ZC} - 2Ab\,[0] - 2aA\,[0] + h^2A$

$= I_{ZC} + h^2A \qquad [\because h^2 = a^2+b^2]$

$\therefore I_{ZP} = I_{ZC} + h^2A$

i.e., the M.I. about the axis through 'P' perpendicular to the plane.

= M.I. about the axis through C (the centroid of the plane area) and perpendicular to the plane (i.e., parallel to each other) + (the distance between the axes)^2 × total area.

Second method: ....(6.19)

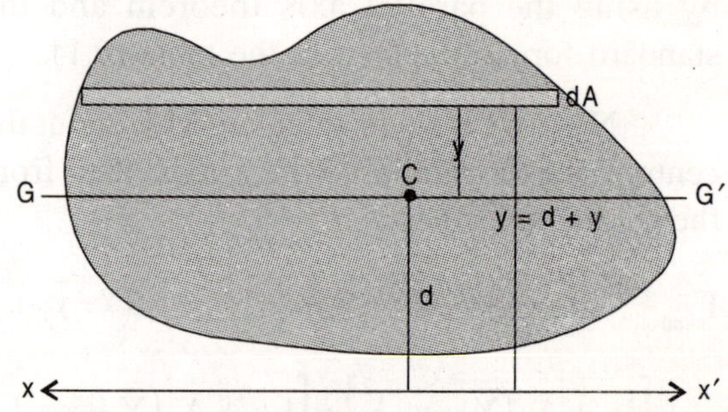

Fig. 6.27

Let the moment of inertia of the given area A with respect to the x-axis be $I_{xx'}$ refer Fig. (6.27); now taking y as the distance of an elementary of area dA from axis xx′, we get,

$$I_{xx'} = \int_A Y^2 dA$$

Now draw axis GG′ parallel to x-axis and passes through the centroid 'C' of the given area; this axis is called the centroidal axis. Taking y as the distance of the elementary area dA to from axis GG′, we get $Y = y + d_{xx'}$; where $d_{xx'}$ is the distance between x-axis and axis-GG′, substituting for y in the integral representing $I_{xx'}$, we can write

$$I_{xx'} = \int_A Y^2 dA$$

$$= \int_A (d_{xx'} + y)^2 dA$$

$$= \int_A (d_{xx'}{}^2 + y^2 + 2d_{xx'} y) dA$$

$$= \int_A d_{xx'}{}^2 dA + \int_A y^2 dA + \int_A 2d_{xx'} y dA$$

$$= d_{xx'}{}^2 \int_A dA + \int_A (y^2) dA + 2\, d_{xx'} y\, dA$$

$$= d_{xx'}{}^2 A + I_{GG'} + 2d_{xx'} \frac{\int_A y dA}{A} A$$

$= Ad_{xx'}{}^2 + I_{GG'} + 2d \times$ distance of centroid from centroidal axis along y-axis) $\times$ A

$$= Ad_{xx'}{}^2 + I_{GG'} + 2Ad \times 0$$

$$I_{xx'} = I_{GG'} + Ad^2$$

Similarly for the figure shown below, we have

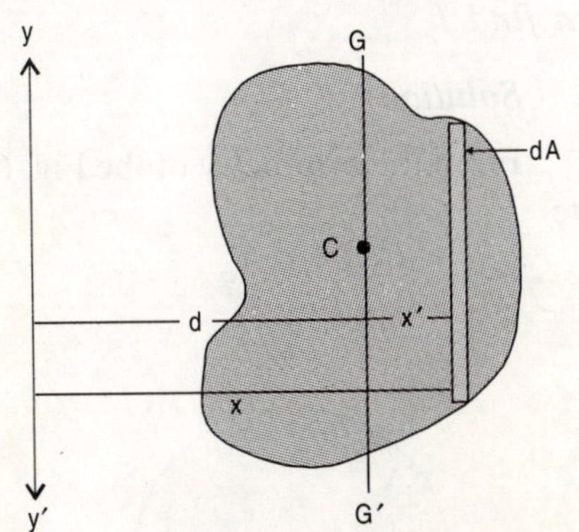

Fig. 6.28

$$I_{yy'} = \int_A x^2 dA$$

$$= \int_A (d_{yy'} + x')^2 dA$$

$$= \int_A (d_{yy'}{}^2 + (x')^2 + 2d_{yy'} x') dA$$

$$= \int_A d_{yy'}{}^2 dA + \int_A (x')^2 dA + \int_A 2d_{yy'} x' dA$$

$$= d_{yy'}^2 \int_A dA + \int_A (x')^2 dA + 2d_{yy'} \left(\frac{\int x' dA}{A} \right) A$$

$= Ad_{yy'}{}^2 + I_{GG'} + 2d_{yy'} A \times$ distance of centroid from centroidal axis.

$$= Ad_{yy'}{}^2 + I_{GG'} + 2Ad \times 0$$

$$= Ad^2 + I_{GG'}$$

$$\therefore I_{yy} = I_{GG'} + Ad^2$$

Now, from the perpendicular axis theorem, we get

$$I_{zz'} = I_{xx'} + I_{yy'}$$

$$= I_{GG' \parallel xx'} + I_{GG' \parallel yy'} + Ad^2_{xx'} + Ad^2_{yy'}$$

$$= I_{GG' \parallel zz'} + A(d^2_{xx'} + d^2_{yy'})$$

$$= I_{GG' \parallel zz'} + Ad^2 \quad \text{where, } d^2 = d^2_{xx'} + d^2_{yy'}$$

Here it is observed that the term Ad^2 is subtracted from the given moment of inertia in order to find the centroidal moment of inertia, while the term is added for getting M.I. about an axis parallel to centroidal axis. In other words, the centroidal moment of inertia is smaller than that of an axis parallel to centroidal axis.

i.e., $I_{\text{centroidal axis}} < I_{\text{any other parallel axis}}$

or $\left| I_{\text{centroidal axis}} - I_{\text{any other parallel axis}} \right| = Ad^2$

***Example 3:** Find the moment of inertia of a circle about AB when I_x = 25 mm^4, refer Fig. (6.29).*

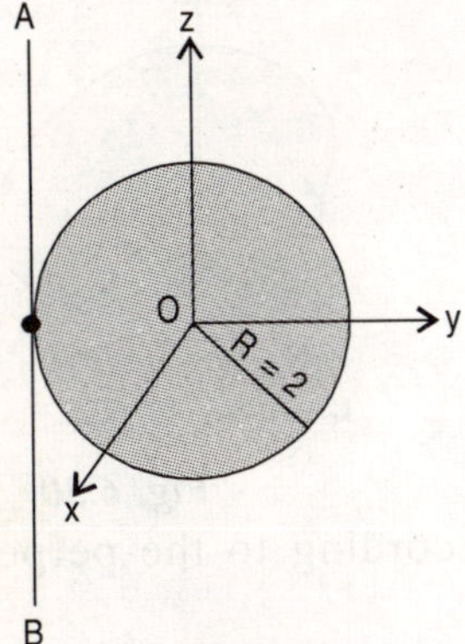

Fig. 6.29

Solution:

From the symmetry of the figure, we have

$$I_x = I_y$$
$$= 25 \text{ mm}^4$$

Now, from the perpendicular axis theorem

$$I_z = I_x + I_y$$
$$= 25 \text{ mm}^4 + 25 \text{ mm}^4$$
$$= 50 \text{ mm}^4$$

Since AB and the z-axis are parallel to each other then from the parallel axis theorem, we have

$$I_{AB} = I_{GG\|AB} + Ad^2$$

or

$$I_{AB} = I_{z\|AB} + Ad^2$$

Since z-axis passing through centroid.

$$= 50 + (\pi R^2)\, R^2 \quad [\because d = R]$$
$$= 50 + \pi R^4$$
$$= 50 + \pi \times 2^4$$
$$= 50 + \pi \times 16$$
$$= 100.24 \text{ mm}^4.$$

6.13 PERPENDICULAR AXIS THEOREM

The moment of inertia (polar moment of inertia) of an area about an axis which is perpendicular to given area at a point is equal to the sum of the moment of inertias about two mutually perpendicular axes which pass through the same point and lie in the plane of the area.

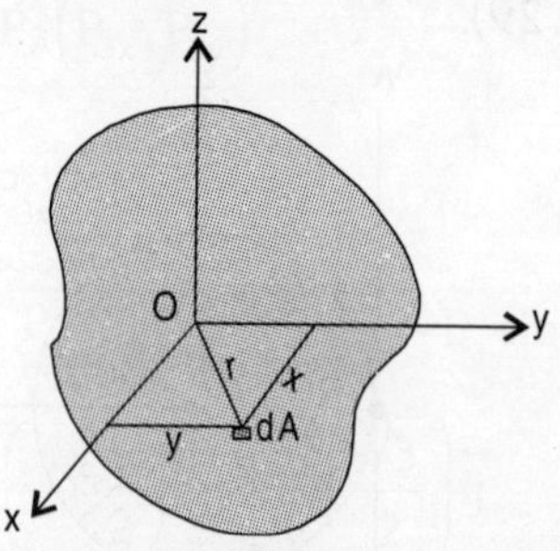

Fig. 6.30

According to the perpendicular axis theorem,

$$I_z = I_x + I_y$$

This theorem could be easily proved. Let us consider an elementary area dA at a distance 'r' from origin. Consider its coordinates are x and y, then from the definition of the polar moment of inertia,

$$I_z = \int_A r^2 dA$$
$$= \int_A \left(x^2 + y^2\right) dA$$
$$[\because r^2 = x^2 + y^2]$$
$$= \int_A x^2 dA + \int_A y^2 dA$$
$$= I_y + I_x$$

Thus $\qquad I_z = I_x + I_y \qquad$ ...(6.20)

If origin is the centroid of the given area then we can write,

$$J_0 = I_x + I_y$$

Example 4: *If the polar moment of inertia of a circle of radius (r) is* $J_o = \frac{\pi}{2} r^4$, *then find* $I_{diameter}$.

Solution:

From the symmetry of the Fig. (6.31), we have

$$I_x = I_y$$

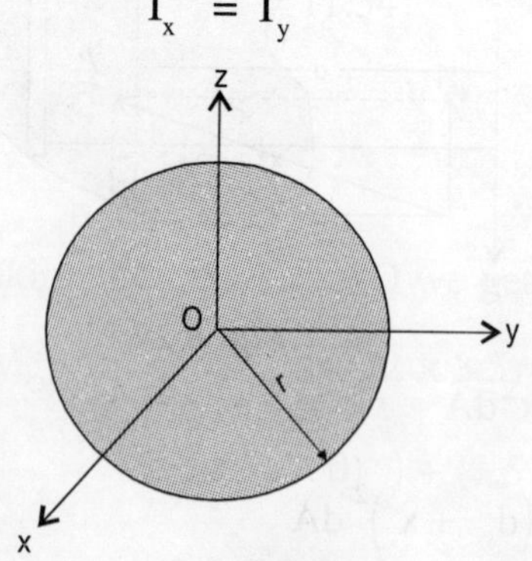

Fig. 6.31

Now from the perpendicular axis theorem, we have

$$I_z = I_x + I_y$$
$$= 2I_x$$

or $I_x = \frac{I_{zz'}}{2} = \frac{\pi}{4} r^4$

$\therefore$ $I_x = I_y$

$= I_{diameter}$

$= \frac{\pi}{4} r^4.$

6.14 PRODUCT OF INERTIA

The integral $I_{xy} = \int xy\, dA$ is obtained by multiplying each element dA of an area A by its coordinates x and y and integrating it over the entire area. This is known as the product of inertia of the area A with respect to x and y axes. Unlike the moment of inertias about the x and y axes, the product of inertia I_{xy} may be either positive or negative depending upon the signs of the x and y coordinates. When one or both of the x and y axes is (are) axis (axes) of symmetry for the area A the product of inertia I_{xy} is equal to zero.

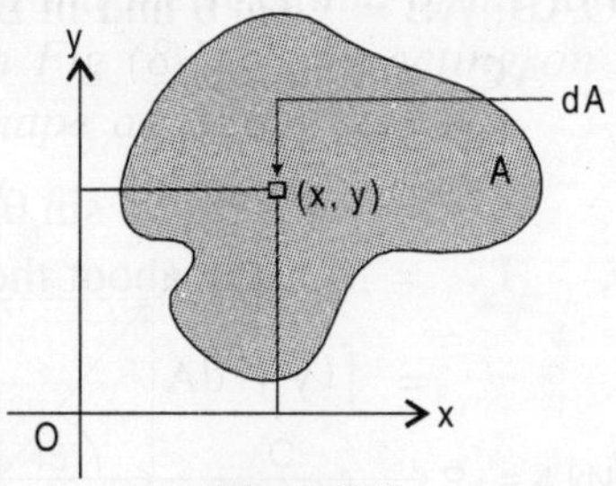

Fig. 6.32

To prove this statement let us consider an elementary area dA having coordinates (x, y) on one side of axis (axes) of symmetry , then on other side of this axis there must exist an elementary area dA having coordinates (–x, y) therefore corresponding elementary product of areas xy dA cancel each other i.e., xy dA+(–xy dA) = 0. Hence the integral

$$\int xy\, dA = \sum_{i=1}^{\infty} xy\, \Delta A = 0.$$

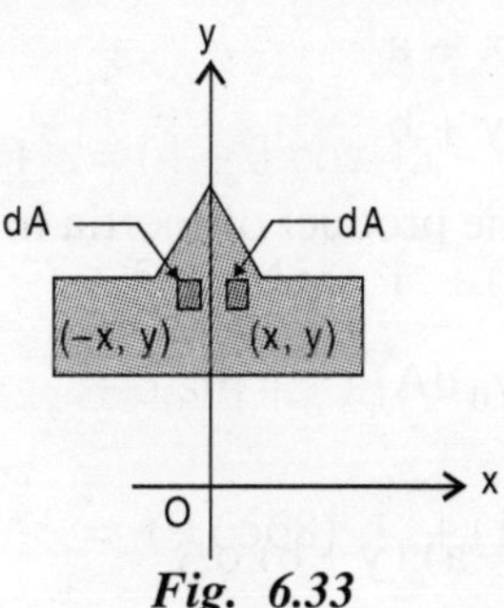

Fig. 6.33

Axis to which product of inertia (I_{xy}) vanishes is known as the principal axes. If I_{xy} = 0 then axes x and y are called as the principal axes and if the origin is the centroid then the axes are called as the centroidal principal axes.

Like the moment of inertia, we can apply the parallel axis theorem for the product of inertia too. If $\bar{I}_{xy}$ is the product of inertia for the axes which pass through the centroid and I_{xy} is the product of inertia parallel to the centroidal axis then the parallel axis theorem gives,

$$I_{xy} = \bar{I}_{xy} + abA$$

where a and b are the coordinates of the centroid 'C' with respect to the parallel axes x and y. Refer the Fig. (6.34).

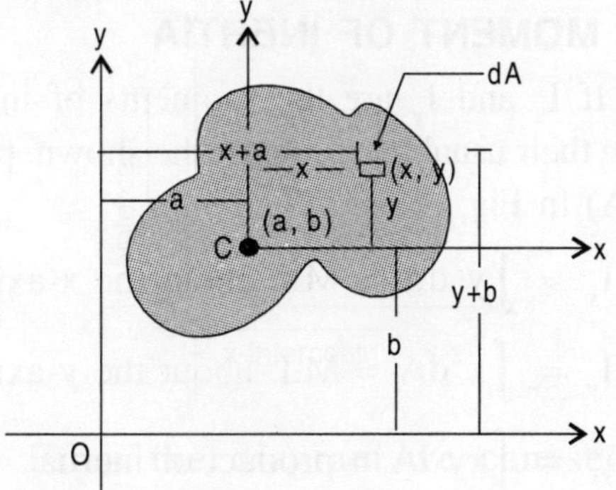

Fig. 6.34

Let (x, y) be the coordinates of the elementary area (dA) with respect to the centroidal axes x-C-y. Therefore the co-ordinates of elementary area dA in xy-axes are,

$x_0 = x + a$

$y_0 = y + b$

Now the product of inertia in x-O-y-axes is,

$$I_{xy} = \int_A x_0 y_0 \, dA$$

$$= \int_A (x+a).(y+b)\, dA$$

$$= \int_A (xy + ay + bx + ab)\, dA$$

$$= \int_A xy\,dA + aA\frac{\int_A y\,dA}{A} + bA\frac{\int_A x\,dA}{A} + abA$$

$= \bar{I}_{xy}$ + aA×(y-coordinate of centroid in centroidal axes) + bA×(x-coordinate of centroid in centroidal axes) + abA

$$= \bar{I}_{xy} + aA \times (0) + bA \times (0) + ab\,A$$

$$= \bar{I}_{xy} + abA$$

Therefore,

$$I_{xy} = \bar{I}_{xy} + abA.$$

6.15 PRINCIPAL AXES AND PRINCIPAL MOMENT OF INERTIA

If I_x and I_y are the moments of inertia having their usual meaning for the shown plane area(A) in Fig. (6.35). Then,

$$I_x = \int y^2 dA = \text{M.I. about the x-axis}$$

$$I_y = \int x^2 dA = \text{M.I. about the y-axis}$$

and $I_{xy} = \int xy\,dA$ = product of inertia about the x and y axes.

Now rotate the axes in an anticlockwise direction through θ and say the new axes are the x′-axis and the y′-axis. Let (x′, y′) be the coordinates of the same elementary area (dA) with respect to x′y′-axes. From the figure,

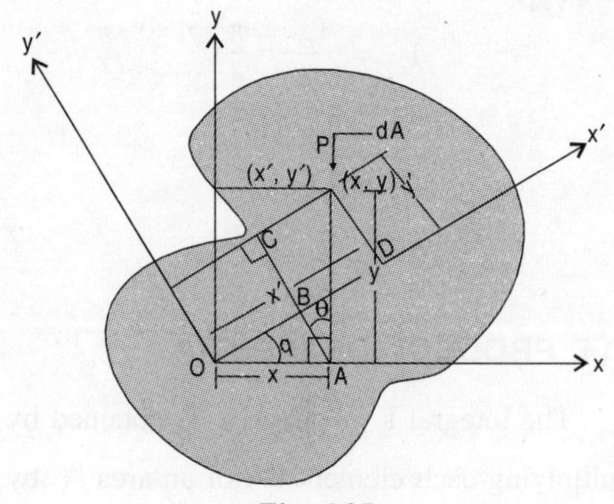

Fig. 6.35

OA = x

AP = y

Now in Δ OAB, OB = x cos θ and in Δ ACP, CP = y sin θ.

∴ OD = x′ = OB + BD

= x cos θ + CP [∵ CP = BD]

∴ x′ = x cos θ + y sin θ

and in Δ OAB, AB = x sin θ and in Δ ACP,

AC = y cos θ

Now DP = y′ = BC = AC − AB

∴ y′ = y cos θ − x sin θ

Now, $I_{x'}$ = M.I. the about the x′-axis

$$= \int (y')^2 dA$$

$$= \int (y\cos\theta - x\sin\theta)^2 dA$$

$$= \int (y^2\cos^2\theta + x^2\sin^2\theta - 2xy\sin\theta\cos\theta)dA$$

$$= \int y^2\cos^2\theta\,dA + \int x^2\sin^2\theta\,dA - \int 2xy\sin\theta\cos\theta\,dA$$

$$= \cos^2\theta \int y^2 dA + \sin^2\theta \int x^2 dA - \sin 2\theta \int xy\,dA$$

$$= I_x(\cos^2\theta) + I_y\sin^2\theta - I_{xy}(\sin 2\theta)$$

$$= I_x\cos^2\theta + I_y\sin^2\theta - I_{xy}\sin 2\theta$$

$$= I_x(1 - \sin^2\theta) + I_y\sin^2\theta - I_{xy}\sin 2\theta$$

$$= I_x - I_x\sin^2\theta + I_y\sin^2\theta - I_{xy}\sin 2\theta$$

$$= I_x + (I_y - I_x)\sin^2\theta - I_{xy}\sin 2\theta$$

$$= I_x + (I_y - I_x)\left(\frac{1-\cos 2\theta}{2}\right) - I_{xy}\sin 2\theta$$

$$= I_x + \frac{I_y - I_x}{2} - \left(\frac{I_y - I_x}{2}\right)\cos 2\theta - I_{xy}\sin 2\theta$$

$$\therefore \quad I_{x'} = \frac{I_x + I_y}{2} + \frac{I_x - I_y}{2}\cos 2\theta - I_{xy}\sin 2\theta \quad \text{...(i)}$$

Similarly, $I_{y'}$ = the M.I. about the y′-axis

$$= \int (x')^2\, dA$$

$$= \int (x\cos\theta + y\sin\theta)^2\, dA$$

$$= \int (x^2\cos^2\theta + y^2\sin^2\theta + 2xy\sin\theta\cos\theta)\, dA$$

$$= \cos^2\theta \int x^2 dA + \sin^2\theta \int y^2 dA + 2\sin\theta\cos\theta \int xy\, dA$$

$$= I_y\cos^2\theta + I_x\sin^2\theta + I_{xy}\sin 2\theta$$

$$= I_y(1 - \sin^2\theta) + I_x\sin^2\theta + I_{xy}\sin 2\theta$$

$$= I_y - I_y\sin^2\theta + I_x\sin^2\theta + I_{xy}\sin 2\theta$$

$$= I_y + (I_x - I_y)\sin^2\theta + I_{xy}\sin 2\theta$$

$$= I_y + (I_x - I_y)\left(\frac{1-\cos 2\theta}{2}\right) + I_{xy}\sin 2\theta$$

$$= I_y + \frac{I_x - I_y}{2} - \frac{I_x - I_y}{2}\cos 2\theta + I_{xy}\sin 2\theta$$

$$\therefore I_{y'} = \frac{I_x + I_y}{2} - \frac{I_x - I_y}{2}\cos 2\theta + I_{xy}\sin 2\theta \quad \text{...(ii)}$$

The product of inertia about the x′y′ axes,

$$I_{x'y'} = \int x'y'\, dA$$

$$= \int (x\cos\theta + y\sin\theta)(y\cos\theta - x\sin\theta)\, dA$$

$$= \int (xy\cos^2\theta + y^2\sin\theta\cos\theta - x^2\sin\theta\cos\theta - xy\sin^2\theta)\, dA$$

$$= \cos^2\theta \int xy\, dA + \sin\theta\cos\theta \int y^2 dA - \sin\theta\cos\theta \int x^2 dA - \sin^2\theta \int xy\, dA$$

$$= \cos^2\theta\, I_{xy} + (I_x - I_y)\sin\theta\cos\theta - \sin^2\theta\, I_{xy}$$

$$= (\cos^2\theta - \sin^2\theta)\, I_{xy} + (I_x - I_y)\sin\theta\cos\theta$$

$$\therefore \quad I_{x'y'} = I_{xy}\cos 2\theta + \frac{I_x - I_y}{2}\sin 2\theta$$

$$= \frac{I_x - I_y}{2}\sin 2\theta + I_{xy}\cos 2\theta \quad \text{...(iii)}$$

Now adding equations (i) and (ii), we get

$$I_{x'} + I_{y'} = \left(\frac{I_x + I_y}{2} + \frac{I_x - I_y}{2}\cos 2\theta - I_{xy}\sin 2\theta\right) + \left(\frac{I_x + I_y}{2} - \frac{I_x - I_y}{2}\cos 2\theta + I_{xy}\sin 2\theta\right)$$

$$= 2\left(\frac{I_x + I_y}{2}\right)$$

or $I_{x'} + I_{y'} = I_x + I_y$ = the polar moment of inertia about the origin is constant for all rectangular axes lie in the plane and pass through origin. That is the sum of the moment of inertias is constant for the x, y axes and x′, y′ axes as they pass through the origin.

Now we have to know the principal axes in which $P_{x'}$ and $P_{y'}$ have to be either maximum or minimum value. Further we should know the critical angle(s). Since $I_{x'}$ and $I_{y'}$ are functions of θ then for the critical angle(s).

$$\frac{dI_{x'}}{d\theta} = 0 \quad \text{...(iv)}$$

and

$$\frac{dI_{y'}}{d\theta} = 0 \quad \text{...(v)}$$

Therefore by differentiating the equation (i) w.r.t. θ, then we have

$$\frac{dI_{x'}}{d\theta} = \frac{d}{d\theta}\left(\frac{I_x + I_y}{2}\right) + \frac{d}{d\theta}\left(\frac{I_x - I_y}{2}\right)\cos 2\theta - \frac{d}{d\theta}\left(I_{xy}\right)\sin 2\theta$$

$$= 0 + \frac{I_x - I_y}{2}\frac{d}{d\theta}(\cos 2\theta) - I_{xy}\frac{d}{d\theta}(\sin 2\theta)$$

$$= -(I_x - I_y)\sin 2\theta - 2I_{xy}\cos 2\theta$$

or $0 = (I_y - I_x)\sin 2\theta - 2I_{xy}\cos 2\theta$

$$\therefore \tan 2\theta = \frac{2I_{xy}}{I_y - I_x} \quad \text{...(vi)}$$

Now by differentiating the equation (ii) w.r.t. θ, we have,

$$\frac{dI_{y'}}{d\theta} = -\left(\frac{I_x - I_y}{2}\right)\frac{d(\cos 2\theta)}{d\theta} + I_{xy}\frac{d}{d\theta}(\sin 2\theta)$$

$$= (I_x - I_y)\sin 2\theta + 2I_{xy}\cos 2\theta$$

or $0 = -(I_y - I_x)\sin 2\theta + 2I_{xy}\cos 2\theta$

or $(I_y - I_x)\sin 2\theta = 2I_{xy}\cos 2\theta$

$$\therefore \quad \tan 2\theta = \frac{2I_{xy}}{I_y - I_x} \quad \text{...(vii)}$$

Therefore solution(s) of equations (vi) or (vii) gives value(s) of θ corresponding to that $I_{x'}$ or $I_{y'}$ will be either maximum or minimum. In other way if $I_{x'}$ is maximum then $I_{y'}$ will be minimum.

Now, from the equation (vi) or (vii) further we get,

$$\sin 2\theta = \frac{2I_{xy}}{\sqrt{(I_y - I_x)^2 + 4I_{xy}^2}}$$

and

$$\cos 2\theta = \frac{I_y - I_x}{\sqrt{(I_y - I_x)^2 + 4I_{xy}^2}} \quad \text{...(viii)}$$

Now using equation (viii), we get,

$$I_{x'y'} = \frac{I_x - I_y}{2}\sin 2\theta + I_{xy}\cos 2\theta$$

$$= \frac{I_x - I_y}{2} \times \frac{2I_{xy}}{\sqrt{(I_y - I_x)^2 + 4I_{xy}^2}} + I_{xy} \times \frac{I_y - I_x}{\sqrt{(I_y - I_x)^2 + 4I_{xy}^2}}$$

$$= \frac{I_{xy}(I_x - I_y)}{\sqrt{(I_y - I_x)^2 + 4I_{xy}^2}} - \frac{I_{xy}(I_x - I_y)}{\sqrt{(I_y - I_x)^2 + 4I_{xy}^2}} = 0$$

Therefore, principal axes are those about which if $I_{x'}$ is maximum then $I_{y'}$ will be minimum and vice-versa and product of inertia will be zero.

Now equation $\tan 2\theta = \frac{2I_{xy}}{I_y - I_x}$ gives two values 2θ and 2θ + π, because tan 2θ = tan (2θ + π). That is difference between two solutions of equations (vii) and (viii) is at right angle. Thus two such axes are called the principal axes. By definition, therefore, the principal axes about the origin are mutually perpendicular sets of axes, one about which the moment of inertia is a maximum and the other about which the moment of inertia is minimum. The maximum and minimum moments of inertia about the axes are called the principal moment of inertia about that axis.

Now from equations (i) and (iii), we get

$$\left(I_{x'} - \frac{I_x + I_y}{2}\right)^2 + I_{x'y'}^2 = \left(\frac{I_x - I_y}{2}\right)^2 \cos^2 2\theta$$

$$+ I^2_{xy}\sin^2 2\theta - 2I_{xy}\left(\frac{I_x - I_y}{2}\right)\sin 2\theta \cos 2\theta$$

$$+\left(\frac{I_x - I_y}{2}\right)^2 \sin^2 2\theta + I^2_{xy} \cos^2 2\theta + 2I_{xy}$$

$$\left(\frac{I_x - I_y}{2}\right) \times \sin 2\theta \cos 2\theta$$

$$= \left(\frac{I_x - I_y}{2}\right)^2 + I^2_{xy}$$

or $$\left(I_{x'} - \frac{I_x + I_y}{2}\right)^2 + I^2_{x'y'} = \left(\frac{I_x - I_y}{2}\right)^2 + I^2_{xy}$$

or $$(I_{x'} - I_{av})^2 + I^2_{x'y'} = \left(\frac{I_x - I_y}{2}\right)^2 + I^2_{xy}$$

or $$(I_{x'} - I_{av})^2 + I^2_{x'y'} = R^2 \quad \text{...(6.21)}$$

where, $$R = \sqrt{\left(\frac{I_x - I_y}{2}\right)^2 + I^2_{xy}}$$

and $$I_{av} = \frac{I_x + I_y}{2}$$

or $(I_{x'} - I_{av})^2 + I^2_{x'y'} = R^2$ is the equation of a circle of radius R centered at a point located by $(I_{av}, 0)$ on abscissa $I_{x'}$ and ordinate $I_{x'y'}$. This circle is called Mohr's circle.

A typical point P on the circle denotes that the moment of inertia about an axis represented by radius vector CP is $I_{x'}$ and the product of inertia with reference to axes is $I_{x'y'}$ as shown in the Fig. (6.36). In particular, the product of inertia at A and B is zero. At point A, I_{max} and at point B, I_{mm}, at points A and B, $I_{x'y'} = 0$.

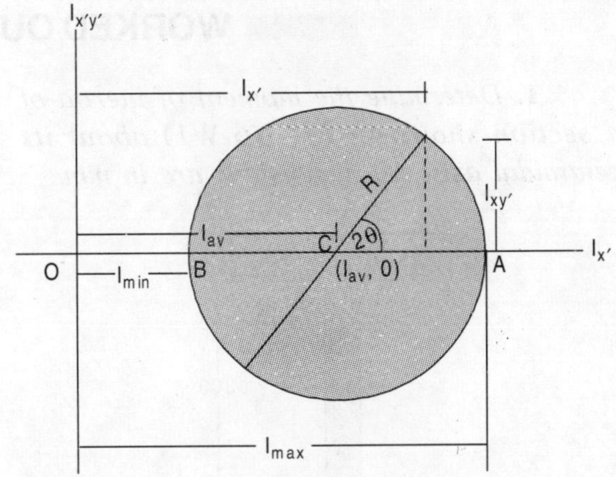

Fig. 6.36

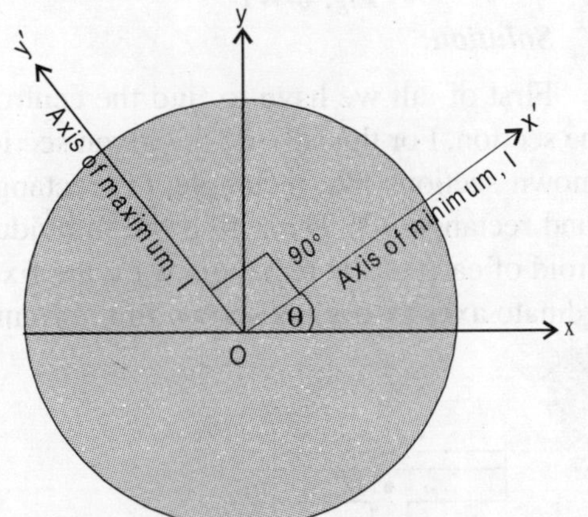

Fig. 6.37

As we know that, the product of inertia for a set of axes vanishes if the area is symmetrical about any of the axes. It follows that if an area possesses an axis of symmetry through a point, this axis must be the principal axis through that point. It could be understood that axes other than the axes of symmetry can be the principal axes. An area may or may not possess one or more axes of symmetry but it must have a set of principal axes.

WORKED OUT EXAMPLES

1. *Determine the moment of inertia of a section shown in Fig. (6-W1) about its centroidal axis. All dimensions are in mm.*

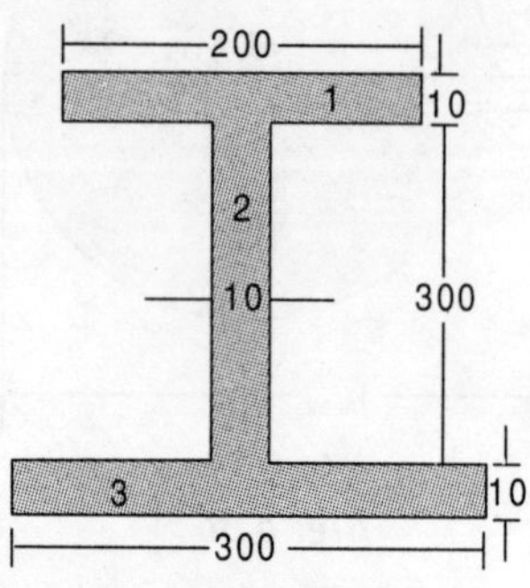

Fig. 6-W1

Solution:

First of all we have to find the centroid of the section. For this, divide the given section in known sections like rectangle, (1) rectangle (2) and rectangle (3) Then, find the individual centroid of each of the rectangle w.r.t. the fixed coordinate axes (x-o-y) as shown in the figure.

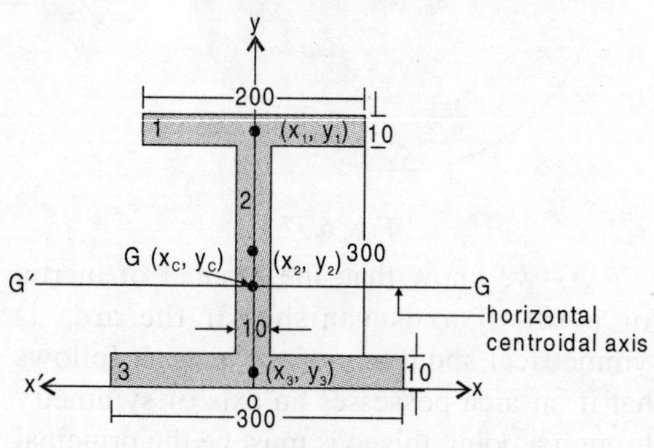

Fig. 6-W2

Now rectangle 1,

$$A_1 = 200 \times 10$$
$$= 2000 \text{ mm}^2$$
$$x_1 = 0$$
$$y_1 = 10 + 300 + \frac{10}{2}$$
$$= 315 \text{ mm}$$

rectangle 2,

$$A_2 = 300 \times 10$$
$$= 3000 \text{ mm}^2$$
$$x_2 = 0$$
$$y_2 = 10 + \frac{300}{2}$$
$$= 160 \text{ mm}$$

and rectangle 3

$$A_3 = 300 \times 10$$
$$= 3000 \text{ mm}^2$$
$$x_3 = 0$$
$$y_3 = \frac{10}{2} = 5 \text{ mm}$$

Therefore,

$$x_c = \frac{x_1A_1 + x_2A_2 + x_3A_3}{A_1 + A_2 + A_3}$$

$$= \frac{0 \times 2000 + 0 \times 3000 + 0 \times 3000}{2000 + 3000 + 3000} = 0$$

because given section is symmetrical about y-axis

and $$y_c = \frac{y_1A_1 + y_2A_2 + y_3A_3}{A_1 + A_2 + A_3}$$

$$= \frac{315 \times 2000 + 160 \times 3000 + 5 \times 3000}{2000 + 3000 + 3000}$$

$$= 140.625 \text{ mm}$$

Now $I_{GG'}$ = moment of inertia about horizontal centroidal axis.

= moment of inertia of A_1 + moment of inertia of A_2 + moment of inertia of A_3

$$= \frac{200 \times 103}{12} + 2000(315 - 14.325)2 + \frac{10 \times 300^3}{12}$$

$$= + 3000\,(160 - 140.625)^2 + \frac{300 \times 10^2}{12}$$

$$+ 3000 \times (140.625 - 5)^2$$

$$= \frac{200 \times 10^3}{12} + \frac{10 \times 300^3}{12} + \frac{300 \times 10^3}{12}$$

$$+ 3000\,(160 - 140.625)^2 + 2000\,(315 - 140.625)^2 + 3000\,(140.325 - 5)^2$$

$$= 22541666.67 + 117121875$$

$$= 139663541.7 \text{ mm}^4.$$

$I_{yy'}$ = moment of inertia about a vertical centroidal axis

= moment of inertia of A_1 + moment of inertia of A_2 + moment of inertia of A_3

$$= \frac{10 \times 200^3}{12} + \frac{300 \times 10^3}{12} + \frac{10 \times 300^3}{12}$$

$$= 29191666.67 \text{ mm}^4$$

2. *Find the moment of inertia of the plane figure shown in the Fig. (6-W3) about its horizontal centroidal x-axis. All the dimensions are in mm.*

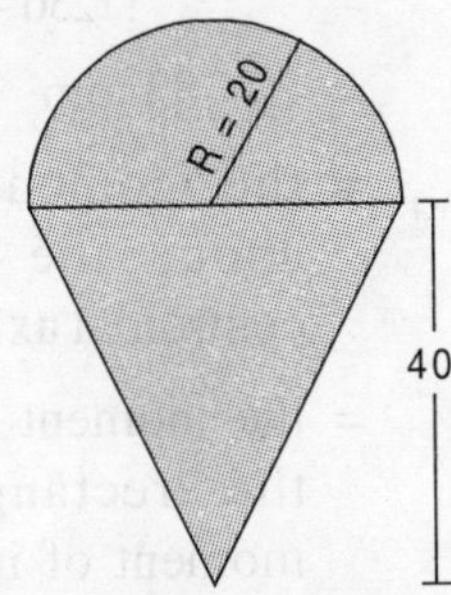

Fig. 6-W3

Solution:

First of all we have to find the centroid of the given plane figure. For this, divide the given section in known regular sections like a semicircle and a triangle. Find their individual centroid and therefore the centroid of the given figure w.r.t. the fixed axis x-O-y.

Now semicircle,

$$A_1 = \frac{1}{2} \times \pi \times 20^2$$

$$= 628 \text{ mm}^2$$

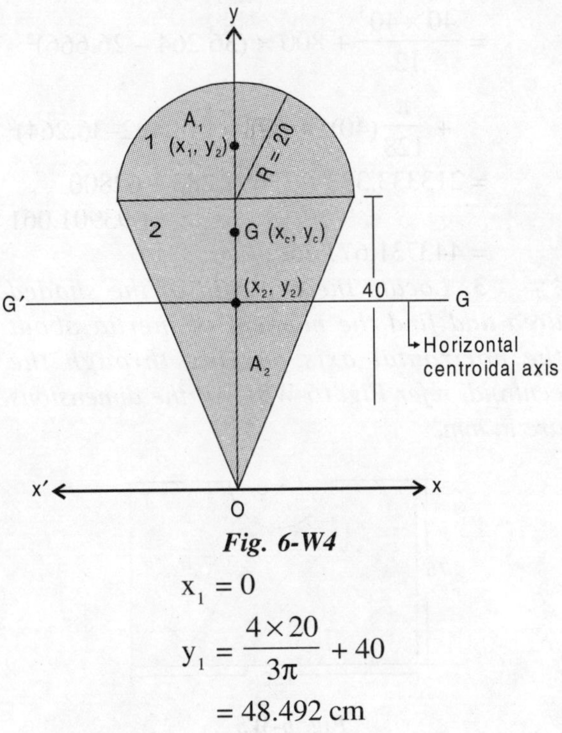

Fig. 6-W4

$$x_1 = 0$$

$$y_1 = \frac{4 \times 20}{3\pi} + 40$$

$$= 48.492 \text{ cm}$$

Triangle,

$$A_2 = \frac{1}{2} \times 40 \times 40$$

$$= 800 \text{ cm}^2$$

$$x_2 = 0$$

$$y_2 = \frac{2 \times 40}{3}$$

$$= 26.666 \text{ cm}$$

If the centroid of the plane figure is (x_c, y_c).

Then, $x_c = 0$

and
$$y_c = \frac{A_1 y_1 + A_2 y_2}{A_1 + A_2}$$

$$= \frac{628 \times 48.492 + 800 \times 26.666}{628 + 800}$$

$$= 36.264 \text{ cm.}$$

$I_{GG'}$ = the moment of inertia about the centroidal horizontal axis

= the moment of inertia of the semicircle + the moment of inertia of the triangle about the same axis

$$= \frac{40 \times 40^3}{12} + 800 \times (36.264 - 26.666)^2$$

$$+ \frac{\pi}{128}(40)^4 + 628 \times (48.492 - 36.264)^2$$

$$= 213333.333 + 73697.283 + 62800 + 93901.061$$

$$= 443731.677 \text{ cm}^4.$$

3. *Locate the centroid of the shaded area and find the moment of inertia about the horizontal axis passing through the centroid, refer Fig. (6-W5). All the dimensions are in mm.*

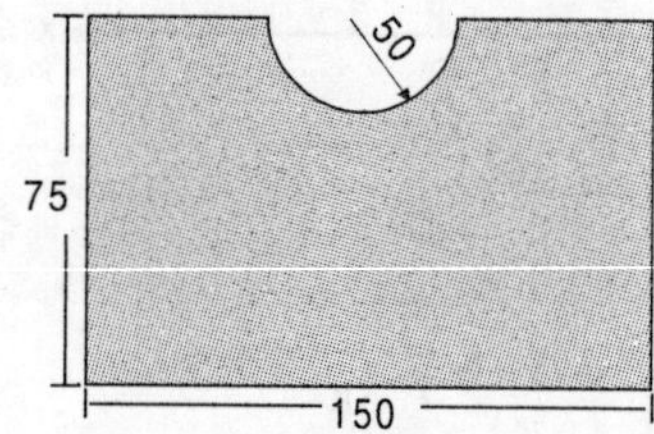

Fig. 6-W5

Solution:

To find the centroid of the given composite section divide it in known sections viz semicircle 1. rectangle, 2. axis is fixed as shown in Fig. (6-W6).

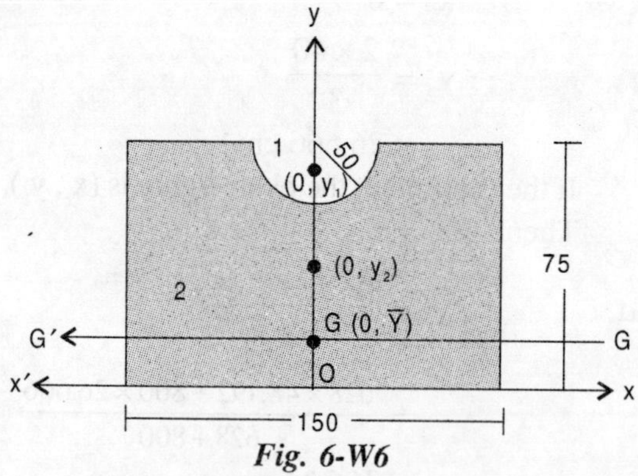

Fig. 6-W6

Semicircle,

$$A_1 = \frac{1}{2} \times \pi \times 50^2$$

$$= 3925 \text{ mm}^2$$

$$x_1 = 0$$

$$y_1 = 75 - \frac{4 \times 50}{3 \times \pi}$$

$$= 53.768 \text{ mm}$$

Rectangle,

$$A_2 = 150 \times 75$$

$$= 11250 \text{ mm}^2$$

$$x_2 = 0 \text{ mm}$$

$$y_2 = \frac{75}{2}$$

$$= 37.5 \text{ mm}$$

∴ The centroid of the composite section

$$\overline{X} = 0$$

[As section is symmetrical about the y-axis.]

$$\overline{Y} = \frac{A_2 y_2 - A_1 y_1}{A_2 - A_1}$$

$$= \frac{11250 \times 37.5 - 3925 \times 53.768}{11250 - 3925}$$

and $= 28.783$ mm

Now $I_{GG'}$ = the moment of inertia about the horizontal centroidal axis

= the moment of inertia of the rectangle – the moment of inertia of the semicircle

$$= \frac{150 \times 75^3}{12} + 11250 \times (37.5 - 28.783)^2$$

$$- \left[\frac{\pi \times 100^4}{128} + 3925 \times (53.768 - 28.783)^2 \right]$$

$$= 5273437.5 + 854843.5013 - 2453125 - 2450182.133$$

$$= 1224973.868 \text{ mm}^4.$$

4. *Determine the moment of inertia of the shaded area about its horizontal centroidal axis. Take all the dimensions in mm. Refer Fig. (6-W7).*

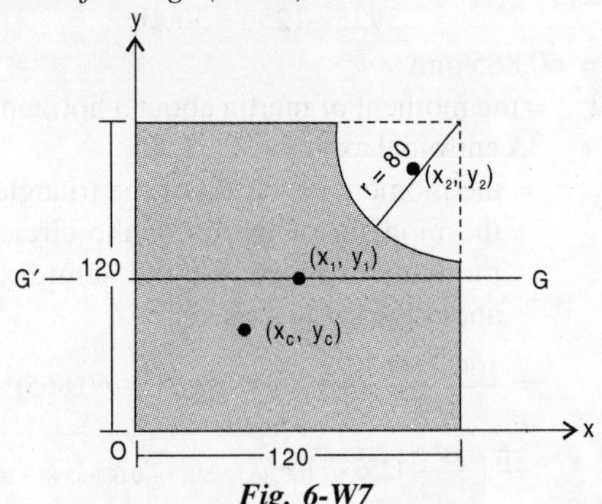

Fig. 6-W7

Solution:

Divide the given shaded area in:

1. a rectangle:

$$A_1 = 120 \times 120$$
$$= 14400 \text{ mm}^2$$
$$x_1 = \frac{120}{2}$$
$$= 60 \text{ mm}$$
$$y_1 = \frac{120}{2}$$
$$= 60 \text{ mm}$$

2. a quarter circle

$$A_2 = \frac{1}{4} \times \pi \times r^2$$
$$= \frac{1}{4} \times 3.14 \times 80^2$$
$$= 5024 \text{ mm}^2$$
$$x_2 = 120 - \frac{4 \times 80}{3 \times 3.14} = 86.029 \text{ mm}$$
$$y_2 = 86.029 \text{ mm}$$

Therefore the centroid of the composite (section),

$$x_c = \frac{A_1x_1 - A_2x_2}{A_1 - A_2}$$
$$= \frac{14400 \times 60 - 5024 \times 86.029}{14400 - 5024}$$
$$= 46.052 \text{ mm}$$

and $y_c = 46.052$ mm

$\therefore$ G is a point (46.052, 46.052)

Now $I_{GG'}$ = the moment of inertia of the shaded area about the horizontal centroidal axis

= the moment of inertia of the rectangle – the moment of inertia of the quarter circle about the same axis

$$= \frac{120 \times 120^3}{12} + 14400 \times (120 - 46.052)^2$$
$$- \left[0.00343 \times 160^2 + 5029 \times (86.507 - 46.052)^2\right]$$
$$= 17280000 + 78743616.54 - 87.808 - 8037149.3$$
$$= 87986379.43 \text{ mm}^4.$$

5. *Determine the second moment of area about the horizontal centroidal axis for the shaded area shown in Fig. (6-W8). Also find the radius of gyration about the same axis where R_1 = 50 mm and R_2 = 20 mm. Take π = 3.14. All the dimensions are given in mm.*

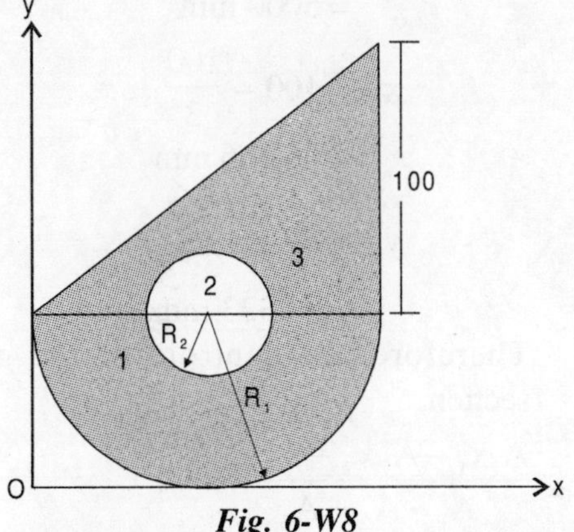

Fig. 6-W8

Solution:

Divide the given section into the following sections:

1. a semicircle

$$A_1 = \frac{\pi}{2}R_1^2$$

$$= \frac{\pi}{2}\times 50^2$$

$$= 3925 \text{ mm}^2$$

$$y_1 = R_1 - \frac{4R_1}{3\pi}$$

$$= 50 - \frac{4\times 50}{3\times 3.14}$$

$$= 28.768 \text{ mm}$$

$$x_1 = R_2$$

$$= 50 \text{ mm.}$$

2. a circle

$-A_2 = \pi R_2^2 = \pi \times 20^2$ [Negative sign because it is a cut section.]

$$= 1256 \text{ mm}^2$$

$$x_2 = R_1$$

$$= 50 \text{ mm}$$

$$y_2 = R_2$$

$$= 50 \text{ mm.}$$

3. a triangle

$$A_3 = \frac{1}{2}\times 100\times 100$$

$$= 5000 \text{ mm}^2$$

$$x_3 = 100 - \frac{100}{3}$$

$$= 66.666 \text{ mm}$$

$$y_3 = 50 + \frac{100}{3}$$

$$= 83.333 \text{ mm.}$$

Therefore, the centroid of the given section,

$$\overline{X} = \frac{A_1x_1 - A_2x_2 + A_3x_3}{A_1 - A_2 + A_3}$$

$$= \frac{3925\times 50 - 1256\times 50 + 5000\times 66.666}{3925 - 1256 + 5000}$$

$$= 66.865 \text{ mm.}$$

$$\overline{Y} = \frac{A_1y_1 - A_2y_2 + A_3y_3}{A_1 - A_2 + A_3}$$

$$= \frac{3925\times 28.768 - 1256\times 50 + 5000\times 83.333}{3925 - 1256 + 5000}$$

$$= 60.865 \text{ mm}$$

$I_{GG'}$ = the moment of inertia about a horizontal centroidal axis

= the moment of inertia of the triangle – the moment of inertia of the circle + moment of inertia of the semicircle about the same axis

$$= \frac{100\times 100^3}{12} + 5000\times(83.333 - 60.865)^2$$

$$-\frac{\pi\times 40^4}{64} - 1256\times (60.865-50)^2 + 0.0068598\times 100^4$$

$$+3925\times(60.865 - 28.768)^2$$

$$= 10857388.45 - 125600 - 148268.5706 + 685980 + 4043603.33$$

$$= 15313103.21 \text{ mm}^4$$

$\therefore$ The radius of gyration $= K_{GG'}$

$$= \sqrt{\frac{I_{GG'}}{A}}$$

$$= \sqrt{\frac{15313103.21}{7669}}$$

$$= 44.68 \text{ mm.}$$

6. *Calculate the moment of inertia of a pre-stressed concrete beam section as shown in Fig. (6-W9) about the horizontal and the vertical axes passing through the centroid. Take all dimensions in mm.*

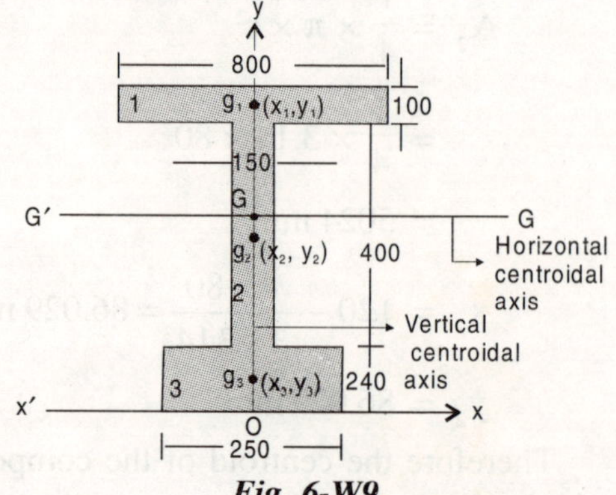

Fig. 6-W9

Solution:

Divide the pre-stressed concrete beam section in

(a) rectangle (1)

$$A_1 = 800 \times 100 = 80000 \text{ mm}^2$$

$$x_1 = 0$$

$$y_1 = 240 + 400 + \frac{100}{2} = 690 \text{ mm.}$$

(b) rectange (2)

$$A_2 = 150 \times 400 = 60000 \text{ mm}^2$$

$$x_2 = 0$$

$$y_2 = 240 + \frac{400}{2} = 440 \text{ mm.}$$

(c) rectangle (3)

$$A_3 = 240 \times 250 = 60000 \text{ mm}^2$$

$$x_3 = 0$$

$$y_3 = \frac{240}{2} = 120 \text{ mm.}$$

If G $(\overline{X}, \overline{Y})$ is the centroid of the composite section about the shown axes then,

$\overline{X} = 0$ [as the composite section is symmetrical about the y-axis]

and
$$\overline{Y} = \frac{A_1y_1 + A_2y_2 + A_3y_3}{A_1 + A_2 + A_3} = \frac{\Sigma A_1y_1}{\Sigma A_1}$$

$$= \frac{80000 \times 690 + 60000 \times 440 + 60000 \times 120}{80000 + 60000 + 60000}$$

$$= 444 \text{ mm.}$$

Therefore, the centroid of the section is G (0, 444). Now, the moment of inertia about the horizontal centroidal axis is,

$$I_{GG'} = \left[\frac{800 \times 100^3}{12} + 80000 \times (690 - 444)^2\right]$$

rectangle (1) $+\left[\frac{150 \times 400^3}{12} + 60000 \times (444 - 440)^2\right]$

rectangle (2) $+\left[\frac{250 \times 240^3}{12} + 60000 \times (444 - 120)^2\right]$

rectangle (3)

$$= 4907946667 + 800960000 + 6586560000$$

$$= 1.23 \times 10^{10} \text{ mm}^4$$

And the moment of inertia about the vertical centroidal axis is,

$$I_{yy'} = \frac{100 \times 800^3}{12} + \frac{400 \times 150^3}{12} + \frac{240 \times 250^3}{12}$$

$$= 4.69 \times 10^9 \text{ mm}^4.$$

7. *Determine the second moment of area about the x-axis. All the dimensions in Fig. (6-W10) are given in millimeters.*

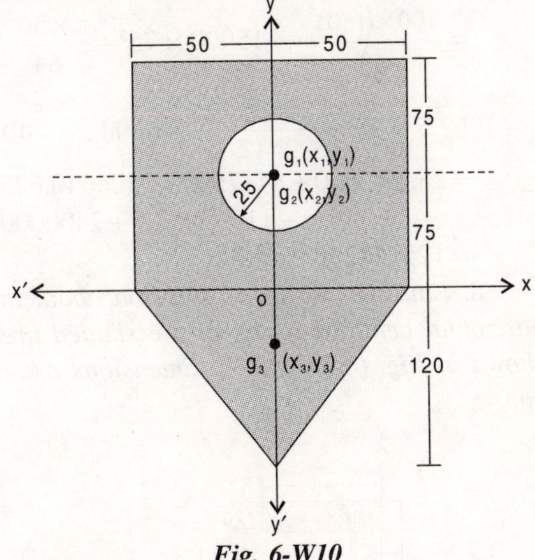

Fig. 6-W10

Solution:

Since, in this problem, the axis is mentioned therefore there is no need to find the centroid to get centroidal axis. We can directly use the derived formula for finding moment of inertia.

Divide the given composite section into

(a) a rectangle;

$A_1 = 100 \times 150$

$= 15000 \text{ mm}^2$

$x_1 = 0$

$y_1 = \frac{150}{2} = 75 \text{ mm.}$

(b) a circle;

$A_2 = \pi(25)^2$

$= 1962.5 \text{ mm}^2$

$x_2 = 0$

$y_2 = 75.$

(c) a triangle;

$A_3 = \frac{1}{2} \times 100 \times 120$

$= 6000 \text{ mm}^2$

$x_3 = 0$

$y_3 = -\frac{120}{3} = -40 \text{ mm.}$

Therefore,

$I_{xx} = I_{xx'}$, rectangle $+ I_{xx'}$, circle $+ I_{xx'}$, triangle

$$= \frac{100 \times (150)^3}{12} + 15000 \times 75^2 - \frac{\pi \times 50^4}{64}$$

$$- 1962.5 \times 75^2 + \frac{100 \times 120^3}{12} + 6000 \times (-40)^2$$

$$= 28{,}125{,}000 + 84375000 - 306640.625$$
$$- 11039062.5 + 24000000$$

$= 125154296.9 \text{ mm}^4.$

8. *Find the radius of gyration about the horizontal centroidal axis of the shaded area shown in Fig. (6-W11). All dimensions are in mm.*

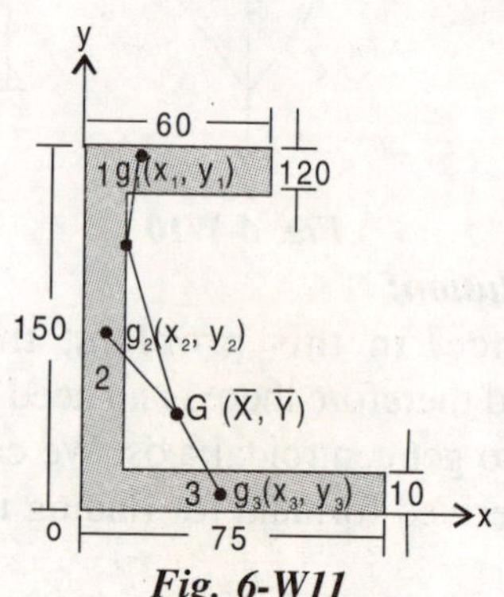

Fig. 6-W11

Solution:

Divide the given section into the following sections:

(a) rectangle (1)

$A_1 = 60 \times 12$

$= 720 \text{ mm}^2$

$x_1 = \frac{60}{2}$

$= 30 \text{ mm}$

$y_1 = 10 + \frac{(150 - 10 - 12)}{2} + 12$

$= 86 \text{ mm}$

(b) rectangle (2)

$A_2 = 128 \times (75 - 60)$

$= 1920 \text{ mm}^2$

$x_2 = \left(\frac{75 - 60}{2}\right)$

$= 7.5 \text{ mm}$

$y_2 = 10 + \frac{128}{2}$

$= 74 \text{ mm}$

(c) rectangle (3)

$A_3 = 75 \times 10$

$= 750 \text{ mm}^2$

$x_3 = \frac{75}{2}$

$= 37.5 \text{ mm}$

$y_3 = \frac{10}{2}$

$= 5 \text{ mm}$

Now, the centroid G $(\overline{X}, \overline{Y})$ of the given shaded area is,

$$\overline{X} = \frac{\Sigma x_1 A_1}{\Sigma A_1}$$

$$= \frac{x_1 A_1 + x_2 A_2 + x_3 A_3}{A_1 + A_2 + A_3}$$

$$= \frac{30 \times 720 + 1920 \times 7.5 + 750 \times 37.5}{720 + 1920 + 750}$$

$= 18.915$ mm

and $\overline{Y} = \frac{\Sigma y_1 A_1}{\Sigma A_1}$

$$= \frac{y_1 A_1 + y_2 A_2 + y_3 A_3}{A_1 + A_2 + A_3}$$

$$= \frac{720 \times 86 + 1920 \times 74 + 750 \times 5}{720 + 1920 + 750}$$

$= 61.283$ mm

$\therefore G(\overline{X}, \overline{Y}) = (18.915, 61.283)$

Now, the moment of inertia about the horizontal centroidal axis $= I_{cent}$

$$= \frac{60 \times (12)^3}{12} + 720 \times (86 - 61.283)^2$$

$$+ \frac{15 \times (128)^3}{12} + 1920 \times (74 - 61.283)^2$$

$$+ \frac{75 \times (10)^3}{12} + 750 \times (61.283 - 5)^2$$

$= 2636330 + 3126208.142$

$= 5762538.142$ mm^4

$\therefore$ Radius of gyration about the same axis

$$= \sqrt{\frac{5762538.142}{3390}} = 41.229 \text{ mm.}$$

9. *Determine the coordinates of the centroid of the composite area with respect to origin as shown in Fig. (6-W12) and find the moment of inertia about the vertical centroidal axis. All dimensions are in mm.*

Solution:

Divide the composite area into the followings sections:

(a) a rectangle (1)

$A_1 = 40 \times 40$

$= 1600$ mm^2

$x_1 = 20 + \frac{40}{2} = 40$ mm

$y_1 = 60 + 40 + 20 + \frac{40}{2}$

$= 140$ mm

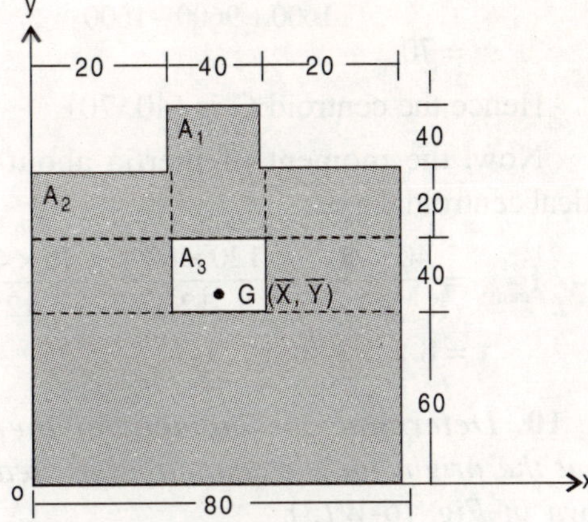

Fig. 6-W12

(b) a rectange (2)

$A_2 = 80 \times 120$

$= 9600$ mm^2

$x_2 = \frac{80}{2}$

$= 40$ mm

$y_2 = \frac{120}{2}$

$= 60$ mm

(c) a rectange (3)

$A_3 = 40 \times 40$

$= 1600$ mm^2

$x_3 = 20 + \frac{40}{2} = 40$ mm

$y_3 = 60 + \frac{40}{2} = 80$ mm

Therefore, the centroid of the composite area w.r.t. origin is

$$\overline{X} = \frac{A_1 x_1 + A_2 x_2 - A_3 x_3}{A_1 + A_2 - A_3}$$

$$= \frac{1600 \times 40 + 9600 \times 40 - 1600 \times 40}{1600 + 9600 - 1600}$$

$= 40$

$$\overline{Y} = \frac{A_1y_1 + A_2y_2 - A_3y_3}{A_1 + A_2 - A_3}$$

$$= \frac{1600 \times 140 + 9600 \times 60 - 1600 \times 80}{1600 + 9600 - 1600}$$

$$= 70$$

Hence the centroid G is (40, 70).

Now, the moment of inertia about the vertical centroidal axis,

$$I_{cent} = \frac{40 \times 40^3}{12} + \frac{120 \times 80^3}{12} - \frac{40 \times 40^3}{12}$$

$$= 8 \times 10^4 \text{ mm}^4.$$

10. *Determine the moment of inertia about the origin for the semicircular area as shown in Fig. (6-W13).*

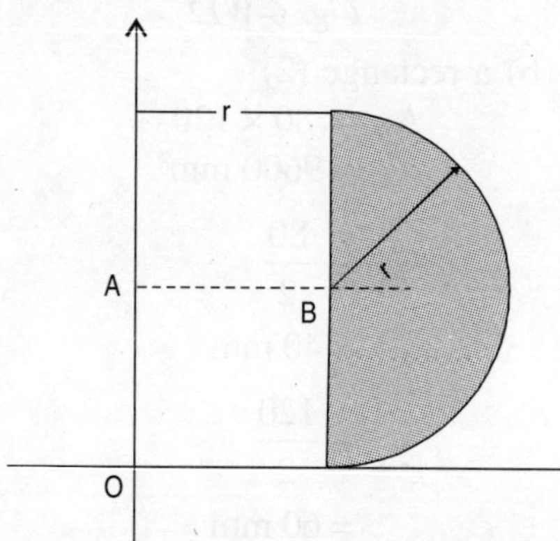

Fig. 6-W13

Solution:

Let us consider an elementary area dA at a distance of r′ from the origin. Let dr′ is the thickness of the area when angle increases by dθ.

therefore, $dA = r'\,dr'\,d\theta$

The moment of inertia of this elementary area about the origin is,

$$dI_O = (r')^2\,dA$$

By integrating it over the proper limits, we get,

$$I_O = \int_r^{\sqrt{5}r} \int_0^{\tan^{-1}2} (r')^2 r'dr'd\theta$$

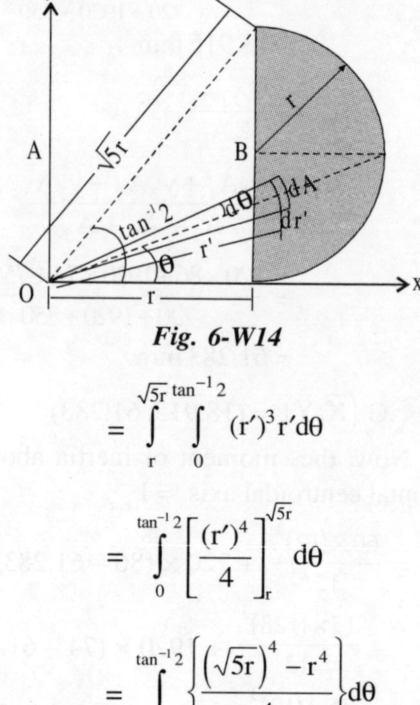

Fig. 6-W14

$$= \int_r^{\sqrt{5}r} \int_0^{\tan^{-1}2} (r')^3 r'd\theta$$

$$= \int_0^{\tan^{-1}2} \left[\frac{(r')^4}{4}\right]_r^{\sqrt{5}r} d\theta$$

$$= \int_0^{\tan^{-1}2} \left\{\frac{\left(\sqrt{5}r\right)^4 - r^4}{4}\right\} d\theta$$

$$= 6r^4 \left[\theta\right]_0^{\tan^{-1}2} = 6r^4 \tan^{-1} 2$$

$$= 6.642\ r^4 \text{ unit}^4.$$

11. *Calculate the moment of inertia of the shaded area about the x-axis.*

Solution:

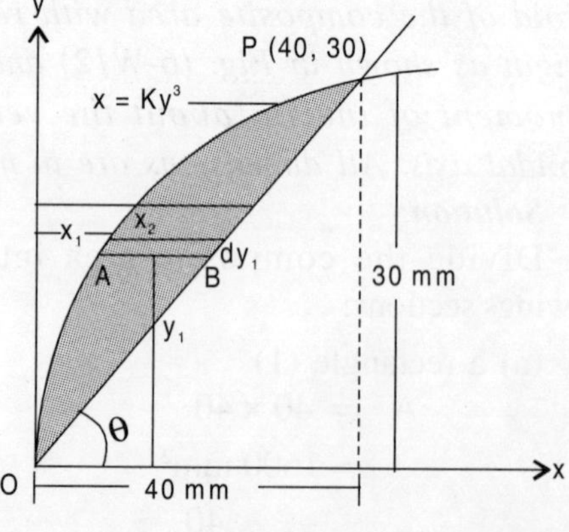

Fig. 6-W15

From Fig. (6-W15), we have

$$\tan\theta = \frac{30}{40} = \frac{3}{4}$$

∴ The equation of the straight line is,

$y = mx$ where $m = \tan\theta$

$$\Rightarrow \quad y = \left(\frac{3}{4}\right)x$$

Consider an elementary rectangular strip as shown in the figure.

∴ dA = an elementary area of the strip

$$= (x_2 - x_1)\,dy = \left(Ky^3 - \frac{4}{3}y\right)dy$$

Since (x_1, y) and (x_2, y) satisfy the equations $x = Ky^3$ and $y = \left(\frac{3}{4}\right)x$ respectively.

Now, since the strip is at a distance y from the x-axis, therefore,

dI_x = M.I. of the strip about the x-axis

$= y^2\,dA$

Integrating dI_x with the proper limits, we get,

$$I_x = \int_{A_1}^{A_2} y^2 dA$$

$$= \int_0^{30} y^2\left(Ky^3 - \frac{4}{3}y\right)dy$$

$$= \int_0^{30}\left(Ky^5 - \frac{4}{3}y^3\right)dy$$

$$= \int_0^{30} Ky^5 dy - \frac{4}{3}\int_0^{30} y^3 dy$$

$$= K\left[\frac{y^6}{6}\right]_0^{30} - \frac{4}{3}\left[\frac{y^4}{4}\right]_0^{30}$$

$$= K\left(\frac{30^6}{6}\right) - \frac{4}{3}\left(\frac{30^4}{4}\right)$$

Since P (40, 30) satisfies $x = Ky^3$

$$\therefore \quad K = \frac{40}{(30)^3}$$

$$\therefore \quad I_x = \frac{40}{(30)^3} \times \frac{(30)^6}{6} - \frac{4}{3} \times \frac{(30)^4}{4}$$

$$= \frac{40}{6} \times (30)^3 - \frac{(30)^4}{3}$$

$$= \left(\frac{4\times 3^3}{6} - \frac{3^4}{3}\right) \times 10^4$$

$$= \left(\frac{4\times 3^3 - 2\times 3^4}{6}\right) \times 10^4$$

$$= \frac{2\times 3^3}{3\times 2} \times 10^4$$

$$= 9 \times 10^4.$$

12. *Find the moment of inertia of an arc of a semicircle about the z-axis passing through the centre of the semicircle.*

Solution:

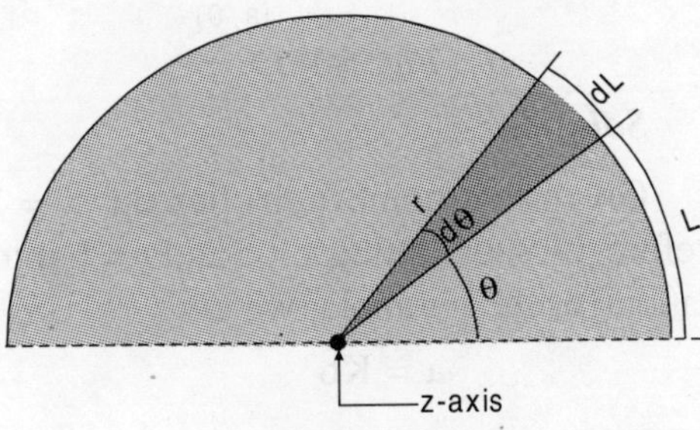

Fig. 6-W16

Let us consider an elementary arc of length dL, when θ increases to θ + dθ.

then, $dL = r\,d\theta$

Therefore, the moment of inertia of an elementary arc of length dL about z-axis is,

$$dI_z = r^2\,dL$$

since, the distance of elementary arc length from the z-axis is the radius of the semicircle.

Integrating with the proper limits, we get

$$I_z = \int_{L_1}^{L_2} r^2\, dL$$

$$= \int_0^{\pi} r^2 r\, d\theta$$

$$= r^3 \int_0^{\pi} d\theta \quad \text{[since 'r' is constant]}$$

$$= \pi r^3 \text{ unit}^3.$$

13. *Determine the moment of inertia (MOI) of the area under the parabola, $x = Ky^2$ about the x-axis as shown in Fig. (6-W17).*

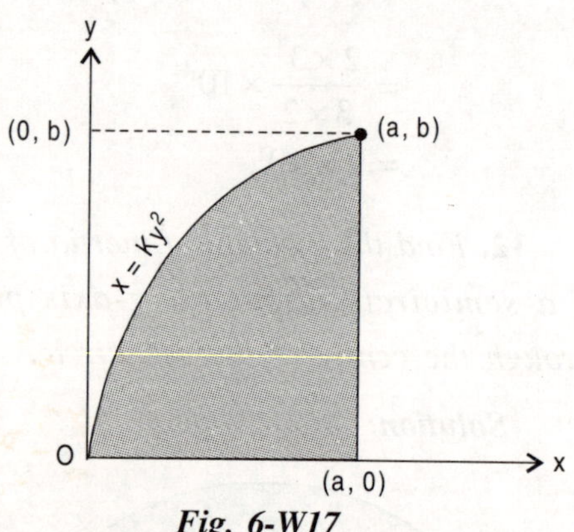

Fig. 6-W17

Solution:

Given that point (a, b) lies on $x = Ky^2$, therefore, by substituting x = a and y = b into the equation for the parabola, we get

$$a = Kb^2$$

$$\therefore \quad K = \frac{a}{b^2}$$

First method:

Let us consider a horizontal strip of area dA at a perpendicular distance y from the x-axis as shown in Fig. (6-W18).

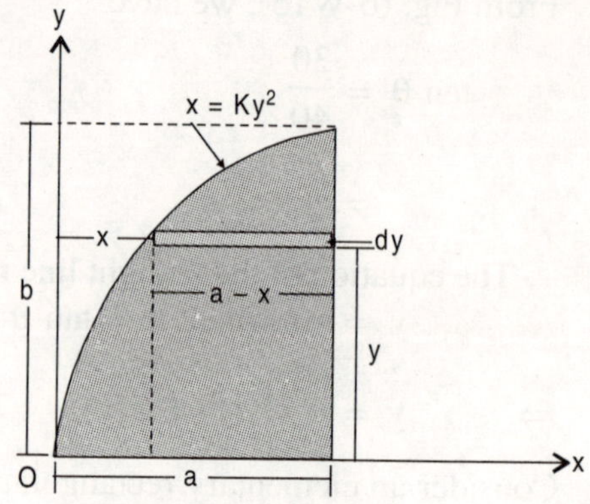

Fig. 6-W18

then, area of elementary strip,

$$dA = \text{width} \times \text{length}$$

$$= dy\,(a - x)$$

$$= (a - x)\, dy$$

Since appoximately all points of the horizontal strip are at the same distance 'y' from the x-axis, therefore the moment of inertia (dI_x) of the elementary strip about the x-axis is $y^2 dA$.

$$\text{i.e.,} \quad dI_x = y^2\, dA$$

$$= \frac{x}{K}(a - x)\, dy$$

Integrating it with the proper limit, we get

$$I_x = \int dI_x$$

$$= \int_0^b \frac{x}{K}(a - x)\, dy$$

$$= \frac{1}{K}\int_0^b (ax - x^2)\, dy$$

$$= \frac{1}{K}\int_0^b (a.Ky^2 - K^2 y^4)\, dy$$

$$= \frac{1}{K}\left[aK\left(\frac{y^3}{3}\right)_0^b - K^2\left(\frac{y^5}{5}\right)_0^b\right]$$

$$= \frac{1}{K}\left[\frac{aKb^3}{3} - \frac{K^2 b^5}{5}\right]$$

$$= \frac{1}{K}\left[\frac{5aKb^3 - 3K^2b^5}{15}\right]$$

$$= \frac{1}{K}\left[\frac{5ab^3 \times \frac{a}{b^2} - 3\times\frac{a^2}{b^4}\times b^5}{15}\right]$$

$$[\because K = \frac{a}{b^2}]$$

$$= \frac{1}{K}\left[\frac{5a^2b - 3a^2b}{15}\right]$$

$$= \frac{1}{K}\left[\frac{2a^2b}{15}\right]$$

$$= \frac{b^2}{a}\times\frac{2}{15}\times a^2b$$

$$= \frac{2ab^3}{15} \text{ unit}^4.$$

***Second method*:**

Let us consider a vertical strip having elementary area dA at a perpendicular distance x from y-axis as shown in Fig. (6-W19)

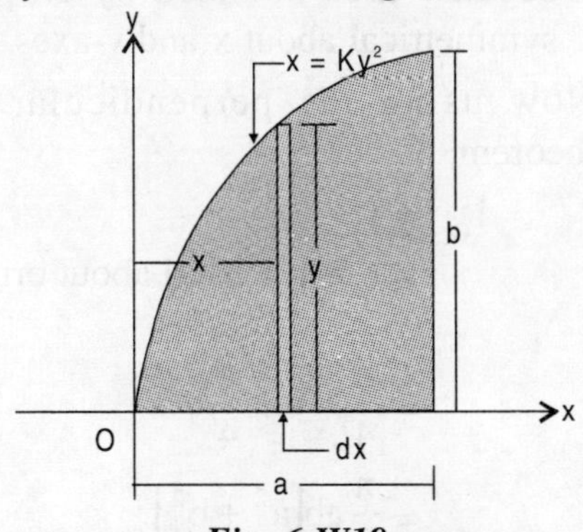

Fig. 6-W19

then, area of the elementary strip,

$$dA = \text{width} \times \text{height}$$
$$= dx \times y$$
$$= y\,dx$$

As we know that, the moment of inertia of a rectangle about its base is $\frac{bh^3}{3}$ then for the elementary rectangular b = dx and h = y, therefore,

$$dI_x = \frac{1}{3}y^3\,dx$$
$$= \frac{y^3dx}{3}$$

Integrating with respect to x over lower to upper limits, we get

$$I = \int dI_x$$

$$= \int_0^a \frac{y^3}{3}dx$$

$$= \frac{1}{3}\int_0^a\left(\sqrt{\frac{x}{K}}\right)^3 dx$$

$$= \frac{1}{3K\sqrt{K}}\int_0^a x^{\frac{3}{2}}dx$$

$$= \frac{1}{3K\sqrt{K}}\left[\frac{x^{\frac{3}{2}+1}}{\frac{3}{2}+1}\right]_0^a$$

$$= \frac{1}{3K\sqrt{K}}\times\frac{2}{5}\left[a^{\frac{5}{2}} - 0^{\frac{5}{2}}\right]$$

$$= \frac{1}{3K\sqrt{K}}\times\frac{2}{5}\times a^{\frac{5}{2}}$$

$$= \frac{2}{15}\times\frac{1}{K\sqrt{K}}\times a^{\frac{5}{2}}$$

$$= \frac{2}{15}\times\frac{b^2}{a}\times\frac{b}{\sqrt{a}}\times a^2\times\sqrt{a}$$

$$= \frac{2ab^3}{15} \text{ unit}^4.$$

14. *Determine the moment of inertia of an elliptical area about the y-axis and find the polar radius of gyration about the origin of the coordinates.*

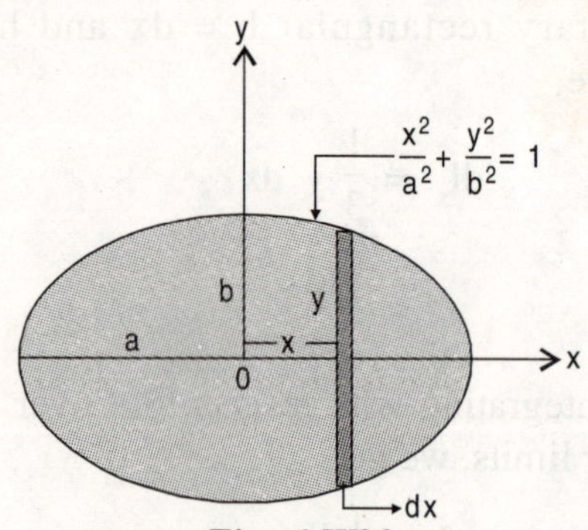

Fig. 6-W20

Solution:

Let us consider a vertical rectangular strip having an elementary area dA at a distance x from the y-axis.

$$\therefore \quad dA = 2y\,dx$$

$$= 2b\sqrt{\left(1-\frac{x^2}{a^2}\right)}\,dx$$

Now since the strip is at a distance x from the y-axis, therefore, dI_y is the moment of inertia of dA about the y-axis is,

$$dI_y = x^2\,dA$$

$\therefore$ The total moment of inertia about the y-axis of elliptical area is,

$$I_y = \int dI_y$$

$$= \int x^2 dA$$

$$= \int_{-a}^{+a} x^2 2b\sqrt{1-\frac{x^2}{a^2}}\,dx$$

$$= 2b\int_{-a}^{+a} x^2\sqrt{1-\frac{x^2}{a^2}}\,dx$$

$$= \frac{2b}{a}\int_{-a}^{a} x^2\sqrt{a^2-x^2}\,dx$$

$$= \frac{4b}{a}\int_{0}^{a} x^2\sqrt{a^2-x^2}\,dx$$

[Using properties of integration since integral is an even function of x.]

Substituting $x = a\sin\theta$, we have

$$dx = a\cos\theta\, d\theta \text{ and } \theta \to 0 \to \frac{\pi}{2}$$

$$= \frac{4b}{a}\int_0^{\frac{\pi}{2}} a^2\sin^2\theta\sqrt{a^2-a^2\sin^2\theta}\,a\cos\theta\,d\theta$$

$$= \frac{4b}{a}\int_0^{\frac{\pi}{2}} a^4\sin^2\theta\sqrt{1-\sin^2\theta}\cos\theta\,d\theta$$

$$= 4ba^3\int_0^{\frac{\pi}{2}} \sin^2\theta\cos^2\theta\,d\theta$$

$$= 4ba^3 \times \frac{1}{4}\times\frac{1}{2}\times\frac{\pi}{2}$$

[Using reduction formula]

$$= \frac{\pi a^3 b}{4}$$

Similarly, $I_x = \dfrac{\pi b^3 a}{4}$

= the moment of inertia about the x-axis because area bounded by ellipse is symmetrical about x and y-axes.

Now using the perpendicular axis theorem,

$$I_O = I_z$$

$$= I_x + I_y = \text{MOI about origin.}$$

$$= \frac{\pi}{4}a^3b + \frac{\pi}{4}b^3a$$

$$= \frac{\pi}{4}ab\left[a^2+b^2\right]$$

$\therefore \quad K_O$ = the radius of gyration about the origin.

$$= \sqrt{\frac{I_z}{A}}$$

$$= \sqrt{\frac{\frac{\pi}{2}ab(a^2+b^2)}{\pi ab}}$$

[$\because$ area of the ellipse is πab]

$$= \sqrt{\frac{a^2 + b^2}{4}}$$

$$= \frac{\sqrt{a^2 + b^2}}{2}.$$

15. *Determine the moment of inertia of the shaded area about the x and y-axes.*

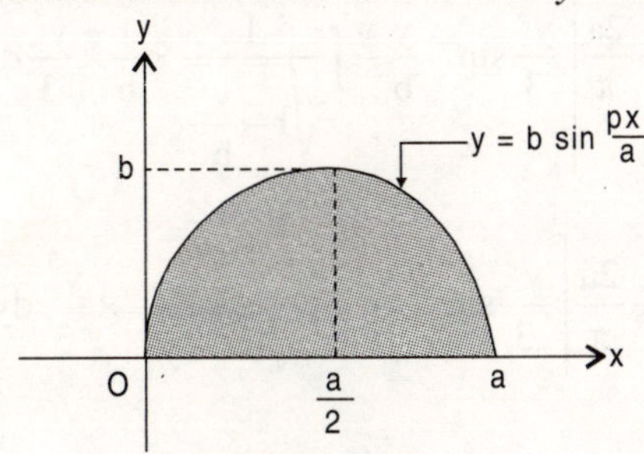

Fig. 6-W21

Solution:

Let us consider a rectangle of an elementary area dA having width dx and height y, therefore,

$$dA = y\,dx$$

$$= b \sin \frac{\pi x}{a}\,dx$$

Now, since the strip of area dA is at a distance x from the y-axis so,

$$dI_y = x^2\,dA$$

By integrating it over the limit $x \to 0 \to a$, we get

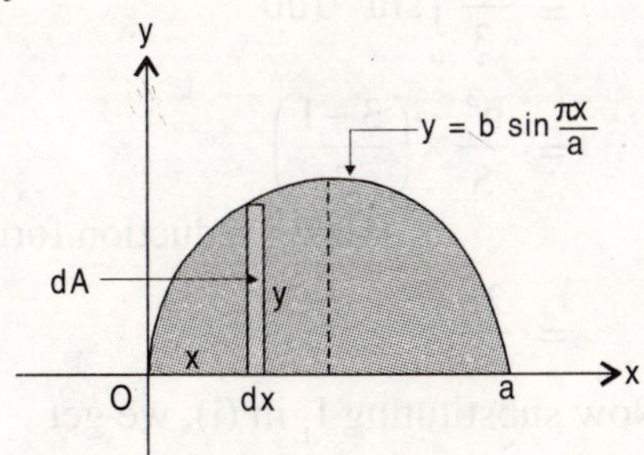

Fig. 6-W22

$$I_y = \int_0^a x^2 dA$$

$$= \int_0^a x^2 b \sin\left(\frac{\pi x}{a}\right) dx$$

$$= b \int_0^a x^2 \sin\left(\frac{\pi x}{a}\right) dx$$

$$= b\left[x^2\left(-\cos\frac{\pi x}{a}\right).\frac{a}{\pi} - \int\left(2x\int \sin\frac{\pi x}{a}dx\right)dx\right]_0^a$$

[Using integration by parts]

$$= b\left[x^2\left(-\cos\frac{\pi x}{a}\right).\frac{a}{\pi} - 2\int x\left(-\cos\frac{\pi x}{a}\right)\frac{a}{\pi}dx\right]_0^a$$

$$= b\left[-x^2 \times \cos\left(\frac{\pi x}{a}\right) \times \frac{a}{\pi} + 2\int x.\frac{a}{\pi}\cos\frac{\pi x}{a}dx\right]_0^a$$

$$= \frac{ab}{\pi}\left[-x^2\cos\frac{\pi x}{a} + 2\int x\cos\left(\frac{\pi x}{a}\right)dx\right]_0^a$$

$$= \frac{ab}{\pi}\left[-x^2\cos\left(\frac{\pi x}{a}\right) + 2\left\{x\int\cos\frac{\pi x}{a} - \int\left(\frac{dx}{dx}\int\cos\frac{\pi x}{a}dx\right)dx\right\}\right]_0^a$$

$$= \frac{ab}{\pi}\left[-x^2\cos\left(\frac{\pi x}{a}\right) + 2\left\{x.\frac{a}{\pi}\sin\frac{\pi x}{a} - \int\frac{a}{\pi}\sin\frac{\pi x}{a}dx\right\}\right]_0^a$$

$$= \frac{ab}{\pi}\left[-x^2\cos\left(\frac{\pi x}{a}\right) + 2\frac{a}{\pi}\left\{x\sin\left(\frac{\pi x}{a}\right) + \frac{a}{\pi}\cos\left(\frac{\pi x}{a}\right)\right\}\right]_0^a$$

$$= \frac{ab}{\pi}\left[-a^2\cos\left(\frac{\pi a}{a}\right) + 2\left\{a \times \frac{a}{\pi}\sin\frac{\pi a}{a} + \frac{a^2}{\pi^2}\cos\frac{\pi a}{a}\right\} - 2\left(-0 + 0 + \frac{a^2}{\pi^2}\cos 0\right)\right]$$

$$= \frac{ab}{\pi}\left[a^2 + 2\left(\frac{a^2}{\pi^2}\right) \times (-1) - 2\left(\frac{a^2}{\pi^2}\right)\right]$$

$$= \frac{ab}{\pi}\left[a^2 - 2\left(\frac{a^2}{\pi^2}\right) - 2\left(\frac{a^2}{\pi^2}\right)\right]$$

$$= \frac{ab}{\pi}\left[a^2 - 4\left(\frac{a^2}{\pi^2}\right)\right]$$

$$= \frac{a^3 b}{\pi}\left[1 - \frac{4}{\pi^2}\right]$$

$$= \frac{ba^3}{\pi}\left[1 - \frac{4}{\pi^2}\right]$$

$$\therefore \quad I_y = \frac{ba^3}{\pi}\left[1 - \frac{4}{\pi^2}\right]$$

Now, consider a rectangular strip of area dA parallel to the x-axis at a distance y from the x-axis then the elementary area,

$$dA = (a - 2x_1)\,dy$$

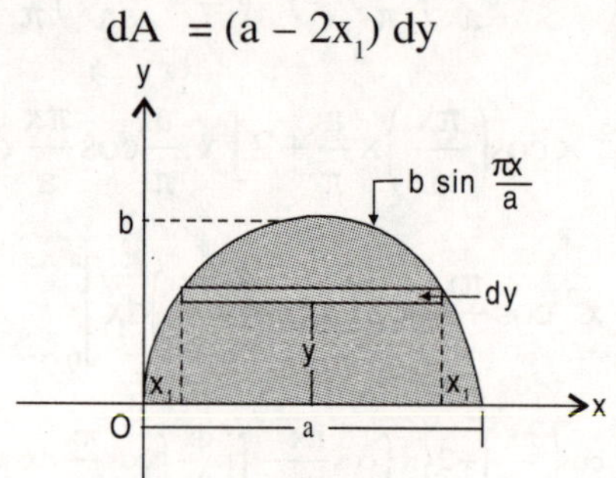

Fig. 6-W23

where $\quad y = b \sin \frac{\pi x_1}{a}$

$$= \left\{a - 2.\frac{a}{\pi}\sin^{-1}\left(\frac{y}{b}\right)\right\}dy$$

$$= \left\{a - \frac{2a}{\pi}\sin^{-1}\left(\frac{y}{b}\right)\right\}dy$$

Now the moment of inertia of the elementary area dA of rectangle is,

$$I_x = \int y^2\, dA$$

$$= \int_0^b y^2\left\{a - \frac{2a}{\pi}\sin^{-1}\left(\frac{y}{b}\right)\right\}dy$$

$$= \int_0^b y^2 a\,dy - \frac{2a}{\pi}\int_0^b y^2 \sin^{-1}\left(\frac{y}{b}\right)dy$$

$$= a\left[\frac{y^3}{3}\right]_0^b - \frac{2a}{\pi}\int_0^{\pi} y^2 \sin^{-1}\left(\frac{y}{b}\right)dy$$

$$= \frac{ab^3}{3} - \frac{2a}{\pi}\left[\sin^{-1}\left(\frac{y}{b}\right)\int y^2 dy - \int\left(\frac{d}{dy}\sin^{-1}\left(\frac{y}{b}\right)\int y^2 dy\right)dy\right]_0^b$$

$$= \frac{ab^3}{3} - \frac{2a}{\pi}\left[\frac{y^3}{3}\sin^{-1}\frac{y}{b} - \int \frac{1}{\sqrt{1-\frac{y^2}{b^2}}} \times \frac{1}{b} \times \frac{y^3}{3}dy\right]_0^b$$

$$= \frac{ab^3}{3} - \frac{2a}{\pi}\left[\frac{y^3}{3}\sin^{-1}\frac{y}{b} - \int \frac{1}{\sqrt{b^2 - y^2}} \times \frac{y^3}{3}dy\right]_0^b \quad \text{....(i)}$$

Let $\int_0^b \frac{1}{\sqrt{b^2-y^2}}\frac{y^3}{3}dy = I_1$

By substituting $y = b \sin \theta$, we have

$$dy = b \cos \theta \, d\theta$$

where $\quad \theta \to 0 \to \frac{\pi}{2}.$

$$\therefore \; I_1 = \int_0^{\frac{\pi}{2}} \frac{1}{b\cos\theta} \times \frac{b^3 \sin^3\theta}{3} \times b\cos\theta\, d\theta$$

$$= \frac{b^3}{3}\int_0^{\frac{\pi}{2}} \sin^3\theta\, d\theta$$

$$= \frac{b^3}{3} \times \left(\frac{3-1}{3}\right)$$

[Using reduction formula]

$$= \frac{2b^3}{9}$$

Now substituting I_1 in (i), we get

$$I_x = \frac{ab^3}{3} - \frac{2a}{\pi}\left[\frac{b^3}{3} \times \frac{\pi}{2} - \frac{2b^3}{9}\right]$$

$$= \frac{ab^3}{3} - \frac{2a}{\pi} \times \frac{b^3}{3} \times \frac{\pi}{2} + \frac{2a}{\pi} \times \frac{2b^3}{9}$$

$$= \frac{ab^3}{3} - \frac{ab^3}{3} + \frac{4ab^3}{9\pi}$$

$$= \frac{4ab^3}{9\pi}$$

$$\therefore \quad I_x = \frac{4ab^3}{9\pi}.$$

16. *A narrow strip of area of constant width b has the form of a spiral $r = K\theta$. After one complete turn i.e., from $\theta = 0$ to $\theta = 2\pi$, the end radius of the spiral is R. Determine the polar moment of inertia and the radius of gyration of the area about origin.*

Solution:

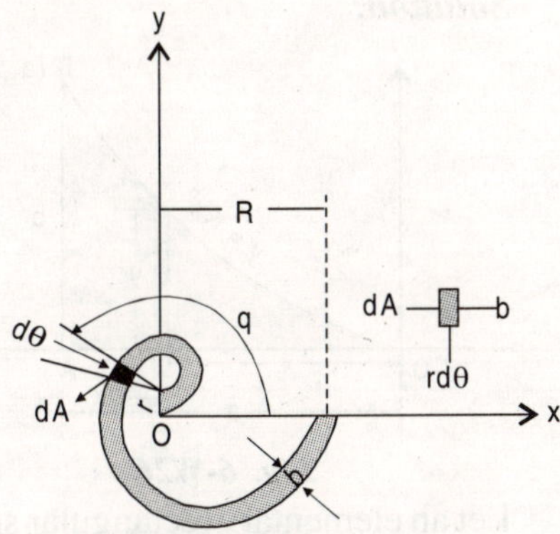

Fig. 6-W24

Consider an elementary area dA as shown in Fig. (6-W24)

$$\therefore \quad dA = r\, d\theta\, b$$

$$= r\, b\, d\theta$$

$\therefore \quad I_z$ = the polar moment of inertia

$$= \int_{\theta=\theta_1}^{\theta=\theta_2} r^2 dA$$

$$= \int_0^{2\pi} (K\theta)^2 . rb\, d\theta$$

$$= \int_0^{2\pi} (K\theta)^2 . (K\theta)\, b\, d\theta$$

$$= bK^3 \int_0^{2\pi} \theta^3 d\theta$$

$$= bK^3 \left[\frac{\theta^4}{4}\right]_0^{2\pi}$$

$$= bK^3 \frac{(2\pi)^4}{4}$$

$$= b\left(\frac{R}{2\pi}\right)^3 \frac{(2\pi)^4}{4}$$

$$= (2\pi)\frac{bR^3}{4}$$

$$= \frac{\pi}{2} bR^3$$

$$= 1.571\; bR^3 \text{ unit}^4$$

Now, K_0 = the radius of gyration about origin

$$= \sqrt{\frac{I_z}{A_T}}$$

Since, $dA = rb\, d\theta$

$$\therefore \quad A = \int_0^{2\pi} rb\, d\theta$$

$$= \int_0^{2\pi} K\theta b\, d\theta$$

$$= Kb\left[\frac{\theta^2}{2}\right]_0^{2\pi}$$

$$= \frac{Kb}{2}\{(2\pi)^2 - 0^2\}$$

$$= \frac{Kb}{2}\{(2\pi)^2\}$$

$$= \frac{4\pi^2 Kb}{2}$$

$$= 2\pi^2 Kb$$

$$= 2\pi^2 \times b \times \frac{R}{2\pi}$$

$$= \pi bR$$

$$\therefore \; K_O = \sqrt{\frac{I_z}{A_T}}$$

$$= \sqrt{\frac{1.571 bR^3}{\pi bR}}$$

$$= \sqrt{\frac{1.571}{3.142}}\, R$$

$$= 0.707 \text{ R.}$$

17. *Determine the polar moment of inertia of the semicircular area about A. The point A is as shown in Fig, (6-W25).*

Solution:

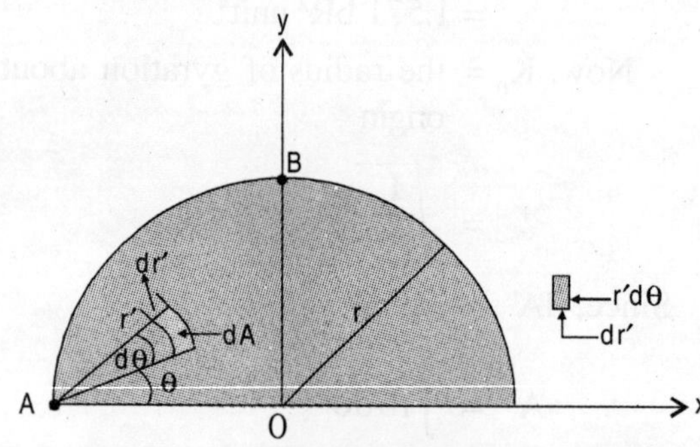

Fig. 6-W25

Let us consider an elementary area dA as shown in Fig. (6-W25), therefore

$$dA = dr'\,(r'd\theta)$$

$$= r'\,dr'\,d\theta$$

Hence, the moment of inertia about point A,

$$I_A = \int_{r'=0}^{r'=2r}\int_0^{\frac{\pi}{2}} r'^2\, r'\, dr'\, d\theta$$

$$= \int_0^{2r}\int_0^{\frac{\pi}{2}} (r')^3 dr'\, d\theta$$

$$= \int_0^{\frac{\pi}{2}} \left[\frac{(r')^4}{4}\right]_0^{2r} d\theta$$

$$= \frac{1}{4}(2r)^4 \int_0^{\frac{\pi}{2}} d\theta$$

$$= 4r^4\,[\theta]_0^{\frac{\pi}{2}}$$

$$= 4r^4 \times \left(\frac{\pi}{2} - 0\right)$$

$$= 2\pi r^4$$

$$\therefore \; I_A = 2\pi r^4.$$

18. *Determine the moment of inertia of the area enclosed within $y = Kx^n$, the x-axis and ordinate at x = a, with respect to the coordinate axes.*

Solution:

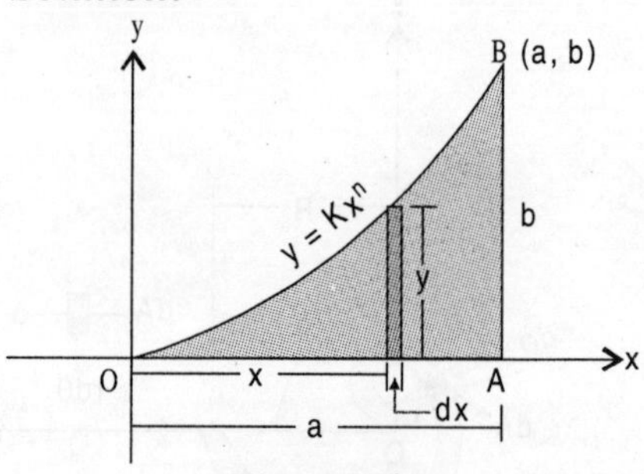

Fig. 6-W26

Let an elementary rectangular strip of area dA then,

$$dA = y\,dx$$

By integrating dA over the proper limits, we get

$$A = \int_{x_1}^{x_2} y\,dx$$

$$= \int_0^a Kx^n\,dx$$

$$= K\int_0^a x^n\,dx$$

$$= K\left[\frac{x^{n+1}}{n+1}\right]_0^a$$

$$= \frac{K}{n+1}\left[x^{n+1}\right]_0^a$$

$$= \frac{K}{n+1}\left[a^{n+1} - 0^{n+1}\right]$$

$$= \frac{K}{n+1}a^{n+1}$$

$$= \frac{ba}{n+1} \qquad [\because b = Ka^n]$$

$$= \frac{ab}{n+1} \text{ unit}^2$$

Now, I_y = M.I. about y-axis

$$= \int_{x_1}^{x_2} x^2\, dA$$

$$= \int_{x_1}^{x_2} x^2\, y\, dx$$

$$= \int_0^a x^2 . Kx^n\, dx$$

$$= K\int_0^a x^{n+2}\, dx$$

$$= K\left[\frac{x^{n+2+1}}{n+2+1}\right]_0^a$$

$$= K\left[\frac{x^{n+3}}{n+3}\right]_0^a$$

$$= \frac{K}{n+3}\left[a^{n+3} - 0^{n+3}\right]$$

$$= \frac{Ka^{n+3}}{n+3}$$

$$= \frac{Ka^n . a^3}{n+3}$$

$$= \frac{ba^3}{n+3} \text{ unit}^4.$$

Now as we know that, the M.I. of a rectangle about the base axis is $\frac{bd^3}{3}$, then the moment of inertia of a rectangular strip about the base axis i.e., x-axis is,

$$dI_x = \frac{1}{3}(dx)\, y^3$$

$$= \frac{1}{3}y^3 dx$$

$$\therefore \quad I_x = \int dI_x = \text{M.I. about x-axis}$$

$$= \int_0^a \frac{1}{3} y^3\, dx$$

$$= \frac{1}{3}\int_0^a (Kx^n)^3\, dx$$

$$= \frac{1}{3}K^3 \int_0^a x^{3n}\, dx$$

$$= \frac{K^3}{3}\left[\frac{x^{3n+1}}{3n+1}\right]_0^a$$

$$= \frac{K^3}{3}\left[\frac{a^{3n+1} - 0^{3n+1}}{3n+1}\right]$$

$$= \frac{K^3}{3}\left[\frac{a^{3n+1}}{3n+1}\right]$$

$$= \frac{K^3 a^{3n+1}}{3(3n+1)}$$

$$= \frac{(Ka^n)^3 . a}{3(3n+1)}$$

$$= \frac{ab^3}{3(3n+1)} \text{ unit}^4$$

19. *Find the moment of inertias about x and y-axes of the enclosed area within the* *curve* $y = b\left[1-\left(\frac{x}{a}\right)^{\frac{1}{2}}\right]$ *and axes. Also find the radius of gyrations.*

Solution:

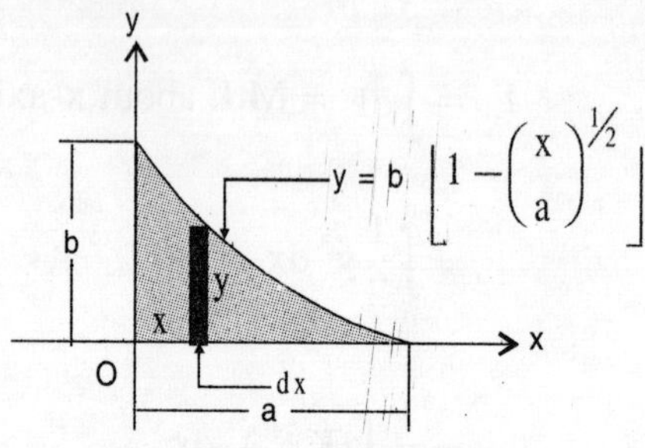

Fig. 6-W27

Let an elementary rectangular strip, at a distance x from y-axis and having width dx and height y, therefore, dA = y dx.

By integrating dA over the proper limits, we get,

$$A = \int_0^a y\,dx$$

$$= \int_0^a b\left[1-\left(\frac{x}{a}\right)^{\frac{1}{2}}\right]dx$$

$$= b\int_0^a \left[1-\left(\frac{x}{a}\right)^{\frac{1}{2}}\right]dx$$

$$= b\int_0^a \left(1-\frac{\sqrt{x}}{\sqrt{a}}\right)dx$$

$$= b\left[x-\frac{1}{\sqrt{a}}\frac{x^{\frac{3}{2}}}{\frac{3}{2}}\right]_0^a$$

$$= b\left[x-\frac{2}{3}\frac{1}{\sqrt{a}}x^{\frac{3}{2}}\right]_0^a$$

$$= b\left[a-\frac{2}{3}\times\frac{1}{\sqrt{a}}\times a\sqrt{a}\right]$$

$$= b\left[a-\frac{2}{3}a\right] = \frac{ab}{3}$$

Now, the moment of inertia about the y-axis,

$$I_y = \int x^2 dA$$

$$= \int_0^a x^2 y\,dx$$

$$= \int_0^a x^2 b\left(1-\left(\frac{x}{a}\right)^{\frac{1}{2}}\right)dx$$

$$= b\int_0^a \left(x^2-\frac{x^{\frac{5}{2}}}{\sqrt{a}}\right)dx$$

$$= b\left[\frac{x^3}{3}-\frac{1}{\sqrt{a}}\times x^{\frac{7}{2}}\times\frac{2}{7}\right]_0^a$$

$$= b\left[\frac{a^3}{3}-\frac{1}{\sqrt{a}}\times\frac{2}{7}\times a^3\times\sqrt{a}\right]$$

$$= b\left[\frac{a^3}{3}-\frac{2a^3}{7}\right]$$

$$= b\left[\frac{7a^3-6a^3}{21}\right]$$

$$= \frac{a^3 b}{21}$$

Similarly, moment of inertia of the enclosed area about x-axis,

$$I_x = \int_0^a \frac{1}{3} y^3\, dx$$

$$= \frac{1}{3}\int_0^a y^3\, dx$$

$$= \frac{1}{3}\int_0^a b^3\left(1-\left(\frac{x}{a}\right)^{\frac{1}{2}}\right)^3 dx$$

$$= \frac{b^3}{3}\int_0^a \left\{1-\left(\frac{x}{a}\right)^{\frac{3}{2}} + 3\left(\frac{x}{a}\right) - 3\left(\frac{x}{a}\right)^{\frac{1}{2}}\right\} dx$$

$$= \frac{b^3}{3}\left[x - \frac{1}{a\sqrt{a}} x^{\frac{5}{2}} \times \frac{2}{5} + \frac{3}{a}\frac{x^2}{2} - \frac{3}{\sqrt{a}}\frac{3}{2} \times \frac{2}{3}\right]_0^a$$

$$= \frac{b^3}{3}\left[x - \frac{1}{a\sqrt{a}} \times \frac{2}{5} \times x^{\frac{5}{2}} + \frac{3}{a} \times \frac{x^2}{2} - \frac{3}{\sqrt{a}} \times \frac{2}{3} x^{\frac{3}{2}}\right]_0^a$$

$$= \frac{b^3}{3}\left[a - \frac{1}{a\sqrt{a}} \times \frac{2}{5} \times \sqrt{a} \times a^2 + \frac{3}{a} \times \frac{a^2}{2} - \frac{3}{\sqrt{a}} \times \frac{2}{3} \times \sqrt{a} \times a\right]$$

$$= \frac{b^3}{3}\left[a - \frac{2a}{5} + \frac{3a}{2} - 2a\right]$$

$$= \frac{b^3}{3}\left[\frac{10a - 4a + 15a - 20a}{10}\right]$$

$$= \frac{ab^3}{30}$$

$\therefore\ K_x = \sqrt{\frac{I_x}{A}}$ = radius or gyration about x-axis

$$= \sqrt{\frac{ab^3}{30} \times \frac{3}{ab}}$$

$$= \sqrt{\frac{b^2}{10}}$$

$$= \frac{b}{\sqrt{10}}$$

$$= 0.316\, b$$

and $K_y = \sqrt{\frac{I_y}{A}}$

= radius of gyration about y-axis

$$= \sqrt{\frac{a^3 b}{21} \times \frac{3}{ab}}$$

$$= \sqrt{\frac{a^2}{7}}$$

$$= \frac{a}{\sqrt{7}}$$

$$= 0.3779\, a.$$

20. *Locate the axes of the minimum and the maximum moments of inertia at the right angle corner of an isosceles triangle and determine the values of the moments of inertia about those axes with b = 4 mm and a = 3 mm.*

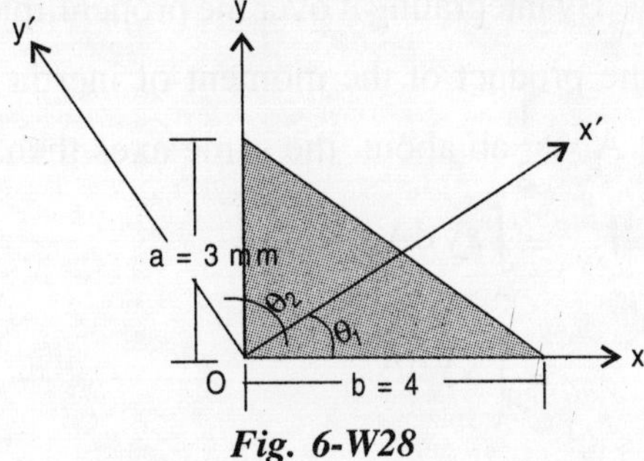

Fig. 6-W28

Solution:

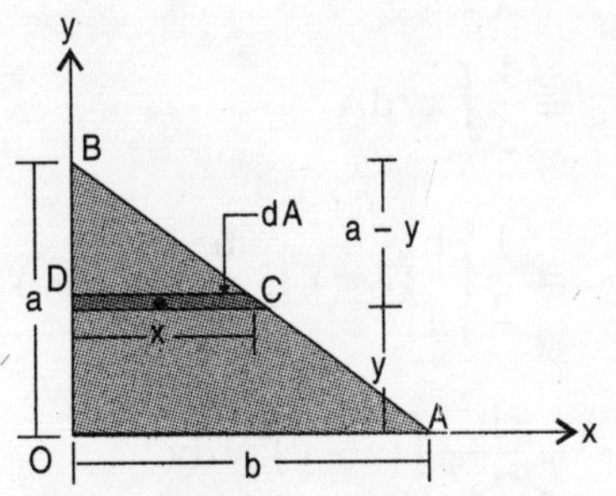

Fig. 6-W29

with reference to figure (6-W29), from similar triangles OAB and DCB, we have

$$\frac{a-y}{x} = \frac{a}{b}$$

or $$x = \frac{b}{a}(a-y)$$

and area of the elementary strip,

$$dA = x\,dy = \frac{b}{a}(a-y)\,dy$$

Let c $(\bar{x}, \bar{y})$ be the centroid of the elementary area.

Then, $$\bar{x} = \frac{x}{2}, \quad \bar{y} = y$$

Therefore, the product of inertia of the elementary strip about the x and y-axes is

$$dI_{xy} = xy\,dA\ ;\ \text{where } x = \bar{x} \text{ and } y = \bar{y}.$$

By integrating it over the proper limits, we get the product of the moment of inertia of an area, $A = \frac{1}{2}ab$ about the same axes then,

$$I_{xy} = \int_A xy\,dA$$

$$= \int_A \bar{x}\bar{y}\,dA$$

$$= \int_A \frac{x}{2}.y\,dA \quad [\because\ \bar{x} = \frac{x}{2} \text{ and } \bar{y} = y]$$

$$= \frac{1}{2}\int_A xy\,dA$$

$$= \frac{1}{2}\int_0^a \frac{b}{a}(a-y)y\frac{b}{a}(a-y)\,dy$$

$$= \frac{b^2}{2a^2}\int_0^a (a-y)^2 y\,dy$$

$$= \frac{b^2}{2a^2}\int_0^a \left(a^2+y^2-2ay\right)y\,dy$$

$$= \frac{b^2}{2a^2}\int_0^a \left(a^2y+y^3-2ay^2\right)dy$$

$$= \frac{b^2}{2a^2}\left[a^2\left[\frac{y^2}{2}\right]_0^a + \left[\frac{y^4}{4}\right]_0^a - 2a\left[\frac{y^3}{3}\right]_0^a\right]$$

$$= \frac{b^2}{2a^2}\left[\frac{a^4}{2}+\frac{a^4}{4}-\frac{2a^4}{3}\right]$$

$$= \frac{b^2}{2a^2}\left[\frac{6a^4+3a^4-8a^4}{12}\right]$$

$$= \frac{a^2b^2}{24}$$

and as we know that, $I_x = \dfrac{ba^3}{12}$

and $I_y = \dfrac{ab^3}{12}$

Now the inclination of the axes of the maximum and the minimum moments of inertia is located by,

$$\tan 2\theta = \frac{2I_{xy}}{I_y - I_x}$$

$$= \frac{2\times\frac{a^2b^2}{24}}{\frac{ab^3}{12}-\frac{ba^3}{12}}$$

$$= \frac{\frac{a^2b^2}{12}}{\frac{ab\left(b^2-a^2\right)}{12}}$$

$$= \frac{ab}{b^2-a^2}$$

$$= \frac{4 \times 3}{4^2 - 3^2}$$

$$= \frac{12}{7}$$

$$= 1.714$$

$$\Rightarrow \quad 2\theta \cong 60 \text{ or } (180 + 60)$$

or $\theta = 30° \text{ or } 120°$

$$\Rightarrow \quad \theta_1 = 30° \text{ and } \theta_2 = 120°$$

As we know that,

$$R = \sqrt{\left(\frac{I_x - I_y}{2}\right)^2 + I_{xy}^2}$$

$$= \sqrt{\frac{1}{4}\left(\frac{ba^3}{12} - \frac{ab^3}{12}\right)^2 + \left(\frac{a^2b^2}{24}\right)^2}$$

$$= \sqrt{\frac{1}{4} \times \frac{a^2b^2}{144}(a^2 - b^2)^2 + \frac{a^4b^4}{576}}$$

$$= \sqrt{\frac{1}{4} \times \frac{9 \times 16}{144} \times 49 + \frac{81 \times 256}{576}}$$

$$= \frac{1}{24}\sqrt{27792}$$

$$= \frac{166.709}{24}$$

$$= 6.94$$

Therefore, we have

$$I_{max} = I_{av} + R$$

$$= \frac{I_x + I_y}{2} + 6.94$$

$$= \frac{1}{24}(3^2 + 4^2) \times 12 + 6.94$$

$$= \frac{25}{2} + 6.94$$

$$= 19.44 \text{ mm}^4$$

and $I_{min} = I_{av} - R$

$$= 12.5 - 6.94$$

$$= 5.56 \text{ mm}^4$$

21. *Determine the product of inertia about the centroidal axis. If it is negative, then explain it.*

Solution: Do it by yourself.

22. *For the skin-stringer cross-section as shown in Fig. (6-W30), determine the*

(*a*) *moment of inertia about the centroidal x-axis.*

(*b*) *moment of inertia about the centroidal y-axis.*

(*c*) *product of inertia.*

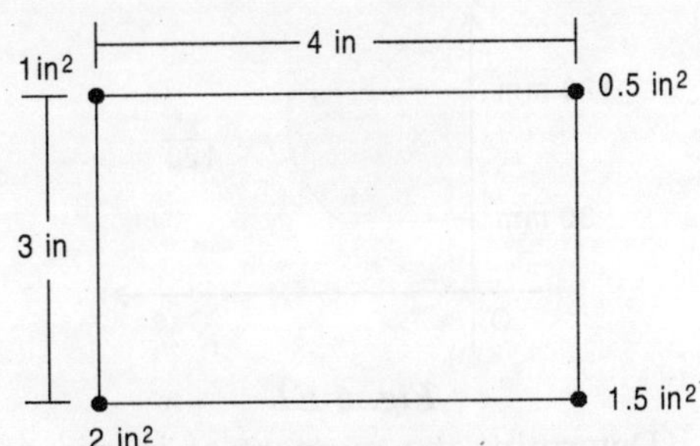

Fig. 6-W30

Solution: With respect to the 2 in^2 stiffener, the centroidal is located at

$$\bar{x} = \frac{\Sigma Ax}{\Sigma A}$$

$$= \frac{2 \times 0 + 1.5 \times 4 + 0.5 \times 4 + 1 \times 0}{2 + 1.5 + 0.5 + 1}$$

$$= 1.6 \text{ in}$$

$$\bar{y} = \frac{\Sigma Ay}{\Sigma A}$$

$$= \frac{2 \times 0 + 1.5 \times 0 + 0.5 \times 3 + 1 \times 3}{2 + 1.5 + 0.5 + 1}$$

$$= 0.9 \text{ in}$$

The rectangular moments and product of inertia are,

$$I_x = \Sigma Ad^2$$

$$= 2 \times (0 - 0.9)^2 + 1.5 \times (0 - 0.9)^2 + 0.5 \times (3 - 0.9)^2 + 1 \times (3 - 0.9)^2$$

$$= 9.45 \text{ in}^4$$

$$I_y = \Sigma Ad^2$$

$$= 2 \times (0 - 1.6)^2 + 1.5 \times (4 - 1.6)^2 + 0.5 \times (4 - 1.6)^2 + 1(0 - 1.6)^2$$

$$= 9.2 \text{ in}^4$$

$$I_{xy} = \Sigma Axy$$

$$= 2 \times (0 - 0.9) \times (0 - 1.6) + 1.5 \times (0 - 0.9) \times (4 - 1.6) + 0.5 \times (5 - 0.9) \times (4 - 1.6) + 1 \times (3 - 0.9) \times (0 - 1.6)$$

$$= -1.2 \text{ in}^4.$$

***Note:** Because the cross-section is not symmetrical about x or y-axes, the product of inertia is not zero.*

EXERCISE

1. Calculate the moment of inertia of the shaded area about the y-axis.

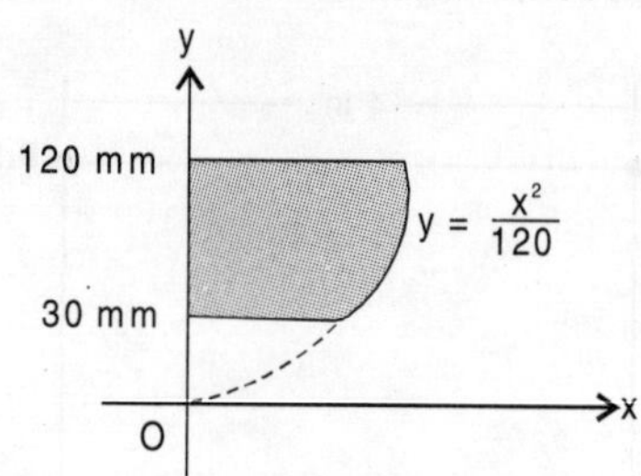

Fig. 6-E1

2. Determine the moment of inertia of the shaded area about the x-axis.

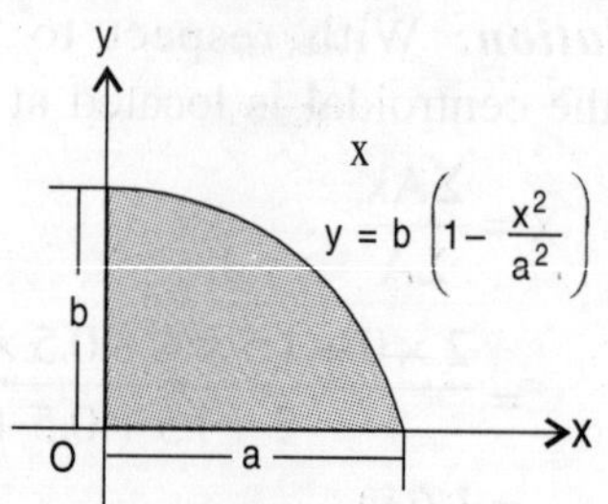

Fig. 6-E2

3. Determine the second moment of inertia of the shaded area about the x-axis and the y-axis.

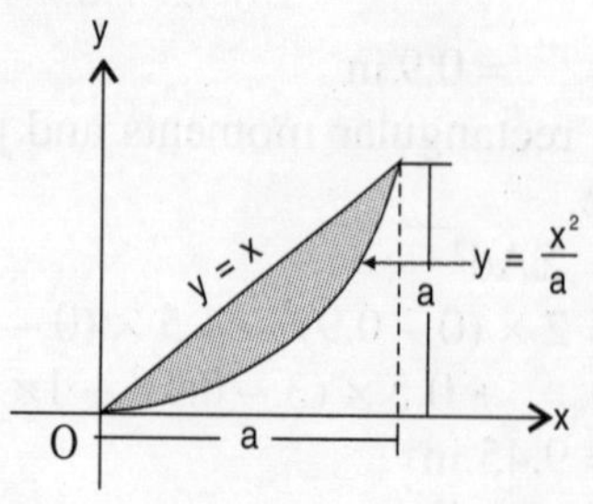

Fig. 6-E3

4. Determine the moment of inertia of the shaded area about the y-axis.

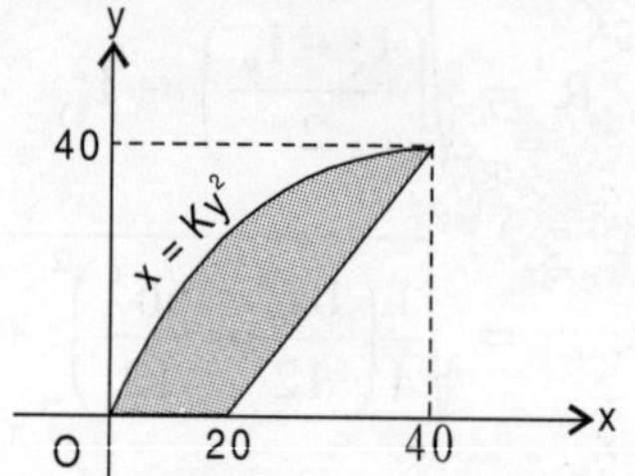

Fig. 6-E4

5. Calculate the moment of inertia of the overlapping shaded area of the two circles about the x-axis.

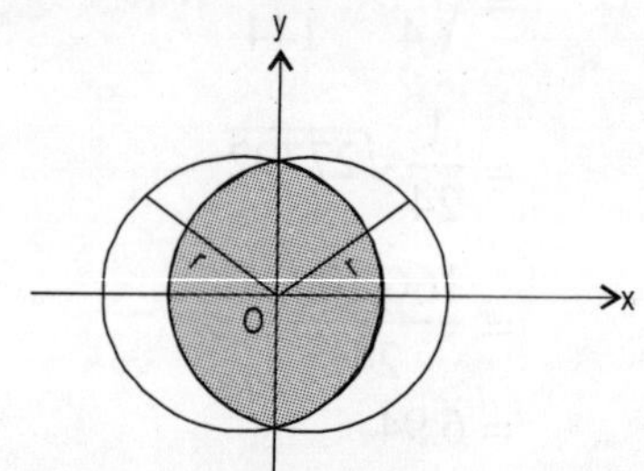

Fig. 6-E5

6. Find the moment of inertia about the x-axis of the area bounded by two parabolas with the dimensions indicated in Fig. (6-E6).

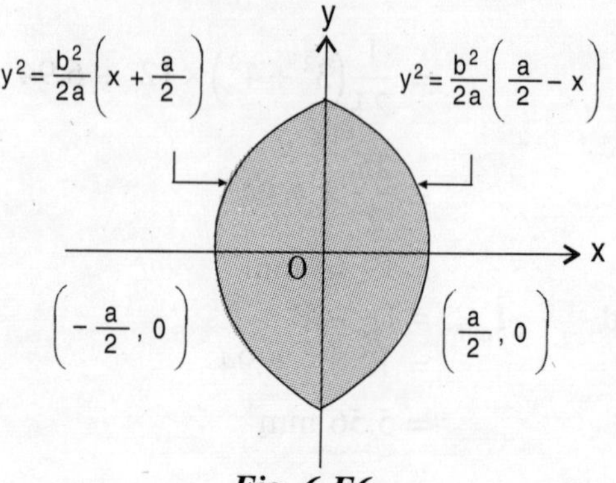

Fig. 6-E6

7. Determine the moment of inertia and the radius of gyration of the shaded area shown with respect to the y-axis.

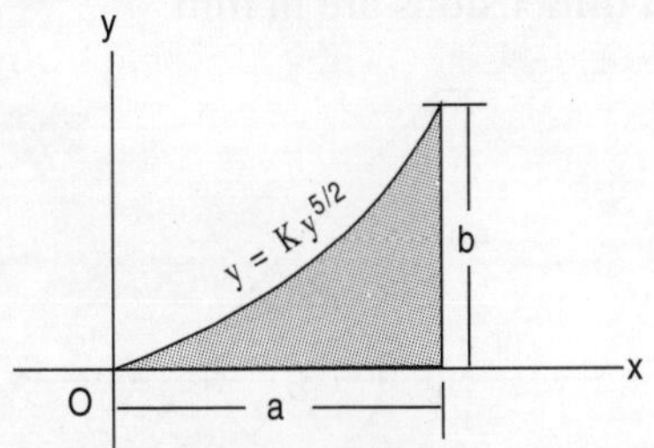

Fig. 6-E7

8. Find I_{yy} for the area between the curves $y = 2 \sin x$ and $y = \sin 2x$ from $x = 0$ to $x = \pi$.

9. Find I_x and I_y for the area between the axes and $y = |x|$ when $-1 \le x \le 1$.

10. Find the M.I. about the x-axis of the area bounded by the curve $y = x - [x]$, $0 < x < 1$ and x-axis.

11. Calculate the least radius of gyration of the section shown in Fig. (6-E8). All dimensions are in mm.

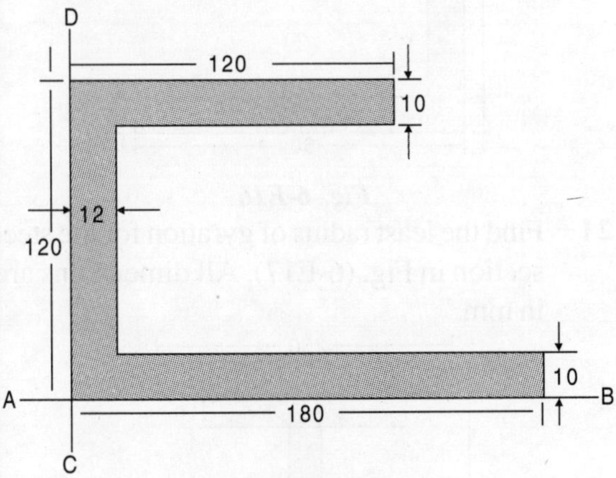

Fig. 6-E8

12. Find the moment of inertia of the shaded area shown in Fig. (6-E9) about the horizontal centroidal axis. And calculate the radius of gyration about the same axis. All the dimensions are in mm.

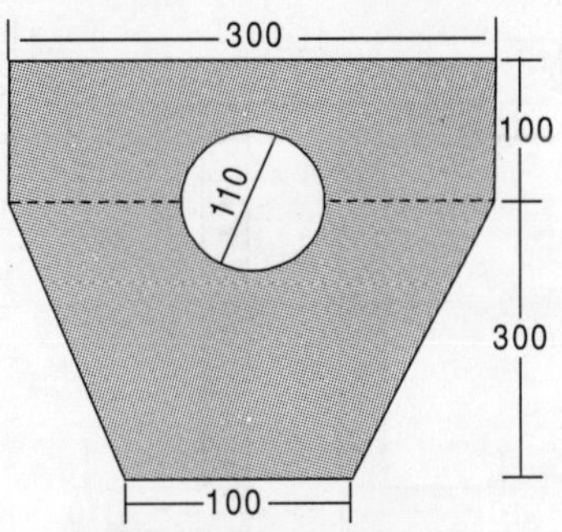

Fig. 6-E9

13. Calculate the moment of inertia of the plane section shown in Fig. (6-E10) about the horizontal axis through the centroid of the section. All dimensions are in mm.

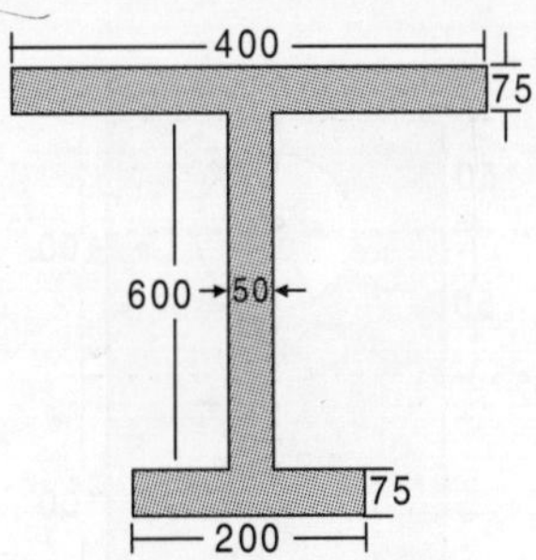

Fig. 6-E10

14. State and prove,
   (a) the parallel axis theorem and
   (b) the perpendicular axis theorem.

15. Compute the least radius of gyration of the lamina shown in Fig. (6-E11). All dimensions are in mm.

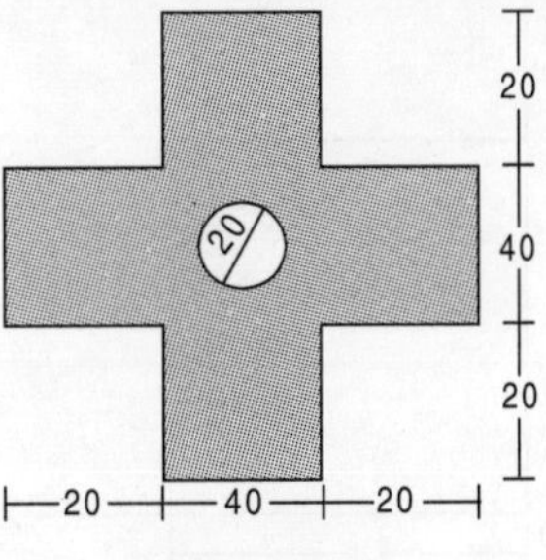

Fig. 6-E11

16. Calculate the least moment of inertia. All dimensions are in mm.

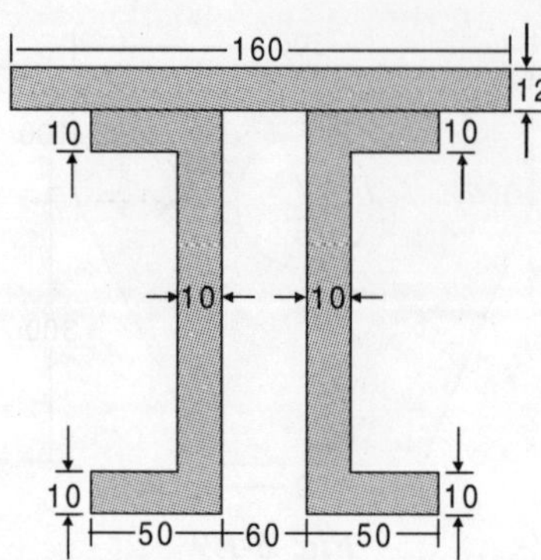

6-E12

17. Compute the least radius of gyration of the plate with a circular hole removed as shown in Fig. (6-E13). All dimensions are in mm.

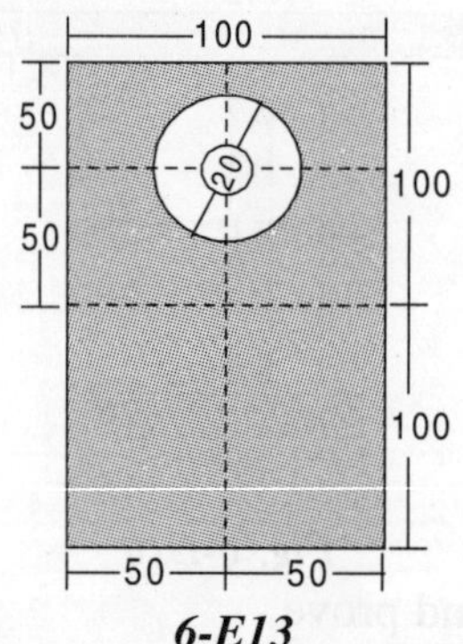

6-E13

18. Calculate the moment of inertia of the shaded area in Fig. (6-E14) with respect to the centroidal axis parallel to the x-axis.

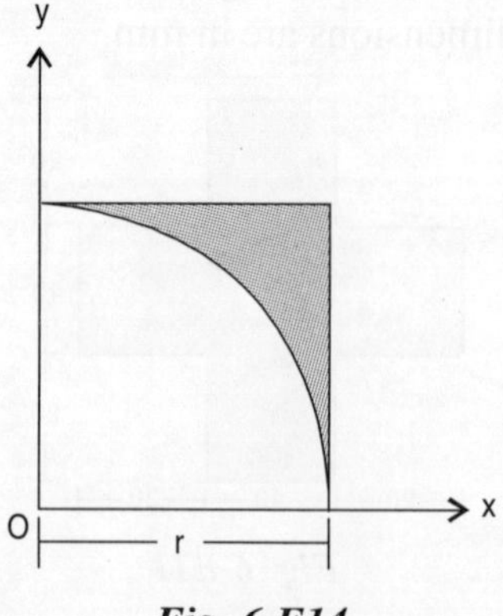

Fig. 6-E14

19. Calculate the moment of inertia of the shaded area in Fig. (6-E15) with respect to the centroidal axis parallel to the x-axis. All dimensions are in mm.

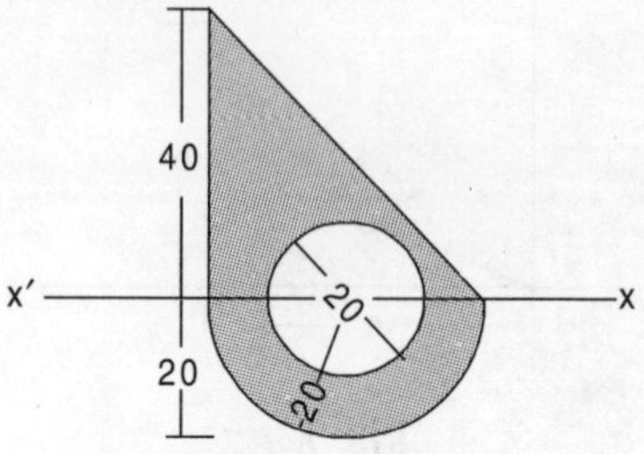

Fig. 6-E15

20. Determine the moment of inertia of the L-section shown in Fig. (6-E16) about its centroidal axes parallel to the x and y-axes. Also find out the polar moment of inertia passing through the centroid. All dimensions are in mm.

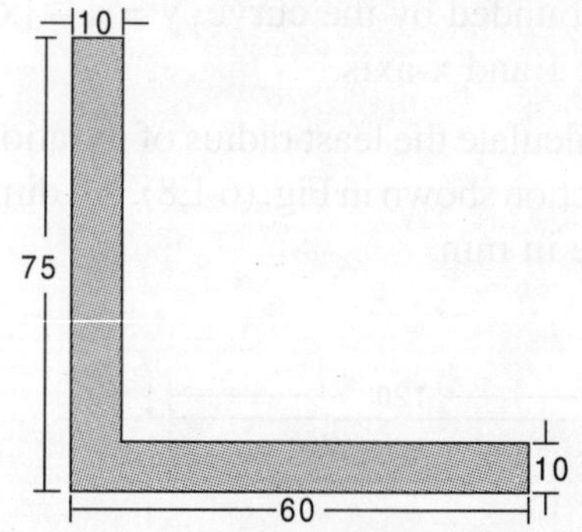

Fig. 6-E16

21. Find the least radius of gyration for the steel section in Fig. (6-E17). All dimensions are in mm.

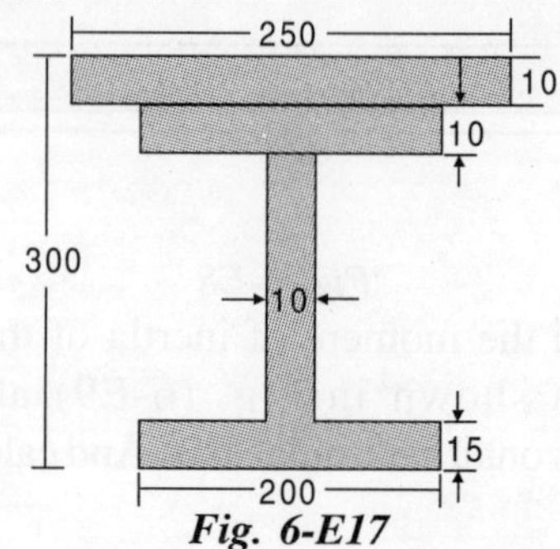

Fig. 6-E17

22. Find the value of $K_{xx'}$ and K_{yy} for the section shown in Fig. (6-E18). All dimensions are in mm.

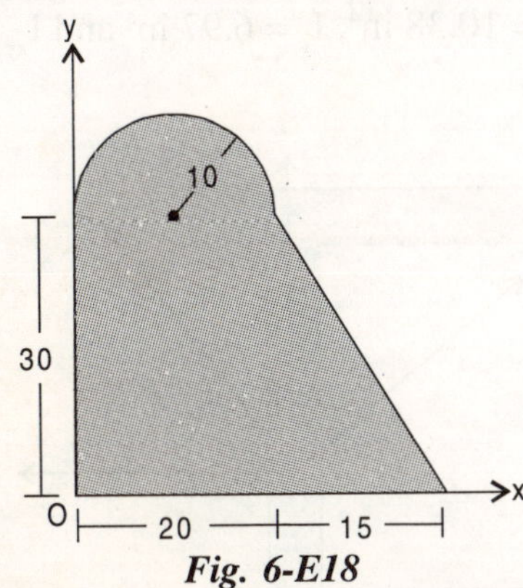

Fig. 6-E18

23. The cross-section of a plain concrete culvert is as shown in Fig. (6-E19). Determine the moment of inertia about the horizontal centroidal axis. All dimensions are in mm.

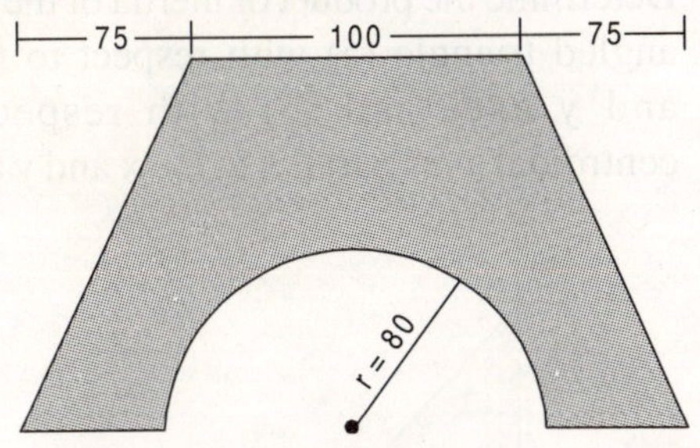

Fig. 6-E19

24. Find the moment of inertia of the section shown in Fig. (6-E20) about x-axis. Take all dimensions in mm.

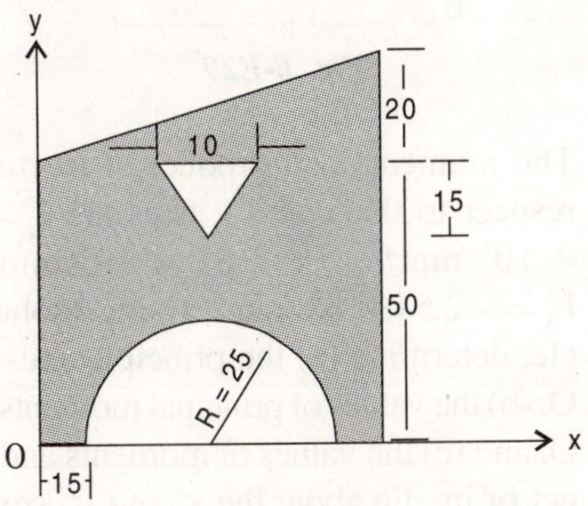

Fig. 6-E20

25. A semi-circular cut is made in a rectangular wooden beam as shown in Fig. (6-E21). Determine the polar moment of inertia about the centroidal axis. All dimensions are in mm.

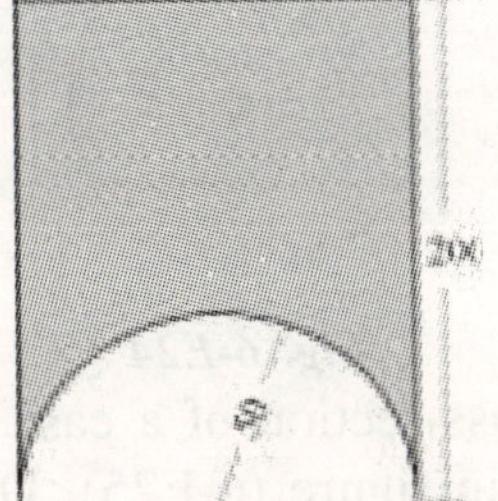

Fig. 6-E21

26. For the lamina shown in Fig. (6-E22), find the moment of inertia about the centroidal x-axis parallel to the base. All dimensions are in mm.

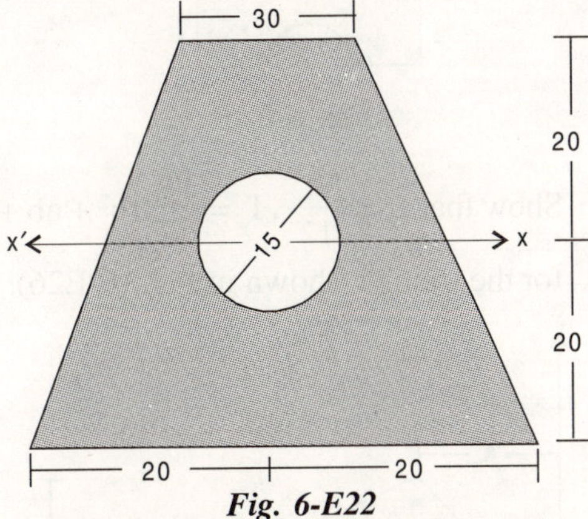

Fig. 6-E22

27. Find the moment of inertia about the centroidal x-axis for the lamina shown in Fig. (6-E23). All dimensions are in mm.

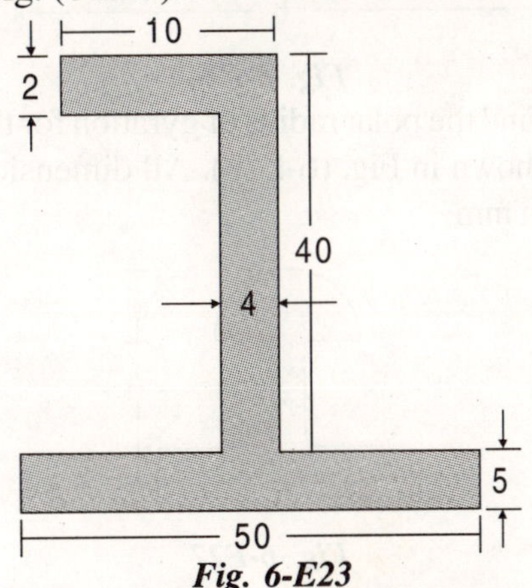

Fig. 6-E23

28. A semi-circle of radius r has been cut out from a circle of radius R as shown in Fig. (6-E24). Calculate the polar moment of inertia about 'O' for the resulting section.

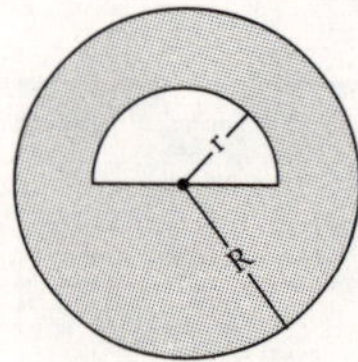

Fig. 6-E24

29. The cross-section of a cast iron beam is shown in figure (6-E25). Determine the moment of inertia about the centroidal axes.

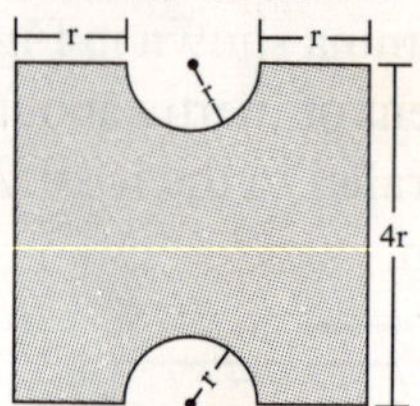

Fig. 6-E25

30. Show that $I_x = \dfrac{bh^3}{12}$, $I_y = \dfrac{bh}{12}(b^2 + ab + a^2)$

for the triangle shown in Fig. (6-E26).

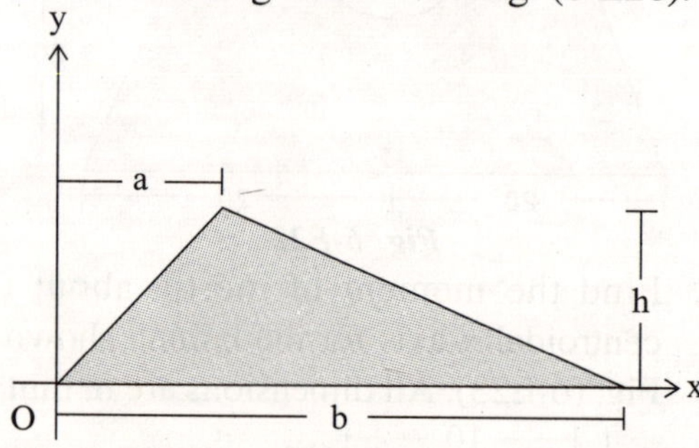

Fig. 6-E26

31. Find the polar radius of gyration for the area shown in Fig. (6-E26). All dimensions are in mm.

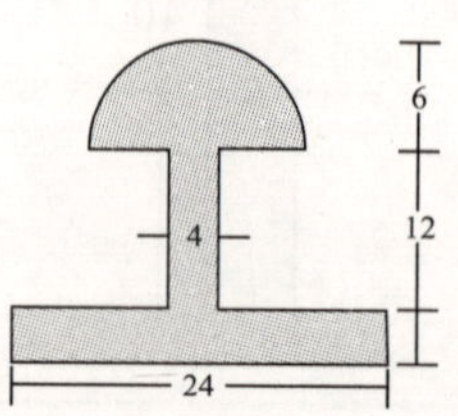

Fig. 6-E27

32. Determine the orientation of the principal axes and principal moments of inertia for the section shown in Fig. (6-E28) where $I_x = 10.38$ in^4, $I_y = 6.97$ in^4 and $I_{xy} = -6.56$ in^4.

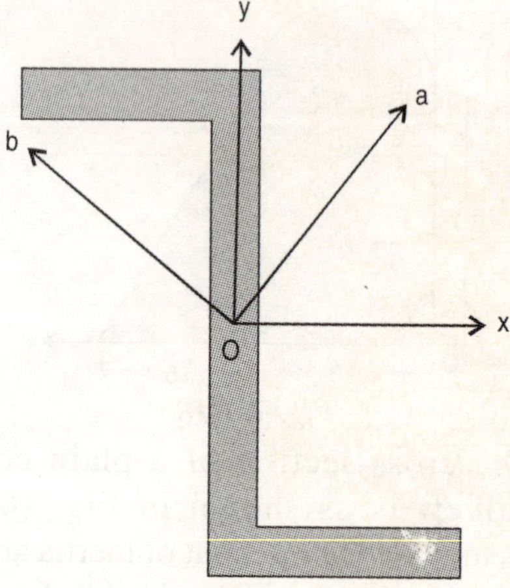

Fig. 6-E28

33. Determine the product of inertia of the right angled triangle (a) with respect to the x and y axes and (b) with respect to centroidal axes parallel to the x and y axes.

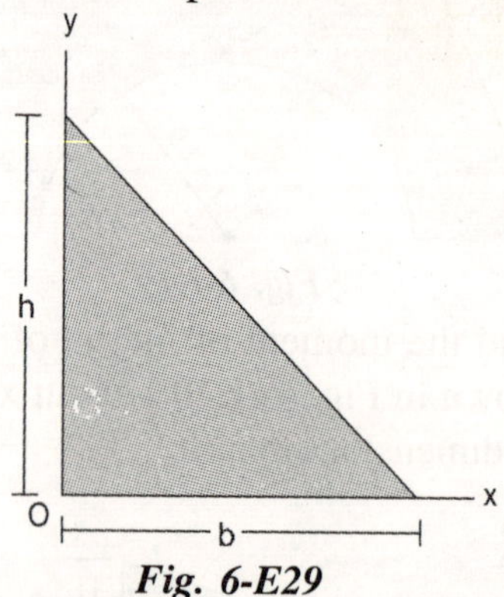

Fig. 6-E29

34. The moments and product of inertia with respect to the x and y axes are $I_x = 7.24 \times 10^6$ mm^4, $I_y = 2.61 \times 10^6$ mm^4 and $I_{xy} = -2.54 \times 10^6$ mm^4. Using Mohr's circle, determine (a) the principle axes about O, (b) the values of principal moments about O, and (c) the values of moments and product of inertia about the x′ and y′ axes.

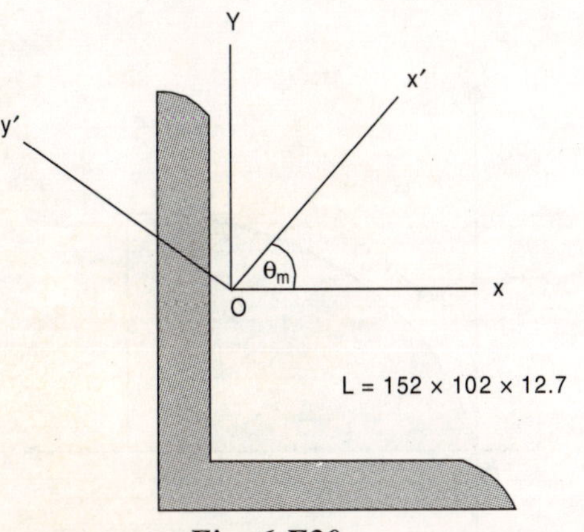

Fig. 6-E30

35. For the cross section shown in Fig. (6-E31), determine the product of inertia.
Hints. $I_{xy} = \Sigma Axy$.

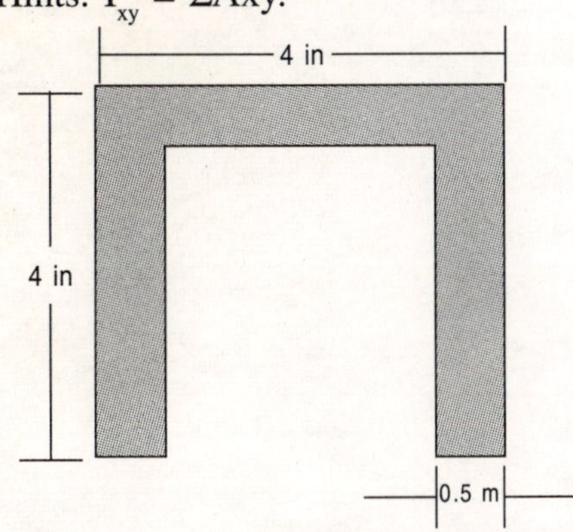

Fig. 6-E31

36. For the skin-stringer cross section shown in Fig. (6-E32), determine the product of inertia. Given that $A = 0.5$ in^2, $I_x = 0.1$ in^4 and $I_y = 0.15$ in^4.

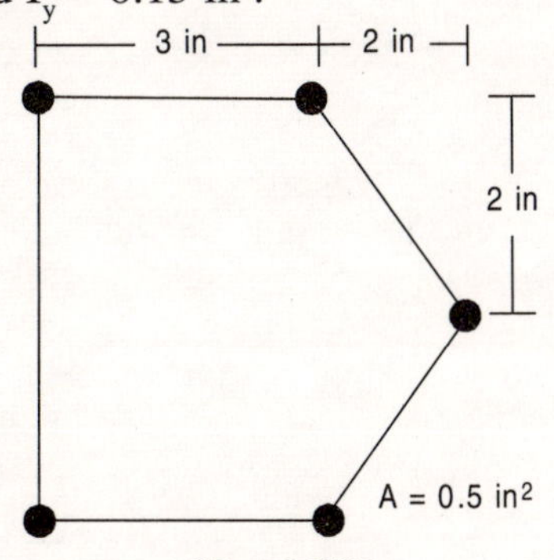

Fig. 6-E32

37. For the skin-stringer cross section as shown in Fig. (6-E33) determine the product of inertia.

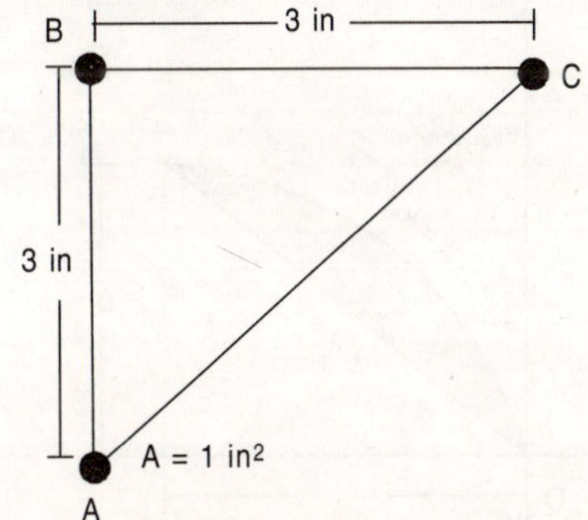

Fig. 6-E33

38. Compute the area M.I. in xy-plane and find the corresponding radius of gyration of the shaded area in Fig. (6-E34).

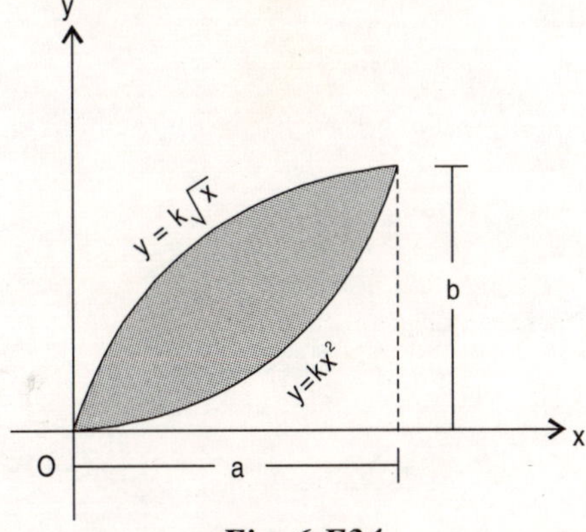

Fig. 6-E34

39. Compute the area M.I. in xy-plane and find the corresponding radius of gyration of the shaded area in Fig. (6-E35).

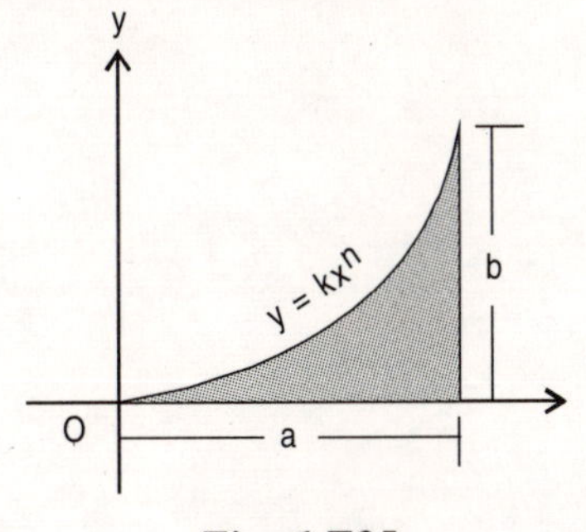

Fig. 6-E35

40. Find the area M.I. of the shaded areas shown in Figs. (6-E36 a & b).

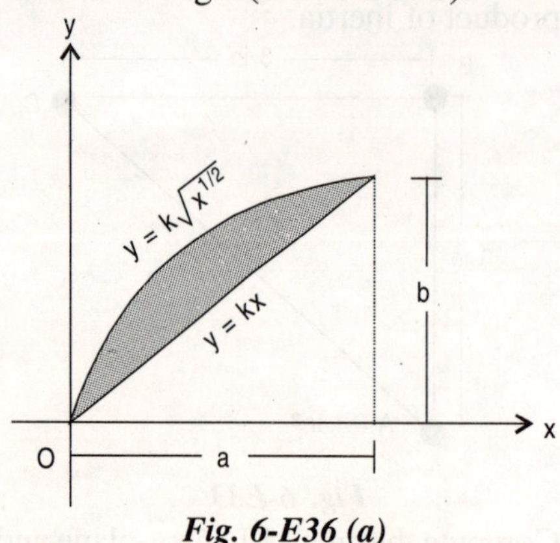

Fig. 6-E36 (a)

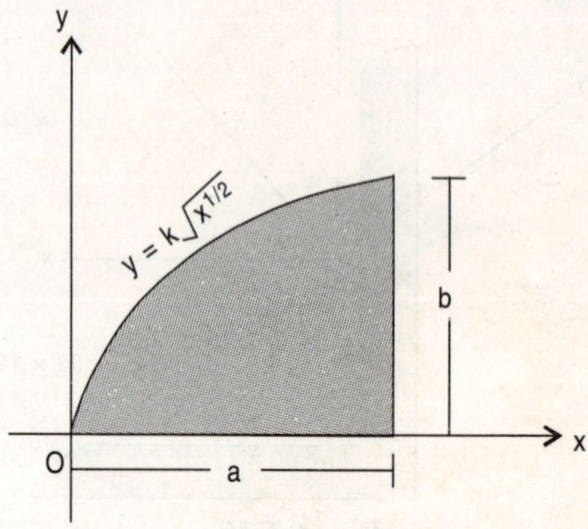

Fig. 6-E36 (b)

7 FRICTION

7.1 FRICTION

Consider a block of mass m placed on a smooth horizontal table as shown in Fig. 7.1. Attach a string to it to measure the force required to set the block in motion. We find that the block will move even though we apply an infinitesimal small force (dF). We say that there is no opposing force which will oppose the motion with respect to the table.

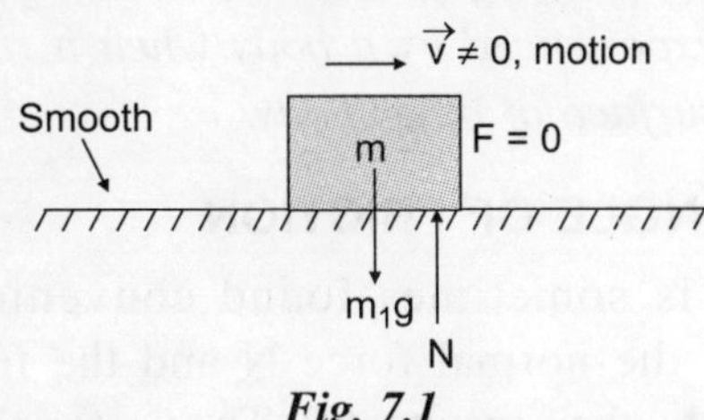

Fig. 7.1

Now place same block on an another horizontal table which is not smooth as in Fig. 7.2. Applying a force F, we find that the block does not move then from Newton's second law, we say that our applied force F is balanced by an opposite frictional force (F_s) exerted on the block by the table, acting along the surface of contact.

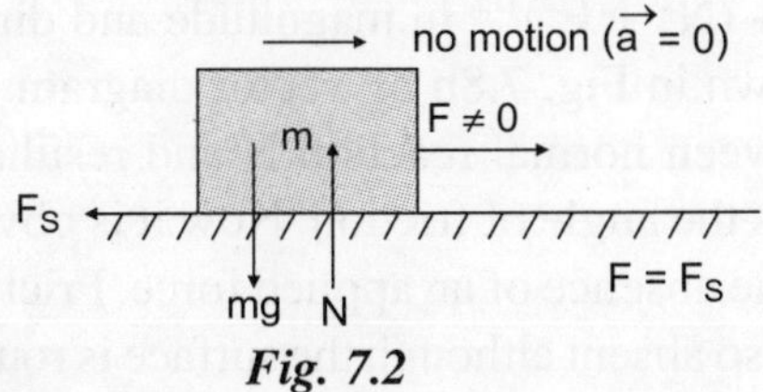

Fig. 7.2

Now we increase the applied force F to F_0, still we observe that the block is not moving, then we can say that the opposing force (F_0) is also increased so that it balances the applied force F_0; this state is shown in Fig. 7.3.

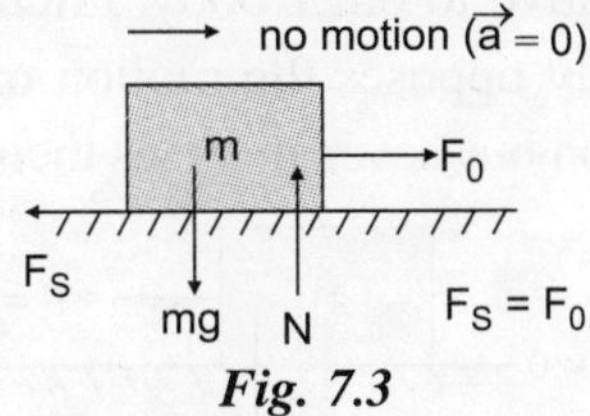

Fig. 7.3

Now if we increase the applied force to F_{max}, we find that block is in impending motion i.e., the block is about to move. By further increasing the applied force to P block moves with uniform acceleration.

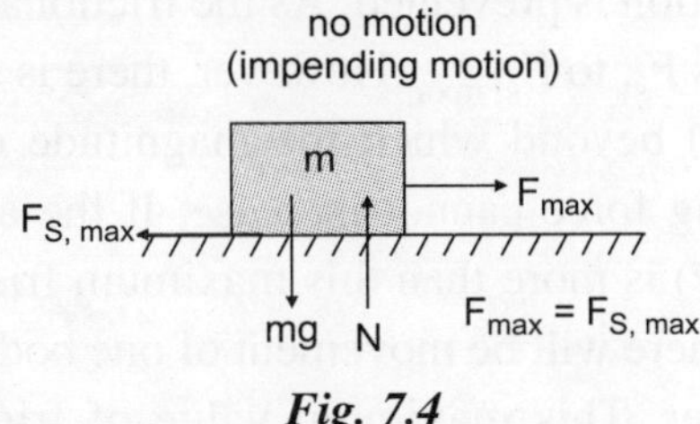

Fig. 7.4

These states are shown in Fig. 7.4 and Fig. 7.5 respectively. Note that force F_{max} increases to P then block starts moving, it does not mean that there is no opposing force at all, rather it is there, call it F_k. Because by the experiment we can observe that acceleration of block is not equal to $\frac{P}{m}$ but it is $\frac{P - F_k}{m}$.

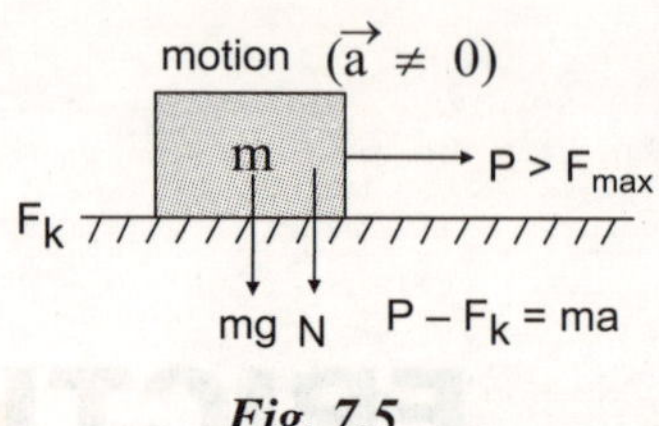

Fig. 7.5

Including all the above observations, we can define friction as follow:

Actually, whenever the surface of one body slides or is about to slide over that of another, parallel to the surfaces. The frictional force on each body is, in direction, opposite to the motion, relative to other body. Frictional force automatically opposes the motion or tendency of the motion but never reverses their tendency.

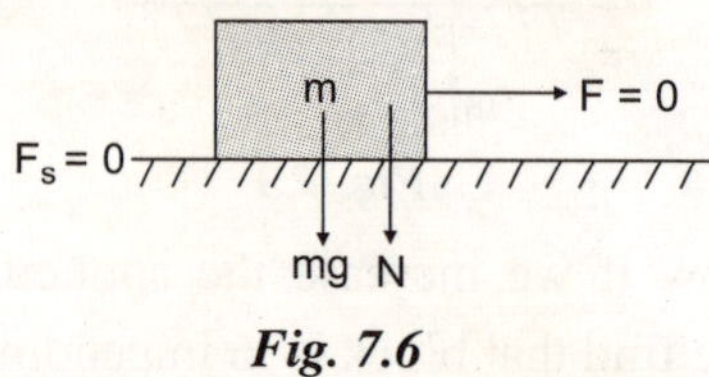

Fig. 7.6

Frictional force has the remarkable property of adjusting itself in magnitude to the force producing or tending to produce the motion so that motion is prevented. As the frictional force changes F_s, to $F_{s, mxa}$. However, there is a limit ($F_{s, max}$) beyond which the magnitude of this opposing force cannot increase. If the applied force (P) is more than this maximum frictional force, there will be movement of one body over the other. This maximum value of frictional force, which comes into the picture, when the motion is impending (Fig. 7.4) is said to be limiting static frictional force.

We observe that when the applied force is between zero to $F_{s, max}$ or $0 \leq P_{applied} \leq F_{s, max}$; block is in static state for this limit, either block is at rest or about to move but there is no motion at all. When the applied force is increased from $F_{s, max}$ the block is in motion and the opposing force (Fk) is called kinetic frictional force. All the experiments or observations can be put in pictorial form as.

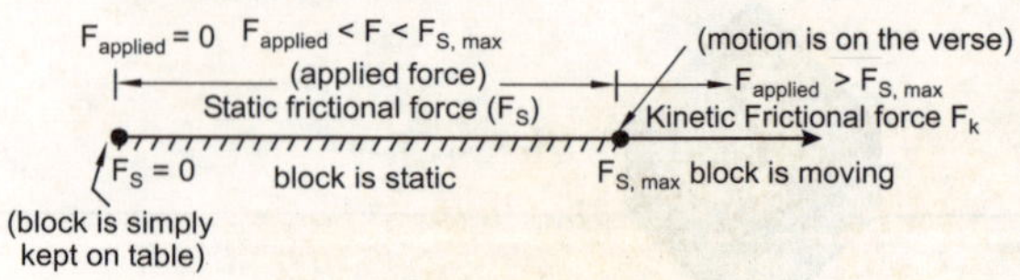

Fig. 7.7

Notes:

1. *When we just keep the block on the horizontal table, then the table does not exert frictional force as shown in Fig. 7.6*
2. *There are two types of kinetic friction.*
   (a) Sliding friction: *It is the friction experienced by a body when it slides (skids) over the other body.*
   (b) Rolling friction: *It is the friction experienced by a body when it rolls on surface of other body.*

7.2 ANGLE OF FRICTION

It is sometimes found convenient to replace the normal force N and the friction force F by their resultant R. The vectorial angle between the normal force and the resultant force is called as an angle of friction (ϕ). Let us consider a block of weight W rests on a horizontal plane surface. If the applied force P (horizontally) tries to move the block then the friction force comes in the picture and tries to oppose the applied force P. Since N and F are perpendicular to each other their resultant R will be + $(N^2 + F^2)^{1/2}$ in magnitude and direction is shown in Fig. 7.8b by vector diagram. Angle ϕ between normal reaction N and resultant force R, is the angle of friction. Now it is obvious that in the absence of an applied force. Friction force is also absent although the surface is rough, then, the resultant force will be only R = N. Now on increasing the applied force, the frictional force also increases till the maximum i.e., till limiting frictional force; or on increasing frictional force, angle ϕ is also increasing up to maximum angle ϕ_{max} corresponding to the limiting frictional force ($F_{max} = \mu_S N$).

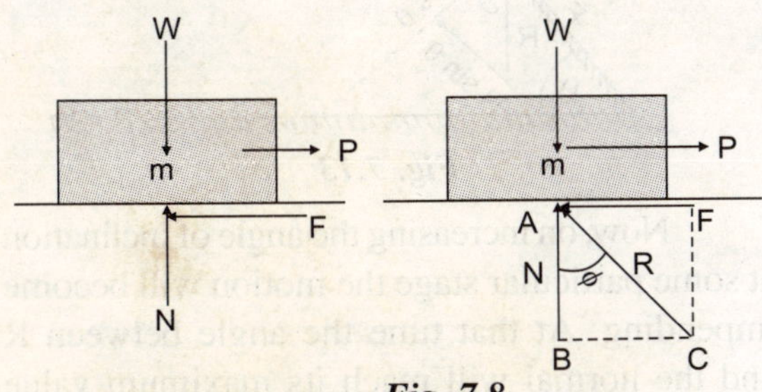

Fig. 7.8

Now in ΔABC_4, we have

$$\tan\theta_{max} = \frac{BC_4}{BA} = \frac{F_{max}}{N}$$

$$= \frac{\mu_S N}{N} \quad [\because F_{max} = \mu_S N]$$

$= \mu_s$, where μ_s is coefficient of static friction

Therefore, $\theta_{max} = \tan^{-1}\mu_s$ ... (7.1)

in other words, the maximum static angle of friction

= $\tan^{-1}$ (coefficient of static friction)

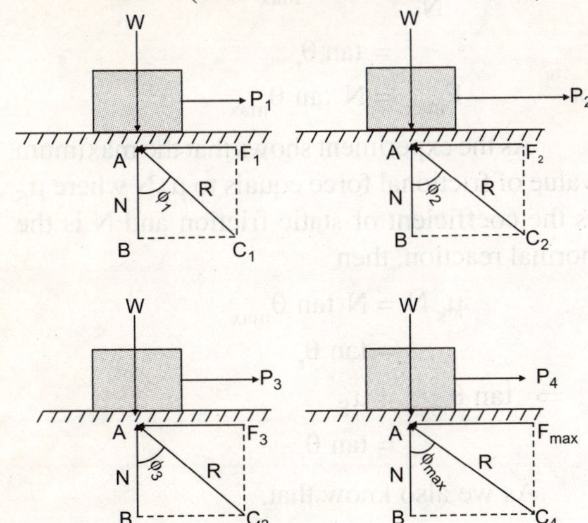

Fig. 7.9

Now, by further increasing the applied force ($p > p_4$) from the state of impending block, will be set in motion; experimental evidences show that the maximum frictional force F_{max} drops to f_k, naturally, ϕ_{max} decreases to ϕ_k.

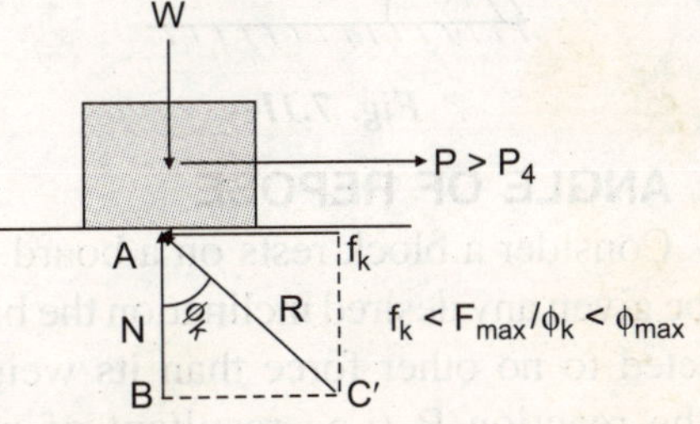

Fig. 7.10

Now in $\Delta ABC'$, we have

$$\tan\phi_k = \frac{BC'}{BA}$$

$$= \frac{f_k}{N}$$

$$= \frac{\mu_k N}{N}$$

$$= \mu_k$$

$$\Rightarrow \quad \phi_k = \tan^{-1}\mu_k \quad ....(7.2)$$

In other words, angle of kinetic friction = $\tan^{-1}$ (coefficient of kinetic friction).

Example 7.1: *Fig. 7.11 shows that a coin is resting on a book that has been tilted at an angle θ with the horizontal. By trial and error you find that when angle 'θ' is increased to 13°, the coin begins to roll down the book. What is the coefficient of state friction μ_s between the coin and the book?*

Solution: By trial and error we find that 13° is that maximum angle of friction at which coin is about to roll over the book then,

$$\mu_s = \tan\phi_{max}$$

$$= \tan 13°$$

$$= 0.23$$

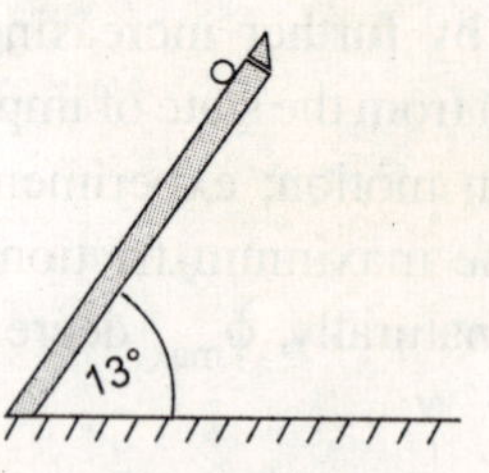

Fig. 7.11

7.3 ANGLE OF REPOSE

Consider a block rests on a board which may be given any desired inclination the block is subjected to no other force than its weight W and the reaction R (i.e., resultant of normal reaction and frictional force) of the board. If the board is horizontal, the reaction R exerted by the board on the block is perpendicular to the block and is actually only a normal reaction because there is no tendency of motion or motion. It is shown in Fig. (7.12).

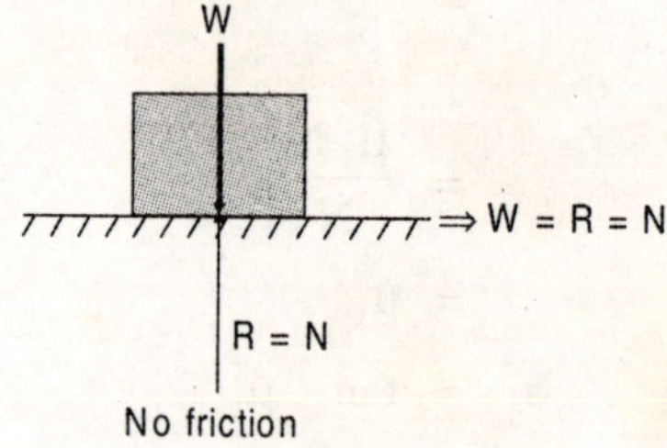

Fig. 7.12

If the board is given a small angle of inclination θ, the reaction R will deviate from the perpendicular direction to the board by the same angle θ and will keep balancing w. Here we have a normal component N of magnitude $N_{\perp} = R\cos\theta$ and a tangential component (frictional force) F of magnitude $F = R \sin\theta$.

Considering the equilibrium of the block we have

$$\Sigma F_x = 0$$

or $$F = R\sin\theta = W\sin\theta \quad \text{...(i)}$$

and $$\Sigma F_y = 0$$

or $$N = R\cos\theta = W\cos\theta \quad \text{...(ii)}$$

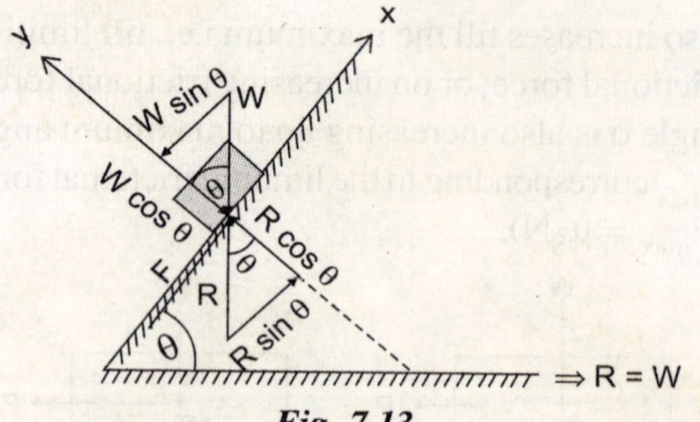

Fig. 7.13

Now, on increasing the angle of inclination at some particular stage the motion will become impending. At that time the angle between R and the normal will reach its maximum value θ_{max}. The value of the angle of inclination corresponding to the impending motion is called the angle of repose. Denoted by $\theta_r = \theta_{max}$. Now corresponding to θ_{max} frictional force F will reach its maximum value.

$$F_{max} = R\sin\theta_{max}$$

$$= W\sin\theta_{max} \quad \text{...(iii)}$$

and $$N = R\cos\theta_{max}$$

$$= W\cos\theta_{max} \quad \text{...(iv)}$$

By dividing (iii) and (iv), we get

$$\frac{F_{max}}{N} = \tan\theta_{max}$$

$$= \tan\theta_r$$

or $$F_{max} = N\tan\theta_{max}$$

As the experiment shows that the maximum value of frictional force equals to $\mu_S N$ where μ_S is the coefficient of static friction and N is the normal reaction, then

$$\mu_s N = N\tan\theta_{max}$$

$$= \tan\theta_r$$

$$\Rightarrow \tan\theta_{max} = \mu_S$$

$$= \tan\theta_r$$

As we also know that,

$$\tan(\theta_{\text{maximum angle of friction}}) = \mu_S$$

Therefore, we get a fundamental result where the angle of repose and the maximum angle of friction are equal in magnitude result shown in Fig. 7.14.

i.e., $\theta_{repose} = \theta_{\text{maximum angle of static friction}}$

or $\theta_r = \theta_{\text{maximum angle of static friction}}$ ....(7.3)

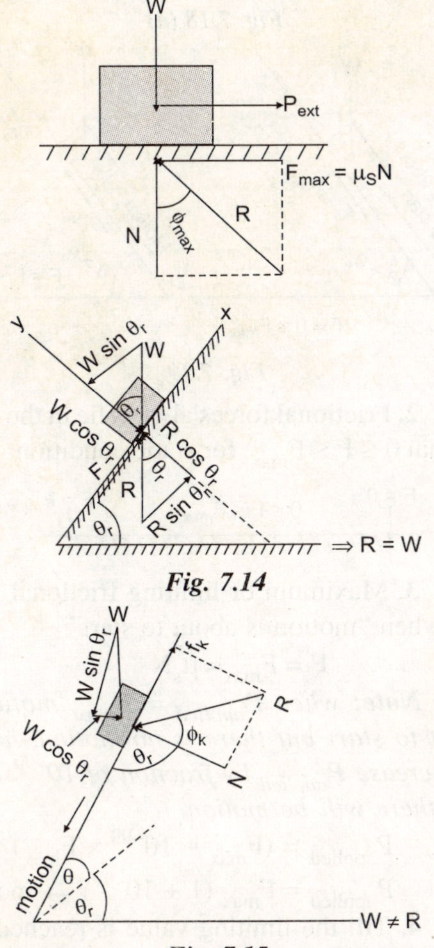

Fig. 7.14

Fig. 7.15

Now, by further increasing θ_{max} the motion will be started and F_{max} will drop to $f_k = \mu_k N$. And the angle between R and the normal drops to a lower value ϕ_k, the reaction R is not vertical any more and forces acting on the block are unbalanced.

7.4 CONE OF FRICTION

Consider a block of weight W which is placed on a horizontal surface and a horizontal applied force P which is acted in such a way that it sets the block on the verse of motion, then the resultant reaction R of normal force (N) and maximum frictional force $F_{max} = \mu_s N$ makes an maximum angle ϕ_S with the normal, this is angle of friction.

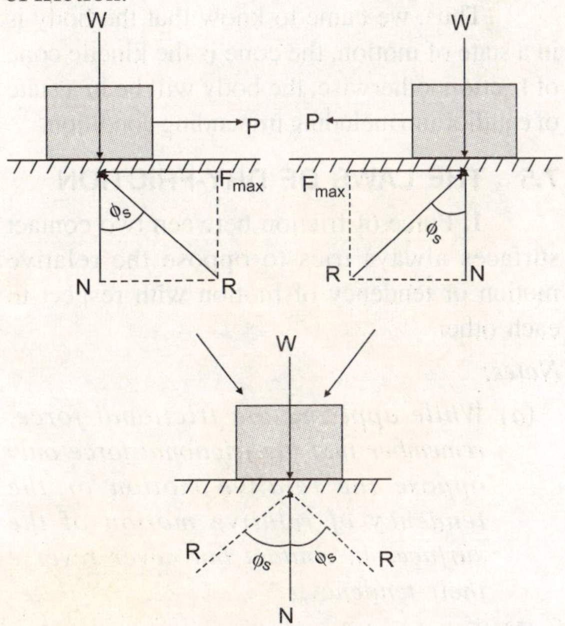

Fig. 7.16

Now if we change the direction of applied force p gradually through 360°, the resultant R generates a right circular cone of semi-central angle ϕ_S.

If the motion is impending, R must be one element of this inverted right circular cone of the vertex angle $2\phi_S$. This is known as the cone of static friction and represents the locus of R corresponding to the impending motion. If the resultant is within this cone the body will be either stationary (impending) or in motion. As we know that the kinetic angle of friction is less than that of the static angle of friction, this cone is known as cone of kinetic friction.

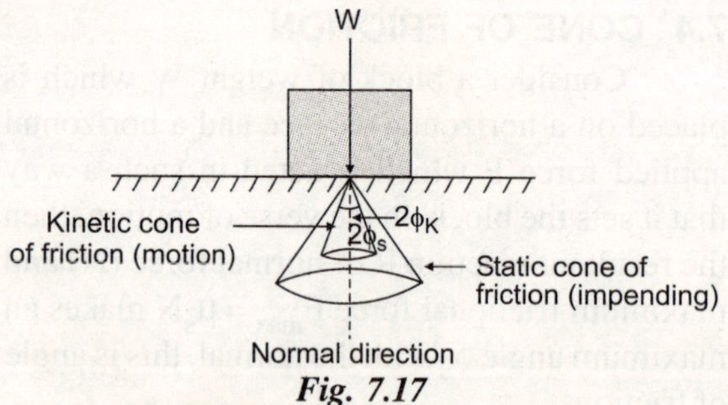

Fig. 7.17

Thus, we came to know that the body is in a state of motion, the cone is the kinetic cone of friction, otherwise, the body will be in a state of equilibrium including impending condition.

7.5 THE LAWS OF DRY-FRICTION

1. Force of friction between two contact surfaces always tries to oppose the relative motion or tendency of motion with respect to each other.

Notes:

(*a*) *While applying the frictional force, remember that the frictional force only oppose the relative motion or the tendency of relative motion of the surfaces in contact but never reverse their tendency.*

(*b*) *Frictional forces always act parallel to contact surfaces.*

(*c*) *If there is no applied force, which can move the block, or tend to move the block, or which is responsible for "motion is about to start of block", then there is no frictional force. As shown in Fig. 7.18.*

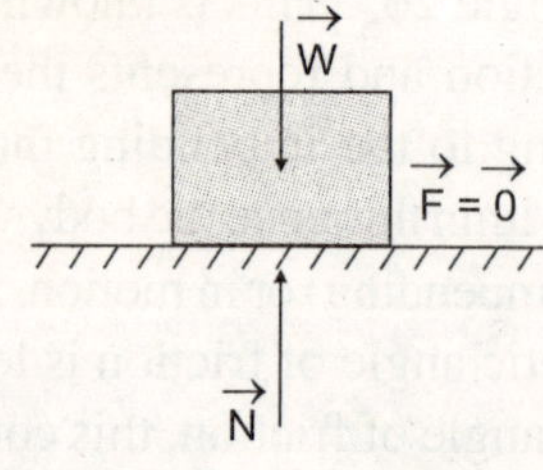

Fig. 7.18

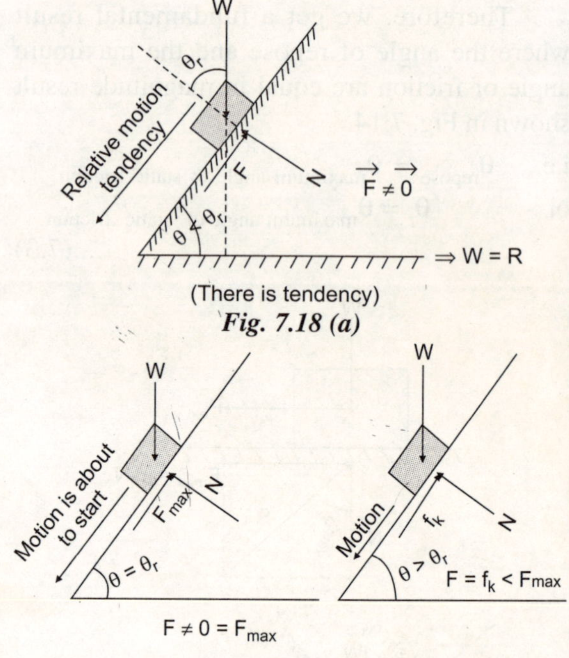

Fig. 7.18 (a)

Fig. 7.19

2. Frictional forces always lie in the close internal $0 \le F \le F_{max}$, for static condition

3. Maximum or limiting frictional force acts when "motion is about to start".

i.e., $F = F_{max} = \mu_s N$

Note: *when $P_{applied} = F_{max}$, motion is about to start but there is no motion, now if we increase $P_{applied}$ by fraction of 10^{-99} F_{max}, then there will be motion.*

i.e., $P_{applied} = (F_{max} + 10^{-99} \times F_{max})$

or $P_{applied} = F_{max}\,(1 + 10^{-99}) \rightarrow$ motion

4. Till the limiting value is reached, the magnitude of frictional force is exactly equal to the applied force which tends to move (but no motion) the body.

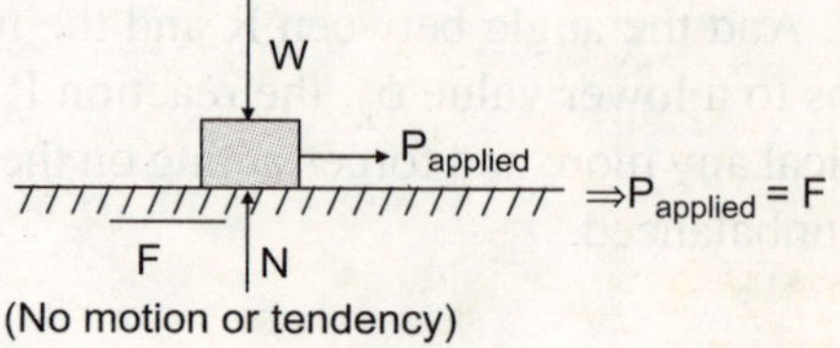

Fig. 7.20

5. Experimental evidences show that the magnitude of maximum frictional force, $F_{s,\ max}$ is proportional to the normal reaction (N) between contact surfaces.

Mathematically, $F_{s,\ max} \propto N$

or $F_{s,\ max} = \mu_s N$; where μ_s is proportionally constant, for a given pair of surfaces.

Thus, we get $F_{s,\ max} = \mu_s N$; where 's' stands for static condition.

In other words, "when motion of block is about to start frictional force $F_{s,\ max} = \mu_s N$ acts on the block."

6. μ_s is known as the coefficient of static friction. It is a dimensionless quantity because of the just ratio of two same physical quantities. μ_s depends strongly on the nature of the contact surfaces in contact. It will be less than one or it may be greater than one, normally it is less than one. If $\mu_s = 0$, the surface is said to be smooth but it is not practically possible to get perfect smooth surfaces.

$\mu_S = 0$ $0 < \mu_S < 1$ $\mu_S > 1$ 0   1   Normal case

Notes:

(*a*) *That there is nothing like "μ_s for one surface is 'k'," but it should be always between two surfaces, means we always use "between" word.*

(*b*) *For experimental process, we can use the result* $\mu_s = \left(\frac{F_{max}}{N}\right)$ *to determine the value of* μ_s *between two given surfaces.*

7. The actual microscopic area of contact is much less than that of the apparent area of contact. In particular the ratio is $1 : 10^4$. The frictional force depends strongly on the actual microscopic area of contact and is independent of the apparent area of contact.

i.e., $F \propto A_{actual}$. It is also a fact that the actual area of contact is proportional to the normal between reaction of contact surfaces so frictional force depends on actual microscopic area of contact.

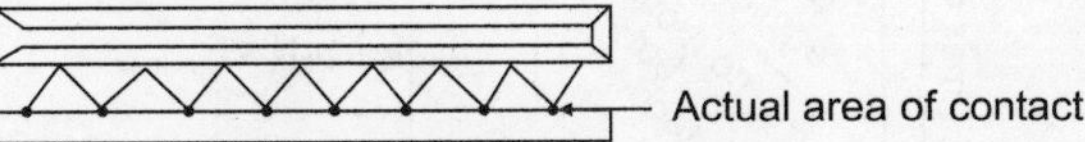

8. Frictional forces are assumed to be due to irregularities of the surfaces in contact and also to a certain extent because of the molecular attraction. It is also depends on the roughness of surfaces. If the roughness is more than μ is more and it is less when roughness changes to smoothness of the surfaces.

9. Experiments show that when we increase the applied force from maximum frictional force (limiting frictional force) the block starts moving in the direction of the applied force and then the frictional force decreases by some amount from the limiting frictional force, this frictional force is known as the kinetic frictional force (f_k) thus, we get

$$f_k = F_{s,max} - \Delta f$$

i.e., $$f_k < F_{s,max}.$$

Like static maximum frictional force, kinetic frictional force also bears a constant ratio with normal reaction, this constant also depends on the roughness of the surfaces in contact and the nature of both surfaces in contact. It is denoted by μ_k.

i.e., $$\mu_k N < \mu_s N$$

or $$\mu_k < \mu_s$$

or $$|\mu_S - \mu_k| \neq 0$$

It is found that for moderate (low) speed of sliding, the total frictional force is practically independent of speed, although the experiment shows that the force F necessary to start sliding is greater than that necessary to maintain sliding. All these things can be summarized in the figure below.

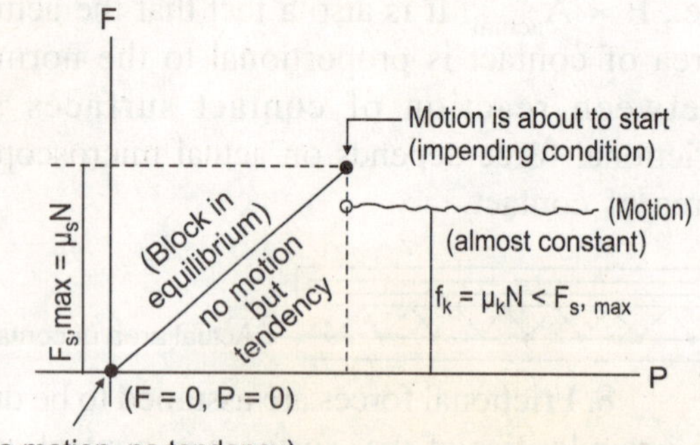

Fig. 7.21

10. $|\mu_s - \mu_K|$ = f (velocity, nature of surfaces in contact)

i.e., the change in coefficients depends on high relative velocity and nature of surfaces in contact.

11. μ_K is also dimensionless quantity and it could be greater than one or less than one. for smooth surfaces $\mu_S = \mu_k = 0$.

7.6 D'ALEMBERT'S PRINCIPLE

The French mathematician Jean le Rond D'Alembert proved in 1743 that the Newton's second law of motion is applicable not only to the motion of a particle but also to the motion of a body, and all accelerated bodies (particles) can be treated as in dynamic equilibrium in the presence of the inertia force of the body; apart from the actual real forces acting on the body (particle).

If a net resultant force $\vec{F}_{net}$ is acting on a particle of mass m and the particle has an acceleration $\vec{a}$ in the direction of the net force then from Newton's second law,

$$\vec{F}_{net} = m\vec{a}$$

or $\quad \vec{F}_{net} - m\vec{a} = 0$

or $\quad \vec{F}_{net} + (-m\vec{a}) = m \times \vec{0}$

or $\quad \vec{F}_{net} + \vec{F}_{inertia} = 0$ i.e., equilibrium condition. ....(7.4)

The term '$-m\vec{a}$' may be considered as a force of magnitude ma, applied in the opposite direction of acceleration and is termed as the inertia force or reverse effective force. To understand clearly see the diagram in Fig. (7.22).

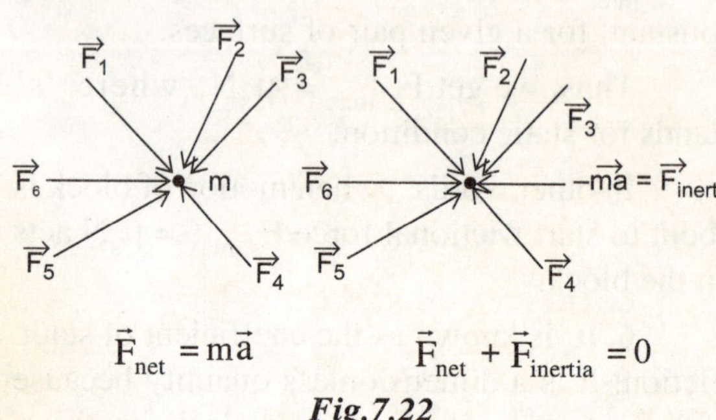

Fig.7.22

The inertia force '$-m\vec{a}$' has a physical meaning. According to Newton's first law of motion, a body continues to be in the state of rest or of uniform motion in a straight line unless acted upon by an external force. This means that every body has a tendency to continue in its state of rest or of uniform motion. This tendency is called inertia. Hence inertia force seems to be the resistance offered by a body to the change in its state of rest or of uniform motion.

Actually inertia force is a technique by which we are in a position to handle accelerated motion body problems by $\Sigma\vec{F} = 0$ instead of $\Sigma\vec{F} = m\vec{a}$. It should be a technique only because a body cannot apply force on itself. Note that there is a difference between inertia force and pseudo force, pseudo force is the non-real force having magnitude $m\vec{a}_0$ where 'm' is the mass of the particle and $\vec{a}_0$, is the acceleration of non-inertial frame of reference in inertial frame of reference. The inertial force $m\vec{a}_0$ is the product of the mass of particle and acceleration of the particle. We can extend this concept to a group of discrete particles or continuous body by taking the product of acceleration of the centre of mass ($\vec{a}_{cm}$) and M the total mass of bodies (group of bodies).

i.e., $\vec{F}_{inertia} = M\vec{a}_{cm}$,

where $M = \Sigma m$;

Example 7.2: *The coefficients of friction between blocks A and C and the horizontal surface are μ_s = 0.24 and μ_k = 0.20, knowing that mass of A = 5 kg, mass of B = 10 kg and mass of C = 10 kg. Determine (a) the tension in the strings (b) the acceleration of each block (use D'Alembert principle).*

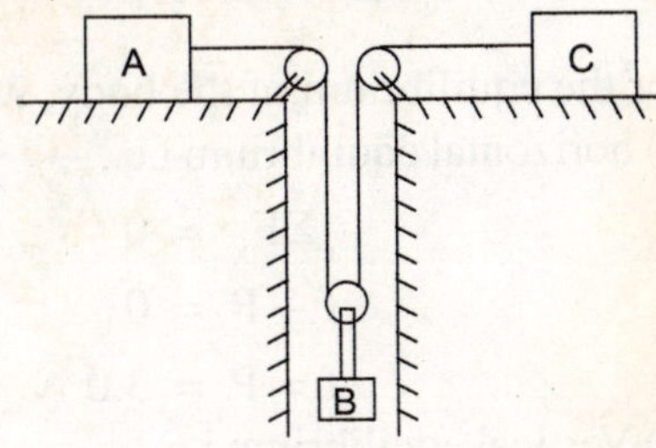

Fig. 7.23

Solution: Since the strings are unstretchable, therefore, the acceleration of each of the blocks will be the same, let it be $\vec{a}$. If the tensions in the strings are T_1 and T_2. Forces acting on blocks and free body diagrams are shown below.

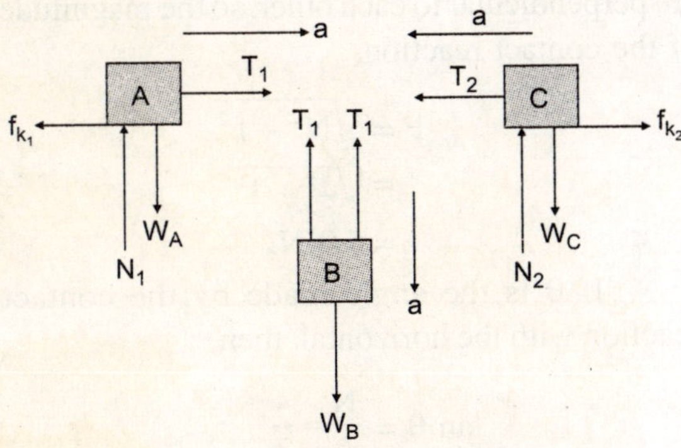

Fig. 7.24

Applying D'Alembert principle we get the following free body diagrams.

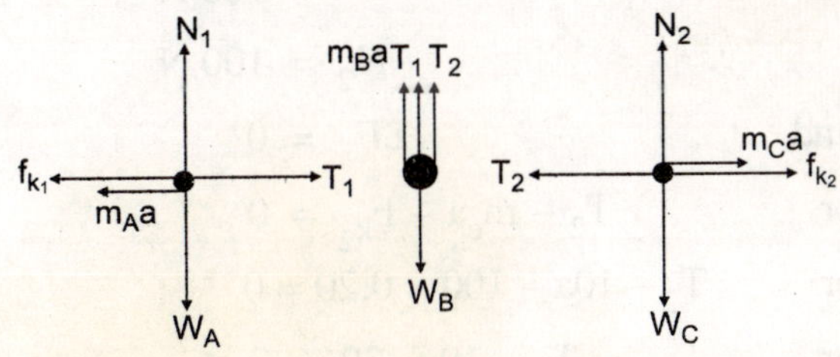

Fig. 7.25

For block A

Block-A is in equilibrium along the y-axis (vertical axis).

i.e., $\Sigma F_y = 0$

$N_1 - W_A = 0$

$N_1 = W_A$

$= 5 \times 10$

$g = 10 \text{ ms}^{-2}$

$\therefore$ $N_1 = 50$ N

Now, after applying the D'Alembert principle, block A is in equilibrium along the horizontal direction

i.e., $\Sigma F_H = 0$

or $\Sigma F_x = 0$

or $T_1 - m_A a - F_{k_1} = 0$

or $T_1 - 5a - N_1\mu_k = 0$

or $T_1 - 5a - 50 \times 0.20 = 0$

or $T_1 - 5a - 10 = 0$

$\therefore$ $T_1 - 5a - 10 = 0$ ....(i)

For block B

Similarly, we have $\Sigma F_y = 0$; as block B is in equilibrium under D'Alembert consideration.

or $T_1 + T_2 + m_B a - W_B = 0$

or $T_1 + T_2 + 10a - 100 = 0$; $g = 10 \text{ ms}^{-2}$

$\therefore T_1 + T_2 + 10a - 100 = 0$ ....(ii)

For block C

We have got $\Sigma F_y = 0$; as there is no motion along the vertical direction.

or $N_2 - W_c = 0$

or $N_2 = 10 \times 10$

$= 100\text{ N}$

$\therefore \quad N_2 = 100\text{ N}$

and $\Sigma F_x = 0$

or $T_2 - m_c a - F_{k_2} = 0$

or $T_2 - 10a - 100 \times 0.20 = 0$

or $T_2 - 10a - 20 = 0$

$\therefore \quad T_2 - 10a - 20 = 0 \quad \text{....(iii)}$

By adding equations (i) and (iii), we get

$$T_1 + T_2 - 15a - 30 = 0$$

From (ii)

$$T_1 + T_2 = 100 - 10a$$

$\therefore \quad 100 - 10a - 15a - 30 = 0$

or $70 = 25a$

$\therefore \quad a = 2.8\text{ ms}^{-2}$

From (i) and (iii), we get

$$T_1 = 10 + 5a = 10 + 5 \times 2.8 = 24\text{ N}$$

and

$$T_2 = 10a + 20 = 10 \times 2.8 + 20 = 48\text{ N.}$$

WORKED OUT EXAMPLES

1. *A body of weight 4.0 N has an impending motion towards the left caused by applied force P as shown in Fig. (7-W1) on a rough horizontal surface. If the frictional force is 3.0 N, find (a) the angle made by the contact reaction on the body with the horizontal (b) the magnitude of the contact reaction.*

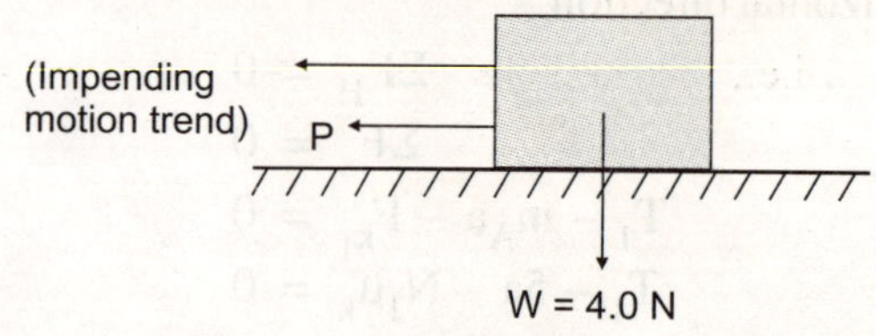

Fig. 7-W1

Solution:

Free body diagram of the body is shown in Fig. (7-W2)

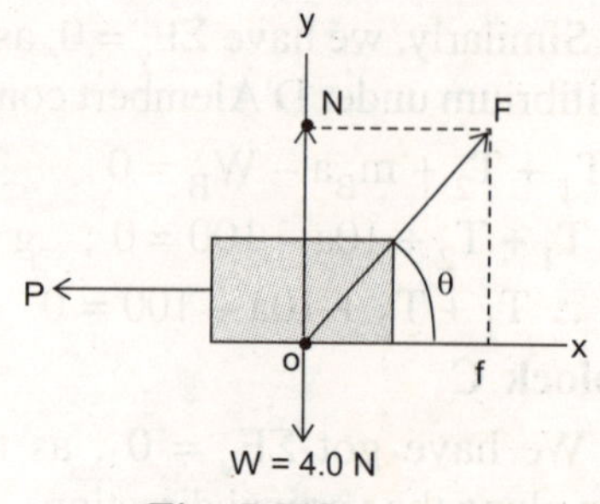

Fig. 7-W2

For the equilibrium of the body, we have

(a) horizontal equilibrium i.e.,

$$\Sigma F_x = 0$$

or $f - P = 0$

or $f = P = 3.0\text{ N.}$

(b) Vertical equilibrium i.e.,

$$\Sigma F_y = 0$$

or $N - W = 0$

or $N = W = 4.0\text{ N.}$

Now, the contact reaction on the body by contact surface is defined by the vector sum of normal reaction and frictional force. Since they are perpendicular to each other, so the magnitude of the contact reaction,

$$F = \sqrt{N^2 + f^2} = \sqrt{4^2 + 3^2} = 5.0\text{ N.}$$

If θ is the angle made by the contact reaction with the horizontal, then,

$$\tan\theta = \frac{N}{f} = \frac{4}{3}$$

$\therefore \quad \theta = 53°$

2. *The coefficient of static friction between a block of mass M kg and the table shown in Fig. (7-W3) is $\mu_s = 0.2$. What should be the maximum value of m so that the block m cannot move? Take $g = 10\ ms^{-2}$. The string and the pulley are light and smooth. Also find the total force on the pulley.*

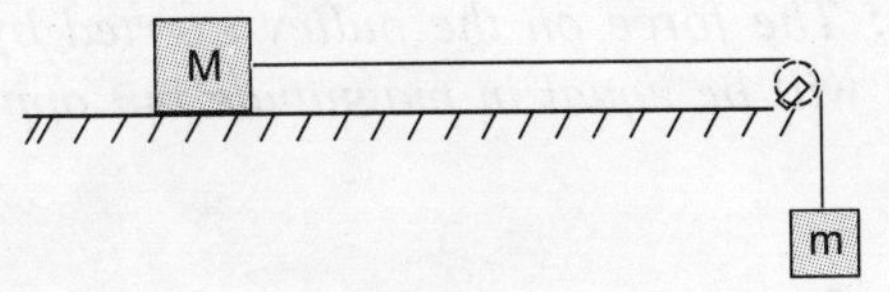

Fig. 7-W3

Analytical method

Solution:

Free body diagram shown in Fig. (7-W4).

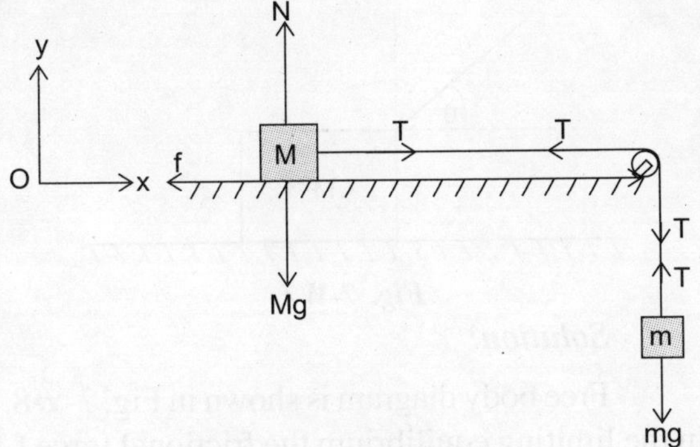

Fig. 7-W4

Consider the equilibrium of mass m. The forces on this block are:

(a) mg downward applied by the Earth.

(b) T, upward by the string.

Hence $\Sigma F_y = 0$

or $T - mg = 0$

or $T = mg$ ...(i)

Now consider the equilibrium of the mass M. The forces on this block are:

(a) Tension, T, towards the right by the string,

(b) Frictional force (f) by table towards the left,

(c) Mg weight downward applied by the Earth, and

(d) N, normal reaction upward by the table.

For vertical equilibrium of this block,

$$\Sigma F_y = 0$$

or $N - Mg = 0$

or $N = Mg$ ...(ii)

According to the problem friction is limiting

Thus, $f = \mu_s N$

$= \mu_s Mg$ ...(iii)

$[\because N = Mg : \text{equation (ii)}]$

For horizontal equilibrium of the block M,

$$\Sigma F_x = 0$$

or $T - f = 0$

or $T = f$

or $T = \mu_s Mg$

[from equations (iii) and (iv)]

Using equations (i) and (iv), equate tension, T.

i.e., $mg = \mu_s Mg$

$\therefore$ $m = \mu_s M$ kg

Now the net force on the pulley is,

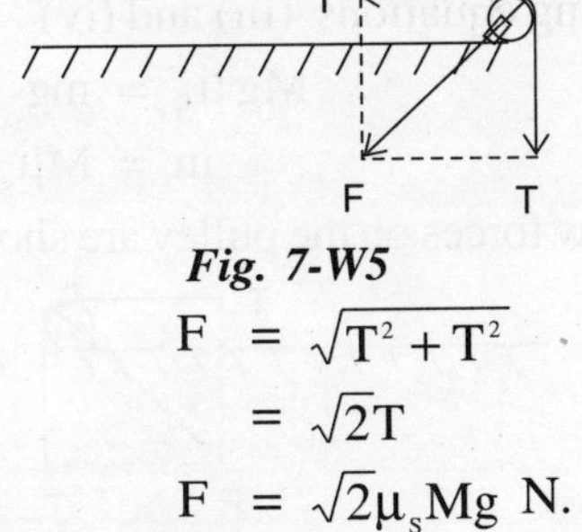

Fig. 7-W5

$$F = \sqrt{T^2 + T^2}$$

$$= \sqrt{2}T$$

$$F = \sqrt{2}\mu_s Mg \text{ N.}$$

Vector method

Consider the equilibrium of block of mass M.

i.e., $$\sum \vec{F} = 0$$

or $$\vec{T} + \vec{f} + \vec{N} + M\vec{g} = 0$$

or $$T\vec{i} + f(-\vec{i}) + N\vec{J} + Mg(-\vec{j}) = 0$$

or $$T\vec{i} - f\vec{i} + +N\vec{J} - Mg\vec{j} = 0$$

or $$(T - f)\vec{i} + (N - Mg)\vec{j} = 0$$

$$\Rightarrow \quad T - f = 0$$

or $$T = f \quad ...(i)$$

and $$N - Mg = 0$$

or $$N = Mg \quad ...(ii)$$

Now consider the equilibrium of block of a mass m.

i.e., $$\Sigma \vec{F} = 0$$

or $$\vec{T} + m\vec{g} = 0$$

or $$T\vec{j} + mg(\vec{j}) = 0$$

or $$(T - mg)\vec{j} = 0$$

$$\Rightarrow \quad T - mg = 0$$

or $$T = mg \quad ...(iii)$$

Since friction is limiting therefore

$$f = N\mu_S$$

$$= Mg\mu_S \quad ....(iv)$$

Using equations (iii) and (iv)

$$Mg\mu_S = mg$$

or $$m = M\mu_s \text{ kg}$$

Now forces on the pulley are shown below

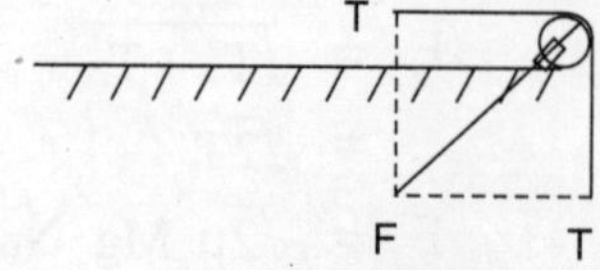

Fig. 7-W6

then $$\vec{F} = \text{net force on the pulley}$$

$$= T(-\vec{i}) + T(-\vec{j})$$

$$= T\vec{i} - T\vec{J}$$

$$\therefore \quad |\vec{F}| = \sqrt{(-T)^2 + (-T)^2}$$

$$= \sqrt{2}\,T$$

$$= \sqrt{2}\,mg$$

or $$F = \sqrt{2}\,M\mu_s g \text{ N.}$$

***Note:** The force on the pulley exerted by the table will be equal in magnitude but opposite to $\vec{F}$.*

3. *A block placed on a horizontal surface is being pushed by a force F making an angle θ with the horizontal. If the coefficient of friction is μ, how much force is needed to get the block just started. Discuss the condition for minimum force (F).*

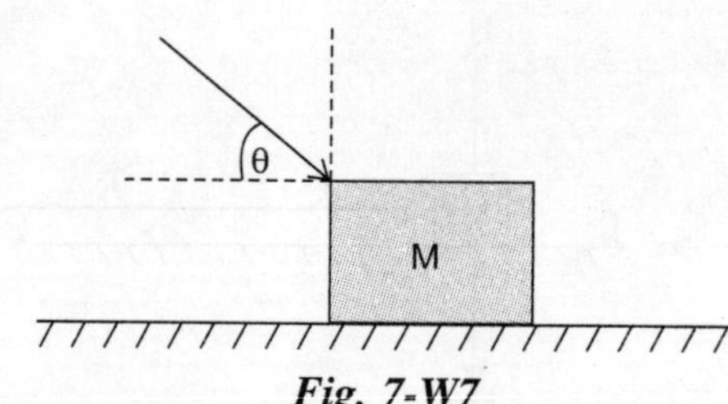

Fig. 7-W7

Solution:

Free body diagram is shown in Fig. 7-W8. In the limiting equilibrium the frictional force f_s will be equal to μN.

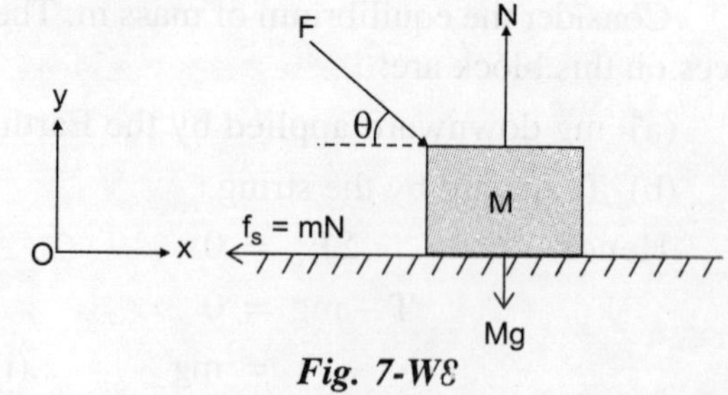

Fig. 7-W8

For horizontal equilibrium,

$$\Sigma F_x = 0$$

or $$F\cos\theta - \mu N = 0$$

or $$F\cos\theta = \mu N \quad ...(i)$$

For vertical equilibrium,

$$\Sigma F_y = 0$$

or $$Mg + F \sin\theta - N = 0$$

or $$Mg + F \sin\theta = N \quad ...(ii)$$

Eliminating N using these two equations, we get

$$\frac{F\cos\theta}{\mu} = Mg + F\sin\theta$$

or $$F\cos\theta = \mu Mg + \mu F\sin\theta$$

or $$F(\cos\theta - \mu\sin\theta) = \mu Mg$$

or $$F = \frac{\mu Mg}{\cos\theta - \mu\sin\theta}$$

For F_{min}, assumed $\cos\theta - \mu\sin\theta = f(\theta)$ should be maximum,

Now, $$F'(\theta) = -\sin\theta - \mu\cos\theta$$

or $$F''(\theta) = -\cos\theta + \mu\sin\theta$$

$$= \mu\sin\theta - \cos\theta$$

For $[f(\theta)]_{max}$, $F'(\theta) = 0$

or $$\sin\theta + \mu\cos\theta = 0$$

or $$\sin\theta = -\mu\cos\theta$$

or $$\tan\theta = -\mu$$

$$\Rightarrow \quad 90° < \theta < 180°$$

Now, $$f''(\theta) = \mu\sin\theta - \cos\theta$$

$$= -\tan\theta\sin\theta - \cos\theta$$

$$= \frac{-\sin^2\theta - \cos^2\theta}{\cos\theta}$$

$$= \frac{-1}{\cos\theta} > 0;$$

as $\cos\theta < 0$ in interval $90° < \theta < 180°$.

$\Rightarrow f(\theta)$ is maximum at $\tan\theta = -\mu$

Therefore, $$F_{min} = \frac{\mu Mg}{\cos\theta + \tan\theta\sin\theta}$$

$$= \mu Mg\cos\theta$$

$$= \mu Mg\frac{1}{\sqrt{\mu^2+1}}$$

$$= Mg\sqrt{\frac{\mu^2}{\mu^2+1}}$$

Vector method:

Since block in equilibrium

i.e., $$\Sigma\vec{F} = 0$$

or $$\vec{F} + \vec{N} + \vec{f_s} + M\vec{g} = 0$$

or $$(F\cos\theta)\,\vec{i} + N(\vec{J}) + f_s(-\vec{i})$$

$$+ Mg(-\vec{j}) - (F\sin\theta)\,\vec{j} = 0$$

or $$F\cos\theta\,\vec{i} - f_s\,\vec{i} + N\,\vec{j} - Mg\,\vec{j} - (F\sin\theta)\,\vec{j} = 0$$

or $$(F\cos\theta - f_s)\,\vec{i} + (N - Mg - F\sin\theta)\,\vec{j} = 0$$

$$\Rightarrow \quad F\cos\theta - F = 0 \quad ....(i)$$

$$N - Mg - F\sin\theta = 0 \quad ....(ii)$$

and according to question $f_s = N\mu$ ....(iii)

Using (i) and (iii), we get

$$F\cos\theta - N\mu = 0$$

or $$N = \frac{F\cos\theta}{\mu}$$

Putting N in equation (iii)

$$\frac{F\cos\theta}{\mu} - Mg - F\sin\theta = 0$$

or $$F\cos\theta - F\mu\sin\theta = Mg\mu$$

or $$F = \frac{\mu Mg}{\cos\theta - \mu\sin\theta}.$$

Further proceed in similar manner as done in analytical method.

4. *Figure 7-W9 shows two blocks connected by a light string placed on the two inclined sides of a triangular structure. The coefficient of static friction is 0.30 at each of the surface. Find the ratio $\frac{m}{M}$ so that the system remains at rest.*

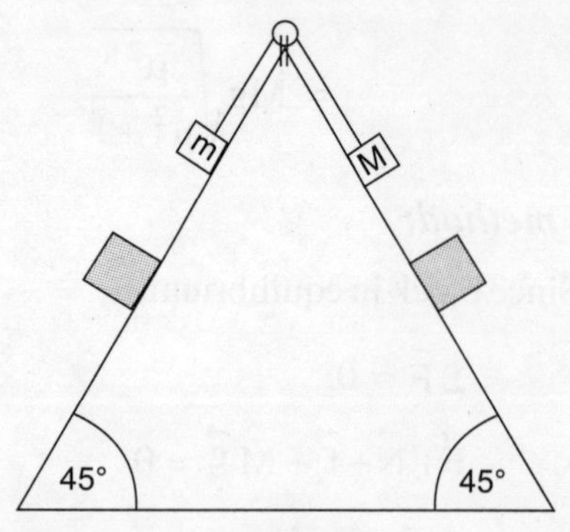

Fig. 7-W9

Solution:

Let us suppose that block of mass M has tendency to slip down (reverse can also be assumed). Free body diagrams of mass m and M are shown in Fig. [(7-W10 (a)] and in Fig. [(7-W10(b)] respectively.

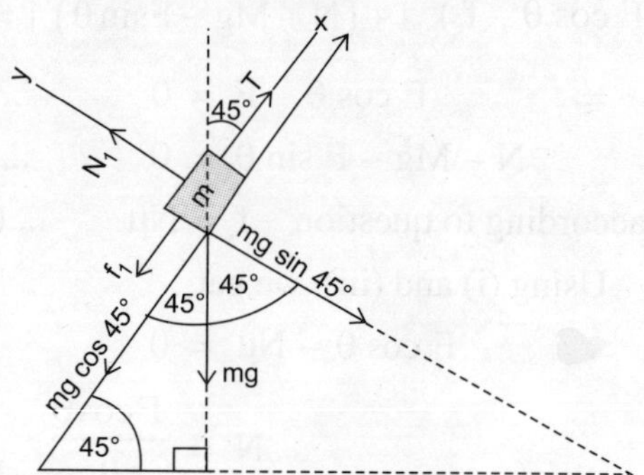

Fig. 7-W10 (a)

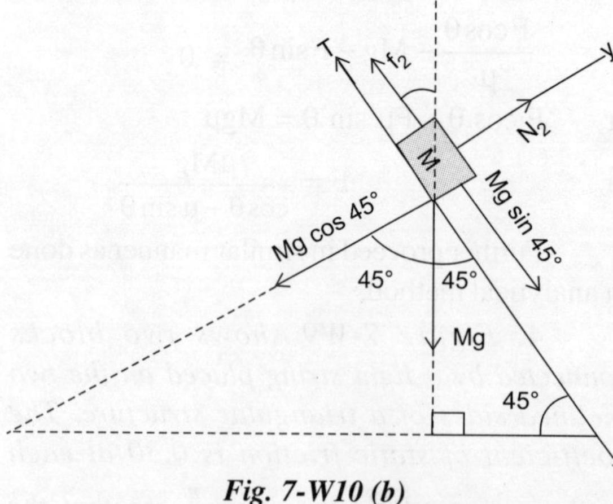

Fig. 7-W10 (b)

As the block(m) is in equilibrium, the resultant force parallel and perpendicular to the incline surface should be zero. Taking component along perpendicular to inclined plane, we get

$$\Sigma F_{\perp} = 0$$

or $$N_1 - mg\cos 45° = 0$$

or $$N_1 = mg\cos 45° = \frac{mg}{\sqrt{2}} \qquad \text{...(i)}$$

Now taking component parallel to the inclined plane we get,

$$\Sigma F_{||ar} = 0$$

or $$T - f_1 - mg\cos 45° = 0$$

or $$T = f_1 + mg\cos 45° \qquad \text{...(ii)}$$

As it is the case of limiting equilibrium, so,

$$f_1 = N_1\mu_s$$

or $$f_1 = \frac{mg}{\sqrt{2}}\times 0.3$$

$$= \left(\frac{0.3}{\sqrt{2}}\right) mg \qquad \text{....(iii)}$$

Using equations (ii) and (iii)

$$T = \frac{0.3}{\sqrt{2}} mg + mg\cos 45°$$

$$= \frac{0.3}{\sqrt{2}} mg + \frac{1}{\sqrt{2}} mg$$

or $$T = \frac{1.3}{\sqrt{2}} mg \qquad \text{...(iv)}$$

As the block (M) is in equilibrium, the resultant force should be zero in parallel direction to incline plane as well as perpendicular to incline plane. Taking component along perpendicular to inclined plane, we get

$$\Sigma F_{\perp} = 0$$

or $$N_2 - Mg\cos 45° = 0$$

or $$N_2 = \frac{Mg}{\sqrt{2}} \qquad \text{...(v)}$$

Now taking component parallel to inclined plane, we get

i.e., $\Sigma F_{|||ar} = 0$

or $T + f_2 - Mg \sin 45° = 0$

or $T = Mg \sin 45° - f_2$

Now as it is the case of limiting friction then,

$$f_2 = \mu N_2$$
$$= 0.3\ Mg \cos 45°$$

$\therefore$ $T = Mg \sin 45° - 0.3\ Mg \cos 45°$
$= (1 - 0.3)\ Mg \cos 45°$

or $$T = \frac{0.7}{\sqrt{2}} Mg \quad ...(vi)$$

We could compare T from equations (iv) and (vi), because we are using same string so tension should be same in both parts.

i.e., $$\frac{1.3}{\sqrt{2}} mg = \frac{0.7}{\sqrt{2}} Mg$$

or $$\frac{M}{m} = \frac{1.3}{0.7} = \frac{13}{7}$$

$\therefore$ $$\frac{M}{m} = \frac{13}{7}$$

5. *Find the maximum value of θ for the situation shown in Fig. (7-W11) so that the system remains at rest. The coefficient of friction at both the contacts is 0.3. Take $\frac{M}{m} = 2$.*

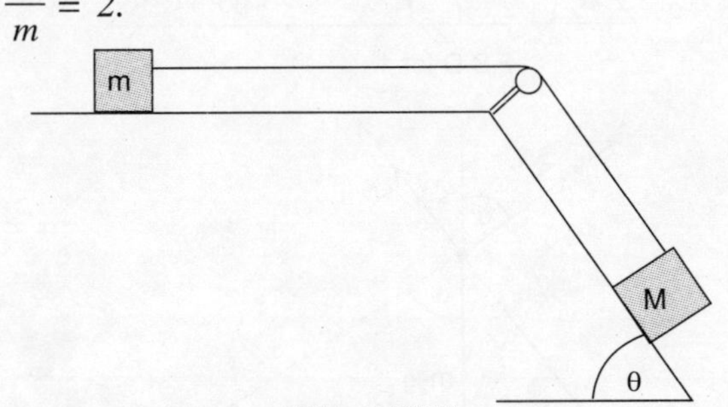

Fig. 7-W11

Solution:

Fig. (7-W12) shows the forces acting on the two blocks. As we are looking for maximum value of θ, therefore, the equilibrium is limiting.

Hence, the frictional forces are equal to 0.3 times the corresponding normal reactions.

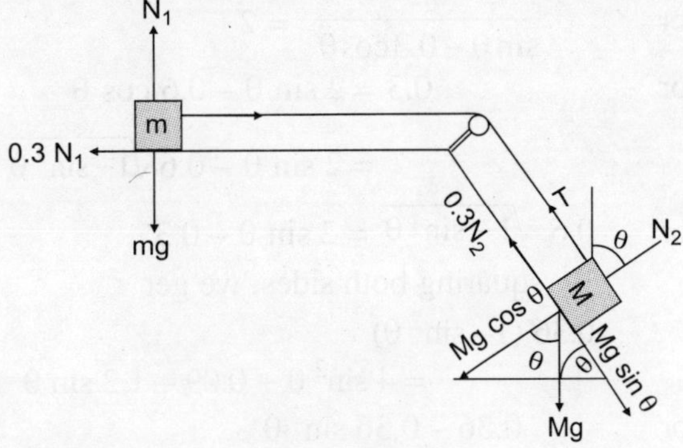

Fig. 7-W12

Equilibrium of the block m gives,

$\Sigma F_{|||ar} = 0$

or $T - 0.3\ N_1 = 0$

or $T = 0.3\ N_1 \quad ....(i)$

and $\Sigma F_{\perp} = 0$

or $N_1 - mg = 0$

or $N_1 = mg \quad ...(ii)$

Now equations (i) and (ii) give

$$T = 0.3\ mg \quad ...(iii)$$

Now, consider the equilibrium of block M. Taking the components parallel to the inclined plane, we get

$\Sigma F_{|||ar} = 0$

or $T + 0.3N_2 - Mg \sin \theta = 0$

or $T + 0.3N_2 = Mg \sin \theta \quad ...(iv)$

Taking the component perpendicular to the inclined plane, we get

$\Sigma F_{\perp} = 0$

or $N_2 - Mg \cos \theta = 0$

or $N_2 = Mg \cos \theta \quad ...(v)$

From equations (iv) and (v)

$$T = Mg (\sin \theta - 0.3 \cos \theta) \quad ...(vi)$$

Comparing (iii) and (vi), we get

$$0.3\ mg = Mg (\sin \theta - 0.3 \cos \theta)$$

$$\text{or} \quad \frac{0.3}{\sin\theta - 0.3\cos\theta} = \frac{M}{m}$$

$$\text{or} \quad \frac{0.3}{\sin\theta - 0.3\cos\theta} = 2$$

$$\text{or} \quad 0.3 = 2\sin\theta - 0.6\cos\theta$$

$$= 2\sin\theta - 0.6\sqrt{1-\sin^2\theta}$$

$$\text{or} \quad 0.6\sqrt{1-\sin^2\theta} = 2\sin\theta - 0.3$$

By squaring both sides, we get

$$0.36(1 - \sin^2\theta) = 4\sin^2\theta + 0.09 - 1.2\sin\theta$$

$$\text{or} \quad 0.36 - 0.36\sin^2\theta = 4\sin^2\theta + 0.09 - 1.2\sin\theta$$

$$\text{or} \quad 4.36\sin^2\theta - 1.2\sin\theta - 0.27 = 0$$

$$\text{or} \quad \sin\theta = \frac{1.2 \pm \sqrt{1.2^2 + 4\times 4.36\times 0.27}}{2\times 4.36}$$

$$= \frac{1.2 \pm 2.48}{2\times 4.36}$$

$$= \frac{1.2 + 2.48}{2\times 4.36} \qquad [\because \sin\theta \nless 0]$$

$$= 0.42 \text{ (approx)}$$

$$\therefore \quad \theta = 25°.$$

6. *Two touching bars 1 and 2 are placed on an inclined plane forming an angle α with the horizontal Fig. (7-W13). The masses of the bars are equal to m_1 and m_2, and the coefficients of friction between the inclined plane and these bars are equal to K_1 and K_2 respectively, assume $K_1 > K_2$. Find the minimum value of the angle α at which the bar starts sliding down.*

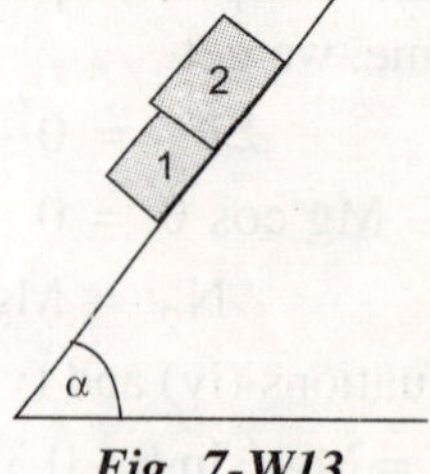

Fig. 7-W13

Solution:

Figures show the forces on these two bars and free body diagrams. (To draw the free body diagrams we have to treat bars as point). Given m_1 and m_2 are the masses of these bars 1 and 2 respectively.

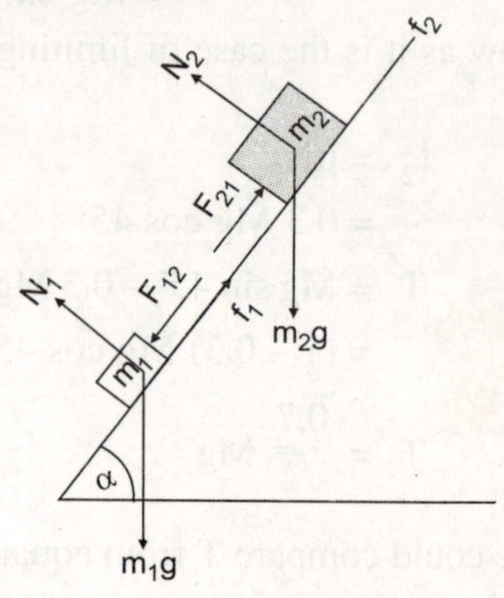

Fig. 7-W14

Here F_{12} and F_{21} are the two contact reactions following Newton's third law and others have their usual meanings.

According to the question, friction is limiting thus, $f_1 = K_1 N_1$ and $f_2 = K_2 N_2$.

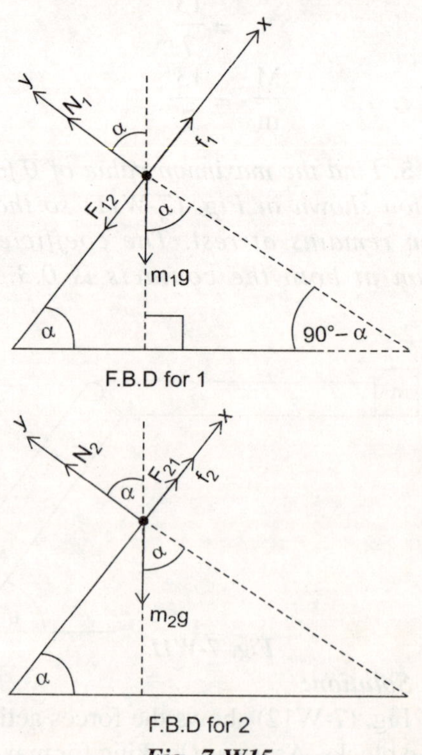

Fig. 7-W15

Now, consider the equilibrium of bar1. The forces on bar 1 are:

(a) Weight, mg,

(b) Contact reaction F_{12} applied by bar 2,

(c) Frictional force, f_1 by inclined surface, and

(d) Normal reaction, N_1 by inclined surface.

Since the bar –1 is in equilibrium (impending motion) then the net force perpendicular to the inclined plane should be zero.

i.e., $\Sigma F_{net, \perp} = 0$

or $N_1 - m_1 g \cos\alpha = 0$

or $N_1 = m_1 g \cos\alpha$ ...(i)

Using (i) and $f = N_1 K_1$, we get

$$f_1 = K_1 m_1 g \cos\alpha \quad ...(ii)$$

And the net force parallel to inclined plane should be zero.

i.e., $\Sigma F_{net, ||ar} = 0$

or $f_1 - F_{12} - m_1 g \sin\alpha = 0$

or $F_{12} = (K_1 \cos\alpha - \sin\alpha)\, m_1 g$ ...(iii)

[Putting $f_1 = K_1 m_1 g \cos\alpha$ from (ii)]

Similarly, consider the equilibrium of bar –2, forces on bar 2 are:

(a) Weight, $m_2 g$,

(b) Contact reaction, f_{21} applied by bar 1,

(c) Frictional force, f_2 by inclined surface, and

(d) Normal reaction, N_2 by inclined surface.

Since the bar 2 is in equilibrium then the net force perpendicular to the inclined plane should be zero.

i.e., $\Sigma F_{net,\perp} = 0$

or $N_2 - m_2 g \cos\alpha = 0$

or $N_2 = m_2 g \cos\alpha$ ...(iv)

By using equations $f_2 = K_2 N_2$ and (iv), we get

$$f_2 = K_2 m_2 g \cos\alpha \quad ...(v)$$

And the net force parallel to the inclined plane should be zero.

i.e., $\Sigma F_{net, ||ar} = 0$

or $F_{21} + f_2 - m_2 g \sin\alpha = 0$

or $F_{21} = m_2 g \sin\alpha - K_2 m_2 g \cos\alpha$

$= (\sin\alpha - K_2 \cos\alpha)\, m_2 g$ ...(vi)

Now by comparing equations (iii) and (vi), we get

$$(K_1 \cos\alpha - \sin\alpha)\, m_1 g = (\sin\alpha - K_2 \cos\alpha)\, m_2 g$$

or $$\frac{m_1}{m_2} = \frac{\sin\alpha - K_2\cos\alpha}{K_1\cos\alpha - \sin\alpha}$$

or $$\frac{m_1 + m_2}{m_1 - m_2} = \frac{\sin\alpha - K_2\cos\alpha + K_1\cos\alpha - \sin\alpha}{\sin\alpha - K_2\cos\alpha - K_1\cos\alpha + \sin\alpha}$$

or $$\frac{m_1 + m_2}{m_1 - m_2} = \frac{(K_1 - K_2)\cos\alpha}{2\sin\alpha - (K_1 + K_2)\cos\alpha}$$

or $$\frac{m_1 + m_2}{m_1 - m_2} = \frac{1}{\dfrac{2\tan\alpha}{K_1 - K_2} - \dfrac{K_1 + K_2}{K_1 - K_2}}$$

or $$\frac{2}{K_1 - K_2}\tan\alpha - \frac{K_1 + K_2}{K_1 - K_2} = \frac{m_1 - m_2}{m_1 + m_2}$$

or $$\frac{2}{K_1 - K_2}\tan\alpha = \frac{m_1 - m_2}{m_1 + m_2} + \frac{K_1 + K_2}{K_1 - K_2}$$

or $$\tan\alpha = \left(\frac{m_1 - m_2}{m_1 + m_2} + \frac{K_1 + K_2}{K_1 - K_2}\right)\frac{K_1 - K_2}{2}$$

or $$\alpha_{min} = \tan^{-1}\left[\left(\frac{m_1 - m_2}{m_1 + m_2} + \frac{K_1 + K_2}{K_1 - K_2}\right)\frac{K_1 - K_2}{2}\right]$$

$$= \tan^{-1}\left[\frac{(m_1K_1 - m_2K_1 - m_1K_2 + m_2K_2 + m_1K_1 + m_1K_2 + m_2K_1 + m_2K_2)}{2(m_1 + m_2)}\right]$$

$$= \tan^{-1}\left[\frac{2(m_1K_1 + m_2K_2)}{2(m_1 + m_2)}\right]$$

$$\therefore \quad \alpha_{min} = \tan^{-1}\left[\frac{m_1K_1 + m_2K_2}{m_1 + m_2}\right].$$

7. *The following parameters of the arrangement of Fig. (7-W16) are available: The angle α which the inclined plane makes with the horizontal, and the coefficient of friction K between the body m_1 and the inclined plane. The mass of the pulley is negligible. Assuming both bodies to be motionless at initial moment, find the mass ratio $\frac{m_2}{m_1}$ at which m_2 will be at rest.*

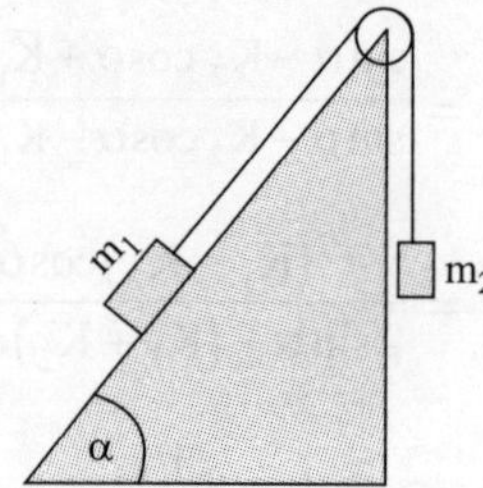

Fig. 7-W16

Solution:

Forces on masses m_1 and m_2 are shown in the Fig. (7-W17).

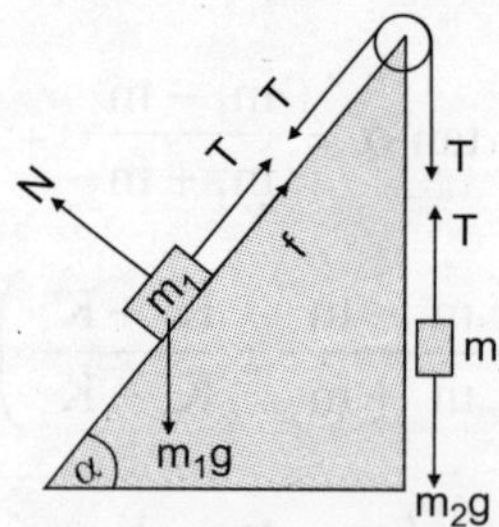

Fig. 7-W17

Consider block (m_1) : Forces on the block m_1 are:

(a) Tension in string (T) towards inclined plane as shown,

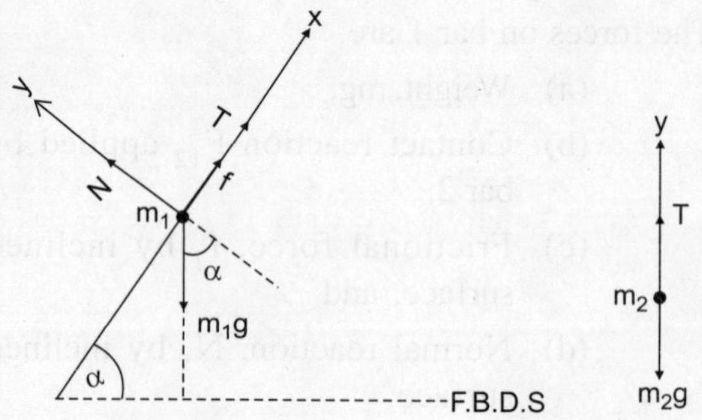

Fig. 7-W18a *Fig. 7-W18b*

(b) Weight, m_1g, vertically downward,

(c) Frictional force (f) opposite to the impending motion trend, and

(d) normal reaction perpendicular to the plane.

Since the block (m_1) is in equilibrium, then the net force perpendicular to the plane should be zero.

i.e., $\Sigma F_{net, \perp} = 0$

or $N - m_1g \cos \alpha = 0$ ...(i)

and that of parallel to plane will be zero.

i.e., $\Sigma F_{net, |||ar} = 0$

or $T + f - m_1g \sin \alpha = 0$

or $T + f = m_1g \sin \alpha$ ...(ii)

Since the block (m_1) is at rest then

$f \le NK = m_1g K \cos \alpha$ ...(iii)

[Putting $N = m_1g \cos \alpha$ from equation (i).]

Now the forces acting on the block (m_2) are:

(a) Tension (T) vertically upward, and

(b) Weight (m_2g) vertically downward.

Since the block (m_2) is in equilibrium then the net force on the block should be zero

i.e., $\Sigma F_{net, y} = 0$

or $T - m_2g = 0$

or $T = m_2g$ ...(iv)

By using equations (ii) and (iii), we get

$$T \le m_1 g \sin \alpha - m_1 g K \cos \alpha$$

$$\le m_1 g (\sin \alpha - K \cos \alpha) \quad ...(v)$$

Now by comparing equation (iv) and (v), we get

$$m_2 g \le m_1 g (\sin \alpha - K \cos \alpha)$$

or $$\frac{m_2}{m_1} \le \sin \alpha - K \cos \alpha \quad ...(vi)$$

Similarly, if we consider the opposite tendency then,

$$T - f - m_1 g \sin \alpha = 0 \quad ...(vii)$$

Since block (m_1) is in a state of rest then,

$$f \le N K \le K m_1 g \cos \alpha$$

From (vii), $T = f + m_1 g \sin \alpha$

or $$T \ge K m_1 g \cos \alpha + m_1 g \sin \alpha$$

or $$m_2 g \ge (K \cos \alpha + \sin \alpha) m_1 g$$

or $$\frac{m_2}{m_1} \ge (K \cos \alpha + \sin \alpha) \quad ...(viii)$$

From (vii) and (viii), we get

$$(K \cos \alpha + \sin \alpha) \le \frac{m_2}{m_1} \le (\sin \alpha - K \cos \alpha).$$

8. *A block (m) is at rest on an inclined plane making an angle θ with the horizontal, as shown in Fig. (7-W19a). As the angle of incline is raised, it is found that slipping just begins at an angle θ_s. What is the coefficient of static friction between the block and incline plane ?*

Solution:

The forces acting on the block (considered to be a particle) are shown in Fig. (7-W19b). W is the weight, N the normal force exerted by the inclined surface on the block and f_s the tangential force of friction exerted by the inclined surface on the block.

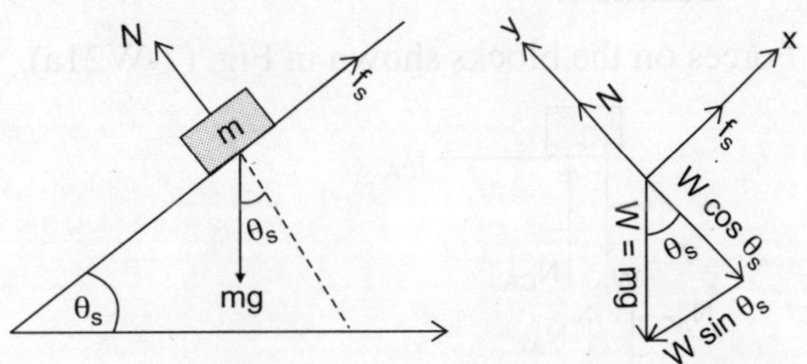

Fig. 7-W19a *Fig. 7-W19b : FBD*

The block is at rest, so that

$$\Sigma F_x = 0$$

or $$f_s - W \sin \theta = 0$$

or $$f_s = W \sin \theta \quad ...(i)$$

and $$\Sigma F_y = 0$$

or $$N - W \cos \theta = 0$$

or $$N = W \cos \theta \quad ...(ii)$$

If we increase the angle of inclination slowly until the slipping just begins, then for that, $\theta = \theta_s$, and we can use $f_s = \mu_s N$.

Substituting this into (i), we obtain

$$\mu_s N = W \sin \theta_s \quad ...(iii)$$

and $$N = W \cos \theta_s \quad ...(iv)$$

Dividing (iii) by (iv), we obtain $\mu_s = \tan \theta_s$.

9. *In Fig. (7-W20) A is a 45.3 kg block and B is a 2.26 kg block (a). Determine the minimum weight of block C which must be placed on A to keep it from sliding, if μ_s between A and the table is 0.20.*

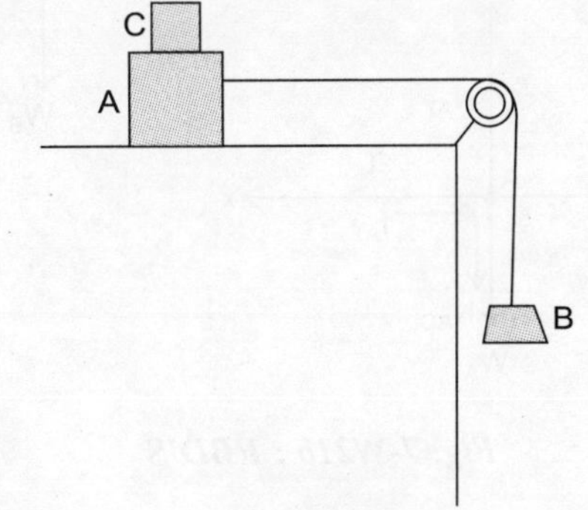

Fig. 7-W20

Solution:

Forces on the blocks shown in Fig. (7-W21a).

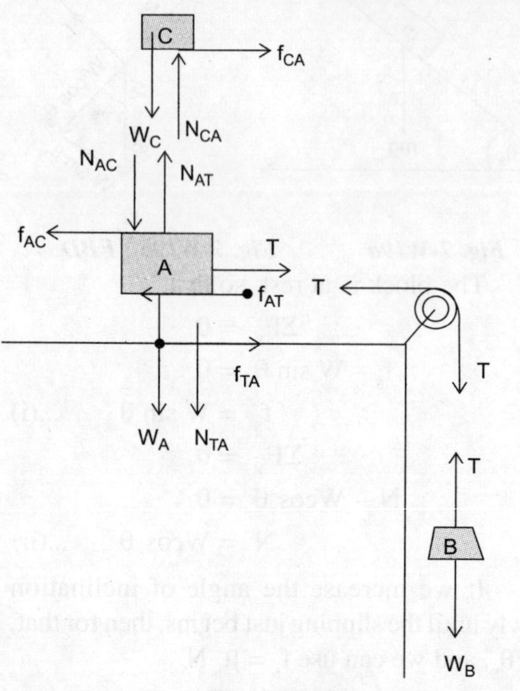

Fig. 7-W21a

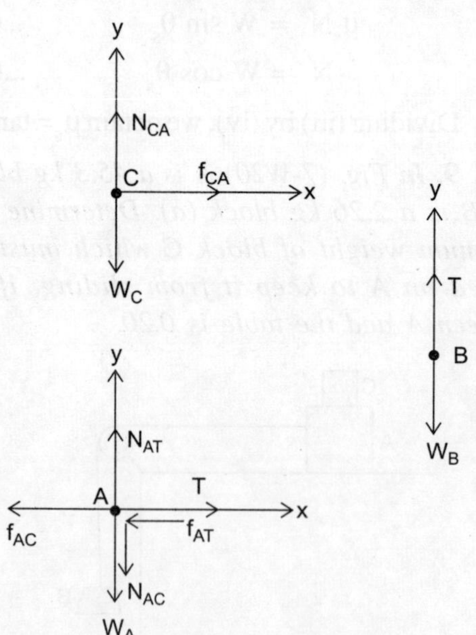

Fig. 7-W21b : FBD'S

Now consider the equilibrium of block B,

i.e., $\Sigma F_{net,\, y} = 0$

or $T - W_B = 0$

or $T = W_B$

$= 2.26 \times 10$

$= 22.6$ N (take $g = 10$ ms^{-2})

$\therefore \quad T = 22.6$ N ...(i)

Now consider the equilibrium of block C

i.e., $\Sigma F_{net,\, y} = 0$

or $N_{CA} - W_C = 0$

or $N_{CA} = W_C$ ...(ii)

and $\Sigma F_{net,\, x} = 0$

or $f_{CA} = 0$ ...(iii)

Similarly, equilibrium of block (A) gives

i.e., $\Sigma F_{net,\, x} = 0$

or $T - f_{AT} - 0\ (= f_{CA}) = 0$

or $T = f_{AT}$

$= N_{AT}\mu$

$= N_{AT} \times 0.20$

$\therefore \quad N_{AT} = \dfrac{T}{0.20}$

$= \dfrac{22.6}{0.20}$

$= 113$ N

and $\Sigma F_{net,\, y} = 0$

$N_{AT} - N_{AC} - W_A = 0$

or $N_{AC} = N_{AT} - W_A$

$= 113 - 45.3$

$= 67.7$ N

Therefore from equation (ii), we get

$W_C = 67.7.$

10. *Block B in Fig. 7-W22 weighs 72.54 kg. The coefficient of static friction between the block and the table is 0.25. Find the maximum weight of the block A for which the system will be in equilibrium.*

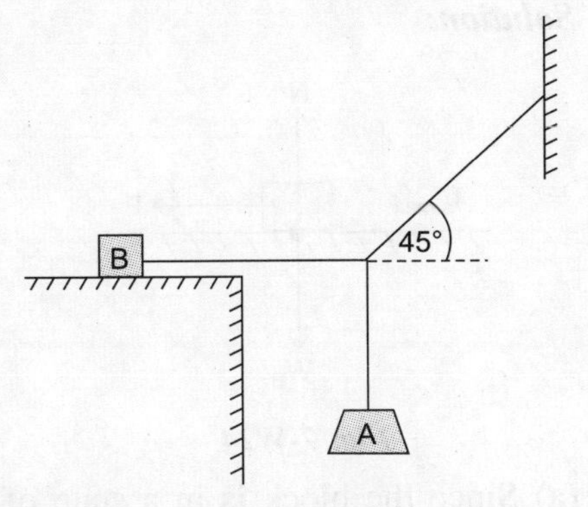

Fig. 7-W22

Solution:

Forces on the blocks are shown in Fig.(7-W22a and b) and FBD'S are shown in Fig.(7-W22c, d and e).

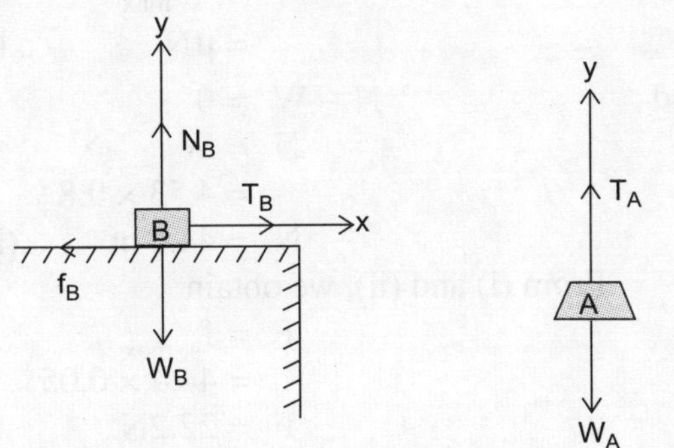

Fig. 7-W22a *Fig. 7-W22b: FBD'S*

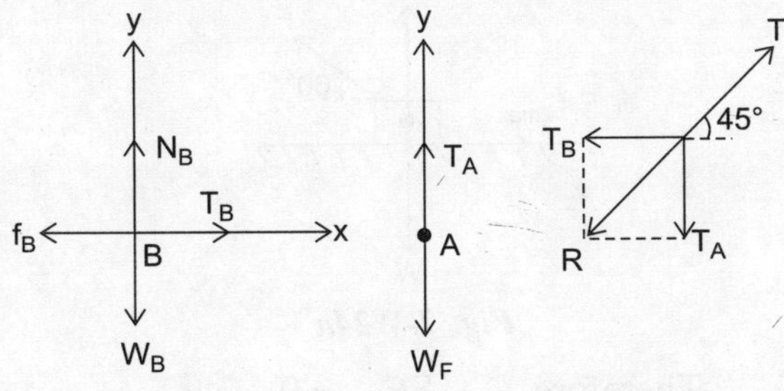

Fig. 7-W22c *Fig. 7-W22d* *Fig. 7-W22e*

Consider the equilibrium of block B.

i.e., the net force along vertically upward should be zero.

or $\Sigma F_{net,\, y} = 0$

$N_B - W_B = 0$

or $N_B = W_B$

$= 72.54 \times 9.8$ N

$= 710.9$ N

$\therefore \quad N_B = 710.9 \text{ N} \quad ...(i)$

Now, equilibrium of block A gives

$\Sigma F_{net} = 0$

or $T_B - f_B = 0$

or $T_B - N_B \times 0.25 = 0$

or $T_B = N_B \times 0.25$

$= 710.9 \times 0.25$

$= 177.7 \text{ N} \quad ....(ii)$

Since T is the resultant of T_A and T_B then,

$T \cos 45 = T_B$

and $T \sin 45 = T_A$

or $\left(\dfrac{\sin 45°}{\cos 45°}\right) = \dfrac{T_A}{T_B}$

or $1 = \dfrac{T_A}{T_B}$

$\Rightarrow \quad T_A = T_B$

$= 177.7 \text{ N} \quad ...(iii)$

Now the equilibrium of block A gives

$T_A = W_A$

$= 177.7$ N. [From iii]

11. *A horizontal force F of 54.4 N pushes a block weighing 2.26 kg, against a vertical wall. The coefficient of static friction between the wall and the block is 0.60, will the block start moving? Take $g = 10\ ms^{-2}$. Refer Fig. (7-W23).*

Fig. 7-W23

Solution:

Forces on block and F.B.D of block are shown in Fig. (7-W23a)

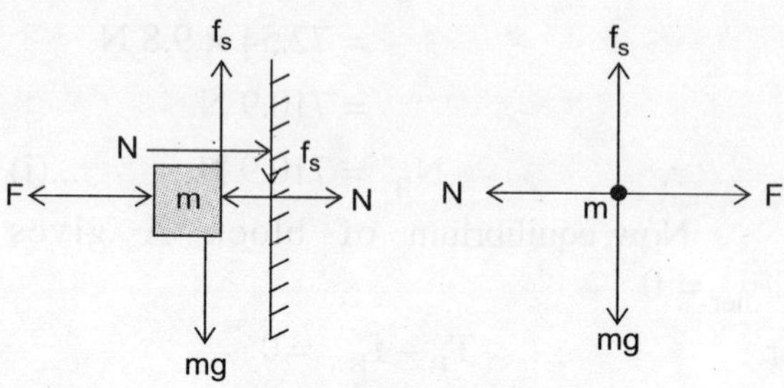

Fig. 7-W23a

Since the block is in rest on the horizontal plane.

Therefore, $\Sigma F_H = 0$

or $F - N = 0$

or $F = N = 54.4$ N.

Now frictional force, $f_{s,\,max} = N\mu$

$= 54.4 \times 0.60$

$= 32.64$ N.

Since $f_{s,\,max}$ > weight of block as we know that, frictional force is an adjustable force so it will adjustable to 22.6 N. Therefore the block remains at rest.

12. *A 4.53 kg block of steel is at rest on a horizontal table. The coefficient of static friction between the block and the table is 0.50.*

(a) What is the magnitude of the horizontal force (F) that will just start the block moving?

(b) What is the magnitude of a force acting along 60° from the horizontal that will just start the block moving? If the force acts down at 60° from the horizontal, how much can it be without causing the block to move?

Solution:

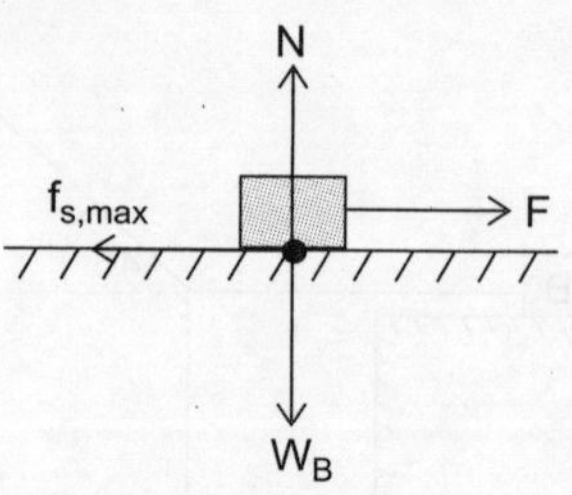

Fig. 7-W24

(a) Since the block is in a state of rest then the net resultant force on the block should be zero.

i.e., $\vec{f}_s + \vec{N} + \vec{W} + \vec{F} = 0$

$\Rightarrow$ $F - f_{s,\,max} = 0$

or $F = f_{s,\,max}$

$= \mu N$ ...(i)

and $N - W = 0$

or $N = W$

$= 4.53 \times 9.8$

$\therefore$ $N = 44.4$ nt ...(ii)

From (i) and (ii), we obtain

$F = f_{s,\,max}$

$= 44.4 \times 0.05$

$\therefore$ $F = 22.2$N.

(b) Since the block just starts moving i.e., friction is limiting.

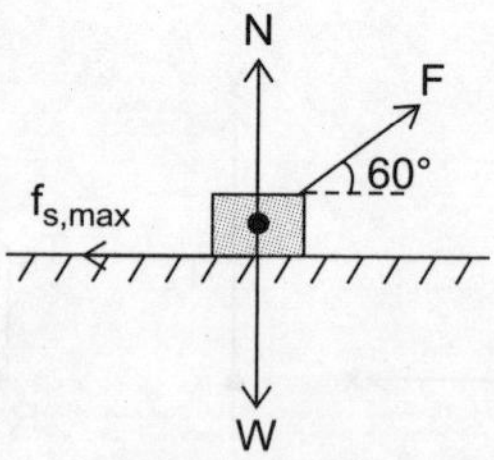

Fig. 7-W24a

Therefore, $\Sigma F_H = 0$

or $F \cos 60° - f_{s,\,max} = 0$

or $f_{s,\,max} = F \cos 60°$

or $N\mu = F \cos 60°$ ...(iii)

and $\Sigma F_V = 0$

or $F \sin 60° + N - W = 0$

or $F \sin 60° = W - N$ ...(iv)

Dividing (iv) by (iii), we obtain

$$\tan 60° = \frac{W-N}{N\mu}$$

or $\sqrt{3}N\mu = W - N$

or $N\left(\mu\sqrt{3}+1\right) = W$

or
$$N = \frac{W}{\mu\sqrt{3}+1} = \frac{4.53\times 9.8}{0.5\times\sqrt{3}+1} = 23.8 \text{ nt}$$

From (iii), we get

$N\mu = F \cos 60°$

or $23.8 \times 0.5 = F \cos 60°$

$\therefore$ $F = 23.8$ nt.

(c) Since the block is in equilibrium then

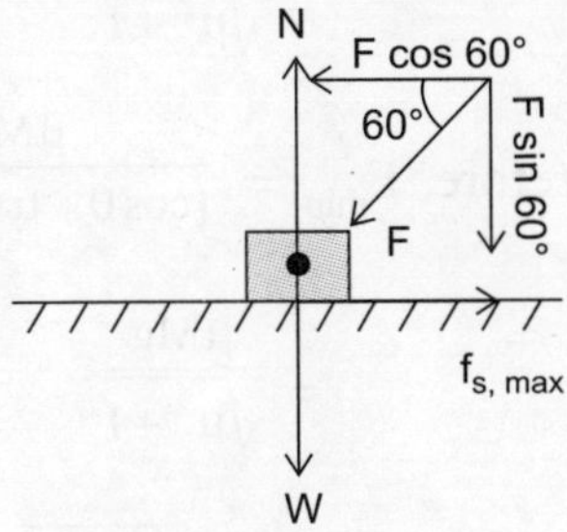

Fig. 7-W24b

$\Sigma F_H = 0$

or $F \cos 60° - f_{s,\,max} = 0$

or $F \cos 60° = f_{s,\,max}$

$= N\mu$ ...(v)

and $\Sigma F_v = 0$

or $N - W - F \sin 60° = 0$

or $N - W = F \sin 60°$ ...(vi)

Dividing (vi) by (v), we obtain

$$\tan 60° = \frac{N-W}{N\mu}$$

or $N\mu\sqrt{3} = N - W$

or $N\left(\sqrt{3}\mu - 1\right) = -W$

or $N\left(1-\sqrt{3}\mu\right) = W$

or
$$N = \frac{W}{1-\sqrt{3}\mu} = \frac{4.53\times 9.8}{1-1.73\times 0.5} = 328.84 \text{ nt}$$

From (v), we get

$F \cos 60° = 328.84 \times 0.5$

$\therefore$ $F = 328.84$ nt.

13. *A block of mass M is kept on a rough horizontal surface. The coefficient of static friction between the block and the surface is μ. The block is to be pulled by applying a force on it, what minimum force is needed to slide the block? In which direction should this force act? Is this inclined direction, if yes, then why?*

Solution:

If the block is being pulled horizontally then the normal reaction is Mg. Consequently the maximum frictional force is Mg μ so the minimum force required to slide the block is Mgμ. If one pull the block making some angle with the horizontal or vertical then one could decrease the normal reaction which consequently decreases the applied force.

Suppose force is being applied at an angle of θ with the horizontal, force diagram and free body diagram are shown in Fig. (7-W25).

Since the block is in equilibrium then, the net force along the x-axis (horizontally right) should be zero.

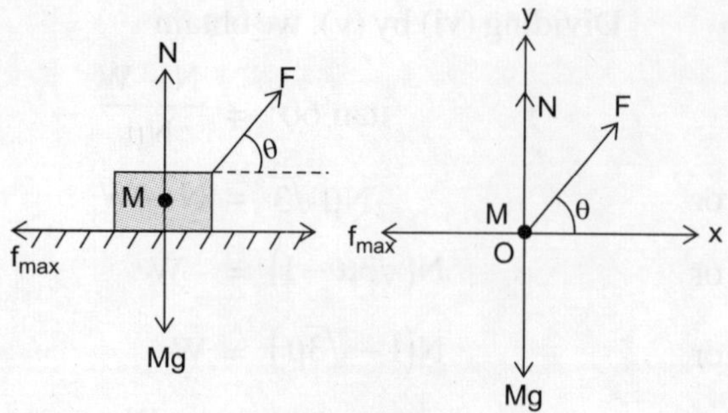

Fig. 7-W25

i.e., $\Sigma F_{net,\,x} = 0$

or $F\cos\theta - f_{max} = 0$ ...(i)

and net force along the y-axis should be zero.

i.e., $\Sigma F_{net,\,y} = 0$

or $N + F\sin\theta - Mg = 0$

$= Mg - F\sin\theta$

$< Mg$ (normal force is decreases by $F\sin\theta$)

Since it is a case of limiting friction (f_{max}) therefore

$$f_{max} = \mu N$$

$$= \mu(Mg - F\sin\theta) \quad \text{...(ii)}$$

From equations (i) and (ii), we get

$F\cos\theta - \mu(Mg - F\sin\theta) = 0$

or $F\cos\theta - \mu Mg + \mu F\sin\theta = 0$

or $F(\cos\theta + \mu\sin\theta) = \mu Mg$

or $$F = \frac{\mu Mg}{\cos\theta + \mu\sin\theta} \quad \text{...(iii)}$$

According to the question F should be minimum. From (iii) we conclude that, the term $\cos\theta + \mu\sin\theta = f(\theta)$ (say) should be maximum.

As $f(\theta) = \cos\theta + \mu\sin\theta$

$$\therefore \quad \frac{d(f(\theta))}{d\theta} = -\sin\theta + \mu\cos\theta$$

At maximum point $f'(\theta) = 0$

i.e., $-\sin\theta + \mu\cos\theta = 0$

or $\mu = \tan\theta$ ...(iv)

Now, $$\frac{d^2(f(\theta))}{d\theta} = -\cos\theta - \mu\sin\theta$$

$$= -(\cos\theta + \mu\sin\theta)$$

$$= -\left[\frac{1}{\sqrt{\mu^2+1}} + \frac{\mu^2}{\sqrt{\mu^2+1}}\right]$$

$$= -\left[\frac{\mu^2+1}{\sqrt{\mu^2+1}}\right]$$

Therefore, $f''(\theta) = -\sqrt{\mu^2+1} < 0$

i.e., at $\theta = \tan^{-1}\mu$, $f(\theta)$ is maximum and maximum value is

$$f(\theta)_{max} = \cos\theta + \mu\sin\theta$$

$$= \frac{1}{\sqrt{\mu^2+1}} + \frac{\mu^2}{\sqrt{\mu^2+1}}$$

$$= \frac{\mu^2+1}{\sqrt{\mu^2+1}}$$

$$= \sqrt{\mu^2+1}$$

Therefore, $$F_{min} = \frac{\mu Mg}{(\cos\theta + \mu\sin\theta)_{max}}$$

$$= \frac{\mu Mg}{\sqrt{\mu^2+1}}$$

$$= Mg\sqrt{\frac{\mu^2}{\mu^2+1}}$$

$$= Mg\sqrt{1 - \frac{1}{\mu^2+1}} < Mg\,\mu.$$

14. *The coefficient of friction between the two blocks shown in Fig. (7-W26) is μ but the floor is smooth, what maximum horizontal force F could be applied without disturbing the equilibrium of the system ?*

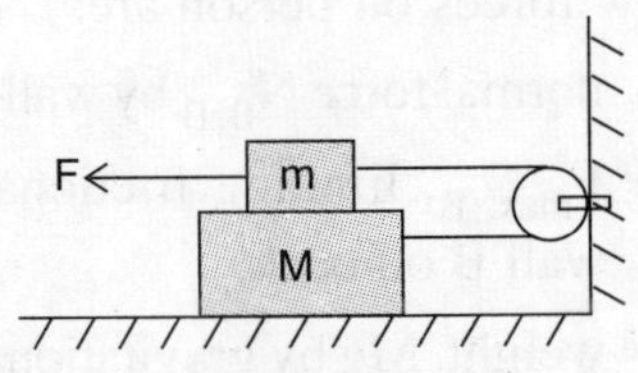

Fig. 7-W26

Solution:

Suppose that tension in the string is T. Figures given below show force diagrams and free body diagrams of block (m) and block (M).

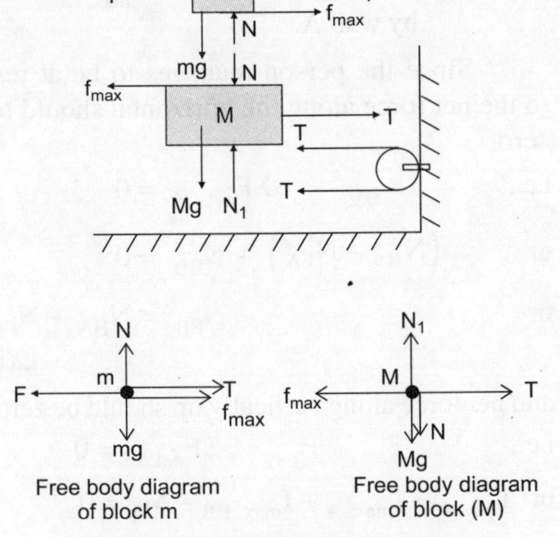

Fig. 7-W27

Forces on block (m) are

(a) tension (T) in string,

(b) weight (mg) in gravitational field of Earth,

(c) normal reaction (N) by block (M),

(d) limiting friction (f_{max}) by block (M), and

(e) F, applied force.

Since the block (m) is in equilibrium therefore the net force along the horizontally right direction should be zero.

i.e., $\Sigma F_{net,\ H} = 0$

or $T + f_{max} - F = 0$

or $F = T + f_{max}$ ...(i)

Since friction is limiting

then, $f_{max} = \mu N$ ...(ii)

And the net force along vertically up should be zero.

i.e., $N - mg = 0$

or $N = mg$ ...(iii)

by combining equations (i), (ii) and (iii), we get

$$F = T + \mu mg \quad \text{...(iv)}$$

Now the forces on block (M) are:

(a) tension (T) in string,

(b) Mg, weight in gravitational field,

(c) normal reaction (N) by block m,

(d) normal reaction (N_1) by floor, and

(e) frictional force (f_{max}) by block (m).

Since the block (M) is in equilibrium then, the net force on the block along horizontally right should be zero

i.e., $\Sigma F_{net,\ H} = 0$

or $T - f_{max} = 0$

or $T = f_{max}$

$= mg\mu$ ...(v)

And net force on block along vertically up should be zero.

i.e., $\Sigma F_{net,\ V} = 0$

or $N_1 - N - Mg = 0$

or $N_1 = N + Mg$

$= mg + Mg$

$= (m + M)g$ ...(vi)

Now from (iv) and (v), we get

$F = T + \mu mg$

$= \mu mg + \mu mg$

$F = 2\,\mu mg.$

15. *A person (40 kg) is managing to be at rest between two vertical walls by pressing one wall A by his hands and feet, and the other wall B by his back Fig. (7-W28). Assuming that the coefficient of friction between his body and walls is 0.8 and the limiting friction acts at all the contacts (a) show that the person pushes the two walls with equal force (b). Find the normal reaction exerted by either of wall on the person. Take* $g = 10\ ms^{-2}$.

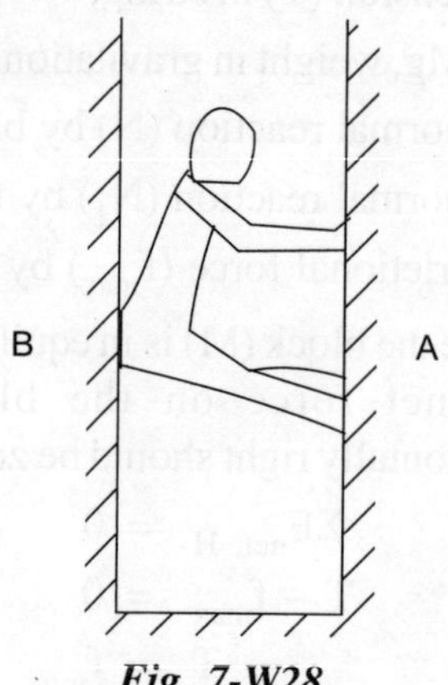

Fig. 7-W28

Solution:

Forces on person are shown in Fig. (7-W29).

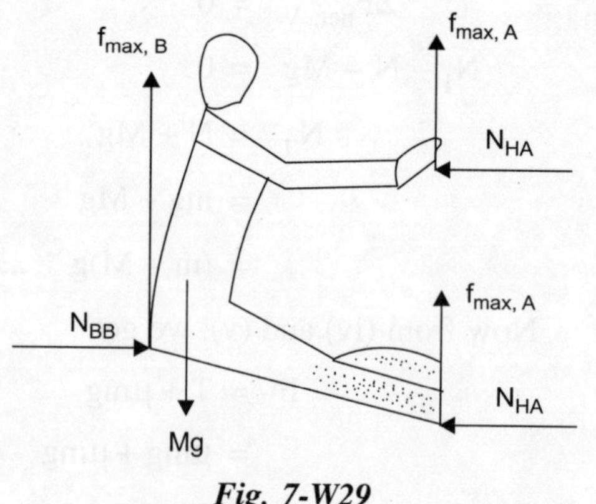

Fig. 7-W29

Now forces on person are:

(a) normal force $N_{B, B}$ by wall B on back,

(b) $F_{max, B}$, limiting frictional force by wall B on back,

(c) weight, Mg by gravitational field,

(d) normal reaction N_{HA} on hands by wall A,

(e) $f_{max, A}$, limiting frictional force on hands by wall A,

(f) normal reaction N_{FA} by wall A on feet,

(g) limiting frictional force $f_{max, A}$ on feet by wall A.

Since the person manages to be at rest so the net force along the horizontal should be zero.

i.e., $\Sigma F_{net, H} = 0$

or $-\left(N_{HA} + N_{FA}\right) + N_{BB} = 0$

or $N_{BB} = N_{HA} + N_{FA}$ ...(i)

and net force along vertically up should be zero.

i.e., $\Sigma F_{net, V} = 0$

or $f_{max, A} + f_{max, A} + f_{max, BB} - Mg = 0$

or $0.8 \times N_{HA} + 0.8 \times N_{FA} + 0.8 \times N_{BB} - 40 \times 10 = 0$

or $N_{HA} + N_{FA} + N_{BB} = 500$

[Since friction is limiting therefore $f_{max, A} = 0.8\ N_{HA}$, similarly $f_{max;\ BB} = 0.8\ N_{BB}$.]

or $N_{HA} + N_{FA} + N_{BB} = 500$ N

or $N_{BB} + N_{BB} = 500$ N [from (i)]

or $N_{BB} = 250$ N

Since $N_{HA} + N_{FA} = N_{BB}$

Therefore, forces exerted by either of wall is 250 N ($N_{BB} = N_B = N_{HA} + N_{FA} = N_A$).

16. *Two blocks with m = 16 kg and M = 88 kg are shown in Fig. (7-W30) are not attached. The coefficient of static friction between the blocks is μ_s = 0.38, but if the surface beneath the larger block is frictionless. What is the minimum magnitude of horizontal force $\vec{F}$ required to keep the smaller block from slipping down the larger block ?*

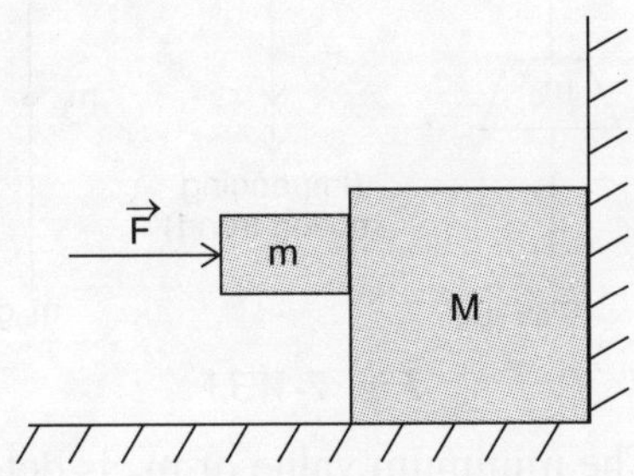

Fig. 7-W30

Solution:

Forces on blocks and free body diagrams are shown in Fig. (7-W30a and 7-W30b).

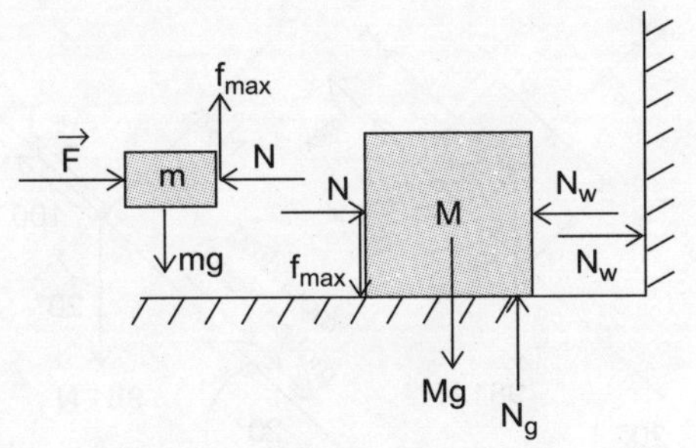

Fig. 7-W30a

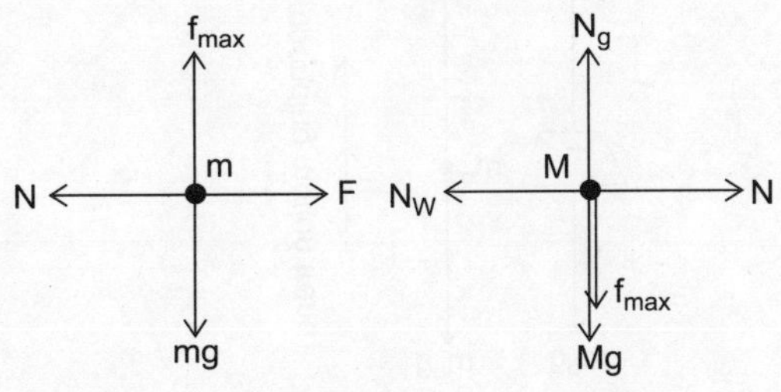

Fig. 7-W30b

Since block (m) is in equilibrium then the net force along the horizontal direction should be zero.

i.e., $\Sigma F_{net,\ mH} = 0$

or $F - N = 0$

or $F = N$ ....(i)

The net force along vertical direction should also be zero.

i.e., $\Sigma F_{net,\ mV} = 0$

or $f_{max} - mg = 0$

or $\mu N - mg = 0$

or $N = \dfrac{mg}{\mu}$

$= \dfrac{16 \times 9.8}{0.38}$

$\therefore = 412.63$ nt

i.e., F = 412.63 N, is the minimum force is required to keep the block in state of rest.

Now equilibrium of block M gives,

$\Sigma F_{net,\ MH} = 0$

or $N - N_w = 0$

or $N = N_w$

$= 412.63$ nt.

and $\Sigma F_{net,\ MV} = 0$

or $Mg + f_{max} = N_g$

or $88 \times 9.8 + 0.38 \times 412.63 = N_g$

$\therefore$ $N_g = 1019.20$ N

Therefore minimum required force, F = 412.63 nt keeps block (m) in equilibrium.

17. *Determine the range of values which the mass m_0 may have so that the 100 kg block shown in the Fig. (7-W31) will neither*

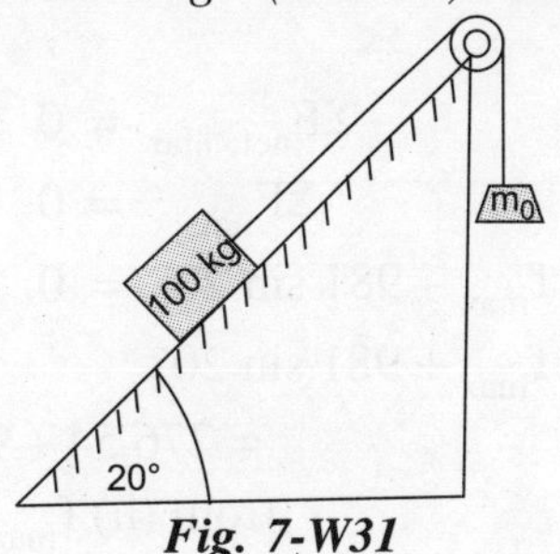

Fig. 7-W31

start moving up the plane nor slip down the plane. The coefficient of static friction for the contact surface is 0.30. Take $g = 9.81\ ms^{-2}$.

Solution:

The maximum value of m_0 come when impending motion is up the plane. The frictional force on the block therefore acts down the plane. Forces acting on the block and free body diagram are shown in Fig. (7-W32). The equilibrium of the block 100 kg, implies that the net force along the perpendicular to the plane should be zero.

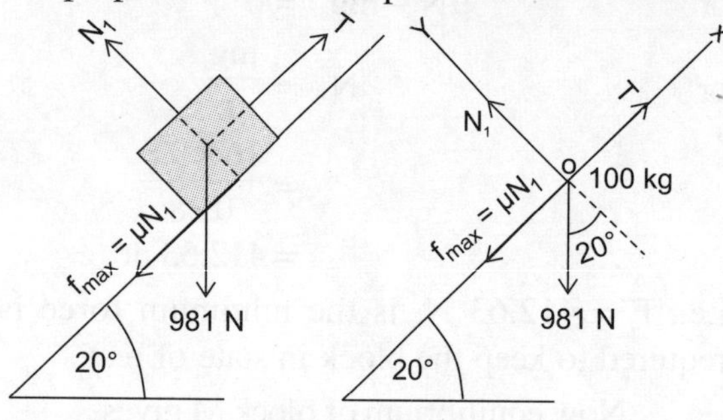

Fig. 7-W32

i.e.,
$$\Sigma F_{net,\ \perp r} = 0$$
or
$$\Sigma F_{net,\ y} = 0$$
or
$$N_1 - 981 \cos 20° = 0$$
or
$$N_1 = 981 \cos 20°$$
$$\therefore \quad = 921.83 \text{ N} \quad ...(i)$$

Now as motion is impending so the frictional force will be maximum that is,

$$f_{max} = \mu N_1$$
$$= 921.83 \times 0.30$$
[from equation (i) $N_1 = 921.83$]
$$= 276.54 \text{ N} \quad ...(ii)$$

Net force along the inclined surface should be zero.

i.e.,
$$\Sigma F_{net,\ ||lar} = 0$$
or
$$\Sigma F_{net,\ x} = 0$$
or
$$T - f_{max} - 981 \sin 20° = 0$$
or
$$T = f_{max} + 981 \sin 20°$$
$$= 276.54 + 981 \sin 20°$$
[from (ii) $f_{max} = 276.54$]
$$= 612.06 \text{ N} \quad ...(iii)$$

Now, the equilibrium of block m_0 gives

$$m_0 g - T = 0$$
or
$$T = m_0 g \quad ...(iv)$$

From (iii) and (iv)

$$9.81\ m_0 = 612.06$$
$$\therefore \quad m_0 = 62.4 \text{ kg.}$$

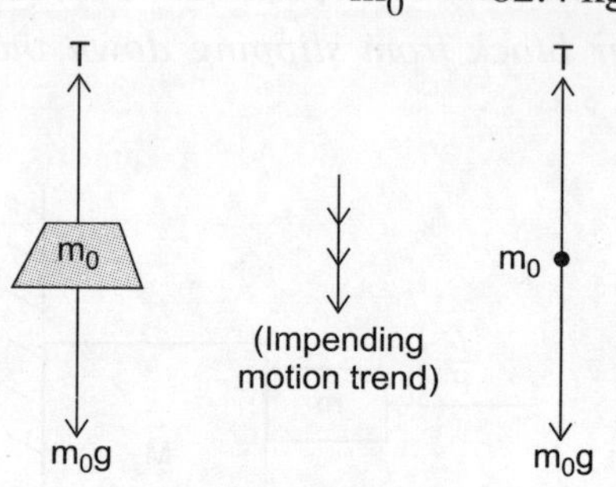

Fig. 7-W33

The minimum value of m_0 is determined when the motion is impending and down the plane. The frictional force on the block will act opposite to the trend of impending motion. The force acting on the block and the free body diagram are shown in the figures below.

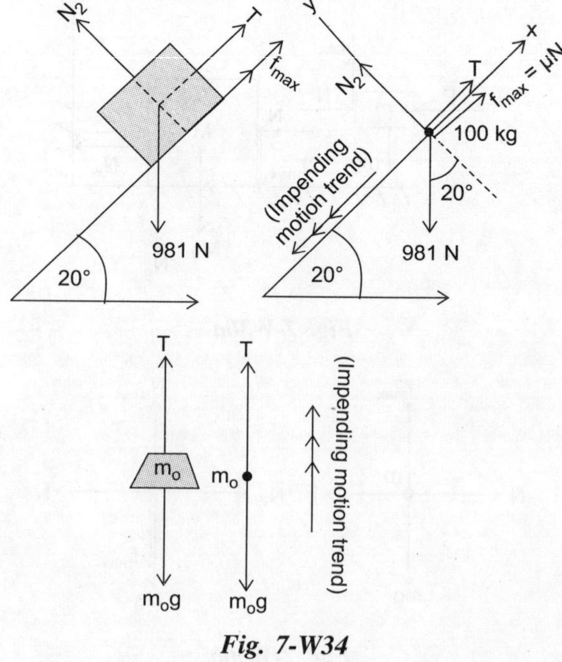

Fig. 7-W34

For equilibrium block of mass 100 kg. The net force perpendicular to the plane should be zero.

i.e., $\Sigma F_{net,\, y} = 0$

or $N_2 - 981 \cos 20° = 0$

or $N_2 = 921.83$ N

$\therefore$ $f_{max} = \mu N_2$

[since motion is impending]

$= 0.30 \times 921.83$

$= 276.54$N ...(v)

Net force parallel to the plane should be zero.

i.e., $\Sigma F_{net,\, x} = 0$

or $T + 276.54 - 981 \sin 20° = 0$

$\therefore$ $T = 981 \sin 20° - 276.83$

$= 58.70$ N ...(vi)

Now equilibrium of block (m_0) gives,

$T - m_0 g = 0$

or $T = m_0 g$ ...(vii)

From equations (vii) and (viii), we get

$m_0 g = 58.70$

$\therefore$ $m_0 = 6$ kg

Thus for the range 6 kg $\leq m_0 \leq$ 62.4 kg block remains at rest.

18. *The three flat blocks are positioned on the 30° inclined as shown in Fig. (7-W35), and a force P parallel on the inclined plane is being applied to the middle block. The upper block is prevented from moving by a wire which attaches it to the fixed support, the coefficient of static friction for each of the three pair of mating surfaces are shown. Determine the maximum value which P could have before slipping takes place.*

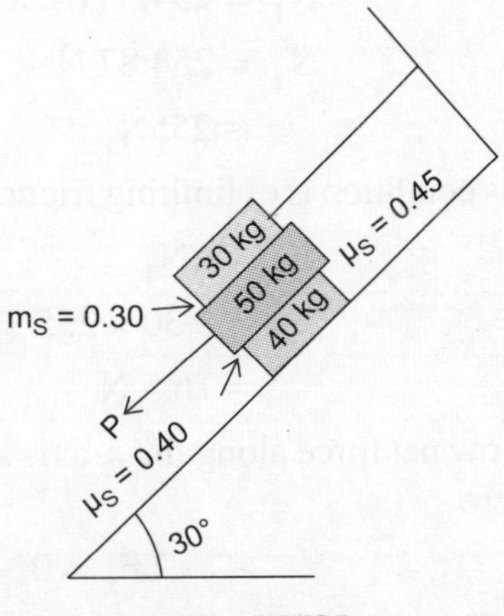

Fig. 7-W35

Solution:

The free body diagram (FBD) of each block is shown below. The forces are assigned in the way to oppose the relative motion which would occur if no friction were present. Here the block of (50 kg) has tendency to slip down, therefore the frictional forces at the surfaces contacts of 30 kg and 40 kg like $\vec{F}_1$ and $\vec{F}_2$ will act up the plane and following Newton's third law of motion frictional forces on 30 kg and 40 kg has been shown along respective directions.

Block (30 kg)

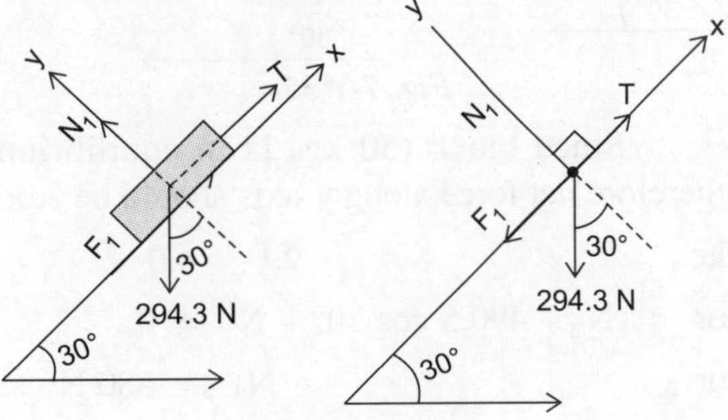

Fig. 7-W35a

Since block (30 kg) is in equilibrium therefore the net force along y-axis should be zero.

i.e., $\Sigma F_y = 0$

or $N_1 - 294.3 \cos 30° = 0$

or $N_1 = 294.3 \cos 30°$

$\therefore$ $N_1 = 254.87$ N

$\cong 255$N

As condition is of limiting friction then,

$F_1 = \mu N_1$

$= 0.30 \times 255$

$= 76.5$ N

Now net force along the x-axis should be zero.

i.e, $\Sigma F_x = 0$

or $-F_1 - 294.30 \sin 30° + T = 0$

or $T = F_1 + 294.30 \sin 30°$

$= 76.5 + 294.30 \sin 30°$

$= 223.65$ N

Block (50 kg)

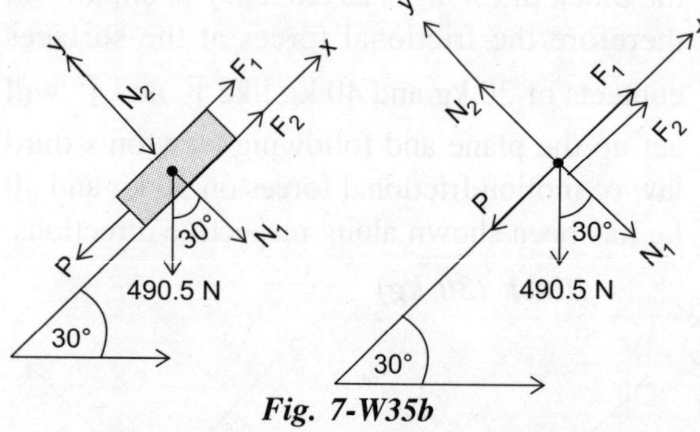

Fig. 7-W35b

Since block (50 kg) is in equilibrium therefore net force along y-axis should be zero.

i.e., $\Sigma F_y = 0$

or $N_2 - 490.5 \cos 30° - N_1 = 0$

or $N_2 \cong 680$ N

Now considering condition of limiting friction then,

$F_2 = \mu N_2$

$= 0.40 \times 680$

$= 272$ N

and net force along x-axis should be zero.

i.e., $\Sigma F_x = 0$

or $F_1 + F_2 - P - 490.5 \sin 30° = 0$

$\therefore$ $P = F_1 + F_2 - 490.5 \sin 30°$

$= 76.46 + 272 - 490.5 \sin 30°$

$= 103.21$ N.

Now block (40 kg)

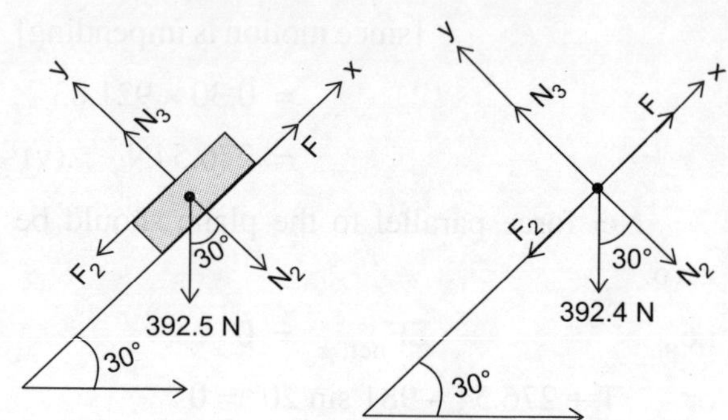

Fig. 7-W35c

Block (40 kg) is in equilibrium, therefore, net force along the y-axis should be zero.

i.e., $\Sigma F_y = 0$

or $N_3 - 392.4 \cos 30° - 680 = 0$

or $N_3 = 1019.82$ N

And net force along the x-axis should be zero.

i.e., $\Sigma F_x = 0$

or $F - F_2 - 392.4 \sin 30° = 0$

or $F = F_2 + 392.4 \sin 30°$

$= 272 + 392.4 \sin 30°$

$= 468.2$ N

But the maximum possible value of $F = \mu_s N_3$

$= 0.45 \times 1019.82$

$= 458.91 \text{ N} < 468.2$ N

$\Rightarrow$ Cannot be in equilibrium, it will slip.

Thus, we conclude that, slipping occurs first of all between the 40 kg block and the incline plane, with the corrected value F = 458.91 N equilibrium of the 40 kg block is exiting.

Therefore, $F - F_2 - 392.4 \sin 30° = 0$

or $458.91 - F_2 - 392.4 \sin 30° = 0$

or $F_2 = 262.71$ N

Therefore, equilibrium of 50 kg block gives final value of P,

$- P + F_1 + F_2 - 490.5 \sin 30° = 0$

or $P = 76.5 + 262.71 - 490.5 \sin 30°$

$\therefore \quad P = 93.96$ N

Now corresponding to P = 93.96 N we can find the value for T for which the system remains in equilibrium.

19. *A horizontal force F_a is being applied to a block which rests on an inclined plane, as shown in Fig. (7-W36). Find the force F_a required to initiate motion of block, up the plane.*

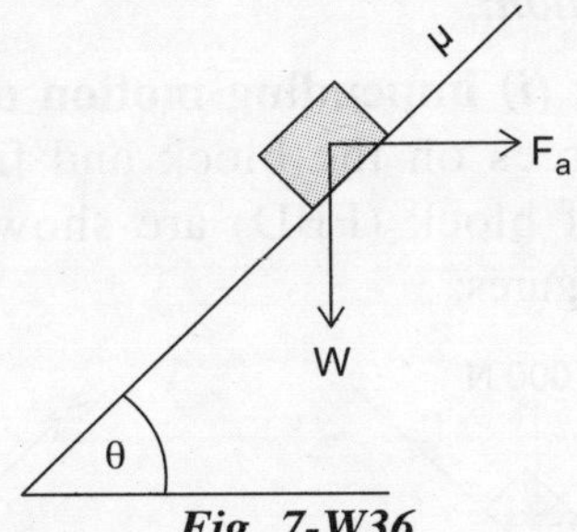

Fig. 7-W36

Solution:

Here, we have to initiate motion up the inclined plane, therefore, the force of limiting friction acts along down the inclined plane. Forces on block and FBD are shown in figures below.

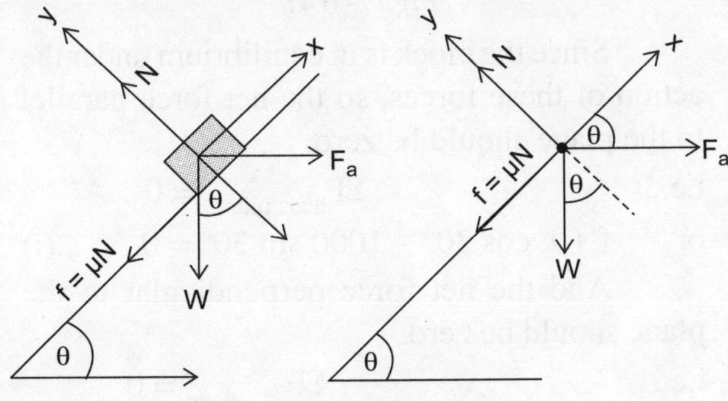

Fig. 7-W37

Since block is on the verge of the motion up the plane, therefore, net force along the x-axis should be zero.

i.e., $\Sigma F_{x,\, block} = 0$

or $F_a \cos\theta - W \sin\theta - \mu N = 0$ ...(i)

Net force perpendicular to plane (along y-axis) should be zero.

i.e., $\Sigma F_{y,\, block} = 0$

or $N - W\cos\theta - F_a \sin\theta = 0$

or $N = W\cos\theta + F_a \sin\theta$ ..... (ii)

By substituting N from (ii) to equation (i), we get

$F_a \cos\theta - W\sin\theta - \mu (W\cos\theta + F_a \sin\theta) = 0$

or $F_a \cos\theta - W\sin\theta - W\mu\cos\theta - \mu F_a \sin\theta = 0$

or $F_a (\cos\theta - \mu\sin\theta) = W\sin\theta + W\mu\cos\theta$

or $$F_a = \frac{W(\sin\theta + \mu\cos\theta)}{\cos\theta - \mu\sin\theta}$$

$$\therefore F_a = W\left(\frac{\tan\theta + \mu}{1 - \mu\tan\theta}\right).$$

20. *A short right circular of weight W rests in a horizontal V-notch having an angle 2α as shown in Fig.(7-W38). If the coefficient of friction is μ, find the horizontal force P necessary for impending motion.*

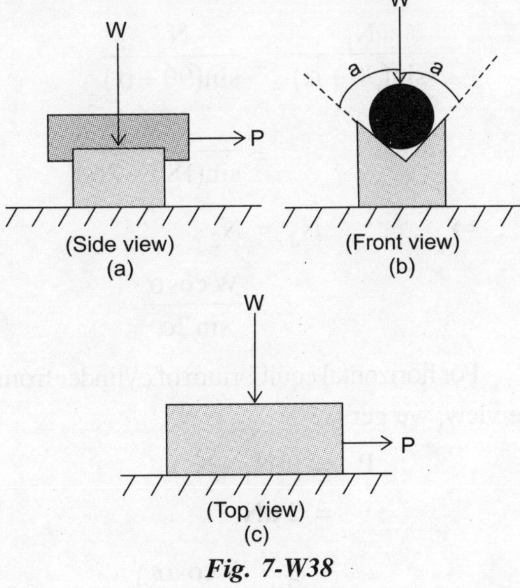

Fig. 7-W38

Solution:

Forces on cylinder and FBD are shown in Figs. (7-W39 a, b, c)

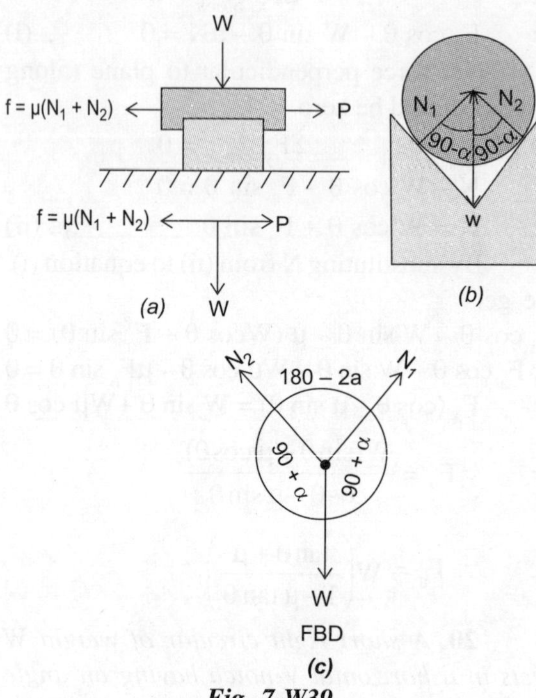

Fig. 7-W39

From FBD, (7-W39c), by applying Lami's theorem, we get

$$\frac{N_1}{\sin(90+\alpha)} = \frac{N_2}{\sin(90+\alpha)}$$

$$= \frac{W}{\sin(180-2\alpha)}$$

$$\Rightarrow \quad N_1 = N_2$$

$$= \frac{W\cos\alpha}{\sin 2\alpha}$$

For horizontal equilibrium of cylinder from side view, we get

$$P = \mu(N_1 + N_2)$$

$$= 2\,\mu N_1$$

$$= 2\mu\left(\frac{W\cos\alpha}{\sin 2\alpha}\right)$$

$$= \frac{\mu W}{\sin\alpha}$$

21. *A small block of weight 1,000 N is placed on a 30° inclined plane having coefficient of friction 0.25 as shown in Fig. (7-W40). Determine, the horizontal force to be applied for (i) the impending motion down the plane and (ii) the impending motion up the plane.*

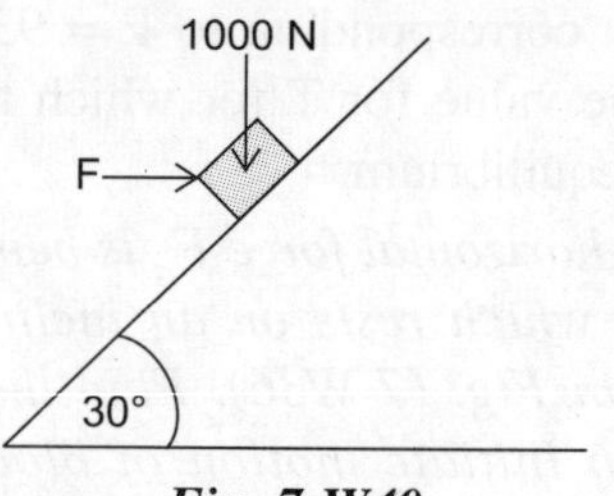

Fig. 7-W40

Solution:

***Case (i)* impending motion down the plane:** Forces on the block and free body diagram of block (FBD) are shown in the following figures:

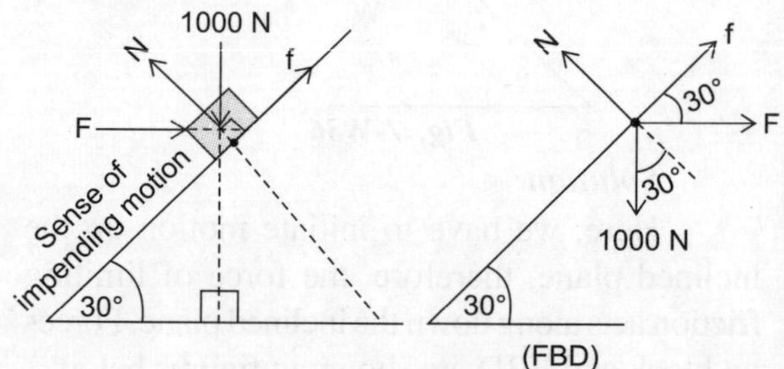

Fig. 7-W41

Since the block is in equilibrium under the action of these forces, so the net force parallel to the plane should be zero.

i.e., $\Sigma F_{\text{|||ar, plane}} = 0$

or $f + F\cos 30° - 1000\sin 30° = 0$ ...(i)

And the net force perpendicular to the plane should be zero.

i.e., $\Sigma F_{\perp\text{r, plane}} = 0$

or $N - 1000\cos 30° - F\sin 30° = 0$

or $\qquad N = 1000 \cos 30° + F \sin 30°$

$= 866.025 + F \sin 30°$

Since the motion of block is impending, then the friction is limiting

$\therefore \qquad f = \mu N$

$= 0.25 \times (866.025 + F \sin 30°)$

$= 216.506 + \dfrac{F}{8}$

Putting value of $\quad f = 216.506 + \dfrac{F}{8}$ in equation (i), we get

$$216.506 + \frac{F}{8} + F \cos 30° - 500 = 0$$

or $\quad \dfrac{F}{8} + F \cos 30° = 283.494$

$\therefore \qquad F = 286.061$ N.

***Case (ii)* Impending motion up to plane**

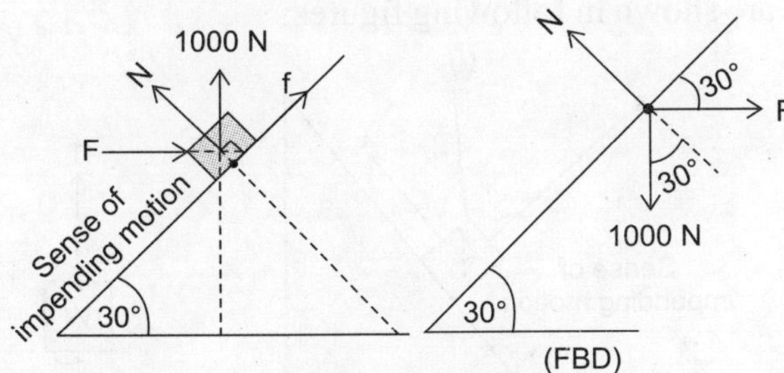

Fig. 7-W42

Using condition of equilibrium, we get

$\Sigma F_{\perp lar,\ plane} = 0$

or $\quad N_2 - 1000 \cos 30° - F \sin 30°$

or $\qquad N_2 = \left(866.025 + \dfrac{F}{2}\right)$nt

and $\qquad \Sigma F_{||lar,\ plane} = 0$

or $\quad F \cos 30° - f - 1000 \sin 30° = 0$

or $\quad F \cos 30° - \left(866.025 + \dfrac{F}{2}\right) 0.25$

$- 1000 \sin 30° = 0$

or $\quad F \cos 30° - \dfrac{1}{8} F = 716.506$

or $\qquad F\left(\cos 30° - \dfrac{1}{8}\right) = 716.506$

$\therefore \qquad F = 966.911$ N.

22. *Two blocks A and B weighing 3 kN and 1.3 kN respectively are connected by a string over a friction less pulley as shown in Fig. (7-W43). Find the minimum value of T to generate an impending motion to the right. Coefficients of friction for surface of contacts for block A and B are 0.20 and 0.30 respectively.*

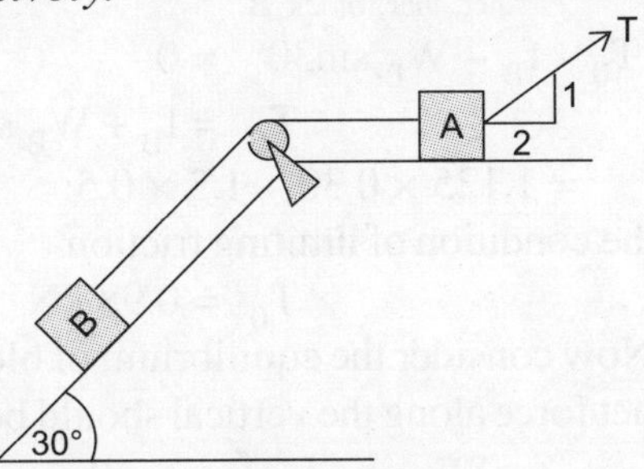

Fig. 7-W43

Solution:

Since block (A) is pulled towards the right so, impending motion of block (B) will be up to the plane. Forces on the blocks and Free diagrams (FBDs) are shown in Fig. (7-W44).

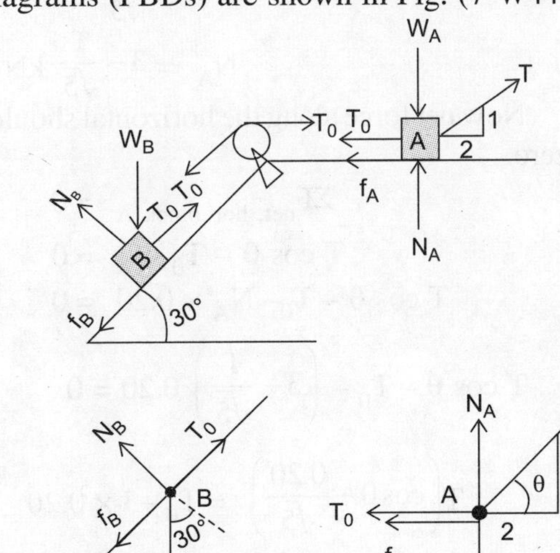

Fig. 7-W44

First let us consider the equilibrium of block B.

Net force along the perpendicular to the plane should be zero.

i.e., $\Sigma F_{\text{net, }\perp\text{lar, block B}} = 0$

or $N_B - W_B \cos 30° = 0$

or $N_B = 1.3 \cos 30°$

$= 1.125 \text{ kN}$

And net force along the plane should also be zero.

i.e., $\Sigma F_{\text{net, }\parallel\text{lar, block B}} = 0$

or $T_0 - f_B - W_B \sin 30° = 0$

or $T_0 = f_B + W_B \sin 30°$

$= 1.125 \times 0.30 + 1.3 \times 0.5;$

Using the condition of limiting friction

$\therefore \quad T_0 = 0.98 \text{ kN}$

Now consider the equilibrium of block A, net force along the vertical should be zero.

i.e., $\Sigma F_{\text{net, ver, block A}} = 0$

or $N_A - W_A + T \sin\theta = 0$

or $N_A = W_A - T\sin\theta$

Where $\tan\theta = \frac{1}{2}$

$\Rightarrow \quad \sin\theta = \frac{1}{\sqrt{5}}$

$\therefore \quad N_A = 3 - \frac{T}{\sqrt{5}} \text{ kN}$

Now net force along the horizontal should be zero.

i.e., $\Sigma F_{\text{net, hori,block A}} = 0$

or $T\cos\theta - T_0 - f_A = 0$

or $T\cos\theta - T_0 - N_A \times 0.20 = 0$

or $T\cos\theta - T_0 - \left(3 - \frac{T}{\sqrt{5}}\right) 0.20 = 0$

or $T\left(\cos\theta + \frac{0.20}{\sqrt{5}}\right) = T_0 + 3 \times 0.20$

$= 0.98 + 3 \times 0.20$

$= 1.58$

or $T\left(\frac{2}{\sqrt{5}} + \frac{0.20}{\sqrt{5}}\right) = 1.58$

$\therefore \quad T = \frac{1.58 \times \sqrt{5}}{2.20}$

$= 1.6 \text{ kN.}$

23. *For the block-pulley-weight system, find the weight W to keep the block A in static equilibrium. The weight of the block A is 100 N. The coefficient of friction between the block and incline plane is 0.25. Neglect friction in the pulley.*

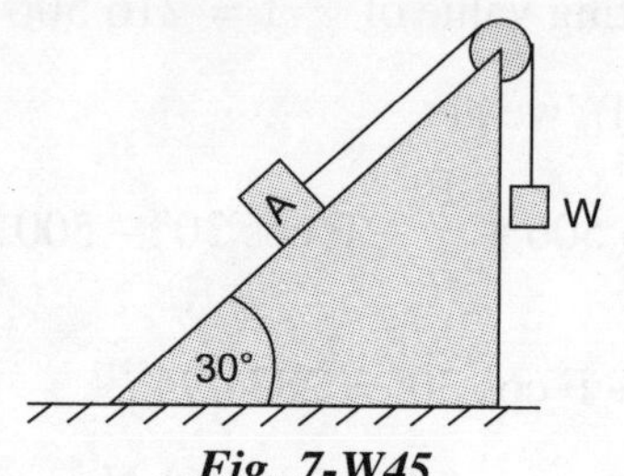

Fig. 7-W45

Solution:

Forces on blocks and free body diagrams are shown in following figures:

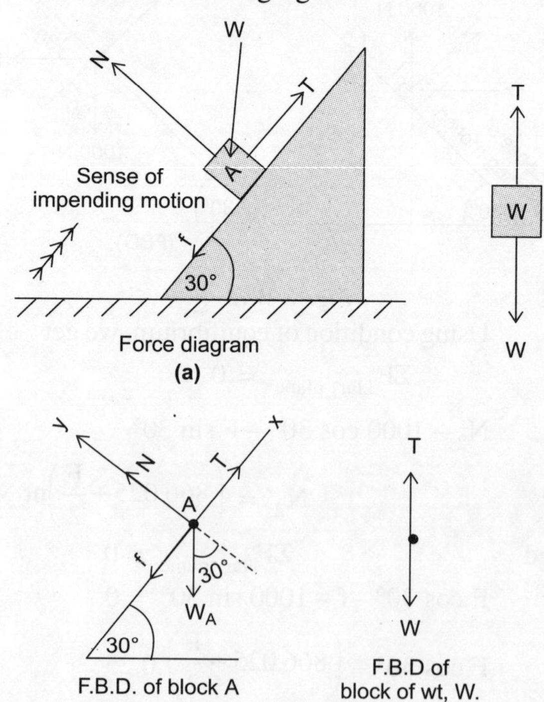

Fig. 7-W46

First consider the equilibrium of weight (W). Since the weight (W) is in vertical equilibrium therefore, the net force acting on W along the vertical should be zero.

i.e., $\Sigma F_{net,\ ver,\ W} = 0$

or $T - W = 0$

or $T = W$ ...(i)

Now, consider the equilibrium of block A, the net force on block (A) along the perpendicular to the plane should be zero, then,

$N - W_A \cos 30° = 0$

or $N = W_A \cos 30°$

$= 100 \times \cos 30°$

$\therefore$ $N = 86.602$ nt

The net force on block (A), parallel up to the plane should be zero,

i.e., $\Sigma F_{net,\ |||ar,\ block\ A} = 0$

or $T - W_A \sin 30° - f = 0$

or $T = W_A \sin 30° + f$

$= W_A \sin 30° + \mu N$ [using limiting friction]

$= 100 \times \sin 30° + 0.25 \times 86.602$

$= 71.65$ N

From equation (i); T = W = 71.65 N is the required weight (W) to keep block(A) in static equilibrium.

24. *A body of weight 200 N is being acted upon by a force P of 40 N shown in Fig. (7-W47). If the coefficient of friction between the inclined plane and the body is 0.3, determine whether the body moves up the plane or down the plane or remains stationary.*

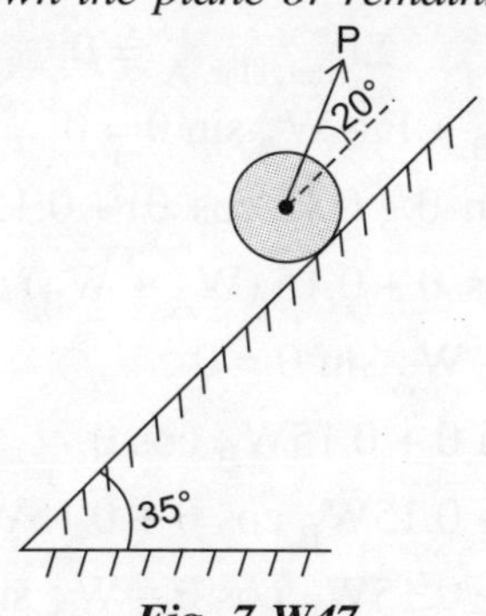

Fig. 7-W47

Solution:

Forces on body and FBD are shown in Fig. (7-W48).

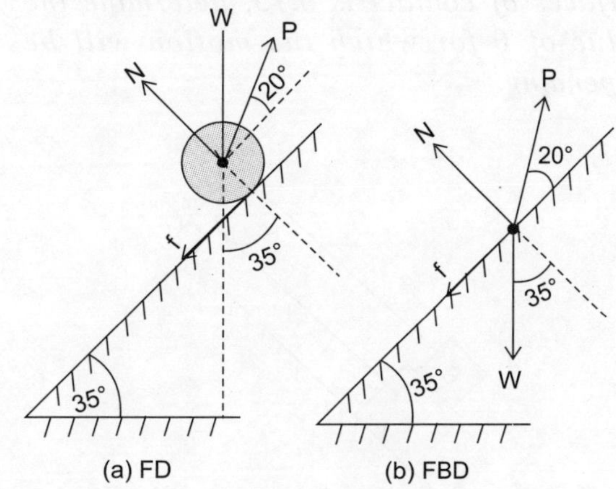

Fig. 7-W48

Since the body is in equilibrium under the action of these forces, then, net force on the body perpendicular to the plane on the body should be zero.

i.e., $\Sigma F_{net,\ \perp lar,\ body} = 0$

or $N - W \cos 35° + P \sin 20° = 0$

or $N = W \cos 35° - P \sin 20°$

$= 200 \cos 35° - 40 \sin 20° = 150.15$ nt

Net force on the body parallel to the plane should be zero

i.e., $\Sigma F_{net,|||ar,\ body} = 0$

or $P \cos 20° - f - W \sin 35° = 0$

or $f = P \cos 20° - W \sin 35°$

$= 40 \cos 20° - 200 \sin 35°$

$= -77.12$ NT

The negative sign shows that the direction of friction force will be opposite to the shown direction.

Now, the limiting frictional force, $f = \mu N$

$= 0.3 \times 150.15$

$= 45.045 < 77.12$

$\Rightarrow$ body will move down the plane.

25. *Block 'A' has a mass of 20 kg and block 'B' has a mass of 10 kg. Knowing that the coefficient of static friction between all surfaces of contact is 0.15, determine the value of θ for which the motion will be impending.*

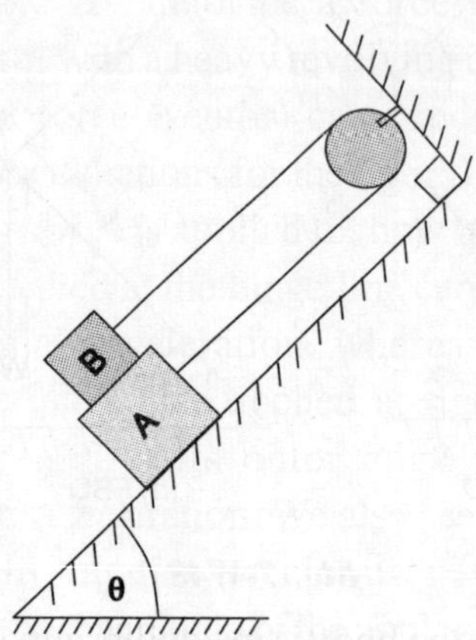

Fig. 7-W49

Solution:

Suppose that the impending motion of block (B) is up the plane, then the friction force should be down the plane. The forces on the blocks and FBD's are shown in the following figures.

Consider the equilibrium of the block B, therefore, the net force perpendicular to the plane should be zero.

i.e., $\Sigma F_{net, \perp r, B} = 0$

or $N_{BA} - W_B \cos\theta = 0$

or $N_{BA} = W_B \cos\theta$ ...(i)

and the net force along the parallel (up) of the plane should be zero.

i.e., $\Sigma F_{net, ||lar, B} = 0$

or $T - W_B \sin\theta - f_{BA} = 0$

or $T = W_B \sin\theta + \mu N_{BA}$

$f_{BA} = \mu N_{BA}$

$= W_B \sin\theta + 0.15\ W_B \cos\theta$

$= W_B (\sin\theta + 0.15 \cos\theta)$ ...(ii)

Now consider the equilibrium of block A, vertical equilibrium of block A gives the net force perpendicular to the plane should be zero.

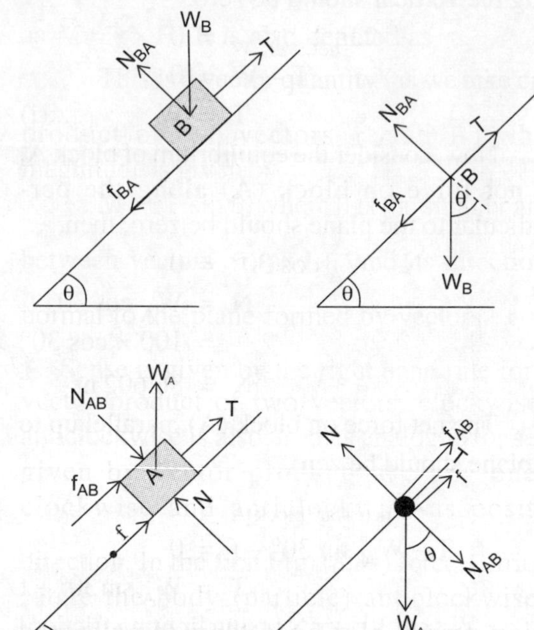

Fig. 7-W50

i.e., $\Sigma F_{net, \perp lar, A} = 0$

or $N - N_{AB} - W_A \cos\theta = 0$

or $N = N_{AB} + W_A \cos\theta$

$= W_B \cos\theta + W_A \cos\theta$

$[\because N_{AB} = N_{BA} = W_B \cos\theta]$

$= (W_A + W_B) \cos\theta$ ....(iii)

Horizontal equilibrium gives the net force parallel (up) the plane, is equal to zero.

i.e., $\Sigma F_{net, ||lar, A,} = 0$

or $T + f_{AB} + F - W_A \sin\theta = 0$

or $W_B (\sin\theta + 0.15 \cos\theta) + 0.15.$

$W_B \cos\theta + 0.15 (W_A + W_B) \cos\theta$

$- W_A \sin\theta = 0$

or $W_B \sin\theta + 0.15 W_B \cos\theta$

$+ 0.15 W_B \cos\theta + 0.15 W_A \cos\theta$

$+ 0.15 W_B \cos\theta - W_A \sin\theta = 0$

or $(W_B - W_A)\sin\theta + (0.15W_B + 0.15W_B + 0.15W_A + 0.15W_B)\cos\theta = 0$

or $\tan\theta = \dfrac{0.15(3W_B + W_A)}{W_A - W_B}$

$= \dfrac{0.15(3\times10+20)}{20-10}$

$= 0.75$

$\therefore \quad \theta \cong 37°.$

26. *A body resting on a horizontal plane required a pull of 10 kN inclined at 60° to the horizontal plane just to move it. It is also found that a push of 12 kN inclined at 30° to the horizontal just moved the body. Find the weight of the body and the co-efficient of friction.*

Solution:

Let us suppose that the weight of the block is W and the coefficient of friction is μ between the body and the horizontal plane.

Consider the condition when the body is pulled by 10 kN inclined at 60° to the horizontal.

Since block is in condition that it is just about to move i.e., impending motion. Forces on the block and free body diagram of the block are shown in the following figures.

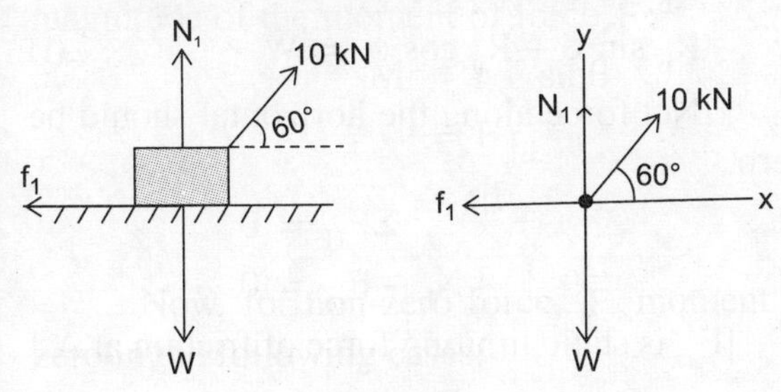

Fig. 7-W51

Now considering vertical equilibrium, we get vertically net force on the body equal to zero.

i.e., $\Sigma F_{net,\,V} = 0$

or $N_1 - W + 10\sin 60° = 0$

or $N = W - 10\sin 60°$

$= (W - 5\sqrt{3})$ ....(i)

Using limiting friction condition, we get

$f_1 = \mu N_1$

$= \mu\,(W - 5\sqrt{3})$

From horizontal equilibrium, we have the net force along the horizontally right, equal to zero.

i.e., $\Sigma F_{net,\,H} = 0$

or $10\cos 60° - f_1 = 0$

or $5 = f_1$

$= \mu\,(W - 5\sqrt{3})$ ...(ii)

Now consider the equilibrium condition when the body is pushed by 12 kN at 30° to the horizontal.

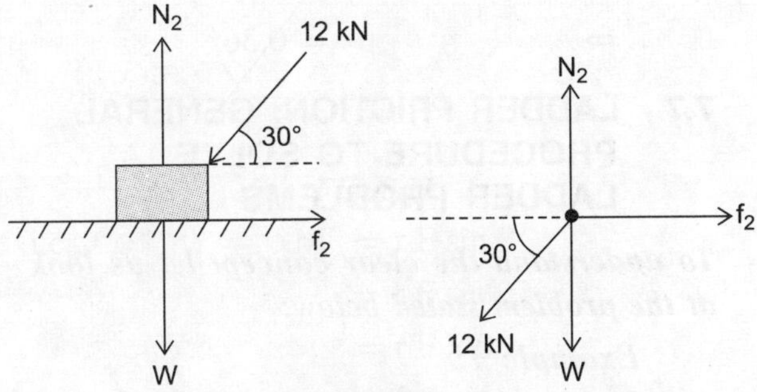

Fig. 7-W52

Now vertical equilibrium of body gives,

$\Sigma F_{net,\,V} = 0$

or $N_2 - W - 12\sin 30° = 0$

or $N_2 = W + 12\sin 30°$

$= W + 6$

Therefore, $f_2 = N_2\,\mu$

$= \mu\,(W + 6)$

[Using condition of limiting friction condition.] ...(iii)

And the horizontal equilibrium of the body gives,

$$\Sigma F_{net,\ H} = 0$$

or $f_2 - 12 \cos 30° = 0$

or $f_2 = 12 \cos 30°$

or $\mu (W + 6) = 6\sqrt{3}$ ...(iv)

$f_2 = \mu (W + 6)$ [From (iii)]

Dividing (ii) by (iv), we get

$$\frac{\mu(W - 5\sqrt{3})}{\mu(W+6)} = \frac{5}{6\sqrt{3}}$$

or $6\sqrt{3}W - 90 = 5W + 30$

or $(6\sqrt{3} - 5)W = 120$

$$\therefore \quad W = \left(\frac{120}{6\sqrt{3}-5}\right)$$

$= 22.25$ kN.

Putting the value of W = 22.25 kN in equation (ii), we get

$$5 = \mu (22.25 - 5\sqrt{3})$$

$\Rightarrow \quad \mu = 0.36.$

7.7 LADDER FRICTION: GENERAL PROCEDURE TO SOLVE LADDER PROBLEMS

To understand the clear concept let us look at the problem stated below:

Example 1:

A ladder AB of length l is supported by a horizontal floor at A and by a vertical wall at B and makes an angle α with the horizontal (Fig. 7-L1). Find the maximum distance x up the ladder at which a man of weight W can stand without causing slipping to occur, if the angle of friction between the floor and ladder and between the wall and ladder is ϕ. Neglect the weight of the ladder itself.

Solution:

When slipping impends, the reactions at A and B will be inclined to the normal by the angle of friction ϕ as shown in Fig. (7-L1) and W is the weight of the man acting downward at a distance x along the ladder.

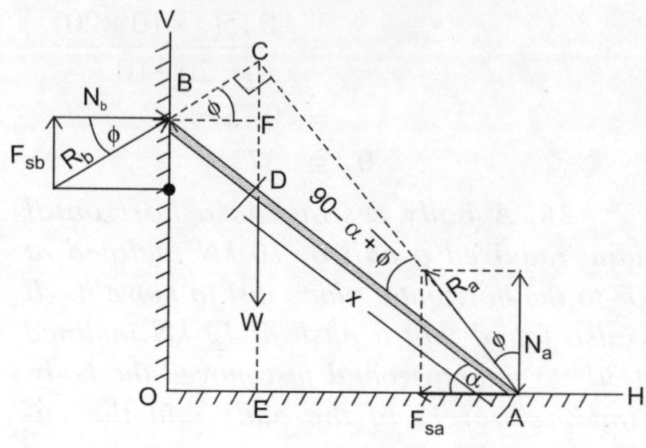

Fig. 7-L1

First method:

Taking the total reaction at points A and B; R_a and R_b respectively making an angle 'ϕ' with the normal at points A and B respectively. Let N_a and N_b are normal reactions.

Since the ladder is in equilibrium then, net force along the vertical should be zero.

i.e., $\Sigma F_V = 0$

or $F_{sb} + N_a - W = 0$

[F_{sb} is static limiting force of friction at B.]

or $R_b \sin \phi + R_a \cos \phi - W = 0$

or $R_b \sin \phi + R_a \cos \phi = W$ ....(i)

Net force along the horizontal should be zero.

i.e., $\Sigma F_H = 0$

$N_b - F_{sa} = 0$

[F_{sa} is static limiting force of friction at A.]

or $R_b \cos\phi - R_a \sin \phi = 0$ ...(ii)

or $R_b \cos \phi = R_a \sin\phi$

or $$\tan \phi = \frac{R_b}{R_a}$$

Net moment about A should be zero.

i.e., $\Sigma M_A = 0$

or $R_b\,\overline{AC} + W\,\overline{AE} \circlearrowright = 0$

$$\left[\because \vec{N}_b + \vec{F}_{sb} = \vec{R}_b\right]$$

or $R_b\, l \cos\{90 - (\alpha + \phi)\} - W\, x \cos\alpha = 0$

or $R_b\, l \sin(\alpha + \phi) - Wx \cos\alpha = 0$ ...(iii)

Now solving equation (ii) and (i), we get

$$R_b = W \sin\phi$$

From equation (iii)

$lW \sin\phi \sin(\alpha + \phi) - Wx \cos\alpha = 0$

or $l \sin\phi \,.\, \sin(\alpha + \phi) = x \cos\alpha$

or $x = l \sin\phi \sin(\alpha + \phi) \sec\alpha$

$\therefore$ $x = l \sin\phi \sin(\alpha + \phi) \sec\alpha$

We conclude that in case of perfectly smooth surfaces ($\phi = 0$) the value of x given by equation (iv) is zero and some sort of support at A will be needed to prevent slipping of the ladder. Again, from equation (iv), we see that, when $\phi \geq 90° - \alpha$, the man may stand safely at the top of the ladder. That is, as long as the ladder makes an angle with the vertical not greater than the angle of friction, then there will be no chance of slipping.

Second method:

If W will pass through the intersection of line of actions of R_a and R_b then the equilibrium condition is automatically satisfied; for that

$$FB = BC \cos\phi$$

$$= l \sin(90 - \overline{\alpha + \phi}) \cos\phi$$

$$= l \cos(\alpha + \phi) \cos\phi$$

$$\therefore EA = l \cos\alpha - l \cos\phi \cos\left(\overline{\alpha + \phi}\right)$$

$$[\because EA = OA - FB]$$

$$= l\left\{\cos\alpha - \cos\phi \cos\left(\overline{\alpha + \phi}\right)\right\}$$

$$\therefore \quad x = EA \sec\alpha$$

$$= l \sec\alpha\, (\cos\alpha - \cos\phi \cos\overline{(\alpha + \phi)})$$

$$= (1 - \cos\phi \cos\overline{\alpha + \phi} \sec\alpha)$$

$$= l \sin\phi \sin(\overline{\alpha + \phi}) \sec\alpha$$

From the above discussion we observe that in any ladder problems there are two normal reactions N_a and N_b and two frictional forces at points A and B (one may be zero). In addition to this, there will be the weight of the ladder or a man or both since these forces are examples of a nonconcurrent coplanar force system and as we know that, for equilibrium caused by this system; we need only three independent equations; two will come when we equate net vertical and horizontal components of forces equal to zero and the third will come, when net moment about any point in the plane, should be zero (take suitable points A or B for simple calculation).

WORKED OUT EXAMPLES

LADDER FRICTION

1. *A ladder of length 4 m weighing 200 N is placed against a vertical wall as shown in Fig. (7-WL1). The coefficient of friction between the wall and the ladder is 0.25 and that between the ladder and the floor is 0.30. Determine the minimum horizontal force to be applied at A to prevent slipping when a man weighing 600 N wants to stand at a distance 3 m from A as shown in the figure.*

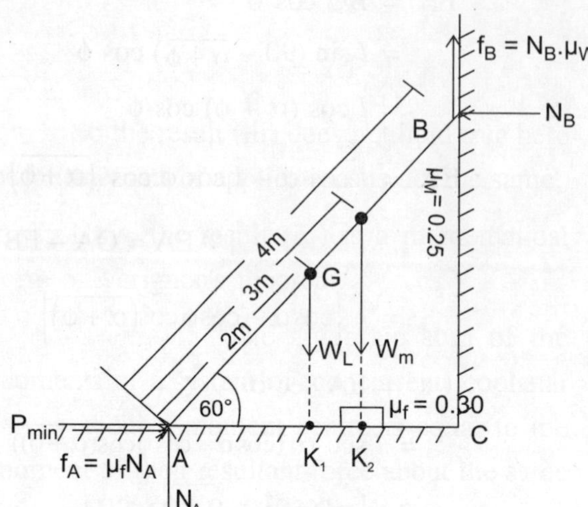

Fig. 7-WL1

Solution:

Forces on the ladder are

(a) W_m, weight of the man stands at a distance 3 m from A,

(b) W_L, weight of ladder passes through G at a distance 2 m from A,

(c) normal force, N_B due to the wall,

(d) f_B, limiting frictional force due to the wall when tendency is down the wall,

(e) N_A, normal reaction due to the floor,

(f) f_A, limiting frictional force due to the floor when tendency is right,

(g) P_{min}, applied minimum force.

Now, since the ladder is in equilibrium, and forces are planar but not concurrent so, for the equilibrium of the ladder we have to consider three conditions.

Net vertical force on the ladder should be zero.

i.e., $\Sigma F_{net,\ V} = 0$

or $f_B - W_B - W_L + N_A = 0$

or $f_B + N_A = W_m + W_L$

$= 600 + 200$

$= 800 \text{ N}$

or $0.25 + N_A = 800\text{N}$ ...(i)

Net horizontally right, force on the ladder should be zero.

i.e., $\Sigma F_{net,\ H} = 0$

or $P_{min} + f_A - N_B = 0$

or $P_{min} + 0.30 \times N_A - N_B = 0$

or $P_{min} + 0.30 \times N_A = N_B$ ....(ii)

Net moment on the ladder should be zero about any convenient point (let. A) [**Note:** This equation comes in picture when forces are not concurrent].

i.e., $\Sigma M_A = 0$

or $P_{min} \times 0 + f_A \times N_A \times 0 - W_L \times \overline{AK_1}$

$- W_m \times \overline{AK_2} + f_B \times AC + N_B \times BC = 0$

Taking anticlockwise direction for positive torque.

or $-200 \times 2 \cos 60° - 600 \times 3 \cos 60° + N_B \times 0.25 \times 4 \cos 60° + N_B \times 4 \sin 60° = 0$

or $-200 \times 2 \times \frac{1}{2} - 600 \times 3 \times \frac{1}{2} + N_B \times 0.25 \times 4 \times \frac{1}{2} + N_B \times 4 \times \frac{\sqrt{3}}{2} = 0$

or $-200 - 900 + 0.5N_B + 2\sqrt{3}\,N_B = 0$

or $N_B\left(2\sqrt{3} + 0.5\right) = 1100$

$\therefore$ $N_B = \frac{1100}{2\sqrt{3} + 0.5}$

$= 277.5 \text{ N}$

Now from equation (i), we get

$0.25N_B + N_A = 800$

or $277.5 \times 0.25 + N_A = 800$

$\therefore$ $N_A = 730.62 \text{ N}$

By substituting the value of N_A and N_B in the equation (ii), we get

$P_{min} = N_B - 0.30\ N_A$
$= 277.5 - 0.30 \times 730.62$
$= 58.314$ N.

Note: *What is the maximum value of the applied force so that the ladder remains in equilibrium?*

Solution:

When maximum force is applied then, as we know that, the frictional force is adjustable so it is adjusted to zero.

i.e., $\Sigma F_{net,V} = 0 \Rightarrow N_A = 800$ ....(i)

and $F_{net,H} = 0$

$\Rightarrow \quad P_{max} = N_B$ ....(ii)

and $200 \times 2 \times \cos 60° + 600 \times 3 \times \cos 60° = 4 \sin 60° N_B$

or $1100 = 2\sqrt{3}\, N_B$

$\therefore \quad P_{max} = N_B$
$= 317.5$ N.

2. *A ladder 5 m long weighing 200 N rests against a verical wall with the angle of inclination of 30°. The coefficient of friction at both the wall and the ground is 0.3. Determine how high a man weighing 800 N can climb before the ladder slips.*

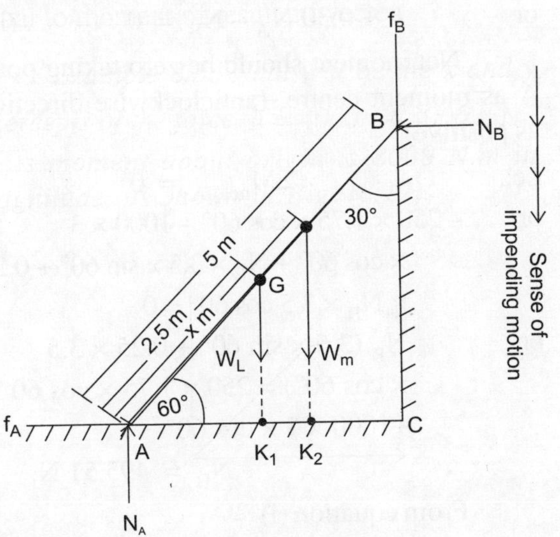

Fig. 7-WL2

Solution :

Forces on the ladder are

(a) W_m, weight of the man climbed at a distance x from point A,

(b) W_L, weight of the ladder passing through centre of gravity of ladder (at a distance 2.5 m from point A),

(c) N_A, normal reaction at point A,

(d) f_A, limiting frictional force due to ground,

(e) N_B, normal reaction at point B by the vertical wall, and

(f) f_B, limiting friction due to wall.

Now since the ladder is in equilibrium and the forces are planar but not concurrent, so apart from equation like $\Sigma\vec{F}_{net} = 0$, we have to consider $\Sigma\vec{M} = 0$ also.

Net force on the ladder along the vertically up should be zero.

i.e., $\Sigma F_{net,V} = 0$

or $N_A + f_B - W_L - W_m = 0$

or $N_A + 0.30 N_B = W_L + W_m = 1000$ ...(i)

Net force on the ladder along the horizontal should be zero.

i.e., $\Sigma F_{net,H} = 0$

or $f_A + (-N_B) = 0$

or $0.30 N_A - N_B = 0$

or $0.30 N_A = N_B$ ...(ii)

From equation (i) and (ii), we get

$N_A + 0.30 \times 0.30 \times N_A = 1000$

or $N_A (1 + 0.30 \times 0.30) = 1000$

or $N_A = \dfrac{1000}{1 + 0.30 \times 0.30}$

$\therefore \quad N_A = 917.43$ N

Therefore, from equation (ii)

$N_B = 0.30 N_A$
$= 917.43 \times 0.30$
$= 275.229$ N

Now considering point A as moment centre and taking anticlockwise moment as positive moment, we get

$$\Sigma M_A = 0$$

or $\quad -W_L \times \overline{AK_1} - W_m \times \overline{AK_2} + N_B \times \overline{BC} + f_B \times \overline{AC} = 0$

or $\quad -200 \times 2.5 \times \cos 60° - 800 \times x \times \cos 60° + 275.229 \times 5 \sin 60° + 275.229 \times 0.30 \times 5 \cos 60° = 0$

or $\quad -250 - 400x + 1191.776 + 206.421 = 0$

or $\quad 400x = 1148.197$

or $\quad x = 2.87 \text{ m}$

Hence at a distance x = 2.87 man can climb without slipping.

3. *A ladder AB 3.5 m in length and weight 250 N is placed against a wall with A at the floor level, B on the wall. AB making 60° with the floor. The coefficient of friction between the wall and ladder is 0.25 and that between floor and ladder is 0.30. In addition, it has to support a load of 1000 N at a distance of 3 m from A along the ladder. To prevent slipping a force F is applied horizontally at A, at the level of the floor. Find the minimum and maximum force F required for this condition.*

Solution:

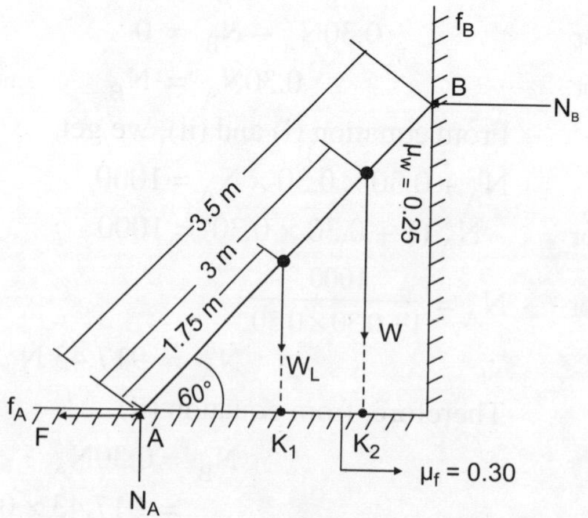

Fig. 7-WL3

Forces on ladder as shown in Fig. are

(a) $W_L = 250$ N, weight of ladder,

(b) W = 1000 N weight of load,

(c) N_B, normal reaction at point B by the wall,

(d) f_B, upward limiting frictional force, by wall as tedency is down,

(e) N_A, normal reaction at point A by floor,

(f) f_A, limiting frictional force by floor due to impending motion towards left, and

(g) F is applied force.

Now considering the equilibrium of ladder we have got net vertically up force on the ladder equal to zero.

i.e., $\quad \Sigma F_{net,\, V} = 0$

or $\quad N_A - W_L - W + f_B = 0$

or $\quad N_A + 0.25 N_B = W_L + W = 250 + 1000$

$\therefore \quad N_A + 0.25\, N_B = 1250 \text{ N} \quad$ ....(i)

Net horizontally right force should be zero.

i.e., $\quad \Sigma F_{net,\, H} = 0$

or $\quad F + f_A - N_B = 0$

or $\quad F + 0.30\, N_A - N_B = 0 \quad$ ....(ii)

Net moment should be zero taking point A as moment centre. (anticlockwise direction as positive)

i.e., $\quad \Sigma M_{net,\, A} = 0$

or $\quad -250 \times 1.75 \times \cos 60° - 1000 \times 3 \times \cos 60° + N_B \times 3.5 \times \sin 60° + 0.25 \times N_B \times 3.5 \cos 60° = 0$

or $\quad N_B\,(3.5 \times \sin 60° + 0.25 \times 3.5 \times \cos 60°) = 250 \times 1.75 \times \cos 60° + 1000 \times 3 \times \cos 60°$

$\therefore \quad N_B = 495.51 \text{ N.}$

From equation (i)

$$N_A + 0.25\, N_B = 1250$$

or $\quad N_A + 0.25 \times 495.51 = 1250$

or $\quad N_A = 1126.12$

$\therefore \quad N_A = 1126.12\text{ N}$

Now substituting the values of N_A and N_B to equation (ii), we get

$$F = N_B - 0.30\, N_A$$
$$= 495.51 - 0.30 \times 1126.12$$

$\therefore \quad F_{min} = F$

$$= 157.67\text{ N}$$

Now when we apply maximum applied force at that instant frictional force will be zero i.e.,

$$F_{max} = F_{min} + f_A = 157.67 + 0.3 \times 1126.12$$
$$= 495.51\text{ N.}$$

4. *A ladder 5 m, long rests on the floor and vertical wall as shown in Fig.(7-WL4). The coefficient of friction at the wall is 0.2 and at the floor is 0.4. The weight of the ladder is 200 N can be considered as acting at the centre of gravity of the ladder. The ladder supports an additional load of 900 N at C, 1 m along the ladder from top. Determine the lowest value of angle θ between ladder and floor at which ladder may be placed without causing any slip.*

Solution:

Forces on the ladder shown in Fig. (7-WL4).

(a) $W_L = 200$ N, weight of the ladder,

(b) $W = 900$ N, weight of the load,

(c) N_A, normal reaction at point A,

(d) Frictional limiting force, f_A, by the floor as impending motion is right

(e) N_B, normal reaction at point B by the floor, and

(f) f_B, limiting frictional force due to the wall as impending motion is down the wall.

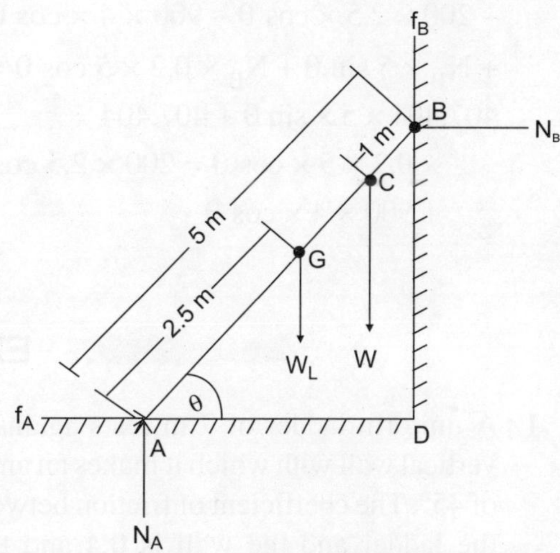

Fig. 7-WL4

Now considering the equilibrium of ladder we have net vertical force on the ladder equal to zero.

i.e., $\quad \Sigma F_{net,\,V} = 0$

or $\quad N_A - 200 - 900 + f_B = 0$

or $\quad N_A + 0.2\, N_B = 1100 \quad$ ...(i)

Net horizontal force on the ladder will be zero.

i.e., $\quad \Sigma F_{net,\,H} = 0$

or $\quad f_A - N_B = 0$

or $\quad 0.4\, N_A - N_B = 0 \quad$ ...(ii)

putting $N_B = 0.4 N_A$ to equation (i) we get,

or $\quad N_A + 0.2\, N_B = 1100$

or $\quad N_A + 0.2 \times 0.4\, N_A = 1100$

or $\quad N_A = \dfrac{1100}{1.08}$

$$= 1018.518\text{ N}$$

$\therefore \quad N_B = 0.4\, N_A$

$$= 0.4 \times 1018.518$$
$$= 407.407\text{ N}$$

Now considering rotational equilibrium and taking point 'A' as moment centre, we get (take as ↺ +ve direction)

$$- 200 \times 2.5 \times \cos\theta - 900 \times 4 \times \cos\theta + N_B \times 5 \sin\theta + N_B \times 0.2 \times 5 \cos\theta = 0$$

or $$407.404 \times 5 \times \sin\theta + 407.404 \times 0.2 \times 5 \times \cos\theta = 200 \times 2.5 \cos\theta + 900 \times 4 \times \cos\theta$$

or $$2037.02 \sin\theta + 407.404 \cos\theta = 500 \cos\theta + 3600 \cos\theta$$

or $$2037.02 \sin\theta = 3692.596 \cos\theta$$

or $$\tan\theta = 1.812$$

$$\therefore \quad \theta \cong 61°10'.$$

EXERCISE

1. A uniform ladder of 7 m rests against a vertical wall with which it makes an angle of 45°. The coefficient of friction between the ladder and the wall is 0.4 and that between the ladder and the floor is 0.5. If a man, whose weight is one-half of that of the ladder, ascends it, how long will it be when the ladder slips.

2. A ladder 3 mts in length and 200 N is placed against a wall with A at the floor level, B on the wall. AB making 60° with the floor, the coefficient of friction between the wall and ladder is 0.25 and between the floor and ladder is 0.35. In addition, it has to support a load of 1000 N at its top at B. To prevent slipping, a force F is applied horizontally at A, at the level of the floor. Find the maximum force required for this condition, find also the angle 'α' above which the above ladder with additional load at the top should be placed to prevent slipping without the force F.

3. A ladder 5 m long is supported by a vertical wall and resting on a horizontal floor, it is inclined at 30° to the vertical. A man of weight 500 N wishes to carry a load of 1000 N, up the ladder, can he reach safely at the top of the ladder? If not, at what position of the man does the ladder tends to slip? The coefficient of friction is 0.2 for all contact surfaces. The weight of ladder is 300 N.

4. A ladder 4 m long and self weight 200 N is supported against the ground and wall as shown in Fig. (7-EL1). If a man of 650 N of climbs to the top of the ladder, determine the inclination with reference to the ground at which the ladder is to be placed to prevent slipping. Take $\mu = 0.25$ for all contact surfaces.

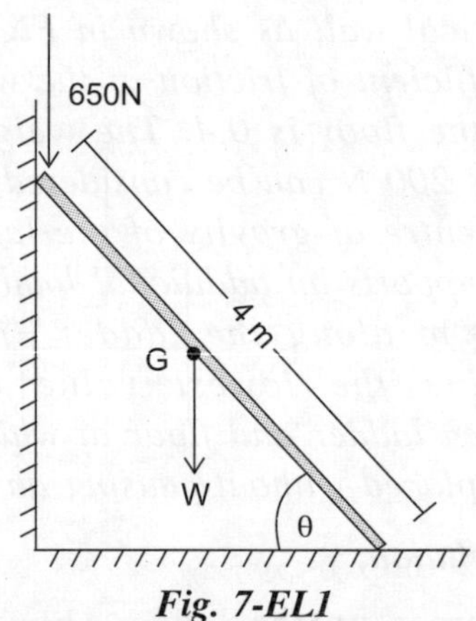

Fig. 7-EL1

5. A ladder of mass 20 kg and length 5m rests on the ground and against a smooth vertical wall as shown in Fig. (7-EL2). The ladder makes an angle of 55° with reference to the ground. If a man of (mass) 70 kg climbs up the ladder through a distance of 4 m. Determine (i) magnitude of frictional force at the ground to avoid slipping (ii) reaction at the ground and the wall and (iii) the coefficient of friction between the ground and ladder.

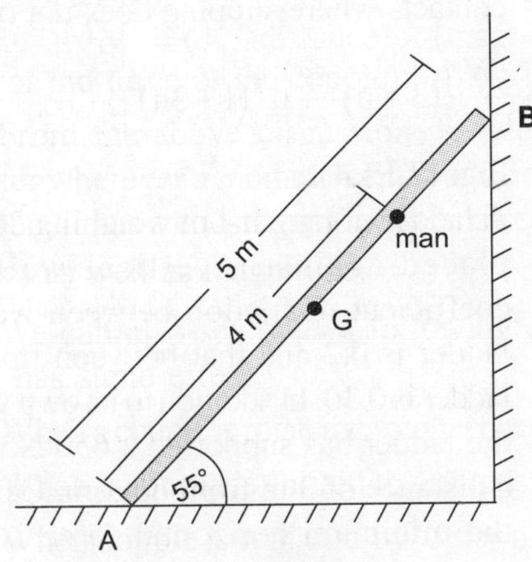

Fig. 7-EL2

6. A 4 m long uniform ladder of mass 70 kg is placed against a smooth vertical wall. The lower end of the ladder is on the floor at a distance of 1.5 m from the wall. Taking the coefficient of friction between the floor and ladder as 0.5, determine the frictional force at the lower end of the ladder for the equilibrium condition. What is the maximum frictional force?

7. A ladder 24 kg mass and 4 m long is supported against a smooth wall and a rough floor. The coefficient of friction between the ladder and the floor is 0.35. Determine the distance through which a man of mass 60 kg can climb the ladder.

8. A ladder rests against a rough vertical wall whereas its lower end is on the rough horizontal floor. Ladder is being inclined at 45° to the horizontal. The coefficient of friction between the ladder and the wall is $\frac{1}{3}$ and that between the ladder and floor is $\frac{1}{2}$. A man whose weight equals one-half of that of the ladder ascends up the ladder till the ladder slips. Determine up to what length of the ladder the man will be able to ascend before the ladder starts slipping.

9. A uniform ladder of length l and weight W rests with its foot on a rough ground, coefficient of friction is μ and its upper end against a vertical smooth wall, its inclination being α. A force P is applied to horizontally at a distance a from the foot, so as to make the foot approach the wall, show that P must exceed

$$\frac{Wl}{l-a}\left(\mu+\frac{1}{2}\cot\alpha\right).$$

10. A ladder AB rests with its lower end A on a rough horizontal surface and other end B against a rough vertical wall, the coefficient of friction being μ_1 and μ_2 at the ground and the wall respectively. If the centre of gravity of the ladder is at a distance a and b from end A and end B respectively. Show that when ladder is in limiting equilibrium, the inclination of the ladder with the horizontal is

$$\tan^{-1}\left\{\frac{a-b\mu_1\mu_2}{(a+b)\mu_1}\right\}.$$

11. Two uniform ladder OA and OB of length l each and weight W each joined at O as shown in Fig. (7-EL3). Rough horizontal floor for which the coefficient of friction is μ. If the ladder are in limiting equilibrium show that the angle AOB is equal to $\sin^{-1}\left\{\frac{4\mu}{1+4\mu^2}\right\}$.

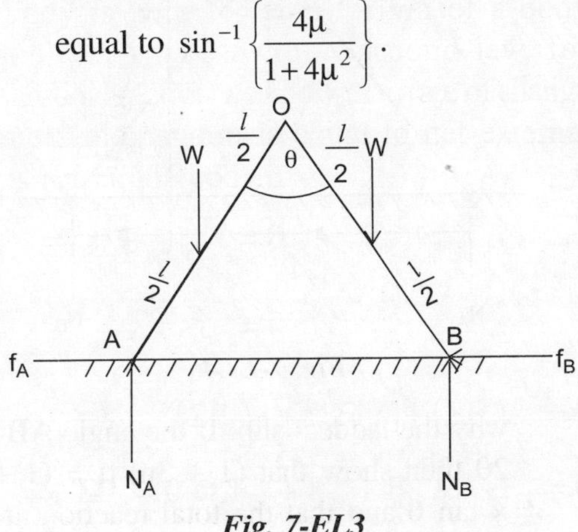

Fig. 7-EL3

12. Two uniform ladder OA and OB of the same length are joined at point O. They started with their lower (ends) A and B on a rough horizontal surface the coefficient of friction between the ladder and floor being μ if each ladder is l unit long and weight W and a man of weight W can start any where on the ladder where the distance between A and B is 2a. Show that μ should not be less than

$$\frac{2a}{3\sqrt{l^2 - a^2}}.$$

13. A ladder resting on a horizontal floor leans against a vertical wall at an angle of 60° with the floor. If the coefficient of friction is 0.025 at all the contact surfaces what % of the length of the ladder a person can ascend without causing the ladder to slip. W is the weight of ladder.

14. Two uniform ladders AB and BC each of length l and weight W and nW (n < 1) are hinged together at the top B and stand on the rough horizontal ground. The coefficient of friction at A and C being μ, the angle ABC is gradually increased until slipping occurs. Then state with reasons

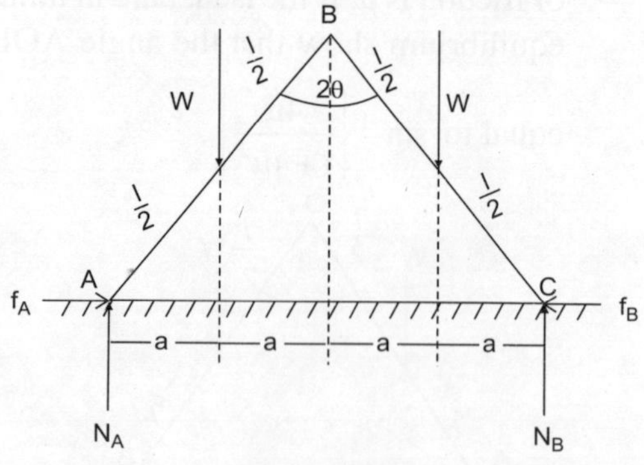

Fig. 7-EL4

why the ladders slip. If the angle ABC is 2θ then show that (1 + 3n) μ = (1 + n) × tan θ and that the total reaction at the contacts where slipping does not occur is

$$\frac{W}{l}\left\{(3+n)^2 + \mu^2(1+3n)^2\right\}^{1/2}.$$ Refer Fig. (7-EL4).

15. A ladder of length 4 m weighing 200 N is placed against a vertical wall. The coefficient of friction between wall and ladder is 0.2 and that between floor and ladder is 0.30. In addition to its own weight, the ladder has supported a 650 N man at a distance of 3 m from bottom. Calculate the minimum horizontal force is to be applied at the bottom to prevent slipping.

16. A uniform ladder of length 10 m and mass 16 kg is resting against a vertical wall making an angle of 37° with it. The vertical wall is friction less but the ground is rough. A mechanic weighing 60 kg climbs up the ladder. If he stays on the ladder at a point 8 m from the lower end what should be the coefficient of friction for the mechanic to work safely.

17. A homogeneous ladder is placed on a flat horizontal surface to rest against a vertical wall. Assuming that the coefficient of friction at each surface is μ, determine the minimum possible inclination of the ladder with the horizontal whether the angle is less than 45° ?

18. A ladder as shown in Fig. (7-EL5) is 6 m long and is supported by a horizontal floor and a vertical wall. The coefficient of friction between the floor and the ladder is 0.4 and between the wall and ladder is 0.25. The weight of the ladder is 200 N and may be considered as concentrated at G. The ladder also supports a vertical load of 900 N at C which is at a distance of 1 m from B. Determine the least value of α at which the ladder may be placed without slipping. Also determine the reactions at this stage.

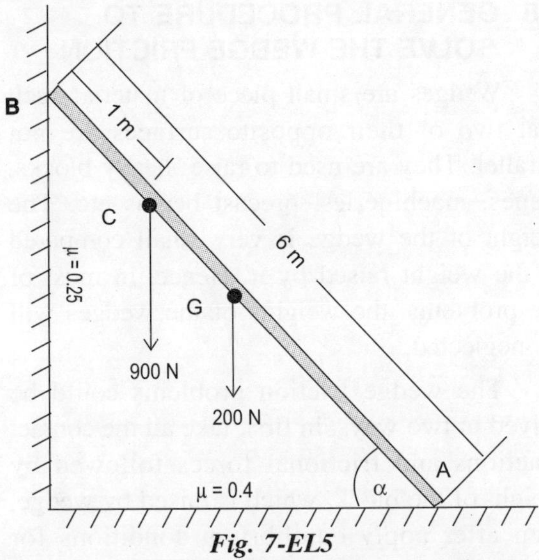

Fig. 7-EL5

19. Show that the ladder shown in the Fig. (7-EL6) is in equilibrium. The length of the ladder is 10 m and has a weight of 250 N. The lower end is at 5 m from the wall. The wall is smooth where the floor is rough with coefficient of friction being 0.35.

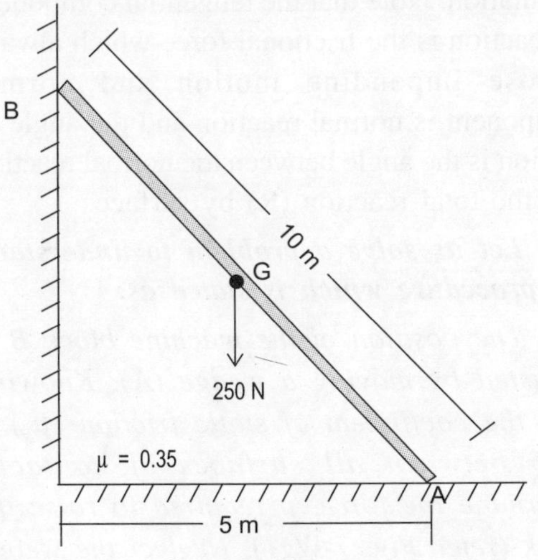

Fig. 7-EL6

20. A 4 m ladder weighing 180 N is placed against a vertical wall as shown in Fig. (7-EL7). A man weighing 750 N reaches a point 2.7 m from A and the ladder is about to slip. Assuming that the coefficient of limiting friction between the ladder and the wall is 0.25, determine the coefficient of friction between the ladder and the floor.

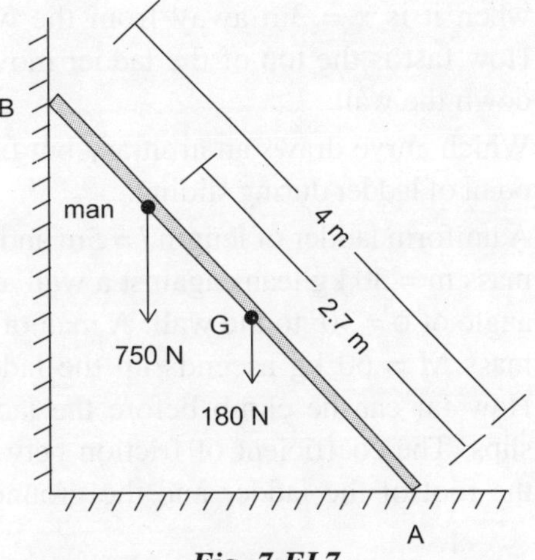

Fig. 7-EL7

21. A ladder of length 4.5 m rests on a horizontal ground and leans against a smooth vertical wall at an angle of 65° with the horizontal, the weight of the ladder is 300 N, the ladder is on the verge of sliping when a man weighing 700 N stands on the rung 1.6 m height. Calculate the coefficient of the friction between the ladder and the floor.

22. There is a ladder with the length c and corridor with width b unfortunately the corridor has a right angled bend. How long may a ladder be so that it may move around the corner ?

23. A box of cube shape with the edge length 1 m stands in front of a wall. A ladder of the length 5 m leans against the wall and just touches the box at an edge. How high on the wall is the top of the ladder ?

24. There are two ladder between two houses. How long is the distance x between the house walls if the ladders A and B and the height c over the ground of crossing are given ?

25. Find the curve of the centre of the ladder during sliding. Length of ladder is c.

26. A ladder c = 5 m long leaning against a wall. The foot of the ladder is pulled away from the wall at a spend of $v_x = 0.3$ just when it is x = 3m away from the wall. How fast is the top of the ladder moving down the wall.

27. Which curve draws an arbitrary but fixed point of ladder during sliding.

28. A uniform ladder of length $l = 5$m and the mass m = 30 kg leans against a wall at an angle of $\phi = 30^o$ to the wall. A man of the mass M = 60 kg ascends up the ladder. How far can he climb before the ladder slips. The coefficient of friction between the foot of the ladder and the ground is $\mu = \frac{1}{3}$. Assume that there is no friction along the wall.

29. Let the weight of the ladder be 16 kg, so that w = 160 N and let it be at 60º to the ground.
Find (i) the forces acting at each end of the ladder (ii) find the minimum coefficient of friction againt the ground.

30. A non-uniform ladder AB of length 15m and mass 40 kg has its centre of gramity at a point 5 m from A. The ladder rests with end A on rough horizontal ground (coefficient of friction 0.25) and end B against a rough vertical wall (coefficient of friction 0.2). The ladder makes an angle α with horizontal such that $\tan \alpha = \frac{9}{4}$. A straight string connects A to a point at base of the wall directly below B. A man of mass 80 kg begins to climb the ladder. How far up the ladder can the man climb without causing tension in the string ? What tension must the string be capable of withstanding if the man is to climb to top of the ladder?

7.8 GENERAL PROCEDURE TO SOLVE THE WEDGE FRICTION

Wedges are small piece of material such that two of their opposite surfaces are not parallel. They are used to raise heavy blocks, stones, machineries, precast beams etc. The weight of the wedge is very small compared to the weight raised by it. Hence, in most of the problems, the weights of the wedges will be neglected.

The wedge friction problems could be solved in two ways, in first, take all the contact reactions and frictional forces followed by weight of the block, which is raised by wedge, then after apply equilibrium conditions for block(s) and wedge, if motion of block is impending. In second, instead of treating normal reaction and frictional force independently, it is good practice to treat their result making an angle equal to angle of friction i.e., $\theta = \tan^{-1} \mu$ with normal (Force). This method decreases the numbers of unknowns and hence ease the calculation. Note that the tangential component of reaction is the frictional force which always oppose impending motion and normal component is normal reaction and the angle of friction is the angle between the normal reaction and the total reaction (R) by surface.

Let us solve a problem to understand the procedure which is stated as:

The position of the machine block B is adjusted by moving a wedge (A). Knowing that the coefficient of static friction (μ_s) is 0.35 between all surfaces of contacts, determine the force, p required to raise the block (refer Fig. 7-Wg1). Neglect the weight of the wedge.

Now we take 1st method in which normal reaction (N) and frictional force (F_s) are taken separately. Then, for the condition of impending upward motion of the block, the free body diagrams for the block and the wedge, respectively, are as shown in Fig. (7-Wg2a and b).

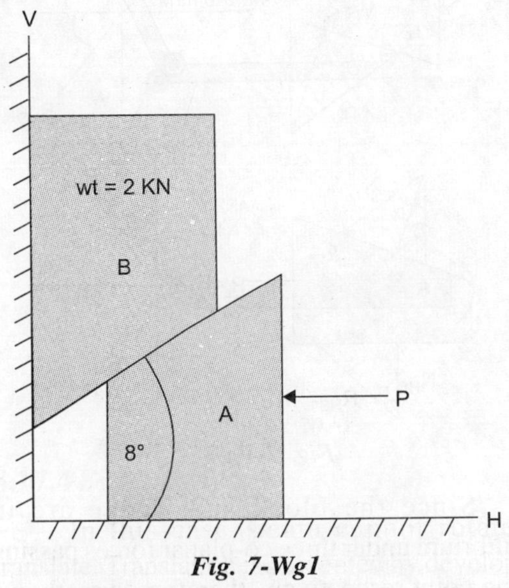

Fig. 7-Wg1

The forces on the block are:

1. Weight of the block acting downward,
2. Normal reaction by vertical wall in horizontal, N_{vs}.
3. As the block has impending motion upward, therefore, the frictional force will be limiting. Frictional force in downward $F_{vs,\ limit} = m_{vs}\ N_{vs}$.
4. Normal reaction (N) by incline surface which is 8° inclined with the horizontal, here the normal (N) makes an angle 8° with the vertical in anticlockwise direction.
5. Frictional force F_{limit} by inclined plane of wedge along the plane such that this opposes the motion and because of impending motion, it is $F_{limit} = \mu N$ where μ is the coefficient of friction between the block and wedge at inclined plane.

Now since block, B is in equilibrium, then net force in horizontal direction = 0

i.e., $N_{vs} - F \cos 8° - N \sin 8° = 0$

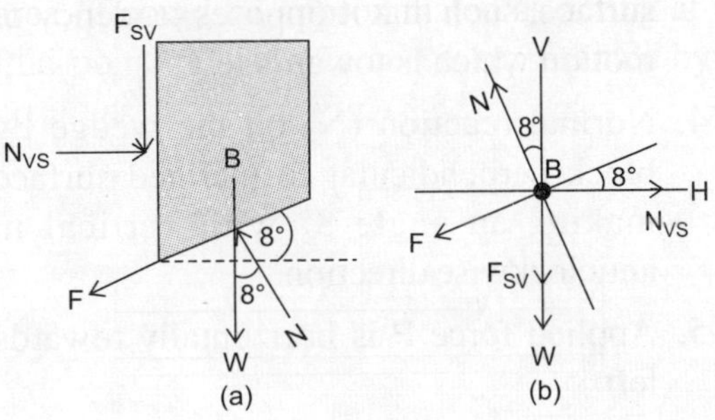

Fig. 7-Wg2

or $N_{vs} = N\mu \cos 8° + N \sin 8°$

$= N (\mu \cos 8° + \sin 8°)$

$= N (0.35 \cos 8° + \sin 8°)$

∴ $N_{vs} = 0.485$ ....(i)

and net force in vertical direction = 0

or $N \cos 8° - F_{sv} - W - F \sin 8° = 0$

or $N \cos 8° - \mu_{sv} N_{vs} - W - \mu N \sin 8° = 0$

or $N \cos 8° - 0.35 \times 0.485 - 2 - 0.35 N \sin 8° = 0$

or $(\cos 8° - 0.35 \times 0.485 - 0.35 \sin 8°) N = 2$

or $0.771 N = 2$

∴ $N = 2.6$ knt

∴ $F = \mu N$

$= 0.35 \times 2.6$

$= 0.91$ knt

From equation (i)

$N_{vs} = 0.485 \times 2.6$ knt

$= 1.261$ knt

∴ $F_{vs} = 1.261 \times 0.35$

$= 0.441$ knt.

Now the forces on the wedge are:

1. Normal reaction by horizontal surface on wedge N_{HS} in upward direction.
2. Limiting frictional force $F_{HS} = m_{SH} N_{HS}$ by horizontal surface towards right.
3. Limiting frictional force, F = mN by block on the wedge along the inclined plane

surface, such that it opposes tendency of motion which is towards left.

4. Normal reaction (N) on the wedge by block perpendicular to inclined surface making an angle 8° with vertical in anticlockwise direction.
5. Applied force P is horizontally towards left.

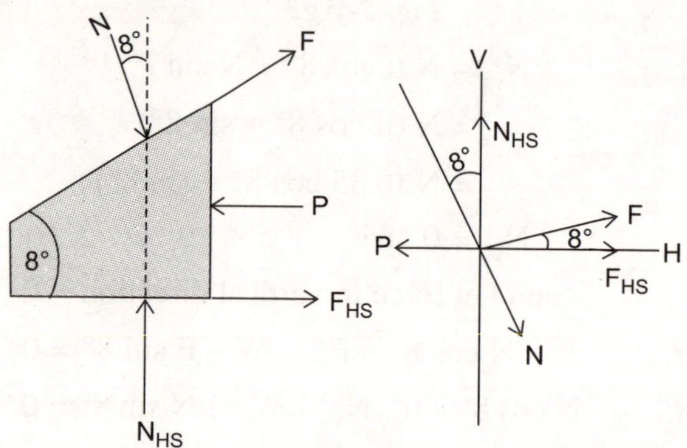

Fig. 7-Wg3

Now since the wedge is in equilibrium, then net force in vertical direction = 0

i.e., $N_{HS} - N \cos 8° + F \sin 8° = 0$

or $N_{HS} = N \cos 8° - F \sin 80°$

$= 2.6 \cos 8° - 0.91 \sin 8°$

$\therefore \quad N_{HS} = 2.448$ knt

$\therefore \quad F_{HS} = \mu_{HS} N_{HS}$

$= 0.35 \times 2.448$ knt

$= 0.856$ knt

and net force in horizontal direction = 0

or $F_{HS} - P + F \cos 8° + N \sin 8° = 0$

or $P = F_{HS} + F \cos 8° + N \sin 8°$

$= 0.856 + 0.91 \cos 8° + 2.6 \sin 8°$

$= 2.118$ kN

$\therefore \quad P = 2118$ N.

Method 2:

In this method, the resultant of normal reaction and frictional force is taken as the contact reaction by the surface.

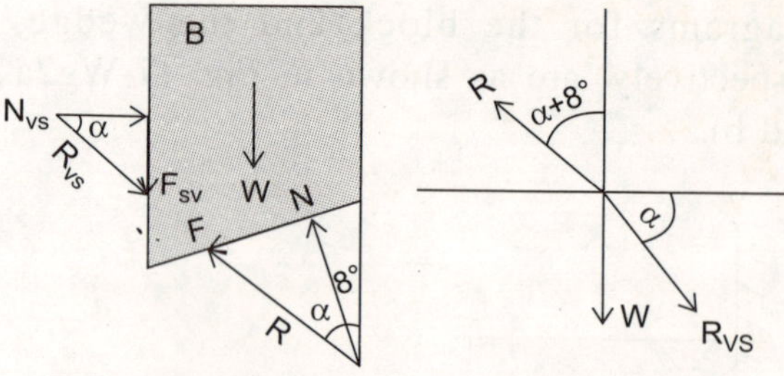

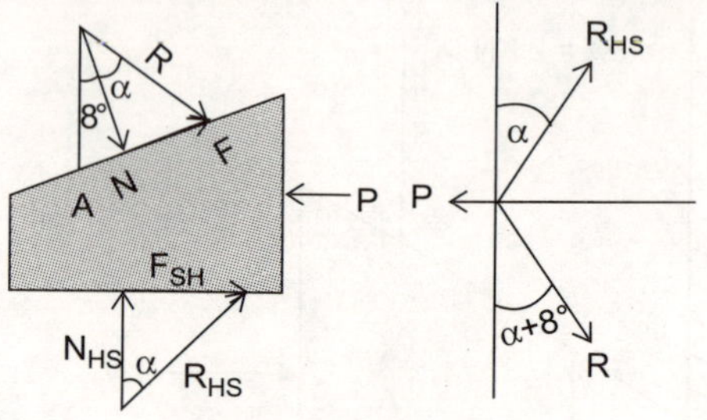

Fig. 7-Wg4

Since the block and wedge are in equilibrium under three co-planar forces passing at one point respectively, therefore, directly we can directly apply Lami's theorem.

For block:

$$\frac{R}{\sin(90-\alpha)} = \frac{W}{\sin(2\alpha+98°)}$$

$$= \frac{R_{vs}}{\sin(172-\alpha)} \quad \text{...(i)}$$

For wedge:

$$\frac{R_{HS}}{\sin(98+\alpha)} = \frac{P}{\sin(180°-2\alpha-8)}$$

$$= \frac{R}{\sin(90+\alpha)} \quad \text{...(ii)}$$

From (i)

$$R = \frac{\sin(90-\alpha)}{\sin(2\alpha+98°)} W \quad \text{...(iii)}$$

and from (ii)

$$R = \frac{\sin(90+\alpha)}{\sin(172-2\alpha)} P \quad \text{....(iv)}$$

Equating equation (iii) and (iv)

$$P = \frac{\sin(172-2\alpha).\sin(90-\alpha)}{\sin(2\alpha+98°).\sin(90+\alpha)} \times W$$

$\because \quad \alpha = \tan^{-1} \mu_s$

$= \tan^{-1} 0.35$

$= 19.29°$

$$\therefore \quad P = \frac{\sin 133.42° \times \cos 19.29°}{\sin 136.58 \times \sin 109.29°} \times W$$

$$= \frac{\sin 133.42}{\sin 136.58} \times 2$$

$= 2.113$ kN

$P = 2113$ N.

We see that there is small difference in P values, because of approximation in sine and cosine values. Thus, we can solve problems by two methods and out of these two the second method is easier than the first.

WORKED OUT EXAMPLES

WEDGE FRICTION:

1. *Determine the force P that must be applied to the 20 kN block B to lift the 100 kN block A. The coefficient of friction for all surfaces of contacts is μ = 0.3.*

Solution: First method:

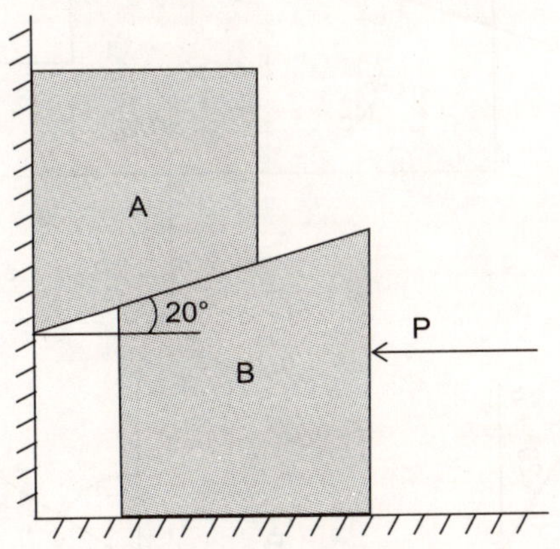

Fig. 7-WWg1

Forces on block (A) and block (B) are shown in following diagrams:

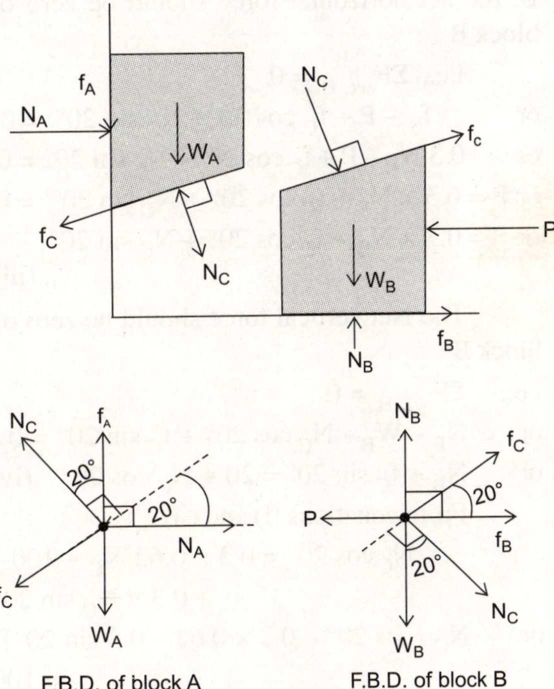

Fig. 7-WWg2

Consider the equilibrium of block A, the net horizontal force should be zero on block A.

i.e., $\Sigma F_{net,\ H} = 0$

or $N_A - f_C \cos 20° - N_C \sin 20° = 0$

or $N_A = \mu N_C \cos 20° + N_C \sin 20°$

$= (0.3 \cos 20° + \sin 20°) N_C$

or $N_A = 0.62\ N_C$ ...(i)

The net vertical force should be zero on block A.

i.e., $\Sigma F_{net,\ V} = 0$

or $N_c \cos 20° - f_A - W_A - f_C \sin 20° = 0$

or $N_C \cos 20° = 0.3\ N_A + W_A + 0.3\ N_C \sin 20°$

or $N_C \cos 20° = 0.3\ N_A + 100$

$+ 0.3\ N_C \sin 20°$ ...(ii)

Now consider the equilibrium of block B, the net horizontal force should be zero on block B

i.e., $\Sigma F_{net,\ H} = 0$

or $f_B - P + f_C \cos 20° + N_C \sin 20° = 0$

or $0.3\ N_B - P + f_C \cos 20° + N_C \sin 20° = 0$

or $P = 0.3 \times N_B + f_C \cos 20° + N_C \sin 20° = 0$

or $P = 0.3 \times N_B + f_C \cos 20° + N_C \sin 20°$ ...(iii)

The Net vertical force should be zero on block B

i.e., $\Sigma F_{net,\ H} = 0$

or $N_B - W_B - N_C \cos 20° + f_C \sin 20° = 0$

or $N_B + f_C \sin 20° = 20 + N_C \cos 20°$ ...(iv)

From equations (i) and (ii)

$N_C \cos 20° = 0.3 \times 0.62\ N_C + 100$

$+ 0.3 \times N_C \sin 20°$

or $N_C \{\cos 20° - 0.3 \times 0.62 - 0.3 \sin 20°\}$

$= 100$

or $N_C \{0.65\} = 100$ [from equation (i)]

or $N_C = 153.84$ kN

$\therefore$ $N_A = 0.62 \times N_C$

$= 0.62 \times 153.84$

$= 95.38$ kN

Now from equation (iv)

$N_B + 0.3 \times N_C \sin 20° = 20 + N_C \cos 20°$

or $N_B = 20 + N_C \cos 20° - 0.3\ N_C \sin 20°$

$= 20 + 153.84 \cos 20° - 0.3 \times 95.38 \sin 20°$

$= 154.77$ kN.

Substituting the values of N_A, N_B and N_C in equation (iii), we get

$P = 0.3 \times 154.77 + 0.3 \times 153.84 \times \cos 20°$

$+ 153.84 \times \sin 20°$

$= 142.12$ kN.

Second method:

In this method we use contact force as sum of the normal reaction and frictional force instead of them. As we know that, this contact force makes an angle $\alpha = \tan^{-1}\mu$ with normal reaction. Free body diagram of these blocks are shown in following figures:

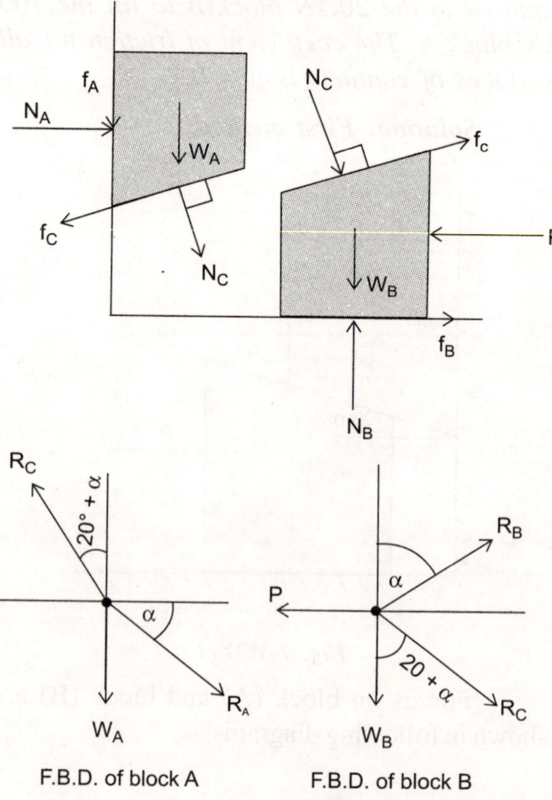

F.B.D. of block A F.B.D. of block B

Fig. 7-WWg3

where $\alpha = \tan^{-1}(0.3)$

$= 16.69°$

Now since block A is in equilibrium under three coplanar concurrent forces, therefore, we can directly use Lami's theorem.

$$\underset{\text{(i)}}{\frac{R_A}{\sin 143.31°}} = \underset{\text{(ii)}}{\frac{R_C}{\sin 73.31°}} = \underset{\text{(iii)}}{\frac{W_A}{\sin 143.38°}}$$

From equation (i) and (ii)

$$R_A = \frac{\sin 143.31°}{\sin 143.38°} \times 100$$

$$= 100.164 \text{ kN}$$

and from equation (ii) and (iii)

$$R_C = \frac{\sin 73.31°}{\sin 143.38°} \times 100$$

$$= 160.580 \text{ kN}$$

Since block B is in equilibrium under four coplanar concurrent forces then $\Sigma F_{net,\,H} = 0$ and $\Sigma F_{net,\,V} = 0$ are sufficient equations to describe the equilibrium of block B.

Here we cannot use Lami's theorem directly because here four coplanar concurrent forces act on block B.

Net horizontal force should be zero.

i.e., $\Sigma F_{net,\,H} = 0$

or $-P + R_C \sin(20 + \alpha) + R_B \sin\alpha = 0$

or $R_C \sin 36.69° + R_B \sin 16.69° = P$

Net vertical force should be zero

i.e., $\Sigma F_{net,\,V} = 0$

or $R_B \cos 16.69° - W_B - R_C \cos 36.69° = 0$

or $R_B \cos 16.69° = 20 + 160.58 \cos 36.69°$

$R_B = 155.308$ kN

$\therefore\ P = R_C \sin 36.69° + R_B \sin 16.69°$

$= 160.580 \times \sin 36.69°$

$+ 155.308 \sin 16.69°$

$= 140.54$ kN.

2. *A block lying over a 10° wedge on a horizontal floor and leaning against a vertical wall and weighing 1800 N is to be raised by applying horizontal force to the wedge. Assuming coefficient of friction between all the surfaces in contacts to be 0.25, determine the minimum horizontal force P to be applied to raise the block shown in Fig. (7-WWg4).*

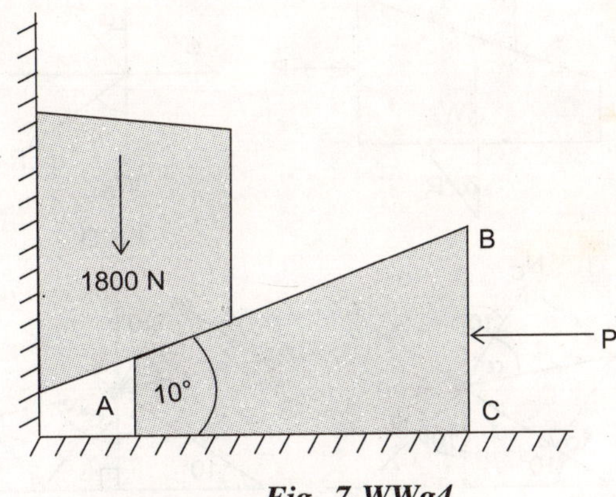

Fig. 7-WWg4

Solution:

Taking contact forces by surfaces in account, forces on blocks and FBDs are shown below.

$\because\ \mu = \tan\alpha$

$\Rightarrow\ 0.25 = \tan\alpha$

$\therefore\ \alpha = 14.036°$

Consider the equilibrium of block (A) then Lami's theorem gives

$$\frac{R_1}{\sin 155.964} = \frac{R_2}{\sin 75.964}$$

$$= \frac{W_A}{\sin 128.072}$$

$$\Rightarrow\ R_1 = \frac{\sin 155.964}{\sin 128.072} \times W_A$$

$$= \frac{\sin 155.964}{\sin 128.072} \times 1800$$

$$= 931.307 \text{ N}$$

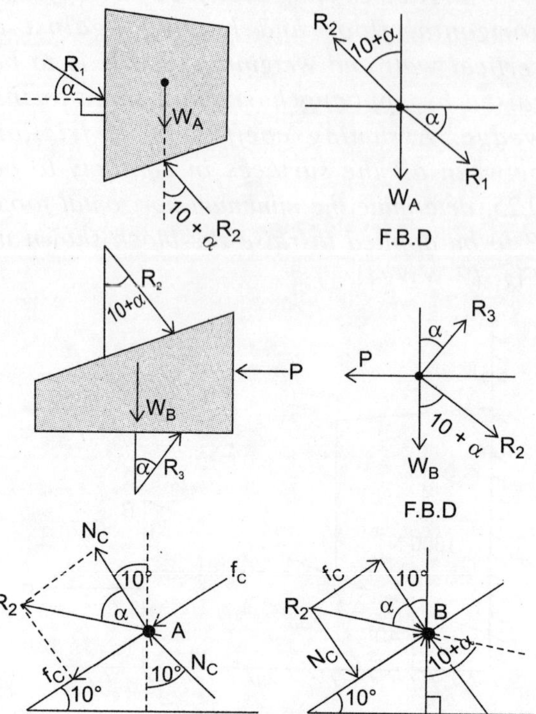

Fig. 7-WWg5

and $$R_2 = \frac{\sin 75.964}{\sin 128.072} \times W_A$$

$$= \frac{\sin 75.964}{\sin 128.072} \times 1800$$

$$= 2218.213 \text{ N.}$$

Similarly for block (B), we get (assume $W_B = 0$)

$$\frac{P}{\sin 141.928} = \frac{R_3}{\sin 114.036} = \frac{R_2}{\sin 104.036}$$

$$\Rightarrow \quad P = \frac{\sin 141.928}{\sin 104.036} \times R_2$$

$$= \frac{\sin 141.928}{\sin 104.036} \times 2218.213$$

$$= 1409.96 \text{ N.}$$

3. *Two blocks each of mass 3000 N resting on horizontal surface as shown in Fig. 7-WWg6, are to be pushed apart by a 75° wedge. The angle of friction is 15° for all contact surfaces. What value of P is required to start the movement of the blocks? How would this answer be changed if the weight of the blocks are increased to 5000 N?*

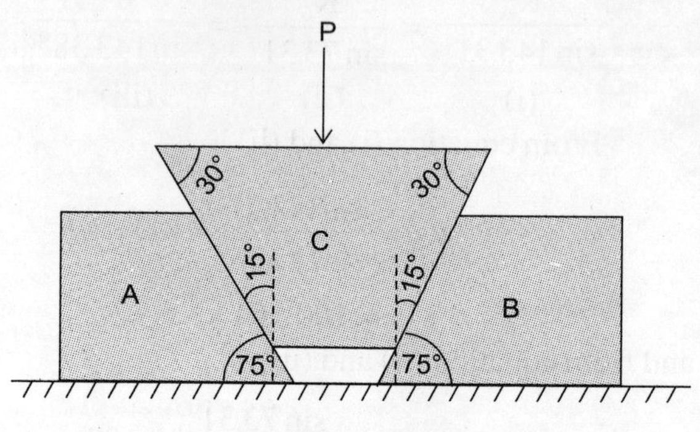

Fig. 7-WWg6

Solution:

Forces on each wedge and FBD's are shown below according to question $\alpha = 15°$ given.

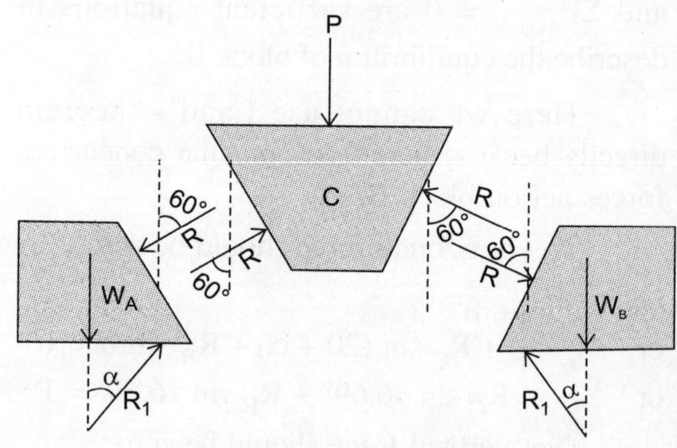

Fig. 7-WWg7

Consider wedge (A):

Since equilibrium of the wedge (A) satisfies all the conditions for Lami's theorem, therefore, by applying Lami's theorem, we get

$$\frac{R_1}{\sin 60°} = \frac{R}{\sin 165°}$$

$$= \frac{3000}{\sin 135°}$$

$$\therefore \quad R = \frac{\sin 165°}{\sin 135°} \times 3000$$

$$= 1098.076 \text{ N}$$

Now consider the wedge (C) applying Lami's theorem, we get

$$\frac{R}{\sin 120°} = \frac{P}{\sin 120} = \frac{R}{\sin 120°}$$

$$\Rightarrow \quad R = P = 1098.076 \text{ N}$$

When $W_A = W_B = W = 5000$ N

then $P = R = \frac{\sin 165°}{\sin 135°} \times 5000 = 1830.127$ N.

PROBLEM

Try it by yourself: *when W_A = 1000 N and W_B = 2000 N (consider blocks are of different material) keeping all datas as given in previous problem. Give reason for your answer.*

[**Ans.** P = 366.025 N]

4. *In Fig. (7-WWg8), C is a stone of unknown weight W. It is being raised slightly by means of two wooden wedges A and B applying force P on the wedge B. magnitude of P is 2.344 kN. The angle between the contacting surfaces of the wedge is 5°. If the coefficient of friction is 0.3 for all contact surfaces compute the value of W required to impend upward motion of the block C. Neglect weight of the wedges.*

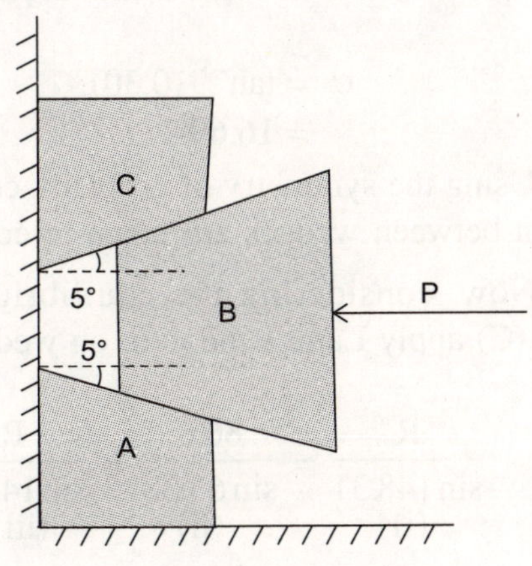

Fig. 7-WWg8

Solution:

Forces on blocks and FBDs are shown in following figures. As the coefficient of friction is 0.3 for all contact surfaces, therefore, angle of friction for all the contact surfaces is $\alpha = \tan^{-1}(0.3) = 16.69°$.

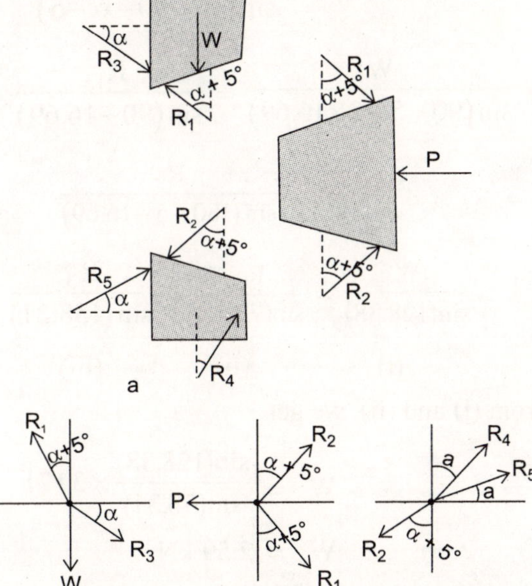

Fig. 7-WWg9

Consider the equilibrium of wedge (B) since there are only three co-planar concurrent forces, under which it is in equilibrium condition so we can apply Lami's theorem.

$$\text{i.e.,} \quad \frac{R_2}{\sin(90+5+\alpha)} = \frac{R_1}{\sin(5+\alpha+90°)} = \frac{P}{\sin(180-2(5+\alpha))}$$

$$\Rightarrow \quad R_1 = R_2$$

$$= \frac{\sin(90+\alpha+5)}{\sin\{180-2(5+\alpha)\}} \times P$$

$$= \frac{\sin(90+16.69+5)}{\sin(180-10-2\times16.69)} \times 2.34$$

$$= 3.171 \text{ kN}$$

Now considering the equilibrium of block A, we get from Lami's theorem.

$$\frac{W}{\sin(90+5+2\alpha)} = \frac{R_1}{\sin(90-\alpha)}$$

$$= \frac{R_3}{\sin(90+90-\alpha-5)}$$

or $$\frac{W}{\sin(90+5+2\times16.69)} = \frac{R_1}{\sin(90-16.69)}$$

$$= \frac{R_3}{\sin(180-5-16.69)}$$

or $$\frac{W}{\sin(128.38)} = \frac{R_1}{\sin(73.31)} = \frac{R_3}{\sin(158.31)}$$

(i) (ii) (iii)

From (i) and (ii), we get

$$W = \frac{\sin(128.38)}{\sin(73.31)} \times 3.171$$

$$\therefore \quad W = 2.59 \text{ kN.}$$

5. *Three wedges A, B and C are arranged as shown in Fig. (7-WWg10). Find the value of P for the impending motion of block C upwards, if the coefficient of friction is 0.3 for all surfaces. The weights of all the wedges may be neglected.*

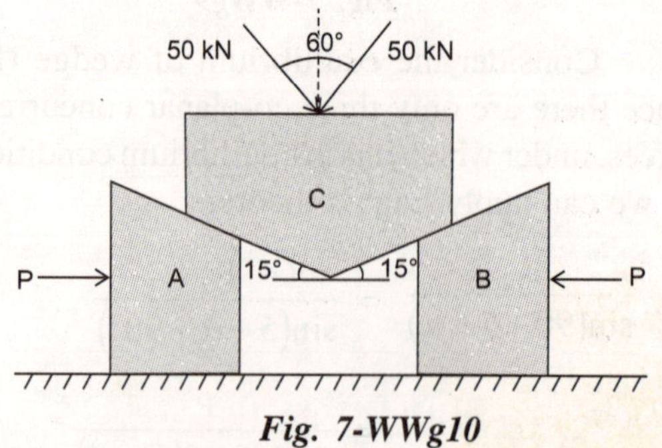

Fig. 7-WWg10

Solution:

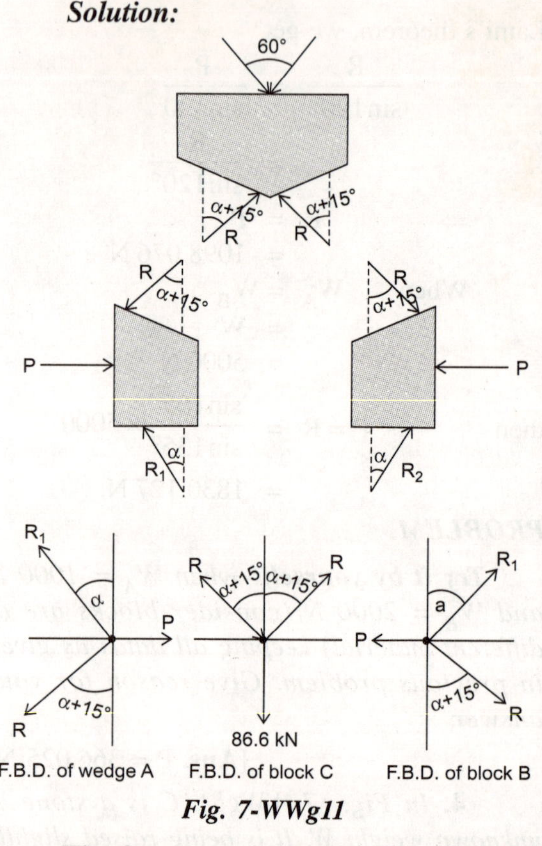

Fig. 7-WWg11

The forces on the wedges are shown in above figures. If α is the angle of limiting friction then

$$\alpha = \tan^{-1}(0.30)$$
$$= 16.69°$$

Using the symmetry of problem, contact reaction between wedges are taken to equal.

Now, considering the equilibrium of wedge (C) apply Lami's theorem on wedge C, we get

i.e., $$\frac{R}{\sin 148.31} = \frac{86.6}{\sin 63.38} = \frac{R}{\sin 148.31}$$

(i) (ii) (iii)

From (i) and (ii), we have

$$R = \frac{\sin 148.31}{\sin 63.38} \times 86.6$$
$$= 50.88 \text{ kN}$$

Similarly, by applying Lami's theorem on wedge (A), we get

$$\underset{(i)}{\frac{P}{\sin 131.62}} = \underset{(ii)}{\frac{R}{\sin 106.69}} = \underset{(iii)}{\frac{R_1}{\sin 121.69}}$$

From (i) and (ii), we have

$$P = \frac{\sin 131.62}{\sin 106.69} \times R$$
$$= \frac{\sin 131.62}{\sin 106.69} \times 50.88$$

$\therefore$ $P = 39.70$ kN.

6. *A 5000 N block rests on the inclined surface of another block of mass 2000 N as shown in the Fig. (7-WWg12). The coefficient of friction between the blocks is 0.20 and that at the supporting surfaces with the blocks is 0.35. Determine the minimum force P required to pull the 2000 N block.*

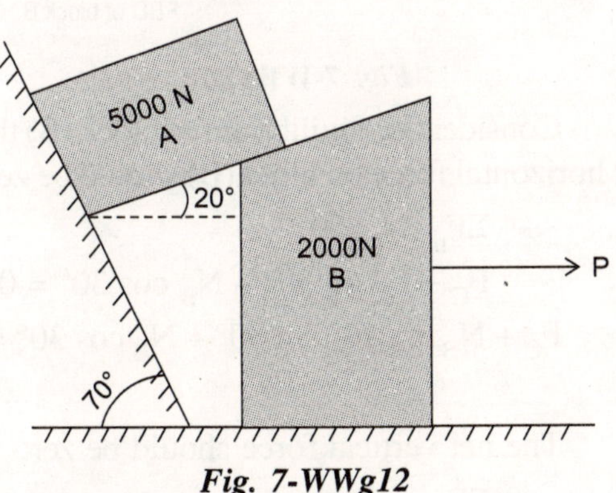

Fig. 7-WWg12

Solution:

If the angle of friction at contact surface between blocks and supporting surfaces are α_1 and α_2 respectively then $\tan \alpha_1 = 0.2$ and $\tan \alpha_2 = 0.35$.

$$\therefore \quad \alpha_1 = 11.3°$$
$$\alpha_2 = 19.3°$$

Forces and free body diagrams of blocks (A) and (B) are shown in Fig. (7-WWg13).

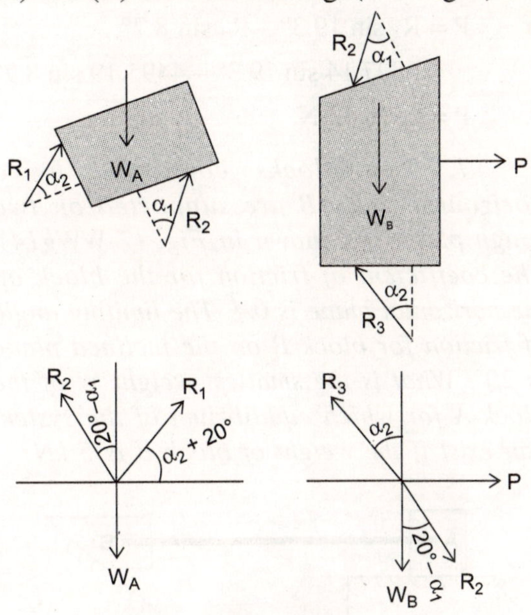

Fig. 7-WWg13

Applying Lami's theorem for block (A), we have

i.e.,
$$\frac{R_1}{\sin 171.3} = \frac{R_2}{\sin 129.3°}$$
$$= \frac{5000}{\sin 59.4}$$

$$\therefore \quad R_1 = \frac{\sin 171.3°}{\sin 59.4} \times 5000$$
$$= 878.66 \text{ N}$$

and
$$R_2 = \frac{\sin 129.3°}{\sin 59.4°} \times 5000$$
$$= 4495.19 \text{ N}$$

Now applying equilibrium equations for block (B), we get

i.e., $\Sigma F_{net,\,V} = 0$

or $R_3 \cos 19.3° - 2000 - R_2 \cos 8.7° = 0$

or $R_3 \cos 19.3° = 2000 + R_2 \cos 8.7°$

$\therefore$ $R_3 = 6827.14$ N

and $\Sigma F_{net,\ H} = 0$

or $P - R_3 \sin 19.3° + R_2 \sin 8.7 = 0$

or $P = R_3 \sin 19.3° - R_2 \sin 8.7°$

$= 6827.14 \sin 19.3° - 4495.19 \sin 8.7°$

$\therefore$ $P = 1576.52$ N.

7. *Two blocks connected by a horizontal link AB are supported on two rough planes as shown in Fig. (7-WWg14). The coefficient of friction for the block on the horizontal plane is 0.4. The limiting angle of friction for block B on the inclined plane is 20°. What is the smallest weight W of the block A for which equilibrium of the system can exist if the weight of block B is 5 kN?*

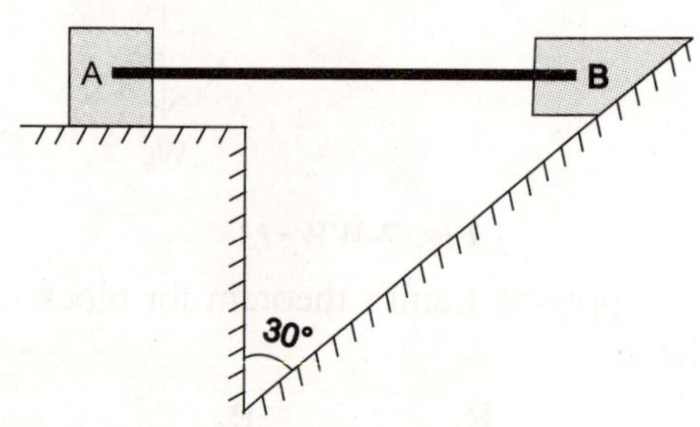

Fig. 7-WWg14

Solution:

Forces on block (A) and block (B) and FBD's are shown in the Figs. (7-WWg15 and 7-WWg16). Taking compression in link as F_C.

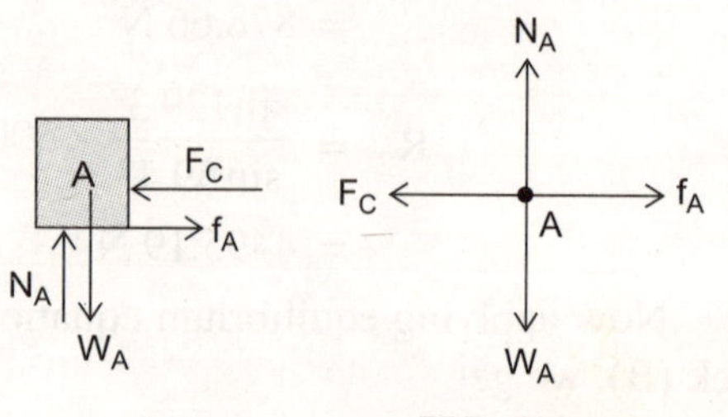

Fig. 7-WWg15

As friction is limiting then,

$$f_A = N_A \mu$$

$$= 0.4 N_A \quad \text{...(i)}$$

Consider the equilibrium of block (A) then horizontal equilibrium gives,

i.e., $\Sigma F_{net,\ H} = 0$

or $f_A - F_C = 0$

or $f_A = F_C \quad \text{....(ii)}$

and vertical equilibrium gives

$\Sigma F_{net,\ V} = 0$

or $N_A - W_A = 0$

or $N_A = W_A \quad \text{....(iii)}$

From (i), (ii) & (iii), we get

$$F_C = 0.4 W_A \quad \text{....(iv)}$$

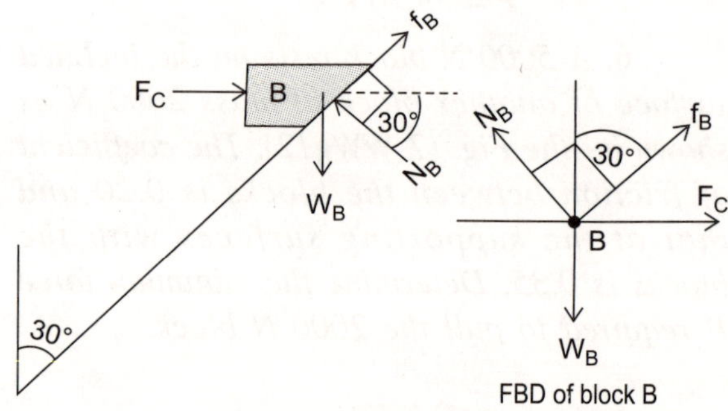

Fig. 7-WWg16

Consider the equilibrium of block (B) then net horizontal force on block(B) should be zero.

i.e., $\Sigma F_{net,\ H} = 0$

or $F_C + f_B \sin 30° - N_B \cos 30° = 0$

or $F_C + N_B \tan 20° \sin 30° - N_B \cos 30° = 0 \quad \text{....(v)}$

The net vertical force should be zero

i.e., $\Sigma F_{net,\ V} = 0$

or $F_B \cos 30° + N_B \sin 30° - W_B = 0$

or $N_B \tan 20° \cos 30° + N_B \sin 30° = W_B$

or $N_B = \dfrac{W_B}{\tan 20°.\cos 30° + \sin 30°}$

$$= \frac{5}{\tan 20°.\cos 30° + \sin 30°}$$

$= 4.19$ kN

From (v), we get

$F_C = N_B \cos 30° - N_B \tan 20°.\sin 30°$
$= N_B (\cos 30° - \tan 20°.\sin 30°)$
$= 6.133 (\cos 30° - \tan 20°.\sin 30°)$
$= 4.19$ kN

and from (iv) $W_A = \frac{F_C}{\tan 20°}$

$$= \frac{4.19}{\tan 20°}$$

$= 11.52$ kN.

7.9 PULLEY FRICTION

A fixed pulley carries a weightless thread over it having masses m_1 and m_2 at its ends. There is friction between the thread and the pulley. It is such that the thread starts slipping when the ratio $m_2/m_1 = \eta_0$. Find,

(a) the coefficient of friction;

(b) the acceleration of the masses when $\frac{m_2}{m_1} = \eta > \eta_0$.

Solution:

(a) Let us examine a small element of the thread in contact with the pulley Fig. (7-PF1) which subtends an angle $\Delta\theta$ at the centre of pulley. If μ_s is the coefficient of static friction between the thread and the pulley and T'_1 and T'_2 are the tensions in two different parts of the thread, when the thread is just about to slide towards right.

Note that $\Delta\theta$ is so small that the length of the thread could be assumed to be a small straight line. The forces on the elementary thread are:

1. T'_1, tension of the thread in one end part.
2. T'_2, tension of the thread in the other end part.

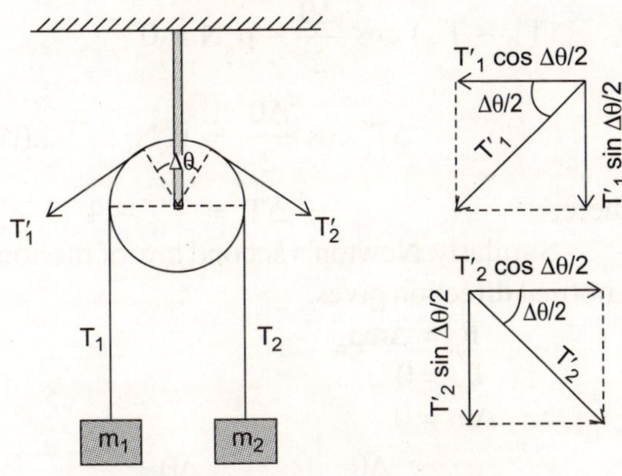

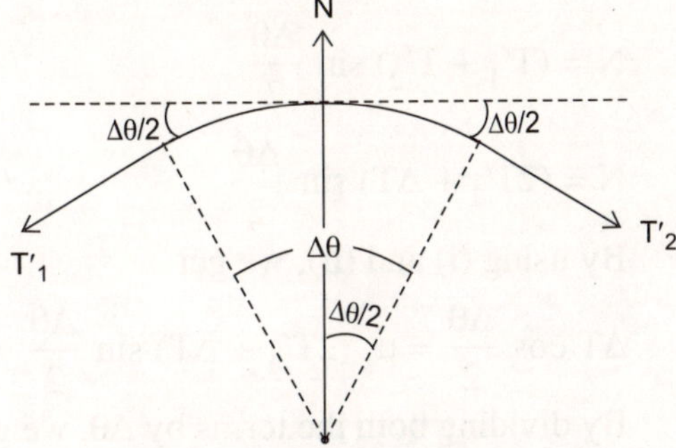

F.B.D. of elementary thread

Fig. 7-PF1

3. N, normal force by the pulley, upward.
4. Limiting frictional force $F_s = \mu_s N$ in tangential direction towards the left as the tendency of motion is assumed to be towards the right.
5. Weight of elementary thread which is approximated to be zero because massless thread is considered.

i.e., $\Delta m = 0$

Now by applying Newton's second of motion law in tangential (horizontal) direction,

$F_t = \Delta m a_t$

or $F_t = 0$ $[\because \Delta m = 0]$

or $$T'_2 \cos \frac{\Delta\theta}{2} - T'_1 \cos \frac{\Delta\theta}{2} - F_s = 0$$

or $$T'_2 \cos \frac{\Delta\theta}{2} - T'_1 \cos \frac{\Delta\theta}{2} - \mu_s N = 0$$

or $\quad (T'_2 - T'_1)\cos\dfrac{\Delta\theta}{2} - \mu_s N = 0$

or $$\Delta T' \cos\frac{\Delta\theta}{2} = \mu_s N \qquad ...(i)$$

where, $\quad \Delta T = T'_2 - T'_1.$

Similarly, Newton's second law of motion in normal direction gives,

$$F_n = \Delta m a_n$$

or $\quad F_n = 0$

$\because \quad \Delta m = 0$

or $\quad N - T'_2 \sin\dfrac{\Delta\theta}{2} - T'_1 \sin\dfrac{\Delta\theta}{2} = 0$

or $\quad N = (T'_1 + T'_2)\sin\dfrac{\Delta\theta}{2}$

or $$N = (2T'_1 + \Delta T)\sin\frac{\Delta\theta}{2} \qquad ...(ii)$$

By using (i) and (ii), we get

$$\Delta T \cos\frac{\Delta\theta}{2} = \mu_s (2T'_1 + \Delta T)\sin\frac{\Delta\theta}{2}$$

By dividing both the terms by $\Delta\theta$, we get

$$\frac{\Delta T}{\Delta\theta}\cos\frac{\Delta\theta}{2} = \frac{\mu_s (2T'_1 + \Delta T)\sin\dfrac{\Delta\theta}{2}}{\Delta\theta}$$

or $$\frac{\Delta T}{\Delta\theta}\cos\frac{\Delta\theta}{2} = \mu_s\left(T_1 + \frac{\Delta T}{2}\right).\left(\frac{\sin\dfrac{\Delta\theta}{2}}{\dfrac{\Delta\theta}{2}}\right)$$

Now by applying $\lim\limits_{\Delta\theta\to 0}$ to both sides, we get

$$\lim_{\Delta\theta\to 0}\left(\frac{\Delta T}{\Delta\theta}\cos\frac{\Delta\theta}{2}\right) = \lim_{\Delta\theta\to 0}\left[\left(T_1 + \frac{\Delta T}{2}\right)\left(\frac{\sin\dfrac{\Delta\theta}{2}}{\dfrac{\Delta\theta}{2}}\right)\mu_s\right]$$

or $$\left(\lim_{\Delta\theta\to 0}\frac{\Delta T}{\Delta\theta}\right).\left(\lim_{\Delta\theta\to 0}\cos\frac{\Delta\theta}{2}\right)$$

$$= \mu_s\left\{\lim_{\Delta\theta\to 0}\left(T'_1 + \frac{\Delta T}{2}\right)\right\}\left\{\lim_{\Delta\theta\to 0}\frac{\sin\dfrac{\Delta\theta}{2}}{\dfrac{\Delta\theta}{2}}\right\}$$

Now, when $\Delta\theta$ (in radian) approaches zero, the tensions in the thread approach to become equal i.e., $\Delta T \to 0$, therefore, we can approximate term as $T'_1 + \dfrac{\Delta T}{2} \cong T'_1$. Also, we know that, $\lim\limits_{\Delta\theta\to 0}\dfrac{\sin\alpha}{\alpha} = 1$ i.e., $\sin\alpha \cong \alpha$ and $\lim\limits_{\Delta\theta\to 0}\cos\alpha = 1$ i.e., $\cos\alpha \cong 1$ when angle α measured in radian and is very small in measure and from the definition of differential coefficient i.e., $\dfrac{dT}{d\theta} = \lim\limits_{\Delta\theta\to 0}\dfrac{\Delta T}{\Delta\theta}$.

Therefore, we can write

$$\frac{dT}{d\theta} = T'_1\mu_s$$

or $$\frac{dT}{T'_1} = \mu_s d\theta$$

Integrating within the limits,

$$\int_{T_1}^{T_2}\frac{dT}{T'_1} = \mu_s\int_0^\theta d\theta$$

or $$[\ln T']_{T_1}^{T_2} = \mu_s\theta$$

or $$\ln\frac{T_2}{T_1} = \mu_s\theta \qquad ...(7.5)$$

or $$\frac{T_2}{T_1} = e^{\mu_s\theta}$$

$\therefore$ $$T_2 = T_1 e^{\mu_s\theta} \qquad ...(7.6)$$

Then from free body diagrams of masses m_1 and m_2 and by applying Newton's second law of motion, we have

$$T_1 - m_1 g = 0$$

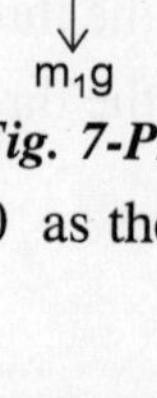

Fig. 7-PF2

[$\because$ a = 0 as there is only tendency]

$\therefore \quad T_1 = m_1 g$ ...(iii)

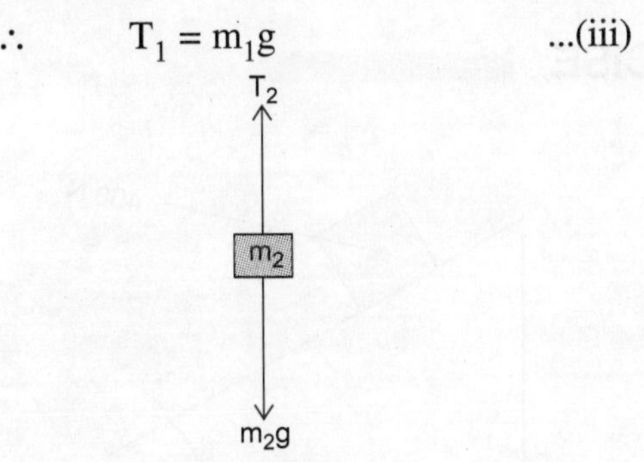

Fig. 7-PF3

$T_2 - m_2 g = 0 \quad a = 0$

$\therefore \quad T_2 = m_2 g$ ...(iv)

By dividing equations (iii) and (iv), we have

$$\frac{T_2}{T_1} = \frac{m_2}{m_1}$$

$$= \eta_0 \quad \text{(given)}$$

Then, from equation (7.5)

$\ln \eta_0 = \mu_s$ [$\because$ $\theta = \pi$ for diametrical ends.]

Therefore, $\frac{\ln \eta_0}{\pi} = \mu_s$ we obtain.

(b) Let a be the acceleration of the blocks when $\frac{m_2}{m_1} = \eta > \eta_0$, and suppose that $\mu_s = \mu_k$.

$$\therefore \quad \frac{T_2}{T_1} = e^{\mu_k \pi}$$

$$= e^{\mu_s \pi}$$

$$= e^{\ln \eta_0}$$

$$= \eta_0$$

$\therefore \quad T_2 = T_1 \eta_0$ ...(v)

From the equations of motion, we get

$T_1 - m_1 g = m_1 a$ ...(vi)

and $\quad m_2 g - T_2 = m_2 a$ ...(vii)

or $\quad m_2 g - T_1 \eta_0 = m_2 a$

By multiplying equation (vi) with η_0 and by adding (vi) and (vii), we get

$$\eta_0 T_1 - m_1 \eta_0 g = m_1 \eta_0 a$$

$$m_2 g - T_1 \eta_0 = m_2 a$$

$$m_2 g - m_1 \eta_0 g = (m_1 \eta_0 + m_2)\, a$$

$$\therefore \quad a = \frac{m_2 - m_1 \eta_0}{m_2 + m_1 \eta_0} g;$$

[divide numerator and denominaor by m_1

$$= \frac{\eta - \eta_0}{\eta + \eta_0} g \quad \frac{m_2}{m_1} = \eta]$$

Now, the following points we should note about the equation (7.5 or 7.6):

1. $T_2 > T_1$; the tension in the pulling side is more than that in the resisting side. The pulling side is called as the tight side and the resisting side is called the slack side.
2. We should measure the angle of contact (θ) in radian only.
3. The angle θ may be larger than 2π; for example, if the thread is wrapped n times, over the pulley then θ is equal to $2n\pi$.
4. Formula-7.5 should be used only if the thread is about to slip, but if the thread is slipping, then we can use similar formula to 7.5 but use μ_k (coefficient of kinetic friction) instead of static coefficient of friction (μ_s).
5. If the rope does not slip, and is not about to slip too then one of these formula may be used.
6. Formula-7.5 holds equally well to problems involving flat massless belts, passing over fixed cylindrical drums.
7. The Difference in tensions is due to the friction between the thread and the pulley and the difference is maximum when the thread is about to slip i.e., in impending condition of motion.
8. Although this equation is derived for a circular pulley, but this relation could be used if the post is non-circular. This is because, every little element can be approximated as a being part of circle, even though the centre of the circle keeps on changing.

EXERCISE

1. (a) State the laws of static friction.

   (b) A block weighing 2000 N is resting on horizontal surface support another block of 4000 N as shown in Fig. (7-E1). Find the horizontal force F required just to move the block to the left. Take coefficient of friction for all contact surfaces as 0.2.

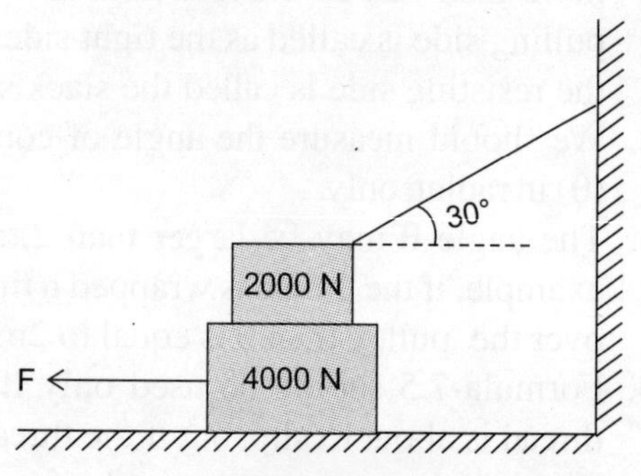

Fig. 7-E1

2. (a) List the laws of dry friction or state the laws of static friction.

   (b) Explain the terms, angle of repose, angle of friction, and cone of friction.

   (c) Explain (i) limiting friction (ii) coefficient of friction.

   (d) State and explain D'Alembert's principle and its application.

   (e) Show that the coefficient of friction is equal to the tangent of the angle of friction.

   (f) Show that the angle of repose is equal to the angle of friction.

3. The coefficient of friction between a 20 kg block and the inclined surface is $\mu_s = 0.4$ and $\mu_k = 0.3$. Determine whether the block is in equilibrium and find the magnitude and direction of the frictional force (refer Fig. 7-E2).

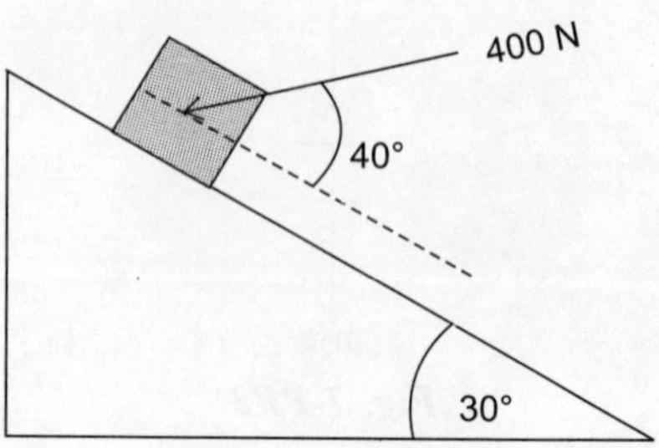

Fig. 7-E2

4. A 300 N box is hold at rest on a smooth inclined plane and acted upon by a horizontal force P as shown in Fig. (7-E3). Determine the value of P.

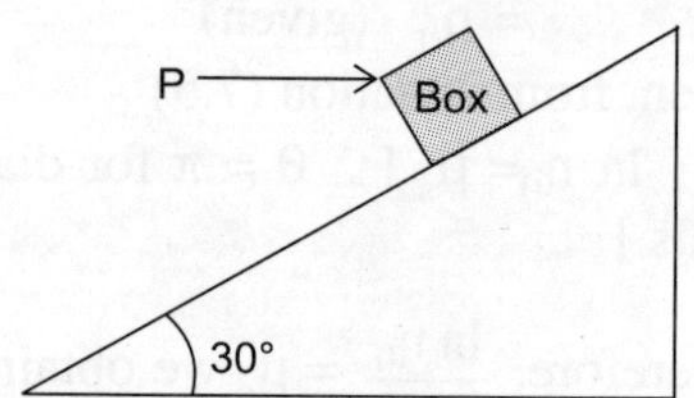

Fig. 7-E3

5. Determine the maximum force P that could be applied on a block of 25 kg mass as shown in Fig. (7-E4) such that 15 kg mass block does not slip. Also find the corresponding acceleration. The coefficient of friction between the two blocks and that between the 25 kg mass block and the plane is 0.35.

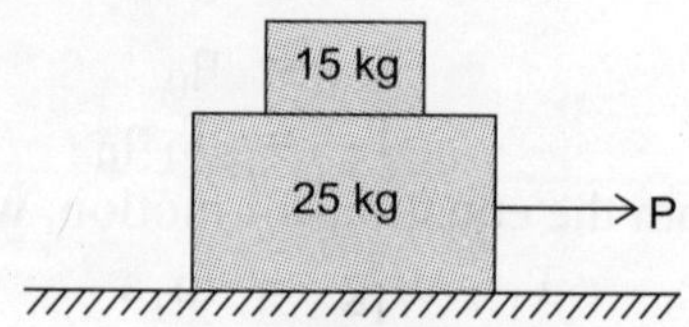

Fig 7-E4

6. Two blocks A and B having weights 8N and 6 N respectively are replaced from rest as shown in Fig. (7-E5). Find the velocity of block B after it moves by 2 m.

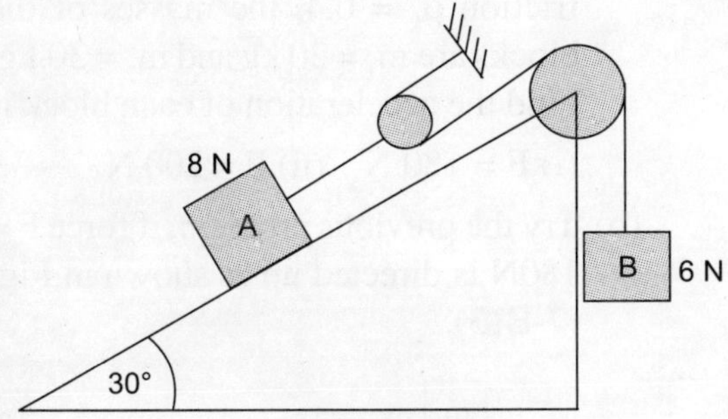

Fig 7-E5

7. A block of weight 1 kN is placed on a 30° incline plane having a coefficient of friction of 0.25 as shown in Fig. (7-E6). Determine the horizontal force to be applied for (i) the impending motion down the plane and (ii) the impending motion up the plane.

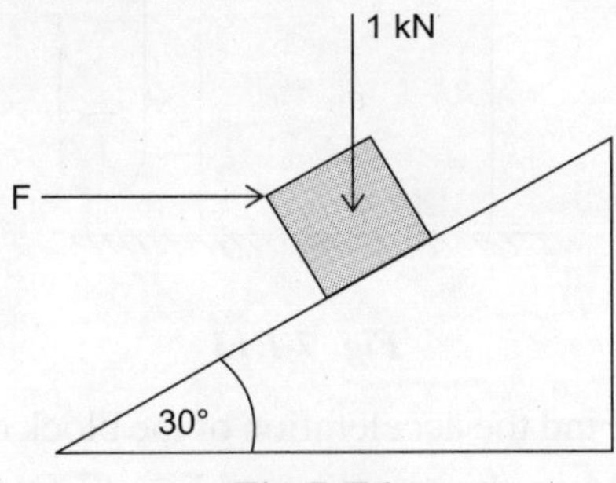

Fig 7-E6

8. A block of weight Q rests on an inclined plane and has attached to it a string that overruns a pulley and carries a weight P at its other end as shown in Fig. (7-E7). If the coefficient of friction between the block Q and the inclined plane is μ, find the limiting values of the ratio P/Q consistent with the equilibrium. Neglect the friction in the pulley and assume that the angle of inclination α of the plane is greater than the angle of friction ϕ = arc tan μ.

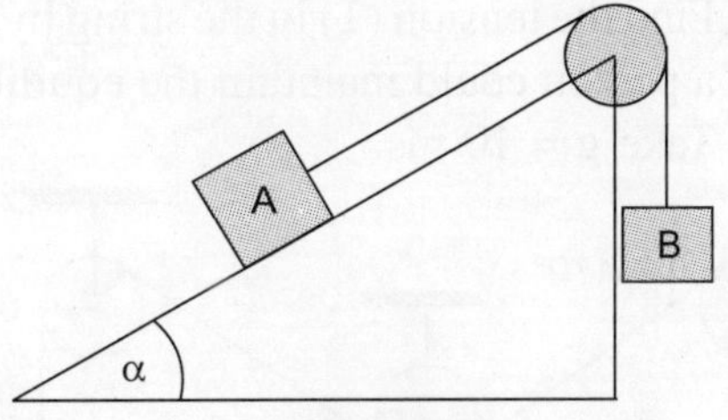

Fig 7-E7

9. Two heavy weighted circular rollers of diameters D and d respectively, rest on a rough horizontal plane as shown in Fig. (7-E8). The larger roller has a string wound around it to which a horizontal force P can be applied as shown. Assuming that the coefficient of friction μ has the same values for all surfaces of contact, determine the necessary condition under which the large roller can be pulled over the small one.

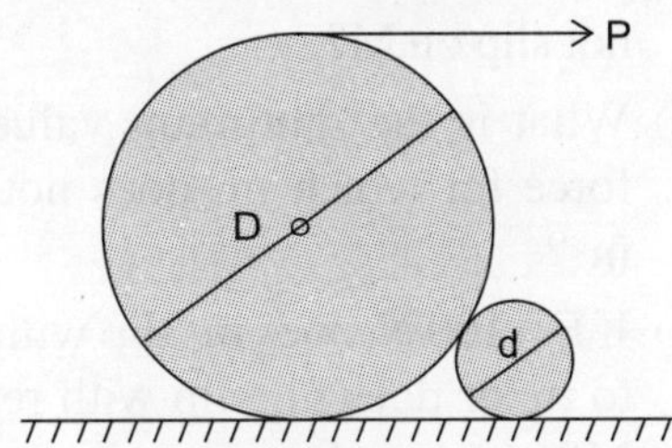

Fig 7-E8

10. Two blocks having weights W_1 and W_2 are connected by a string and rest on horizontal planes as shown in Fig. (7-E9). If the angle of friction for each block is ϕ, find the magnitude and direction of the least force P applied to the upper block that will include sliding.

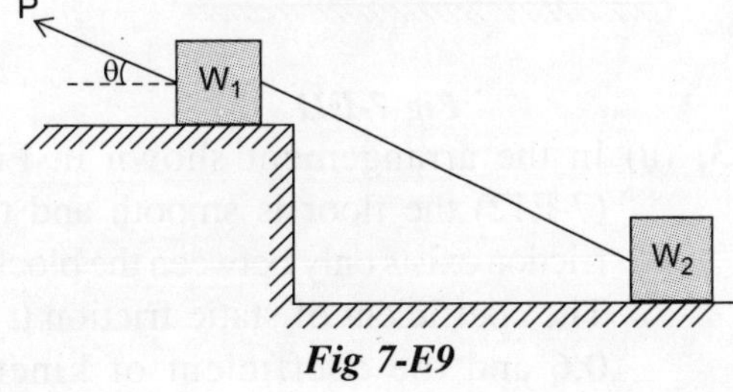

Fig 7-E9

11. Find the tension (T) in the string by which a person could maintain the equilibrium. Take g = 10 ms^{-2}

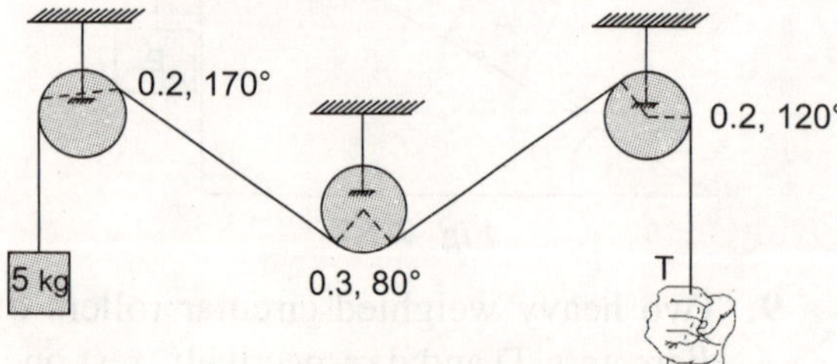

Fig 7-E10

12. The system in Fig. (7-E11) M = 13.4kg, m_1 = 1 kg, m_2 = 2 kg, and m_3 = 3.6 kg. The coefficient of static friction between m_1 and m_2 is 0.75 and that between m_2 and M is 0.6. All other surfaces are frictionless.

(a) What minimum horizontal forces F must be applied to M, so that m_2 does not slip on M?

(b) What is the minimum value of the force for which m_1 does not slip on m_2?

(c) If F = 100 N, does m_1 slip with respect to m_2 or does m_2 slip with respect to M?

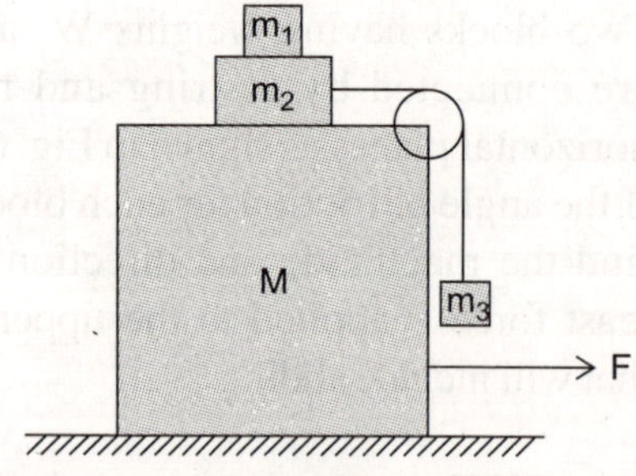

Fig 7-E11

13. (a) In the arrangement shown in Fig. (7-E12) the floor is smooth and the friction exists only between the blocks. The coefficient of static friction μ_s = 0.6 and the coefficient of kinetic friction μ_k = 0.4, the masses of the blocks are m_1 = 20 kg and m_2 = 30 kg. Find the acceleration of each block if

(i) F = 180 N (ii) F = 200 N

(b) Try the previous problem, if force F = 180N is directed up as shown in Fig. (7-E13).

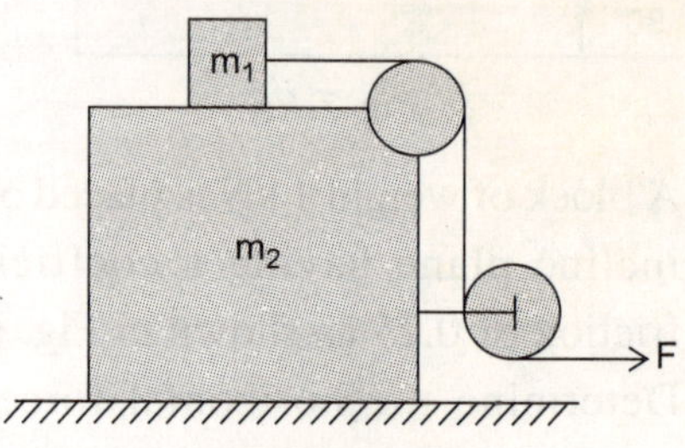

Fig 7-E12

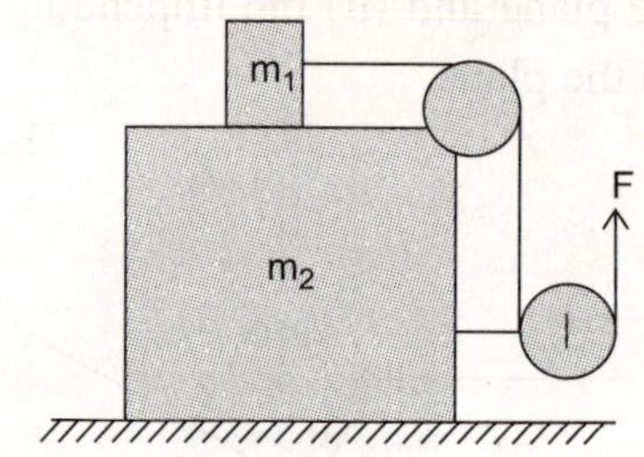

Fig. 7-E13

14. Find the acceleration of the block of mass M in the situation of Fig. (7-E14). The coefficient of friction between the two blocks is μ_1 and that between the bigger block and the ground is μ_2.

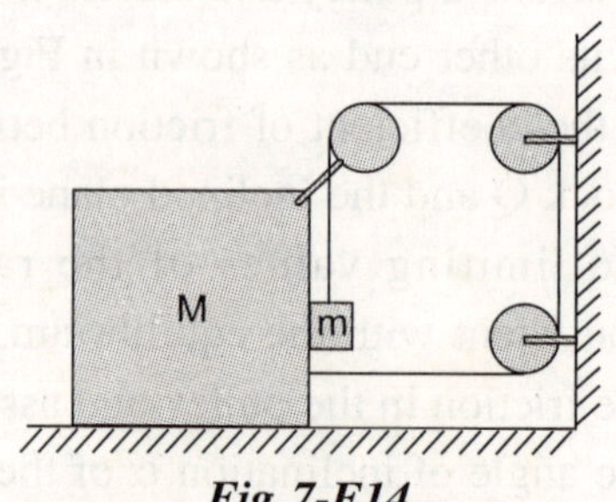

Fig 7-E14

15. A crate slides down an inclined right angled trough as in Fig. (7-E15). The coefficient of kinetic friction between the crate and the trough is μ_k. What is the acceleration of the crate in terms of μ_k, g and θ?

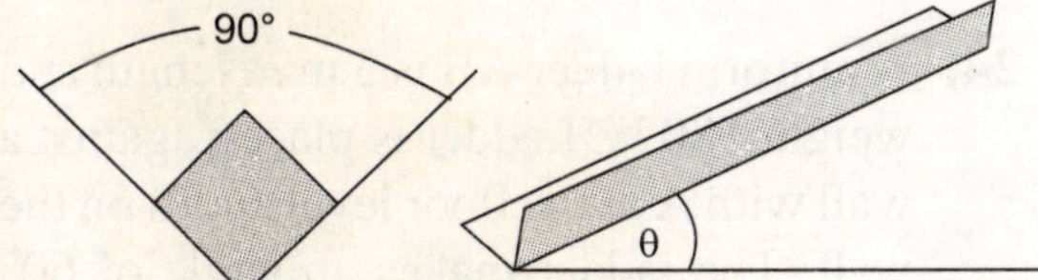

Fig 7-E15

16. Two blocks A and B which are identical, weighing W each, are supported by a rod inclined at 45° to the horizontal as shown in Fig. (7-E16). If the blocks are in limiting equilibrium, find the coefficient of friction assuming it to be the same at both the floors and the wall.

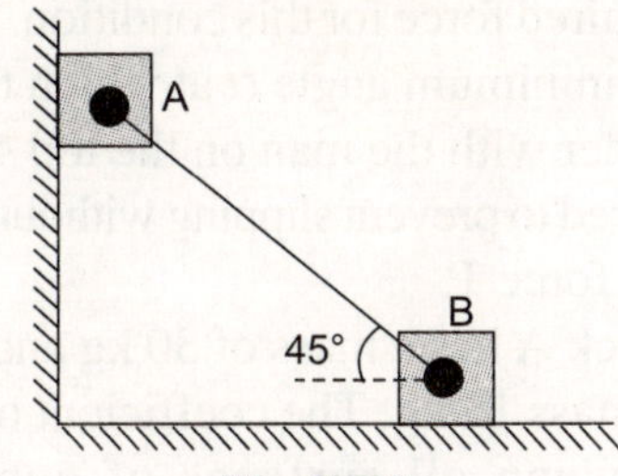

Fig 7-E16

17. A block having weight of 10 kN is to be raised by means of a 20° wedge as shown in Fig. (7-E17) below. Find the horizontal force (P) which will just raise the block if the coefficient of friction for all surfaces of contact is 0.3.

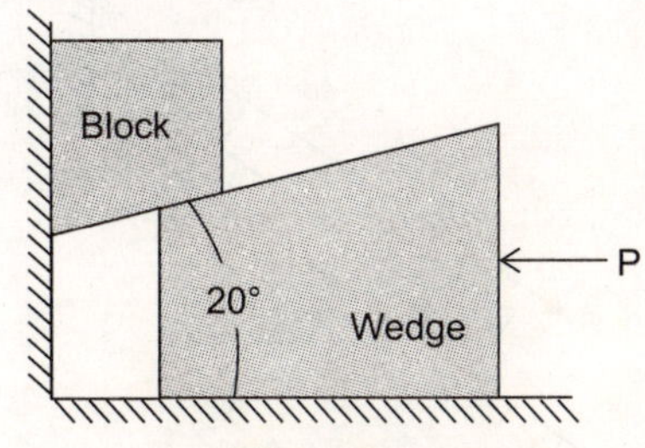

Fig. 7-E17

18. Find the value of P required to lift the crate. Take the angle of friction at all surfaces as 12°.

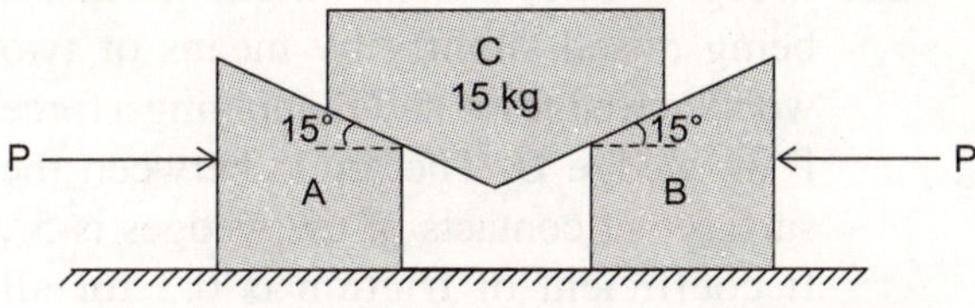

Fig. 7-E18

19. Referring to Fig. (7-E19), the coefficient of frictions are as follows: 0.25 at the floor, 0.30 at the wall and 0.20 between blocks. Find the minimum value of the horizontal force P applied to the lower block that will hold the system in equilibrium.

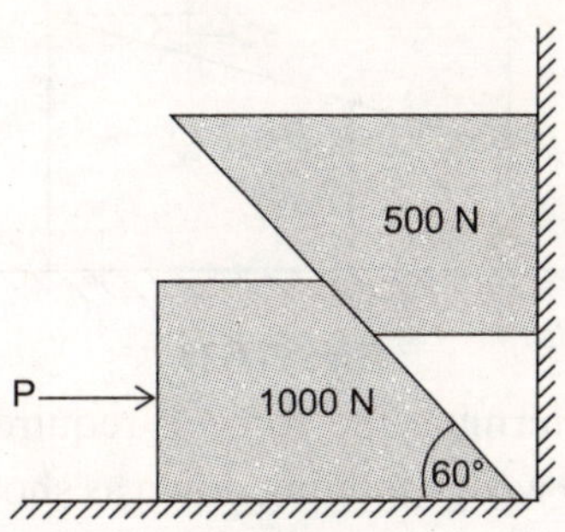

Fig. 7-E19

20. Two blocks A and B weighing 3 kN and 15 kN respectively, are held in position against an inclined plane by applying a horizontal force P as shown in Fig. (7-E20). Find the least value of P which will include motion of the block A upward. The angle of friction for all contact surfaces is 12°.

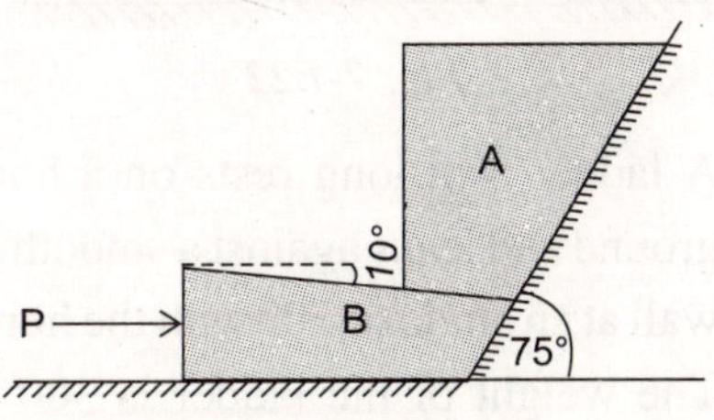

Fig. 7-E20

21. In Fig. (7-E21) a stone weighing 6 kN, is being raised slightly by means of two wooden wedges A and B applying a force P on wedge B. The angle between the surfaces of contacts of the wedges is 5°. If coefficient of friction is 0.3 for all surfaces, compute the value P required to impend motion of the stone C. Neglect the weight of the wedges.

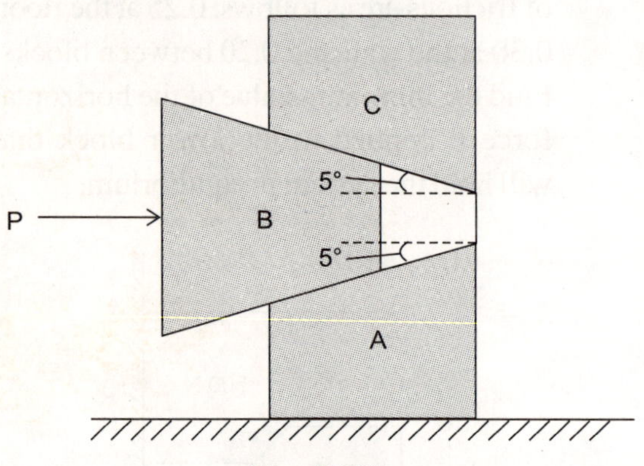

Fig. 7-E21

22. Determine the force P required to start the wedge moving down as shown in Fig. (7-E22). The angle of friction for all surfaces of contact is 15°.

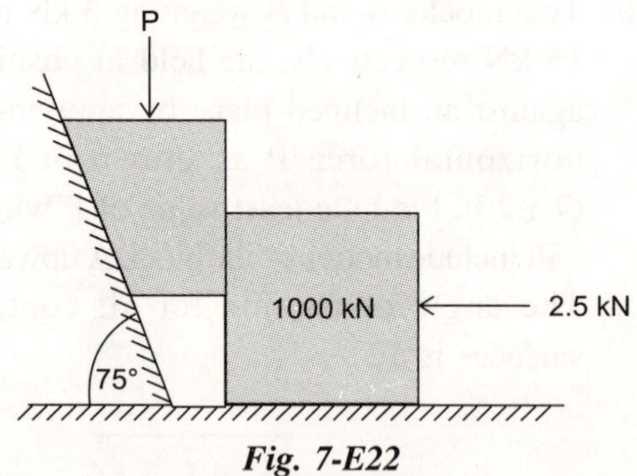

Fig. 7-E22

23. A ladder 5 m long rests on a horizontal ground and leans against a smooth vertical wall at an angle of 70° with the horizontal. The weight of the ladder is 300 N. The ladder is on the verge of sliding when a man weighing 750 N stand on a rung 1.5 m high. Calculate the coefficient of friction between the ladder and the floor.

24. A uniform ladder AB is 3 m in length and weight 180 N. Ladder is placed against a wall with A at the floor level and B on the wall. The ladder makes an angle of 60° with the floor. The coefficient of friction between the wall and the ladder is 0.25 and between the floor and the ladder is 0.35. In addition to the self weight of the ladder, it has to support a man weighing 900 N at its top B. To prevent slipping, a force P is applied horizontally at the level of the floor. Find the minimum force required force for this condition. Find also the minimum angle α at which the above ladder with the man on the top should be placed to prevent slipping without applying the force P.

25. Block A has a mass of 30 kg and block B of mass 15 kg. The coefficient of friction between all surfaces of contact are $\mu_s = 0.15$ and $\mu_k = 0.10$. Knowing that $\theta = 30°$ and the magnitude of the force F applied to block A is 250 N, determine (i) acceleration of block A, (ii) the tension in the rope.

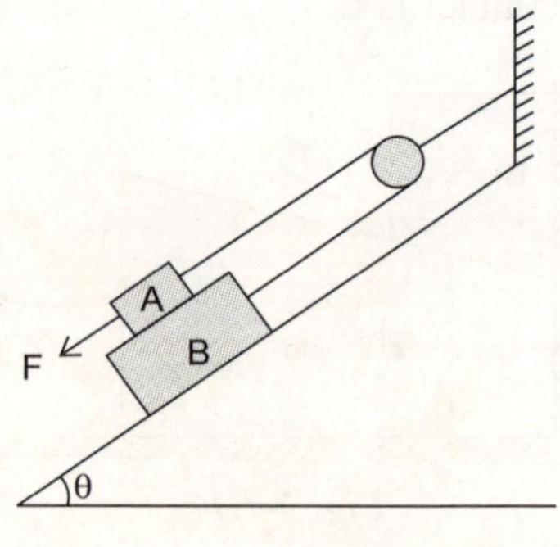

Fig. 7-E23

26. A worker wishes to pile a cone of sand onto a circular area in his yard. The radius of the circle is R and no sand is to

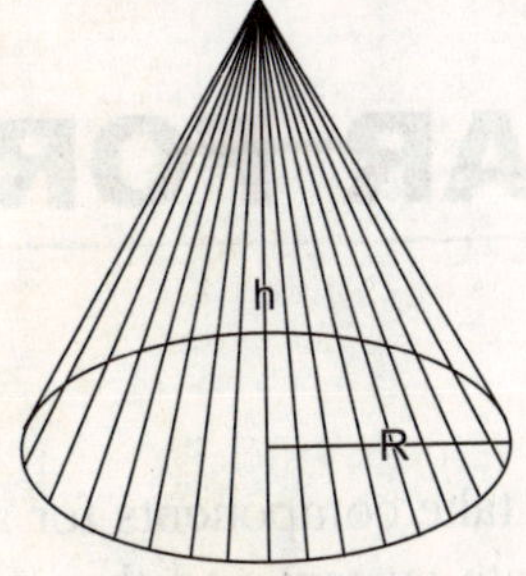

Fig. 7-E24

spill onto the surrounding area as shown in Fig. (7-E24). If μ_s is the static coefficient of friction between each layer of sand along the slope and the sand beneath if (along which it might slip), show that the greatest volume of sand that can be stored in this manner is $\dfrac{\pi\mu_s R^3}{3}$.

27. A small bar starts sliding down an inclined plane forming an angle θ with horizontal. The coefficient of friction depends on the distance x covered as $\mu = ae^{x^2}$, where 'a' is a constant. Find the distance covered by the bar till it stops and its maximum velocity over this distance.

28. The coefficient of friction is 0.30 between the rope and the fixed drum and between other surface of contact, μ = 0.40 determine the minimum weight, w to prevent downward motion of 500 N body.

COPLANAR FORCES

8.1 DEFINITION

If all the forces in a given system lie in a single plane and pass through a single point, then the system of forces is called **coplanar concurrent forces** system. Example: forces $\vec{F}_1$, $\vec{F}_2$, $\vec{F}_4$ and $\vec{F}_4$ lie in a single xy-plane and all pass through a single point P, therefore, these forces constitute system of coplanar concurrent forces.

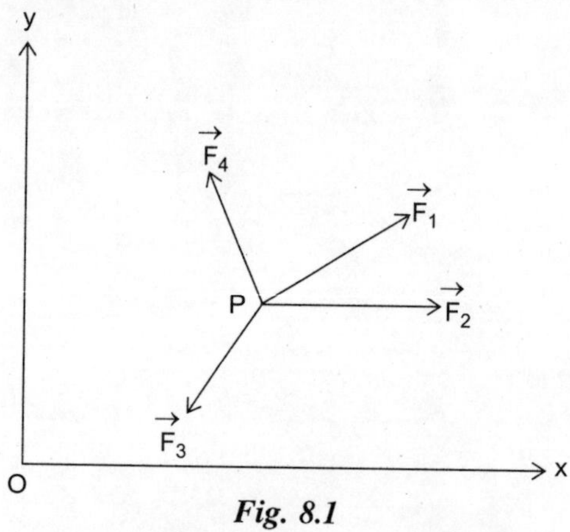

Fig. 8.1

8.2 COMPOSITION OF FORCES

Composition of forces mean finding a single force which will have the same effect as that of a number of forces acting on a body. Such a single force is called resultant force.

Since force is a vector quantity, therefore, resultant can be evaluated geometrically either using parallelogram law of vector addition (subtraction), triangle rule or in general by polygon rule of vector (force) addition. Analytically take components for all the forces in a coordinate system and then find resultant $\vec{R}$ knowing that,

$$R_x = \Sigma x\text{-components} = \Sigma F_x$$

$$R_y = \Sigma y\text{-components} = \Sigma F_y$$

and $$R_z = \Sigma z\text{-components} = \Sigma F_z$$

then, $$\vec{R} = R_x\,\vec{i} + R_y\,\vec{j} + R_z\,\vec{k} \quad ...(8.1)$$

$$= (\Sigma F_x)\,\vec{i} + (\Sigma F_y)\,\vec{j} + (\Sigma F_z)\,\vec{k}$$

where, $$|\vec{R}| = \sqrt{(\Sigma F_x)^2 + (\Sigma F_y)^2 + (\Sigma F_z)^2} \quad ...(8.2)$$

8.3 TYPES OF FORCES ON A BODY

Two types of forces act on a body they are applied forces and non-applied forces.

8.3.1 Applied Forces

Applied forces are the forces applied externally to a body. Each of the forces has got a point of contact with the body. For example, if a cricketer is hitting the ball, at the time of playing his shots, he is applying a force through the bat on the ball, this force is applied force.

When a farmer pulls a stone, then he applies a force and this force is applied force.

8.3.2 Non-Applied Forces

For non-applied force there is no direct contact with the body. Instead of direct interaction, there is indirect interaction. i.e., another body does not take part directly, through their fields they interact, e.g., self weight is a result of gravitational fields set up by Earth. Electric field, magnetic field, weak forces, contact reactions are electromagnetic forces between two contact surfaces. Due to huse uses in engineering problems of self weight and contact reactions we are taking extra care here.

8.3.2.1 Self Weight

Due to gravitational field $\vec{g}$ (gravitational acceleration), Earth applies a force called self weight of a body, is equal to

$$\vec{w} = m\,\vec{g}$$

where near the Earth surface gravitational field is 9.81 m/s^2 in magnitude and vertically downward in direction, it passes through centre of gravity (centroid, for planar body) of the body, if mass (m) of the body is negligible then product 'mg' is also small and for taking care of simplification, it is neglected.

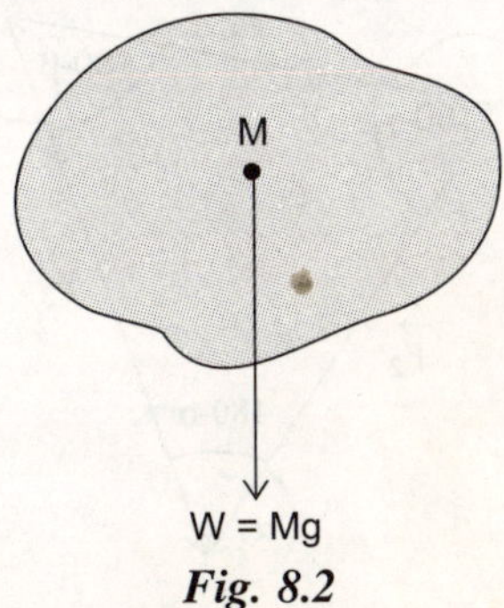

Fig. 8.2

8.4 REACTION

This is a self-adjusting force exerted by the other body which comes in contact with the body under consideration. According to Newton's third law of motion, the reaction is equal and opposite to the action (i.e action-reaction pair). If the surface of contact is smooth, the direction of the reaction is normal to the contact surface if the surface is not smooth, apart from normal reaction there is frictional reaction too, therefore, reaction now will be resultant of these two.

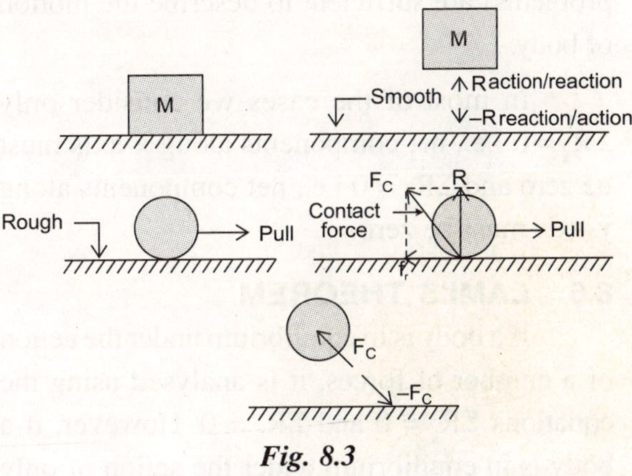

Fig. 8.3

8.5 EQUILIBRIUM OF A BODY

A body is said to be in equilibrium when it is at rest or continues to be in uniform linear velocity or rotates with uniform speed about an axis (point). According to Newton's law of motion it means that the resultant of all the forces acting on the body is zero i.e., $\Sigma\vec{R}_i = 0$ and net moment about a point (axis) will also be zero ($\Sigma\vec{\tau} = 0$), but, since here we are considering only a coplanar concurrent force system, therefore, net moment (torque) must be zero already, so equation $\Sigma\vec{\tau} = 0$ is satisfied.

Now for mathematical point of view

$$\text{Since,}\quad \Sigma\vec{R}_i = 0$$

$$\Rightarrow \quad |\Sigma\vec{R}_i| = 0$$

$$\Rightarrow \quad \sqrt{(\Sigma R_x)^2 + (\Sigma R_y)^2 + (\Sigma R_z)^2} = 0$$

$$\Rightarrow \quad (\Sigma R_x)^2 + (\Sigma R_y)^2 + (\Sigma R_z)^2 = 0$$

$$\left.\begin{aligned} \Rightarrow \quad \Sigma R_x &= 0 \\ \Sigma R_y &= 0 \text{ Simultaneously} \\ \text{and} \quad \Sigma R_z &= 0 \end{aligned}\right\} \quad \text{...(8.3)}$$

But since we are considering only coplanar concurrent force system, therefore, out of three equations only two of them (according to problems) are sufficient to describe the motion of body.

In most of the cases we consider only $\Sigma R_x = 0$ i.e., net components along x-axis must be zero and $\Sigma R_y = 0$ i.e., net components along y-axis must be zero.

8.6 LAMI'S THEOREM

If a body is in equilibrium under the action of a number of forces, it is analysed using the equations $\Sigma R_x = 0$ and $\Sigma R_y = 0$. However, if a body is in equilibrium under the action of only three coplanar concurrent forces, we can use directly Lami's theorem. Note that if a body is in equilibrium under the action of a number of forces then add them vectorially and finally you have to get only three forces to whom Lami's theorem is applied directly. We can also note that Lami's theorem is not an independent theorem, rather it is another form of 'sine rule' only.

Statement: If a body is in equilibrium under the action of three coplanar concurrent forces each force is proportional to the sine of the angle between the other two forces.

If $\vec{F}_1$, $\vec{F}_2$ and $\vec{F}_3$ are three coplanar concurrent forces and under them if a body is in equilibrium, α, β and γ are the angles between

$$\left\{\vec{F}_2 \text{ and } \vec{F}_3\right\}, \left\{\vec{F}_3 \text{ and } \vec{F}_1\right\} \text{ and } \left\{\vec{F}_1 \text{ and } \vec{F}_2\right\}$$

respectively, Ref. Fig. (8.4) then from Lami's theorem.

$$\frac{F_1}{\sin\alpha} = \frac{F_2}{\sin\beta} = \frac{F_3}{\sin\gamma} \qquad \text{...(8.4)}$$

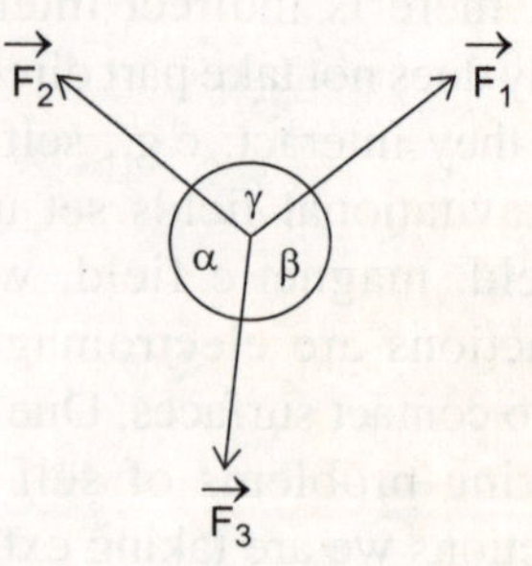

Fig. 8.4

Proof:

Since, body is in equilibrium under the action of $\vec{F}_1$, $\vec{F}_2$ and $\vec{F}_3$ therefore equilibrium gives $\Sigma \vec{F}_i = 0$ i.e., from triangle rule and definition of angle between two vectors we must get a close triangle like Fig. (8.5). Therefore using sine rule in shown triangle, we get

$$\frac{F_1}{\sin(180-\alpha)} = \frac{F_2}{\sin(180-\beta)} = \frac{F_3}{\sin(180-\gamma)}$$

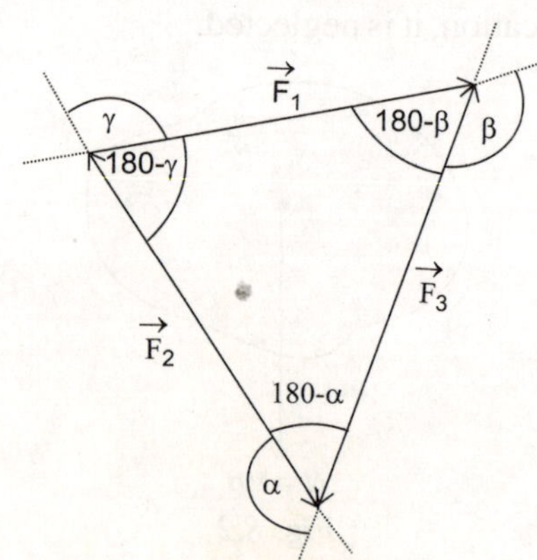

Fig. 8.5

or
$$\frac{F_1}{\sin\alpha} = \frac{F_2}{\sin\beta} = \frac{F_3}{\sin\gamma}$$

$$[\because \sin(180-\theta) = \sin\theta]$$

This equation is equal to statement of Lami's Theorem.

8.7 FREE BODY DIAGRAM

If one wants to know that whether a particular body is in state of rest or in state of motion from a system of bodies then from fundamental laws of motion, we have to consider only those forces which are acting on this particular body, out of the forces those this body is applying on other bodies as shown in the diagram (8.6) forces applying by body itself are not considered the diagram where all the forces acting on the body is shown is called Free Body Diagram (FBD). For example,

Suppose particles P_1, P_2, P_3 and P_4 lie on a table, considering particle P_1, then free body diagram consists of following forces:

1. Force applied by P_2 i.e., $\vec{F}_{21}$.
2. Force applied by P_3 i.e., $\vec{F}_{31}$.
3. Force applied by P_4 i.e., $\vec{F}_{41}$.
4. Force due to gravitational acceleration i.e., $m\vec{g}$, vertically downward, applied by Earth.
5. Normal reaction vertically upward by surface of table.
6. Considering $\vec{F}$ is the force applied by a external agent then free body diagram of particle P_1 is shown in Fig. (8.6).

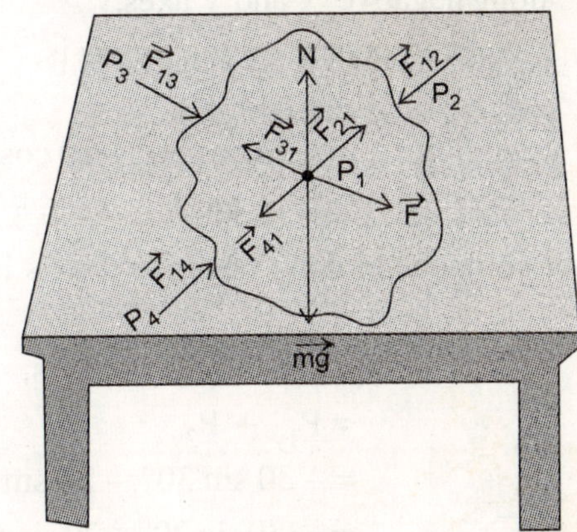

Fig. 8.6

Note:

1. *$\vec{F}_{21}$, $\vec{F}_{31}$, and $\vec{F}_{41}$ are gravitational attractive forces.*
2. *For free body diagram treat finally body as a point to clear visualization.*

WORKED OUT EXAMPLES

1. *The guy wire of an electric pole shown in Fig. (8-WCC1) makes 60° with the horizontal and is subjected to 20 kN and 30 kN forces. Find the horizontal and vertical components of the forces.*

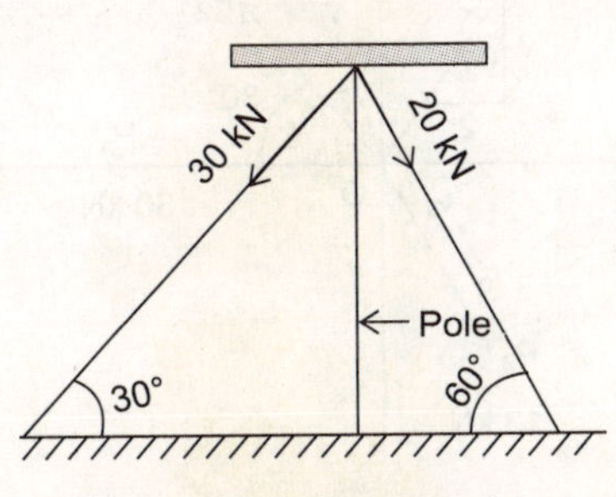

Fig. 8-WCC1

Solution:

Let P_1 = 30 kN and P_2 = 20 kN. P_{1x}, P_{2x}, P_{1y} and P_{2y} are components along x and y-axis respectively, then from triangles shown Fig. 8-WCC2.

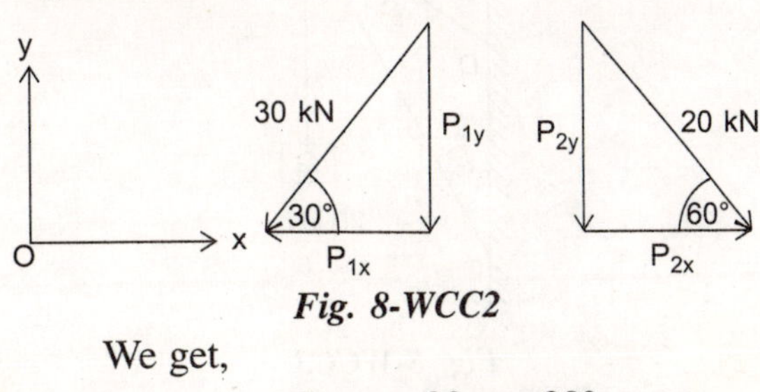

Fig. 8-WCC2

We get,

$$P_{1x} = -30 \cos 30°$$
$$P_{1y} = -30 \sin 30°$$

and $P_{2x} = 20 \cos 60°$

$P_{2y} = -20 \sin 60°$

[Negative sign because components are along negative x and y axes.]

∴ Net components along x-axis is,

$$= P_{1x} + P_{2x}$$
$$= -30 \cos 30° + 20 \cos 60°$$
$$\cong -16 \text{ kN}$$

i.e., net component along negative x-axis is 15 16 kN.

And net component along y-axis is,

$$= P_{1y} + P_{2y}$$
$$= -30 \sin 30° - 20 \sin 60°$$
$$= -(30 \sin 30° + 20 \sin 60°)$$
$$= -32.32 \text{ kN}$$

i.e., net component along negative y-axis is 32.32 kN.

2. *The frictionless pulley A shown in Fig. (8-WCC3) is supported by two bars AB and AC which are hinged at B and C to a vertical wall. The flexible cable DG hinged at D goes over the pulley and supports a load of 20 kN at G. The angles between the various members are shown in Fig. Determine the forces in the bars AB and AC, neglect the weight of the pulley.*

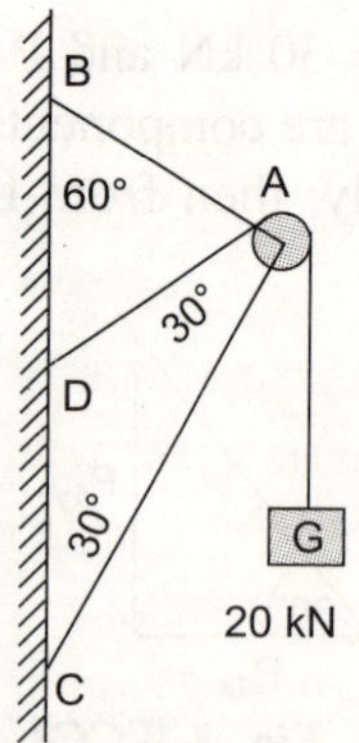

Fig. 8-WCC3

Solution:

Forces in the bars and on the pulley are shown in the figure (8-WCC4). Since GA and AD are same cable therefore 20 kN force will be in both parts AG and AD and they are tensile forces.

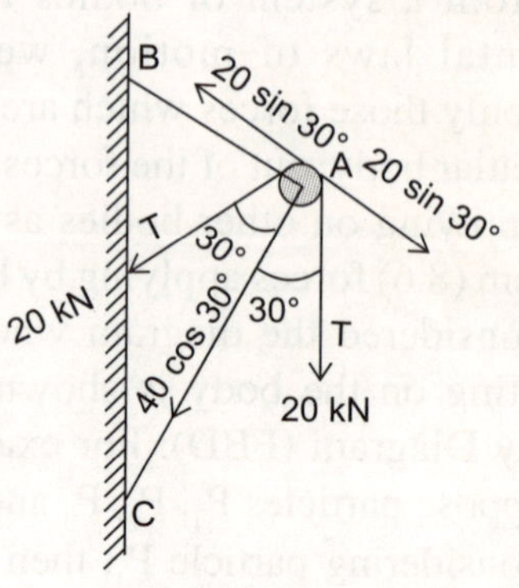

Fig. 8-WCC4

Now to balance the system, force in the bar AC is,

$$F_{AC} = 20 \cos 30° + 20 \cos 30°$$
$$= 2 \times 20 \times \cos 30°$$
$$= 40 \cos 30°$$
$$= 34.641 \text{ kN directed from A to C.}$$

and force in bar AB is,

$$F_{AB} = 20 \sin 20° + (-20 \sin 30°)$$
$$= 0.$$

3. *A system of four forces acting on a body is as shown in Fig. (8-WCC5). Determine the resultant.*

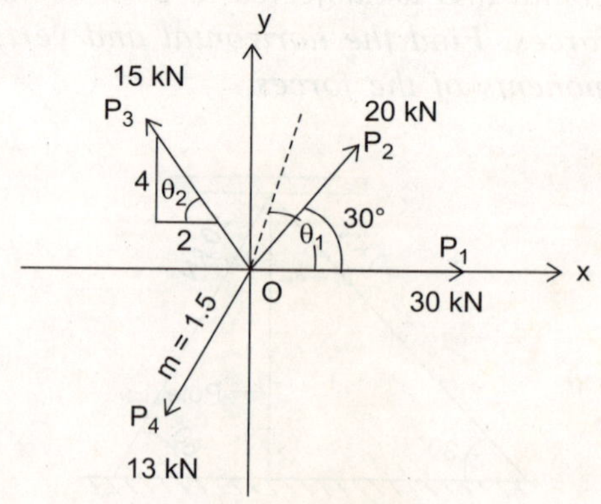

Fig. 8-WCC5

Solution:

Let θ_1 and θ_2 are inclinations of the extended line as shown and for the force 1.5 kN then $\tan\theta_1 = 1.5 = \frac{3}{2}$ and $\tan\theta_2 = \frac{4}{2} = 2.$

$$\therefore \quad \sin\theta_1 = \frac{3}{\sqrt{13}}$$

$$\sin\theta_2 = \frac{4}{\sqrt{20}}$$

and
$$\cos\theta_1 = \frac{2}{\sqrt{13}}$$

$$\cos\theta_2 = \frac{2}{\sqrt{20}}$$

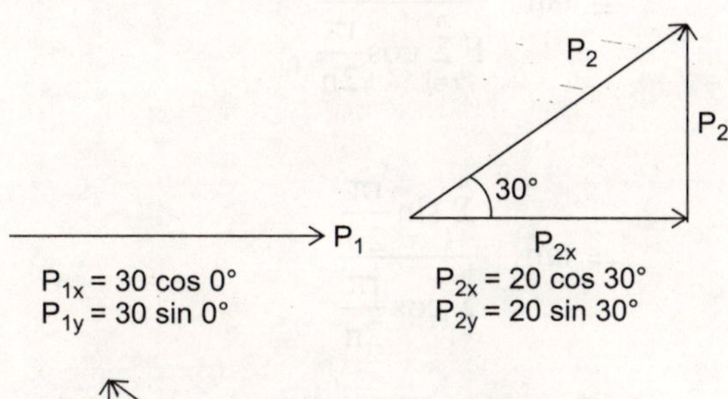

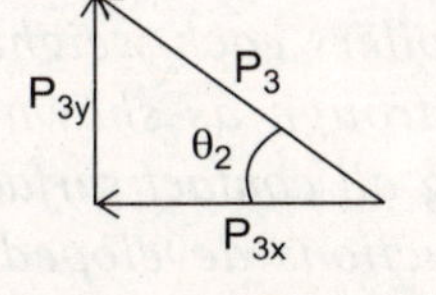

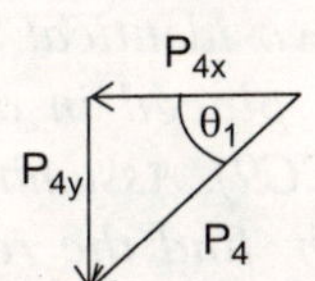

Fig. 8-WCC6

Now sum of components along x and y axes.

$$\therefore \quad R_x = P_{1x} + P_{2x} + P_{3x} + P_{4x}$$

$$= 30\cos 0° + 20\cos 30° - 15\cos\theta_2 - 13\cos\theta_1$$

$$= 30\cos 0° + 20\cos 30° - 15\times\frac{2}{\sqrt{20}} - 13\times\frac{2}{\sqrt{13}}$$

$$= 33.40 \text{ kN}$$

and
$$R_y = 30\sin 0° + 20\sin 30° + 15\sin\theta_2 - 13\sin\theta.$$

$$= 30\sin 0° + 20\sin 30° + 15\times\frac{4}{\sqrt{20}} - 13\times\frac{3}{\sqrt{13}}$$

$$= 0 + 10 + \frac{60}{\sqrt{20}} - \frac{39}{\sqrt{13}}$$

$$= 10 + \sqrt{\frac{60\times 60}{20}} - \sqrt{\frac{39\times 39}{13}}$$

$$= 10 + \sqrt{180} - \sqrt{117}$$

$$= 12.6 \text{ kN}$$

$$\therefore \quad R = \sqrt{R_x^2 + R_y^2}$$

$$= \sqrt{(33.40)^2 + (12.6)^2}$$

$$= 35.7 \text{ kN}$$

and
$$\theta = \tan^{-1}\left(\frac{R_y}{R_x}\right)$$

$$= \tan^{-1}\left(\frac{12.6}{33.4}\right)$$

$$= 22.66°.$$

4. *Determine the resultant of the forces acting on a block as shown in Fig. (8-WCC7).*

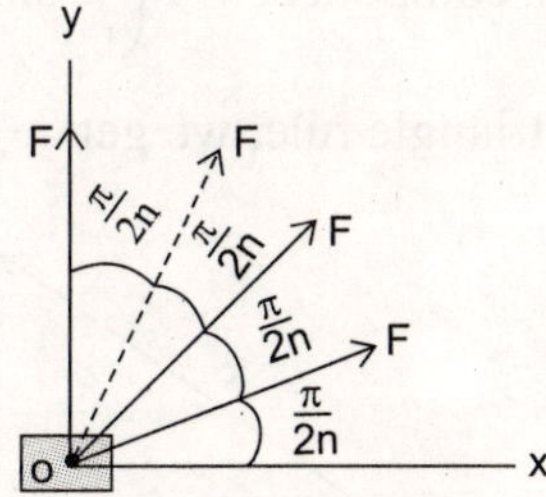

Fig. 8-WCC7

Solution:

	Force	x-component	y-component
1.	F	$F\cos\frac{\pi}{2n}$	$F\sin\frac{\pi}{2n}$
2.	F	$F\cos\frac{2\pi}{2n}$	$F\sin\frac{2\pi}{2n}$
3.	F	$F\cos\frac{3\pi}{2n}$	$F\sin\frac{3\pi}{2n}$
(n –1)	F	$F\cos(n-1)\frac{\pi}{2n}$	$F\sin(n-1)\frac{\pi}{2n}$
n	F	$F\cos\frac{n\pi}{2n}$	$F\sin\frac{n\pi}{2n}$

$$\Sigma\text{x-component} = F\cos\frac{\pi}{2n} + F\cos\frac{2\pi}{2n}$$

$$+ F\cos\frac{3\pi}{2n} + F\cos(n-1)\frac{\pi}{2n} + F\cos\frac{n\pi}{2n}$$

$$= F\left\{\cos\frac{\pi}{2n} + \cos\frac{2\pi}{2n} + \cos\frac{3\pi}{2n} + \cos\frac{(n-1)\pi}{2n} + \cos\frac{n\pi}{2n}\right\}$$

$$= F\left(\sum_{r=1}^{n}\cos\frac{r\pi}{2n}\right)$$

Similarly, $\Sigma\text{y-component} = F\left(\sum_{r=1}^{n}\sin\frac{r\pi}{2n}\right)$

From triangle rule, we get

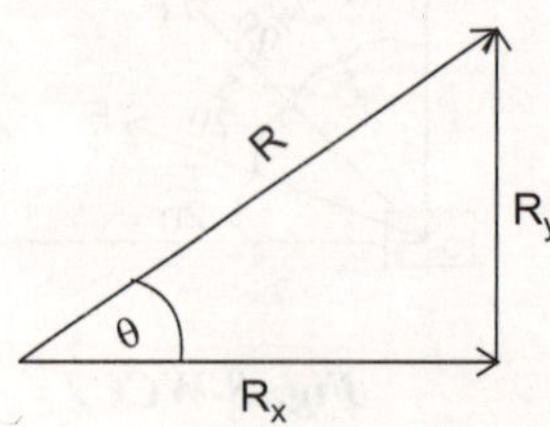

Fig. 8-WCC8

$$\therefore |\vec{R}| = \sqrt{(\Sigma x - \text{component})^2 + (\Sigma y - \text{component})^2}$$

$$= \sqrt{\left(F\sum_{r=1}^{n}\cos\frac{r\pi}{2n}\right)^2 + \left(F\sum_{r=1}^{n}\sin\frac{r\pi}{2n}\right)^2}$$

$$= F\sqrt{\left(\sum_{r=1}^{n}\cos\frac{r\pi}{2n}\right)^2 + \left(\sum_{r=1}^{n}\sin\frac{r\pi}{2n}\right)^2}$$

$$\text{and } \theta = \tan^{-1}\frac{\Sigma\text{y-omponent}}{\Sigma\text{x-component}}$$

$$= \tan^{-1}\frac{F\sum_{r=1}^{n}\sin\frac{r\pi}{2n}}{F\sum_{r=1}^{n}\cos\frac{r\pi}{2n}}$$

$$= \tan^{-1}\frac{\sum_{r=1}^{n}\sin\frac{r\pi}{2n}}{\sum_{r=1}^{n}\cos\frac{r\pi}{2n}}.$$

5. *Two identical rollers each weighing 200 N are placed in a trough as shown in Fig. (8-WCC9). Assuming all contact surfaces are smooth, find the reactions developed at contact surfaces A, B, C and D.*

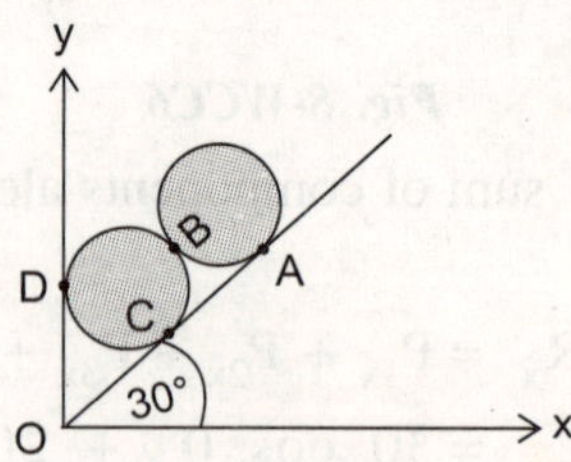

Fig. 8-WCC9

Solution:

Let the reactions at contact surfaces are R_A, R_B, R_C and R_D respectively. The free body diagrams of rollers are shown below.

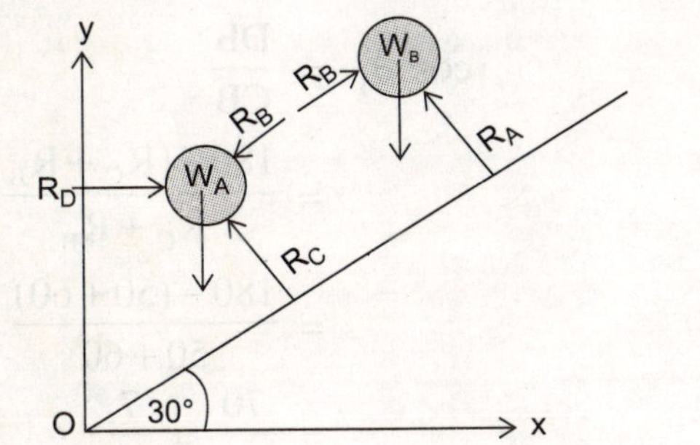

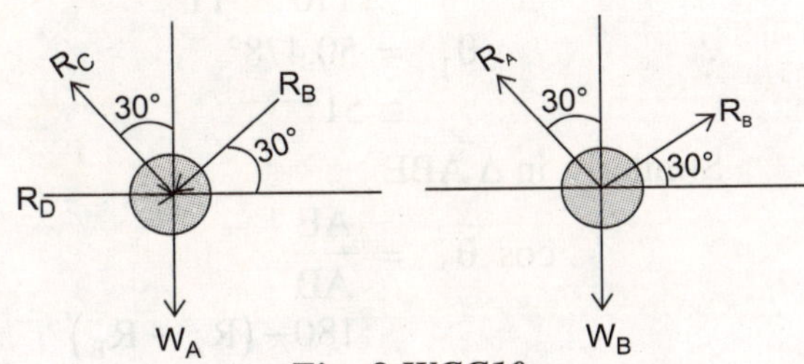

Fig. 8-WCC10

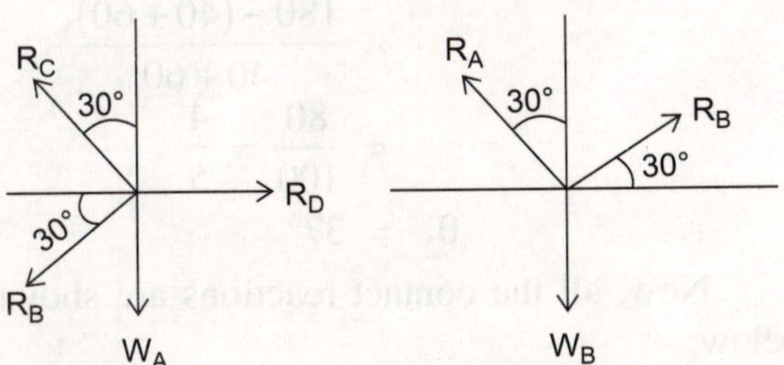

Fig. 8-WCC11

Since both the rollers are in equilibrium, therefore,

Net force along x-axis on roller-1 = 0

or $\Sigma F_x = 0$

or $R_D - R_B \cos 30° - R_C \sin 30° = 0$

or $R_D - \frac{\sqrt{3}}{2} R_B - \frac{1}{2} R_C = 0$ ...(i)

Net force along y-axis on roller-1 should be equal to zero.

$\Sigma F_y = 0$

or $R_C \cos 30° - R_B \sin 30° - W_A = 0$

or $\frac{\sqrt{3}}{2} R_C - \frac{1}{2} R_B - W_A = 0$ ...(ii)

Similarly, net force along x-axis on roller-2 = 0

or $\Sigma F_x = 0$

or $R_B \cos 30° - R_A \sin 30° = 0$

or $\frac{\sqrt{3}}{2} R_B - \frac{1}{2} R_A = 0$

or $\sqrt{3}\, R_B - R_A = 0$

or $R_A = \sqrt{3}\, R_B$ ...(iii)

and net force along y-axis on roller-2 should be zero.

$\Sigma F_y = 0$

or $R_A \cos 30° + R_B \sin 30° - W_B = 0$

or $\frac{\sqrt{3}}{2} R_A + \frac{1}{2} R_B = 200$

or $\sqrt{3} R_A + R_B = 400$

or $3R_B + R_B = 400$

or $4R_B = 400$

$\therefore$ $R_B = 100$ N ...(iv)

$\therefore$ from equation(iii) $R_A = \sqrt{3}\, R_B$

$= 100\sqrt{3}$ N.

Now, from equation (ii)

i.e., $R_C \frac{\sqrt{3}}{2} - \frac{R_B}{2} - W_A = 0$

or $\frac{R_C\sqrt{3}}{2} - 50 - 200 = 0$

or $\frac{R_C\sqrt{3}}{2} = 250$

$\therefore$ $R_C = \frac{500}{\sqrt{3}}$

Using equation (i), we get

$R_D = \frac{\sqrt{3}}{2} R_B + \frac{1}{2} R_C$

$$= \frac{\sqrt{3}}{2} \times 100 + \frac{1}{2} \times \frac{500}{\sqrt{3}}$$

$$= 50\sqrt{3} + \frac{250}{\sqrt{3}}$$

$$= \frac{50 \times 3 + 250}{\sqrt{3}}$$

$$= \frac{150 + 250}{\sqrt{3}}$$

$$= \frac{400}{\sqrt{3}} \text{ N}$$

Therefore, $R_B = 100$ N, $R_A = 173.20$ N, $R_C = 288.67$ N and $R_D = 230.94$ N.

6. *Three cylinders are pied up in a rectangular channel as shown in Fig. (8-WCC12). Determine the reactions at the surfaces of contacts between the bottom cylinder and the vertical face of the channel. Assume smooth surfaces at all contact surfaces.*

Cylinder	Details
Weight	Radius
A .... 150 N	40 mm
B .... 400 N	60 mm
C .... 200 N	50 mm

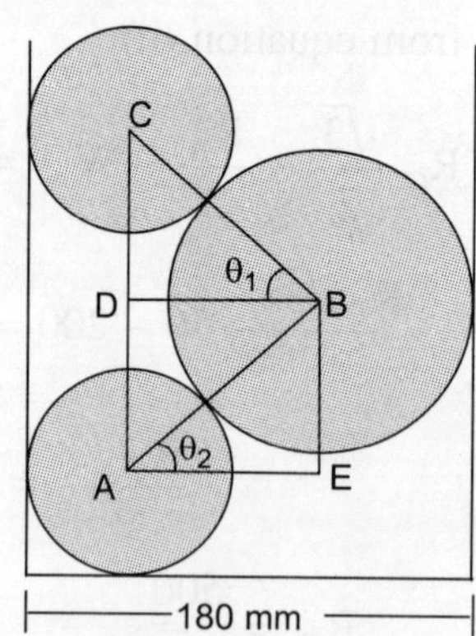

Fig. 8-WCC12

Solution:

Let R_A, R_B and R_C are the radii of cylinders respectively. Then in Δ BCD,

$$\cos \theta_1 = \frac{DB}{CB}$$

$$= \frac{180 - (R_C + R_B)}{R_C + R_B}$$

$$= \frac{180 - (50 + 60)}{50 + 60}$$

$$= \frac{70}{110} = \frac{7}{11}$$

$$\therefore \quad \theta_1 = 50.478°$$

$$\cong 51°$$

Similarly, in Δ ABE

$$\cos \theta_2 = \frac{AE}{AB}$$

$$= \frac{180 - (R_A + R_B)}{R_A + R_B}$$

$$= \frac{180 - (40 + 60)}{40 + 60}$$

$$= \frac{80}{100} = \frac{4}{5}$$

$$\therefore \quad \theta_2 = 37°$$

Now, all the contact reactions are shown below:

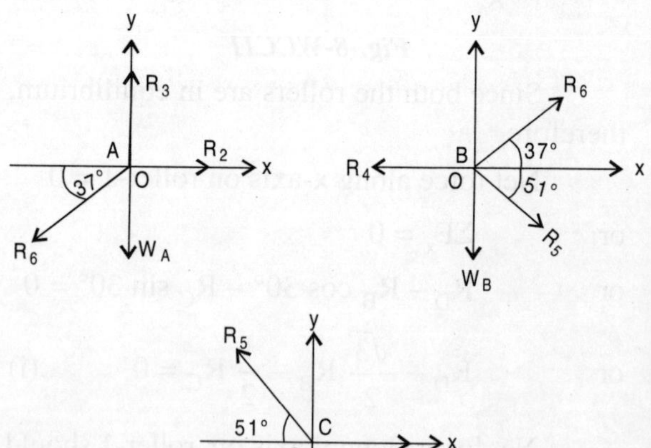

Fig. 8-WCC13

Consider cylinder-C, then from equilibrium of this cylinder we get,

$$\Sigma F_x = 0$$

or $R_1 - R_5 \cos 51° = 0$

or $R_1 = R_5 \cos 51°$

$\therefore \quad R_1 = R_5 \cos 51° \quad$ ...(i)

and $\Sigma F_y = 0$

or $R_5 \sin 51° - W_C = 0$

or $R_5 \sin 51° = W_C$

or $R_5 = \dfrac{W_C}{\sin 51°}$

$\therefore \quad R_5 = \dfrac{200}{\sin 51°}$

$= 257.35$ N.

From equation (i)

$$R_1 = R_5 \cos 51°$$
$$= 257.35 \times \cos 51°$$
$$= 162 \text{ N}$$

Cylinder-B, then from equilibrium of this cylinder, we get

$$\Sigma F_x = 0$$

or $-R_4 + R_5 \cos 51° + R_6 \cos 37° = 0$

or $R_5 \cos 51 + R_6 \cos 37° = R_4 \quad$ ...(ii)

and $\Sigma F_y = 0$

or $R_6 \sin 37° - W_B - R_5 \sin 51° = 0$

or $R_6 \sin 37° = W_B + R_5 \sin 51° = 0$

or $R_6 \sin 37° = 400 + 257.35 \sin 51°$

$= 600$

$\therefore \quad R_6 = \dfrac{600}{\sin 37°}$

$= 997$ N

From equation (ii)

$$R_4 = R_5 \cos 51° + R_6 \cos 37°$$
$$= 257.35 \cos 51° + 997 \cos 37°$$
$$= 958.20 \text{ N}$$

Cylinder-A, then from equilibrium of this cylinder we get,

$$\Sigma F_x = 0$$

or $R_2 - R_6 \cos 37° = 0$

$\therefore \quad R_2 = R_6 \cos 37°$

$= 997 \cos 37°$

$= 796.23$ N

and $\Sigma F_y = 0$

or $R_3 - W_A - R_6 \sin 37° = 0$

or $R_3 = W_A + R_6 \sin 37°$

$= 150 + 997 \sin 37°$

$= 750$ N

Therefore, $R_1 = 162$ N, $R_2 = 796.23$ N, $R_3 = 750$ N, $R_4 = 958.23$, $R_5 = 257.35$ and $R_6 = 997$ N.

7. *Determine the magnitude of a horizontal force P applied at the centre C of a roller of weight Q = 1000 N and radius r = 15 mm, which will be necessary to pull it over a 3 mm curb refer Fig. (8-WCC14).*

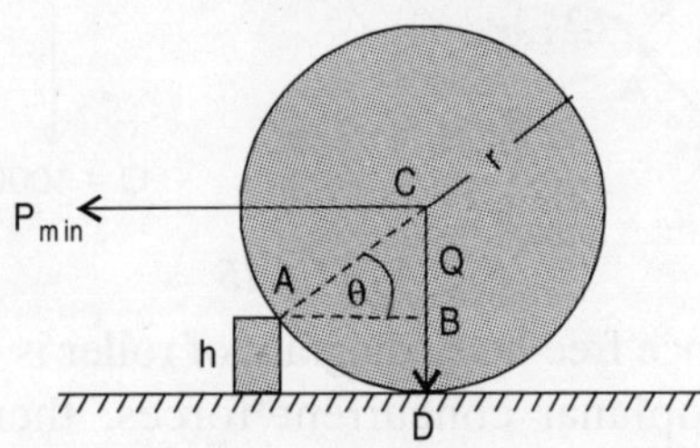

Fig. 8-WCC14

Solution:

In Δ ABC, we get

$$\sin \theta = \frac{BC}{AC}$$
$$= \frac{DC - BD}{AC}$$
$$= \frac{r - h}{r} = 1 - \frac{h}{r}$$
$$= 1 - \frac{3}{15}$$
$$= \frac{12}{15} = \frac{4}{5}$$

$\therefore \quad \theta = 53°$

Let R_A is the reaction at A and consider R_A passes through centre of the roller. R_D is the reaction at D-point of contact on horizontal surface. Let P_{min} is the minimum applied pull necessary to pull it over a 3 mm curb and at this position there will be no contact with horizontal surface i.e., $R_D = 0$. Over this consideration roller is in equilibrium i.e., P_{min} is just sufficient to cause motion of the roller to impend. Since P_{min} and Q = wt. Intersect at the center C of the roller then reaction R_A must pass through this same point also. Here we are not considering friction.

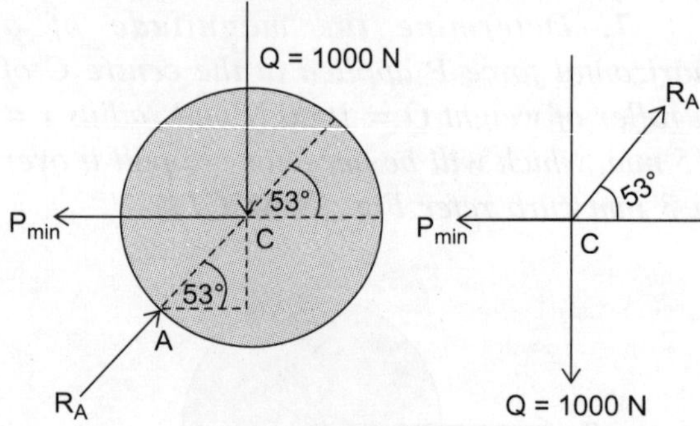

Fig. 8-WCC15

Since free-body diagram of roller is having three coplanar concurrent forces, therefore, straight way we could apply Lami's theorem.

i.e., $$\frac{R_A}{\sin 90°} = \frac{P_{min}}{\sin 143°} = \frac{Q}{\sin 127°}$$
(i) (ii) (iii)

From equations (ii) and (iii)

$$\frac{P_{min}}{\sin 143°} = \frac{1000}{\sin 127°}$$

$$\therefore \quad P_{min} = \frac{\sin 143°}{\sin 127°} \times 1000$$

$$= 753.55 \text{ N}$$

Therefore, $$R_A = \frac{\sin 90}{\sin 127} \times 1000$$

$$= 1252.13 \text{ N.}$$

8. *Determine the reactions at contact points for the system shown in Fig. (8-WCC16).*

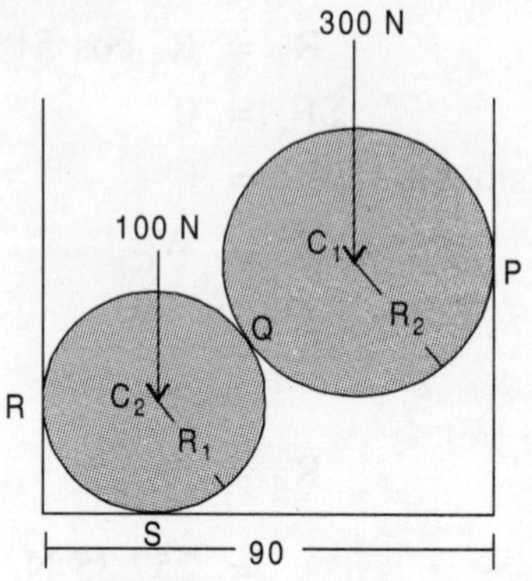

Fig. 8-WCC16

Solution:

Let us draw the following Fig. (8-WCC17).

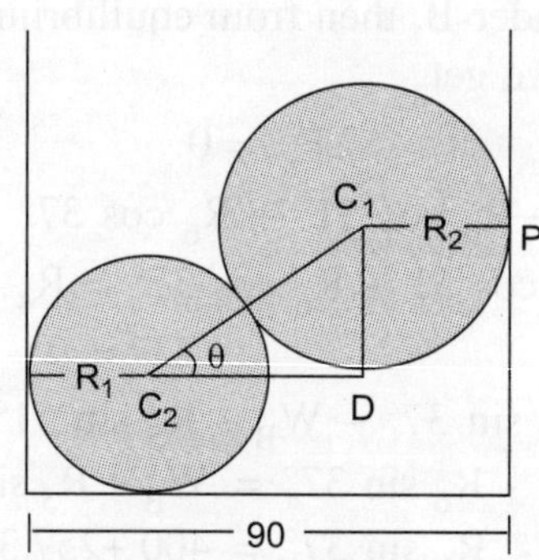

Fig. 8-WCC17

From the figure, it is clear that

$$C_2D = 90 - R_1 - R_2$$
$$= 90 - 20 - 30$$
$$= 40 \text{ mm}$$

and $$C_1C_2 = R_1 + R_2 = 20 + 30$$
$$= 50 \text{ mm}$$

$$\therefore \quad \cos\theta = \frac{C_2D}{C_1C_2}$$

$$= \frac{40}{50} = \frac{4}{5}$$

$$\therefore \quad \theta = 37°$$

Let R_R, R_S, R_Q and R_P are the reactions at points R, S, Q, P respectively. Free body diagrams are shown in Fig. (8-WCC18).

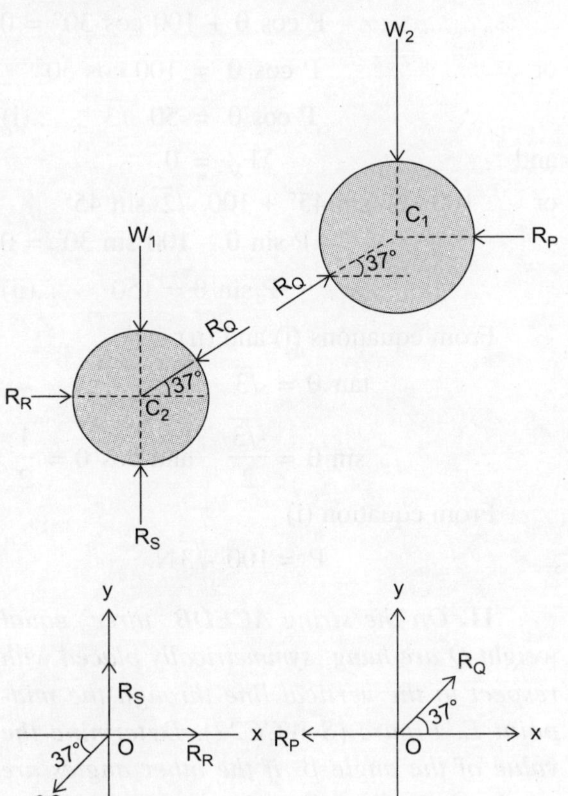

Fig. 8-WCC18

Applying Lami's theorem at C_2, we get

$$\frac{R_P}{\sin 127°} = \frac{R_Q}{\sin 90°} = \frac{W_2}{\sin 143°}$$

(i) (ii) (iii)

From (i) and (iii), we get

$$R_P = \frac{\sin 127°}{\sin 143°} \times W_2$$

$$= \frac{\sin 127°}{\sin 143°} \times 300 = 398.11 \text{ N}$$

From (i) and (iii)

$$R_Q = \frac{\sin 90°}{\sin 143°} \times W_2$$

$$= \frac{300}{\sin 143°} = 498.49 \text{ N.}$$

Since C_2 roller is also in equilibrium, therefore,

$$\Sigma F_x = 0$$

or $\quad R_R - R_Q \cos 37° = 0$

or $\quad R_R = R_Q \cos 37°$

$$= 498.49 \cos 37°$$

$$= 398.11$$

and $\quad \Sigma F_y = 0$

or $R_S - W_1 - R_Q \sin 37° = 0$

or $\quad R_S = W_1 + R_Q \sin 37°$

$$= 100 + 498.49 \sin 37°$$

$$= 400 \text{ N}$$

Thus, reactions are $R_R = 398.11$ N, $R_Q = 498.49$ N, $R_S = 400$ N and $R_P = 398.11$ N.

9. *A 3 kN crate is to be supported by the rope and pulley arrangement as shown in Fig. (8-WCC19). Determine the magnitude and direction of the force F which should be exerted at the free end of the rope.*

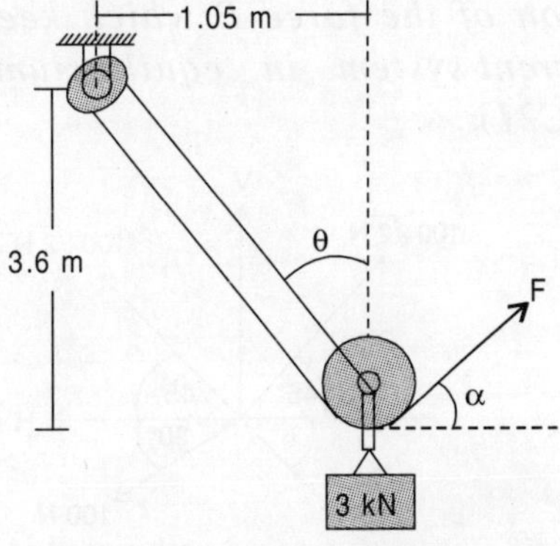

Fig. 8-WCC19

Solution:

From the above figure, $\tan\theta = \frac{1.05}{3.6}$

$\therefore \qquad \theta = 16.26°$

Forces on crate and free body diagram of the crate is shown below.

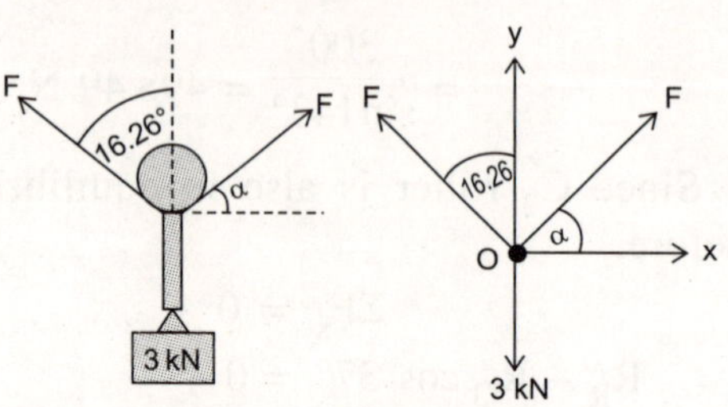

Fig. 8-WCC20

Since crate is in equilibrium, therefore,

$$\Sigma F_x = 0$$

or $\quad F\cos\alpha - F\sin 16.26° = 0$

or $\quad \cos\alpha - \sin 16.26° = 0$

or $\quad \cos\alpha = \sin 16.26$

$\therefore \quad \alpha = \cos^{-1}[\sin(16.26)]$

$= 73.74°$

and $\quad \Sigma F_y = 0$

or $\quad F\cos 16.26 + F\sin\alpha - 3 = 0$

or $\quad F = \frac{3}{\cos 16.26° + \sin 73.74°}$ kN

$= 1.56$ kN

Therefore, α = 73.4° and F = 1.56 kN.

10. *Determine the magnitude and direction of the force P which keeps the concurrent system in equilibrium Fig. (8-WCC21).*

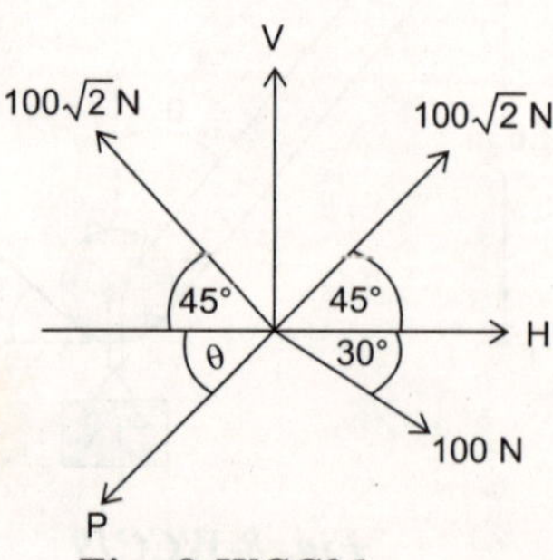

Fig. 8-WCC21

Solution:

Since concurrent system of forces is in equilibrium.

Therefore,

$$\Sigma F_H = 0$$

or $\quad 100\sqrt{2}\cos 45° - 100\sqrt{2}\cos 45° - P\cos\theta + 100\cos 30° = 0$

or $\quad P\cos\theta = 100\cos 30°$

$\therefore \quad P\cos\theta = 50\sqrt{3} \quad$ ...(i)

and $\quad \Sigma F_V = 0$

or $\quad 100\sqrt{2}\sin 45° + 100\sqrt{2}\sin 45° - P\sin\theta - 100\sin 30° = 0$

$\therefore \quad P\sin\theta = 150 \quad$ ...(ii)

From equations (i) and (ii)

$$\tan\theta = \sqrt{3}$$

$\therefore \quad \sin\theta = \frac{\sqrt{3}}{2}$ and $\cos\theta = \frac{1}{2}$

From equation (i)

$$P = 100\sqrt{3}\text{N}.$$

11. *On the string ACEDB three equal weight Q are hung symmetrically placed with respect to the vertical line through the mid-point E. Figure (8-WCC22). Determine the value of the angle β if the other angles are as shown below.*

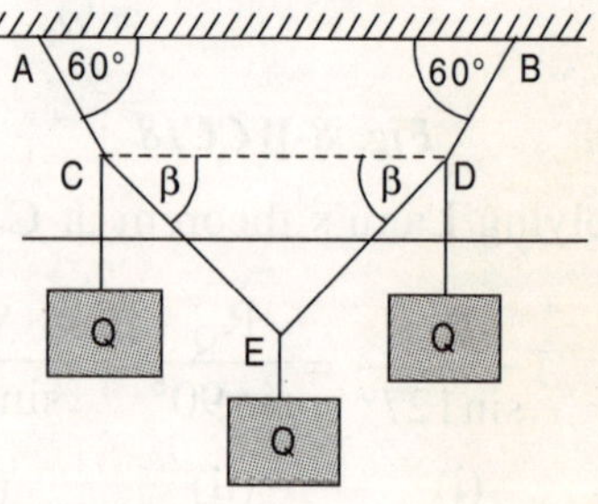

Fig. 8-WCC22

Solution:

Let $T_{AC} = T_{BD} = T_1$ and $T_{CE} = T_{DE} = T_2$ because it is the symmetrical figure.

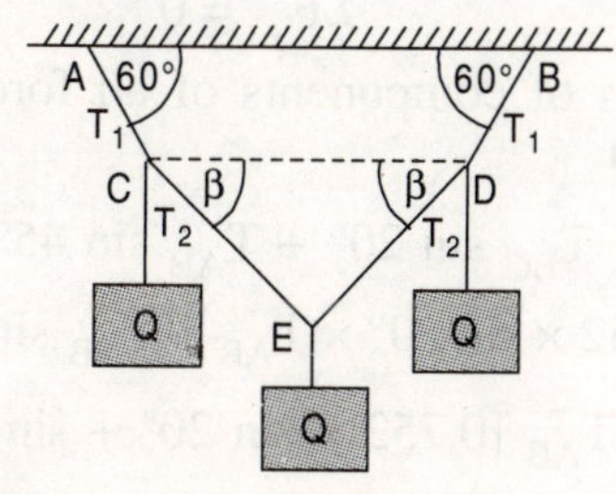

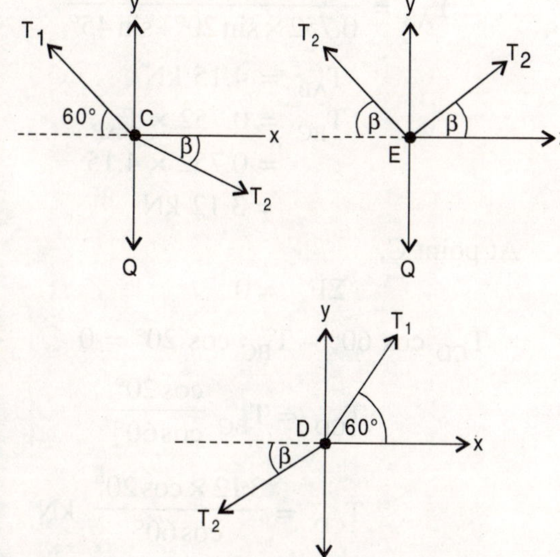

Fig. 8-WCC23

Since the system is in equilibrium, therefore consider,

Point-C

$$\Sigma F_{x,\,C} = 0$$

or $\quad T_1 \cos 60° - T_2 \cos \beta = 0$

or $\quad \dfrac{T_1}{2} = T_2 \cos \beta$

$\therefore \quad T_1 = 2T_2 \cos \beta$ ...(i)

and $\quad \Sigma F_{y,\,C} = 0$

or $\quad T_1 \sin 60° - Q - T_2 \sin \beta = 0$

or $\quad T_1 \sin 60° = Q + T_2 \sin \beta$

or $\quad 2T_2 \cos \beta \times \sin 60° = \theta + T_2 \sin \beta$

or $\quad T_2 \left(\sqrt{3} \cos \beta - \sin \beta\right) = Q$

$\therefore \quad T_2 \left(\sqrt{3} \cos \beta - \sin \beta\right) = Q$ ....(ii)

Now at point E,

$$\Sigma F_{x,\,E} = 0$$

or $\quad T_2 \cos \beta - T_2 \cos \beta = 0$

or $\quad 0 = 0$

$\Rightarrow$ It is a trivial result.

and $\quad \Sigma F_{y,\,E} = 0$

or $\quad T_2 \sin \beta + T_2 \sin \beta - Q = 0$

or $\quad 2T_2 \sin \beta = Q$

or $\quad 2T_2 \sin \beta = Q$ ...(iii)

From equations (ii) and (iii)

$$T_2 \left(\sqrt{3} \cos \beta - \sin \beta\right) = 2T_2 \sin \beta$$

or $\quad \sqrt{3} \cos \beta - \sin \beta = 2 \sin \beta$

or $\quad \sqrt{3} \cos \beta - 3 \sin \beta = 0$

or $\quad \sin \beta = \dfrac{1}{\sqrt{3}} \cos \beta$

or $\quad \tan \beta = \dfrac{1}{\sqrt{3}}$

$\therefore \quad \beta = 30°.$

Note: *From equilibrium at point D we will get again equations (i) and (ii).*

12. *A string is subjected to the forces 4 kN and W as shown in Fig. (8-WCC24). Determine the magnitude of W and the tensions induced in various portions of the string.*

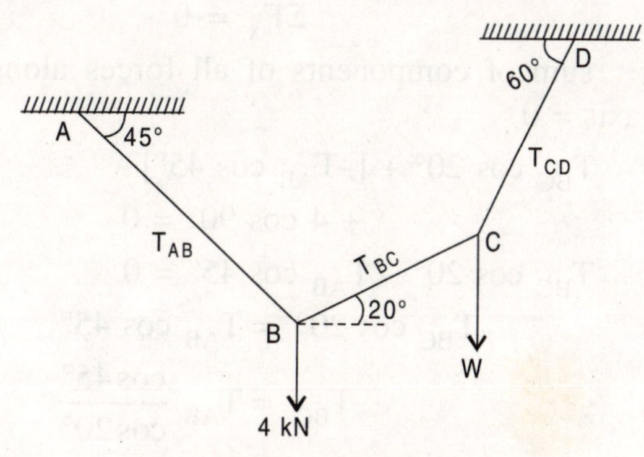

Fig. 8-WCC24

Solution:

Let tensions in AB, BC and CD strings are T_{AB}, T_{BC} and T_{CD} respectively. Forces acting at points B and C are shown in free body diagrams below.

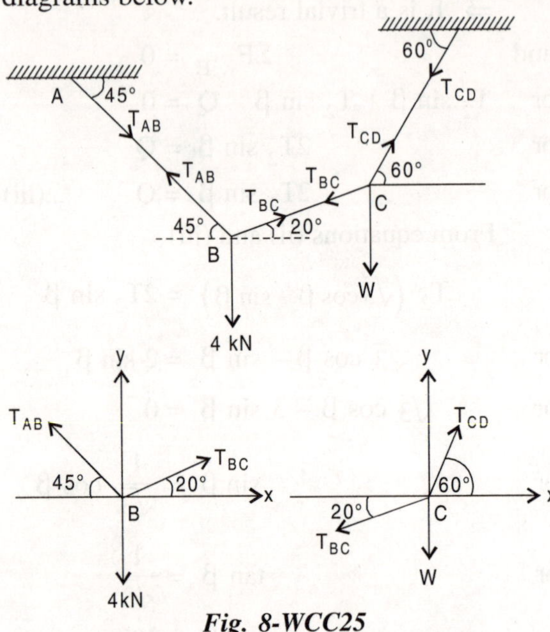

Fig. 8-WCC25

Now since at points B and C all forces are passing through single points B and C respectively i.e., they are coplanar concurrent system of forces then for equilibrium of points B and C there are two equations. $\Sigma F_x = 0$ and $\Sigma F_y = 0$, are sufficient to describe the equilibrium.

At point C,

$$\Sigma F_x = 0$$

or sum of components of all forces along x-axis = 0

or $T_{BC} \cos 20° + [-T_{AB} \cos 45°] + 4 \cos 90° = 0$

or $T_{BC} \cos 20° - T_{AB} \cos 45° = 0$

or $T_{BC} \cos 20° = T_{AB} \cos 45°$

$\therefore$ $T_{BC} = T_{AB} \dfrac{\cos 45°}{\cos 20°}$

$$T_{BC} = 0.752\, T_{AB}$$

and $\Sigma F_y = 0$

i.e., sum of components of all forces along y-axis = 0

or $T_{BC} \sin 20° + T_{AB} \sin 45° - 4 = 0$

or $0.752 \times \sin 20° \times T_{AB} + T_{AB} \sin 45° = 4$

or $T_{AB}\, [0.752 \times \sin 20° + \sin 45°] = 4$

$\therefore$ $T_{AB} = \dfrac{4}{0.752 \times \sin 20° + \sin 45°}$

or $T_{AB} = 4.15$ kN

$\therefore$ $T_{BC} = 0.752 \times T_{AB}$

$= 0.752 \times 4.15$

$= 3.12$ kN

At point C,

$$\Sigma F_x = 0$$

or $T_{CD} \cos 60° - T_{BC} \cos 20° = 0$

or $T_{CD} = T_{BC} \dfrac{\cos 20°}{\cos 60°}$

or $T_{CD} = \dfrac{3.12 \times \cos 20°}{\cos 60°}$ kN

$[\because T_{BC} = 3.12 \text{ kN}]$

$\therefore$ $T_{CD} = 5.863$ kN

and $\Sigma F_y = 0$

or $T_{CD} \sin 60° - T_{BC} \sin 20° - W = 0$

or $W = T_{CD} \sin 60° - T_{BC} \sin 20°$

$= 4.010$ kN

$\therefore$ $W = 4.010$ kN

Hence, tensions in strings are $T_{BC} = 3.120$ kN, $T_{AB} = 4.15$ kN, $T_{CD} = 5.863$ kN and weight W = 4.010 kN respectively.

13. *Two cylinders, A of weight 4000 N and B of weight 2000 N rests on smooth inclined surface as shown in Fig. (8-WCC26). They are connected by a bar of negligible mass hinged to each cylinder at its* *geometrical centre by smooth pins. Find the force P to be applied, as shown in Fig. (8-WCC26) such that it will hold the system in the given position.*

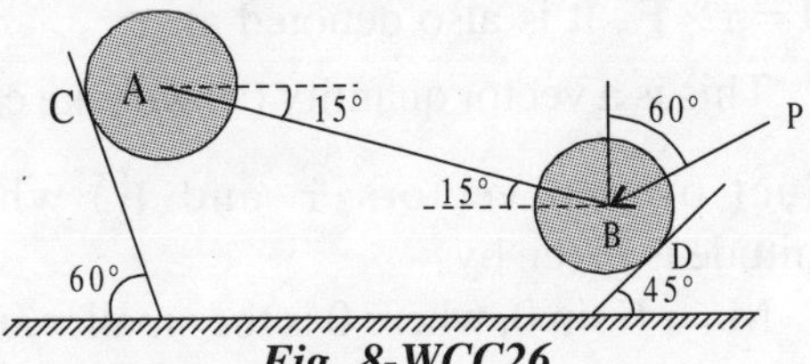

Fig. 8-WCC26

Solution:

Forces and free body diagrams are shown below for both the cylinders. Let us take F_C as the compression in the rod (note that compression force always act towards from the body).

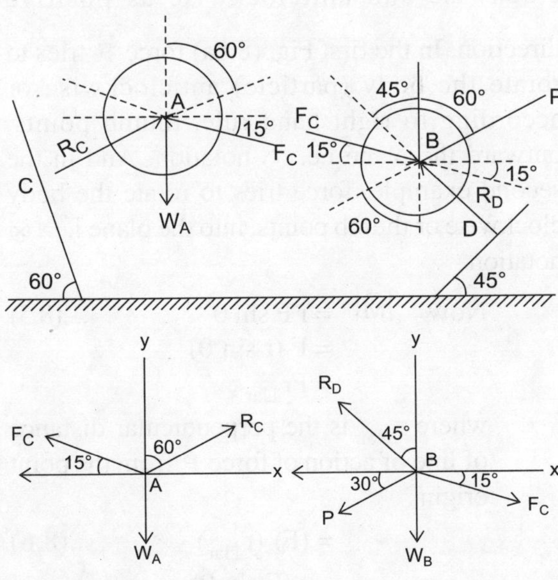

Fig. 8-WCC27

Cylinder-A

Applying Lami's theorem,

$$\frac{R_C}{\sin 105°} = \frac{F_C}{\sin 120°} = \frac{W_A}{\sin 135°}$$
$$\text{(i)} \qquad \text{(ii)} \qquad \text{(iii)}$$

From (i) and (iii), we get

$$R_C = \frac{\sin 105°}{\sin 135°} \times W_A$$
$$= \frac{\sin 105°}{\sin 135°} \times 4000$$
$$= 5464.10 \text{ N}$$

and
$$F_C = \frac{\sin 120°}{\sin 135°} \times 4000$$
$$= 4898.98 \text{ N}$$

Cylinder-B

$$\Sigma F_{x,\,B} = 0$$

or $F_C \cos 15° - P \cos 30° - R_D \cos 45° = 0$

or $P \cos 30° + R_D \cos 45° = F_C \cos 15°$

$= 4732.04$ ...(i)

and $\Sigma F_{y,\,B} = 0$

or $R_D \cos 45° - P \sin 30° = F_C \sin 15°$

$+ 2000 = 3267.94$ ...(ii)

Solving (i) and (ii) simultaneously, we get

$R_D = 5379.43$ N

and $P = 1071.8$ N.

8.8 COPLANAR NONCONCURRENT FORCES

If all the forces of a system lie in a (single) plane and the lines of action of all the forces do not pass through a (single) point, the system is said to be coplanar nonconcurrent force system. Some such systems are shown below.

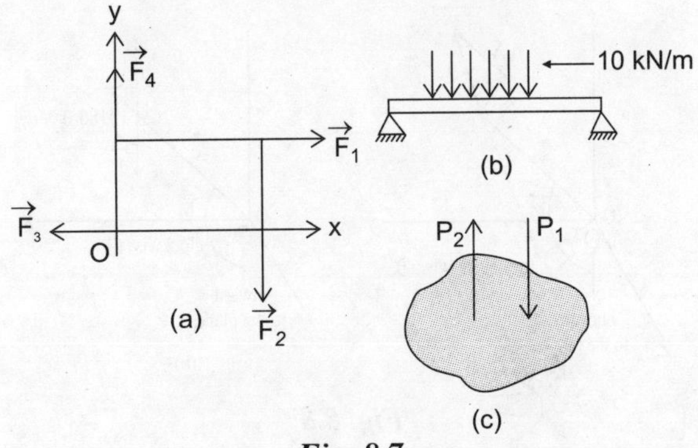

Fig. 8.7

8.9 MOMENT OF A FORCE

In translational motion we associate a force with a linear acceleration of a body (particle). In rotational motion, which quantity shall we associate with an angular acceleration of a body? It cannot be a force because, an experiment with a heavy revolving door teaches us, given force (vector) can produce various angular accelerations for the door depending on where the force is applied to, how it is directed; a force applied at the hinge line cannot produce any angular acceleration, whereas a force of the same magnitude applied at right angles to the door and at its outer edge produces a maximum acceleration. We also see that, for the maximum angular acceleration we have to increase the magnitude of the applied force when we move from outer edge to the nearest edge of hinge, the force will be maximum at infinitesimally nearest to hinge. Thus, we conclude that now only force is not sufficient to describe the rotational motion. So we define a new physical vector quantity torque or moment of a force which is rotational analogue of the force. Thus, moment of a force about a point is the measure of its rotational effect.

8.10 MOMENT ABOUT A POINT

If a force $\vec{F}$ acts on a particle (body) whose position vector with respect to the origin

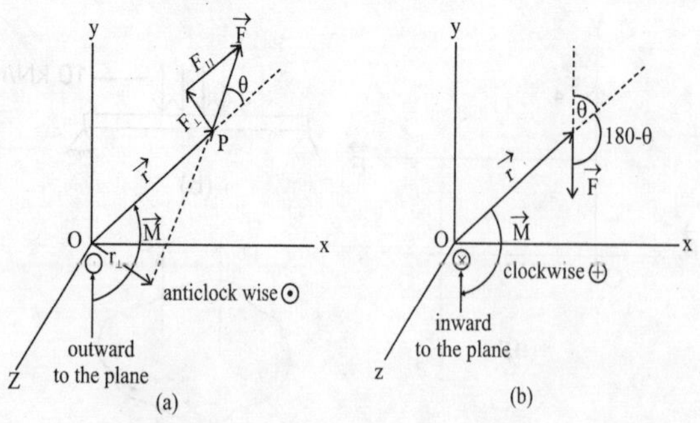

Fig. 8.8

of a frame (inertial) is given by the displacement vector $\vec{r}$, the torque (moment) acting on the particle (body) with respect to origin is defined as $\vec{M} = \vec{r} \times \vec{F}$. It is also denoted as $\vec{\tau}$.

This is a vector quantity (as we take cross product of two vectors $\vec{r}$ and $\vec{F}$) whose magnitude is given by

M = r F sin θ, where θ is the smaller angle between vectors $\vec{r}$ and $\vec{F}$ and its direction is normal to the plane formed by vectors $\vec{r}$ and $\vec{F}$. Sense is given by the right hand rule for the vector product of two vectors; clockwise or anticlockwise is also used as analogue for sense given by vector product take any one of clockwise and anticlockwise as positive direction. In the first Fig. (8.8a) force $\vec{F}$ tries to rotate the body (particle) anticlockwise or according to right hand rule, thumb points, outward the plane i.e, ⊙ notation. And in the second example, force tries to rotate the body clockwise or thumb points, into the plane i.e., ⊗ notation.

$$\text{Now} \quad |\vec{M}| = r F \sin\theta \qquad \text{....(8.5)}$$

$$= F (r \sin\theta)$$

$$= F r_{\perp lar};$$

where $r_{\perp lar}$ is the perpendicular distance of line of action of force $\vec{F}$ from the point origin.

$$= (F)\,(r_{\perp lar}) \qquad \text{....(8.6)}$$

$$= r\,(F \sin\theta)$$

$$= r F_{\perp lar}$$

where $F_{\perp lar}$ is component of force which is perpendicular to the vector $\vec{r}$

$$= (r)\,(F_{\perp lar}) \qquad \text{....(8.7)}$$

Thus, from equation (8.6) we can say that the magnitude of the moment is equal to the simple product of magnitude of the force and perpendicular distance of line of action of force from point origin, which is also called as moment centre and the perpendicular distance ($r_{\perp lar}$) is called as moment arm.

Similarly, from equation (8.7) we conclude that only that component of force cause turning effect (moment of the force) which is perpendicular to vector $\vec{r}$ i.e., $F_{\perp lar}$; effect of the component which is parallel to vector $\vec{r}$ is zero. Unit of moment of a force is Nm or kNm.

Some Special Cases:

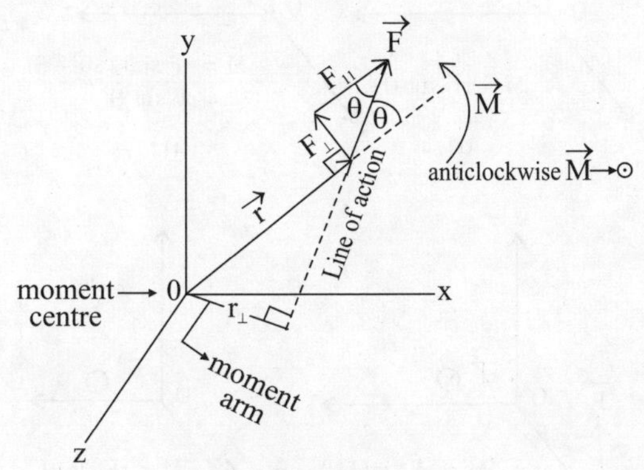

Fig. 8.9

From the above Fig. (8.9), $\vec{F}$ is a force which is at a displacement of $\vec{r}$ from the moment centre origin. Angle between $\vec{r}$ and $\vec{F}$ is θ, then magnitude of the moment of force $\vec{F}$;

$$
\begin{aligned}
M &= r\,F\,\sin\theta \\
&= F\,r_{\perp} \\
&= r\,F_{\perp}
\end{aligned}
$$

Now, for non-zero force, $\vec{F}$ moment is zero in the following cases.

(a) From equation (8.5) we observe that when $\vec{r} = \vec{0}$ i.e., when force $\vec{F}$ passes through moment centre then moment of force is zero.

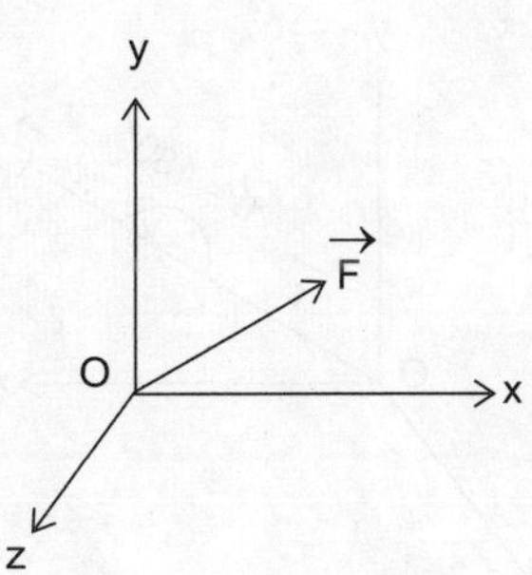

Fig. 8.10

i.e., $|\vec{M}_0| = 0 \times F \times \sin\theta$
$= 0.$

(b) From equation (8.5) we also observe that, when θ = 0 i.e., force $\vec{F}$ and $\vec{r}$ are parallel then, $\vec{M} = 0$

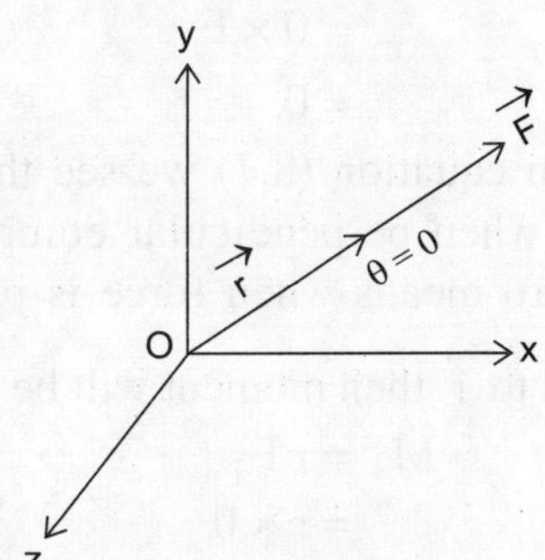

Fig. 8.11

i.e.,
$$
\begin{aligned}
M &= r\,F\sin\theta \\
&= r\,F\sin 0 \\
&= r\,F \times 0 \\
&= 0.
\end{aligned}
$$

(c) We also see that when force, $\vec{F}$ and displacement $\vec{r}$ are antiparallel to each other then $\vec{M} = 0$

i.e.,
$$
\begin{aligned}
M &= r\,F\sin\theta \\
&= r\,F\sin 180^\circ \\
&= r\,F \times 0 \\
&= 0
\end{aligned}
$$

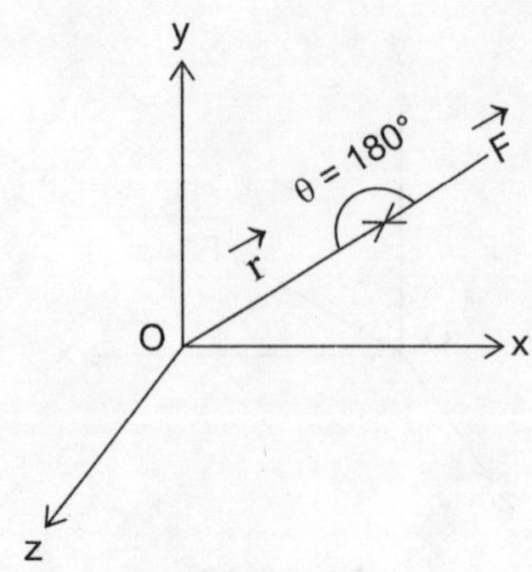

Fig. 8.12

(d) From equation (8.6) we see that when moment arm ($r_{\perp lar}$) is equal to zero then moment is zero. Now $r_{\perp lar} = 0$, means $\vec{F}$ is either parallel to or antiparallel to $\vec{r}$ which is matching case (b) and (c),

i.e., $$M = r_{\perp}F = 0 \times F = 0.$$

(e) From equation (8.7) we see that where $F_{\perp lar}$, i.e., when perpendicular component of force is zero means when force is parallel or antiparallel to $\vec{r}$ then moment will be zero.

i.e., $$M = r F_{\perp} = r \times 0 = 0.$$

Thus, finally we conclude that "when force passes through moment centre (case (a)) or line of action of force passes through moment centre i.e., cases (b), (c), (d), (e), then moment of the force is equal to zero".

(f) As in Figs. (ii) and (iii), we reverse the direction of $\vec{F}$ and $\vec{r}$ respectively. We see that, the magnitude of moment remains unchanged due to the fact that, $\sin\theta = \sin(180 - \theta)$, $|\vec{F}| = |-\vec{F}|$ and $|\vec{r}| = |-\vec{r}|$. But the direction of moment is reversed with respect to Fig. (i). i.e., (⊙) to (⊗), due to the fact that, $\vec{r} \times \vec{F} = -\left(-\vec{r} \times \vec{F}\right) = -\left(\vec{r} \times \left(-\vec{F}\right)\right)$. When reversing the direction of both vectors we get same magnitude and same direction since $\vec{r} \times \vec{F} = \left(-\vec{r}\right) \times \left(-\vec{F}\right)$.

Note: *Student should always use right hand rule to find out the direction of the moment.*

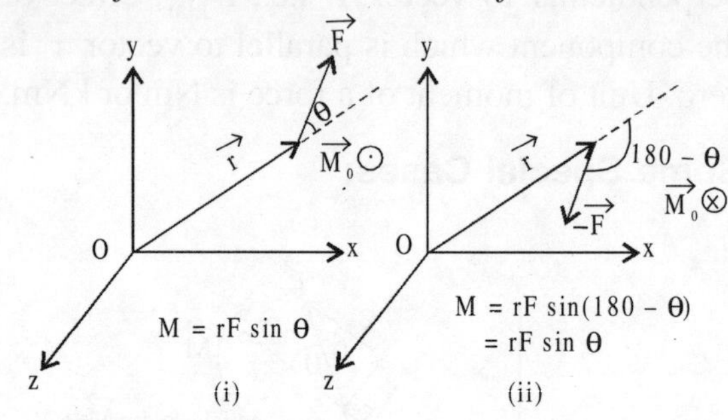

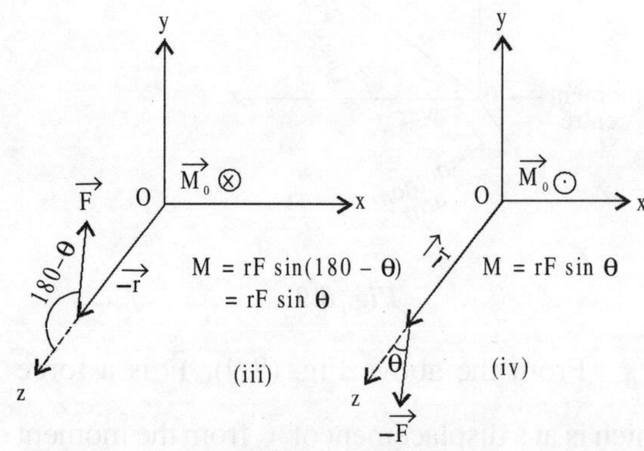

Fig. 8.13

Example 8.1: *Find the moment about point A of the force $\vec{F} = 2\vec{i} + 3\vec{j} + 5\vec{k}$ as shown in Fig. (8.14), where A and B are points (1, 1, 1) and (2, 2, 2) respectively.*

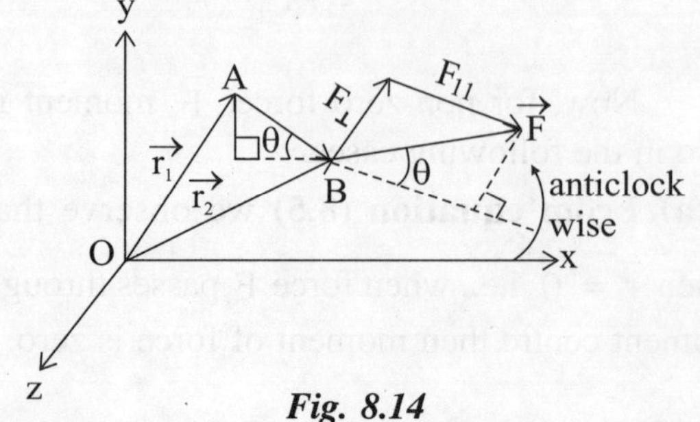

Fig. 8.14

Solution:

We know that

$$\vec{M}_A = \vec{r}_{AB} \times \vec{F} = \left(\vec{r}_{OB} - \vec{r}_{OA}\right) \times \vec{F}$$

$$\left[\because \vec{r}_{AB} = \vec{r}_B - \vec{r}_A = \vec{r}_2 - \vec{r}_1\right]$$

$$= \left[\left(2\vec{i}+2\vec{j}+2\vec{k}\right) - \left(\vec{i}+\vec{j}+\vec{k}\right)\right] \times \left(2\vec{i}+3\vec{j}+5\vec{k}\right)$$

$$= \left(\vec{i}+\vec{j}+\vec{k}\right) \times \left(2\vec{i}+3\vec{j}+5\vec{k}\right)$$

$$= \begin{vmatrix} \vec{i} & \vec{j} & \vec{k} \\ 1 & 1 & 1 \\ 2 & 3 & 5 \end{vmatrix}$$

$$= \vec{i}[5-3] - \vec{j}[5-2] + \vec{k}[3-2]$$

Hence, $\vec{M}_A = 2\vec{i} - 3\vec{j} + \vec{k}$

Therefore, $|\vec{M}_A| = \sqrt{2^2 + (-3)^2 + 1^2}$

$= \sqrt{4+9+1} = \sqrt{14}$

= 3.74 unit and directed out ward to the plane i.e., (⊙) and tries to rotate the body in anticlockwise direction.

8.11 VARIGNON'S THEOREM

Let there are more than one forces for e.g., $\vec{F}_1$, $\vec{F}_2$, $\vec{F}_3$,...., $\vec{F}_n$ are acting on a body simultaneously and we have to find total moment of these forces; about a point (say A).

To obtain the total torque (moment), we have to get separately the torques of the individual forces and then add them.

i.e., $\vec{M}_{total,A} = \vec{r}_{1,A} \times \vec{F}_1 + \vec{r}_{2,A} \times \vec{F}_2 + \vec{r}_{3,A} \times \vec{F}_3 + ... + \vec{r}_{n,A} \times \vec{F}_n$

$$= \sum_{i=1}^{n} \left(\vec{r}_{i,A}\right) \times \left(\vec{F}_i\right) \quad ....(i)$$

We could attempt to add the forces $\vec{F}_1$, $\vec{F}_2$,....., $\vec{F}_n$ ..... vertically and then obtain the torque of resultant force about point 'A'.

i.e., $\vec{M}_{total,A} = \vec{r} \times \vec{R}$ ....(ii)

Where $\vec{R}$ is resultant of $\vec{F}_1$, $\vec{F}_2$, $\vec{F}_3$ ..... $\vec{F}_n$ and $\vec{r}$ is position vector of resultant from point A.

Combining these equations and we get

$$\sum_{i=1}^{n} \vec{r}_{i,A} \times \vec{F}_i = \vec{r} \times \vec{R} \quad ...(iii)$$

But this won't always hold true. Even if $\vec{R} = 0$; $\vec{r}_{1,A} \times \vec{F}_1 + \vec{r}_{2,A} \times \vec{F}_2 + ... + \vec{r}_{n,A} \times \vec{F}_n$ may not be equal to zero. However, if the forces act on the same point (concurrent) one can use the result (iii). Let a force system shown in Fig. (8.15).

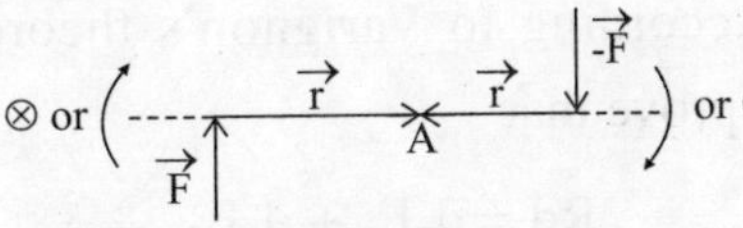

⊗ or ($\vec{r}$ A $\vec{r}$ $-\vec{F}$ $\vec{F}$) or ⊗

Fig. 8.15

Here $\vec{R} = \vec{F} - \vec{F}$

$= \vec{0}$

i.e., $\vec{M}_{total,A} = \vec{r} \times \vec{R}$

$= \vec{r} \times \vec{0}$

$= 0$

But $\vec{M}_{total,A} = \{rF \otimes\} + \{rF \otimes\}$

$= rF \circlearrowright + rF \circlearrowright$

$= 2rF \otimes$

$= 2rF$)

So the result (iii) does not hold true here because points of applications is not the same.

Now, the result (iii) is a mathematical form of Varignon's theorem.

Statement: "The algebraic sum of the moments of a system of (concurrent) coplanar forces about a moment centre is equal to the moment of their resultant force about the same moment centre."

Proof:

For the sake of simplicity we consider two forces $\vec{F_1}$ and $\vec{F_2}$ and their resultant as $\vec{R}$. Let 'B' is the moment centre in the same plane, as shown in Fig. (8.16) where d_1, d_2, d are the perpendiculars from the moment centre B to the $\vec{F_1}$, $\vec{F_2}$ and $\vec{R}$ respectively.

According to Varignon's theorem, we have to prove that

$$Rd = d_1F_1 + d_2F_2 \quad ....(i)$$

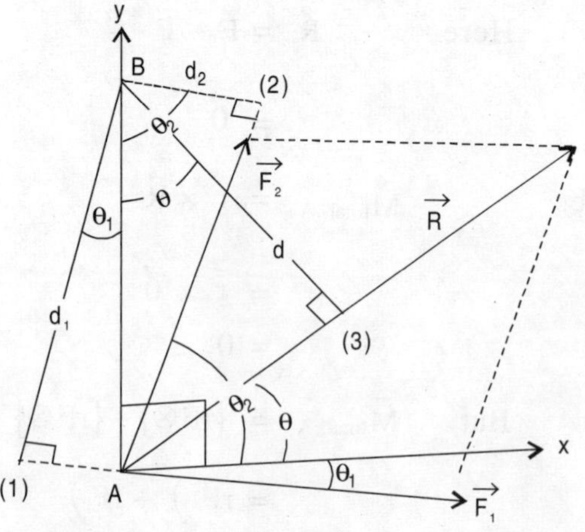

Fig. 8.16

Join AB and consider it as y-axis and draw x-axis at right angle through A.

Suppose θ_1, θ_2 and θ are the angles made by $\vec{F_1}$, $\vec{F_2}$ and $\vec{R}$ with x-axis respectively then from geometry, angles between perpendiculars are same as that between two lines (Here, x-axis , $\vec{F_1}$, $\vec{F_2}$, $\vec{R}$ and perpendiculars to them are AB (y-axis), d_1, d_2 and d respectively as shown).

Now R.H.S. of equation (i) :

$d_1F_1 + d_2F_2 = F_1$ (AB cos θ_1) + F_2 (AB cos θ_2)

[Since in triangles BA1 and AB2, we get d_1 = AB cos θ_1 and d_2 = AB cos θ_2]

$= AB (F_1 \cos \theta_1) + AB (F_2 \cos \theta_2)$

$= AB\ F_{1x} + AB\ F_{2x}$

where F_{1x} and F_{2x} are x-components of $\vec{R}1$ and $\vec{F}2$ respectively,

$= AB (F_{1x} + F_{2x}) = AB\ R_x \quad ...(ii)$

As we know that , $R_x = F_{1xp} + F_{2x}$

where R_x is the x-component of resultant $\vec{R}$.

L.H.S. of equation (i): Rd

$= R(AB \cos \theta)$

[see right angle triangle AB3]

$= AB (R \cos\theta)$

$= ABR_x \quad ...(iii)$

So from equations (ii) and (iii), we see that

$$Rd = d_1F_1 + d_2F_2$$

$$= ABR_x$$

$$\therefore \quad Rd = d_1F_1 + d_2F_2$$

This result can be extended for a number of forces as given below.

Let $\vec{F_1}$, $\vec{F_2}$, $\vec{F_3}$,......,$\vec{F_n}$ be 'n' concurrent coplanar forces and $\vec{R}$ is the resultant. Let d_1,

$d_2, d_3, ..., d_n$ and d are the perpendicular distances, respectively, if about point B we are taking moments then,

$$d_1F_1 + d_2F_2 = d'_1R_1 \quad ....(i)$$

where $\vec{R_1}$ is resultant of $\vec{F_1}$ and $\vec{F_2}$ and d_1' is perpendicular distance of $\vec{R_1}$ from the moment centre B.

Again,

$$d_1'R_1 + d_3F_3 = d_2' \quad ...(ii)$$

where $\vec{R_2}$ is the resultant of $\vec{R_1}$ and $\vec{F_3}$ and d_2' is perpendicular distance of $\vec{R_2}$ from the moment centre B.

Now using (i) and (ii)

$$d_1F_1 + d_2F_2 + d_3F_3 = d_2''R_2$$

Proceeding in similar way we can proof that,

$$d_1F_1 + d_2F_2 + d_3F_3 + ... + d_nF_n = Rd$$

i.e.,

$$R.d = \sum d_1F_1 \quad ....(8.8)$$

⇒ algebraic sum of individual moments is equal to moment of resultant vector.

Example 8.2: *What will be the x and y intercepts of the force $\vec{F} = 4000\,\vec{i} + 3000\,\vec{k}\,N$, if its moment about origin is 8000 N.m in magnitude, as shown in Fig. (8.17).*

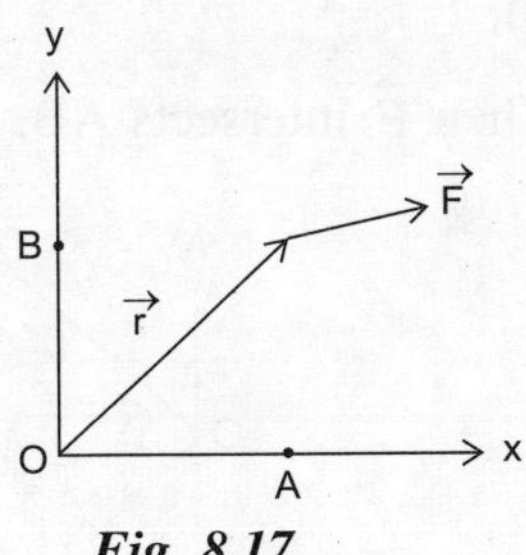

Fig. 8.17

Vector method:

Solution:

Let $\vec{r} = x\,\vec{i} + y\,\vec{j}$

But according to question

$$\left|\vec{r} \times \vec{F}\right| = 8000$$

or $\left|\left(x\,\vec{i} + y\,\vec{j}\right) \times \left(4000\,\vec{i} + 3000\,\vec{j}\right)\right| = 8000$

or $\left|3000x\left(\vec{i} \times \vec{j}\right) + 4000y\left(\vec{j} \times \vec{i}\right)\right| = 8000$

or $\left|3000x\left(\vec{k}\right) - 4000y\left(\vec{k}\right)\right| = 8000$

or $\left|(3000x - 4000y)\vec{k}\right| = 8000$

or $3x - 4y = \pm 8$

Now for x-intercept, $y = 0$

or $3x = \pm 8$

or $x = \pm \dfrac{8}{3}$

But according to figure $x > 0$ for point A

∴ $x = \dfrac{8}{3}$ m

and for y-intercept, $x = 0$

or $-4y = \pm 8$

or $y = \pm 2$

But according to Figure, $y > 0$ for point B

∴ $y = 2$ m.

Analytical method:

To find the x-intercept of force $\vec{F} = 4000\,\vec{i} + 3000\,\vec{j}$, $\vec{F}$ is shifted to point A and that for y-intercept is to be shifted to point B.

By Varignon's Theorem, we know that moment of $\vec{F}$ about origin is equal to sum of moments of its components about the origin.

i.e., 8000 $= y \times F_x + 0 \times F_y$;

[applying Varignon's theorem at point B]

$= y \times 400 + 0$

$= 400y$

$\therefore \quad y = 2$ m.

Similarly, $8000 = x \times F_y + 0 \times F_x$

[applying Varignon's Theorem at point A]

$= x \times 300 + 0$

$\therefore \quad x = \frac{8}{3}$ m

8.12 TORQUE OF A FORCE ABOUT THE AXIS OF ROTATION

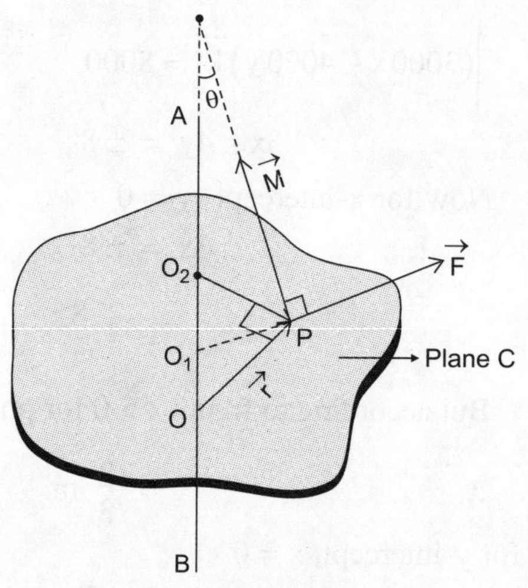

Fig. 8.18

Fig. (8.18) shows a plane C. AB is a line passing through point O_1 on the plane and point O which is below the plane and P is a point on the plane at which a force $\vec{F}$ is acts as shown. Let $\vec{OP} = \vec{r}$ is the position vector of point P with respect to O. Therefore, $\vec{M} = \vec{r} \times \vec{F}$ will be perpendicular to both $\vec{r}$ and $\vec{F}$ simultaneously. Now moment $\vec{M}$ about point O makes some angle with the line AB, say θ, then M cos θ is the component along the line AB, 'M cos θ' is called moment about line AB.

The moment of a force about a line is independent of the choice of the origin on which it is chosen. This can be proved as given below. Let $O_{||}$ is any point on the line AB then moment about $O_{||}$ is

$$\vec{r} \times \vec{F} = \vec{O_{||}P} \times \vec{F} = \left(\vec{O_{||}O} + \vec{OP}\right) \times \vec{F}$$

$$= \vec{O_{||}O} \times \vec{F} + \vec{OP} \times \vec{F}$$

$$= \vec{O_{||}O} \times \vec{F} + \vec{r} \times \vec{F}$$

But $\vec{O_{||}O} \times \vec{F}$ is perpendicular to $\vec{O_{||}O}$. So $\vec{O_{||}P} \times \vec{F}$ does not have component along AB.

$\therefore$ Component of $\vec{O_{||}P} \times \vec{F}$ along the line AB is equal to $\vec{r} \times \vec{F}$. So moment about a line is independent of choice of origin on the line.

Some Special Cases:

Case (a): $\vec{F} \parallel \vec{AB}$

When $\vec{F} \parallel \vec{AB}$, then $\vec{r} \times \vec{F}$ will be perpendicular to $\vec{F}$. Therefore, $\vec{r} \times \vec{F}$ will be perpendicular to $\vec{AB}$. Thus, component of it along AB will be zero as derived.

$$\vec{r} \times \vec{F} \cos 90° = 0$$

Case (b):

When $\vec{F}$ intersects AB:

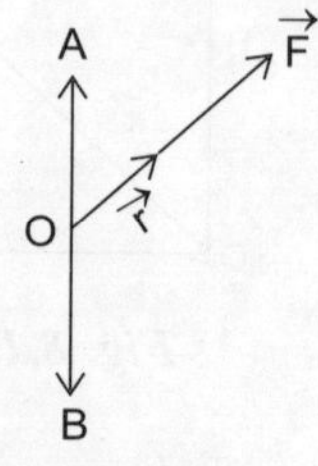

Fig. 8.19

Since angle between $\vec{r}$ and $\vec{F}$ is zero then $\vec{r} \times \vec{F}$ is zero itself. So component along AB will be also zero.

Case (c):

When $\vec{F} \perp \vec{AB}$, but $\vec{F}$ and $\vec{AB}$ do not intersect each other.

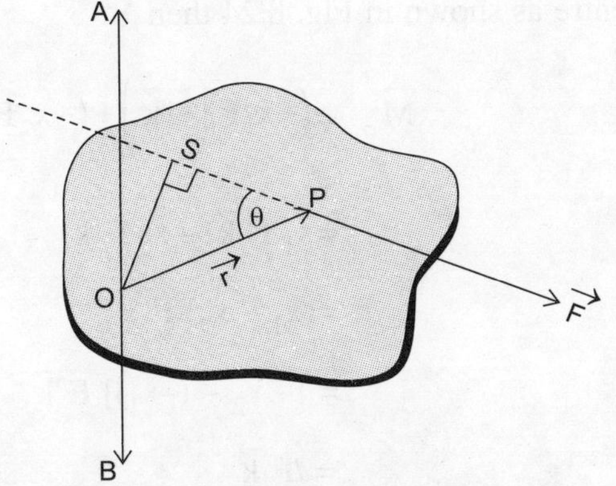

Fig. 8.20

In three dimensions two lines may be perpendicular without intersecting each other. For example, a vertical line on the surface of a wall of your room and a horizontal line on the opposite wall are mutually perpendicular but they never intersect. These two non-parallel and non-intersecting lines are called skew lines.

Fig. (8.20) shows the plane through the particle P and it is perpendicular to the axis of rotation AB. Suppose the plane intersects the axis at the point 'O'. The force $\vec{F}$ lies in the plane. Considering the origin at 'O'. We can write,

$$\vec{M} = \vec{r} \times \vec{F}$$

$$= \vec{OP} \times \vec{F}$$

and $$|\vec{M}| = r F \sin\theta$$

$$= F \times (OS)$$

where OS is the perpendicular from O to the line of action of the force $\vec{F}$, the line OS is also perpendicular to the axis of rotation. Thus, it is the length of the common perpendicular to the force and the axis of rotation.

The direction of $\vec{M}$ is along the axis AB because $\vec{AB} \perp \vec{OP}$ as well as $\vec{AB} \perp \vec{F}$. The moment about AB is, therefore, equal to the magnitude of $\vec{M}$ that is (F) × (OS).

$\therefore\ |\vec{M}| = |\vec{F}| \times$ length of common perpendicular

$=$ magnitude of force × length of common perpendicular.

Case (d):

When $\vec{F}$ and $\vec{r}$ are skew but not perpendicular. Take components of $\vec{F}$, parallel and perpendicular to the axis.

The moment of the parallel component is zero as proved in case (a) and that of perpendicular part can be calculated by 'magnitude × common perpendicular length'.

Example: 8.3: Find the moment of weight of mass 10 kg about the axis of rotation.

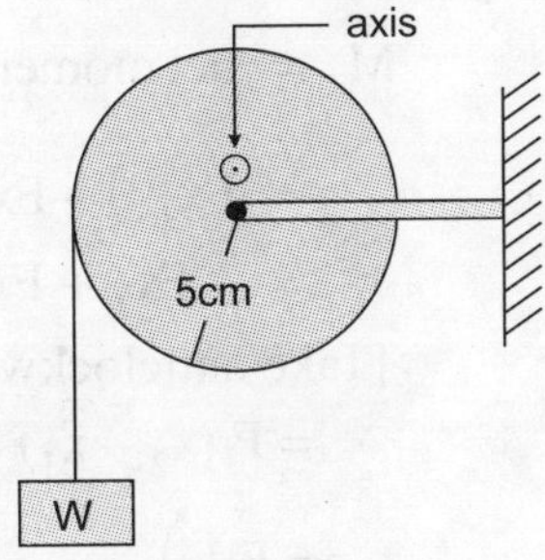

Fig. 8.21

Solution:

Axis and weight are perpendicular but not intersecting therefore, M = magnitude of force × common perpendicular

$$= 10 \times 9.8 \text{ N} \times 5 \times 10^{-2} \text{ m}$$

$$= 10 \times 9.8 \times 5 \times 10^{-2} \text{ N.m}$$

$$= 4.9 \text{ N.m}$$

8.13 COUPLE

Definition : Two parallel forces equal in magnitude and opposite in direction and separated by a constant distance is said to form a couple.

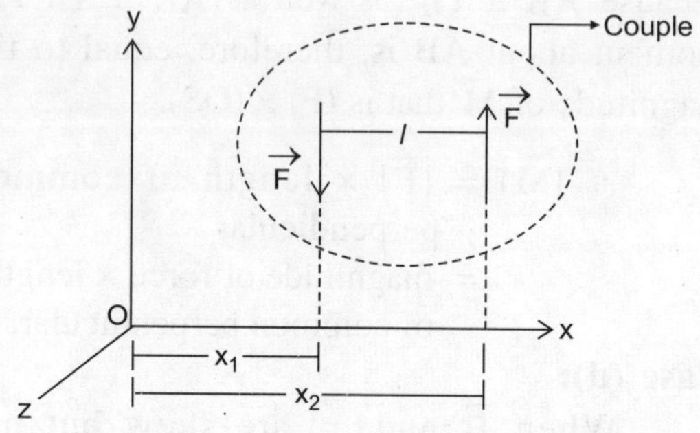

Fig. 8.22

Fig. (8.22) shows a couple of two forces, having magnitude F and separated by a perpendicular distance l, then the resultant is $\vec{R} = \vec{F} - \vec{F} = \vec{0}$. The fact that the resultant of a couple is zero that implies a couple has no translational effect.

If 'O' is moment centre, than

$$\vec{M}_o = \text{net moment about point O}$$

$$= Fx_1 \circlearrowright + Fx_2 \circlearrowleft$$

$$= -Fx_1 + Fx_2$$

[Take anticlockwise as positive]

$$= F(x_2 - x_1)$$

$$= F\,l \circlearrowleft$$

$$= Fl\,\vec{k} \qquad \text{...(i)}$$

$\therefore \quad |\vec{M}_o| = Fl$ = magnitude of one of the force of the couple × perpendicular distance between parallel forces of the couple.

Let O be moment centre, as shown in Fig. (8.23) then,

$$\vec{M}_o = \left(\frac{l}{2} \times F\right)\circlearrowleft + \left(\frac{l}{2} \times F\right)\circlearrowleft$$

$$= \left(\frac{lF}{2} + \frac{lF}{2}\right)\circlearrowleft$$

$$= Fl \circlearrowleft$$

$$= Fl\,\vec{k} \qquad \text{...(ii)}$$

$\therefore |\vec{M}_o| = lF$ = perpendicular distance × Magnitude of one the force. Now take moment centre as shown in Fig. 8.24 then

$$\vec{M}_o = (-x_1 F)\left(-\vec{k}\right) + (-x_2 F)\vec{k}$$

$$= x_1 F\,\vec{k} - x_2 F\,\vec{k}$$

$$= (x_1 - x_2) F\,\vec{k}$$

$$= [-x_2 - (-x_1)] F\,\vec{k}$$

$$= lF\,\vec{k} \qquad \text{....(iii)}$$

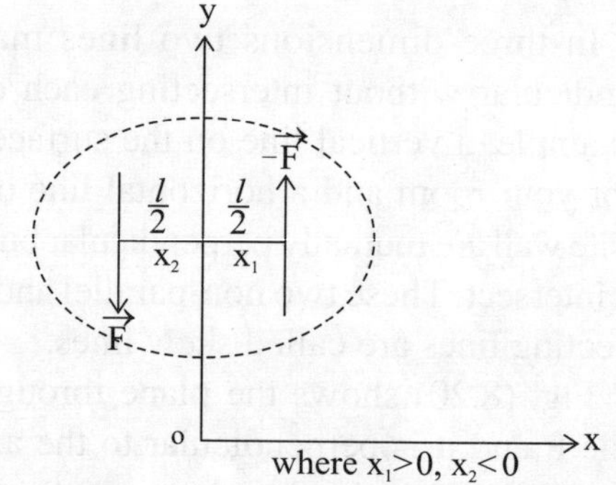

Fig 8.23

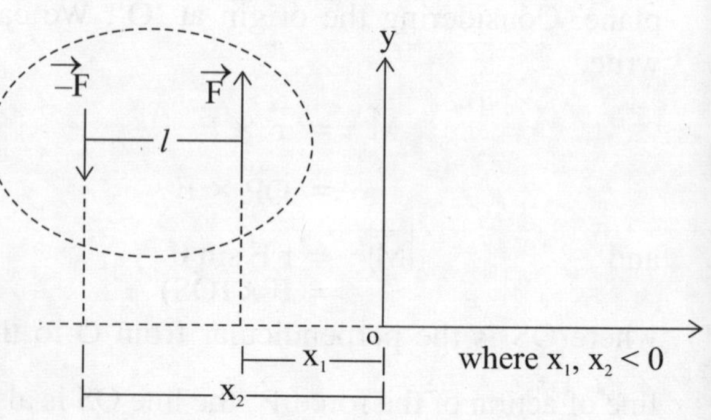

Fig 8.24

$\therefore \left|\vec{M}_O\right| = (\perp_{lar}$ distance) × (magnitude of one of the force of the couple). .... (8.9)

From the above discussion we see that no matter wherever a moment centre of a couple is chosen, it has a constant moment in direction (here $\vec{k}$) as well as magnitude.

We can also see that in following cases couple has same effect.

1. When a couple is rotated through any angle.
2. When a couple is shifted to any position.
3. When couple is replaced by other pair of forces whose rotational effect is same.

8.14 TYPES OF COUPLE

8.14.1 Anticlockwise Couple

A Couple which tries to rotate a body in anticlockwise direction. Refer example in Fig. (8.25). A couple which has net moment(Fd) points towards positive z-axis, if forces lie in xy-plane, in right handed coordinate system.

y
$\vec{F}$ d $-\vec{F}$ anticlockwise $\vec{M}\odot$
$\vec{M} = (Fd)\,\vec{k}$
0 x

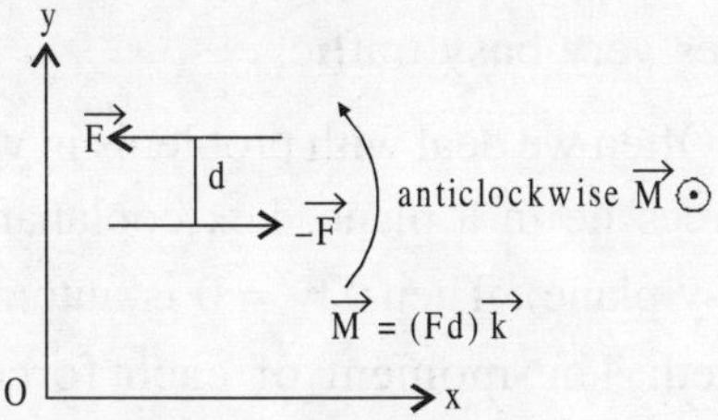

Fig. 8.25

8.14.2 Clockwise Couple

A couple which tries to rotate the body in clockwise direction. Refer example in Fig. (8.26). A couple which has net moment(Fd) points towards negative z-axis in right handed coordinate system if forces lie in xy-plane.

y
$\vec{F}$ d $-\vec{F}$ anticlockwise $\vec{M}\odot$
$\vec{M} = Fd\,(-\vec{k})$
0 x

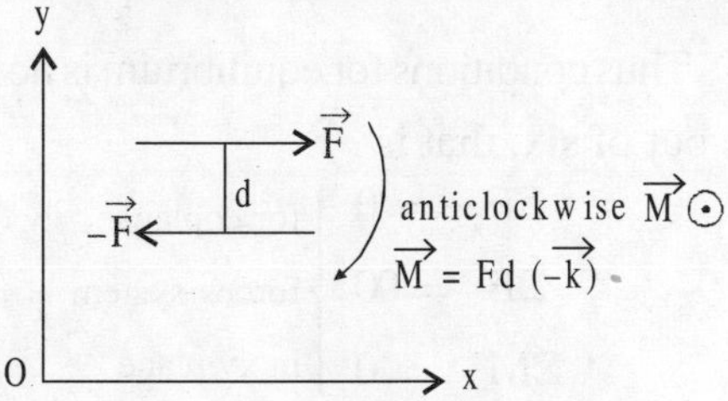

Fig. 8.26

8.15 EQUILIBRIUM CONDITIONS

A body is said to be in translational equilibrium when a body is in state of rest (static equilibrium) or it moves with constant velocity (dynamic) equilibrium.

Mathematically,

$\vec{v} = 0$ or $\vec{v}$ = constant

Combinly we can say that the rate of the change of velocity is zero with respect to time.

i.e., $$\frac{d\vec{v}}{dt} = 0$$

or $$m\frac{d\vec{v}}{dt} = 0;$$ where m is mass of body and assumed that it is not changing.

or $$\frac{d\left(m\vec{v}\right)}{dt} = 0;$$

M is constant of time.

or $$\frac{d\vec{p}}{dt} = 0,$$

where $\vec{p} = m\vec{v}$ = linear momentum of a body. That is rate of change of linear momentum of the body is zero for equilibrium of a body (Particle). From Newton's second law, for translational motion we know that, rate of change of linear momentum is equal to net external forces act on the body.

$$\Rightarrow \quad \vec{F}_{net,\,ext} = 0$$

or $$\Sigma\vec{F}_{ext} = 0$$

or $$(\Sigma F_x)\,\vec{i} + (\Sigma F_y)\,\vec{j} + (\Sigma F_z)\,\vec{k} = 0$$

$$\Rightarrow \Sigma F_x = 0 \text{ or } F_{1x} + F_{2x} + F_{3x} + ... + F_{nx} = 0$$

$$\Sigma F_y = 0 \text{ or } F_{1y} + F_{2y} + F_{3y} + ... + F_{ny} = 0$$

$$\text{and } \Sigma F_z = 0 \text{ or } F_{1z} + F_{2Z} + F_{3z} + ... + F_{nz} = 0 \quad ...(8.10)$$

These three simultaneous equations are sufficient to describe the translational equilibrium for a body.

Now body is said to be in rotational equilibrium when either body is in state of rest or it rotates with uniform speed with respect to the axis (inertial frame of reference).

Mathematically,

either $\omega = 0$

or $\omega = \text{constant}$

i.e., $\frac{d\vec{\omega}}{dt} = 0$

i.e., rate of change of angular velocity is equal to zero.

or $\vec{\alpha} = 0$

where $\vec{\alpha}$ = angular acceleration

or $I\vec{\alpha} = 0$,

where I = moment of inertia of the body about axis of rotation.

or $\vec{M} = 0$ where $\vec{M} = I\vec{\alpha} = \vec{\tau}$

Newton's law of rotational motion states that "moment of forces act on a body about axis of rotation is equal to moment of inertia of the body about the axis of rotation multiply by its angular acceleration".

or $(M_x)\vec{i} + (M_y)\vec{j} + (M_z)\vec{k} = 0$

or $(\Sigma M_x)\vec{i} + (\Sigma M_y)\vec{j} + (\Sigma M_z)\vec{k} = 0$

$$\Rightarrow \quad M_x = 0 \text{ or } \Sigma M_x = 0 \text{ or } M_{1x} + M_{2x} + M_{3x} + ... + M_{nx} = 0$$

$$M_y = 0 \text{ or } \Sigma M_y = 0, \text{ or } M_{1y} + M_{2y} + M_{3y} + ... + M_{ny} = 0$$

$$\text{and} \quad M_z = 0 \text{ or } \Sigma M_Z = 0, \text{ or } M_{1z} + M_{2z} + M_{3z} + ... + M_{nz} = 0 \quad ....(8.11)$$

These three simultaneous equations are sufficient to describe the rotational equilibrium of the body. Thus, when we say body is in equilibrium that means body is in rotational equilibrium as well as in translational equilibrium.

Hence, we have six independent conditions on forces acting on body, to be in equilibrium. The analysis of specially static equilibrium is very important in an engineering practice. A designing engineer must isolate and identify all the external forces and moments that may act on a structure and, by good design and wise choice of materials, he makes sure that the structure can tolerate the loads. Such analysis is necessary to make sure, example Mahatma Gandhi Setu (largest in Asia) that connect Patna and Hajipur over Ganga river tolerates very busy traffic.

Often we deal with problems in which all the forces lie in a plane, i.e., coplanar forces. (say x-y plane). Then $\Sigma F_z = 0$ is automatically satisfied. The moment of each force about x-axis is identically zero because either the force intersects the axis or it is parallel to axis [case (a) and case (b) in section 8.12]. Similarly, the moment of each force about y-axis is automatically zero. Therefore,

$$\Sigma M_x = 0 \text{ and } \Sigma M_y = 0 \text{ are satisfied.}$$

Thus conditions for equilibrium is now only three out of six, that is

$$\left.\begin{array}{l} \Sigma F_x = 0 \\ \Sigma F_y = 0 \\ \Sigma M_z = 0 \end{array}\right\} \begin{array}{l}\text{for coplanar} \\ \text{forces system} \\ \text{in xy plane}\end{array} \quad ....(8.12)$$

Similarly,

$$\left.\begin{aligned}\Sigma F_y &= 0\\ \Sigma F_z &= 0\\ \Sigma M_x &= 0\end{aligned}\right\}\text{for coplanar force system in } yz \text{ plane} \qquad ...(8.13)$$

$$\left.\begin{aligned}\Sigma F_z &= 0\\ \Sigma F_x &= 0\\ \text{and} \quad \Sigma M_y &= 0\end{aligned}\right\}\text{for coplanar force system in } xy \text{ plane} \qquad ...(8.14)$$

Notice that while taking moment about z-axis, the moment centre can be chosen either on axis of rotation or any line perpendicular to plane. For simplicity we can take point on the plane, i.e., moment can be taken about any line perpendicular to the plane of the forces. In general we can prove that moment is different about different perpendiculars on plane but if resultant is zero then about any perpendicular line it will be same. If about any one is zero, then for such perpendiculars it will be zero also. It could be proved that if a body is in equilibrium under the action of three forces those forces must be concurrent.

Example 8.4: *Figure (8.27) shows an overhead view of a uniform rod in static equilibrium. Find the value of F_1 and F_2.*

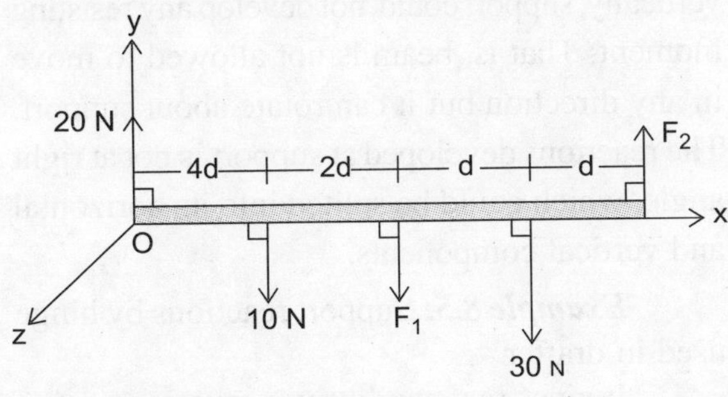

Fig. 8.27

Solution:

We are considering the shown figure, is in xy-plane, then for equilibrium first condition is,

$$\Sigma F_x = 0$$

or $\quad 0 = 0$

[already satisfied according to problem]

and second condition is

$$\Sigma F_y = 0$$

or $\quad 20 - 10 - F_1 - 30 + F_2 = 0$

or $\quad -F_1 + F_2 - 20 = 0$

or $\quad F_2 - F_1 = 20 \quad ...(i)$

For third condition, take point origin on the z-axis and on the plane also then,

$$\Sigma M_z = 0$$

becomes $\quad \Sigma M_o = 0$

or $\quad (4d \times 10) + (6d \times F_1) + (7d \times 30)$

$- (8d \times F_2) = 0$; taking clockwise moment as positive.

or $\quad 40d + 6F_1d + 210\,d - 8F_2d = 0$

or $\quad 6F_1d - 8F_2d + 250d = 0$

or $\quad d\,(6F_1 - 8F_2 + 250) = 0$

since $\quad d \neq 0$

$\therefore \quad 6F_1 - 8F_2 + 250 = 0$

or $\quad 4F_2 - 3F_1 = 125 \quad ....(ii)$

Solving equation (i) and (ii) simultaneously, we get,

$$F_1 = 45 \text{ N and } F_2 = 65 \text{ N}.$$

8.16 REACTIONS AT SUPPORTS OF A BEAM

8.16.1 Beam

A beam is element of structure which has one dimension larger than other two dimensions and is supported at few points. Beam is usually loaded in vertical direction. It is a fact that due to self weight of beam and applied loads, reactions developed at supports, and the system of forces consisting loads, reactions and self-weight of beam keep the beam in state of static equilibrium. Further beam is classified in statically determinate beam and statically indeterminate beam.

8.16.2 Statically determinate beam

The beam which can be analysed using only on the basis of equilibrium equations i.e., number of unknowns are equal to number of independent equilibrium equations.

8.16.3 Statically indeterminate beam

The beam which cannot be analysed only by equilibrium equations (numbers of independent equilibrium equations are less than number of unknowns), this type of beam can be analysed using the conditions of continuity in deformations in addition to equilibrium equations. Such type of beams are not considered in this book. Now while writing equilibrium equations, we see type of supports upon which nature of reactions depends.

8.17 TYPES OF SUPPORTS

8.17.1 Simple support

If the beam rests simply on a support it is called a simple support.

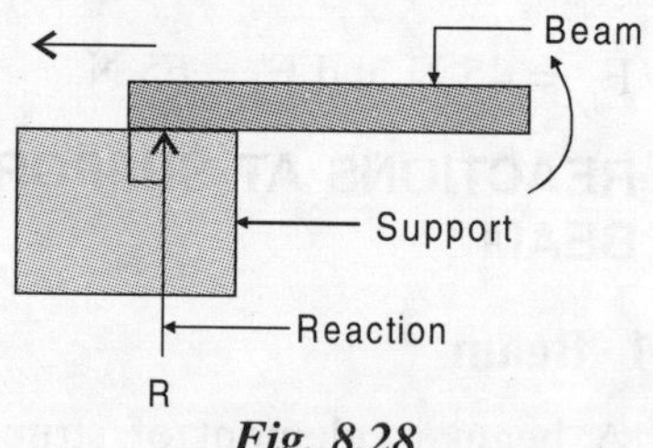

Fig. 8.28

In such type of support, the reactions developed would be at right angle to the support and the beam may be free to move horizontally and also it may be free to rotate about the support; example: A rod on the table i.e., simple support resists vertically only.

Analytically this support gives vertical equilibrium equations only. Example,

R – Weight – lead (s) = 0

or Σy-components = 0

8.17.2 Roller support

Beam end is supported on rollers. In such cases, reaction is normal to the support.

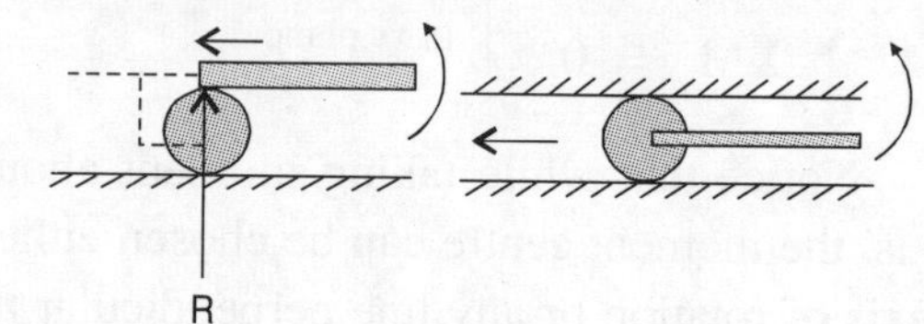

Fig. 8.29

At roller support beam is free to more along the support (horizontally) and also free to rotate about the support, i.e., support resists vertically only.

Analytically, this support gives vertical equilibrium equation only. Example,

R – self weight – lead (s) = 0

or Σy-components = 0

8.17.3 Hinged support

In this case, beam is supported by a hinge which can resist motion along horizontally and vertically, support could not develop any resisting moment. That is, beam is not allowed to move in any direction but it can rotate about support. The reactions developed at support is not at right angle, which could be splited into its horizontal and vertical components.

Example 8.5: Support reactions by hinge used in drafter.

It gives two equilibrium equations out of three necessary independent equations.

i.e., Σx-components = 0

and Σy-components = 0

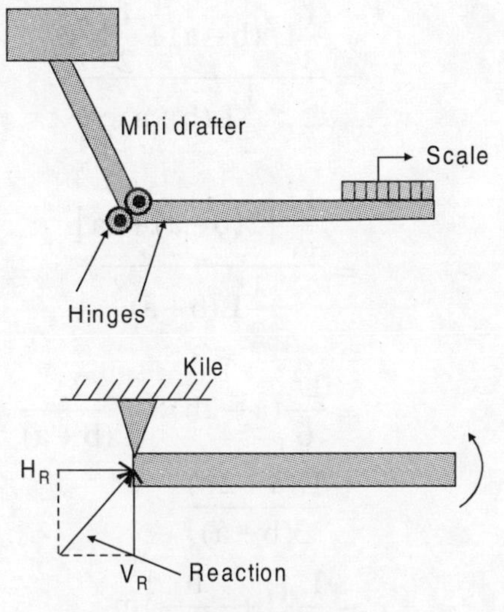

Fig. 8.30

8.17.4 Fixed support

In this case, beam cannot rotate or translate. Translation is prevented by developing support reaction in any required direction. Rotation is prevented by developing support moment.

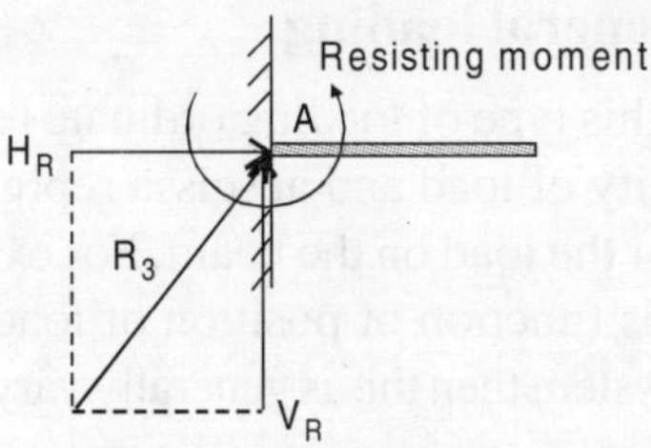

Fig. 8.31

Analytically: It gives all equilibrium equations necessary to describe equilibrium of beam (planar section) i.e.,

$$\Sigma\text{x-component} = 0$$

$$\Sigma\text{y-components} = 0$$

and $$\Sigma M_A = 0$$

8.18 TYPES OF LOADING

8.18.1 Concentrated loads

If a load acts on a beam over a very small area, it could be approximated as acting at the mid-point of that area and is represented by an arrow as shown in Fig. (8.32).

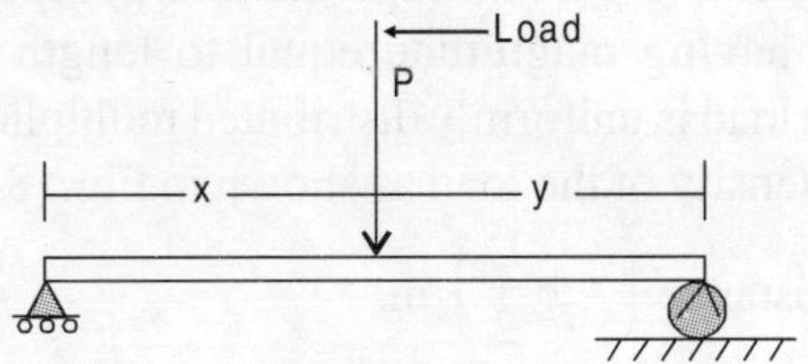

Fig. 8.32

8.18.2 Uniformly distributed load (UDL)

Over considerably long distance such load has got uniform intensity (means load per unit length, area or volume). It is represented as shown in Fig. (8.33a) or as in Fig. (8.33b). For finding reaction, this load may be assumed as total load acts at centre of gravity, for plane section acting at centroid of area formed by rectangle with length as the distance over which load is distributed and height as intensity of load; over length only total load acts at mid-point of distance over which the loads are distributed.

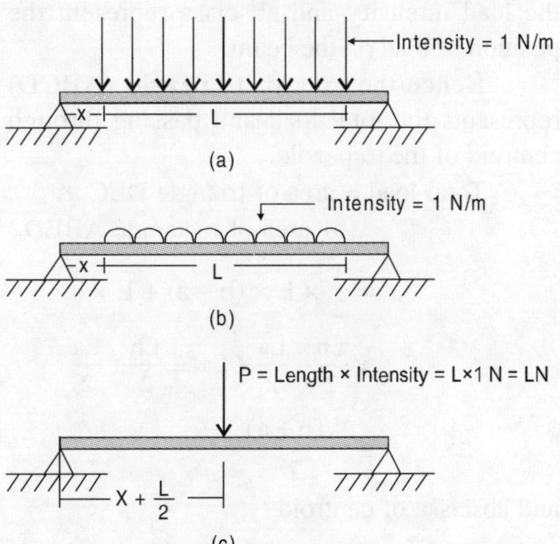

Fig. 8.33

For example, load shown in Fig. (8.33) acts having intensity of 1 N/m and is distributed over beam uniformly on the length L. This load is assumed to pass through the mid-point of the beam having magnitude equal to length over which load is uniformly distributed multiplied by the intensity of the load as shown in Fig. (8.33c) at a distance $\left(x+\frac{L}{2}\right)$ m.

8.18.3 Uniformly varying load (UVL)

The load has got uniformly varing intensity is known as uniformly varing load.

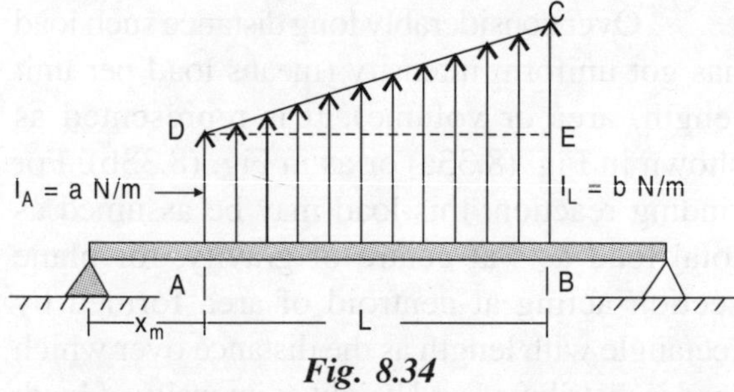

Fig. 8.34

The load shown in Fig. (8.34) varing uniformly from A (intensity I_o = a N/m) to B (intensity I_L = b N/m). The ordinate represents the load intensity and abscissa represent the position of load on the beam.

Hence the area of trapezoid (ABCD) represents the total load and passing through centroid of the trapezoid.

Total load = area of triangle DEC + area of rectangle ABED

$$= \frac{1}{2} \times L \times (b-a) + L \times a$$

$$= \frac{Lb}{2} - \frac{La}{2} + La = \frac{Lb}{2} + \frac{La}{2}$$

$$= \frac{L(b+a)}{2}$$

and abscissa of centroid

$$= \frac{\frac{1}{2}L(b-a) \times \frac{2}{3}L + La \times \frac{L}{2}}{\frac{1}{2}L(b-a) + La}$$

$$= \frac{\frac{1}{3}L^2(b-a) + \frac{1}{2}L^2 a}{\frac{1}{2}L(b+a)}$$

$$= \frac{\frac{L^2}{6}[2(b-a)+3a]}{\frac{1}{2}L(b+a)}$$

$$= \frac{L^2}{6}(a+2b) \times \frac{2}{L(b+a)}$$

$$= \frac{L(a+2b)}{3(b+a)}$$

$$= \frac{L}{3}\left\{1 + \frac{b}{a+b}\right\} m$$

Therefore, net load $\frac{1}{2}L(b+a)$Nm passing through the point $\left[x + \frac{1}{3}\left(1 + \frac{b}{a+b}\right)\right]$ from left support.

8.18.4 General loading

In this type of loading ordinate represents the intensity of load and abscissa represents the position of the load on the beam. For example, if intensity is function of position of load in a co-ordinate system then this is generally vary loading.

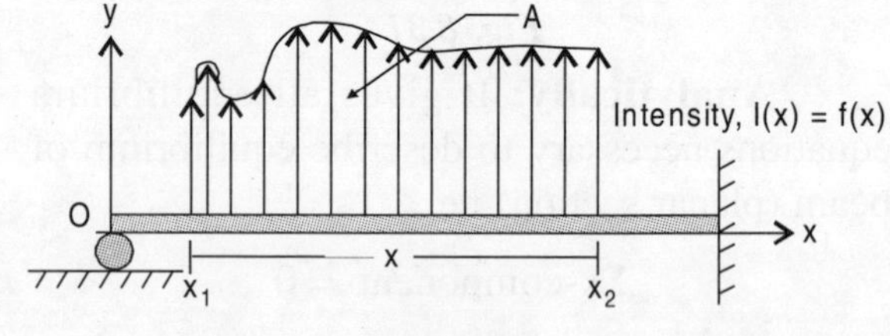

Fig. 8.35

Therefore, total load, L = area A

$$= \int_{x_1}^{x_2} I(x)\, d_x$$

and load passing through the centroid of the shaded area.

8.18.5 External moment

A beam may be subjected to external moment at a point. While writing moment equilibrium equations this moment we have to add.

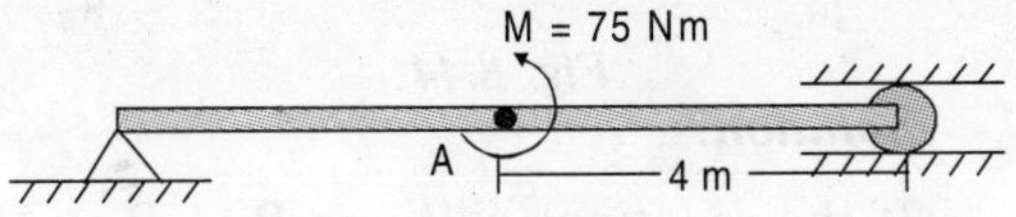

Fig. 8.36

Example in Fig. (8.36) shows a external moment of 75 Nm acting in anticlockwise direction at a distance 4 m from right support.

8.19 TYPES OF BEAMS

8.19.1 Simply supported beam

When both ends of a beam are simply supported it is called simply supported, beam Fig. (8.37). Such a beam support load in the normal direction to its axis. Any other load may give translational motion or rotational motion i.e., beam resist motion perpendicular to axis of beam.

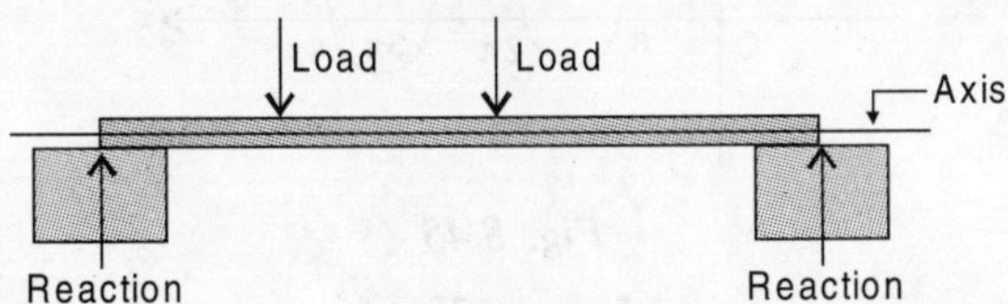

Fig. 8.37

8.19.2 Beam with one end hinged and the other on rollers

If one end of a beam is hinged and other end is on roller. This type of beam can resist loads in any other direction as shown in Fig. (8.38). This beam cannot resist load which could give rotational tendency to move.

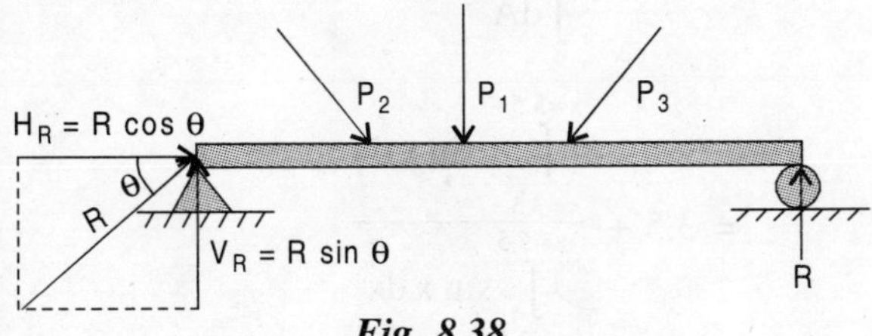

Fig. 8.38

8.19.3 Over-hanging beam

If length of a beam is greater than the distance between two supports, then beam is called over hanging beam as shown in Fig. (8.39).

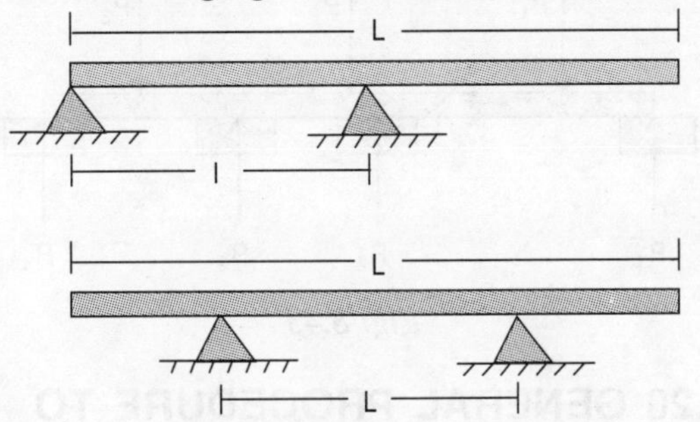

Fig. 8.39

8.19.4 Cantilever beam

If a beam is fixed at one end and other end is free, it is called cantilever beam Fig. (8.40).

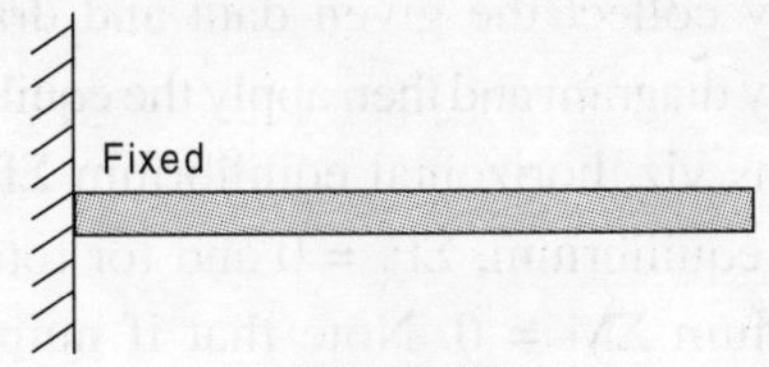

Fig. 8.40

8.19.5 Propped cantilever

It is a beam which one end is fixed and other end is simply supported Fig. (8.41).

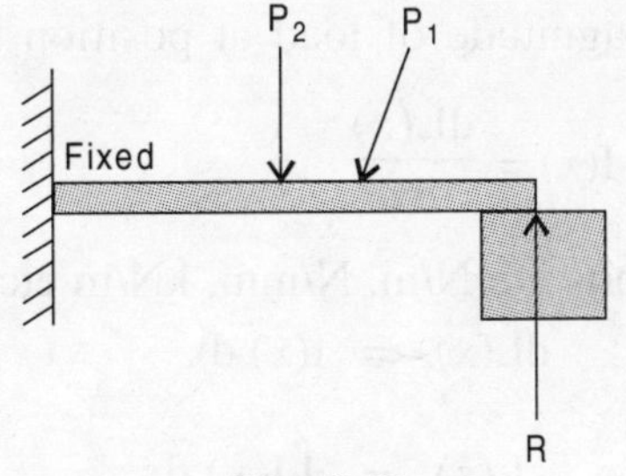

Fig. 8.41

8.19.6 Both end hinged

Both ends hinged at supports

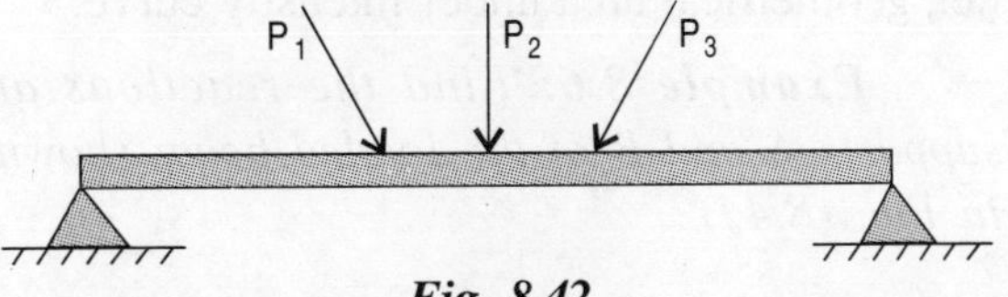

Fig. 8.42

8.19.7 Continuous beam

If beam is supported at more than two points it is called continuous beam (Fig. 8.43).

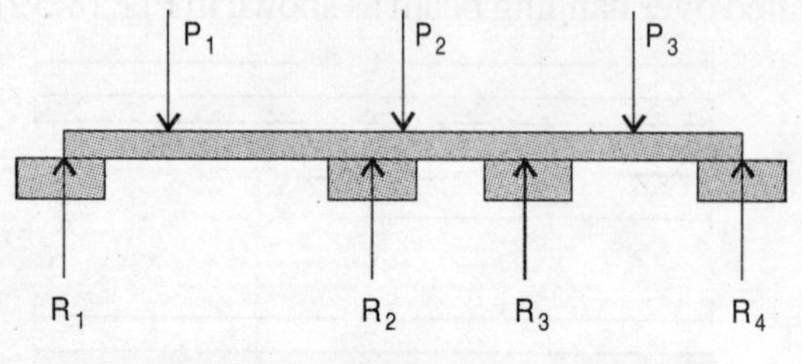

Fig. 8.43

8.20 GENERAL PROCEDURE TO ANALYSE PROBLEMS RELATED TO BEAMS

There is no hard and fast rule to solve this type of problems, read the problems carefully collect the given data and draw the free body diagram and then apply the equilibrium equations viz, horizontal equilibrium $\Sigma F_H = 0$, vertical equilibrium, $\Sigma F_v = 0$ and for rotational equilibrium $\Sigma M = 0$. Note that if number of independent equations are not equal to number of unknown then we cannot analyse. Further mathematically we define "intensity as rate of change of load w.r.t. position". Therefore, if L(x) is the magnitude of load at position (x) then

$$\text{intensity } I(x) = \frac{dL(x)}{dx}$$

where units are N/m, N/mm, kN/m etc.

$$\text{or} \qquad dL(x) = I(x)\,dx$$

$$\therefore \qquad L(x) = \int I(x)\,dx$$

In general, total load, $L(x) = \int I(x)dx$

i.e., geometrical area under intensity curve.

Example 8.6: *Find the reactions at supports A and B of the loaded beam shown in Fig. (8.44).*

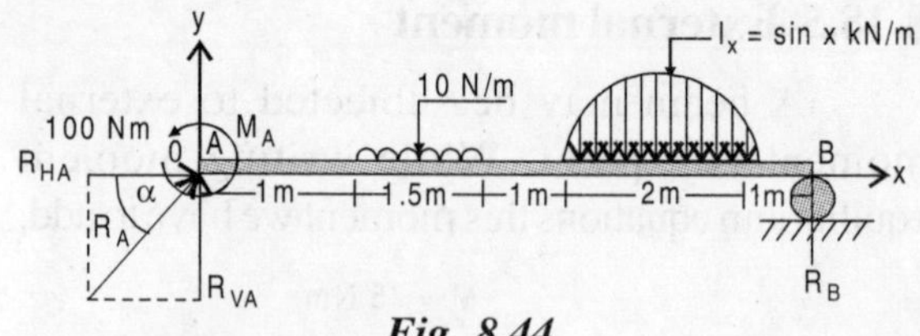

Fig. 8.44

Solution:

If the reactions at A are R_{HA}, R_{VA}, M_A = 100 N-m and at B is R_B. Now for UDL of intensity 10 N/m total load (L_1) of magnitude $1.5 \times 10 = 15$ N acting at point $1 + \frac{1.5}{2}$ = 1.75m from support A.

And for general loading of intensity $I_x = \sin(x)$ the magnitude of total load will be

$$L_2 = \int_{x=3.5}^{x=5.5} \sin x\,dx$$

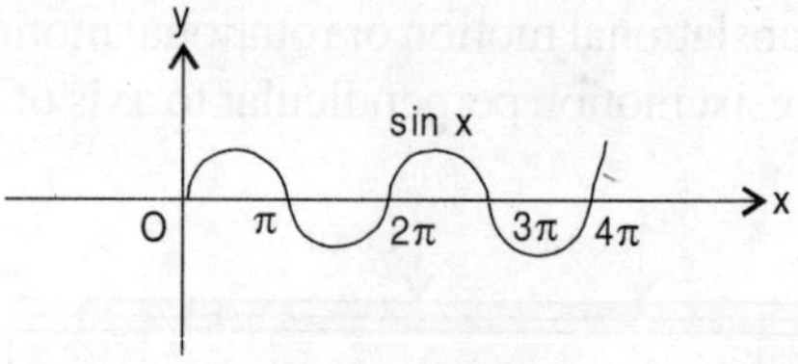

Fig. 8.45

$$= [-\cos x]_{x=3.5}^{x=5.5}$$

$$= -[\cos 5.5 - \cos 3.5]$$

$$= \cos 3.5 - \cos 5.5$$

$$= 2.738 \text{ N}$$

and acting at centroid of area i.e., $x = 3.5 + x_C$ from support A.

$$= 3.5 + \frac{\int x\,dA}{\int dA}$$

$$= 3.5 + \frac{\int_{x=3.5}^{x=5.5} x \sin x\,dx}{\int_{x=3.5}^{x=5.5} \sin x\,dx}$$

$$= 3.5 + \frac{1}{2.738}\left|\int_{x=3.5}^{x=5.5} x \sin x \, dx\right|$$

$$= 3.5 + \left|\frac{1}{2.738}\left[x\int \sin x \, dx - \int\left(\frac{dx}{dx} f \sin x dx\right)dx\right]_{x=3.5}^{x=5.5}\right|$$

$$= 3.5 + \left|\frac{1}{2.738}\left[-x\cos x + \sin x\right]_{x=3.5}^{x=5.5}\right|$$

$$= 3.5 + \left|\frac{1}{2.738}\left[\sin x - x\cos x\right]_{x=3.5}^{x=5.5}\right|$$

$$= 3.5 + \left|\frac{1}{2.738}\left[(\sin 5.5 - 5.5\cos 5.5) - (\sin 3.5 - 3.5\cos 3.5)\right]\right|$$

$$= 3.5 + \left|\frac{1}{2.738}\left[(\sin 5.5 - \sin 3.5) + (3.5\cos 3.5 - 5.5\cos 5.5)\right]\right|$$

$$= 3.5 + \left|\frac{1}{2.738}\left[34.797\times10^{-3} + (3493.471\times10^{-3} - 5474.673\times10^{-3})\right]\right|$$

$$= 3.5 + \frac{1}{2.738}\left[34.797\times10^{-3} + 3493.471\times10^{-3} - 5474.673\times10^{-3}\right]$$

$$= \left|\frac{1}{2.738}[34.797 + 3493.471 - 5474.673]\times10^{-3}\right|$$

$\cong 4.5$ m

We can get same using symmetry i.e., $x = 3.5 + \frac{2}{2} = 4.5$ m. Therefore, finally free body diagram of beam is shown in Fig. (8.46).

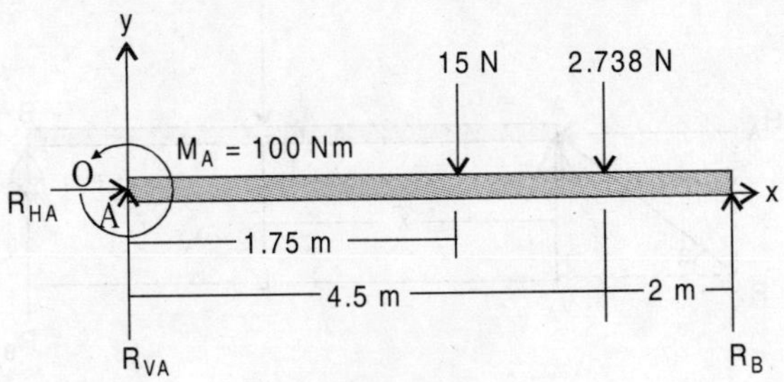

Fig. 8.46

Now for vertical equilibrium Σy-component = 0

or $R_{VA} - 15 - 2.738 + R_B = 0$

or $R_{VA} + R_B = 17.738$ ...(i)

for horizontal equilibrium Σx-component = 0

or $R_{HA} = 0$ ...(ii)

∴ $R_{HA} = 0$

and for rotational equilibrium, taking anticlockwise as positive and moment centre as A, we get

$$\Sigma M_A = 0$$

or $100 + (-1.75 \times 15) + (-4.5 \times 2.738) + 6.5 \times R_B = 0$

or $100 - 1.75 \times 15 - 4.5 \times 2.738 + 6.5 \times R_B = 0$

or $61.429 + 6.5\, R_B = 0$

or $R_B = -\frac{61.429}{6.5}$

$= -9.45$ N

Negative sign shows direction of reaction at B will be opposite to shown direction.

Therefore from equation (i)

$R_{VA} + R_B = 17.738$

or $R_{VA} + (-9.45) = 17.738$

or $R_{VA} = 17.738 + 9.45$

∴ $R_{VA} = 27.188$ N.

8.21 GENERAL PROCEDURE TO SOLVE DISTRIBUTED LOADING PROBLEMS

1. *First of all load distribution may be of two types (i) uniformly distributed load (ii) general distribution (may be function of x as W = f (x)). Now UDL may be triangular, rectangular or any other known distribution. For net load of UD loading, find out area of triangle or rectangle or of any other known figure, notice that some time a complex figure could be divided into simpler figures. And total load will pass through their centroid. For general loading net load will be given by*

$$W_{net} = \int_{x_1}^{x_2} w(x)dx\,; \quad where\ W(x) = f(x)$$

i.e., area under the curve.

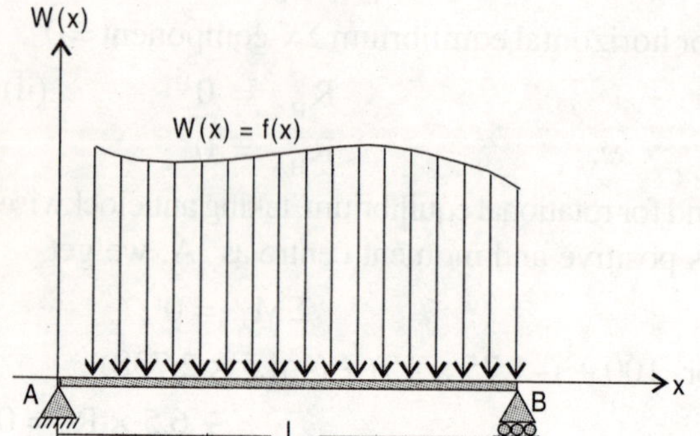

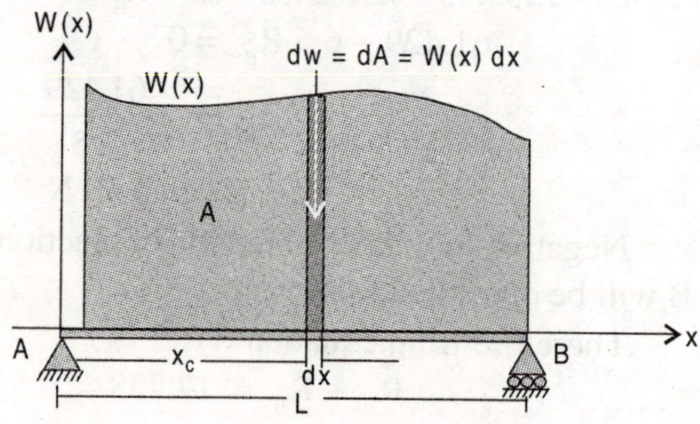

Fig. 8.47

$$\therefore \quad W_{net} = \int_0^L W(x)dx$$

$$= \int dA$$

$$= A$$

i.e., $\quad W_{net} = A \quad$ ....(8.15)

We shall now determine where a single concentrated load W, of the same magnitude W as the total distributed load, should be applied on the beam if it is to produce the same reactions at the supports. This position is centroid of the given area under distributed load.

where, $\quad \overline{X} = \dfrac{\int_0^L x\,dA}{A_{Total}}$

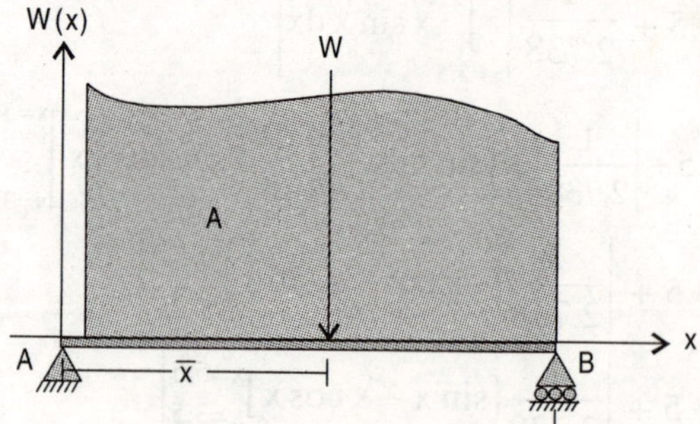

Fig. 8.48

$$= \frac{\int_0^L x\,w(x)dx}{A_{Total}} \quad ....(8.16)$$

Thus distributed load on a beam may be replaced by a concentrated load; the magnitude of this single load equal to the area under the load curve, and its line of action passes through the centroid of this area.

2. *Now assume reactions as per given support (hinged support, fixed support, roller support etc.)*
3. *Apply the equilibrium conditions like $\Sigma F_x = 0$, $\Sigma F_y = 0$ and $\Sigma M = 0$; for coplanar concurrent forces assumed to lie in xy-plane. For given beam in Fig. (8.49) finally we get FBD as shown in Fig. (8.49).*

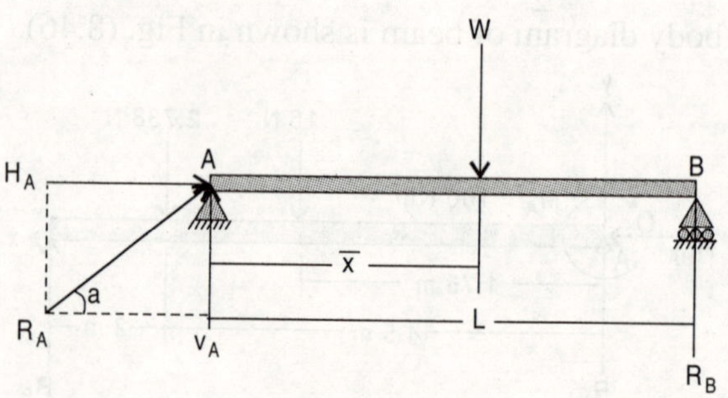

Fig. 8.49

For horizontal equilibrium;

$$\Sigma F_x = 0$$

or $$H_A = 0$$

$$\therefore \quad H_A = 0$$

Vertical equilibrium;

$$\Sigma F_y = 0 \quad \text{...(i)}$$

or $$R_B + V_A - W = 0$$

or $$W = V_A + R_B$$

$$\therefore \quad V_A = W + R_B \quad \text{...(ii)}$$

$$\Sigma M_A = 0$$

Rotational equilibrium;

or $$W\overline{X} = R_B L$$

$$\therefore \quad R_B = \frac{W\overline{X}}{L}$$

Finally, we will get unknown support reactions, V_A, R_B and H_A by solving equations (i), (ii) and (iii).

8.22 RESOLUTION OF A FORCE INTO A FORCE AND A COUPLE

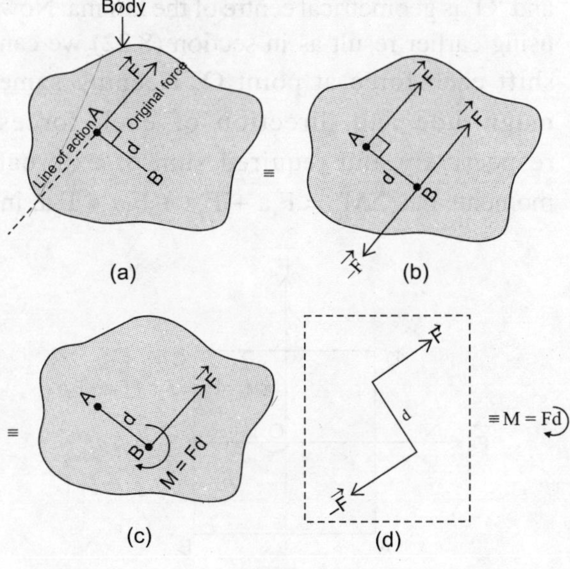

Fig. 8.50

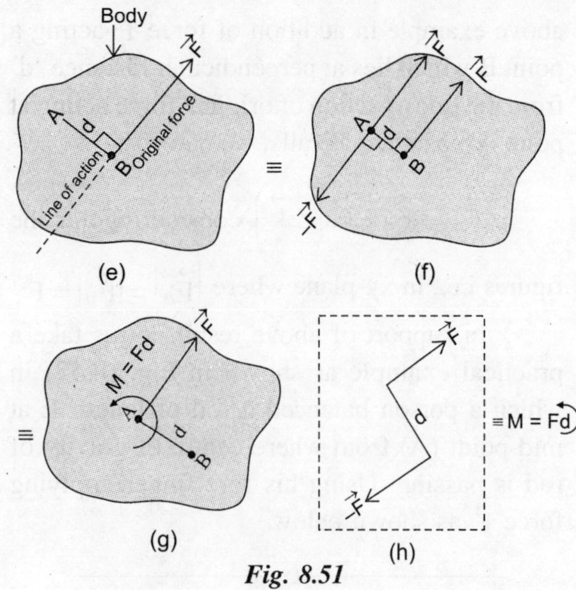

Fig. 8.51

Resolution of a force into its components is a common process and it makes problem easy. But sometimes resolution of a force into a force and a couple (external moment) also solves problem easily. Suppose a force of magnitude $\vec{F}$ and is directed along the directions shown in Fig. (a or e) acting at points A and B on a body respectively. Take a point B which lies at perpendicular distance d from the line of action of force $\vec{F}$. Now if body is in state of equilibrium under this force then without disturbing equilibrium we can apply two equal and opposite forces having magnitude F as shown in Fig. (b) at point B. Now original force $\vec{F}$ and $-\vec{F}$ form a couple of magnitude Fd and directed along clockwise as shown in boxes. As we know that about any point or orientations the couple has same effect, therefore instead of these couple we can apply external torque (moment) of magnitude Fd in respective directions.

Thus we see that 'for a force $\vec{F}$ acting a point A of body, equivalently we can take a external moment of magnitude Fd and directed along the direction of the couple as formed in above example in addition of force $\vec{F}$ acting a point B which lies at perpendicular distance 'd' from the line of action of original force acting at point A". Mathematically, we can write

$$\vec{F}_A \equiv \vec{F}_B + Fd\left(\pm\vec{k}\right)$$ considering that the figures i.e., in xy-plane where $|\vec{F}_A| = |\vec{F}_B| = F$

In support of above result, let us take a practical example as shown in Fig. (8.52) in which a person balanced a rod of length L at mid-point (A) from where centre of gravity of rod is passing. Using his fore finger applying force $\vec{F}$ as shown below.

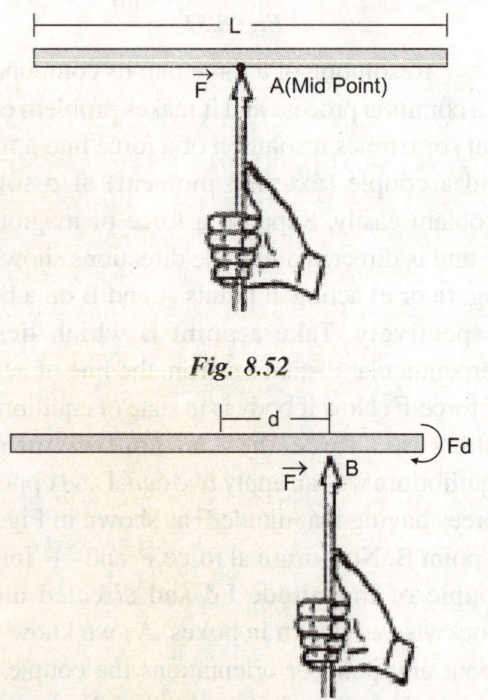

Fig. 8.52

Fig. 8.53

Now if he wants to replace his forefinger at point B at a distance d from A as shown in Fig. (8.53), then he cannot balance the rod until and unless he applies a clockwise external moment of magnitude Fd.

8.23 RESULTANT OF FORCE SYSTEM

8.23.1 Definition

The resultant of a force system is the one which has the same rotational and translational effect as the given system of forces, it may be a single force, a pure moment or a force and a moment.

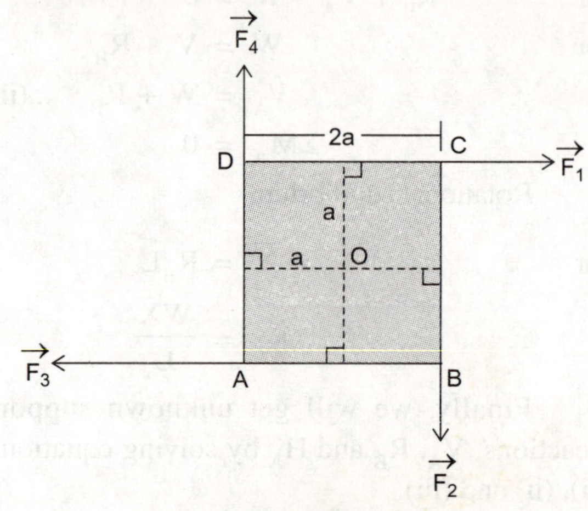

Fig. 8.54

Let $\vec{F}_1$, $\vec{F}_2$, $\vec{F}_3$ and $\vec{F}_4$ constitute a system of forces acting on a rectangular lamina ABCD and 'O' is geometrical centre of the lamina. Now using earlier result as in section (8.22) we can shift each force at point O. Keeping same magnitude and direction of each forces respectively and required sum of external moments i.e., $\Sigma M_O = F_1a + F_2a + F_3a + F_4a$, in

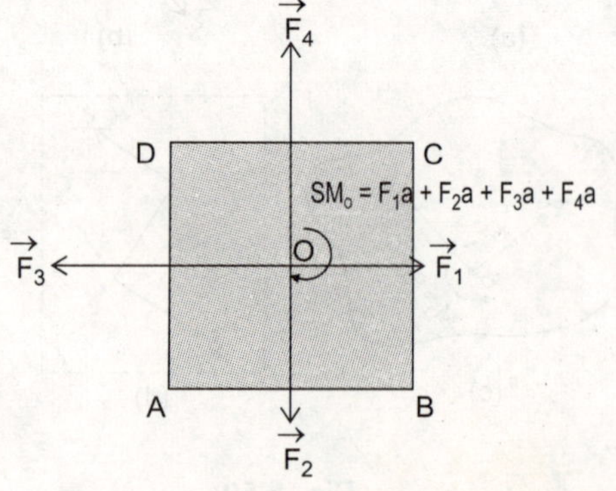

Fig. 8.55

clockwise direction as shown in Fig. (8.55). Now, since $\vec{F}_1$, $\vec{F}_2$, $\vec{F}_3$ and $\vec{F}_4$ are passing through single point O then instead of all these forces we can use their resultant $= \vec{R}$ $= \vec{F}_1 + \vec{F}_2 + \vec{F}_3 + \vec{F}_4$.

Analytically :

$$|\vec{R}| = \sqrt{\Sigma x\text{-components}^2 + \Sigma y\text{-components}^2} \quad ...(8.17)$$

Direction = tan α = tan(resultant makes angle with x-axis)

$$= \frac{\Sigma y\text{-component}}{\Sigma x\text{-component}} \quad ...(8.18)$$

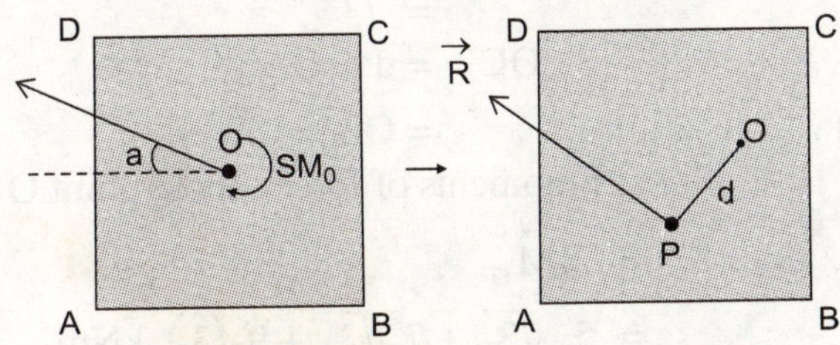

Fig. 8.56

Since at point O resultant $\vec{R}$ and sum of external moment ΣM_o are acting, therfore using the reverse of result obtained in section (8.22), we can pass resultant $\vec{R}$ through point P which has perpendicular distance d from O such that

$$\Sigma M_o = Rd. \quad ...(8.19)$$

Therefore, using three results (8.17), (8.18) and (8.19) we can obtain resultant which has same rotational and translational effect.

Notes:

(i) *From result $\Sigma M_o = Rd$ or, $F_1a + F_2a + F_3a + F_4a = Rd$ we cannot mean that we are applying Varignon's theorem directly for nonconcurrent system of forces.*

(ii) *If Σx-components = 0 and Σy-components = 0 the resultant is a moment.*

(iii) *If $\vec{F}_1$, $\vec{F}_2$, $\vec{F}_3$ and $\vec{F}_4$ are concurrent force system initially, then resultant will be a force.*

(iv) *Otherwise resultant could be a force and a moment.*

8.23.2 x and y Intercepts of a Resultant

Let d is the distance of the resultant $\vec{R}$ from origin and the resultant $\vec{R}$ is inclined at an angle α with y-axis. We have to find x-intercept and y-intercept of the resultant force. Consider the Fig. (8.57).

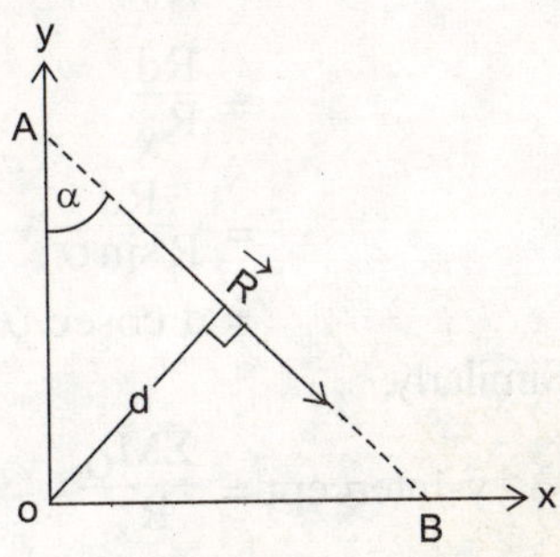

Fig. 8.57

Now to find the x-intercept shift the resultant at point A on x-axis. Similarly, find y-intercept shift the resultant on y-axis at point B.

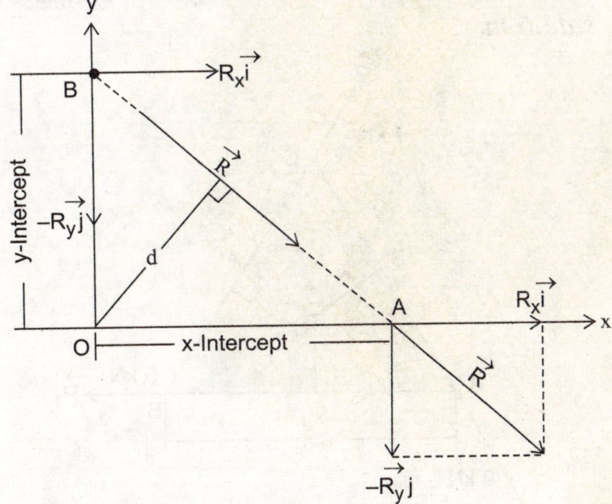

Fig. 8.58

Since R_x and R_y are components so we can apply Varignon's theorem i.e., moment of resultant about origin.

M_O = Sum of moments of components of the resultant.

or $$Rd = 0 \times R_x + (\text{x-intercept}) \times R_y$$
$$= 0 + (\text{x-intercept}) \times R_y$$

or $$Rd = (\text{x-intercept}) \times R_y$$

or $$R_y = \frac{dR}{\text{x-intercept}}$$

or $$\text{x-intercept} = \frac{\Sigma M_O}{R_y}$$
$$= \frac{Rd}{R_y}$$
$$= \frac{Rd}{R\sin\alpha}$$
$$= d \operatorname{cosec} \alpha \qquad ...(8.20)$$

Similarly,

$$\text{y-intercept} = \frac{\Sigma M_O}{R_x}$$
$$= \frac{Rd}{R_x}$$
$$= d \sec \alpha. \qquad ....(8.21)$$

Example 8.7: *Find the resultant of the force system shown in Fig. (8.59) acting on a lamina of equilateral triangular shape of side 6 m.*

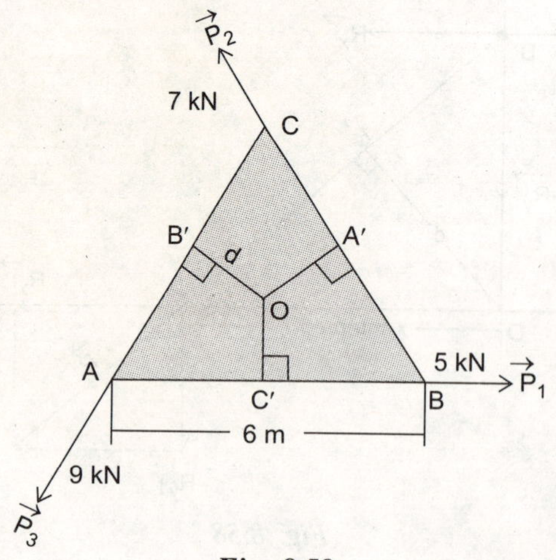

Fig. 8.59

Solution: We see that

area (Δ ABC) = area (Δ AOB) + area (Δ BOC) + area (Δ COA)

or $$\frac{\sqrt{3}}{4} \times 6^2 = 3 \times \text{area } (\Delta \text{ AOB})$$

[∵ area (Δ AOB) = area (Δ BOC) = area (Δ COA) from symmetry of lamina]

or $$\frac{\sqrt{3}}{4} \times 6^2 = 3 \times \left(\frac{1}{2} \times AB \times OC'\right)$$

or $$\frac{\sqrt{3}}{4} \times 6^2 = \frac{3}{2} \times 6 \times OC'$$

or $$OC' = d = \frac{\sqrt{3}}{4} \times 6^2 \times \frac{2}{3} \times \frac{1}{6}$$
$$= \sqrt{3}$$

∴ $$OC' = d = OA'$$
$$= OB' = \sqrt{3} \text{ m.}$$

Sum of moments of forces about point O

$$= \Sigma \vec{M}_O$$
$$= 5\ \sqrt{3}\circlearrowright + 7\ \sqrt{3}\circlearrowright + 9\sqrt{3}\circlearrowright \text{ kNm}$$
$$= 21\sqrt{3}\circlearrowright \text{ kNm}$$
$$= 36.373 \text{ kNm}\circlearrowright$$

Therefore, using result obtained in section (8.23) we can shift forces at O with external moment $\Sigma \vec{M}_O$ = 36.373 kNm↻. as shown in Fig. (8.60) below.

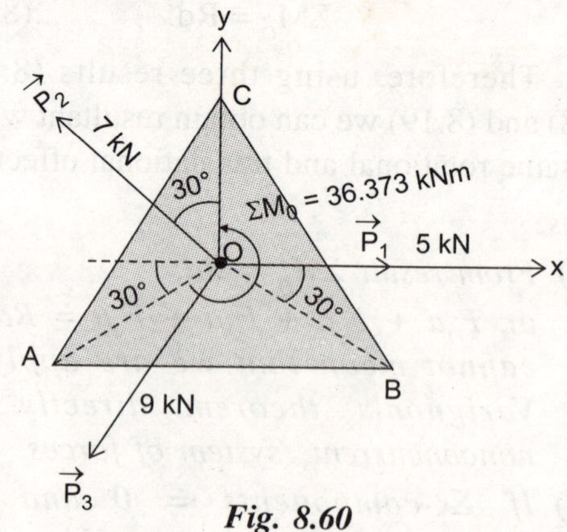

Fig. 8.60

Now suppose that $\vec{R}$ is resultant of $\vec{P}_1$, $\vec{P}_2$ and $\vec{P}_3$ in the coordinate system shown in Fig. (8.61).

where, $\vec{R} = R_x \vec{i} + R_y \vec{j}$

$$= (p_{1x} + p_{2x} + p_{3x})\vec{i} + (p_{1y} + p_{2y} + p_{3y})\vec{j}$$

$$= \{5 + (-7 \sin 30°) + (-9 \sin 30°)\}\vec{i} + \{0 + 7 \cos 30° + (-9 \cos 30°)\}\vec{j}$$

$$= (5 - 7 \sin 30° - 9 \sin 30°)\vec{i} + \{0 + 7 \cos 30° - 9 \cos 30°)\vec{j}$$

$$= -3\vec{i} + \left(-\sqrt{3}\right)\vec{j}$$

$\therefore$ $\vec{R} = R_x \vec{i} + R_y \vec{j}$

$$= -3\vec{i} - \sqrt{3}\,\vec{j} \text{ kN.}$$

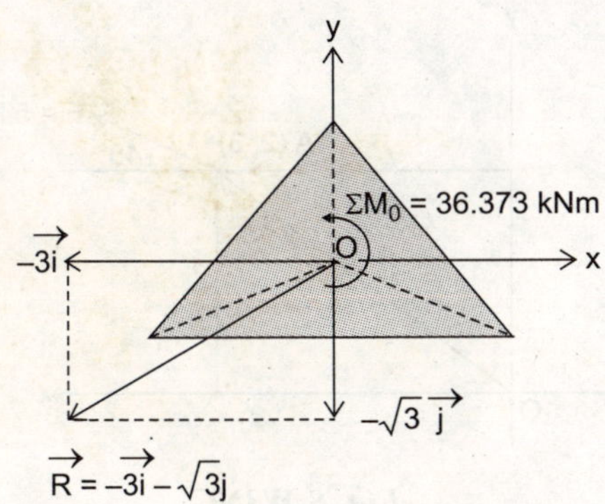

Fig. 8.61

Using reverse of the result obtained in section (8.23) we can shift the resultant at point B such that $\Sigma M_O = Rd'$ where d is the perpendicular distance of resultant from point O.

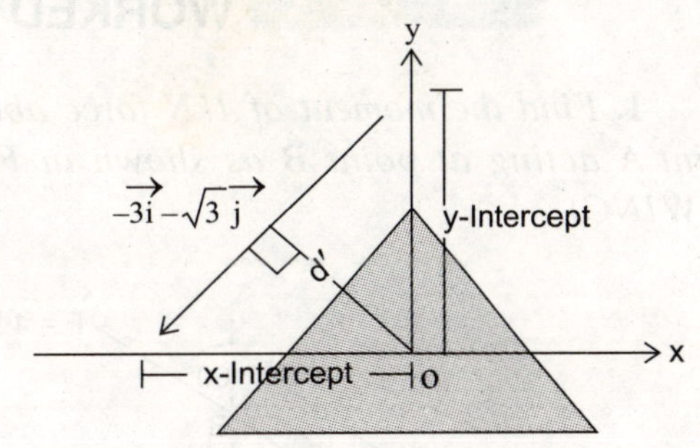

Fig. 8.62

then $\Sigma M_O = Rd'$

or $36.373 = \sqrt{3^2 + \left(\sqrt{3}\right)^2}\, d'$

or $36.373 = \sqrt{12}\, d'$

or $d' = \dfrac{36.373}{\sqrt{12}}$

$\therefore$ $d' = 10.5$ m

Further, x-intercept $= \left|\dfrac{\Sigma M_O}{\Sigma y\text{-component}}\right|$

$$= \left|\frac{\Sigma M_O}{R_y}\right|$$

$$= \frac{36.373}{\sqrt{3}}$$

$$= 21 \text{ m}$$

and y-intercept $= \left|\dfrac{\Sigma M_O}{R_x}\right|$

$$= \frac{36.373}{3}$$

$$= 12.12 \text{ m}$$

WORKED OUT EXAMPLES

1. *Find the moment of 1kN force about point A acting at point B as shown in Fig. (8-W1NC).*

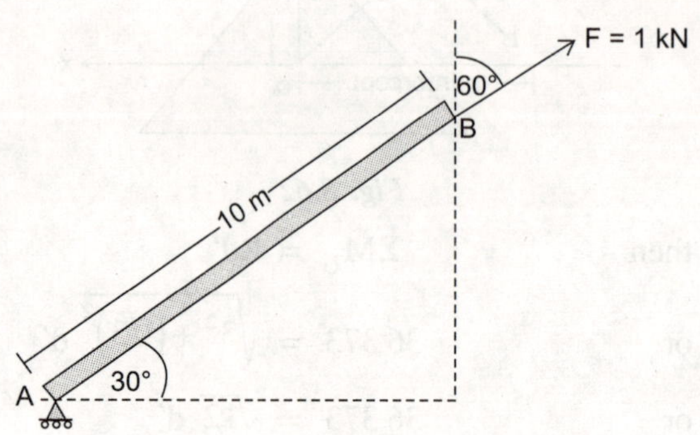

Fig. 8-W1NC

Solution:

Taking rectangular component of the force we get F_x = 866 N and F_y = 500 N.

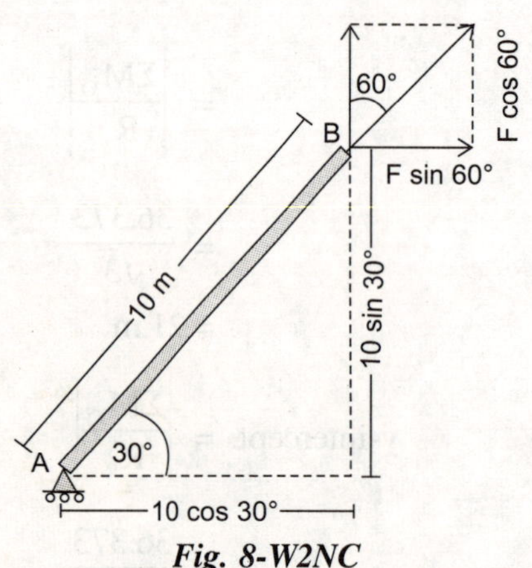

Fig. 8-W2NC

From Fig. (8-W2NC) we observe that about moment centre A, distance of line of action of F cos 60° = 0.5 kN is 10 cos 30° and that of F sin 60° = 0.866 kN is 10 sin 30°.

Thus applying Varignon's theorem moment of force F about A,

i.e.,
$$M_A = (F\cos 60°) \times (10\cos 30°) - (F\sin 60°) \times (10\sin 30°)$$
$$= 0.5 \times 8.660 - 0.866 \times 5$$
$$= 0;\ \text{taking +for } \circlearrowright$$

Note: *Since line of action of F passing through point A therefore moment should be zero, this problem is also proving that the fundamental result.*

2. *What will be the y-intercept of 5 kN force if its moment about A is 3.535 kN.m in Fig. (8-W3NC).*

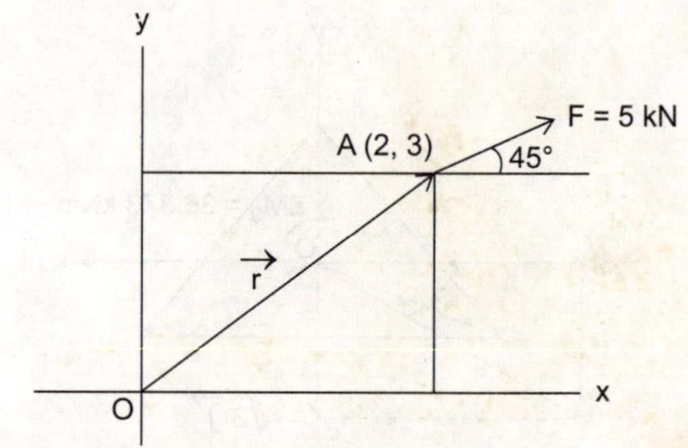

Fig. 8-W3NC

Solution:

First method:

Now we decompose F = 5 kN into F_x = F cos 45° = 3.535 kN and F_y = F sin 45° = 3.535 kN.

then
$$\vec{F} = F_x\,\vec{i} + F_y\,\vec{j}$$
$$= 3.535\,(\vec{i} + \vec{j})$$

and position vector is
$$\vec{r} = 2\,\vec{i} + 3\,\vec{j}$$

then $\vec{\tau}_O = \vec{r} \times \vec{F}$

$$= (2\vec{i} + 3\vec{j}) \times 3.535\ (\vec{i} + \vec{j})$$

$$= 3.535\{(2\vec{i} + 3\vec{j}) \times (\vec{i} + \vec{j})\}$$

$$= 3.535(3\vec{j} \times \vec{i} + 2\vec{i} \times \vec{j})$$

$$= 3.535(-3\vec{i} \times \vec{j} + 2\vec{i} \times \vec{j})$$

$$= -3.535\ \vec{i} \times \vec{j}$$

$$= -3.535\ \vec{k}$$

$$\Rightarrow |\vec{\tau}_O| = 3.535 \text{ kN-m.}$$

Now as we know that,

$$\text{y-intercept} = \frac{\tau_O}{\Sigma F_x}$$

$$= \frac{3.535}{3.535}$$

$$= 1 \text{ m.}$$

Second method:

Extend line of action such that it passes through y-axis.

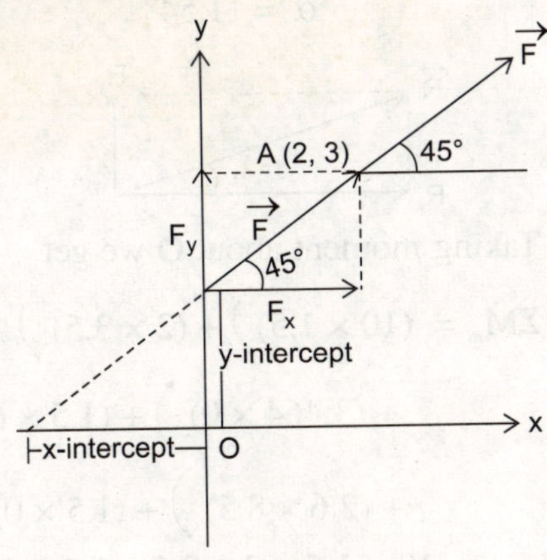

Fig. 8-W4NC

Taking O as moment centre then,

$$M_O = y \times F_x = y \times 3.535$$

or $3.535 = y \times 3.535$

or $y = 1$ m.

3. *A system of loads act on a horizontal beam as shown in Fig. (8-W5NC). Determine the position of the resultant of the loads.*

Solution:

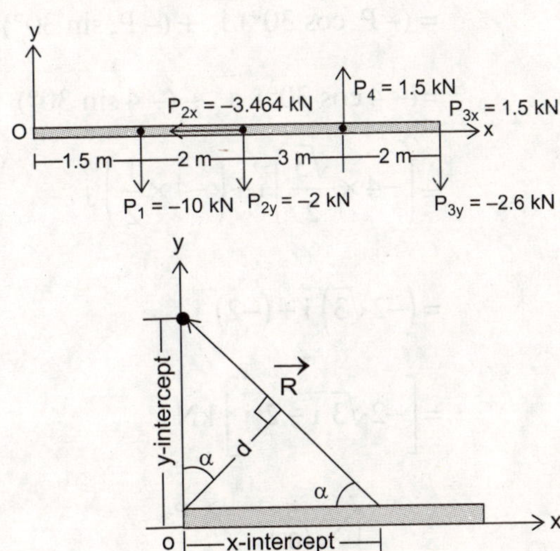

Fig. 8-W5NC

Taking x-components and y-components of each load (force), we can write

$$\vec{R} = (\Sigma F_x)\ \vec{i} + (\Sigma F_y)\ \vec{j}$$

or $\vec{L} = (\Sigma L_x)\ \vec{i} + (\Sigma L_y)\ \vec{j}$

Now $P_1 : \downarrow \Rightarrow P_1\ (0_x - P_1)$

$$= P_1$$

$$= 0\ \vec{i} + (-P_1)\ \vec{j}$$

$$= [0\ \vec{i} + (-10)\ \vec{j}] \text{ kN}$$

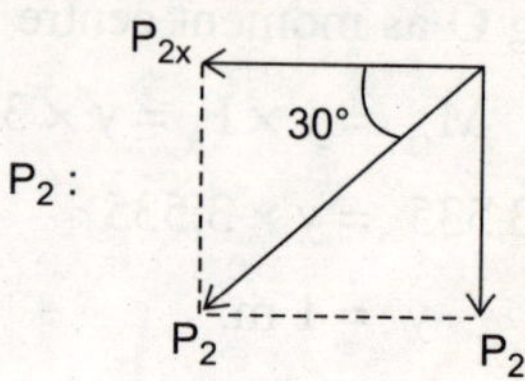

$$\vec{P}_2 = \vec{P}_2\,(P_{2x}, P_{2y})$$

$$= (-P_{2x})\,\vec{i} + (-P_{2y})\,\vec{j}$$

$$= (-P_2 \cos 30°)\,\vec{i} + (-P_2 \sin 30°)\,\vec{j}$$

$$= (-4 \cos 30°)\,\vec{i} + (-4 \sin 30°)\,\vec{j}$$

$$= \left(-4 \times \frac{\sqrt{3}}{2}\right)\vec{i} + \left(-4 \times \frac{1}{2}\right)\vec{j}$$

$$= \left(-2\sqrt{3}\right)\vec{i} + (-2)\,\vec{j}$$

$$= \left[-2\sqrt{3}\,\vec{i} - 2\,\vec{j}\right] \text{kN}$$

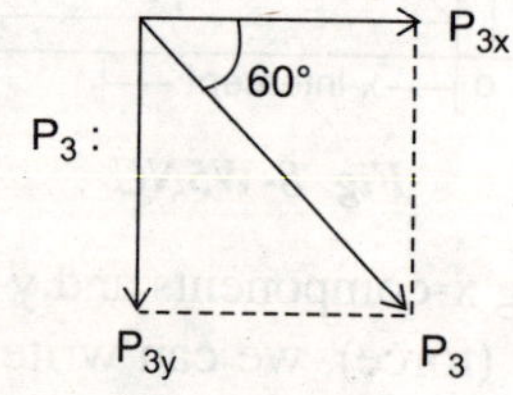

$$\vec{P}_3 = \vec{P}_3\left(P_{3x}, P_{3y}\right)$$

$$= P_{3x}\,\vec{i} - P_{3y}\,\vec{j}$$

$$= (P_3 \cos 60°)\,\vec{i} + (-P_3 \sin 60°)\,\vec{j}$$

$$= \left(3 \times \frac{1}{2}\right)\vec{i} - \left(3 \times \frac{\sqrt{3}}{2}\right)\vec{j}$$

$$= 1.5\,\vec{i} - 1.5\sqrt{3}\,\vec{j}$$

P_4 : $\overset{P_4}{\uparrow}$

$$\Rightarrow \vec{P}_4 = \vec{P}_4\,(0, P_{4y})$$

$$= \vec{P}_4\,(0, P_4)$$

$$= \vec{P}_4\,(0, 15)$$

$$\equiv \vec{P}_4$$

$$= [0\,\vec{i} + 15\,\vec{j}]\,\text{kN}$$

Therefore resultant load,

$$\vec{R} = \left[0 + \left(-2\sqrt{3}\right) + 1.5 + 0\right]\vec{i} + \left[(-10) + (-2) + 1.5\sqrt{3} + 15\right]\vec{j}$$

$$= (1.5 - 2\sqrt{3})\,\vec{i} + (15 - 10 - 2 - 1.5\sqrt{3})\,\vec{j}$$

$$\vec{R} = -1.964\,\vec{i} + 0.401\,\vec{j}$$

$$\therefore \quad |\vec{R}| = 2.004 \text{ kN.}$$

and $\tan \alpha = \left|\frac{R_y}{R_x}\right| = \frac{0.401}{1.964}$

or $\alpha = 11.54°$

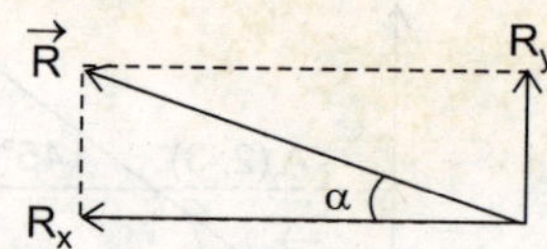

Taking moment about O we get

$$\Sigma M_O = (10 \times 1.5) + (2 \times 3.5) + (3.464 \times 0) + (1.5 \times 6.5) + (2.6 \times 8.5) + (1.5 \times 0)$$

$$= 10 \times 1.5 + 2 \times 3.5 - 1.5 \times 6.5 + 2.6 \times 8.5$$

[Taking ↻ as positive direction]

= 34.35 kNm in anticlockwise direction.

Resultant load does the same effect as quit that of the system of forces.

i.e., $Rd = 34.35$

or $2.004\,d = 34.35$

or $d = \dfrac{34.35}{2.004}$

$= 17.14$ m.

Therefore, x-intercept $= \dfrac{d}{\sin\alpha}$

$= \dfrac{17.14}{\sin 11.54}$

$= 85.68$ m

and y-intercept $= \dfrac{d}{\cos\alpha}$

$= \dfrac{17.14}{\cos 11.54}$

$= 17.50$ m.

4. *Find the resultant of the force system shown in Fig (8-W6NC), acting on a lamina of the shape as shown.*

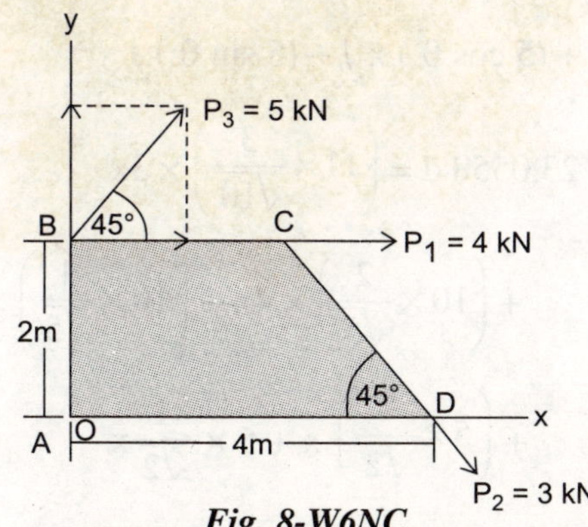

Fig. 8-W6NC

Solution:

Taking 2D-components of each load, we have

$P_1 = P_{1x} = 4$ kN

$P_2 = P_2\,(3\cos 45°, -3\sin 45°)$

and $P_3 = P_3\,(5\cos 45°, 5\sin 45°)$

Therefore,

resultant $(\vec{R}) = (4 + 3\cos 45° + 5\cos 45°)\,\vec{i} + (0 - 3\sin 45° + 5\sin 45°)\,\vec{j}$

$= (9.6568)\,\vec{i} + (1.4142)\,\vec{j}$

$= 9.6568\,\vec{i} + 1.4142\,\vec{j}$

$\therefore\ |\vec{R}| = 9.7598$ kN

Applying Varignon's theorem,

$Rd = P_{3y} \times 0 + (P_{3x} \times 2) + (P_{1x} \times 2) + P_{2x} \times 0 + P_{2y} \times 4$

Taking clockwise moment as positive moment, we have

$Rd = P_{3x} \times 2 + P_{1x} \times 2 + P_{2y} \times 4$

$= 5\cos 45° \times 2 + 4 \times 2 + 3\cos 45° \times 4$

or $9.6568\,d = 23.5563$ kNm

or $d = \left(\dfrac{23.5563}{9.6568}\right)$ m

$\therefore\ d = 2.4393$ m

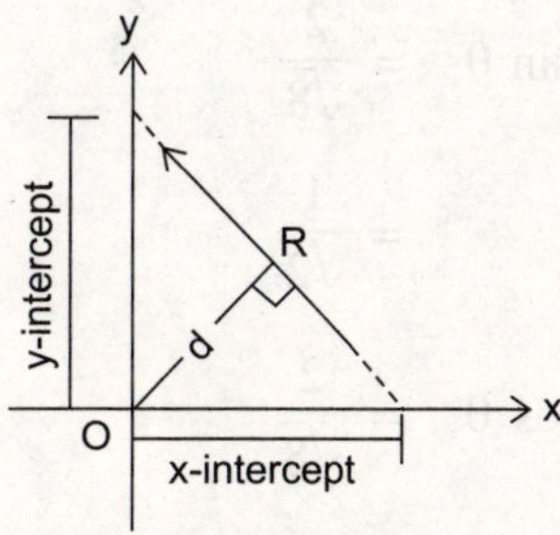

From the diagram we can see the x-intersect and y-intersect of resultant.

5. *Find the resultant of a set of coplanar forces acting on a lamina as shown in Fig. (8-W7NC). The smallest uniform sub-lamina is square.*

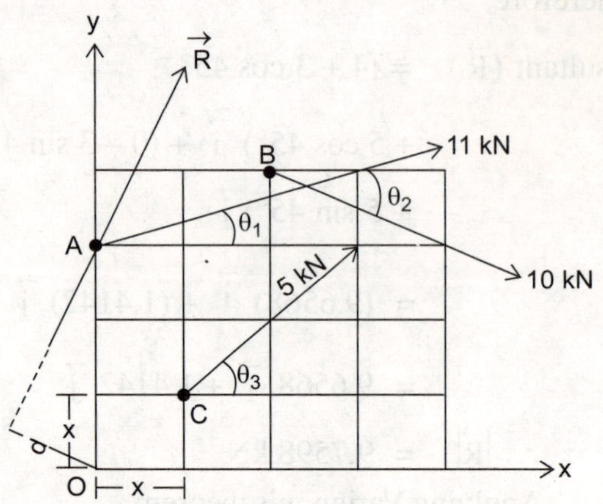

Fig. 8-W7NC

Solution:

If square has the side of length 'x' m, then we can write

$$\sin\theta_1 = \frac{1}{\sqrt{10}}$$

$$\cos\theta_1 = \frac{3}{\sqrt{10}}$$

$$\sin\theta_2 = \frac{1}{\sqrt{5}}$$

$$\cos\theta_2 = \frac{2}{\sqrt{5}}$$

$$\sin\theta_3 = \frac{2}{\sqrt{8}} = \frac{1}{\sqrt{2}}$$

$$\cos\theta_3 = \frac{2}{\sqrt{8}} = \frac{1}{\sqrt{2}}.$$

Taking 2D-components of each load, we have

$$\Sigma F_x = 11\cos\theta_1 + 10\cos\theta_2 + 5\cos\theta_3$$

$$= 11\times\frac{3}{\sqrt{10}} + 10\times\frac{2}{\sqrt{5}} + 5\times\frac{1}{\sqrt{2}}$$

$$= 22.9153 \text{ kN}$$

and $\Sigma F_y = 11\sin\theta_1 - 10\cos\theta_2 + 5\cos\theta_3$

$$= 11\times\frac{1}{\sqrt{10}} - 10\times\frac{1}{\sqrt{5}} + 5\times\frac{1}{\sqrt{2}}$$

$$= 2.5419 \text{ kN.}$$

∴ magnitude of resultant load

$$R = \sqrt{\Sigma F_x^2 + \Sigma F_y^2}$$

$$= \sqrt{22.9153^2 + 2.5419^2}$$

$$= 23.0558 \text{ kN}$$

Let us suppose that resultant passes through the line of action having a perpendicular distance d from the origin, then we can write from Varignon's theorem,

$$Rd\circlearrowright = (11\cos\theta_1)\,3x\circlearrowright + (11\sin\theta_1)\times 0$$

$$+ (10\cos\theta_2)\,4x\circlearrowright + (10\sin\theta_2)\times 2x\circlearrowright$$

$$+ (5\cos\theta_3)\,x\circlearrowright + (5\sin\theta_3)\circlearrowright$$

or $$23.0558\,d = \left(11\times\frac{3}{\sqrt{10}}\right)\times 3x$$

$$+\left(10\times\frac{2}{\sqrt{5}}\right)\times 4x + \left(10\times\frac{1}{\sqrt{5}}\right)\times 2x$$

$$+\left(5\times\frac{1}{\sqrt{2}}\right)x + 5\times\frac{1}{\sqrt{2}}x$$

$$= \left(\frac{11\times3\times3}{\sqrt{10}} + \frac{10\times2\times4}{\sqrt{5}} + \frac{10\times1\times2}{\sqrt{5}} + \frac{5\times1}{\sqrt{2}} + \frac{5\times1}{\sqrt{2}}\right)x$$

$$= 83.0989\,x \text{ kNm.}$$

or d = 3.6042 m when x = 1 m

∴ d = 3.6042 m as shown in Fig. (8-W7NC).

6. *Find the resultant on the lamina of non-uniform dimensions as shown in Fig. (8-W8NC).*

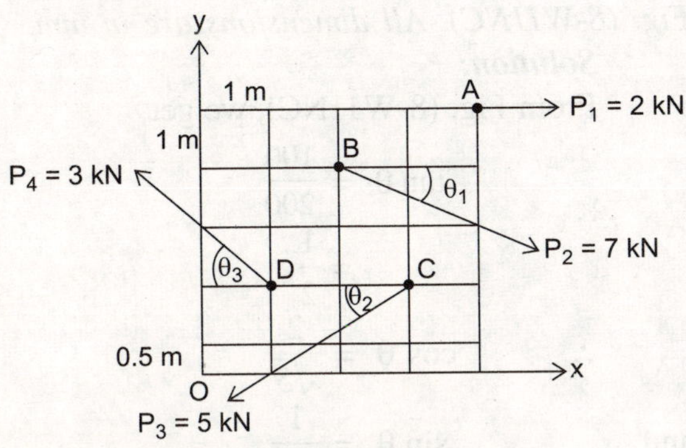

Fig. 8-W8NC

Solution:

From the above figure, we have

$$\cos\theta_1 = \frac{2}{\sqrt{5}} = 0.89$$

$$\sin\theta_1 = \frac{1}{\sqrt{5}} = 0.44$$

$$\cos\theta_2 = \frac{2}{\sqrt{2^2+1.5^2}} = \frac{2}{2.5} = 0.8$$

$$\sin\theta_2 = \frac{1.5}{2.5} = 0.6$$

and $$\cos\theta_3 = \frac{1}{\sqrt{2}} = 0.7$$

$$\sin\theta_3 = \frac{1}{\sqrt{2}} = 0.7$$

Taking rectangular component of loads, we get component of resultant load,

$$R_x = \text{x-component of the resultant}$$
$$= P_{1x} + P_{2x} + P_{3x} + P_{4x}$$
$$= P_1 + P_2 \cos\theta_1 - P_3 \cos\theta_2 - P_4 \cos\theta_3$$
$$= 2 + 7\times(0.89) - 5\times 0.8 - 3\times 0.7$$
$$= 2 + 6.23 - 4 - 2.1$$
$$= 2.13 \text{ kN.}$$

and $$R_y = \text{y-component of the resultant}$$
$$= 0 + P_{2y} + P_{3y} + P_{4y}$$
$$= 0 + (-P_2 \sin\theta_1) + (-P_3 \sin\theta_2) + (P_4 \sin\theta_3)$$
$$= (-7\times 0.44) + (-5\times 0.6) + (3\times 0.7)$$
$$= (-3.08) + (-3) + 2.1$$
$$= -3.98 \text{ kN.}$$

Therefore, magnitude of resultant,

$$R = \sqrt{R_x^2 + R_y^2}$$
$$= \sqrt{2.13^2 + (-3.98)^2}$$
$$= \sqrt{2.13^2 + 3.98^2}$$
$$= 4.51 \text{ kN.}$$

and $$\tan\alpha = \left|\frac{R_y}{R_x}\right| = \left|\frac{3.98}{2.83}\right|$$

$$\Rightarrow \quad \alpha = 61.84°$$

Fig. 8-W9NC

Taking moment about point O we get,

$$M_0 = 4.5 \times P_1 + 3.5 \times P_{2x} + 2 \times P_{2y} + 1.5 \times P_{3x} + 3 \times P_{3y} + 1 \times P_{4y} + 1.5 \times P_{4x}$$

$$= 4.5 \times 2 + 3.5 \times 6.23 + 2 \times 3.08 - 1.5 \times 4 + 3 \times 3 - 1 \times 2.1 - 1.5 \times 2.1$$

taking anticlockwise direction as positive direction.

$$= 34.715 \text{ kNm}$$

Let us suppose that d_x and d_y is the x-intercept and y-intercept of resultant load then,

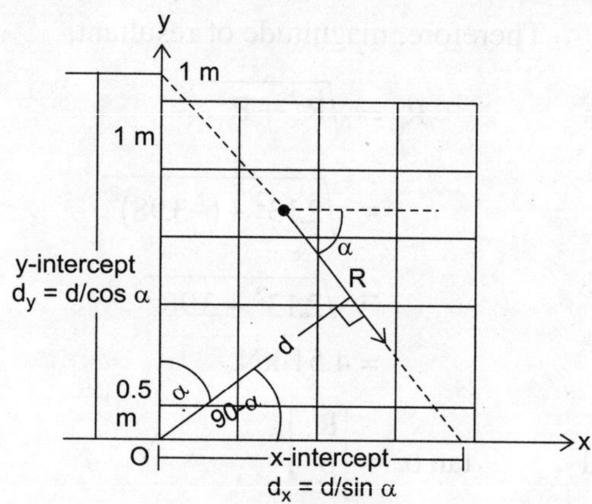

Fig. 8-W10NC

Applying Varignon's theorem,

i.e., $Rd = 34.715$

or $4.51\, d = 34.715$

or $d = \dfrac{34.715}{4.51}$

$= 7.69$ m

$\therefore$ d_x = x-intercept

$= \dfrac{d}{\sin \alpha}$

$= \dfrac{7.69}{\sin 61.84}$

$= 8.72$ m

and d_y = y-intercept

$= \dfrac{7.69}{\cos 61.84°}$

$= 16.29$ m.

7. *Determine the magnitude and y-intercept of the resultant of force system acting on the lamina as shown in Fig. (8-W11NC). All dimensions are in mm.*

Solution:

From Fig. (8-W11NC), we get

$$\tan \theta = \frac{100}{200} = \frac{1}{2}$$

$\therefore$ $\cos \theta = \dfrac{2}{\sqrt{5}}$

and $\sin \theta = \dfrac{1}{\sqrt{5}}$

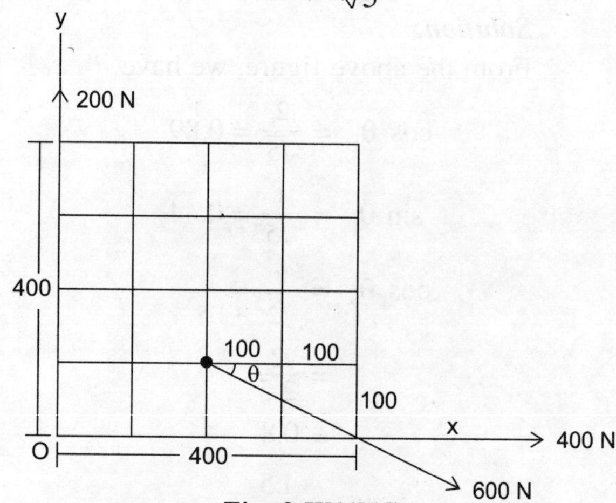

Fig. 8-W11NC

Taking rectangular component of each load, we get x-component of resultant.

$$R_x = 400 + 0 + 600 \cos \theta$$

$$= 400 + 600 \times \frac{2}{\sqrt{5}}$$

$$= 400 + 240\sqrt{5}$$

$$= 400 + 536.656$$

$$= 936.656 \text{ N}$$

and

$$R_y = 200 - 600 \times \sin \theta$$

$$= 200 - 600 \times \frac{1}{\sqrt{5}}$$

$$= 200 - 120\sqrt{5}$$

$$= -68.328 \text{ N}$$

∴ Magnitude of resultant lead is

$$R = \sqrt{R_x^2 + R_y^2}$$

$$= \sqrt{(936.636)^2 + (-68.328)^2}$$

$$= \sqrt{881955.7121}$$

$$= 939.124 \text{ N}$$

Considering point O as moment centre, net moment about origin is

$$\Sigma M_o = (600 \sin\theta) \times 200 \curvearrowright + (600 \cos\theta) \times 100 \curvearrowright + 400 \times 0 \curvearrowright + 200 \times 0 \curvearrowright$$

Taking clockwise moment as positive, we get

$$\Sigma M_o = 600 \times \frac{1}{\sqrt{5}} \times 200 + 600 \times \frac{2}{\sqrt{5}} \times 100$$

$$= 107331.2629 \text{ Nmm}$$

As we know that y-intercept $= \dfrac{\Sigma M_o}{R_x}$

$$= \frac{107331.2629 \text{ Nmm}}{936.656 \text{ N}}$$

$$= 114.592 \text{ mm.}$$

8. *Find out the point of application of the resultant force on Y-shape rod as shown in Fig. (8-W12NC).*

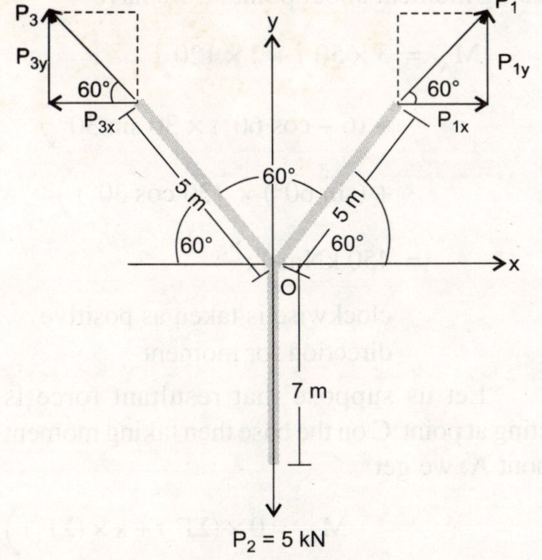

Fig. 8-W12NC

Solution:

Taking rectangular components of P_1, P_2 and P_3, the component of resultant is

$$R_x = P_{1x} + P_{2x} + P_{3x}$$
$$= P_1 \cos 60° + P_2 \cos 90° - P_3 \cos 60°$$
$$= P_1 \cos 60° - P_3 \cos 60°$$
$$= (P_1 - P_3) \cos 60°$$
$$= (3 - 4) \cos 60°$$
$$= -\cos 60° = -0.5 \text{ kN.}$$

and

$$R_y = P_{1y} + P_{2y} + P_{3y}$$
$$= 3 \sin 60° - 5 \sin 90° + 4 \sin 60°$$
$$= 3 \sin 60° - 5 + 4 \sin 60°$$
$$= 7 \sin 60° - 5$$
$$= 7 \times \frac{\sqrt{3}}{2} - 5$$
$$= 3.5 \times \sqrt{3} - 5$$
$$= 1.0621 \text{ kN.}$$

$$\therefore R = \sqrt{R_x^2 + R_y^2}$$
$$= \sqrt{(-0.5)^2 + (1.0621)^2}$$
$$= \sqrt{1.378}$$
$$= 1.1739 \text{ kN.}$$

Taking O as moment centre, then net moment will be

$$M_O = (P_{1x}) 5 \sin 60° \curvearrowright + (P_{1y}) 5 \cos 60° \curvearrowleft + (P_{3x}) 5 \sin 60° \curvearrowleft + (P_{3y}) 5 \cos 60° \curvearrowright + (0)P \curvearrowright$$

$$= (3 \cos 60°) \times (5 \sin 60°) - (3 \sin 60°) \times (5 \cos 60°) - (4 \cos 60°) \times (5 \sin 60°) + (4 \sin 60°)(5 \cos 60°) + 0 = 0$$

From Varignon's theorem

$$Rd = M_O$$

or

$$Rd = 0$$

$$\Rightarrow \quad d = 0 \qquad [\because R \neq 0]$$

i.e., Resultant passes through moment centre.

where $\tan\alpha = \frac{R_y}{R_x}$

$= \frac{1.0621}{0.5}$

$\Rightarrow \alpha = \mathbf{64.79°}$

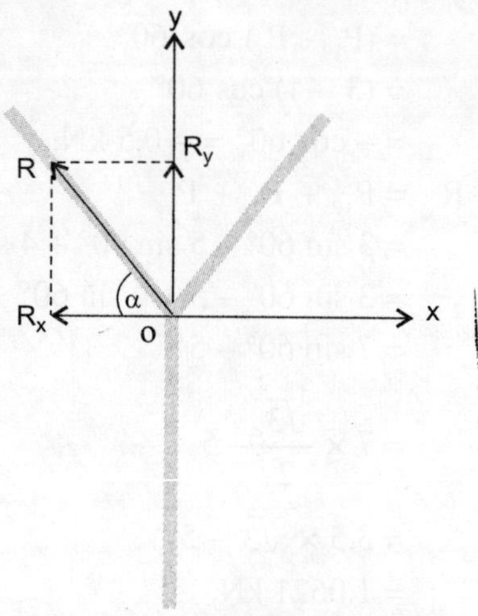

Fig. 8-W13NC

9. *Determine the resultant of the three forces acting on the dam section as shown in Fig. (8-W14NC) and its intersection with the base AB. For a safe design this intersection should occur within the middle third. Is it a safe design? Consider α = 30° and β = 60°.*

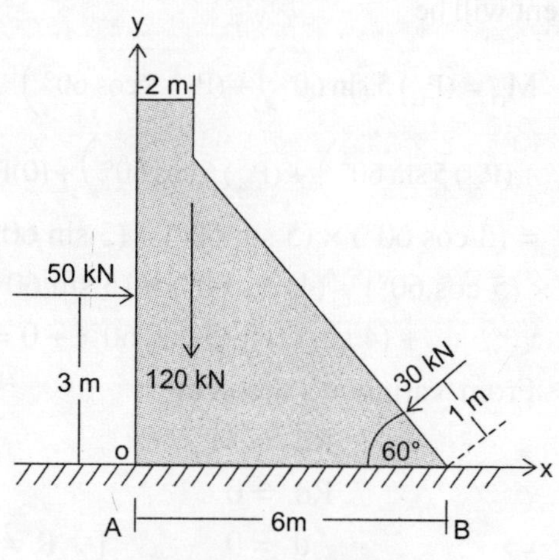

Fig. 8-W14NC

Solution:

Taking two dimensional components, we get components of resultant,

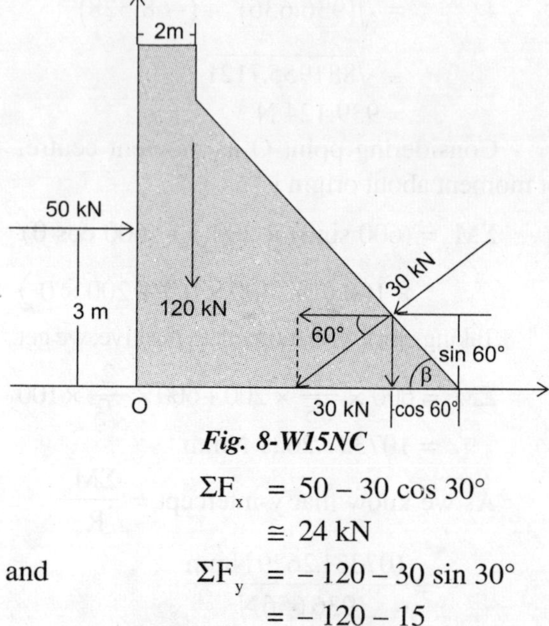

Fig. 8-W15NC

$$\Sigma F_x = 50 - 30\cos 30° \cong 24 \text{ kN}$$

and

$$\Sigma F_y = -120 - 30\sin 30° = -120 - 15 = -135 \text{ kN}$$

(Negative sign because ΣF_y pointing along negative direction of y-axis.)

Taking moment about point A, we have

$$M_A = 3 \times 50 + 2 \times 120 + (6 - \cos 60°) \times 30\sin 30° + (\sin 60°) \times (30\cos 30°) = 450 \text{ kNm};$$

clockwise is taken as positive direction for moment

Let us suppose that resultant force is acting at point C on the base then taking moment about A, we get

$$M_A = 0 \times (\Sigma F_x) + x \times (\Sigma F_y)$$

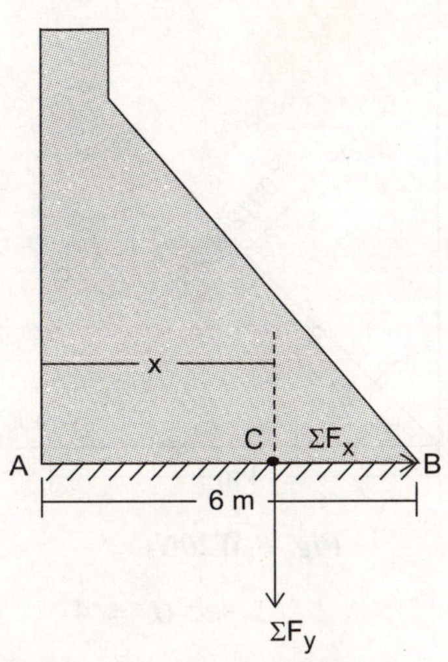

Fig. 8-W16NC

Applying Varignon's theorem, we have

$$x \times (\Sigma F_y) = 450$$

or $$x \times 135 = 450$$

or $$x = \frac{450}{135}$$

$$= 3.33 \text{ m.}$$

From the Fig. (8-W17NC) it is clear that resultant passes through middle third of the base, therefore, dam is in safe position.

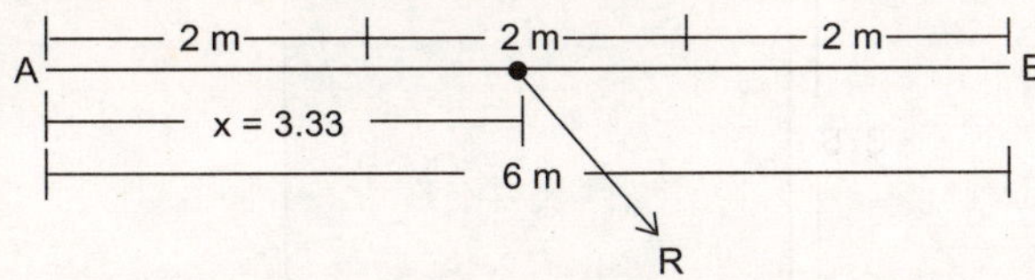

Fig. 8-W17NC

10. *For a nonconcurrent coplanar system of forces shown in Fig. (8-W18NC), determine the magnitude, direction and position of resultant force with reference to A.*

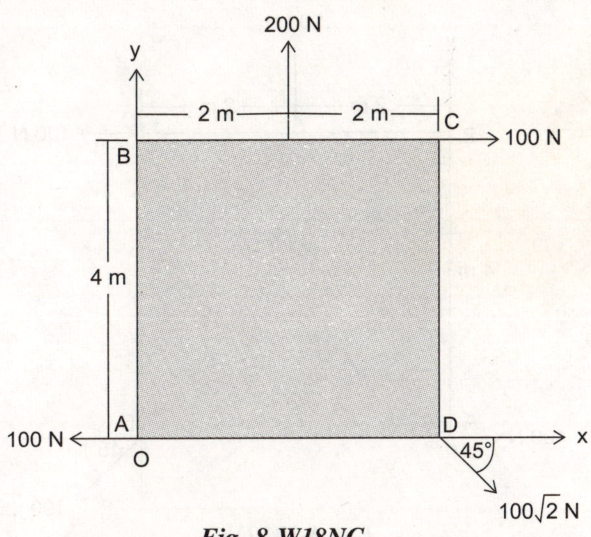

Fig. 8-W18NC

Solution:

Net x-component = ΣF_x

$$= 100 - 100 + 100\sqrt{2}\cos 45° + 200\cos 90°$$

$$= 100 \text{ N}$$

and net y-component = ΣF_y

$$= 200 - 100\sqrt{2}\sin 45° + 100\sin 0° + 100\sin 0°$$

$$= 100 \text{ N}$$

Therefore, net force = R

$$= \sqrt{(\Sigma F_x)^2 + (\Sigma F_y)^2}$$

$$= \sqrt{100^2 + 100^2}$$

$$= 100\sqrt{2} \text{ N}$$

Now let us draw the rectangular components of each force.

Now considering the point A as moment centre and clockwise moment as positive moment. Then net moment about A is

$$\Sigma M_A = 100 \times 0 + 100 \times 4 \circlearrowright + 100 \times 4 \circlearrowright + 200 \times 2 \circlearrowleft$$

$$= 400 + 400 - 400$$

$$= 400 \text{ Nm}$$

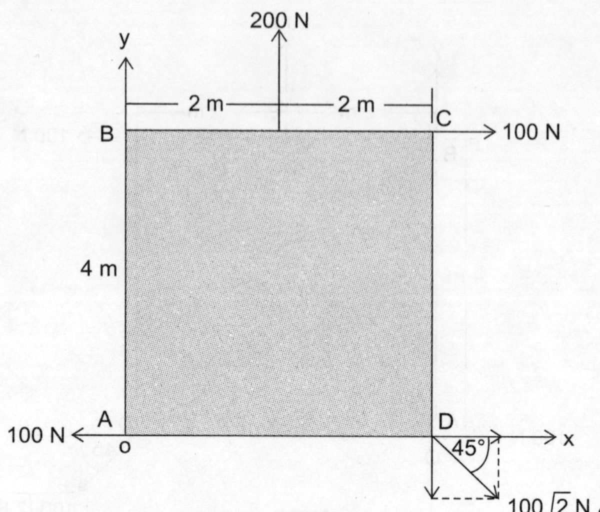

Fig. 8-W19NC

Suppose 'd' is the perpendicular distance of the resultant from point A then

$$\Sigma M_A = Rd$$

or $$400 = 100\sqrt{2}\ d$$

$\therefore$ $$d = \frac{400}{100\sqrt{2}}$$

$$= 2\sqrt{2}\ \text{m}$$

As we know that,

$$\text{x-intercept} = \frac{\Sigma M_A}{R_y}$$

$$= \frac{400}{100}$$

$$= 4\ \text{m}$$

and $$\text{y-intercept} = \frac{\Sigma M_A}{R_x}$$

$$= \frac{400}{100}$$

$$= 4\ \text{m}$$

where, $$d \operatorname{cosec} \alpha = \text{x-intercept}$$

or $$2\sqrt{2} \operatorname{cosec} \alpha = 4$$

or $$\operatorname{cosec} \alpha = \sqrt{2}$$

or $$\alpha = 45°$$

or $$d \sec \alpha = \text{y-intercept}$$

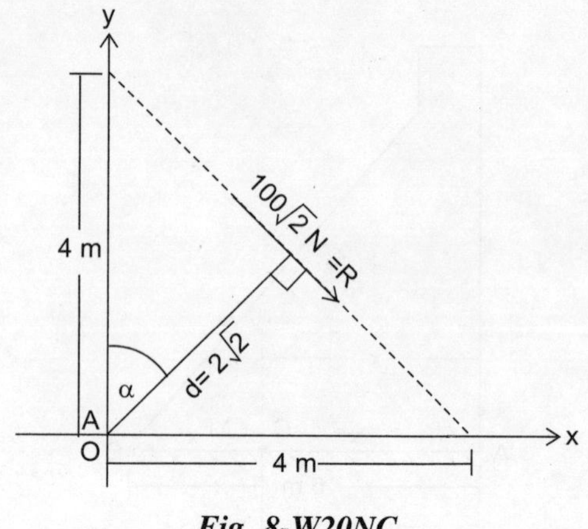

Fig. 8-W20NC

or $$2\sqrt{2} \sec \alpha = 4$$

or $$\sec \alpha = \sqrt{2}$$

or $$\alpha = 45°.$$

11. *Four forces act on a 700 mm × 375 mm plate (a) Find the resultant of these forces (b) Locate the point where the line of action of the resultant intersects the edge AB of the plate. Refer Fig. (8-W21NC).*

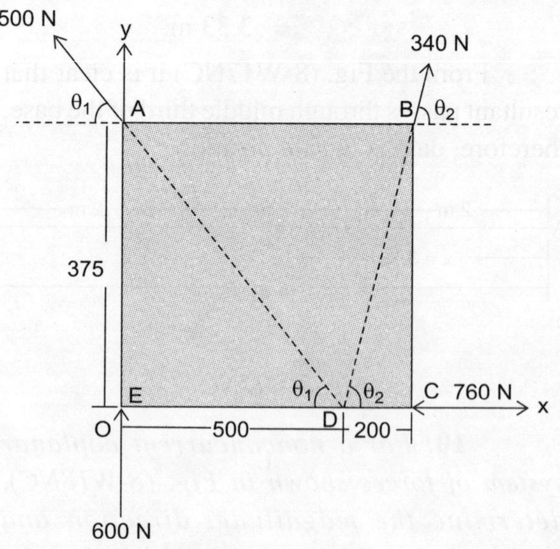

Fig. 8-W21NC

Solution:

From the Fig. (8-W21NC), we have

$$\tan\theta_1 = \frac{AE}{ED} = \frac{375}{500}$$

$$\therefore \quad \theta_1 \cong 37°$$

and

$$\tan\theta_2 = \frac{BC}{DC} = \frac{375}{200}$$

$$\therefore \quad \theta_2 \cong 62°$$

Now taking rectangular components of each force, we get components of resultant as

R_x = x-component of resultant

$$= 340\cos\theta_2 - 760 - 500\cos\theta_1$$
$$= 340\cos 62° - 760 - 500\cos 37°$$
$$\cong -1000 \text{ N}$$

and

$$R_y = 600 + 500\sin\theta_1 + 340\sin\theta_2$$
$$= 600 + 500\sin 37° + 340\sin 62°$$
$$\cong 1201 \text{ N.}$$

Therefore, magnitude of resultant is

$$R = \sqrt{R_x^2 + R_y^2} = \sqrt{(-1000)^2 + (1201)^2} = \sqrt{2442401}$$

and

$$\tan\alpha = \frac{R_y}{R_x};$$

where α is inclination of resultant from negative x-axis.

$$= \frac{1201}{1000}$$

$$\therefore \quad \alpha \cong 50°$$

$$\therefore \quad R = 1562.81 \text{ N and } \alpha = 50°$$

Consider point E as moment centre and take clockwise moment as positive moment then net moment about point E is

$$\Sigma M_E = 600\times 0 + 760\times 0 + 500\sin\theta_1\times 0 + 500\cos\theta_1\times 375 + 340\cos\theta_2 \times 375 + 340\sin\theta_2\times 700$$

$$= (500\sin 37°)\times 0 + 600\times 0 + 760\times 0 - (500\cos 37°)\times 375 + (340\cos 62°)\times 375 - (340\sin 62°)\times 700$$

$$= (340\cos 62°)\times 375 - (340\sin 62°)\times 700 - (500\cos 37°)\times 375$$

$$= -300028.06 \text{ Nmm}$$

$$\therefore \quad |\Sigma M_E| = 300028.06 \text{ Nmm}$$

If d is the perpendicular distance of resultant from A

then $Rd = |\Sigma M_E|$; [Varignon's theorem]

or $d\,(1562.81) = 300028.06$

or $$d = \frac{300028.06}{1562.81}$$

$$\therefore \quad d \cong 192 \text{ mm.}$$

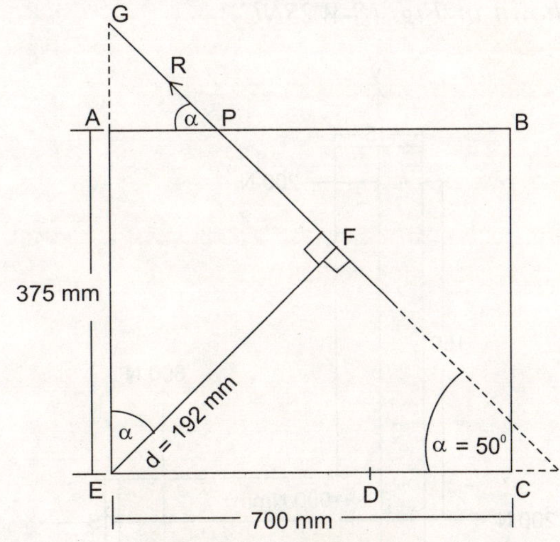

Fig. 8-W22NC

From Δ EFG

$$\cos\alpha = \frac{EF}{EG}$$

$$= \frac{192}{EG}$$

$$\therefore \quad EG = \frac{192}{\cos\alpha}$$

$$= \frac{192}{\cos 50^\circ}$$

$$\cong 299 \text{ mm}$$

Therefore, GA = – 76 m

$$\therefore \quad \tan\alpha = \frac{GA}{AP}$$

$$\therefore \quad AP = \frac{-76}{\tan 50^\circ}$$

$$\cong -63 \text{ mm}$$

Negative sign shows that point p divides AB externally at distance 63 m from point A.

12. *Find the equilibrium force with respect to O as origin for the system of forces shown in Fig. (8-W23NC).*

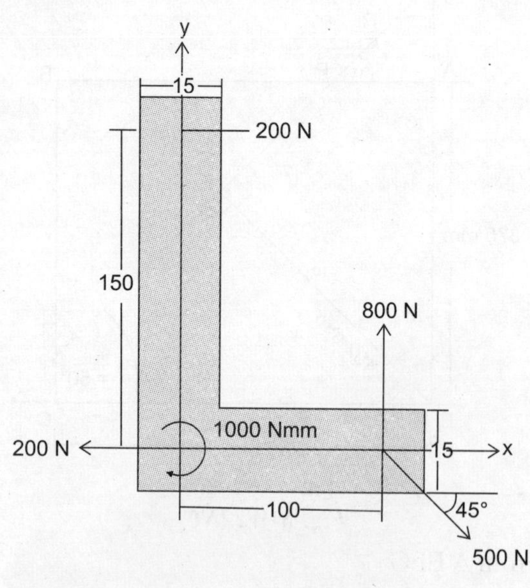

Fig. 8-W23NC

Solution:

Now, $R_x = \Sigma F_x$

= net force along x-axis

$= 200 - 200 + 500 \cos 45^\circ + 800 \cos 90^\circ$

$= 353.55$ N

and $R_y = \Sigma F_y$ = net force along y-axis

$= 800 - 500 \sin 45^\circ + 200 \sin 0^\circ + 200 \sin 0^\circ$

$= 446.44$ N

$\therefore$ R = magnitude of resultant force

$$= \sqrt{R_x^2 + R_y^2}$$

$$= \sqrt{353.55^2 + 446.44^2}$$

$$= 569.46 \text{ N}$$

Take O as moment centre and anti-clockwise moment as positive moment, we get net moment of forces about point O is

$$\Sigma M_0 = 200 \times 150 \text{ (clockwise)} + \text{external moment} + 200 \times 0 + (500 \cos 45^\circ) \times 0 + (-500 \sin 45^\circ + 800) \times 100 \text{ (anticlockwise)}$$

$$= 200 \times 150 \text{ (clockwise)} + 1000 \text{ (clockwise)} + (800 - 500 \sin 45^\circ) \times 100 \text{ (anticlockwise)}$$

$$= -200 \times 150 - 1000 + (800 - 500 \sin 45^\circ) \times 100$$

$$= -30000 - 1000 + 44644.66$$

$$= 13644.66 \text{ Nmm}$$

Suppose resultant force R passes at a distance d from the point O, then from Varignon's theorem, we get

$$Rd = \Sigma M_A$$

or $$d\,(569.46) = 13644.66$$

or $$d = \frac{13644.66}{569.47}$$

$$\therefore \quad d \cong 24 \text{ mm.}$$

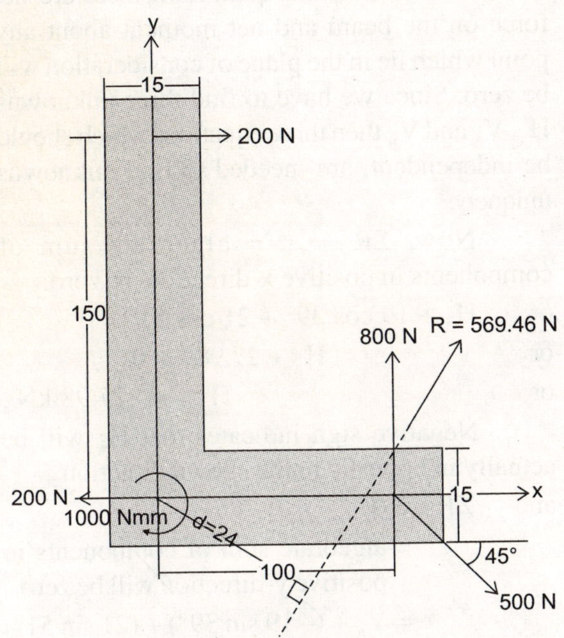

Fig. 8-W24NC

13. *Two 60 N forces are applied as shown at the corners A and C of a 200 mm × 200 mm square plate. Determine the moments of the couple formed by these two forces (a) by multiplying their magnitude with their perpendicular distance (b) by resolving each force into horizontal and vertical components and by adding the moments of the two resulting couples.*

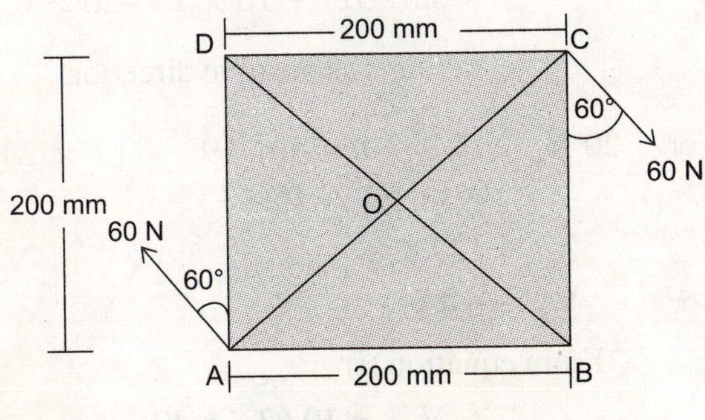

Fig. 8-W25NC

Solution:

First method:

From Pythagorous theorem,

$$AC = 200\sqrt{2} \text{ mm}$$

$$\therefore \quad OC = OA = 100\sqrt{2} \text{ mm}$$

Therefore, distance of couple is

$$AC = 200\sqrt{2} \text{ mm}$$

then magnitude of couple is

$$|M_O| = Fd$$

$$= 60 \times 200\sqrt{2} \text{ N mm}$$

$$\cong 17 \text{ kN mm}$$

Therefore, M_O = 17 kNmm is in anticlockwise direction.

Second method:

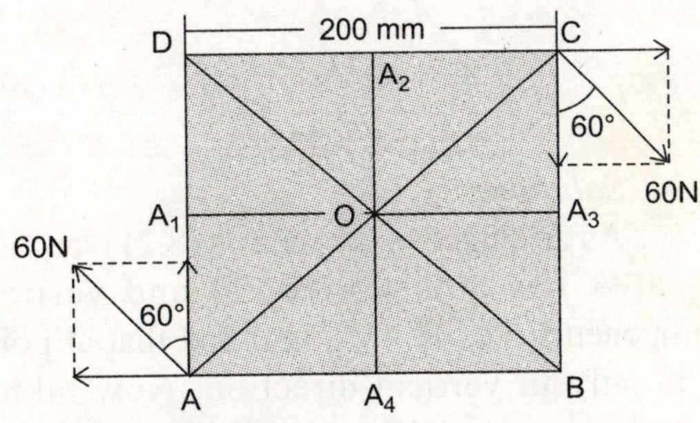

Fig. 8-W26NC

Consider O as a moment centre, then net moment about O is

$$M_O = 100 \times 60 \sin 60° + 100 \times 60 \cos 60°$$

$$+ 100 \times 60 \sin 60° + 100 \times 60 \cos 60°$$

$$= 60 \times 100 (\sin 60° + \cos 60° + \sin 60° + \cos 60°)$$

$$= 60 \times 100 (2 \sin 60 + 2 \cos 60°)$$

$$= 60 \times 100 \times 2\left(\frac{\sqrt{3}}{2}+\frac{1}{2}\right)$$

$$= 60 \times 100\left(\sqrt{3}+1\right)$$

$$= 6 \times \left(\sqrt{3}+1\right) \text{ kN mm}$$

$\cong$ 17 kN mm in anticlockwise direction.

14. *The beam AB of span 20 m shown in Fig. (8-W27NC) is hinged at A and is on roller at B. Determine the reactions at A and B for the loading shown in the figure.*

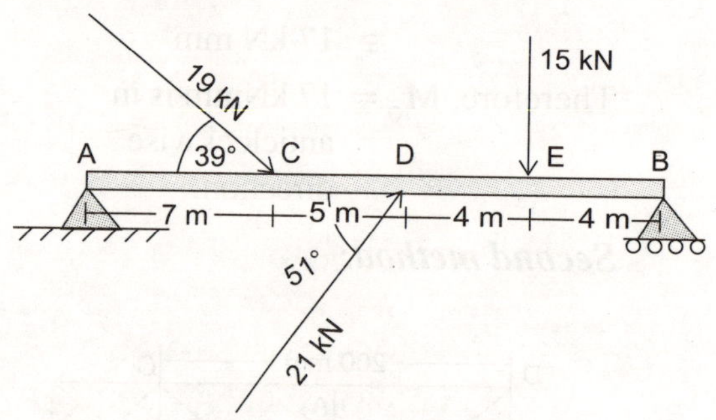

Fig. 8-W27NC

Solution:

As discussed in the section (8.21) reaction R_A at A has both horizontal and vertical components i.e., R_A (V_A, H_A) and that at point B is only in vertical direction. Now taking rectangular components of each load we get the geometrical representation is as below:

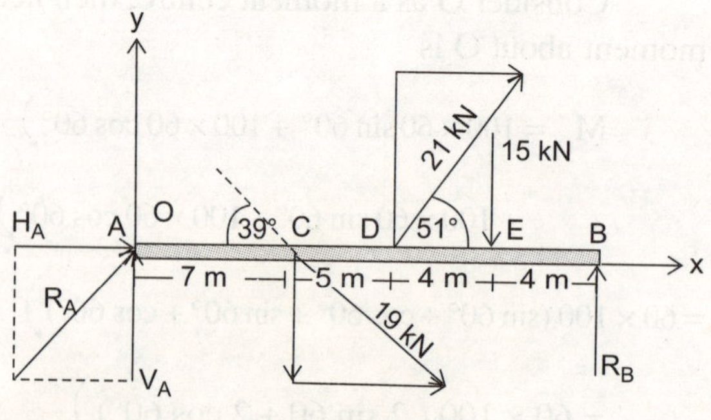

Fig. 8-W28NC

Since beam is in equilibrium, therefore, net force on the beam and net moment about any point which lie in the plane of consideration will be zero. Since we have to find three unknowns H_A, V_A and V_B then three equations which should be independent, are needed to find unknowns uniquely.

Now, $\Sigma R_x = 0$ = algebraic sum of components in positive x-direction is zero.

or $H_A + 19 \cos 39° + 21 \cos 51° = 0$

or $H_A + 27.98 = 0$

or $H_A = -27.98$ kN

Negative sign indicates that H_A will be actually in opposite to the shown direction

and $\Sigma R_y = 0$

= algebraic sum of components in positive y-direction will be zero.

$= V_A + (-19 \sin 39°) + (21 \sin 51°) - 15 + V_B$

or $V_A + V_B = 19 \sin 39° - 21 \sin 51° + 15$

$= 10.63$ kN

or $V_A + V_B = 10.63$ kN ...(i)

Consider point 'A' as moment centre then net moment about it is

$M_A = (7 \times 19 \sin 39°) + (12 \times 21 \sin 51°) + (16 \times 15) + 20 \times R_B$

or $0 = 7 \times 19 \times \sin 39° - 12 \times 21 \times \sin 51° + 16 \times 15 - 20 \times R_B$

taking ↻ as positive direction.

or $20 R_B = 7 \times 19 \times \sin 39° - 12 \times 21 \times \sin 51° + 16 \times 15$

$= 127.858$

or $V_B = 6.4$ kN

From equation (i)

$V_A = 10.63 - 6.40$

$= 4.23$ kN.

15. *Find the reactions at supports A and B and α for the beam AB as shown in Fig. (80W29NC).*

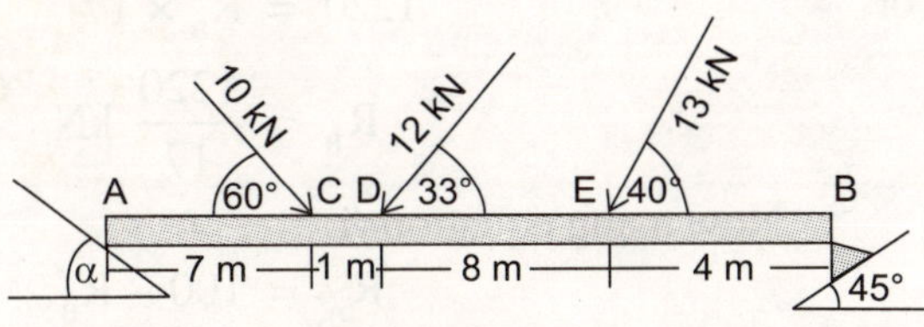

Fig. 8-W29NC

Solution:

If reactions at point A and point B are R_A and R_B respectively, then R_A and R_B should be in shown directions.

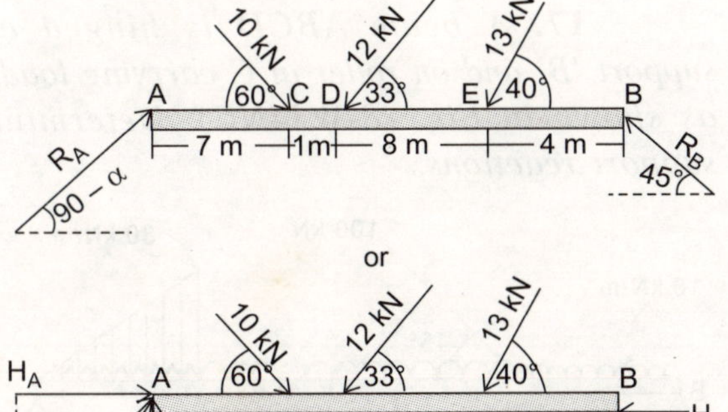

Fig. 8-W30NC

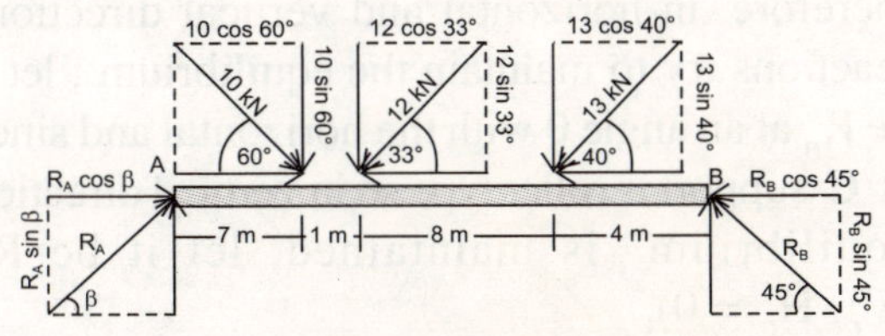

Fig. 8-W31NC

Since beam is in equilibrium then,

Σx-components = 0

or $R_A \cos\beta + 10 \cos 60° - 12 \cos 33° - 13 \cos 40° - R_B \cos 45° = 0$

or $R_A \cos\beta - R_B \cos 45° = 15.02$ ...(i)

and Σy-components = 0

or $R_A \sin\beta - 10 \sin 60° - 12 \sin 33° - 13 \sin 40° + R_B \sin 45° = 0$

or $R_A \sin\beta + R_B \sin 45° = 23.55$ ...(ii)

Considering B as moment centre, we get net moment about B as

$$M_B = (4 \times 13 \sin 40°) + (12 \sin 33°) \times 12 + (10 \sin 60°) \times 13 + (R_A \sin\beta \times 20) = 0$$

or $52 \sin 40° + 144 \sin 33° + 130 \sin 60° - 20 R_A \sin\beta = 0$; take as positive direction of moments.

or $224.43 = 20 R_A \sin\beta$

or $$R_A \sin\beta = \frac{224.43}{20} = 11.22 \text{ kN}$$

Putting $R_A \sin\beta$ in equation (ii), we have

$11.22 + R_B \sin 45° = 23.55$

or $R_B \sin 45° = 12.32$

or $R_B = 17.43$ kN

$\therefore$ $R_B = 17.43$ kN

Further from equation (i), we have

$R_A \cos\beta = 15.02 + R_B \cos 45° = 27.34$

$\therefore$ $R_A \cos\beta = 27.34$

Dividing equation (iii) by equation (iv), we have

$$\frac{R_A \sin\beta}{R_A \cos\beta} = \frac{11.22}{27.34}$$

or $\tan\beta = 0.41$

or $\beta = \tan^{-1} 0.41 \cong 22.29°$

$\therefore$ $\alpha = 90° - 22.29° = 67.70°$.

16. *Find out the reactions at supported points of a beam of uniform cross-section of length 17 m is loaded by a load of uniform intensity 15 kN/m as shown in Fig. (8-W32NC).*

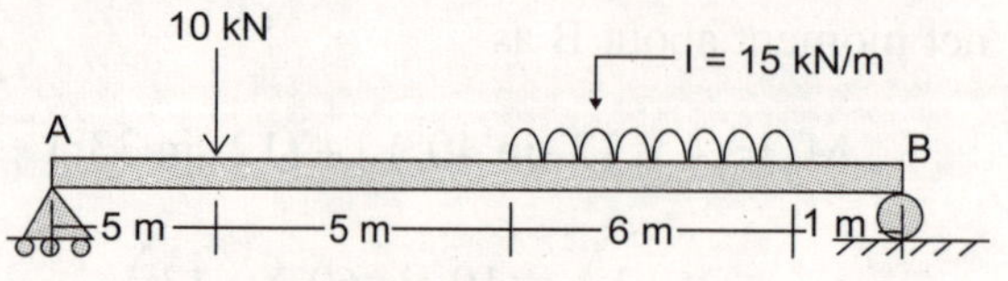

Fig. 8-W32NC

Solution:

Let reactions at A and B are R_A and R_B. As both supports are of roller support type, therefore reactions are in perpendicular directions.

Since, intensity of load is I = 15 kN/m and it has breadth of 6 m then total load is 15 × 6 kN = 90 kN acts at middle of 6 m all reactions are shown in Fig. (8-W33NC).

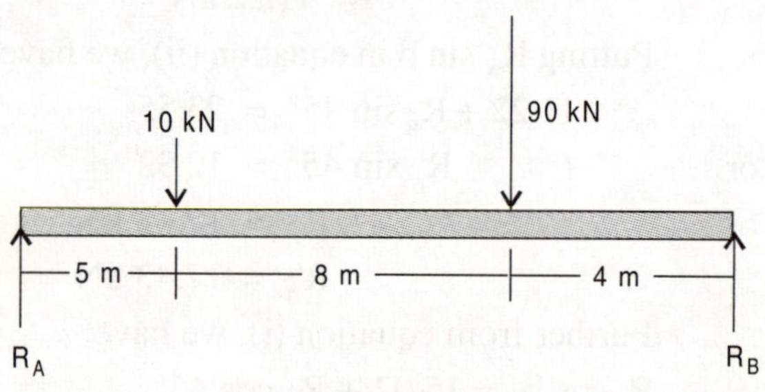

Fig. 8-W33NC

Since, beam is in equilibrium then, from conditions of equilibrium, we have

$$\Sigma F_x = 0$$

$$\Rightarrow \quad 0 = 0$$

[obviously satisfied the condition]

and $$\Sigma F_y = 0$$

or $$R_A + R_B = 10 + 90$$

or $$R_A + R_B = 100 \text{ kN} \quad \text{...(i)}$$

and $\Sigma M_A = 0$; point 'A' as moment centre and clockwise moment as positive moment

or $$5 \times 10 + 13 \times 90 + R_B \times 17 = 0$$

or $$50 + 1170 - R_B \times 17 = 0$$

or $$1220 = R_B \times 17$$

$$\therefore \quad R_B = \frac{1220}{17} \text{ kN}$$

Now using equation (i)

$$R_A = 100 - R_B$$

$$= 100 - \frac{1220}{17}$$

$$= \frac{1700 - 1220}{17}$$

$$= \frac{480}{17} \text{ kN.}$$

17. *A beam ABCD is hinged at support 'B' and on roller at C carrying loads as shown in Fig. (8-W34NC). Determine support reactions.*

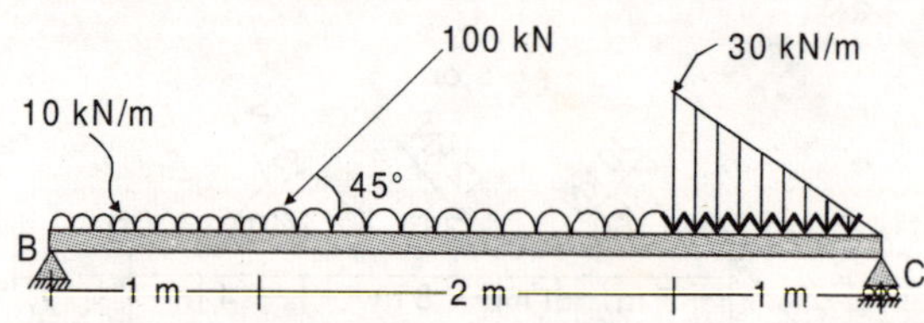

Fig. 8-W34NC

Solution:

Since at 'B' support is hinge type, therefore, in horizontal and vertical directions reactions try to maintain the equilibrium ; let it be R_B at an angle θ with the horizontal and since at C support is roller type so in vertical direction equilibrium is maintained, let it be R_C (V_C, $H_C = 0$).

Now, net rectangular uniformly distributed load = intensity × distance over which it is distributed

$$= (10 \text{ kN/m}) \times (1\text{m} + 2\text{m})$$

$$= 30 \text{ kN}$$

$$\text{Position} = \frac{3}{2}$$

$= 1.5$ m [from point B]

and net triangular uniformly distributed load

= area of triangle

$= \frac{1}{2} \times 1\text{m} \times 30$ kN/m

$= 15$ kN

and its position $= 3 + \frac{1}{3}h$ [From the base]

$= 3 + \frac{1}{3} \times 1$

$= 3 + \frac{1}{3}$

$= \frac{10}{3}$ m [from point B]

Now drawing all loads and reactions on beam we see following force diagram (FBD) or load diagram.

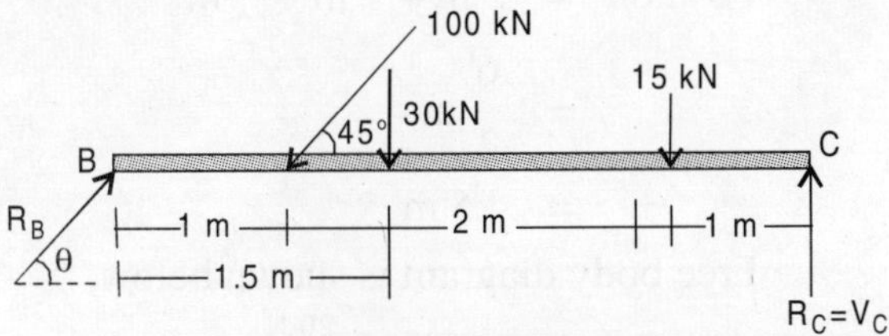

Fig. 8-W35NC

Since beam is in equilibrium, therefore, net force along horizontal direction = 0

i.e., $\Sigma F_H = 0$

or $R_B \cos\theta - 100\cos 45° = 0$

or $R_B \cos\theta = 100 \cos 45°$

$\therefore$ $R_B \cos\theta = 70.71$ ...(i)

and net force along vertical = 0

or $\Sigma F_V = 0$

or $R_B \sin\theta - 100 \sin 45° - 30 - 15 + V_C = 0$

or $R_B \sin\theta + V_C = 115.71$

$\therefore$ $R_B \sin\theta + V_C = 115.71$ ...(ii)

Taking point B as moment centre and anticlockwise moment as positive, we get

$\Sigma M_B = 0$

or $(100 \sin 45° \times 1) \curvearrowright + 30 \times 1.5 \curvearrowright + 15\times \frac{10}{3} \curvearrowright + V_C \times 4 \curvearrowleft = 0$

or $100 \sin 45° + 30 \times 1.5 + 15 \times \frac{10}{3} - V_C \times 4 = 0$

or $165.71 = 4V_C$

$\therefore$ $V_C = 41.42$ kN

From (ii)

$R_B \sin\theta = 115.71 - 41.42$ kN

$= 74.29$ ....(iii)

Dividing equation (iii) by equation (i), we get

$\tan\theta = \frac{74.29}{70.71}$

$= 1.05$

$\therefore$ $\theta \cong 47°$

From (i) $R_B = \frac{70.71}{\cos 47°}$

$= 103.68$

$\cong 104$ kN.

18. *Determine the reactions at A and B for the loaded beam shown in Fig. (8-W36NC).*

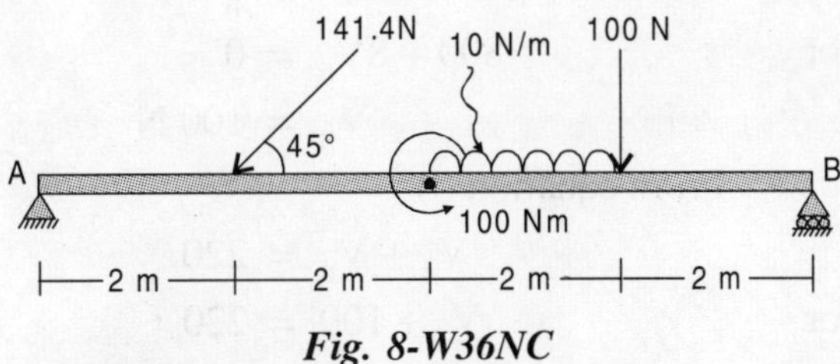

Fig. 8-W36NC

Solution:

Magnitude of uniformly distributed load = intensity × distance over which it is distributed

$= (10 \text{ N/m}) \times 2\text{m}$

$= 20$ N.

$$\text{Position} = 2\text{ m} + 2\text{ m} + \frac{2}{2}\text{ m}$$
$$= 5\text{ m}$$

Let R_A (H_A, V_A), and R_B (0, V_B) are reactions at A and B respectively.

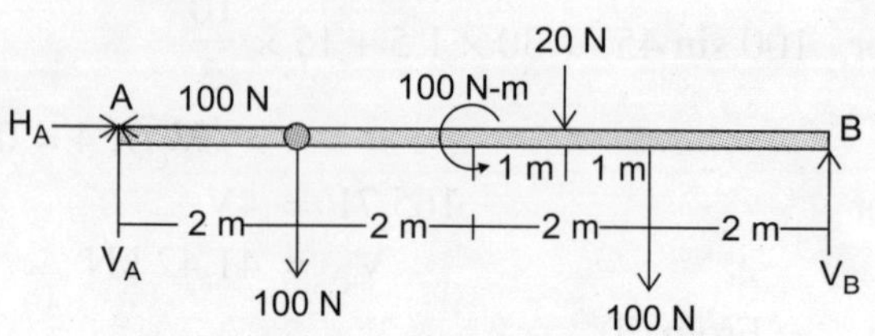

Fig. 8-W37NC

Since beam is in equilibrium then,

$$\Sigma F_H = 0$$

or $$H_A - 100 = 0$$

$\therefore$ $$H_A = 100\text{ N}$$

and $$\Sigma F_V = 0$$

or $$V_A - 100 - 20 - 100 + V_B = 0$$

or $$V_A + V_B = 220 \quad \text{...(i)}$$

Choose point A as moment centre and anticlockwise direction as positive direction, therefore,

$$\Sigma M_A = 0$$

or $$100 \times 2 \circlearrowright + 100 \circlearrowleft + 5 \times 20 \circlearrowright + 100 \times 6 \circlearrowright + 8 \times V_B \circlearrowleft = 0$$

or $$200 - 100 + 100 + 600 - 8V_B = 0$$

or $$800 - 8V_B = 0$$

$\therefore$ $$V_B = 100\text{ N}$$

From equation (i)

$$V_A + V_B = 220$$

or $$V_A + 100 = 220$$

or $$V_A = 220 - 100$$

$\therefore$ $$V_A = 120\text{ N}$$

Thus, $$V_A = 120\text{ N},$$
$$V_B = 100\text{ N}$$

and $$H_A = 100\text{ N}.$$

19. *For the beam shown in Fig. (8-W38NC) determine the reactions at the supports.*

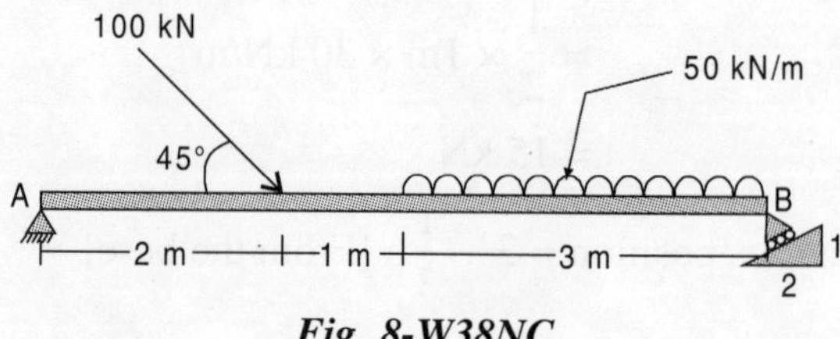

Fig. 8-W38NC

Solution :

Let R_A (H_A, V_A) and R_B are the reactions at the supports A and B respectively.

Then, net load of uniformly distributed load = intensity × distance over which it is distributed

$$= 50 \times 3$$
$$= 150\text{ kN}$$

$$\text{Position} = 2\text{ m} + 1\text{ m} + \frac{3}{2}\text{m}$$
$$= \frac{9}{2}\text{ m}$$
$$= 4.5\text{ m}$$

Free body diagram is shown below,

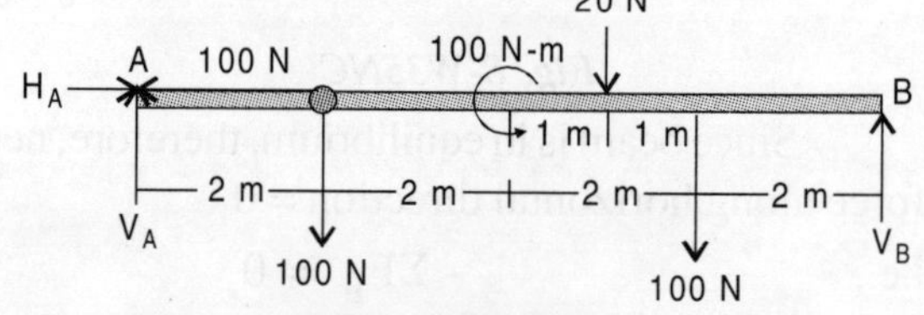

Fig. 8-W39NC

Since, beam is in equilibrium, then

$$\Sigma F_H = 0$$

or $$H_A + 100 \cos 45° - R_B \sin\theta = 0$$

or $$H_A + 100 \cos 45° - R_B \times \frac{1}{\sqrt{5}} = 0$$

or $$H_A - \frac{R_B}{\sqrt{5}} + 70.71 = 0$$

$\therefore$ $$H_A + 70.71 = \frac{R_B}{\sqrt{5}} \quad \text{....(i)}$$

and $\Sigma F_V = 0$

or $V_A - 100 \sin 45° - 150 + R_B \cos \theta = 0$

or $V_A + R_B \cos \theta = 150 + 100 \sin 45°$

$= 220.71$

or $$V_A + R_B \times \frac{2}{\sqrt{5}} = 220.71$$

or $$V_A + \frac{2R_B}{\sqrt{5}} = 220.71$$

$$\therefore \quad V_A + \frac{2R_B}{\sqrt{5}} = 220.71 \quad \text{...(ii)}$$

Since all forces i.e., in a plane and beam is in equilibrium, therefore, taking point 'A' as moment centre, we get

$$\Sigma M_A = 0$$

or $100 \sin 45° \times 2 + 150 \times 4.5$

$+ R_B \cos \theta \times 6 = 0$

or $200 \sin 45° + 150 \times 4.5 - R_B \cos\theta \times 6 = 0$

[Clockwise moment considered as positive.]

or $$200 \sin 45° + 150 \times 4.5 - R_B \times \frac{2}{\sqrt{5}} \times 6 = 0$$

or $$\frac{12R_B}{\sqrt{5}} = 150 \times 4.5 + 200 \sin 45°$$

or $$R_B = \frac{(150 \times 4.5 + 200 \times \sin 45°)\sqrt{5}}{12}$$

$\therefore \quad R_B = 152.13$ kN

From equation (ii)

$$V_A + \frac{2R_B}{\sqrt{5}} = 220.71$$

or $$V_A = 220.71 - \frac{2R_B}{\sqrt{5}}$$

$$= 220.71 - \frac{2 \times 152.13}{\sqrt{5}}$$

$= 84.64$ kN

From equation (i)

$$H_A + 70.71 = \frac{R_B}{\sqrt{5}}$$

$$H = \frac{152.13}{\sqrt{5}} - 70.71$$

$= -2.67$ kN

Negative sign shows that direction of H_A should be opposite to that of shown direction.

Therefore, reaction at A is R_A (– 2.67 kN, 84.64 kN) and reaction at B is

$R_B = 152.13$ kN.

20. *Determine the distance x in Fig. (8-W40NC) such that the reactions R_A and R_B are equal (consider beam is in equilibrium).*

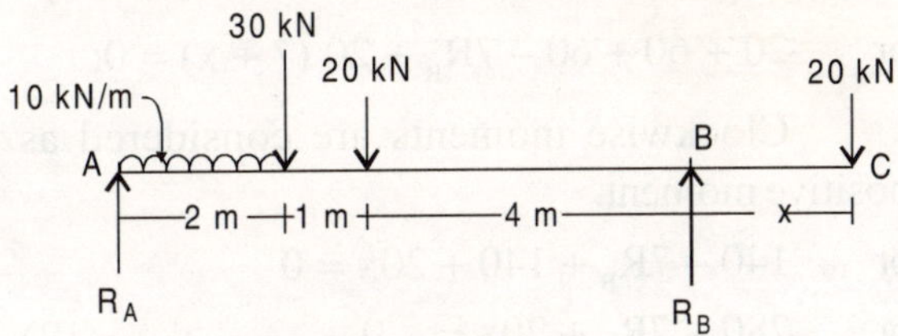

Fig. 8-W40NC

Solution:

Net load due to uniformly distributed load of intensity 10 kN/m

= intensity × distance over which it is distributed

= (10 kN/m) × 2m

= 20 kN

and at position $= \frac{2}{2}$

= 1 m from point A.

All reactions and loads are shown below.

Fig. 8-W41NC

Since beam is in equilibrium and all reactions and loads lie in vertical plane only, so $\Sigma F_y = 0$ and $\Sigma M_A = 0$ are sufficient to describe the equilibrium of beam because forces are coplanar but not concurrent.

Now $\Sigma F_y = 0$

or $R_A - 20 - 30 - 20 + R_B - 20 = 0$

$\therefore$ $R_A + R_B = 90$ ...(1)

and $\Sigma M_A = 0$

i.e., net moment about point A = 0

or $20 \times 1 + 30 \times 2 + 20 \times 3$

$+ R_B \times 7 + 20 \times (7 + x) = 0$

or $20 + 60 + 60 - 7R_B + 20\,(7 + x) = 0;$

Clockwise moments are considered as positive moment.

or $140 - 7R_B + 140 + 20x = 0$

or $280 - 7R_B + 20x = 0$ ...(2)

It is given that,

$$R_A = R_B$$

Therefore, from equation (i)

$$R_A + R_A = 90$$

or $2R_A = 90$

$\therefore$ $R_A = R_B$

$= 45$ kN

Now, putting the value of $R_B = 45$ kN in equation (ii), we get

$$280 - 7 \times 45 + 20\,x = 0$$

or $2\,x = 7 \times 45 - 280$

$= 35$

$\therefore$ $x = 17.5$ m.

21. *Beam AB shown in Fig. (8-W42NC) has hinged at support A and roller support at B. Determine the reactions developed at the supports, where the loads are act in their respective directions.*

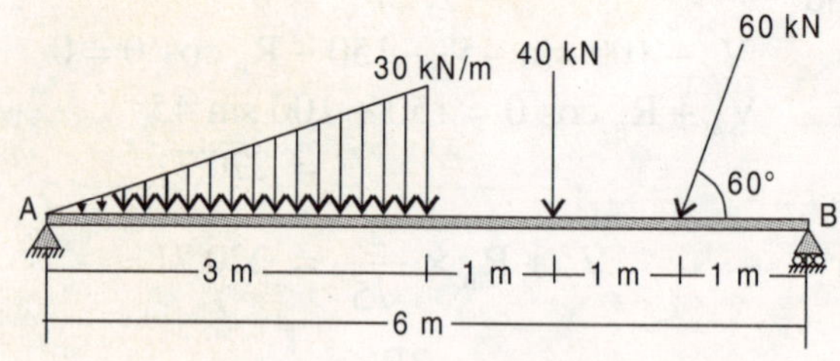

Fig. 8-W42 NC

Solution:

Let reactions at A and B are R_A (H_A, V_A), R_B $(0, V_B)$ respectively.

Uniformly distributed load

Triangular load $= \frac{1}{2} \times 3 \times 30$

$= 45$ kN

and acts at position $= \frac{2h}{3}$

$= \frac{2 \times 3}{3}$

$= 2$ m

Now all forces on the beam are shown in Fig. (8-W43NC).

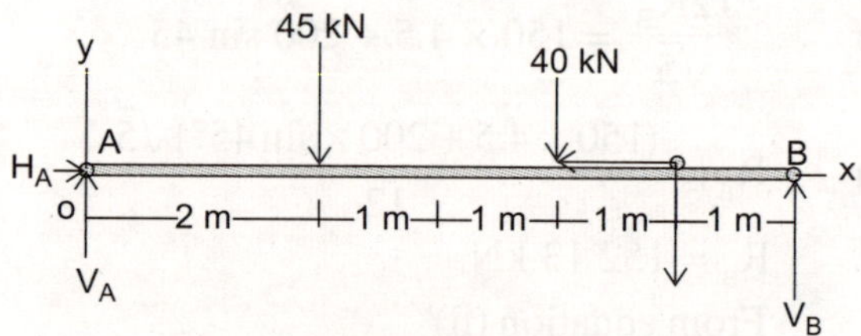

Fig. 8-W43NC

Since beam is in equilibrium, therefore

$\Sigma F_x = 0$

or $H_A - 60 \cos 60° = 0$

or $H_A = 60 \cos 60°$

$= 30$ kN ...(i)

and $\Sigma F_y = 0$

or $V_A - 45 - 40 - 60 \sin 60° + V_B = 0$

or $V_A + V_B = 45 + 40 + 60 \sin 60°$

$= 136.96$

$\cong 137$ kN ....(ii)

Taking point A as moment centre and since all the loads lie in xy-plane

Therefore,

$$\Sigma M_A = 0$$

or $45 \times 2 + 40 \times 4 + (60 \cos 60°) \times 0$

$+ (60 \sin 60°) \times 5 + V_B \times 6 = 0$

or $90 + 160 + 0 + 259.80 - V_B \times 6 = 0;$

Clockwise moments are taken as positive moments.

or $90 + 160 + 260 = 6V_B$

or $510 = 6V_B$

$\therefore$ $V_B = 85$ kN

From equation (ii)

$V_A = 137 - 85$
$= 52$ kN

From equation (i)

$H_A = 60 \cos 60°$
$= 30$ kN

Therefore, $H_A = 30$ kN, $V_A = 52$ kN and $V_B = 85$ kN .

22. *Determine the reactions at the supports A, B, C and D for the arrangement of the beams shown in Fig. (8-W44NC). Neglect the self-weights of the beams.*

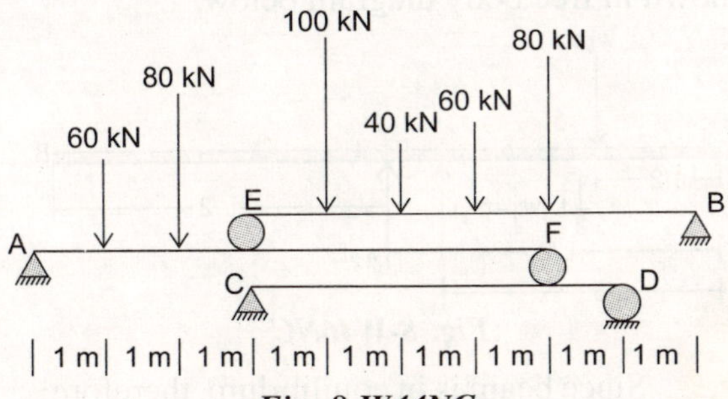

Fig. 8-W44NC

Solution:

Let R_A, R_B, R_C, R_D, R_E and R_F are the reactions at respective points respectively. Note that all the loads are in vertical only, therefore, all the reactions act vertically only. Forces on each beam are shown below.

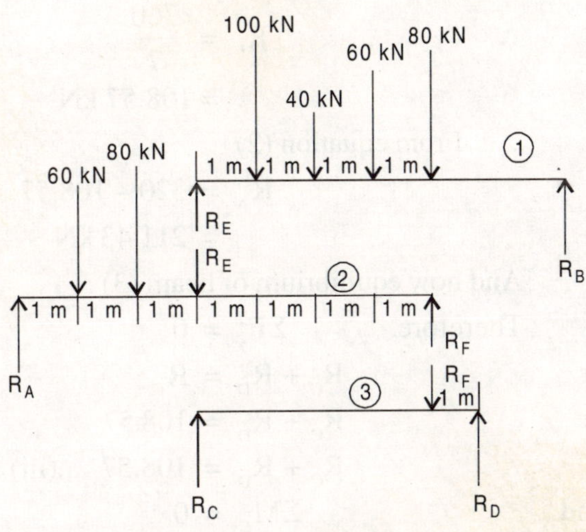

Fig. 8-W44(a)NC

Let us first consider equilibrium of beam (1).

Therefore, $\Sigma F_y = 0$

or $R_E + R_B = 280$

$\therefore$ $R_E + R_B = 280$ ...(i)

and $\Sigma M_E = 0$

or $100 \times 1 + 40 \times 2 + 60 \times 3$

$+ 80 \times 4 + R_B \times 6 = 0$

or $100 + 80 + 180 + 240 = 6R_B$

or $R_B = 100$ kN

$\therefore$ $R_E = 280 - 100$
$= 180$ kN.

Now equilibrium of beam (2) :

Therefore, $\Sigma F_y = 0$

or $R_A + R_F - R_E - 60 - 80 = 0$

or $R_A + R_F = 80 + 60 + 180$

[$\because R_E = 180$ kN]

or $R_A + R_F = 320$

$\therefore$ $R_A + R_F = 320$ ...(ii)

and $\Sigma M_A = 0$

or $60 \times 1 + 80 \times 2 + 180 \times 3 = 7 \times R_F$

or $60 + 160 + 540 = 7R_F$

or $760 = R_F$

$\therefore \quad R_F = \dfrac{760}{7}$

$= 108.57$ kN

$\therefore$ From equation (2)

$R_A = 320 - 108.57$

$= 211.43$ kN

And now equilibrium of beam (3) :

Therefore, $\Sigma F_y = 0$

or $R_C + R_D = R_F$

or $R_C + R_D = 108.57$

$\therefore \quad R_C + R_D = 108.57 \quad$ ...(iii)

and $\Sigma M_C = 0$

or $R_F \times 4 = R_D \times 5$

or $108.57 \times 4 = R_D \times 5$

$\therefore \quad R_D = \dfrac{108.57 \times 4}{5}$

$= 86.85$ kN

Then from equation (3)

$R_C = 108.57 - 86.85$

$= 21.71$ kN

Therefore calculated values are

$R_A = 211.43$ kN, $R_B = 100$ kN,

$R_C = 21.71$ kN, $R_D = 86.85$ kN,

$R_E = 180$ kN and $R_F = 108.57$ kN.

23. *A beam AB supports a uniformly distributed load of intensity w_0 and rests on*

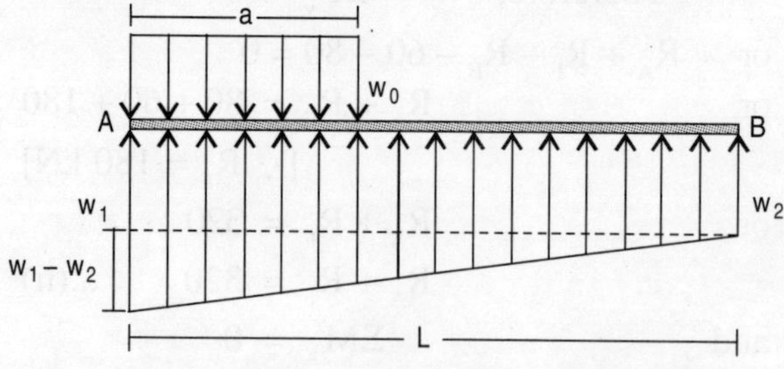

Fig. 8-W45NC

soil which exerts a uniformly varying upward reactions as shown in Fig. (8-W45NC). Determine w_1 and w_2 corresponding to equilibrium. Note that the soil can exert only an upward reaction at any point on the beam. Hence, state for what range of values of $\frac{a}{L}$ the results obtained are valid (a < L)?

Solution:

Net load by UDL $= aw_0$

and acts at position from A= $\dfrac{a}{2}$ from A.

Divide uniformly varying upward load in two parts.

(i) Uniformly distributed load of intensity w_2 and

(ii) triangular load vary from intensity 0 to $w_1 - w_2$.

Centroid of area covered by uniformly distributed load of intensity w_2 is $\dfrac{L}{2}$ from 'A' and that of triangular varying load is $\dfrac{L}{3}$ from the point, 'A' respectively, and their net loads are $w_2 \times L$ and $\dfrac{1}{2}L(w_1 - w_2)$ respectively are shown in free body diagram below.

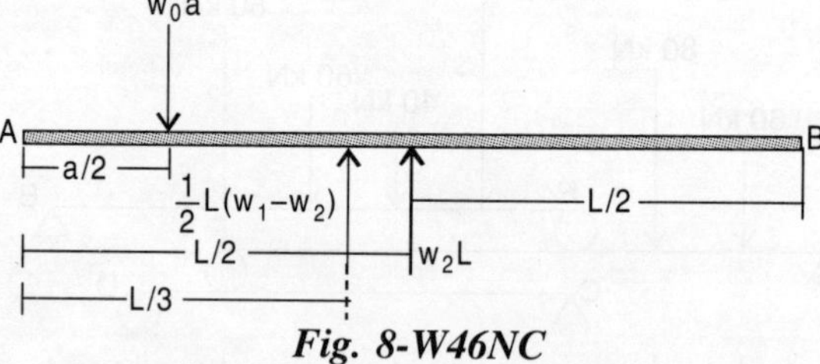

Fig. 8-W46NC

Since beam is in equilibrium, therefore,

$$\Sigma F_y = 0$$

or $\quad w_2L + \dfrac{1}{2}L(w_1 - w_2) - w_0a = 0$

or $\quad L\left[w_2 + \dfrac{w_1}{2} - \dfrac{w_2}{2}\right] = w_0a$

or $L\left[\frac{w_1 + w_2}{2}\right] = w_0 a$

or $L(w_1 + w_2) = 2w_0 a$

or $w_1 + w_2 = \frac{2w_0 a}{L}$ ....(i)

Note: *All loads are in vertical direction. Therefore, there is no question of $\Sigma F_x = 0$, it is obvious.*

Now, $\Sigma M_A = 0$

or $(w_0 a)\frac{a}{2} + (w_2 L) \times \frac{L}{2} + \frac{1}{2}L$

$(w_1 - w_2)\frac{L}{3} = 0$

take anticlockwise moment as positive then

$$-\frac{w_0 a^2}{2} + \frac{w_2 L^2}{2} + \frac{(w_1 - w_2)L^2}{6} = 0$$

or $L^2\left[\frac{w_2}{2} + \frac{w_1 - w_2}{6}\right] = \frac{w_0 a^2}{2}$

or $L^2\left[\frac{3w_2 + w_1 - w_2}{6}\right] = \frac{w_0 a^2}{2}$

or $L^2\left[\frac{2w_2 + w_1}{6}\right] = \frac{w_0 a^2}{2}$

or $L^2[w_1 + 2w_2] = 3w_0 a^2$

or $w_1 + 2w_2 = \frac{3w_0 a^2}{L^2}$ ....(ii)

Subtracting (i) from (ii), we get

$$w_1 + 2w_2 = \frac{3w_0 a^2}{L^2}$$

$$-w_1 - w_2 = -\frac{2w_0 a}{L}$$

$$w_2 = \frac{3w_0 a^2}{L^2} - \frac{2w_0 a}{L}$$

$$= \frac{w_0 a}{L}\left[\frac{3a}{L} - 2\right]$$

Similarly, from $(i \times 2) - (ii)$

$$2w_1 + 2w_2 = \frac{4w_0 a}{L}$$

$$-w_1 - 2w_2 = -\frac{3w_0 a^2}{L^2}$$

We get $w_1 = \frac{4w_0 a}{L} - \frac{3w_0 a^2}{L^2}$

$$= \frac{w_0 a}{L}\left[4 - \frac{3a}{L}\right]$$

Therefore, $w_2 = \frac{w_0 a}{L}\left[\frac{3a}{L} - 2\right]$

and $w_1 = \frac{w_0 a}{L}\left[4 - \frac{3a}{L}\right]$

Since, w_1 and w_2 cannot be negative physically (may be zero).

Therefore, $w_2 \geq 0$

or $\frac{w_0 a}{L}\left[\frac{3a}{L} - 2\right] \geq 0$

or $\frac{3a}{L} - 2 \geq 0$

or $\frac{3a}{L} \geq 2$

or $\frac{a}{L} \geq \frac{2}{3}$ ....(iii)

Similarly, $w_1 \geq 0$

or $\frac{w_0 a}{L}\left[4 - \frac{3a}{L}\right] \geq 0$

or $4 - \frac{3a}{L} \geq 0$

or $4 \geq \frac{3a}{L}$

or $\frac{3a}{L} \leq 4$

or $\frac{a}{L} \leq \frac{4}{3}$ ....(iv)

Therefore, $\frac{2}{3} \le \frac{a}{L} \le \frac{4}{3}$

but, $a < L$

$\therefore \quad \frac{a}{L} < 1$

So, finally valid interval for

$\frac{a}{L}$ is $\frac{2}{3} \le \frac{a}{L} < 1$

24. *Force $\vec{F}$ acting away from orgin has components 500 N and – 150 N along x and y axes respectively as shown in Fig. (8W47NC). If the inclination θ_y of the force is 30° with respect to y-axis, obtain the magnitude of the force vector $\vec{F}$. Find its moment also about point C (12, 4, 3) and moment arm.*

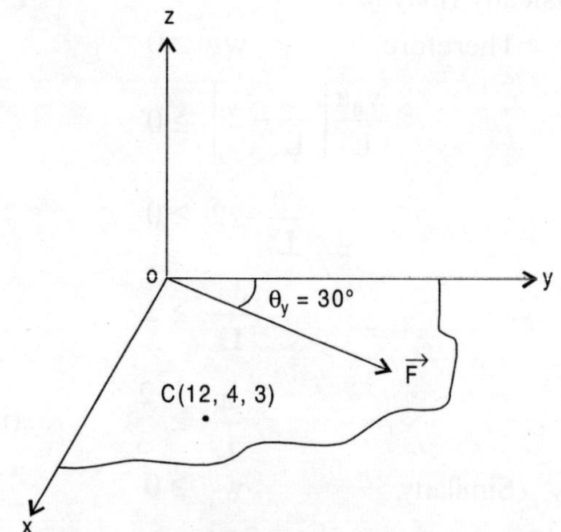

Fig. 8-W47NC

Solution:

Since force has rectangular components $F_x = 500$ N and $F_y = -150$ N then we can write,

$$\vec{F} = F_x \vec{i} + F_y \vec{j}$$

or
$$\vec{F} = 500\,\vec{i} - 150\,\vec{j}$$

Since, vector $\vec{F}$ makes an angle 30° with y-axis, let one unit length along $\vec{F}$; then unit vector along $\vec{F}$.

$$\hat{F} = \vec{i}\,(1 \sin 30°) + (1 \cos 30°)\,\vec{j}$$

$$= (\sin 30°)\,\vec{i} + (\cos 30°)\,\vec{j}$$

$$= \frac{1}{2}\vec{i} + \frac{\sqrt{3}}{2}\vec{j}$$

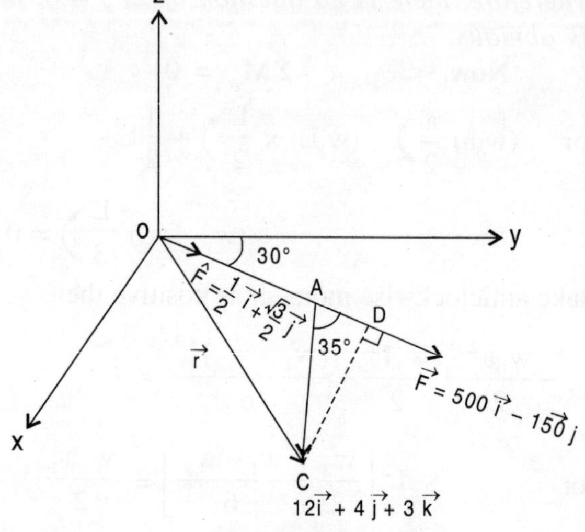

Fig. 8-W48NC

Let, $\hat{F} = \vec{OA}$

$$= \frac{1}{2}\vec{i} + \frac{\sqrt{3}}{2}\vec{j} \text{ and } \vec{OC}$$

$$= 12\vec{i} + 4\vec{j} + 3\vec{k}$$

$$= \vec{r}$$

Therefore, from triangle rule of vector addition

$$\vec{OC} = \vec{OA} + \vec{AC}$$

$$\therefore \quad \vec{AC} = \vec{OC} - \vec{OA}$$

$$= (12\vec{i} + 4\vec{j} + 3\vec{k}) - \left(\frac{1}{2}\vec{i} + \frac{\sqrt{3}}{2}\vec{j}\right)$$

$$= \left(12 - \frac{1}{2}\right)\vec{i} + \left(4 - \frac{\sqrt{3}}{2}\right)\vec{j} + 3\vec{k}$$

$$= 11.5\,\vec{i} + 3.133\,\vec{j} + 3\vec{k}$$

Now, $\vec{F} \cdot \vec{AC} = (500\vec{i} - 150\vec{j})$

$\cdot (11.5 + 3.133\vec{j} + 3\vec{k})$

$= 500 \times 11.5 - 150 \times 3.133 + 0 \times 3$

$= 5280.05$

If θ is the angle between $\vec{F}$ and $\vec{AC}$,

Therefore, $\cos\theta = \dfrac{\vec{F} \cdot \vec{AC}}{|\vec{F}||\vec{AC}|}$

$$= \frac{5280.05}{\sqrt{500^2 + 150^2} \times \sqrt{11.5^2 + 3.133^2 + 3^2}}$$

$= 0.823$

or $\theta = \cos^{-1} 0.823$

$\therefore \theta \cong 35°$

From triangle ACD,

$\sin 35° = \dfrac{CD}{|\vec{AC}|}$

or $CD = |\vec{AC}| \sin 35°$

$= \sqrt{11.5^2 + 3.133^2 + 3^2} \times \sin 35°$

$= 12.3 \times \sin 35°$

$\cong 7$

$\therefore$ CD = moment arm = 7 m

Therefore moment about C

= moment arm × $|\vec{F}|$

$= 7 \times 522.01$

$= 3654.1$ Nm.

25. *The transition between the loads of 10 kN/m and 37 kN/m is accomplished by means of a cubic function of the form $w = k_0 + k_1x + k_2x^2 + k_3x^3$, the slope of which is zero at its end points x = 1m and x = 4 m. Determine the reactions at A and B.*

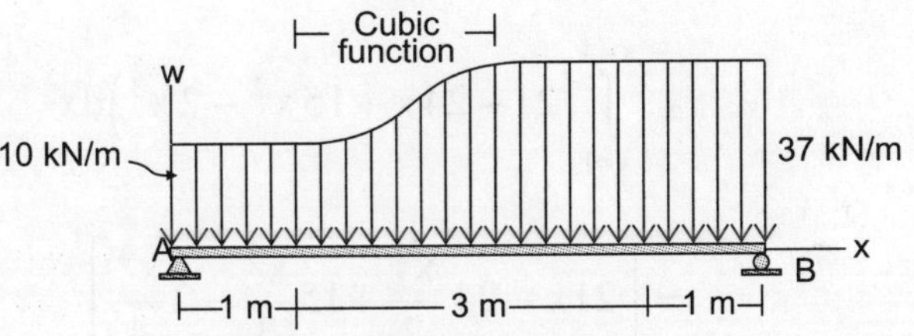

Fig. 8-W49NC

Solution:

Given intensity is function of x,

i.e., $w = k_0 + k_1x + k_2x^2 + k_3x^3$

or $\dfrac{dw}{dx} = k_1 + 2x\,k_2 + 3x^2 k_3$

= slope at x

Now, $\left.\dfrac{dw}{dx}\right|_{x=1} = 0$ (given)

or $k_1 + 2k_2 + 3k_3 = 0$

$\therefore k_1 + 2k_2 + 3k_3 = 0$ ...(i)

and $\left.\dfrac{dw}{dx}\right|_{x=4} = k_1 + 8k_2 + 48k_3$

= 0 (given)

$\therefore k_1 + 8k_2 + 48k_3 = 0$ ...(ii)

According to question it is given that,

$(w)_{x=1} = 10$ and $(w)_{x=4} = 37$

Therefore,

$k_0 + k_1 + k_2 + k_3 = 10$ ...(iii)

and $k_0 + 4k_1 + 16k_2 + 64k_3 = 37$ ...(iv)

Solving equations (i), (ii), (iii) and (iv) simultaneously, we get

$w = 21 - 24x + 15x^2 - 2x^3$

Now, w_{net}^{cubic} = net load due to cubic function of intensity.

$$= \int_{x=1}^{x=4} w\,dx$$

$$= \int_{x=1}^{x=4} \left(21 - 24x + 15x^2 - 2x^3\right) dx$$

$$= \left[21x - 24\frac{x^2}{2} + 15\frac{x^3}{3} - 2\frac{x^4}{4}\right]_1^4$$

$$= 21 \times 4 - 12 \times 4^2 + 5 \times 4^3 - \frac{1}{2} \times 4^4$$

$$- 21 + 12 - 5 + \frac{1}{2}$$

$$= 70.5 \text{ kN}$$

and $\bar{x}$ = position of centroid from where this cubic load will act

$$= \frac{\int_{x=1}^{x=4} x_c \, dA}{\int_{x=1}^{x=4} w dx}$$

$$= \frac{\int_{x=1}^{x=4} x(w dx)}{70.5}$$

$$= \frac{\int_{x=1}^{x=4} (21x - 24x^2 + 15x^3 - 2x^4) dx}{70.5}$$

$$= \frac{\left[21\frac{x^2}{2} - 24\frac{x^3}{3} + 15\frac{x^4}{4} - 2\frac{x^5}{5}\right]_1^4}{70.5}$$

$$= 2.84 \text{ m}$$

Net load due to constant distributed load of intensity 10 kN/m

= 10 × 1

= 10 kN and acts at position

x = 0.5 and due to constant distributed load of intensity 37 kN/m

= 37 × 1

= 37 and acts at position

$4 + \frac{1}{2}$ = 4.5 from point A. If R_A and R_B are the reactions at A and B respectively, then free body diagram of beam is as shown below.

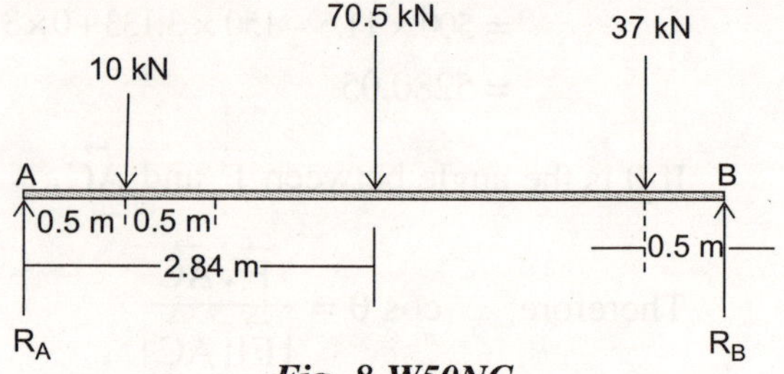

Fig. 8-W50NC

Now, $\Sigma F_y = 0$

or $R_A + R_B = 117.5$ ....(v)

and $\Sigma M_A = 0$

or $10 \times 0.5 + 70.5 \times 2.84 + 37 \times 4.5 = R_B \times 5$

or $371.72 = R_B \times 5$

∴ $R_B = 74.34$ kN.

From equation (v)

$$R_A = 117.5 - 74.34 = 43.15 \text{ kN.}$$

26. *A roller of radius r = 12 mm and weight Q = 500 N is to be pulled over a curb of height h = 6 mm by a horizontal force P applied to the end of a string wound around the circumference of 1the roller (Fig. 8-W51NC). Find the magnitude of P required to start the roller rolling over the curb:*

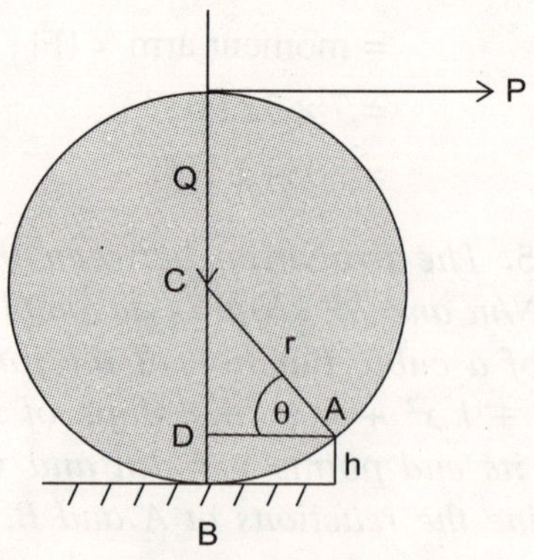

Fig. 8-W51NC

Solution:

From the Fig.(8-W52NC); in Δ ADC, we get

$$\sin\theta = \frac{CD}{AC}$$

$$= \frac{BC - BD}{AC}$$

$$= \frac{r - h}{r}$$

$$= \frac{12 - 6}{12}$$

$$= \frac{1}{2}$$

$$\therefore \quad \theta = \sin^{-1}(0.5)$$

$$= 30°$$

Let R_a be the contact reaction at point A and is passing through centre of the roller, making an angle $\theta = 30°$ with the horizontal, weight Q = 500 N is also passing through centre. We have to find P such that roller starts rolling over the curb, therefore, at that moment reaction at B will become zero.

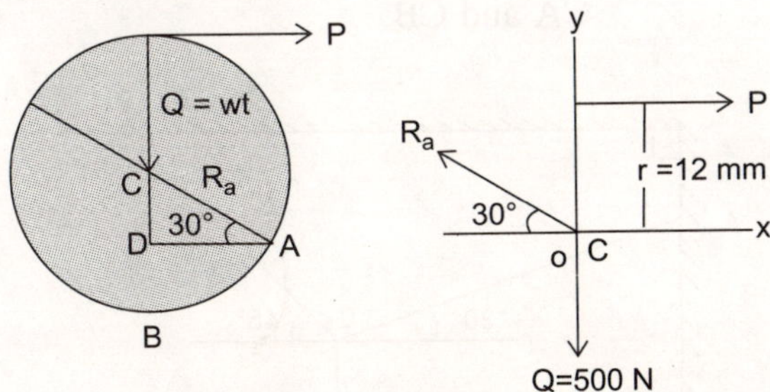

Fig. 8-W52NC

Since, all forces are coplanar but not concurrent, we have to consider $\Sigma M = 0$ also apart from $\Sigma F_x = 0$ and $\Sigma F_y = 0$.

Now, $\quad \Sigma F_x = 0$

or $\quad P - R_a \cos 30° = 0$

or $\quad P = R_a \cos 30°$

$\therefore \quad P = R_a \cos 30° \quad$ ...(i)

and $\quad \Sigma M_A = 0$

or $\quad r\, Q \cos 30 = P \times (r + r - h)$

or $\quad Q\,(r \cos 30°) = P \times (2r - h)$

or $$500\left[12 \times \frac{\sqrt{3}}{2}\right] = P \times [2 \times 12 - 6]$$

or $$500 \times 6\sqrt{3} = P \times 18$$

$$\therefore \quad P = \frac{500 \times \sqrt{3} \times 6}{18}$$

$$= \frac{500}{\sqrt{3}}$$

$$= 288.67 \text{ N}$$

Since, $\quad \Sigma F_y = 0$

or $\quad R_a \sin 30° = Q$

or $\quad R_a = 2Q$

$\therefore \quad R_a = 2 \times 500$

$\quad = 1000$ N.

Therefore, from equation (i)

$$P = R_a \cos 30°$$

$$= 1000 \times \cos 30°$$

$$= 1000 \times \frac{\sqrt{3}}{2}$$

$$= 500\sqrt{3}$$

$$= 866.02 \text{ N}$$

But since forces are not concurrent, so we cannot consider conditions $\Sigma F_x = 0$ and $\Sigma F_y = 0$ simultaneously, consideration of $\Sigma M_A = 0$ along with one of $\Sigma F_x = 0$ and $\Sigma F_y = 0$ is correct. Therefore, P = 288.67 N not 866.02 N.

EXERCISE

Coplanar concurrent system

1. The force P is applied to a small wheel which rolls on the cable ACB as shown in Fig. (8-ECC1) knowing that the tension in the cable is 600 N, determine the magnitude and direction of P.

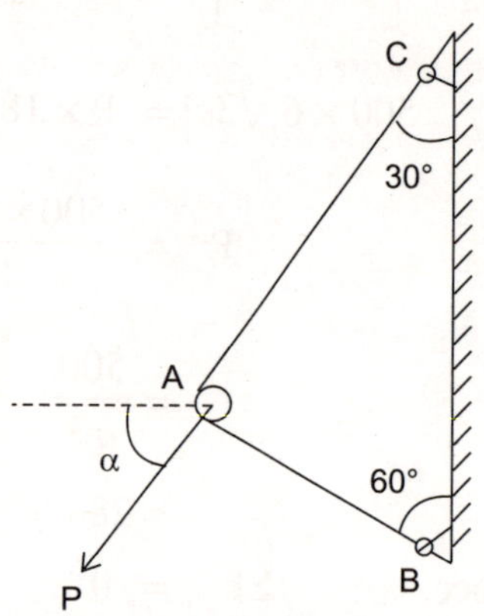

Fig. 8-ECC1

2. One end of the rod AB rests at the corner 'A' and the other end is attached to the cord BC. If the rod supports 200 N load at D. Find the reaction at A and tension in the chord. (Refer Fig. 8-ECC2).

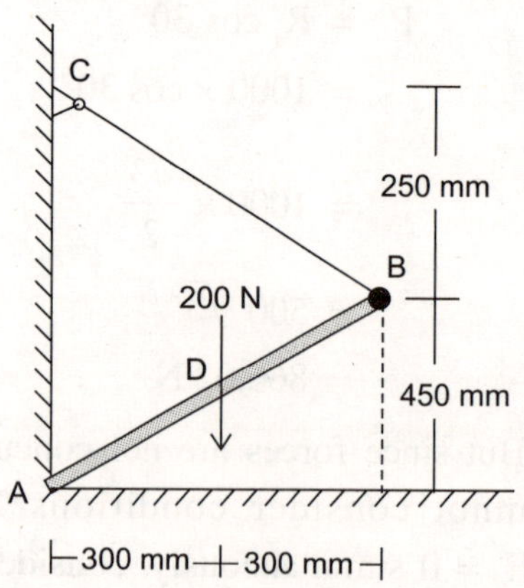

Fig. 8-ECC2

3. Explain different systems of forces with examples.

4. A system of forces is shown in Fig. (8-ECC3). Find the magnitude of 'F' if the resultant of the system R = 100 kN is acting horizontally.

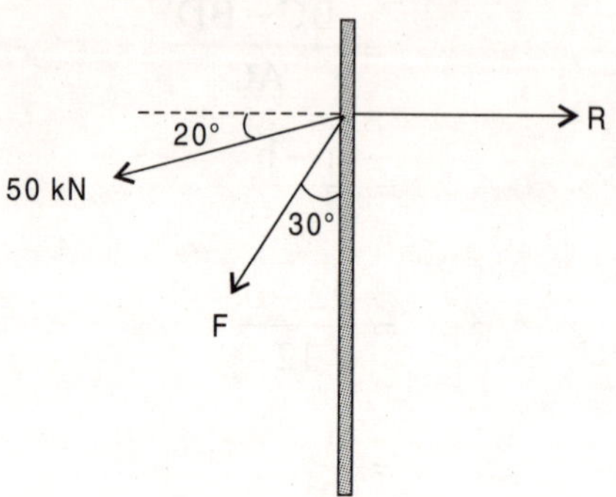

Fig. 8-ECC3

5. Distinguish between (i) coplanar and noncoplanar force system; (ii) resultant and equilibrium of the force system.

6. (a) What is the significance of a free body diagram and give two examples?
   (b) Two cables are connected at A and B as shown in Fig. (8-ECC4), and a force of 20 kN is applied at C. Determine forces on the cables along CA and CB.

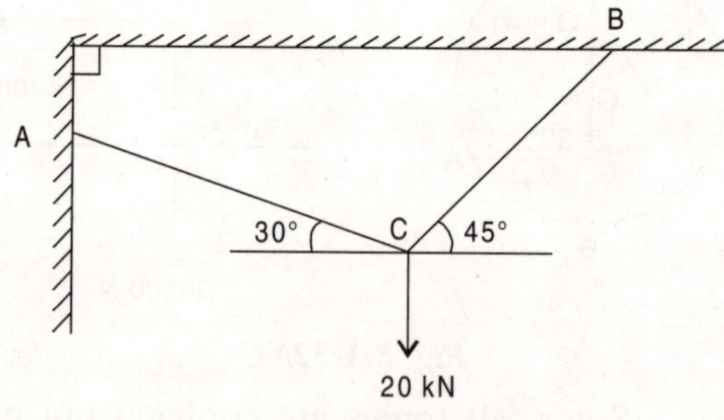

Fig. 8-ECC4

7. (a) State and explain the principle of transmissibility of a force.
   (b) Mention the characteristics of a force and a couple.

8. State the conditions for equilibrium of a rigid body subjected to forces in (i) concurrent three-dimensional space oriented and (ii) nonconcurrent three-dimensional space oriented.

9. Find the value of W which is required to maintain equilibrium of system shown in Fig. (8-ECC5).

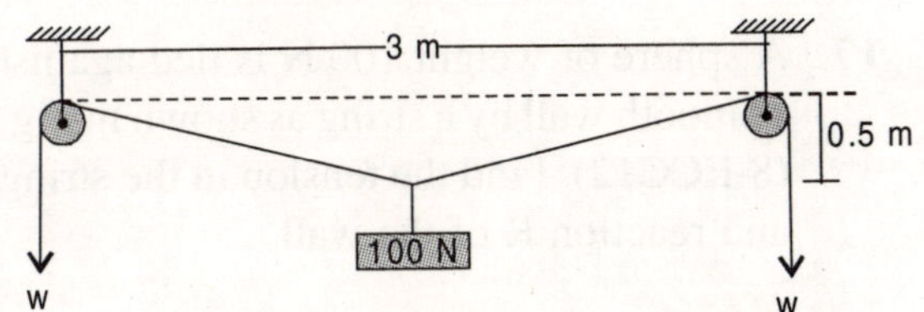

Fig. 8-ECC5

10. Find the resultant of system of forces shown and locate the resultant.

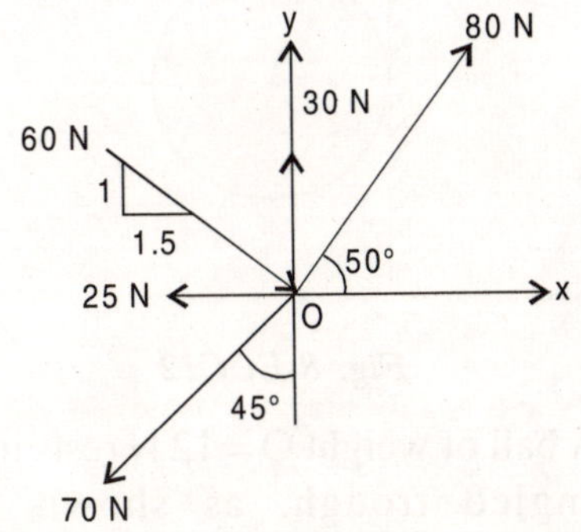

Fig. 8-ECC6

11. Three spheres A, B and C having their diameters 500 mm, 500 mm and 800 mm respectively, are placed in a trench with smooth side walls and floor as shown in Fig. (8-ECC7). The centre-to-centre distance of spheres A and B is 600 mm. The weights of the spheres A, B and C are 4 kN, 4 kN and 8 kN respectively. Determine the reactions at P, Q, R and S.

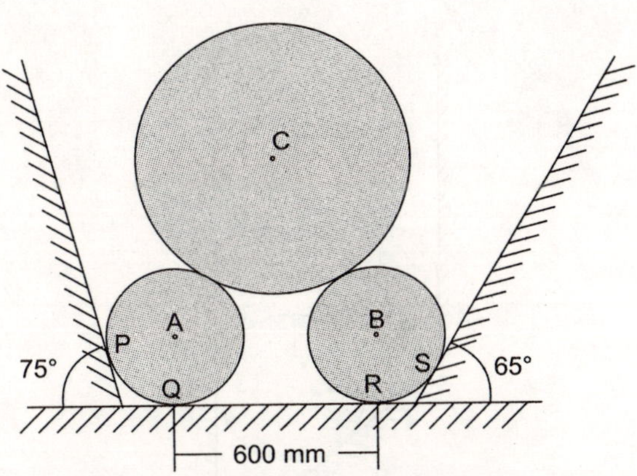

Fig. 8-ECC7

12. Three uniform, homogeneous and smooth spheres A, B and C weighing 300 N, 600 N and 300 N, respectively and having diameters 800 mm, 1200 mm and 800 mm respectively, are placed in a trench as shown in Fig. (8-ECC8). Determine the reactions at the contact points P, Q, R and S.

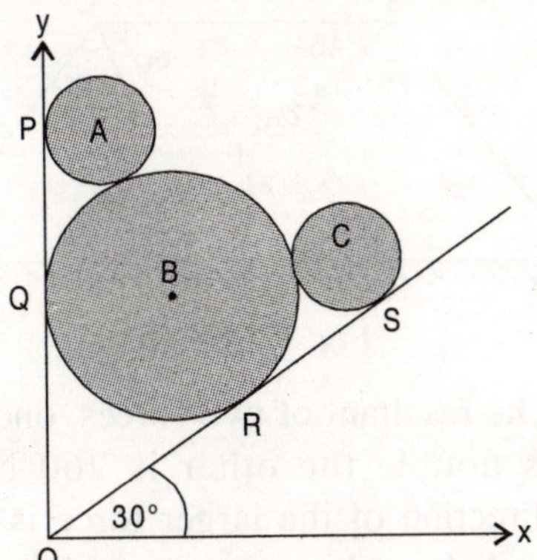

Fig. 8-ECC8

13. Two cylinders of diameters 100 mm and 50 mm, weighing 200 N and 50 N respectively are placed in a trough as shown in Fig. (8-ECC9). Neglecting friction, find the reactions at contact surfaces 1, 2, 3 and 4.

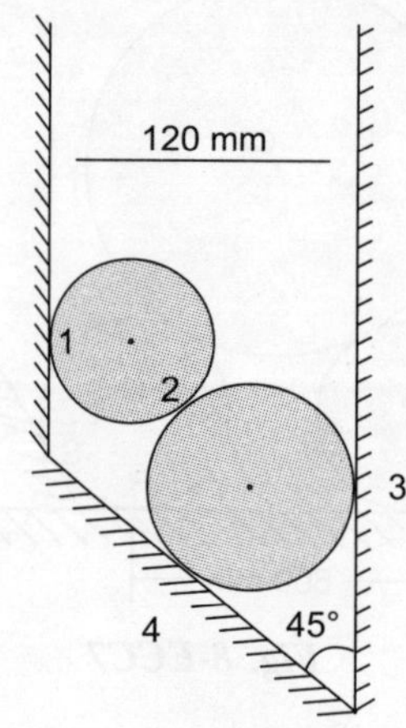

Fig. 8-ECC9

14. Three bars, hinged at A and D and pinned at B and C as shown in Fig. (8-ECC10) form a four-linked mechanism. Determine the value of P that will prevent movement of bars.

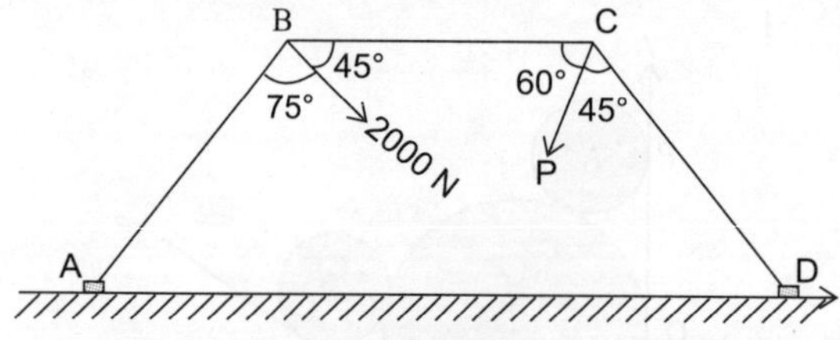

Fig. 8-ECC10

15. The resultant of two forces, one of them is double the other is 260 N. If the direction of the larger force is reversed and the other remains unaltered, the resultant reduces to 180 N. Determine the magnitude of the forces and the angle between the forces.

16. Determine the horizontal force P to be applied to a block of weight 1500 N to hold it in position on a smooth inclined plane AB which makes an angle of 30° with the horizontal Fig. (8-ECC11).

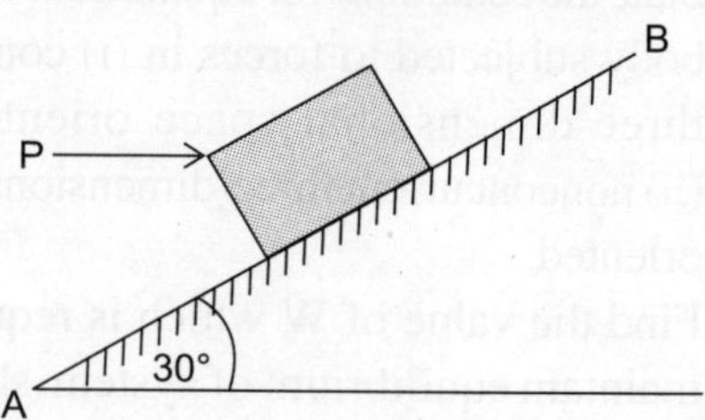

Fig. 8-ECC11

17. A sphere of weight 100 N is tied against a smooth wall by a string as shown in Fig. (8-ECC12). Find the tension in the string and reaction R of the wall.

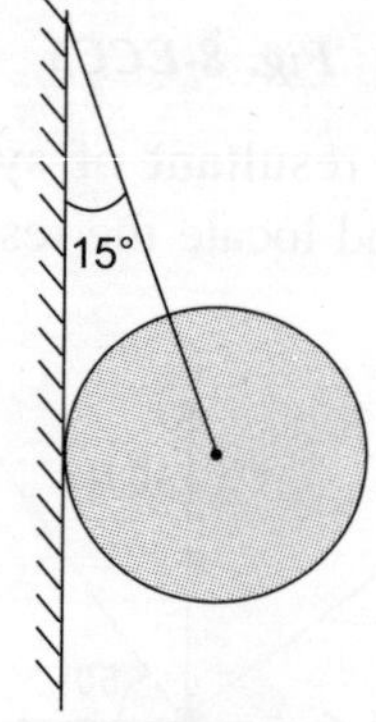

Fig. 8-ECC12

18. A ball of weight Q = 12 N rests in a right-angled trough, as shown in Fig. (8-ECC13). Determine the forces exerted on the sides of the trough at D and E if all surfaces are perfectly smooth.

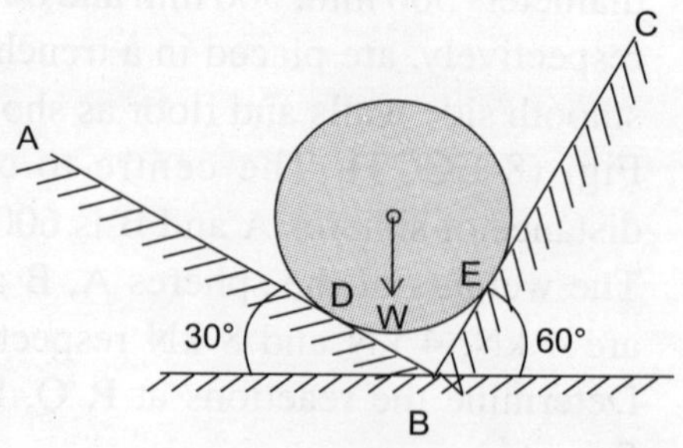

Fig. 8-ECC13

19. What axial force does the vertical load P induce in the members of system shown in Fig. (8-ECC14). Neglect the weights of the members themselves and assume an ideal hinge at A and a perfectly flexible sting BC.

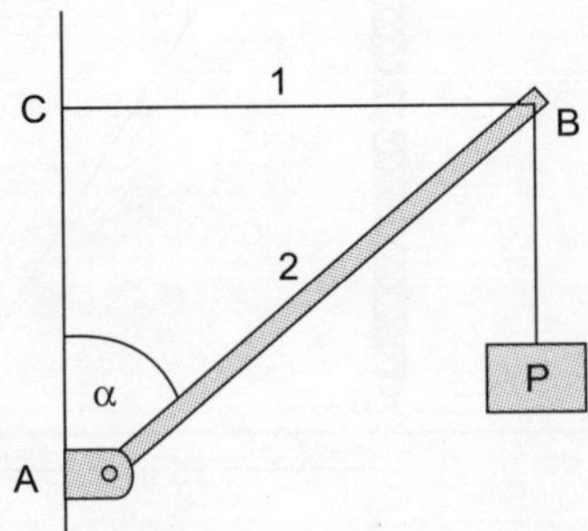

Fig. 8-ECC14

20. A right circular roller of weight W rests on a smooth horizontal plane and is held a in position by an inclined bar AC as shown in Fig. (8-ECC15). Find the tension S in the bar AC and vertical reaction R_b at B if there is also a horizontal force P acting at C.

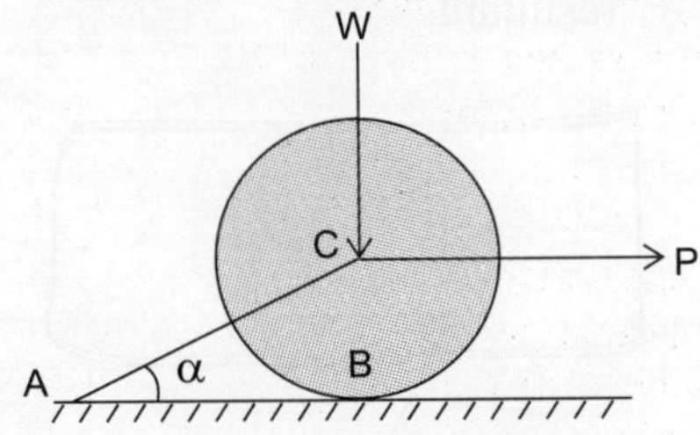

Fig. 8-ECC15

21. Two smooth circular cylinders each of weight W = 100 N and radius r = 6 mm are connected at their centres by a string AB of length l = 16 mm, and rest on a horizontal plane, suporting a third cylinder of weight Q = 200 N and radius r = 6 mm above them as shown in Fig. (8-ECC16) and radius r = 6 mm. Find the force 'S' in the string AB and the pressures produced on the points of contact D and E.

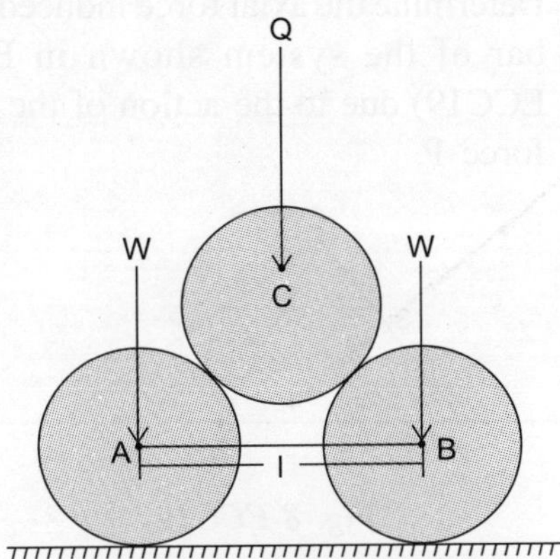

Fig. 8-ECC16

22. A weight Q is suspended to point B of a cord ABC, the ends of which are pulled by equal weights P overhanging small pulleys A and C which are on the same level Fig. (8-ECC17). Neglecting the radii of the pulleys, determine the sag BD if l = 12 cm, P = 20 N, and Q = 10 N.

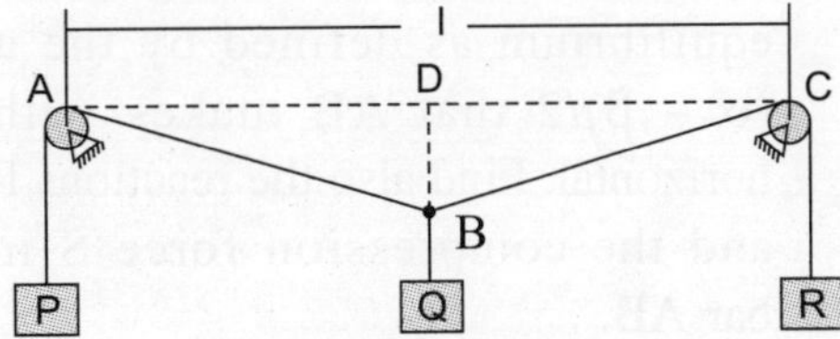

Fig. 8-ECC17

23. A hinged square ABCD Fig. (8-ECC18) with diagonal BD is subjected to the action of two equal and opposite forces P applied as shown. Determine the forces induced in all bars.

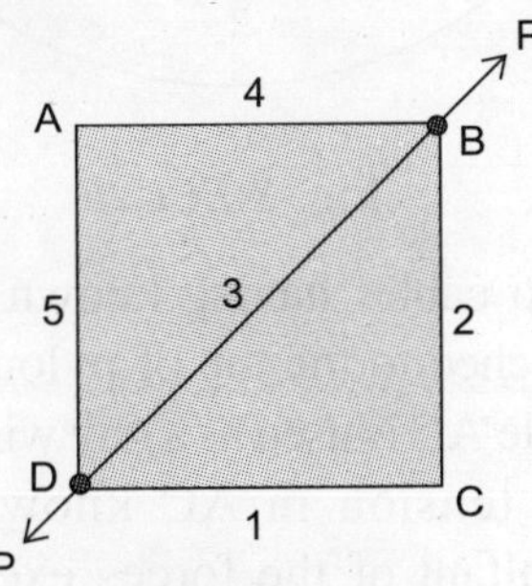

Fig. 8-ECC18

24. Determine the axial force induced in each bar of the system shown in Fig. (8-ECC19) due to the action of the applied force P.

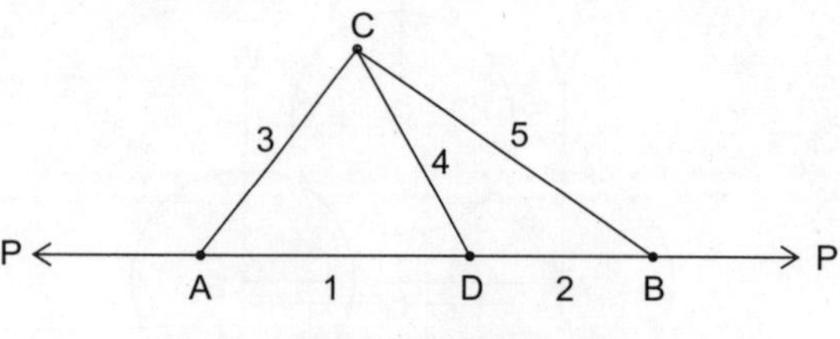

Fig. 8-ECC19

25. A rigid bar AB with rollers of weights P = 50 N and Q = 100 N at it's ends is supported inside a circular ring in a vertical plane as shown in Fig. (8-ECC20). The radius of the ring and the length AB are such that AC and BC form a right angle at C; that is, $\alpha + \beta = 90°$. Neglect the friction and the weight of the bar AB. Find the configuration of equilibrium as defined by the angle $(\alpha - \beta)/2$ that AB makes with the horizontal. Find also the reactions R_a, R_b and the compression force S in the bar AB.

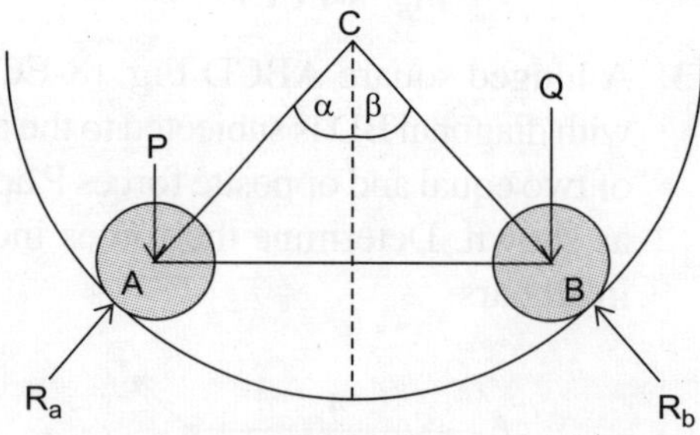

Fig. 8-ECC20

26. Two cables having known tensions, are attached to the top of pylon AB. A third cable AC is used as a guy wire. Determine the tension in AC knowing that the resultant of the forces exerted at A by the three cables must be vertical.

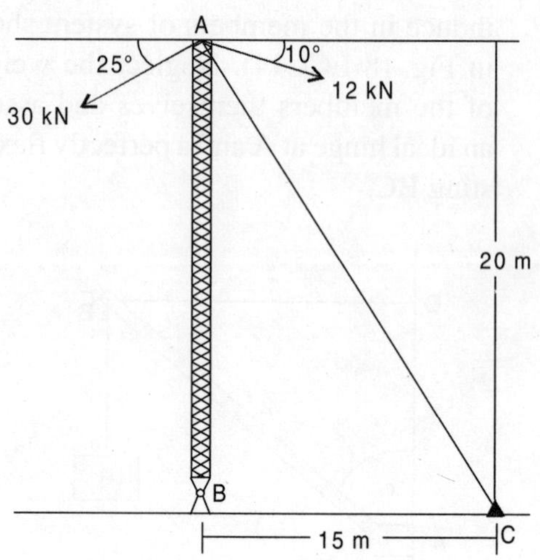

Fig. 8-ECC21

27. A hoist trolley is subjected to the three forces as shown in Fig. (8-ECC22). Determine

(a) the value of the angle α for which the resultant of the forces is vertical.

(b) the corresponding magnitude of the resultant.

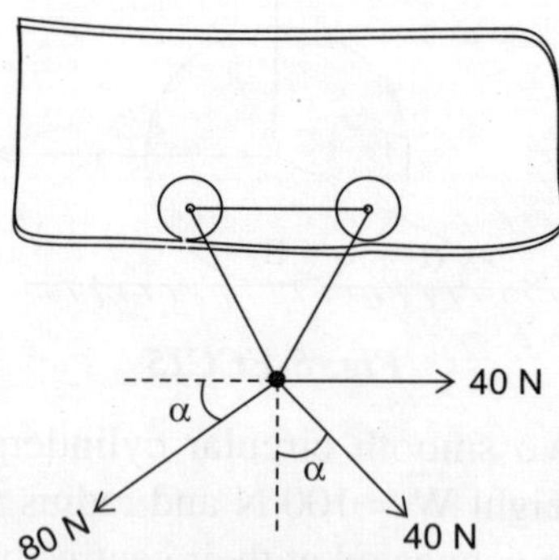

Fig. 8-ECC22

28. A part of the design of a new sail boat, it is desired to determine the drag force which may be expected at a given speed. To do so, a model of proposed hull is placed in a test channel and three cables are used to keep its bow on the center line of channel. Dynamometer reading indicates that for a given speed, the tension is 40 N in the cable AB and 60 N in cable AE. Determine the drag force exerted on the hull and the tension in cable AC.

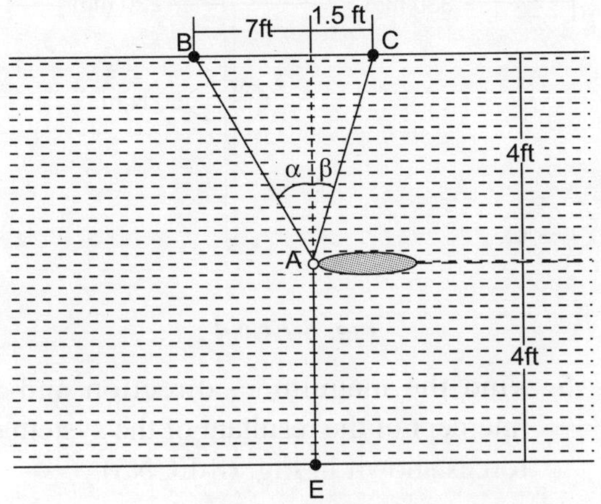

Fig. 8-ECC23

29. Two cables are tied together at C and loaded as shown in Fig. (8-ECC24). Knowing P = 400 N and α = 75°, determine the tensions in AC and BC.

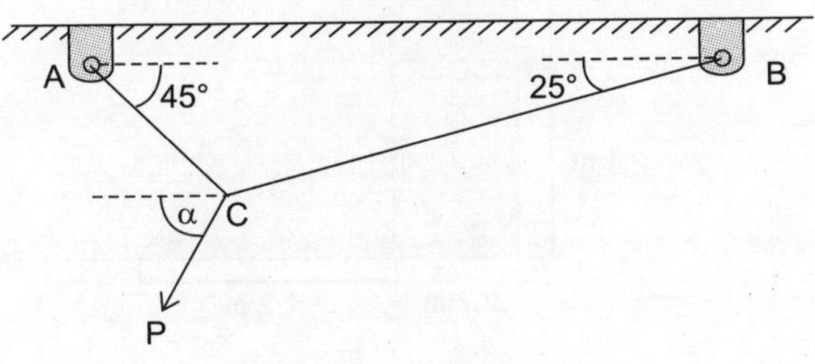

Fig. 8-ECC24

30. A string ABCD attached to two fixed points A and D has two equal weights of 500 N attached to it at B and C. The weights rest with portions AB and CD inclined at angles of 30° and 60° respectively to vertical as shown in Fig. (8-ECC25). Find the tensions in the portions AB, BC and CD of the string. The inclination of the portion BC with vertical is 120°.

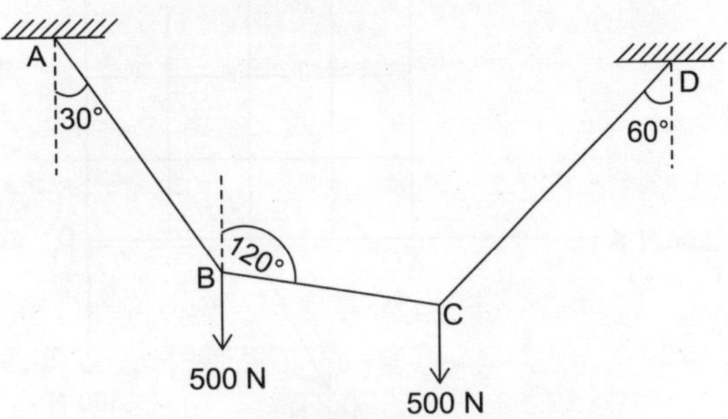

Fig. 8-ECC25

31. A 26 kN force is the resultant of two forces. One of them is as shown in Fig. (8-ECC26). Determine the other force.

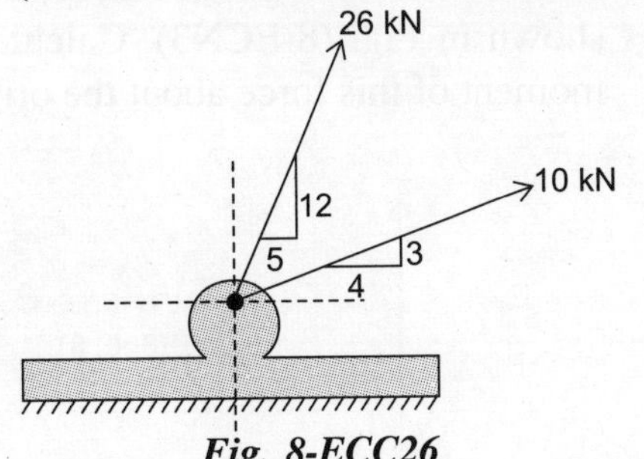

Fig. 8-ECC26

EXERCISE

Coplanar Nonconcurrent system

1. A coplanar parallel force system consisting of three forces act on a rigid bar as shown in Fig. (8-ECN1). Determine the single force and its location to give equivalent action for the system of forces.

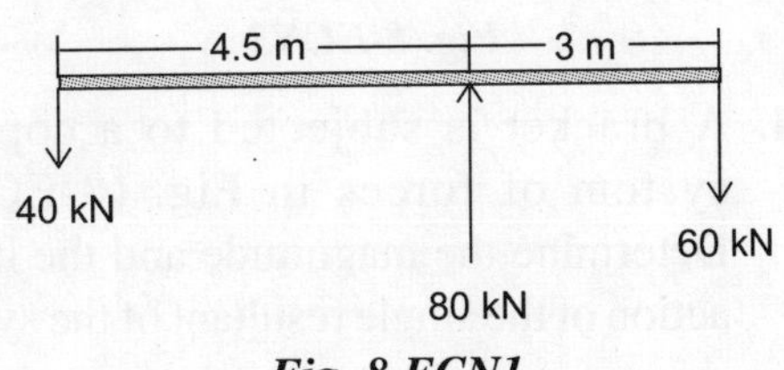

Fig. 8-ECN1

2. A flat plate is subjected to the coplanar system of forces as shown in Fig. (8-ECN2). Each square of the inscribed grid is having length of 1 m. Determine the resultant by vector approach.

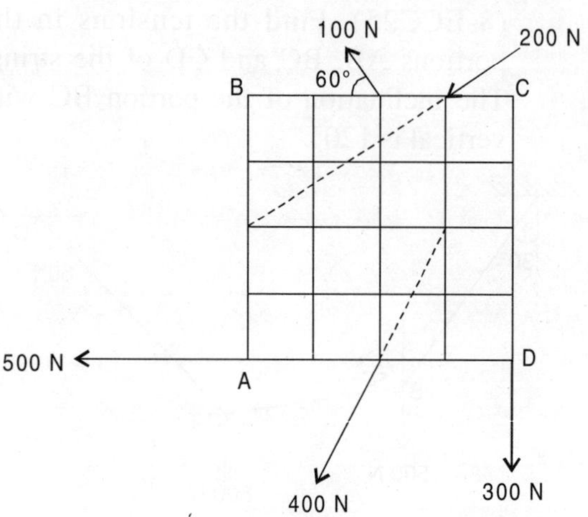

Fig. 8-ECN2

3. A force of 140 N passes through a point C (– 6, 2, 2) and goes to B (6, 6, 8) as shown in Fig. (8-ECN3). Calculate the moment of this force about the origin .

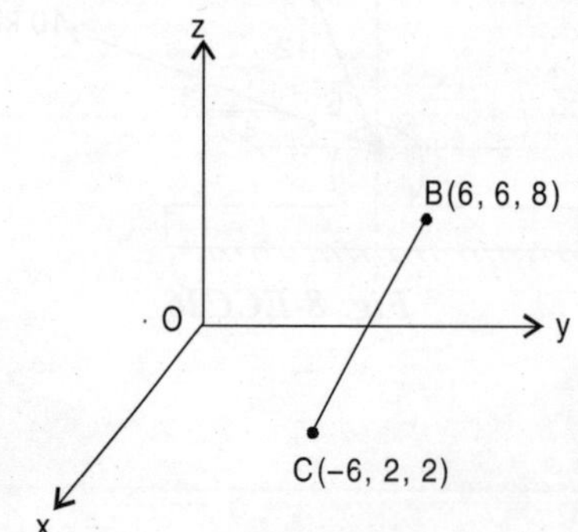

Fig. 8-ECN3

4. A bracket is subjected to a coplanar system of forces in Fig. (8-ECN4). Determine the magnitude and the line of action of the single resultant of the system.

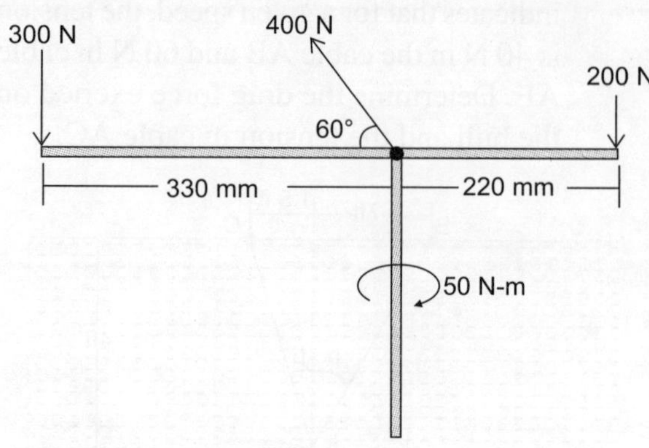

Fig. 8-ECN4

5. Find the magnitude, direction and x-intercept of the resultant of the system of forces shown in Fig. (8-ECN5).

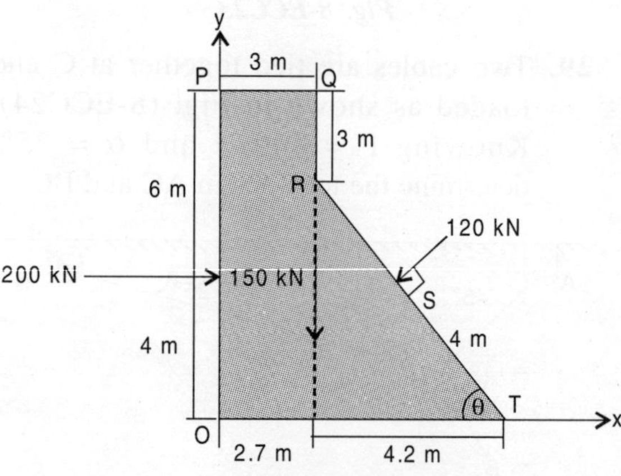

Fig. 8-ECN5

6. Find the sum of moments of the three forces shown in figure (8-ECN6) about A.

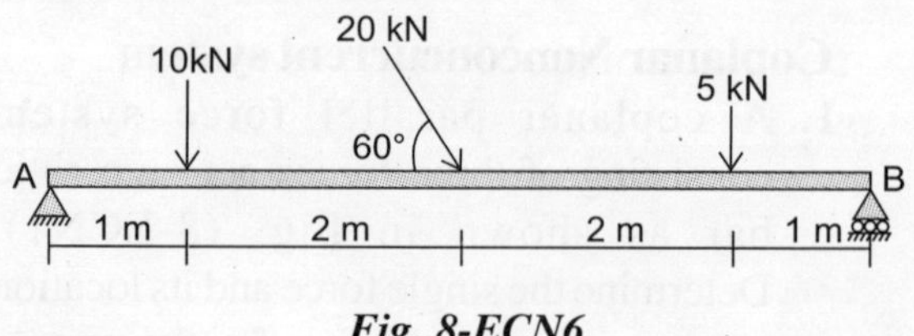

Fig. 8-ECN6

7. Replace the 80 N force at A by an equivalent force and couple moment at the origin.

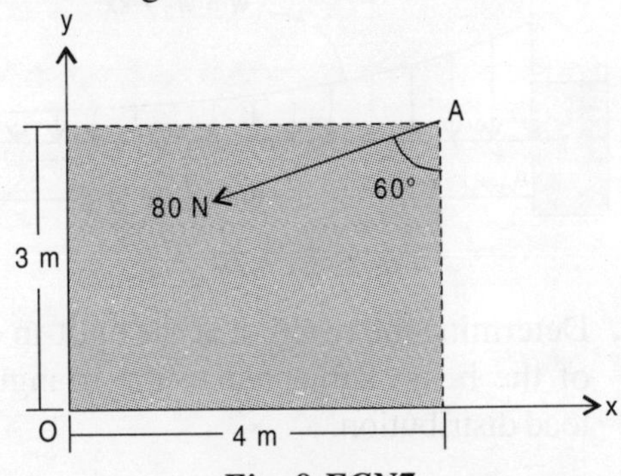

Fig. 8-ECN7

8. A force and couple act as shown in Fig. (8-ECN8) on a square plate of side a = 625 mm, knowing that p = 300 N, Q = 200 N and α = 50°, replace the given force and couple by a single force applied at a point located on line AB.

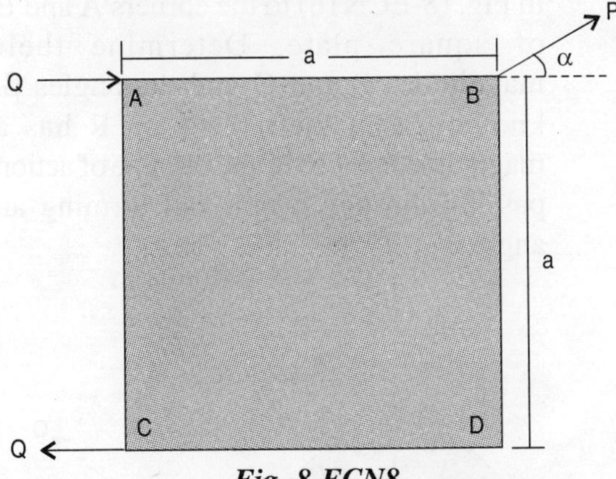

Fig. 8-ECN8

9. Find the equilibrium with respect to point A for the system of forces shown in Fig. (8-ECN9). On a flat plate, the square may be assumed to be of unit size.

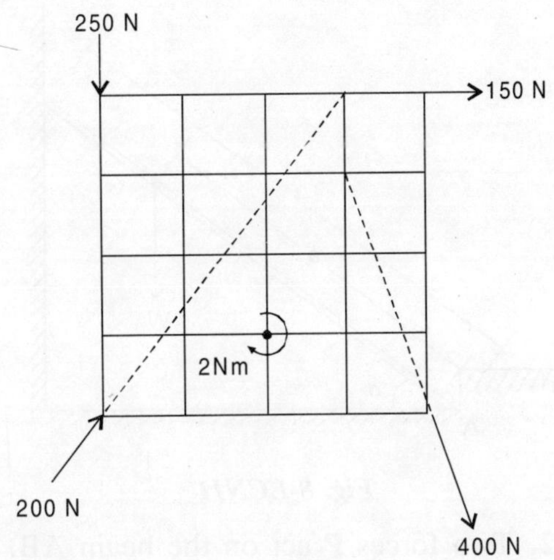

Fig. 8-ECN9

10. Find the magnitude and position of resultant for the coplanar system of forces shown in Fig. (8-ECN10)

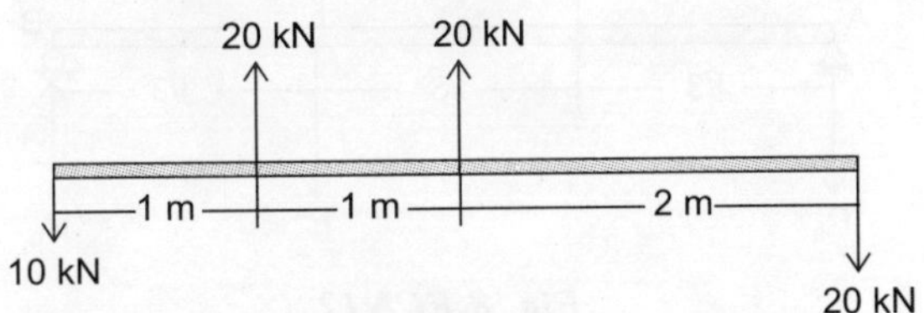

Fig. 8-ECN10

11. Along a ladder supported at A and B as shown in Fig. (8-ECN11). A vertical load W can have any position as defined by the distance 'a' from the bottom. Neglecting friction, determine the magnitude of reaction R_b at B. Neglect the weight of ladder.

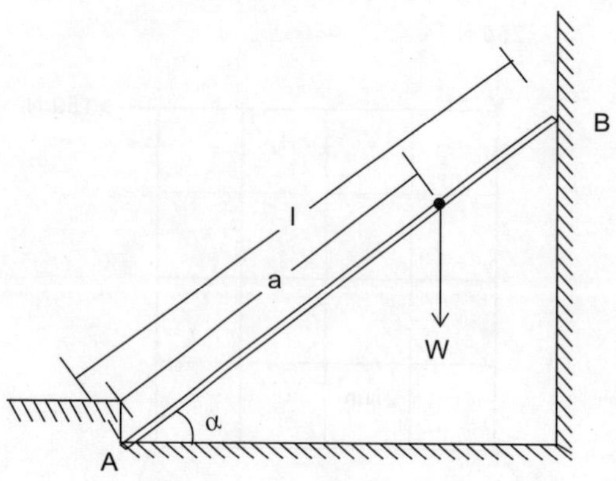

Fig. 8-ECN11

12. Two forces P act on the beam AB, as shown in Fig. (8-ECN12). Determine graphically the reaction R_a and R_b at supports.

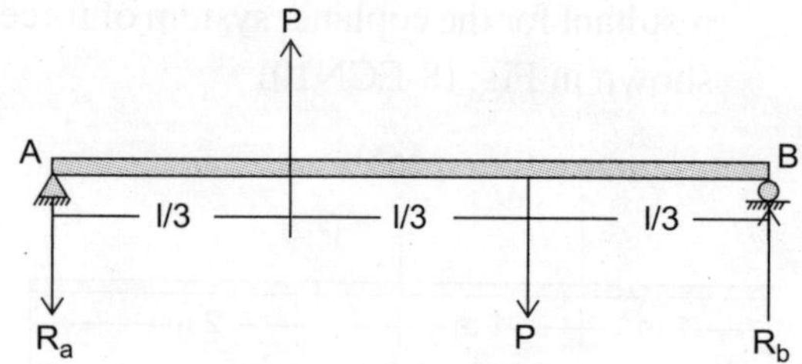

Fig. 8-ECN12

13. A beam is subjected to the variable loading as shown in Fig. (8-ECN13), calculate the support reactions at A and B.

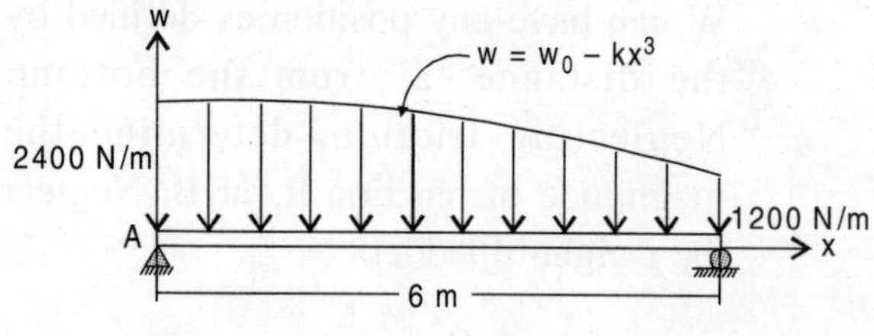

Fig. 8-ECN13

14. A cantilever beam supports the variable load as shown in Fig. (8-ECN14), calculate the supporting force R_A and moment M_A at A.

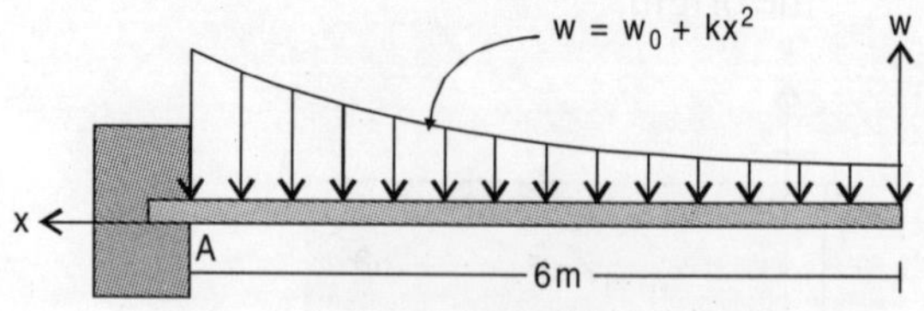

Fig. 8-ECN14

15. Determine the reaction at the built-in end of the beam subjected to the triangular load distribution.

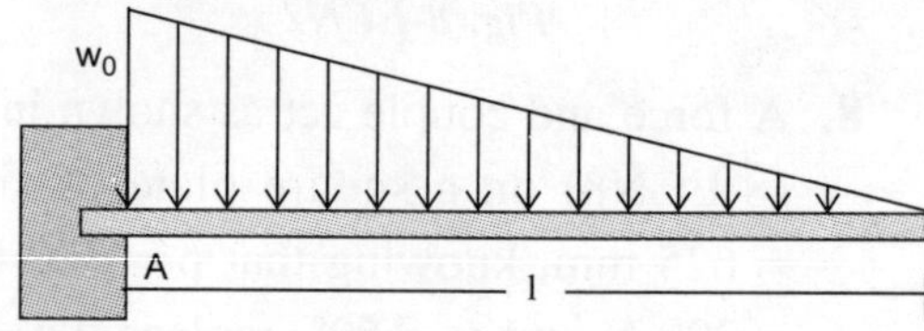

Fig. 8-ECN15

16. Two forces P and Q are applied as shown in Fig. (8-ECN16) to the corners A and B of square plate. Determine their magnitudes P and Q and the angles β, knowing that their resultant R has a magnitude R = 120 N and the line of action passing through origin and forming an angle θ = 30° with the x-axis.

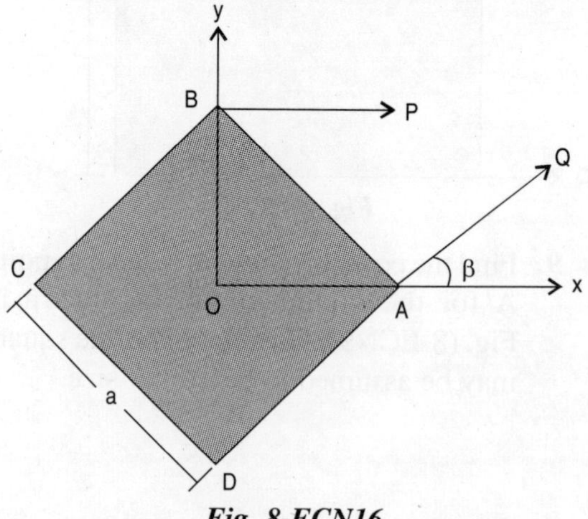

Fig. 8-ECN16

17. A man raises a 10 kg joist of length 4 m, by pulling with a rope. Find the tension T in the rope and the reaction at A.

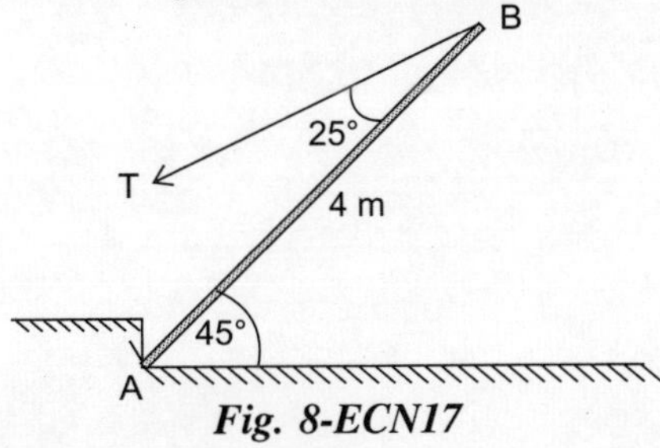

Fig. 8-ECN17

18. Determine the reaction at the supports of the beam for the given loading condition.

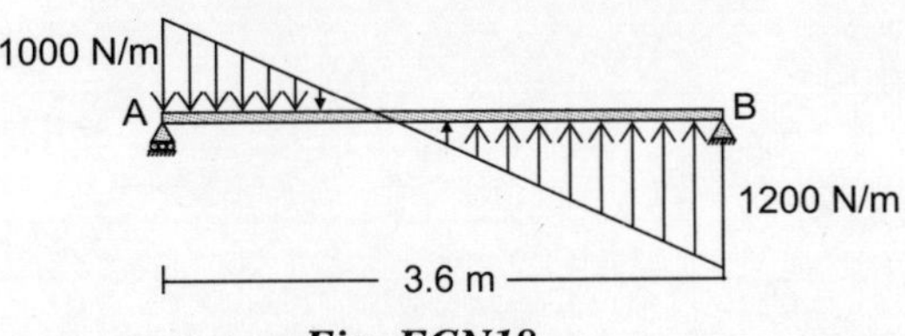

Fig. ECN18

19. Determine the resultant of the three forces on the dam section shown in Fig. (8-ECN19) and locate its intersection with the base AB. For a safe design this intersection should occur within the middle third, is it a safe design?

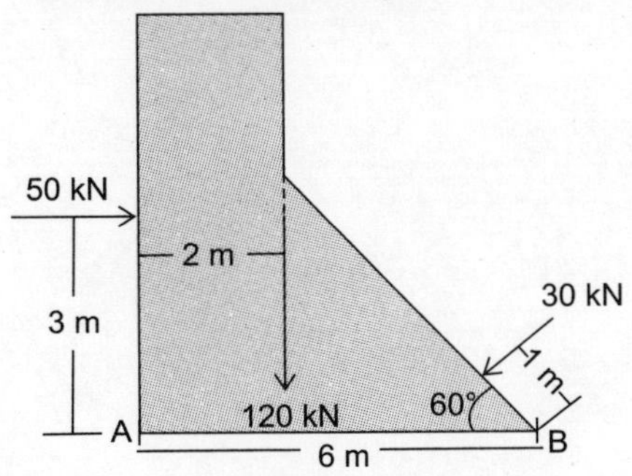

Fig. 8-ECN19

20. Beam AB 8.5 m long, is hinged at A and is supported on roller at B, the roller support is inclined at 45° to the horizontal. Find the reaction at A and B, if the loads on it are as shown in Fig. (8-ECN20).

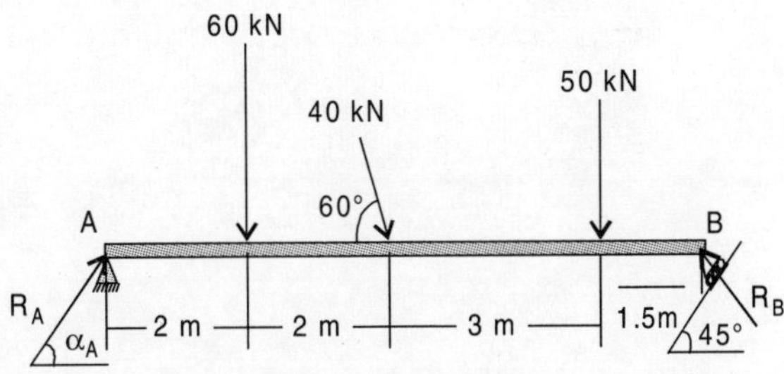

Fig. 8-ECN20

21. Determine the tension in the string BC and the reaction at the hinged support D for the beam ABD shown in Fig. (8-ECN21) is in equilibrium.

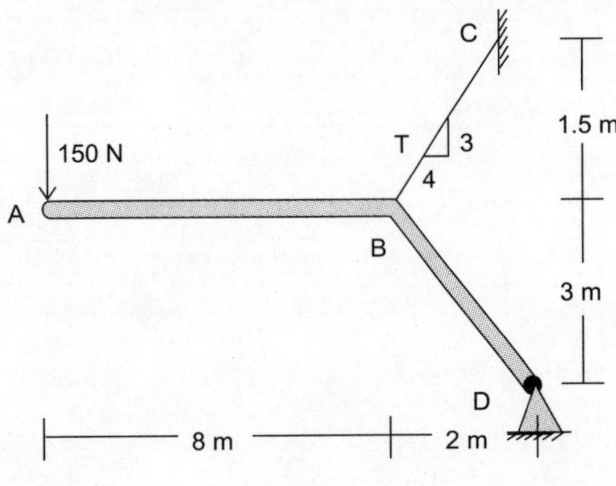

Fig. 8-ECN21

APPENDICES

1. SOME VECTOR FORMULAE

(i) $\mathbf{a} \cdot (\mathbf{b} + \mathbf{c}) = \mathbf{a} \cdot \mathbf{b} + \mathbf{a} \cdot \mathbf{c}$

(ii) $\mathbf{a} \times (\mathbf{b} + \mathbf{c}) = \mathbf{a} \times \mathbf{b} + \mathbf{a} \times \mathbf{c}$

(iii) $\mathbf{a} \cdot \mathbf{b} = a_x b_x + a_y b_y + a_z b_z$

(iv) $$\mathbf{a} \times \mathbf{b} = \begin{vmatrix} \mathbf{i} & \mathbf{j} & \mathbf{k} \\ a_x & a_y & a_z \\ b_x & b_y & b_z \end{vmatrix} = -\mathbf{b} \times \mathbf{a}$$

Note : That the constant unit vectors **i, j** and **k** obey the circular permutation.

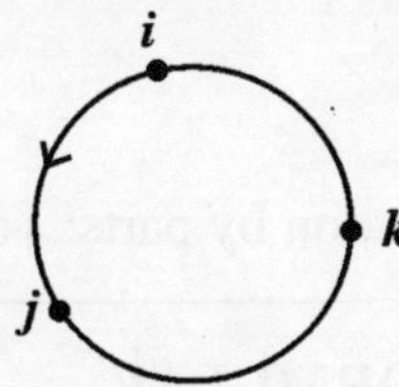

(v) $$\mathbf{a} \cdot (\mathbf{b} \times \mathbf{c}) = \begin{vmatrix} a_x & a_y & a_z \\ b_x & b_y & b_z \\ c_x & c_y & c_z \end{vmatrix}$$
$$= \mathbf{b} \cdot (\mathbf{c} \times \mathbf{a}) = \mathbf{c} \cdot (\mathbf{a} \times \mathbf{b})$$

Note: *That the scalar triple product obeys the circular permutation.*

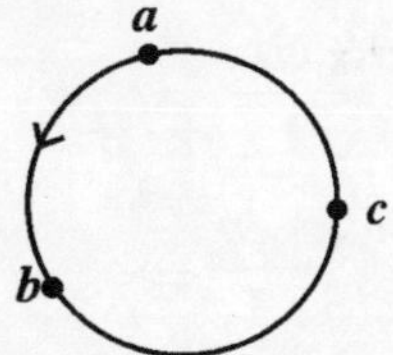

(vi) $\mathbf{a} \times (\mathbf{b} \times \mathbf{c}) = (\mathbf{a} \cdot \mathbf{c})\, \mathbf{b} - (\mathbf{a} \cdot \mathbf{b})\, \mathbf{c}$

(vii) $$\frac{d(\mathbf{a}+\mathbf{b})}{dt} = \frac{d\mathbf{a}}{dt} + \frac{d\mathbf{b}}{dt}$$

(viii) $$\frac{d}{dt}(k\mathbf{a}) = k\frac{d\mathbf{a}}{dt}$$

(ix) $$\frac{d}{dt}(\mathbf{a} \cdot \mathbf{b}) = \mathbf{b} \cdot \frac{d\mathbf{a}}{dt} + \mathbf{a} \cdot \frac{d\mathbf{b}}{dt}$$
$$= \frac{d\mathbf{a}}{dt} \cdot \mathbf{b} + \mathbf{a} \cdot \frac{d\mathbf{b}}{dt}$$

(x) $$\frac{d}{dt}(\mathbf{a} \times \mathbf{b}) = \frac{d\mathbf{a}}{dt} \times \mathbf{b} + \mathbf{a} \times \frac{d\mathbf{b}}{dt}$$

2. DERIVATIVES AND INTEGRALS

Functions	Derivatives	Functions	Integrals
x^n	$n\,x^{n-1}$	$x^n\,dx$	$\frac{x^{n+1}}{n+1}$ $(n \neq -1)$
e^{nx}	ne^{nx}	$\frac{dx}{x}$	$\ln\lvert x \rvert$
a^x	$a^x \ln a$	$\sin x\,dx$	$-\cos x$
$\ln x$	$\frac{1}{x}$	$\cos x\,dx$	$\sin x$
$\sin x$	$\cos x$	$\tan x\,dx$	$\ln\lvert\sec x\rvert$
$\cos x$	$-\sin x$	$\cot x\,dx$	$\ln\lvert\sin x\rvert$
$\tan x$	$\sec^2 x$	$\frac{dx}{\cos^2 x}$	$\tan x$
$\cot x$	$-\operatorname{cosec}^2 x$	$\operatorname{cosec}^2 x\,dx$	$-\cot x$

(Contd.)

Functions	Derivatives	Functions	Integrals
$\sin^{-1} x$	$\frac{1}{\sqrt{1-x^2}}$	$e^x\,dx$	e^x
$\cos^{-1} x$	$-\frac{1}{\sqrt{1-x^2}}$	$\frac{dx}{1+x^2}$	$\tan^{-1} x$
$\tan^{-1} x$	$\frac{1}{1+x^2}$	$\frac{dx}{\sqrt{1-x^2}}$	$\sin^{-1} x$
$\cot^{-1} x$	$-\frac{1}{1+x^2}$	$-\frac{dx}{\sqrt{1-x^2}}$	$\cos^{-1} x$
$\sqrt{u}$	$\frac{u'}{2\sqrt{u}}$	$\frac{dx}{\sqrt{x^2-1}}$	$\ln\left(x+\sqrt{x^2+1}\right)$
$\ln u$	$\frac{u'}{u}$		
$\frac{u}{v}$	$\frac{vu'-v'u}{v^2}$	Integration by parts: $\int u\,dv = uv - \int v\,du$	

3. BASIC TRIGONOMETRICAL FORMULAE

$$\sin(\alpha \pm \beta) = \sin\alpha\cos\beta \pm \cos\alpha\sin\beta$$

$$\cos(\alpha \pm \beta) = \cos\alpha\cos\beta \mp \sin\alpha\sin\beta$$

$$\tan(\alpha \pm \beta) = \frac{\tan\alpha \pm \tan\beta}{1 \mp \tan\alpha\tan\beta}$$

$$\cot(\alpha \pm \beta) = \frac{\cot\alpha\cot\beta \mp 1}{\cot\beta \pm \cot\alpha}$$

$$\sin\alpha = \frac{1}{\sqrt{1+\cot^2\alpha}}$$

$$\cos\alpha = \frac{1}{\sqrt{1+\tan^2\alpha}}$$

$$\sin 2\alpha = 2\sin\alpha\cos\alpha$$

$$\cos 2\alpha = \cos^2\alpha - \sin^2\alpha = 2\cos^2\alpha - 1 = 1 - 2\sin^2\alpha$$

$$\tan 2\alpha = \frac{2\tan\alpha}{1-\tan^2\alpha}$$

$$\cot 2\alpha = \frac{\cot^2\alpha - 1}{2\cot\alpha}$$

$$\sin\alpha + \sin\beta = 2\sin\frac{\alpha+\beta}{2}\cos\frac{\alpha-\beta}{2}$$

$$\sin\alpha - \sin\beta = 2\cos\frac{\alpha+\beta}{2}\sin\frac{\alpha-\beta}{2}$$

$$\cos\alpha + \cos\beta = 2\cos\frac{\alpha+\beta}{2}\cos\frac{\alpha-\beta}{2}$$

$$\cos\alpha - \cos\beta = -2\sin\frac{\alpha+\beta}{2}\sin\frac{\alpha-\beta}{2}$$

$$\tan\alpha \pm \tan\beta = \frac{\sin(\alpha \pm \beta)}{\cos\alpha\cos\beta}$$

$$\cot\alpha \pm \cot\beta = \pm\frac{\sin(\alpha \pm \beta)}{\sin\alpha\sin\beta}$$

$$\sin\frac{\alpha}{2} = \sqrt{\frac{1-\cos\alpha}{2}}$$

$$\cos\frac{\alpha}{2} = \sqrt{\frac{1+\cos\alpha}{2}}$$

$$2\sin\alpha\sin\beta = \cos(\alpha-\beta) - \cos(\alpha+\beta)$$

$$2\cos\alpha\cos\beta = \cos(\alpha-\beta) + \cos(\alpha+\beta)$$

$$2\sin\alpha\cos\beta = \sin(\alpha-\beta) + \sin(\alpha+\beta)$$

$$2\sin\beta\cos\alpha = \sin(\alpha+\beta) - \sin(\alpha-\beta)$$

4. NUMERICAL CONSTANTS AND APPROXIMATIONS

Numerical constants	Approximate formulae (for $\alpha << 1$)
$\pi = 3.1416$	$(1 \pm \alpha)^n \approx 1 \pm n\alpha$
$e = 2.7183$	$e^{\alpha} \approx 1 + \alpha$
$\log e = 0.4343$	$\ln(1 + \alpha) \approx \alpha$
$\ln 10 = 2.3026$	$\sin\alpha \approx \alpha,\ \cos\alpha \approx 1 - \frac{\alpha^2}{2},\ \tan\alpha \approx \alpha$

5. GREEK ALPHABETS

α	alpha	ν	nu
β	beta	ξ	xi
γ	gamma	o	omicron
δ, Δ	delta	π	pi
ε	epsilon	ρ	rho
ζ	zeta	σ	sigma
η	eta	τ	tau
θ	theta	ϕ, Φ	phi
ι	iota	χ	chi
κ	kappa	ψ	psi
λ	lambda	ω, Ω	omega
μ	mu		

6. ASTRONOMICAL DATA (the Sun, the Earth and the Moon)

Property	Unit	Sun[a]	Earth	Moon
Mass	kg	1.99×10^{30}	5.98×10^{24}	7.36×10^{22}
Mean radius	m	6.96×10^{8}	6.37×10^{6}	1.74×10^{6}
Mean density	kg/m^3	1410	5520	3340
Surface gravity	m/s^2	274	9.81	1.67
Escape velocity	km/s	618	11.2	2.38
Period of rotation[c]	–	37 days-poles[b] 26 days-equator	23 hr 56 min.	27.3 days

[a]the Sun radiates energy at the rate of 3.90×10^{26} W; assuming normal incidence, just outside the Earth's atmosphere, at the rate of 1340 W/m^2.

[b]the Sun, a ball of gas, does not rotate as a rigid body.

[c]Measured with respect to the distant stars.

Some distances from the Earth:

To the moon*	3.82×10^{8} m
To the sun	1.50×10^{11} m
To the nearest star (Proxima Centauri)	4.04×10^{16} m
To the center of our galaxy	2.2×10^{20} m

* Mean distance

7. DENSITY OF SUBSTANCES (in increasing order)

Gases at S.T.P.			
	g/cm³		
Hydrogen	0.09	Sodium	0.97
Methane	0.72	Graphite	1.6
Ammonia	0.77	Porcelain	2.3
Nitrogen	1.25	Aluminium	2.7
Air	1.293	Diamond	3.5
Oxygen	1.43	Titanium	4.5
Carbon dioxide	1.98	Zinc	7.0
Chlorine	3.21	Tin	7.4
Liquids	**g/cm³**	Iron (steel)	7.8
Ether	0.72	Cadmium	8.65
Alcohol	0.79	Cobalt	8.9
Kerosene oil	0.80	Copper	8.9
Benzene	0.88	Nickel	8.9
Castor oil	0.90	Molibdenum	10.2
Water	1.00	Silver	10.5
Heavy water	1.10	Lead	11.3
Glycerine	1.26	Uranium	19.0
Mercury	13.60	Tungsten	19.1
Solids	**g/cm³**	Gold	19.3
Cork	0.20	Platinum	21.5
Ice	0.916		

8. THERMAL EXPANSION COEFFICIENTS (at room temperature) in increasing order

Solids	**Linear expansion coefficient (α) 10^{-6} K^{-1}**
Common glass	8.5
Steel (Iron)	11
Copper	16.7
Brass	18.9
Aluminium	22.9
Liquids	
Mercury	1.8
Water	2.1
Glycerine	5.0
Kerosene	10.0
Ethyl alcohol	11.0

9. PERMITTIVITY (relative values) in increasing order

Air	1.00058 ≅ 1	Glass	6.0
Kerosene	2.0	Porcelain	6.0
Paraffin	2.0	Mica	7.5
Polyethylene	2.3	Alcohol	26
Ebonite	2.7	Water	81
Plexiglas	3.5		

10. RESISTIVITIES OF CONDUCTORS (in increasing order)

Conductor	Resistivity (at 20°)	Temperature coefficient
	ρ, n Ω m	α, K^{-1}
Silver	15	4.1
Copper	16	4.3
Gold	20	4.0
Aluminium	25	4.5
Tungsten	50	4.8
Iron	90	6.5
Lead	190	4.2

11. REFRACTIVE INDICES (in increasing order)

Substance	Indices
Air	1.00029 ≅ 1
Water	1.33
Glass	1.50
Diamond	2.42

12. DECIMAL PREFIXES

Factor	Prefix	Symbol
10^{12}	tera	T
10^{9}	giga	G
10^{6}	mega	M
10^{3}	kilo	k
10^{2}	hecto	h
10	deca	da
10^{-1}	deci	d
10^{-2}	centi	c
10^{-3}	milli	m
10^{-6}	micro	μ
10^{-9}	nano	n
10^{-12}	pico	p
10^{-15}	femto	f
10^{-18}	atto	a

13. SOME EXTRA SYSTEM UNITS

1 year	$= 3.11 \times 10^7$ s	1 Å	$= 10^{-10}$ m
1 atm	= 101.3 1 k Pa = 760 mm Hg	1 b	$= 10^{-28}$ m^2
		1 e V	$= 1.6 \times 10^{-19}$J
1 bar	= 100 k Pa		$= 1.6 \times 10^{-24}$ erg
1 mm Hg	= 133.3 Pa	1 a.m.u	$= 1.66 \times 10^{-24}$ g
1 Cal	= 4.18 J		$= 9.314 \times 10^2$ MeV
1 atm	= 101.3J	1 Ci (Curie)	$= 3.70 \times 10^{10}$ dis/s

14. CONVERSION FACTORS

Solid angle

1 sphere = 4π steradians = 12.57 steradians

Length

1 meter = 100 cm = 10^{-3} km = 39.37 inch = 3.281 ft = 6.214×10^4 mi

1 nautical mile = 1852 m = 1.151 miles = 6076 ft

1 light year = 9.460×10^{12} km

1 par sec = 3.084×10^{13} km

1 fathom = 6 ft

1 Bohr radius = 5.292×10^{-11} m

1 yard = 3 ft

1 rod = 16.5 ft

1 mil = 10^{-3} in

1 fermi = 10^{-15} m

Mass

1 kilogram = 1000 g = 6.852×10^{-2} slug = 2.205 lb
= 1.102×10^{-3} ton

Density

1 kg/m^3 = 6.243×10^{-2} lb/ft^3

Time

1 second = 3.169×10^{-8} year = 1.15×10^{-5} days
= 2.778×10^{-4} hour = 1.667×10^{-2} min

Force

1 N = 10^5 dyne = 0.2248 lb = 102 gf

Pressure

1 atm = 1.013×10^6 dyne/cm^2 = 76 cm Hg
= 1.013×10^5 Pa = 14.70 lb/in^2 = 216 lb/ft^2

Energy, Work, Heat

1 Btu = 1.055×10^{10} erg = 777.9 ft.lb = 1055 J = 252 Cal.
= 6.585×10^{15} MeV = 1.174×10^{-14} kg = 7.070×10^{12} u

Power

1 Btu/h = 3.929×10^{-4} hp = 6.998×10^{-2} Cal/s = 2.93×10^{-4} kW = 0.293 W

Magnetic flux

1 maxwell = 10^{-8} weber

Magnetic field

1 gauss = 10^{-4} Tesla = 1000 milligauss

15. UNITS AND DIMENSIONS OF PHYSICAL QUANTITIES

Quantity	Common symbol	SI unit	Demension
Mass	m, M	kg	M
Distance	d D	m	L
Time	t	s	T
Area	S, A	m^2	L^2
Volume	V	m^3	L^3
Density	ρ	kg/m^3	$M L^{-3}$
Velocity	u, ν	m/s	$L T^{-1}$
Acceleration	a	m/s^2	$L T^{-2}$
Force	F	N	$M L T^{-2}$
Work	W	J	$M L^2 T^{-2}$
Power	P	W	$M L^2 T^{-3}$
Momentum	p	kg. m/s	$M L T^{-1}$
Angle	θ, α	radian	–
Angular velocity	ω	radian/s	T^{-1}
Angular acceleration	α	$radian/s^2$	T^{-2}
Angular momentum	L	kg. m^2/s	$M L^2 T^{-1}$
Moment of inertia	I	kg. m^2	$M L^2$
Torque	τ	N-m	$M L^2 T^{-2}$
Angular frequency	ω	radian/s	T^{-1}
Frequency	*ν*, f	Hz	T^{-1}
Period	T	s	T
Young's modulus	Y	N/m^2	$M L^{-1} T^{-2}$
Bulk modulus	β	N/m^2	$M L^{-1} T^{-2}$
Shear modulus	η	N/m^2	$M L^{-1} T^{-2}$
Surface tension	T	N/m	$M T^{-2}$

Quantity	Common symbol	SI unit	Demension
Coefficient of viscosity	η	$N.s/m^2$	$ML^{-1}T^{-1}$
Pressure	p	N/m^2, Pa	$M L^{-1} T^2$
Wavelength	λ	m	L
Intensity of wave	I	w/m^2	$M T^{-3}$
Temperature	T	K	K
Specific heat capacity	c	J/kg.K	$L^2 T^{-2} K^{-1}$
Stefan's constant	σ	$W/m^2.K^4$	$M T^3 K^{-4}$
Current	Q	J	$M L^2 T^{-2}$
Heat	k	W/m.K	$M L T^{-3} K^{-1}$
Thermal conductivity	I, i	A	I
Charge	q, Q	C	I T
Current density	j	A/m^2	$I L^{-2}$
Electric conductivity	σ	$1/\Omega.m$	$I^2 T^3 M^{-1} L^{-3}$
Dielectric constant	k		
Electric dipole moment	P	C.m	L I T
Electric field	E	V/m	$M L T^{-1} T^{-3}$
Potential (voltage)	V	V (= J/C)	$M L^2 T^{-1} T^{-3}$
Resistivity	ρ	Ω.m	$L^2 M^2 T^{-3} I^{-2}$
Flux of a vector field	Φ	V.m	$M L^3 I^{-1} T^{-3}$
Capacitance	C	F	$I^2 T^4 M^{-1} T^{-2}$
Electromotive force	ξ	V	$M L^2 I^{-1} T^{-3}$
Resistance	R	Ω	$M L^2 I^{-2} T^{-3}$
Permittivity of a space	ε_0	$C^2/N.m^2 = F/m$	$I^2 T^4 M^{-1} L^{-3}$
Permeability of space	μ_0	N/A^2	$M L I^{-2} T^{-3}$
Magnetic field	B	$T = Wb/m^2$	$M I^{-1} T^{-2}$
Magnetic dipole moment	μ	N.m/T	$I L^2$
Inductance	L	H	$M L^2 I^{-2} T^{-2}$
Luminous intensity	I	Cd	$M L^2 T^{-3}$
Luminous flux	Φ	lm	$M L^2 T^{-3}$
Illumination	E	lux	$M T^{-3}$
Luminance or brightness	B	Cd/m^2	$M T^{-3}$
Radiant intensity	I	W/sr	$M L^2 T^{-3}$

16. UNIVERSAL CONSTANT

Quantity	Symbol	Value	Unit
Gravitational constant	G	6.67259×10^{-11}	$N.m^2/kg^2$
Speed of light in vacuum	c	2.99792458×10^{8}	m/s
Avogadro constant	N_0	6.0221367×10^{23}	mol^{-1}
Gas constant	R	8.314510	J/K.mol
Boltzmann constant	k	1.380658×10^{-23}	J/K
		8.617385×10^{-5}	eV/K
Stefan-Boltzmann constant	σ	5.67051×10^{-8}	$W/m^2.K^4$
Wien's displacement law constant	b	2.897756×10^{-3}	m.K
Elementary charge	e	$1.60217733 \times 10^{-19}$	C
Rest mass of electron	m_e	$9.1093897 \times 10^{-31}$	kg
		$5.48579903 \times 10^{-4}$	
Rest mass of proton	m_p	$1.6726231 \times 10^{-27}$	kg
		1.007276470	μ
Rest mass of neutron	m_n	$1.6749286 \times 10^{-27}$	kg
		1.008664904	μ
Permeability of vacuum	μ_0	$4\pi \times 10^{-7}$	N/A^2
Permittivity of vacuum	ε_0	$\frac{1}{\mu_0 C^2} = 8.8554187817 \times 10^{-12}$	$C^2/N.m^2 = F/m$
Faraday constant	F	96485.3029	C/mol
Planck constant	h	$6.6260755 \times 10^{-34}$	J.s
		$4.1356692 \times 10^{-15}$	eV.s
Rydberg constant	R	$1.097371534 \times 10^{-7}$	m^{-1}
Ground state energy of hydrogen atom	–	13.605698	eV
First Bohr radius	a_0	$5.29177249 \times 10^{-11}$	m
Molar volume of ideal gas at S.T.P.	V_0	22.4	1/mol
Atomic mass unit	a.m.u.	1 a.m.u. = 4.66×10^{-24}	g
e/m_c ratio of electron		1.76×10^{11}	C/kg
Crompton wavelength of an electron	h /mc	3.86×10^{-13}	m

TABLES OF BASIC ELEMENTARY FUNCTIONS

1. Trigonometric functions

α°	sin α	tan α	cot α	cos α	(90–α)°
0	0.0000	0.0000	–	1.000	90
1	0.0175	0.0175	57.3	1.000	89
2	0.0349	0.0349	28.6	0.999	88
3	0.0523	0.0524	19.1	0.999	87
4	0.0697	0.0699	14.3	0.998	86
5	0.0872	0.0875	11.4	0.996	85
6	0.1045	0.1051	9.51	0.995	84
7	0.1219	0.1228	8.14	0.993	83
8	0.139	0.141	7.11	0.990	82
9	0.156	0.158	6.31	0.988	81
10	0.174	0.176	5.67	0.985	80
11	0.191	0.194	5.145	0.982	79
12	0.208	0.213	4.705	0.978	78
13	0.225	0.231	4.331	0.974	77
14	0.242	0.249	4.011	0.970	76
15	0.259	0.268	3.732	0.966	75
16	0.276	0.287	3.487	0.961	74
17	0.292	0.306	3.271	0.956	73
18	0.309	0.325	3.078	0.951	72
19	0.326	0.344	2.904	0.946	71
20	0.342	0.364	2.747	0.940	70
21	0.358	0.384	2.605	0.934	69
22	0.375	0.404	2.475	0.927	68
23	0.391	0.424	2.350	0.921	67
24	0.407	0.445	2.246	0.914	66
25	0.423	0.466	2.145	0.906	65
26	0.438	0.488	2.050	0899	64
27	0.454	0.510	1.963	0.891	63
28	0.469	0.532	1.881	0.883	62
29	0.485	0.554	1.804	0.875	61
30	0.500	0.577	1.732	0.866	60
31	0.515	0.601	1.664	0.857	59

α°	sin α	tan α	cot α	cos α	(90–α)°
32	0.530	0.625	1.600	0.848	58
33	0.545	0.649	1.540	0.839	57
34	0.559	0.675	1.483	0.829	56
35	0.574	0.700	1.428	0.819	55
36	0.588	0.727	1.376	0.809	54
37	0.601	0.754	1.327	0.799	53
38	0.616	0.781	1.280	0.788	52
39	0.629	0.810	1.235	0.777	51
40	0.643	0.839	1.192	0.766	50
41	0.656	0.869	1.150	0.755	49
42	0.669	0.900	1.111	0.743	48
43	0.682	0.933	1.072	0.731	47
44	0.695	0.966	1.036	0.719	46
45	0.707	1.000	1.000	0.707	45
	cos α	cot α	tan α	sin α	α°

2. Inverse quantities, square and cubic roots, logarithms, exponential function

x	$\frac{1}{x}$	$\sqrt{x}$	$\sqrt{10x}$	$\sqrt[3]{x}$	$\sqrt[3]{10x}$	$\sqrt[3]{100x}$	$\ln x$	e^x
1.0	1.000	1.00	3.16	1.00	2.15	4.64	0.000	2.72
1.1	0.909	1.05	3.32	1.03	2.22	4.79	0.095	3.00
1.2	0.833	1.10	3.46	1.06	2.29	4.93	0.182	3.32
1.3	0.769	1.14	3.61	1.09	2.35	5.07	0.262	3.67
1.4	0.714	1.18	3.74	1.12	2.41	5.19	0.336	4.06
1.5	0.667	1.22	3.87	1.14	2.47	5.31	0.405	4.48
1.6	0.625	1.26	4.00	1.17	2.52	5.43	0.470	4.95
1.7	0.588	1.30	4.12	1.19	2.57	5.54	0.530	5.47
1.8	0.556	1.34	4.24	1.22	2.62	5.65	0.588	6.05
1.9	0.526	1.38	4.36	1.24	2.67	5.75	0.642	6.69
2.0	0.500	1.41	4.47	1.26	2.71	5.85	0.693	7.39
2.1	0.476	1.45	4.58	1.28	2.76	5.94	0.742	8.17
2.2	0.455	1.48	4.69	1.30	2.80	6.03	0.789	9.03
2.3	0.435	1.52	4.80	1.32	2.84	6.13	0.833	9.97
2.4	0.417	1.55	4.90	1.34	2.88	6.21	0.875	11.0

x	$\frac{1}{x}$	$\sqrt{x}$	$\sqrt{10x}$	$\sqrt[3]{x}$	$\sqrt[3]{10x}$	$\sqrt[3]{100x}$	ln x	e^x
2.5	0.400	1.58	5.00	1.36	2.92	6.30	0.916	12.2
2.6	0.385	1.61	5.10	1.38	2.96	6.38	0.955	13.5
2.7	0.370	1.64	5.20	1.39	3.00	6.46	0.993	14.9
2.8	0.357	1.67	5.29	1.41	3.04	6.54	1.030	16.4
2.9	0.345	1.70	5.39	1.43	3.07	6.62	1.065	18.2
3.0	0.333	1.73	5.48	1.44	3.11	6.69	1.099	20.1
3.1	0.323	1.76	5.57	1.46	3.14	6.77	1.131	22.2
3.2	0.313	1.79	5.66	1.47	3.18	6.84	1.163	24.5
3.3	0.303	1.81	5.75	1.49	3.21	6.91	1.194	27.1
3.4	0.294	1.84	5.83	1.50	3.24	6.98	1.224	30.0
3.5	0.286	1.87	5.92	1.52	3.27	7.05	1.253	33.1
3.6	0.278	1.90	6.00	1.53	3.30	7.11	1.281	36.6
3.7	0.270	1.92	6.08	1.55	3.33	7.18	1.308	40.4
3.8	0.263	1.95	6.16	1.56	3.36	7.24	1.335	44.7
3.9	0.256	1.98	6.25	1.57	3.39	7.31	1.361	49.4
4.0	0.250	2.00	6.33	1.59	3.42	7.37	1.386	54.6
4.1	0.244	2.03	6.40	1.60	3.45	7.43	1.411	60.3
4.2	0.238	2.05	6.48	1.61	3.48	7.49	1.435	66.7
4.3	0.233	2.07	6.56	1.63	3.50	7.55	1.458	73.7
4.4	0.227	2.10	6.63	1.64	3.53	7.61	1.482	81.5
4.5	0.222	2.12	6.71	1.65	3.56	7.66	1.504	90.0
4.6	0.217	2.15	6.78	1.66	3.58	7.72	1.526	99.5
4.7	0.213	2.17	6.86	1.68	3.61	7.78	1.548	110.0
4.8	0.208	2.19	6.93	1.69	3.63	7.83	1.569	121.0
4.9	0.204	2.21	7.00	1.70	3.66	7.88	1.589	134.0
5.0	0.200	2.24	7.07	1.71	3.68	7.94	1.609	148.0
5.1	0.196	2.26	7.14	1.72	3.71	7.99	1.629	164.0
5.2	0.192	2.28	7.21	1.73	3.73	8.04	1.649	181.0
5.3	0.189	2.30	7.28	1.74	3.76	8.09	1.668	200.0
5.4	0.185	2.32	7.35	1.75	3.78	8.14	1.686	221.0
5.5	0.182	2.35	7.42	1.77	3.80	8.19	1.705	244.0
5.6	0.179	2.37	7.48	1.78	3.83	8.24	1.723	270.0

x	$\frac{1}{x}$	$\sqrt{x}$	$\sqrt{10x}$	$\sqrt[3]{x}$	$\sqrt[3]{10x}$	$\sqrt[3]{100x}$	ln x	e^x
5.7	0.175	2.39	7.55	1.79	3.85	8.29	1.740	299.0
5.8	0.172	2.41	7.62	1.80	3.87	8.34	1.758	330.0
5.9	0.170	2.43	7.68	1.81	3.89	8.39	1.775	365.0
6.0	0.167	2.45	7.75	1.82	3.92	8.43	1.792	403.0
6.1	0.164	2.47	7.81	1.83	3.94	8.48	1.808	446.0
6.2	0.161	2.49	7.87	1.84	3.96	8.53	1.825	493.0
6.3	0.159	2.51	7.94	1.85	3.98	8.57	1.841	545.0
6.4	0.156	2.53	8.00	1.86	4.00	8.62	1.856	602.0
6.5	0.154	2.55	8.06	1.87	4.02	8.66	1.872	665.0
6.6	0.152	2.57	8.12	1.88	4.02	8.71	1.887	735.0
6.7	0.149	2.59	8.19	1.89	4.06	8.75	1.902	812.0
6.8	0.147	2.61	8.25	1.90	4.08	8.79	1.918	898.0
6.9	0.145	2.63	8.31	1.90	4.10	8.84	1.932	992.0
7.0	0.143	2.65	8.37	1.91	4.12	8.88	1.946	1097.0
7.1	0.141	2.67	8.43	1.92	4.14	8.92	1.960	1212.0
7.2	0.139	2.68	8.49	1.93	4.16	8.96	1.974	1339.0
7.3	0.137	2.70	8.54	1.94	4.18	9.00	1.988	1480.0
7.4	0.135	2.72	8.60	1.95	4.20	9.05	2.001	1636.0
7.5	0.133	2.74	8.66	1.96	4.22	9.09	2.015	1808
7.6	0.132	2.76	8.72	1.97	4.24	9.13	2.028	1998
7.7	0.130	2.78	8.78	1.98	4.25	9.17	2.041	2208
7.8	0.128	2.79	8.83	1.98	4.27	9.21	2.054	2440
7.9	0.127	2.81	8.89	1.99	4.29	9.24	2.067	2697
8.0	0.125	2.83	8.94	2.00	4.31	9.28	2.079	2981
8.1	0.124	2.85	9.00	2.01	4.33	9.32	2.092	3294
8.2	0.122	2.86	9.06	2.02	4.34	9.36	2.104	3641
8.3	0.121	2.88	9.11	2.03	4.36	9.40	2.116	4024
8.4	0.119	2.90	9.17	2.03	4.38	9.44	2.128	4447
8.5	0.118	2.92	9.22	2.04	4.40	9.47	2.140	4914
8.6	0.116	2.93	9.27	2.05	4.41	9.51	2.152	5432
8.7	0.115	2.95	9.33	2.06	4.43	9.55	2.163	6003
8.8	0.114	2.97	9.38	2.07	4.45	9.58	2.175	6634

x	$\frac{1}{x}$	$\sqrt{x}$	$\sqrt{10x}$	$\sqrt[3]{x}$	$\sqrt[3]{10x}$	$\sqrt[3]{100x}$	ln x	e^x
8.9	0.112	2.98	9.43	2.07	4.47	9.62	2.186	7332
9.0	0.111	3.00	9.49	2.08	4.48	9.66	2.197	8103
9.1	0.110	3.02	9.54	2.09	4.50	9.69	2.208	8955
9.2	0.109	3.03	9.59	2.10	4.51	9.73	2.219	9897
9.3	0.108	3.05	9.64	2.10	4.53	9.76	2.230	10938
9.4	0.106	3.07	9.69	2.11	4.55	9.80	2.241	12088
9.5	0.105	3.08	9.75	2.12	4.56	9.83	2.251	13360
9.6	0.104	3.10	9.80	2.13	4.58	9.87	2.263	14765
9.7	0.103	3.11	9.84	2.13	4.60	9.90	2.272	16318
9.8	0.102	3.12	9.90	2.14	4.61	9.93	2.282	18034
9.9	0.101	3.13	9.95	2.15	4.63	9.97	2.293	19930
10.0	0.100	3.16	10.00	2.15	4.64	10.0	2.303	22026

Formulae for approximate calculation of roots.

1. $\sqrt[n]{1+x} \approx 1 + \frac{x}{n} + \frac{1-n}{2n^2} x^2$ for $|x| < 1$.

2. $\sqrt[n]{a^n + b} \approx a\left(1 + \frac{b}{na^n} + \frac{1-n}{2n^2} \cdot \frac{b^2}{a^{2n}}\right)$ for $\left|\frac{b}{a^n}\right| < 1$.

LOGARITHMIC AND TRIGONOMETRIC TABLES

Table I: Logarithms

	0	1	2	3	4	5	6	7	8	9	Mean Differences 1	2	3	4	5	6	7	8	9
10	0000	0043	0086	0128	0170	0212	0253	0294	0334	0374	4	8	12	17	21	25	29	33	37
11	0414	0453	0492	0531	0569	0607	0645	0682	0719	0755	4	8	11	15	19	23	26	30	34
12	0792	0828	0864	0899	0934	0969	1004	1038	1072	1106	3	7	10	14	17	21	24	18	31
13	1139	1173	1206	1239	1271	1303	1335	1367	1399	1430	3	6	10	13	16	19	23	26	29
14	1461	1492	1523	1553	1584	1614	1644	1673	1703	1732	3	6	9	12	15	18	21	24	27
15	1761	1790	1818	1847	1875	1903	1931	1959	1987	2014	3	6	8	11	14	17	20	22	25
16	2041	2068	2095	2122	2148	2175	2201	2227	2253	2279	3	5	8	11	13	16	18	21	24
17	2304	2330	2355	2380	2405	2430	2455	2480	2504	2529	2	5	7	10	12	15	17	20	22

	0	1	2	3	4	5	6	7	8	9	Mean Differences								
											1	2	3	4	5	6	7	8	9
18	2553	2577	2601	2625	2648	2672	2695	2718	2742	2765	2	5	7	9	12	14	16	19	21
19	2788	2810	2833	2856	2878	2900	2923	2945	2967	2989	2	4	7	9	11	13	16	18	20
20	3010	3032	3054	3075	3096	3118	3139	3160	3181	3201	2	4	6	8	11	13	15	17	19
21	3222	3243	3263	3284	3304	3324	3345	3365	3385	3404	2	4	6	8	10	12	14	16	18
22	3424	3444	3464	3483	3502	3522	3541	3560	3579	3598	2	4	6	8	10	12	14	15	17
23	3617	3636	3655	3674	3692	3711	3729	3747	3766	3784	2	4	6	7	9	11	13	15	17
24	3802	3820	3838	3856	3874	3892	3909	3927	3945	3962	2	4	5	7	9	11	12	14	16
25	3879	3997	4014	4031	4048	4065	4082	4099	4116	4133	2	3	5	7	9	10	12	14	15
26	4150	4166	4183	4200	4216	4232	4249	4265	4281	4298	2	3	5	7	8	10	11	13	15
27	4314	4330	4386	4362	4378	4393	4409	4425	4440	4456	2	3	5	6	8	9	11	13	14
28	4472	4487	4502	4518	4533	4548	4564	4579	4594	4609	2	3	5	6	8	9	11	12	14
29	4624	4639	4654	4669	4683	4698	4713	4728	4742	4757	1	3	4	6	7	9	10	12	13
30	4771	4786	4800	4814	4829	4843	4857	4871	4886	4900	1	3	4	6	7	9	10	11	12
31	4914	4928	4942	4955	4969	4983	4997	5011	5024	5038	1	3	4	6	7	8	10	11	12
32	5051	5065	5079	5092	5105	5119	5132	5145	5159	5172	1	3	4	5	7	8	9	11	12
33	5185	5198	5211	5224	5237	5250	5263	5276	5289	5302	1	3	4	5	6	8	9	10	12
34	5315	5328	5340	5353	5366	5378	5391	5403	5416	5428	1	3	4	5	6	8	9	10	11
35	5441	5453	5465	5478	5490	5502	5514	5527	5539	5551	1	2	4	5	6	7	9	10	11
36	5563	5575	5587	5599	5611	5623	5635	5647	5658	5670	1	2	4	5	6	7	8	10	11
37	5682	5694	5705	5717	5729	5740	5752	5763	5775	5786	1	2	3	5	6	7	8	9	10
38	5798	5809	5821	5832	5843	5855	5866	5877	5888	5899	1	2	3	5	6	7	8	9	10
39	5911	5922	5933	5944	5955	5966	5977	5988	5999	6010	1	2	3	4	5	7	8	9	10
40	6021	6031	6042	6053	6064	6075	6085	6096	6107	6117	1	2	3	4	5	6	8	9	10
41	6128	6138	6149	6160	6170	6180	6191	6201	6212	6222	1	2	3	4	5	6	7	8	9
42	6232	6243	6253	6263	6274	6284	6294	6304	6314	6325	1	2	3	4	5	6	7	8	9
43	6335	6345	6355	6365	6375	6385	6395	6405	6415	6425	1	2	3	4	5	6	7	8	9
44	6435	6444	6454	6464	6474	6484	6493	6503	6513	6522	1	2	3	4	5	6	7	8	9
45	6532	6542	6551	6561	6571	6580	6590	6599	6609	6618	1	2	3	4	5	6	7	8	9
46	6628	6637	6646	6656	6665	6675	6684	6693	6702	6712	1	2	3	4	5	6	7	7	8
47	6721	6730	6739	6749	6758	6767	6776	6785	6794	6803	1	2	3	4	5	5	6	7	8
48	6812	6821	6830	6839	6848	6857	6866	6875	6884	6893	1	2	3	4	5	5	6	7	8
49	6902	6911	6920	6928	6937	6946	6955	6964	6972	6981	1	2	3	4	4	5	6	7	8
50	6990	6998	7007	7016	7024	7033	7042	7050	7059	7067	1	2	3	3	4	5	6	7	8
51	7076	7084	7093	7101	7110	7118	7126	7135	7143	7152	1	2	3	3	4	5	6	7	8
52	7160	7168	7177	7185	7193	7202	7210	7218	7226	7235	1	2	2	3	4	5	6	7	7
53	7243	7251	7259	7267	7275	7284	7292	7300	7308	7316	1	2	2	3	4	5	6	6	7
54	7324	7332	7340	7348	7356	7364	7372	7380	7388	7396	1	2	2	3	4	5	6	6	7
55	7404	7412	7419	7427	7435	7443	7451	7459	7466	7474	1	2	2	3	4	5	5	6	7
56	7482	7490	7497	7505	7513	7520	7528	7536	7543	7551	1	2	2	3	4	5	5	6	7
57	7559	7566	7574	7582	7589	7597	7604	7612	7619	7627	1	2	2	3	4	5	5	6	7
58	7634	7642	7649	7657	7664	7672	7679	7686	7694	7701	1	2	2	3	4	4	5	6	7

	0	1	2	3	4	5	6	7	8	9	Mean Differences								
											1	2	3	4	5	6	7	8	9
59	7709	7716	7723	7731	7738	7745	7752	7760	7767	7774	1	1	2	3	4	4	5	6	7
60	7782	7789	7796	7803	7810	7818	7825	7832	7839	7846	1	1	2	3	4	4	5	6	6
61	7853	7860	7868	7875	7882	7889	7896	7903	7910	7917	1	1	2	3	4	4	5	6	6
62	7924	7931	7938	7945	7952	7959	7966	7973	7980	7987	1	1	2	3	3	4	5	6	6
63	7993	8000	8007	8014	8021	8028	8035	8041	8048	8055	1	1	2	3	3	4	4	5	6
64	8062	8069	8075	8082	8089	8096	8102	8109	8116	8122	1	1	2	3	3	4	4	5	6
65	8129	8136	8142	8149	8156	8162	8169	8176	8182	8189	1	1	2	3	3	4	4	5	6
66	8195	8202	8209	8215	8222	8228	8235	8241	8248	8254	1	1	2	3	3	4	4	5	5
67	8261	8267	8274	8280	8287	8293	8299	8306	8312	8319	1	1	2	3	3	4	4	5	5
68	8325	8331	8338	8344	8351	8357	8363	8370	8376	8382	1	1	2	3	3	4	4	5	5
69	8388	8395	8401	8407	8414	8420	8426	8432	8439	8445	1	1	2	2	3	4	4	5	5
70	8451	8457	8463	8470	8476	8482	8488	8494	8500	8506	1	1	2	2	3	4	4	5	5
71	8513	8519	8525	8531	8537	8543	8549	8555	8561	8567	1	1	2	2	3	4	4	5	5
72	8573	8579	8585	8591	8597	8603	8609	8615	8621	8627	1	1	2	2	3	4	4	4	5
73	8633	8639	8645	8651	8657	8663	8669	8675	8681	8686	1	1	2	2	3	4	4	4	5
74	8692	8698	8704	8710	8716	8722	8727	8733	8739	8745	1	1	2	2	3	4	4	4	5
75	8751	8756	8762	8768	8774	8779	8785	8791	8797	8802	1	1	2	2	3	3	4	4	5
76	8808	8814	8820	8825	8831	8837	8842	8848	8854	8859	1	1	2	2	3	3	4	4	5
77	8865	8871	8876	8882	8887	8893	8899	8904	8910	8915	1	1	2	2	3	3	4	4	5
78	8921	8927	8932	8938	8943	8949	8954	8960	8965	8971	1	1	2	2	3	3	4	4	5
79	8976	8982	8987	8993	8998	9004	9009	9015	9020	9025	1	1	2	2	3	3	4	4	5
80	9031	9036	9042	9047	9053	9058	9063	9069	9074	9079	1	1	2	2	3	3	4	4	5
81	9085	9090	9096	9101	9106	9112	9117	9122	9128	9133	1	1	2	2	3	3	4	4	5
82	9138	9143	9149	9154	9159	9165	9170	9175	9180	9186	1	1	2	2	3	3	4	4	5
83	9191	9196	9201	9206	9212	9217	9222	9227	9232	9238	1	1	2	2	3	3	4	4	5
84	9243	9248	9253	9258	9263	9269	9274	9279	9284	9289	1	1	2	2	3	3	4	4	5
85	9394	9299	9304	9309	9315	9320	9325	9330	9335	9340	1	1	2	2	3	3	4	4	5
86	9345	9350	9335	9360	9365	9370	9375	9380	9385	9390	1	1	2	2	3	3	4	4	5
87	9395	9400	9405	9410	9415	9420	9425	9430	9435	9440	0	1	1	2	2	3	3	4	4
88	9445	9450	9455	9460	9465	9469	9474	9479	9484	9489	0	1	1	2	2	3	3	4	4
89	9494	9499	9504	9509	9513	9518	9523	9528	9533	9538	0	1	1	2	2	3	3	4	4
90	9542	9547	9552	9557	9562	9566	9571	9576	9581	9586	0	1	1	2	2	3	3	4	4
91	9590	9595	9600	9605	9609	9614	9619	9624	9628	9633	0	1	1	2	2	3	3	4	4
92	9638	9643	9647	9652	9657	9661	9666	9671	9675	9680	0	1	1	2	2	3	3	4	4
93	9685	9689	9694	9699	9703	9708	9713	9717	9722	9727	0	1	1	2	2	3	3	4	4
94	9731	9736	9741	9745	9750	9754	9759	9763	9768	9773	0	1	1	2	2	3	3	4	4
95	9777	9782	9786	9791	9795	9800	9805	9809	9814	9818	0	1	1	2	2	3	3	4	4
96	9823	9827	9832	9836	9841	9845	9850	9854	9859	9863	0	1	1	2	2	3	3	4	4
97	9868	9872	9877	9881	9886	9890	9894	9899	9903	9908	0	1	1	2	2	3	3	4	4
98	9912	9917	9921	9926	9930	9934	9939	9943	9948	9952	0	1	1	2	2	3	3	4	4
99	9956	9961	9965	9969	9974	9978	9983	9987	9991	9996	0	1	1	2	2	3	3	3	4

Table II: Antilogarithms

	0	1	2	3	4	5	6	7	8	9	Mean Differences								
											1	2	3	4	5	6	7	8	9
.00	1000	1002	1005	1007	1009	1012	1014	1016	1019	1021	0	0	1	1	1	1	2	2	2
.01	1023	1026	1028	1030	1033	1035	1038	1040	1042	1045	0	0	1	1	1	1	2	2	2
.02	1047	1050	1052	1054	1057	1059	1062	1064	1067	1069	0	0	1	1	1	1	2	2	2
.03	1072	1074	1076	1079	1081	1084	1086	1089	1091	1094	0	0	1	1	1	1	2	2	2
.04	1096	1099	1102	1104	1107	1109	1122	1114	1117	1119	0	1	1	1	1	2	2	2	2
.05	1122	1125	1127	1130	1132	1135	1138	1150	1143	1146	0	1	1	1	1	2	2	2	2
.06	1148	1151	1153	1156	1159	1161	1164	1167	1169	1172	0	1	1	1	1	2	2	2	2
.07	1175	1178	1180	1183	1186	1189	1191	1194	1197	1199	0	1	1	1	1	2	2	2	2
.08	1202	1205	1208	1211	1213	1216	1219	1222	1225	1227	0	1	1	1	1	2	2	2	3
.09	1230	1233	1236	1239	1242	1245	1247	1250	1253	1256	0	1	1	1	1	2	2	2	3
.10	1259	1262	1265	1268	1271	1274	1276	1279	1282	1285	0	1	1	1	1	2	2	2	3
.11	1288	1291	1294	1297	1300	1303	1306	1309	1312	1315	0	1	1	1	2	2	2	2	3
.12	1318	1321	1324	1327	1330	1334	1337	1340	1343	1346	0	1	1	1	2	2	2	2	3
.13	1349	1352	1355	1358	1361	1365	1368	1371	1374	1377	0	1	1	1	2	2	2	3	3
.14	1380	1384	1387	1390	1393	1396	1400	1403	1406	1409	0	1	1	1	2	2	2	3	3
.15	1413	1416	1419	1422	1426	1429	1432	1435	1439	1442	0	1	1	1	2	2	2	3	3
.16	1445	1449	1452	1455	1459	1462	1466	1469	1472	1476	0	1	1	1	2	2	2	3	3
.17	1479	1483	1486	1489	1493	1496	1500	1503	1507	1510	0	1	1	1	2	2	2	3	3
.18	1514	1517	1521	1524	1528	1531	1535	1538	1542	1545	0	1	1	1	2	2	2	3	3
.19	1549	1552	1556	1560	1563	1567	1570	1574	1578	1581	0	1	1	1	2	2	3	3	3
.20	1585	1589	1592	1596	1600	1603	1607	1611	1614	1618	0	1	1	1	2	2	3	3	3
.21	1622	1626	1629	1633	1637	1641	1644	1648	1652	1656	0	1	1	2	2	2	3	3	3
.22	1660	1663	1667	1671	1675	1679	1683	1687	1690	1694	0	1	1	2	2	2	3	3	3
.23	1698	1702	1706	1710	1714	1718	1722	1726	1730	1734	0	1	1	2	2	2	3	3	4
.24	1738	1742	1746	1750	1754	1758	1762	1766	1770	1774	0	1	1	2	2	2	3	3	4
.25	1778	1782	1786	1791	1795	1799	1803	1807	1811	1816	0	1	1	2	2	2	3	3	4
.26	1820	1824	1828	1832	1837	1841	1845	1849	1854	1858	0	1	1	2	2	3	3	3	4
.27	1862	1866	1871	1875	1879	1884	1888	1892	1897	1901	0	1	1	2	2	3	3	3	4
.28	1905	1910	1914	1919	1923	1928	1932	1936	1941	1945	0	1	1	2	2	3	3	4	4
.29	1950	1954	1959	1963	1968	1972	1977	1982	1986	1991	0	1	1	2	2	3	3	4	4
.30	1995	2000	2004	2009	2014	2018	2023	2028	2032	2037	0	1	1	2	2	3	3	8	4
.31	2042	2046	2051	2056	2061	2065	2070	2075	2080	2084	0	1	1	2	2	3	3	4	4
.32	2089	2094	2099	2104	2109	2113	2118	2123	2128	2133	0	1	1	2	2	3	3	4	4

	0	1	2	3	4	5	6	7	8	9	Mean Differences								
											1	2	3	4	5	6	7	8	9
.33	2138	2143	2148	2153	2158	2163	2168	2173	2178	2183	0	1	1	2	2	3	3	4	4
.34	2188	2193	2198	2203	2208	2213	2218	2223	2228	2234	1	1	2	2	3	3	4	4	5
.35	2239	2244	2249	2254	2259	2265	2270	2275	2280	2286	1	1	2	2	3	3	4	4	5
.36	2291	2296	2301	2307	2312	2317	2323	2328	2333	2339	1	1	2	2	3	3	4	4	5
.37	2344	2350	2355	2360	2366	2371	2377	2382	2388	2393	1	1	2	2	3	3	4	4	5
.38	2399	2404	2410	2415	2421	2427	2432	2438	2443	2449	1	1	2	2	3	3	4	4	5
.39	2455	2460	2466	2472	2477	2483	2489	2495	2500	2506	1	1	2	2	3	3	4	5	5
.40	2512	2518	2523	2529	2535	2541	2547	2553	2529	2564	1	1	2	2	3	4	4	5	5
.41	2570	2576	2582	2588	2594	2600	2606	2612	2618	2624	1	1	2	2	3	4	4	5	5
.42	2630	2636	2642	2649	2655	2661	2667	2673	2679	2685	1	1	2	2	3	4	4	5	6
.43	2692	2698	2704	2710	2716	2723	2729	2735	2742	2748	1	1	2	3	3	4	4	5	6
.44	2754	2761	2767	2773	2780	2786	2793	2799	2805	2812	1	1	2	3	3	4	4	5	6
.45	2818	2815	2831	2838	2844	2851	2858	2864	2871	2877	1	1	2	3	3	4	5	5	6
.46	2884	2891	2897	2904	2911	2917	2924	2931	2938	2944	1	1	2	3	3	4	5	5	6
.47	2951	2958	2965	2972	2979	2985	2992	2999	3006	3013	1	1	2	3	3	4	5	5	6
.48	3020	3027	3034	3041	3048	3055	3062	3069	3076	3083	1	1	2	3	4	4	5	6	6
.49	3090	3097	3105	3112	3119	3126	3133	3141	3148	3155	1	1	2	3	4	4	5	6	6
.50	3162	3170	3177	3184	3192	3199	3206	3214	3221	3228	1	1	2	3	4	4	5	6	7
.51	3236	3243	3251	3258	3266	3273	3281	3289	3296	3304	1	2	2	3	4	5	5	6	7
.52	3311	3319	3327	3334	3342	3350	3357	3365	3373	3381	1	2	2	3	4	5	5	6	7
.53	3388	3396	3404	3412	3420	3428	3436	3443	3451	3459	1	2	2	3	4	5	6	6	7
.54	3467	3475	3483	3491	3499	3508	3516	3524	3532	3540	1	2	2	3	4	5	6	6	7
.55	3548	3556	3565	3573	3581	3589	3597	3606	3614	3622	1	2	2	3	4	5	6	7	7
.56	3631	3639	3648	3656	3664	3673	3681	3690	3698	3707	1	2	3	3	4	5	6	7	8
.57	3715	3724	3733	3741	3750	3758	3767	3776	3784	3793	1	2	3	3	4	5	6	7	8
.58	3802	3811	3819	3828	3837	3846	3855	3864	3873	3882	1	2	3	4	4	5	6	7	8
.59	3890	3899	3908	3917	3926	3936	3945	3954	3963	3972	1	2	3	4	5	5	6	7	8
.60	3981	3990	3999	4009	4018	4027	4036	4046	4055	4064	1	2	3	4	5	6	6	7	8
.61	4074	4083	4093	4102	4111	4121	4130	4140	4150	4159	1	2	3	4	5	6	7	8	9
.62	4169	4178	4188	4198	4207	4217	4227	4236	4256	4256	1	2	3	4	5	6	7	8	9
.63	4266	4276	4285	4295	4305	4315	4325	4335	4345	4355	1	2	3	4	5	6	7	8	9
.64	4365	4375	4385	4395	4406	4416	4426	4436	4446	4457	1	2	3	4	5	6	7	8	9
.65	4467	4477	4487	4498	4508	4519	4529	4539	4550	4560	1	2	3	4	5	6	7	8	9
.66	4571	4581	4592	4603	4613	4624	4634	4645	4656	4667	1	2	3	4	5	6	7	9	10

	0	1	2	3	4	5	6	7	8	9	**Mean Differences**								
											1	2	3	4	5	6	7	8	9
.67	4677	4688	4699	4710	4721	4732	4742	4753	4764	4775	1	2	3	4	5	7	8	9	10
.68	4786	4797	4808	4819	4831	4842	4853	4864	4875	4887	1	2	3	4	6	7	8	9	10
.69	4898	4909	4920	4932	4943	4955	4966	4977	4989	5000	1	2	3	5	6	7	8	9	10
.70	5012	5023	5035	5047	5058	5070	5082	5093	5105	5117	1	2	4	5	6	7	8	9	11
.71	5129	5140	5152	5164	5176	5188	5200	5212	5224	5236	1	2	4	5	6	7	8	10	11
.72	5248	5260	5272	5284	5297	5309	5321	5333	5346	5358	1	2	4	5	6	7	9	10	11
.73	5370	5383	5395	5408	5420	5433	5445	5458	5470	5483	1	3	4	5	6	8	9	10	11
.74	5495	5508	5521	5534	5546	5559	5572	5585	5598	5610	1	3	4	5	6	8	9	10	12
.75	5623	5636	5649	5662	5675	5689	5702	5715	5728	5741	1	3	4	5	7	8	9	10	12
.76	5754	5768	5781	5794	5808	5821	5834	5848	5861	5875	1	3	4	5	7	8	9	11	12
.77	5888	5902	5916	5929	5943	5957	5970	5984	5998	6012	1	3	4	5	7	8	10	11	12
.78	6026	6039	6053	6067	6081	6095	6109	6124	6138	6152	1	3	4	6	7	8	10	11	13
.79	6166	6180	6194	6209	6223	6237	6252	6266	6281	6295	1	3	4	6	7	9	10	11	13
.80	6310	6324	6339	6353	6368	6383	6397	6412	6427	6442	1	3	4	6	7	9	10	12	13
.81	6457	6471	6486	6501	6516	6531	6546	6561	6577	6592	2	3	5	6	8	9	11	12	14
.82	6607	6622	6637	6653	6668	6683	6699	6714	6730	6745	2	3	5	6	8	9	11	12	14
.83	6761	6776	6792	6808	6823	6839	6855	6871	6887	6902	2	3	5	6	8	9	11	13	14
.84	6918	6934	6950	6966	6982	6998	7015	7031	7047	7063	2	3	5	7	8	10	11	13	14
.85	7079	7096	7112	7129	7145	7161	7178	7194	7211	7228	2	3	5	7	8	10	12	13	15
.86	7244	7261	7278	7295	7311	7328	7345	7362	7379	7396	2	3	5	7	8	10	12	13	15
.87	7413	7434	7447	7464	7482	7499	7516	7534	7551	7568	2	3	5	7	9	10	12	14	16
.88	7586	7603	7621	7638	7656	7674	7691	7709	7727	7745	2	4	5	7	9	11	12	14	16
.89	7762	7780	7798	7816	7834	7852	7870	7889	7907	7925	2	4	5	7	9	11	13	14	16
.90	7943	7962	7980	7998	8017	8035	8054	8072	8091	8110	2	4	6	7	9	11	13	15	17
.91	8128	8147	8166	8185	8204	8222	8241	8260	8279	8299	2	4	6	8	9	11	13	15	17
.92	8318	8337	8356	8375	8395	8414	8453	8453	8472	8492	2	4	6	8	10	12	14	15	17
.93	8511	8531	8551	8570	8590	8610	8630	8650	8670	8690	2	4	6	8	10	12	14	16	18
.94	8710	8730	8750	8770	8790	8810	8831	8851	8872	8892	2	4	6	8	10	12	14	16	18
.95	8913	8933	8954	8974	8995	9016	9036	9057	9078	9099	2	4	6	8	10	12	15	17	19
.96	9120	9141	9162	9183	9204	9226	9247	9268	9290	9311	2	4	6	8	11	13	15	17	19
.97	9333	9354	9376	9397	9419	9441	9462	9484	9506	9528	2	4	7	9	11	13	15	17	20
.98	9550	9572	9594	9616	9638	9661	9683	9705	9727	9750	2	4	7	9	11	13	16	18	20
.99	9772	9795	9817	9840	9863	9886	9908	9931	9954	9977	2	5	7	9	11	14	16	18	20

ANSWERS

Chapter-1: MATHEMATICS AND MECHANICS

1. (i) 2

(ii)The limit does not exit.

(iii)∞

(iv)$-\infty$

(v)4

(vi)$-\frac{1}{6}$

(viii) $\frac{108}{7}$

2. $\pm\sqrt{15}$

3. x = 2

4. At x = 0 max and at x = 2 min.

5. x = 1 is neither point of maxima nor minima.

6. 2a + 3b + 6c = 0

7. $-\sqrt{3} \le a \le \sqrt{3}$

8. $2\left(c^2 \log 4 - 8\right)$

10. $\frac{16}{3}$

11. x + y = 3

12. $\left(\frac{7}{2}, \sqrt{\frac{7}{2}}\right)$

14. 11

15. 0.71 hour.

16. $\left[\left(\frac{1}{2}\right)^{\frac{1}{3}}, \pm\left(\frac{1}{2}\right)^{-\frac{1}{6}}\right]$

17. One root is between 0 and 1 and other root is between 1 and 2.

18. y = 3x + 5 and y = 3x – 5

19. $\sqrt{2}\sin^{-1}(\sin x - \cos x) + C$

20. $\log\left(\frac{xe^x}{1+xe^x}\right) + \frac{1}{1+xe^x} + C$

21. 2

22. $\frac{e^2 - 5}{4e}$

23. $\left(-\frac{36}{7}, -\frac{46}{7}\right)$

24. Are concurrent, pass through the fixed point (1, 1) and touch some fixed circle.

25. $\left(\frac{1}{2}, -2^{\frac{1}{2}}\right)$

26. 9

27. $x^2 + x + 1 = 0$

28. 1

29. 41

30. $\frac{1}{5}(n-4)$

31. Divisible by k.

32. Negative

33. –9, 2, 7

34. $\frac{33}{2}$

35. 4

36. 2

37. $\frac{1}{16}$

38. Independent of P.

39. f(x).

40. $\frac{1}{e}$.

Chapter-2: SYSTEMS OF UNITS

1. c
2. c
3. b
4. a
5. d
6. a
7. d
8. e
9. a, c
10. b
11. b
12. 6.0×10^6
13. 31.2 km
14. 8×10^2 km
15. 3.156×10^7 s
16. (a) 10^3 kg/m^3 (b) 158 kg/s
17. (a) 1.18×10^{-29} m^3 (b) 0.282 nm
18. 36×10^5 cm min^{-2}
19. All are dimensionally correct
20. (a) $FL^{-4}T^2$ (b) FL^{-2} (c) FT (d) FL
21. $E = kmc^2$

Chapter-3: VECTOR

1. 50 miles, 120 miles.
2. $\vec{i} - 2\vec{j}$
3. 6950 miles, pointing through the Earth from Washigton to Manila.
5. 110°
6. 54.73°
7. (a) up, unit magnitude (b) zero (c) South, unit magnitude (d) 1.00 (e) 0
8. $a^2 b\left(\hat{b} - \hat{a}\cos\phi\right)$
9. – 24 kJ.
10. $5\left(1+\sqrt{2}\right)$
11. $\frac{1}{\sqrt{1+x^2}}\cos[x]\vec{i} + \sin[x]\vec{j} + (x-[x])\vec{k}$
12. (a) 1.6 m, (b) 0.98 m, 1.3 m (c) 52.85°
13. $12\vec{k}$
14. (a) $\sqrt{29}$, $\sqrt{14}$ (b) $3\vec{i} + \vec{j} + 7\vec{k}$ (c) $\sqrt{59}$

(d) $\vec{i} + 5\vec{j} + \vec{k}$ (e) $\sqrt{27}$; yes
15. $\left(\Delta x^2 + \Delta y^2\right)^{\frac{1}{2}};\tan^{-1}\frac{\Delta y}{\Delta x}$
16. $\frac{\left(\vec{a}\bullet\vec{b}\right)\vec{c} - \left(\vec{a}\bullet\vec{c}\right)\vec{b}}{\left(\vec{a}\bullet\vec{b}\right)}$
19. – 1.
20. 0
35. $R = \sqrt{P^2 + Q^2 + 2PQ\cos\alpha}$

$\beta = \sin^{-1}\left(\frac{Q}{R}\sin\alpha\right)$

$\gamma = \sin^{-1}\left(\frac{P}{R}\sin\alpha\right)$
36. W – Q
37. R = 381.57 N

β = 30°

γ = 27°
38. 10 N inclined by 36.52°
40. 61.45°

Chapter-4: NEWTON'S LAWS OF MOTION

1. $w = \dfrac{m_0 - k(m_1 + m_2)}{m_1 + m_2 + m_3} g$

$T = \dfrac{(1+k)m_0}{m_0 + m_1 + m_2} m_2 g$

2. 0.16

3. (a) $m_2/m_1 > \sin\alpha + k\cos\alpha$ (b) $m_2/m_1 < \sin\alpha - k\cos\alpha$ (c) $\sin\alpha - k\cos\alpha < m_2/m_1 < \sin\alpha + k\cos\alpha$.

4. $\tan 2\alpha = \dfrac{-1}{k}$, $\alpha = 49°$; $t_{min} = 1.0$ s

5. $\beta = \tan^{-1} k$; $T_{min} = \dfrac{mg(\sin\alpha + k\cos\alpha)}{\sqrt{1+k^2}}$

6. $V = \sqrt{\left(\dfrac{2g}{3a}\right)\sin\alpha}$

7. $W = \dfrac{2g(2\eta - \sin\alpha)}{4h+1}$

8. $W_A = \dfrac{g}{1+\eta\cot^2\alpha}$, $W_B = \dfrac{g}{\tan\alpha + \eta\cot\alpha}$

9. $W = \dfrac{g\sqrt{2}}{2+k+\dfrac{M}{m}}$

10. $H = \dfrac{6h\eta}{\eta+4} = 0.6$ m

11. $W_{min} = \dfrac{g(1-k)}{(1+k)}$

12. $W_{max} = \dfrac{g(1+k\cot\alpha)}{(\cot\alpha - k)}$

13. $W = \dfrac{mg\sin\alpha}{M + 2m(1-\cos\alpha)}$

14. (a) $|\langle F \rangle| = 2\sqrt{2}\ mv^2/\pi R$

(b) $|\langle F \rangle| = mwt$

15. (a) $W = g\sqrt{1+3\cos^2\theta}$; $T = 3$ mg cos θ

(b) $T = mg\sqrt{3}$

(c) $\cos\theta = \dfrac{1}{\sqrt{3}}$; θ = 54.7°.

16. (a) $k = (\ln \eta_0)/\pi$

(b) $w = g(\eta - \eta_0)/(\eta + \eta_0)$

17. $T = (\omega + \theta + \omega^2 R/g)\ mg/2\pi$

18. $S = \dfrac{2}{a}\tan\alpha$; $V_{max} = \sqrt{\dfrac{g}{a}\sin\alpha\tan\alpha}$

19. 0

20. 10 N

21. $a \dfrac{Mg\cos\theta}{m + M\cos^2\theta}$

22. $t_{min} = 1$s

24. (a) g sin θ down the plane (b) g sin θ down the plane (c) (g – a) sin θ down the plane (d) (g + a) sin θ down the plane (e) zero.

28. $\left(\dfrac{x}{r_0}\right)^2 - K\left(\dfrac{v}{v_0}\right) = 1$

29. (i) 10 s (ii) 20 s (iii) 15 ms^{-2}, 2.5 ms^{-2}

30. 1.29 m

31. $\dfrac{mm'g}{M(m+m') + m^2 + 2mm'}$

33. $a_1 = 9.02\ ms^{-2}$; $a_2 = 0.39\ ms^{-2}$

34. $a_1 = a_2 = g$

35. v = 24.8 m/s

36. μ = 2.2

37. $a = \dfrac{(M+m)\,g\sin\alpha\cos\alpha}{(M + m\sin^2\alpha)}$

38. $2a_1 + a_2 + a_3 = 0$

39. F = 100 N

40. T = 2 μmg, F = 9 μmg

41. $w = \dfrac{2(2\eta - \sin\alpha)}{(4\eta+1)} g$

42. $a_2 = -7a_1$

Chapter-5: CENTROID

1. $\frac{2}{\pi a}\left(b+\frac{a}{e}\sin^{-1}e\right)$

2. $\frac{\sqrt{2}+\ln\left(1+\sqrt{2}\right)}{2}$

3. $x_c = a - \tan h\left(\frac{a}{2}\right);\ y_c = \frac{a}{2\sinh a}+\frac{\cosh a}{2}$

4. $x_c = \pi a;\ y_c = \frac{4}{3}a$

5. $x_c = \frac{\pi}{6(4-\pi)};\ y_c = \frac{12-\pi^2}{12-3\pi}$

6. $x_c = \frac{b(h_1+2h_2)}{3(h_1+h_2)};\ y_c = \frac{h_1^2+h_2^2+h_1h_2}{3(h_1+h_2)}$

7. $\overline{X} = 0$ and $Y = \frac{4b}{7}$

8. $\overline{X} = 1.44;\ \overline{Y} = 0.361$

9. $\overline{X} = \frac{4}{3\pi}(a+b);\ \overline{Y} = 0$

10. $\overline{X} = 1.797;\ \overline{Y} = 2.34\ (1-\cos\beta)$

11. $\overline{X} = \frac{2r}{3\beta}\sin\beta;\ \overline{Y} = \frac{2r}{3\beta}$

14. (ii) $\overline{X} = h\quad \overline{Y} = \frac{h}{4}$

16. $\bar{x} = 0,\ \bar{y} = 110.95$ mm

17. $\bar{x} = 0,\ \bar{y} = 6.92$ mm

18. (a) [20, 20] (b) 16 mm above the base

20. $G\left(\frac{b}{3},\frac{h}{3}\right)$

21. $\bar{x} = 11.5$ mm; $\bar{y} = 24.9$ mm

22. $\bar{x} = 0;\ \bar{y} = 34$ mm

23. $\bar{x} = 0;\ \bar{y} = 55$ mm

24. $\bar{x} = 25$ mm; $\bar{y} = 35$ mm

25. x = 47.2 mm; y = 62.2 mm

26. $\overline{Y} = 39.19$ mm

28. $\bar{x} = \bar{y} = \frac{7}{12}a$

29. $\bar{x} = \frac{11}{18}a;\ \overline{Y} = \frac{a}{2}a$

30. $\bar{x} = 0,\ \bar{y} = -0.28a$

31 $\bar{x} = 76.516$ mm; $\bar{y} = 63.792$ mm

32. $\bar{x} = 4.195$ m, $3\bar{y} = 2.645$ m

35. $\bar{x} = 50$ mm; $\bar{y} = 40.86$ mm

36. $\bar{x} = 72.692$ mm; $\bar{y} = 59.9$ mm

37. $\bar{x} = 81.6$ mm; $\bar{y} = 29$ mm

38. $\bar{x} = 4.38$ mm; $\bar{y} = 3.385$ mm

43. $\bar{x} = 2.72$ m; $\bar{y} = 1.9218$ m

44. $\bar{x} = 39.44$ mm; $\bar{y} = 28.47$ mm

45. $\bar{x} = 3.523$ m; $\bar{y} = 2.775$ m

47. 1.6, 2.29

48. $\bar{x} = \frac{3}{8}a;\ \bar{y} = \frac{2}{5}b.$

Chapter-6: MOMENT OF INERTIA (AREA)

1 $I_y = 26.8 \times 10^6$ mm^4

2. $I_x = \dfrac{16ab^3}{105}$

4. $I_y = 27.8 \times 10^4$ mm^4

5. $I_x = 0.1988r^4$ unit^4

6. $I_x = \dfrac{ab^3}{30}$

7. $\dfrac{2a^2b}{11}; a\sqrt{\dfrac{7}{11}}$

8. $\dfrac{5}{2}\pi^2 - 8$

18. $0.0071r^4$

19. 243 mm^4

26. 183193.34 mm^4

32. $\theta_n = 37.7^\circ$, θ_m 127.7°, $I_a = I_{max} = 15.45$ in^4; $I_b = I_{min} = 1.897$ in^4

33. $I_{xy} = \dfrac{1}{24}b^2h^2$; $I_{xy} = \dfrac{1}{72}b^2h^2$

34. $q_m = 23.8^\circ$, $I_{max} = 8.36$ ´ 10^6 mm^4, $I_{min} = 1.49 \times 10^6$ mm^4, $I_{x'} = 5.96 \times 10^6$ mm^4, $I_{y'} = 3.89 \times 10^6$ mm^4; $I_{x'y'} = 3.28 \times 10^6$ mm^4.

35. 0

36. 0

37. 3 in^4

38. $I_x = \dfrac{3}{35}ab^3$, $I_y = \dfrac{3}{35}a^3b$, $r_x = \sqrt{\dfrac{3}{35}}b$; $r_y = \sqrt{\dfrac{9}{35}}a$

39. $I_x = \dfrac{1}{21}ab^3$, $I_y = \dfrac{1}{5}a^3b$, $r_x = \sqrt{\dfrac{1}{7}}b$; $r_y = \sqrt{\dfrac{3}{5}}a$

Chapter-7: FRICTION

LADDER

1. 5.25 m
3. 1.73 m
5. $R_A = 882.9$N, $R_B = 453.4$ N; $\mu = 0.514$
6. F = 139 N; $f_{max} = 206$ N
7. 1.8 m.
8. $\left[\left(\dfrac{5}{7}l\right)\text{from foot}\right]$
13. 46.6%
15. 61.76 N
16. $\mu = 0.553$
17. $\tan^{-1}\dfrac{1-2\mu^2}{\mu}$
20. 0.35

WEDGE

8. $\sin\alpha - \mu\cos\alpha \le P/Q \le \sin\alpha \sin\alpha + \mu\cos\alpha$

9. $\mu \ge \sqrt{\dfrac{d}{D}}$

10. $P_{min} = (W_1 + W_2)\sin\phi$

11. 208.11 N

12. (a) 117.7 N (b) 147.15 N (c) m_2 slips on M

13. (a) (i) $a_1 = a_2 = 3.6$ ms^{-2} (ii) $a_1 = 6.08$ ms^{-2}; $a_2 = 2.6$ ms^{-2}

(b) $a_1 = 5.08$ ms^{-2}; $a_2 = 3.4$ ms^{-2}

14. $\dfrac{[2m - \mu_2(m + m)]g}{M + m[5 + 2(\mu_1 - \mu_2)]}$

16. 0.414

19. $P_{min} = 81.2$ N.

20. 14.025 kN

21. 2.344 kN

24. 5.21 ms^{-2}, 215 N

35. 59.22 N

Chapter-8: COPLANAR CONCURRENT FORCES

11. $R_P = 2.1545$ kN, $R_Q = 7.4424$ kN, $R_R = 7.0296$ kN; $R_S = 2.2962$ kN

12. $R_P = 61.24$ N, $R_Q = 631.58$ N, $R_R = 1095$N; $R_S = 290.43$ N

13. $R_1 = 37.5$ N, $R_2 = 62.5$ N, $R_3 = 287.5$ N; $R_4 = 353.5$ N

14. P = 3047.2 N

15. $P_1 = 100$ N, $P_2 = 200$ N; $\theta = 63.8961°$

16. 866.03 N

18. $R_d = 10.4$ N; $R_e = 6.0$ N.

19. $S_1 = P \tan \alpha$ (tension) and $S_2 = P \sec \alpha$ (compression)

20. $S = P \sec \alpha$, $R_b = W + P \tan \alpha$

23. $S_1 = S_2 = S_4 = S_5 = 0$, $S_3 = P$ (tension)

25. $\dfrac{\alpha - \beta}{2} = 18°26'$, $R_a = 67.1$ N, $R_b = 134$ N and S = 63.3 N.

26. 25.6 kN

27. 36.9°; 80 N.

28. $F_D = 19.66$ N; $T_{AC} = 42.9$ N

29. $T_{AC} = 326$ N; $T_{BC} = 369$ N

30. $T_{AB} = 875.133$ N, $T_{BC} = 500$ N; $T_{CD} = 500$ N

Chapter-8: COPLANAR NONCONCURRENT SYSTEM

11. $R_b = \dfrac{wa}{l \tan \alpha}$

12. $R_a = \dfrac{P}{3}$ down, $R_b = \dfrac{P}{3}$ up

15. $R_A = \dfrac{w_0 l}{2}$; $M_A = \dfrac{w_0 l^2}{6}$

16. P = 60 N, Q = 74.4 N; $\beta = 53.8°$

17. T = 81.9 N; $R_A = 147.8$ N at angle of 58.6°

19. Resultant intersect AB 3.333 m from A. It is a safe design.

20. $R_A = 89.4312$ kN, $\alpha_A = 54.76°$; $R_B = 101.2517$ kN.

Bibliography

1. Feidinard P Beer, E. Russell Johnston, Jr. Mechanics for Engineers (static), 4th edition, McGraw -Hill International edition.
2. S. Timoshenko, D.H. Young. Engineering Mechanics, 4th edition, McGraw-Hill International edition.
3. Irving H. Shames. Engineering Mechanics (static and dynamic), Prentice Hall of India Pvt. Ltd.
4. T.L. Meriam, L.G. Kraige. Engineering Mechanics (static), vol. 1, 3rd edition, John Wiley & Sons Pvt. Ltd.
5. Robert Resnick, David Halliday. Physics (Part-1), New Age International (P) Limited, Publishers.
6. David Halliday, Robert Resnick, Jearl Walker. Fundamental of Physics, extended 4th edition.
7. Francis W. Sears. Nark W. Zimansky Hugh D. Young. University Physics, 6th edition, Narosa Publishing House.
8. I.E. Irodov, Problems in General Physics, Mir Publishers, Moscow/CBS Publishers and Distributors, India.